CHRISTOPHER JON BJERKNES

THE MANUFACTURE
AND SALE OF
SAINT EINSTEIN

III

OMNIA VERITAS

Christopher Jon Bjerknes

The Manufacture and Sale of Saint Einstein

Volume III

Published by
Omnia Veritas Ltd

www.omnia-veritas.com

© Omnia Veritas Ltd - Christopher Jon Bjerknes - 2019.

All Rights Reserved. No part of this publication may be reproduced, distributed, or transmitted in any form or by any means, including photocopying, recording, or other electronic or mechanical methods, without the prior written permission of the publisher, except in the case of brief quotations embodied in critical reviews and certain other noncommercial uses permitted by copyright law.

Table of Contents:

5 THE PROTOCOLS OF THE LEARNED ELDERS OF ZION — 7

5.10 The Holocaust as a Zionist Eugenics Program for the Jewish "Remnant": Zionist Nazis Use Natural and Artificial Selection to Strengthen the Genetic Stock of Jews Destined for Forced Deportation to Palestine — 7
5.11 Zionist Lies — 36
5.12 Zionists Declare that Anti-Semitism is the Salvation of the "Jewish Race" — 57
5.13 Communist Jews in America — 79
5.14 The Attempted Assassination of Henry Ford — 84
5.15 How the Zionists Blackmailed President Wilson — 101
   5.15.1 Before the War, the Zionists Plan a Peace Conference After the War—to be Led by a Zionist Like Woodrow Wilson — 105
   5.15.2 "Colonel" Edward Mandell House — 128
   5.15.3 The Balfour Declaration—QUID PRO QUO — 177
5.16 A Newspaper History of Zionist Intrigues During the First World War, which Proves that Jewish Bankers Betrayed Germany — 206
5.17 The Germans' Side of the First World War — 256

6 ZIONISM IS RACISM — 277

6.1 Introduction — 277
6.2 Political Zionism is a Form of Racism — 280
6.3 Most Jews Opposed Zionism — 282
6.4 The Brotherhood of Anti-Semites and Zionists — 303
6.5 Albert Einstein Becomes a Cheerleader for Racist Zionism — 379
   6.5.1 While Zionists and Sycophants Hailed Einstein, Most Scientists Rejected Him and "His" Theories — 379
   6.5.2 Hypocritical and Cowardly Einstein Plays the "Race Card" and Cripples Scientific Progress — 395
   6.5.3 What is Good for the Goose is not Good for the Goyim — 404
      6.5.3.1 Supremacist and Segregationist Jewish "Neo-Messianism" — 408
      6.5.3.2 It is Alright for Jews to Claim that "Einstein's Theories" are "Jewish", but Goyim Dare Not Say It — 424

7 NAZISM IS ZIONISM — 448

7.1 Introduction — 448
7.2 *Blut und Boden*—A Jewish Ideal — 449
7.3 Zionism is Built on Lies and Hatred — 458
7.4 The Hypocritical Vilification of Caligula—Ancient Jewish Historians are not Credible — 489
7.5 All the Best Zionists are Anti-Semites — 518
   7.5.1 Nazism is a Stalking Horse for Zionism and Communism — 567
   7.5.2 Hitler and Goebbels Reveal Their True Motives at War's End — 578

7.5.3 Zionists and Communists Delight in Massive Human Sacrifices to the Jewish Messianic Cause -- *585*

5 THE PROTOCOLS OF THE LEARNED ELDERS OF ZION

5.10 The Holocaust as a Zionist Eugenics Program for the Jewish "Remnant": Zionist Nazis Use Natural and Artificial Selection to Strengthen the Genetic Stock of Jews Destined for Forced Deportation to Palestine

Jews promoted the Eugenics movement. Rabbi Max Reichler published the following paper in 1916:

> "JEWISH EUGENICS
> A PAPER READ BEFORE THE
> NEW YORK BOARD OF JEWISH MINISTERS
> BY
> Rabbi MAX REICHLER, B. A.
> NEW YORK
> BLOCH PUBLISHING COMPANY
> 1916
>
> JEWISH EUGENICS
>
> Who knows the cause of Israel's survival? Why did the Jew survive the onslaughts of Time, when others, numerically and politically stronger, succumbed? Obedience to the Law of Life, declares the modern student of eugenics, was the saving quality which rendered the Jewish race immune from disease and destruction. 'The Jews, ancient and modern,' says Dr. Stanton Coit, 'have always understood the science of eugenics, and have governed themselves in accordance with it; hence the preservation of the Jewish race.' [*Footnote:* Cf. also *Social Direction of Human Evolution*, by Prof. William E. Kellicott, 1911, p. 231.]
>
> I. Jewish Attitude
>
> To be sure eugenics as a science could hardly have existed among the ancient Jews; but many eugenic rules were certainly incorporated in the large collection of Biblical and Rabbinical laws. Indeed there are clear indications of a conscious effort to utilize all influences that might improve the inborn qualities of the Jewish race, and to guard against any practice that might vitiate the purity of the race, or 'impair the racial qualities of future generations' either physically, mentally, or morally.[*Footnote:* Sir Francis

Galton defines eugenics as 'the science which deals with all influences that improve the inborn qualities of the race.'] The Jew approached the matter of sex relationship neither with the horror of the prude, nor with the passionate eagerness of the pagan, but with the sane and sound attitude of the far-seeing prophet. His goal was the creation of the ideal home, which to him meant the abode of purity and happiness, the source of strength and vigor for body and mind.[*Footnote:* Cf. Ps. cxxviii, 3-4. The National Conference on Race Betterment which met recently at Battle Creek declared that 'the core of race betterment consists in promoting more and better homes.'

II. Home of the Pure Bloods

The very founder of the Jewish race, the patriarch Abraham, recognized the importance of certain inherited qualities, and insisted that the wife of his 'only beloved son' should not come from 'the daughters of the Canaanites,' but from the seed of a superior stock.[*Footnote:* Gen. xxiv, 3-4.]

In justifying this seemingly narrow view of our patriarch, one of the Rabbis significantly suggests: ' Even if the wheat of your own clime does not appear to be of the best, its seeds will prove more productive than others not suitable to that particular soil.'[*Footnote:* Ber. Rabbah 59, 11.]

This contention is eugenically correct. Davenport tells of a settlement worker of this city who made special inquiry concerning a certain unruly and criminally inclined section of his territory, and found that the offenders came from one village in Calabria, known as 'the home of the brigands.' [*Footnote: Heredity in Relation to Eugenics*, by Charles B. Davenport, New York, 1911, p. 183.] Just as there is a home of the brigands, so there may be 'a home of the pure bloods.'

Eugenists also claim that though consanguineous marriages are in most cases injurious to the progeny, yet where relatives possess 'valuable characters, whether apparent or not, marriages between them might be encouraged, as a means of rendering permanent a rare and valuable family trait, which might otherwise be much less likely to become an established characteristic.'[*Footnote: Social Direction of Human Evolution*, p. 154; *Heredity in Relation to Eugenics*, p. 185. The Biblical expression 'a bone of my bones' (Gen. ii, 23), refers, according to the Rabbis, to a man who marries one of his relatives. (Bereshith Kabbah 18, 5). The marriage between uncle and niece is also recommended (Yebamoth 63b).] Abraham's servant, Eliezer, so the Midrash states, desired to offer his own daughter to Isaac, but his master sternly rebuked him, saying: 'Thou art cursed, and my son is blessed, and it does not behoove the cursed to mate with the blessed, and thus deteriorate the quality of the race.'[*Footnote:* Ber. Rabbah 59, 12; cf. Gen. ix, 25-26.]

III. Early Marriages

The aim of eugenics is to encourage the reproduction of the good and 'blessed' human protoplasm and the elimination of the impure and 'cursed' human protoplasm. According to Francis Galton, it is 'to check the birthrate of the unfit, and to further the productivity of the fit by early marriages and the rearing of healthful children.'

The Rabbis may or may not have had such a definite purpose in mind, but their Halachic legislation and Haggadic observations naturally tended to bring about the same results. Early marriages were praised as most desirable. Rabbi Ishmael claimed that God was greatly displeased with the man who did not marry before the age of twenty.[*Footnote:* Kiddushin 29b.] Rav Hunah refused to see Rav Hamnuna, a man of great repute (*adam gadol*), after the former discovered that his visitor was a bachelor.[*Footnote:* Ibid.] 'He who is not married,' runs a Talmudic saying, 'is destitute of all joy, blessing, and happiness.'[*Footnote:* Midrash Lekach Tob, Gen. 2, ed. Buber p. 21.] 'He has no conception of the sweetness of life';[*Footnote:* Ber. Rabbah ch. 17.] indeed 'he cannot be regarded as a man at all.'[*Footnote:* Yalkut Gen. ii, 23.]

IV. Reproduction

Among the seven types not acceptable before God are included both the unmarried man and the married man without children.[*Footnote:* Pesachim 113b.] A man without children experiences death in life,[*Footnote:* Nedarim 64b.] and surely deserves our pity when he departs from this earth.[*Footnote:* M. K. 27b.] For only he is dead who leaves no son behind to continue his work, while he who leaves even one worthy son is not really dead but merely sleeps.[*Footnote:* B. B. 110b.] He who does not contribute his share to the reproduction of the race, reduces the divine type,[*Footnote:* Yebamoth 63b.] causes the Shechinah to depart from Israel,[*Footnote:* Ibid. 64a.] and is guilty of murder.[*Footnote:* Ibid 63b, 64a.] The duty of reproduction is incumbent on all, both young and old.[*Footnote:* Ibid 62b. Cf. Koheleth Kabbah 7, 8, also *Social Direction of Human Evolution*, p. 124, concerning pathological defects of first born and earlier members of the family.]

The Rabbis, like the eugenists of to-day, measured the success of a marriage by the number and quality of the offspring. In their judgments the main objects of marriage were the reproduction of the human race (*leshem piryah veribyah*), and the augmentation of the favored stock (*lethikun havlad*).[*Footnote:* Cf. Tur Eben Haezer ch. 25.] Hence they advised that an extremely tall man should not marry an extremely tall woman, lest the children be awkwardly tall; nor should one of short stature marry a woman of the same size, lest their offspring be dwarfed. For the same reason, the intermarriage between blonds or between dark-complexioned people was not countenanced.[*Footnote:* Bechoroth 45b.] A number of precautions in sexual relations were prescribed in order to prevent the birth of defectives, such as lepers,[*Footnote:* Sifra, Mezora ch. 3.] epileptics,[*Footnote:*

Pesachim 112b.] the deaf and the dumb, the lame and the blind.[*Footnote:* Nedarim 20a.]

V. Intelligent Love

Raba advised every young man not to marry a girl before he knew all about her immediate family, especially about her brothers, for 'children usually inherit the traits of their mother's brothers.'[*Footnote:* B. B. 110a] 'Take your time,' counsels a Talmudic proverb, 'before you ask a woman to be your wife';[*Footnote:* Yebamoth 63a.] in other words, 'fall in love intelligently.' Other well- known Rabbinic maxims are: 'a man drinketh not out of a cup which he hath not inspected,'[*Footnote:* Kethuboth 75b.] and 'a bride whose eyes are defective, ought to undergo a general physical examination.'[*Footnote:* Shir Hashirim Rabbah 4, 1-3; cf. Taanith 24a.]

In the opinion of Rabbi Jonathan both Eliezer, the servant of Abraham, and Saul, king of Israel, acted most indiscreetly by treating marriage in a rather frivolous manner. Eliezer said: 'Behold the virgin which will say drink, and I will also draw for the camels, that is the woman whom the Lord hath appointed for my master's son.' Suppose that woman had some physical defects, would she have been a suitable mate for Isaac? Similarly Saul proclaimed: 'The man who killeth Goliath, the king will give him his daughter.' If that man had been a slave or possessed other hereditary defects, would Saul have sanctioned the marriage?[*Footnote:* Taanith 4a.]

VI. Non-Eugenic Marriages

The attempt to limit the multiplication of the undesirable elements in the Jewish race, resulted in three kinds of prohibitions. First, prohibition against the marriage of defectives by reason of heredity (*pesul yochesin*) ; secondly, the prohibition against the marriage of personal defectives (*debar shebagufon*); thirdly, the prohibition against consanguineous marriages (*ervah*).[*Footnote:* Tur Eben Haezer, Piryah Veribyah, ch. 4.]

Besides the prohibition against defective marriages mentioned in the Mosaic code,[*Footnote:* Deuteronomy xxiii, 2.] the Talmud forbade one to marry into a confirmed leprous or epileptic family,[*Footnote:* Yebamoth 64a.] or to marry a woman who had buried three husbands.[*Footnote:* Niddah 64a. It is interesting to note that a late authority insists that the same rule should apply to a man who buried three wives. Cf. Beer Heteb to Eben Haezer, Ishoth 9, 2.] The union between an old man and a young girl was condemned in unequivocal terms.[*Footnote:* Sanhedrin 76a; cf. also Yebamoth 106b and Ruth Rabbah 3, 10.] Persons or families manifesting continuous antagonism to each other were advised not to intermarry.[*Footnote:* Kiddushin 71b. Cf. *Heredity in Relation to Eugenics*, p. 8, where the suggestion is made that the curious antipathy of red-haired persons of the opposite sexes for each other, may be an eugenic antipathy.] Great, in the eyes of the Rabbis, was the offense of him who married a woman from an element classed among the unfit. His act was as reprehensible as if he had dug up every fertile field in existence and sown it with salt. [*Footnote:* Kiddushin 70a.] A quintuple transgression was his, [*Footnote:* Aboth Derabbi Nathan, ch. 26.] for which he will be bound hand and foot by Elijah, the great purifier, [*Footnote:* Cf. Kiddushin 71a.] and flogged by God himself. 'Woe unto him who deteriorates the quality of his children and defiles the purity of his family,' is the verdict of Elijah endorsed by God. [*Footnote:* Kiddushin 70a.] On the other hand, the mating of two persons possessing unique and noble traits cannot but result in the establishment of superior and influential families. [*Footnote:* Bamidbar Rabbah 3, 4.] When God will cause his Shechinah to dwell in Israel, only such which scrupulously preserved the purity of their families, will be privileged to witness the manifestation of the Holy Spirit. [*Footnote:* Kiddushin 70b.]

VII. Psychical Eugenics

The distinctive feature, however, of Jewish eugenics lies in the greater emphasis laid on the psychical well-being of posterity, in contradistinction to the merely physical well-being which is the chief concern of modern eugenists. At the Congress of Eugenics recently held at London, one of our modern eugenists, Professor Samuel C. Smith of the University of Minnesota, exclaimed : 'If I were to choose my own father, I would rather have a robust burglar than a consumptive bishop.' The Rabbis, on the other hand, tell us that when the question came up whether or not the Gibeonites should be permitted to intermarry with the children of Israel, David tested them, in order to ascertain not so much their physical fitness but rather their psychical fitness, and found them wanting. He discovered that they did not possess the three 'unit characters' peculiar to Israel, namely: sympathy, modesty and philanthropy. He therefore thought it eugenically inadvisable to allow their mating with a spiritually better-developed stock. [*Footnote:* Yebamoth 79a.] Rabbi Levi enumerates nine undesirable psychical qualities which ought to be eliminated from amongst the Jewish race. [*Footnote:* Nedarim 20b.]

VIII. Eugenics and Religion

The Jew took his spiritual mission as representing a 'kingdom of priests and a holy kingdom' quite seriously, and used all possible eugenic means to preserve those rare emotional and spiritual qualities developed during centuries of slow progress and unfolding. Intuitively he felt the truth, so well expressed by a modern student of eugenics, that 'Religion would be a more effective thing, if everybody had a healthy emotional nature; but it can do nothing with natures that have not the elements of love, loyalty and devotion.' [*Footnote: Heredity in Relation to Eugenics*, p. 255.] The Rabbis would say: Religion can do nothing with natures that have not the elements of sympathy, modesty and philanthropy. Hence they urged that a man should be willing to offer all his possessions for the opportunity of marrying a member of a psychically well-developed family. [*Footnote:* Pesachim 49b.]

The marriage between the offspring of inferior stock and that of superior stock, such as the marriage between a scholar and the daughter of an *am-haarez*, or between an *am-haarez* and the daughter of a scholar, was considered extremely undesirable, and was condemned very strongly. [*Footnote:* Kiddushin 49b; cf. also Pesachim 49b.] Moreover, no Rabbi or *Talmid Chacham* was allowed to take part in the celebration of such a non-eugenic union. [*Footnote:* Pesachim 49b.]

An historical case is cited by Rabbi Eliezer to prove that one should always select his soul-mate from amongst the spiritually better-developed families. Moses married a daughter of Jethro, a heathen priest, and the result was that one of his grandsons, Jonathan, became an idolatrous priest. Aaron, on the other hand, married the daughter of Abinadab, and history records the name of his grandson Phinehas as the hero who defended the honor and purity of Israel. [*Footnote:* B. B. 109b.]

Parents living normal and righteous lives are not only a blessing to themselves, but also to their children and children's children, until the end of all generations; while parents living abnormal and immoral lives bring ruin and calamity not only on themselves, but also on their children and children's children, to the end of all generations. [*Footnote:* Yoma 87a.]

IX. Heredity

A parallel to the 'rough eugenic ideal' of marrying 'health, wealth and wisdom' [*Footnote: Heredity in Relation to Eugenics*, p. 8.] is found in the words of Rabbi Akiba, who claims that 'a father bequeaths to his child beauty, health, wealth, wisdom and longevity.' [*Footnote:* Eduyoth 2, 9.] Similarly, ugliness, sickness, poverty, stupidity and the tendency to premature death, are transmitted from father to offspring. [*Footnote:* Yer. Kiddushin 1, 7.] Hence we are told that when Moses desired to know why some of the righteous suffer in health and material prosperity, while others prosper and reap success; and again, why some of the wicked suffer, while others enjoy success and material well-being; God explained that the righteous and wicked who thrive and flourish, are usually the descendants of righteous parents, while those who suffer and fail materially are the descendants of wicked parents. [*Footnote:* Berachoth 7a.]

X. Priceless Heritage

Thus the Rabbis recognized the fact that both physical and psychical qualities were inherited, and endeavored by direct precept and law, as well as by indirect advice and admonition, to preserve and improve the inborn, wholesome qualities of the Jewish race. It is true that they were willing to concede that 'a pure-bred individual may be produced by a hybrid mated with a pure bred,' for they found examples of that nature in Ruth the Moabitess, Naamah the Ammonitess, [*Footnote:* Yebamoth 63a.] Hezekiah and Mordecai. [*Footnote:* Bamidbar Kabbah, Chukath ch. 19.] As a general eugenic rule, however, they maintained that one cannot produce 'a clean thing out of an unclean,' and discouraged any kind of intermarriage even with proselytes. [*Footnote:* Pesachim 112b, Kiddushin 70b.] Their ideal was a race healthy in body and in spirit, pure and undefiled, devoid of any admixture of inferior human protoplasm. [*Footnote:* Yer. Kilayim ch. 1.]

Such an ideal, though apparently narrow and chauvinistic, has its eugenic value, as the following suggestive quotation from a well-known eugenist clearly indicates. 'Families in which good and noble qualities of mind and body have become hereditary, form a natural aristocracy; and if such families take pride in recording their pedigrees, marry among themselves, and establish a predominant fertility, they can assure success and position to the majority of their descendants in any political future. They can become the guardians and trustees of a sound inborn heritage, which, incorruptible and undefiled, they can preserve in purity and vigor throughout whatever period of ignorance and decay may be in store for the nation at large. Neglect to hand on undimmed the priceless germinal qualities which such families possess, can be regarded only as a betrayal of a sacred trust. [*Footnote:* See *Social Direction of Human Evolution*, p. 238.]"

The Jewish book of *Isaiah* 48:10 states,

"Behold, I have refined thee, but not with silver; I have chosen thee in the furnace of affliction."

The Jewish book of *Deuteronomy* 4:20 states,

"But the LORD hath taken you, and brought you forth out of the iron furnace, *even* out of Egypt, to be unto him a people of inheritance, as *ye are* this day."

The Jewish book of *Ezekiel* 5:12 states:

"A third *part* of thee shall die with the pestilence, and with famine shall they be consumed in the midst of thee: and a third *part* shall fall by the sword round about thee and I will scatter a third *part* into all the winds, and I will draw out a sword after them."

The Jewish book of *Zechariah* 13:8, 9 calls on Jews to ensure that two thirds of the extant population of Jews in the world is exterminated,

"8 And it shall come to pass, *that* in all the land, saith the LORD, two parts therein shall be cut off *and* die; but the third shall be left therein. 9 And I will bring the third *part* through the fire, and will refine them as silver is refined, and will try them as gold is tried: they shall call on my name, and I will hear them: I will say, It *is* my people: and they shall say, The LORD *is* my God."

Jews do not worry that the pogroms they intentionally cause will completely wipe out the Jewish populations of the Earth. They count on the promise of *Leviticus* to ensure that a "remnant" of Jews will survive the persecutions Jews deliberately cause other Jews, in order to keep Jews segregated, to ensure that some Jewish population will outlive those who are slaughtered by design by other Jews. *Leviticus* 26:44 states,

"And yet for all that, when they be in the land of their enemies, I will not cast them away, neither will I abhor them, to destroy them utterly, and to break my covenants with them: for I *am* the Lord their God."

Racist Zionist Israel Zangwill wrote,

"In the diaspora anti-Semitism will always be the shadow of Semitism. The law of dislike for the unlike will always prevail. And whereas the unlike is normally situated at a safe distance, the Jews bring the unlike into the heart of every milieu and must thus defend a frontier-line as large as the world. The fortunes of war vary in every country, but there is a perpetual tension and friction even at the most peaceful points, which tend to throw back the race on itself. The drastic method of love—the only human dissolvent—has never been tried upon the Jew as a whole, and Russia carefully conserves— even by a ring fence—the breed she designs to destroy. But whether persecution extirpates or brotherhood melts, hate or love can never be simultaneous throughout the diaspora, and so there will probably always be a nucleus from which to restock this eternal type. But what a melancholy immortality! 'To be *and* not to be'—that is a question beside which Hamlet's alternative is crude."[1]

Theodor Herzl wrote in his book *The Jewish State*,

"Oppression and persecution cannot exterminate us. No nation on earth has survived such struggles and sufferings as we have gone through. Jew-baiting has merely stripped off our weaklings; the strong among us were invariably true to their race when persecution broke out against them. This attitude was most clearly apparent in the period immediately following the emancipation

---

[1]. I. Zangwill, *The Problem of the Jewish Race*, Judaen Publishing Company, New York, (1914), p. 18; which was first published as an article, "The Jewish Race", *The Independent*, Volume 71, Number 3271, (10 August 1911), pp. 288-295, at 294.

of the Jews. Later on, those who rose to a higher degree of intelligence and to a better worldly position lost their communal feeling to a very great extent. Wherever our political well-being has lasted for any length of time, we have assimilated with our surroundings. I think this is not discreditable. Hence, the statesman who would wish to see a Jewish strain in his nation would have to provide for the duration of our political well-being; and even Bismarck could not do that. [***] The Governments of all countries scourged by Anti-Semitism will serve their own interests in assisting us to obtain the sovereignty we want. [***] Great exertions will not be necessary to spur on the movement. Anti-Semites provide the requisite impetus. They need only do what they did before, and then they will create a love of emigration where it did not previously exist, and strengthen it where it existed before. [***] I imagine that Governments will, either voluntarily or under pressure from the Anti-Semites, pay certain attention to this scheme; and they may perhaps actually receive it here and there with a sympathy which they will also show to the Society of Jews."[2]

An article in the *Christian Reader*, Volume 3, Number 67, (19 November 1824), page 366, evinces that the Rothschilds were aware that the Jews of Europe did not have the skills, abilities or character needed to successfully farm the fields of Palestine or build the palaces which wealthy Western Jews wanted. Indeed, Lord Sydenham pointed out in 1922 that the Jews emigrating to Palestine under the British Palestine Mandate had no business being in the region and only served to worsen the situation in the Middle East. *The London Times* published a Letter to the Editor from Lord Sydenham of Combe, "British Policy in Palestine. Divergence from Balfour Declaration." on 4 April 1923, on page 6, which stated, *inter alia*,

"Into Palestine we are dumping successive shiploads of impecunious aliens, we are imposing a loan equal to the whole annual revenue, and we have ordained a third official language perfectly useless to the people. All this, together with minor inflictions, we are doing in opposition to the strongly expressed wishes of a huge majority of Palestinians. It would be interesting if the 'Zionist Organization' would explain what 'civil rights' are left to a little people so circumstanced, and how the declaration, 'revised in the Zionist offices in America as well as in England,' can be reconciled with this use of British military forces."

On 7 April 1922, on page 8, *The London Times* published a Letter to the Editor from Lord Sydenham, "Jewish 'National Home.' Lord Sydenham Urges Inquiry.":

---

[2]. T. Herzl, *A Jewish State: An Attempt at a Modern Solution of the Jewish Question*, The Maccabæan Publishing Co., New York, (1904), pp. 5-6, 25, 68, 93.

"Are these colonies or any of them being worked on an economic basis today? Palestine does not lend itself to cheap irrigation; but that aspect of the question needs investigation. My own strong opinion is that the national home must eventually break down on economic grounds, because you cannot indefinitely maintain colonies unable to pay their way. This is also the view of some leading American Jews besides Mr. Morgenthau. If, then, as Dr. Weizmann proposes, 'between 50,000 and 60,000 Jews per annum' are deposited in the Holy Land, we shall soon be confronted with appalling difficulties—partly economic and partly arising from the hostility of the rightful owners of the land, who would find themselves displaced by the growing horde of immigrants. My conclusion is that, in the interests of the Jews as well as the Arabs, immigration must be stopped until a full inquiry has taken place, if serious troubles are to be averted. For moral as well as economic reasons, the 'powerful irritant' must be removed."

On 8 September 1922 on page 9, *The London Times* published correspondence which had taken place between Lord Sydenham and Winston Churchill, "Our Palestine Policy", in which Lord Sydenham wrote,

"3. A 'Jewish National Home' can be interpreted in several ways, and Mr. Balfour's undertaking—that the 'civil rights' of the Palestinians would not be prejudiced naturally reassured me. I never dreamed that a Jewish Government would be set up, and I imagined only a slow immigration of desirable Jews under a purely British Government. In 1917, it was not yet clear that there would be a rush of Russian and Central European Jews to other countries, and that a portion of them would be dumped down in Palestine. I was further reassured in 1918 by General Allenby's Proclamation, which appeared to render impossible what is now happening, while the text of the Treaty with the Hedjaz, which is disputed, was unknown to me at the time. Since 1917 I have given much thought and study to the Jewish problems, and I have been forced to change my opinions. I was, as you suggest, 'mistaken in thinking that the Jews were entitled to regard Palestine as the 'National Home.' I consider that they have no more claim to Palestine than the modern Italians to Britain, or the Moors to Southern Spain. I also think that 'a horde of aliens' correctly describes the immigrants."

Jewish Messianic prophecy called for the expulsion of all non-Jews from Palestine. This necessitated the development of a Jewish workforce suited to fulfill the needs of wealthy Jews. The Jews of Eastern Europe would have to be toughened and trained in construction and agriculture before they would be prepared to build the Palestine the Zionists wanted. The Nazis set about the task of building the Jewish workforce the Zionist Jews had demanded as least as early as 19 November 1824.

Exodus 1:8-14 and 3:2 taught the Jews that oppression strengthened their "race" and ultimately increased their numbers, and note the ancient declaration made by the Jews themselves (the story is a fabrication) that the Jews were a dangerously disloyal nation within a nation, note also the image of enduring a holocaust,

> "8 Now there arose up a new king over Egypt, which knew not Joseph. 9 And he said unto his people, Behold, the people of the children of Israel *are* more and mightier than we: 10 Come on, let us deal wisely with them; lest they multiply, and it come to pass, that, when there falleth out any war, they join also unto our enemies, and fight against us, and *so* get them up out of the land. 11 Therefore they did set over them taskmasters to afflict them with their burdens. And they built for Pharaoh treasure cities, Pithom and Raamses. 12 But the more they afflicted them, the more they multiplied and grew. And they were grieved because of the children of Israel. 13 And the Egyptians made the children of Israel to serve with rigour: 14 And they made their lives bitter with hard bondage, in morter, and in brick, and in all manner of service in the field: all their service, wherein they made them serve, *was* with rigour. [***] 3:2 And the angel of the LORD appeared unto him in a flame of fire out of the midst of a bush: and he looked, and, behold, the bush burned with fire, and the bush was not consumed."

Zionists often stated that Moses saved the Jews by persecuting them. Racist Zionist Jakob Klatzkin stated in 1925,

> "When Moses came to redeem the children of Israel, their leaders said to him, 'You have made our odor evil in the eyes of Pharaoh and in the eyes of his servants, giving them a sword with which to kill us.' Nevertheless, Moses persisted in worsening the situation of the people, and he saved them."[3]

Adolf Hitler was the Zionists' modern Moses.

The *Christian Reader*, Volume 3, Number 67, (19 November 1824), page 366:

> "CHRISTIAN REGISTER.
>
> BOSTON, FRIDAY, NOVEMBER 19, 1824.
>
> THE JEWS. It is stated with much assurance in the Gazette of Spires, that the Sublime Porte has recently made proposals to the House of Rothschild for the loan of a considerable sum of money, and has offered as a security for payment, the entire country of Palestine. It is stated also that in consequence of this proposal a confidential agent had been dispatched by

---

[3]. J. Klatzkin, *Tehumim: Ma'amarim*, Devir, Berlin, (1925). English translation in J. B. Agus, *The Meaning of Jewish History*, Volume 2, Abelard-Schuman, New York, (1963), p. 426.

that House to Constantinople, 'to examine into the validity of the pledge offered by the Turkish Cabinet.'

The editor of the *National Advocate* observes in relation to this report, that he at first supposed it was intended as a satire on the prevailing custom of raising loans for different nations; but on a nearer view of the subject, the proposition might be supposed probable. The Advocate proceeds with some interesting remarks on the subject, tending to show, that if such a proposition had been made it could not be accepted with any prospect, on the part of the Rothschilds, (who are Jews,) of the immediate restoration of their countrymen to Palestine, as it was probably not in the power even of the Turkish government, to guarantee to the Jews the quiet possession of the country against the prejudices and interests of the Egyptians, the Wechabites, the Wandering Arabs, and the Tartar Hordes.

It is also argued that the descrepancy of education, habits, views, and manners, existing between the Jews of different countries, unfit them to amalgamate and become united under one government. They must be prepared for this by the same discipline which their fathers, who went out of Egypt were subjected to under Moses, for forty years in the wilderness, to prepare them for the promised land. 'Our country,' continues the Advocate, 'must be an asylum to the ancient people of God. Here they must reside; here, in calm retirement, study laws, governments, sciences; become familiarly known to their brethren of other religious denominations; cultivate the useful arts; acquire a knowledge of legislation, and become liberal and free. So, that appreciating the blessings of just and salutary laws, they may be prepared to possess permanently their ancient land, and govern righteously.'"

Racist Zionists have long complained that Jewish genes and Jewish mores have been corrupted in the Diaspora by the persecutions of the Middle Ages. Racist Zionists, like Adolf Hitler, believed that Americans constituted superior racial strains with the strongest and most adventurous of the races having migrated to the New World.[4] Racist Zionists, including Adolf Hitler, believed that the Holocaust would cause a process of natural selection that would improve the Jewish blood and undo the damage of the Diaspora, leaving only the strongest and smartest Jews left alive. Racist Zionists believed that American Jews and the improved Jewish remnant of the Holocaust would stock Israel. Jewish prophets predicted that Jews in the end times would be superior to those who had come before them.

The are numerous accounts from Holocaust survivors of *SS* doctors reviewing new arrivals at the concentration camps and selecting out some Jews for a chance at survival—should they be strong enough to survive the poor rations and rampant diseases and be clever enough to outlive their fellows. These

---

[4]. A. Hitler, translated by S. Attanasio, *Hitler's Secret Book*, Bramhall House, New York, (1962).

same doctors allegedly selected out some Jews, or part-Jews, for immediate death. The rationale given for this selection process was that the Nazis only spared the lives of Jews who were fit to work. However, the Nazis had conquered numerous territories and could more easily have used those populations as a slave labor force, with no chance that a select remnant of Jews would survive.

If the accounts of the artificial selection of the fittest Jews are true, and not scripted, this was likely part of a broader eugenics program to improve the genetic stock and "racial purity" of the Jews who survived the war and who were slated by the Zionists to dwell in Palestine. Even if these stories are not true, there is no doubt that the Jews were starved and overworked and faced deadly diseases, which had the effect of natural selection and the survival of the fittest. We know that the Zionists handpicked the best Jews to smuggle out of the Reich. We know further that the Nazis aided the Zionists in training fit recruits for the Zionist cause.

The Zionists wanted young, strong and clever Jews to populate their land. They did not want old, very young or feeble Jews to get in the way of the founding of their "Jewish State". The racist Zionists overlooked such sentimentalities as the innate value of human life. They justified mass murder as the prophesied birth pangs of the Messianic Age, the "hevlei Mashiah". They doubted that Palestine could absorb more than a fraction of the Jews under Hitler's control, so the loss of some assimilatory Jews to a eugenics program that profited the Zionists was an overall gain, in their minds.

We know that the Nazis and Zionists collaborated and practiced human selection of the best Jews slated to survive in Israel. Hannah Arendt wrote in her book *Eichmann in Jerusalem: A Report on the Banality of Evil*,

> "Of greater importance for Eichmann were the emissaries from Palestine, who would approach the Gestapo and the S.S. on their own initiative, without taking orders from either the German Zionists or the Jewish Agency for Palestine. They came in order to enlist help for the illegal immigration of Jews into British-ruled Palestine, and both the Gestapo and the S.S. were helpful. They negotiated with Eichmann in Vienna, and they reported that he was 'polite,' 'not the shouting type,' and that he even provided them with farms and facilities for setting up vocational training camps for prospective immigrants. ('On one occasion, he expelled a group of nuns from a convent to provide a training farm for young Jews,' and on another 'a special train [was made available] and Nazi officials accompanied' a group of emigrants, ostensibly headed for Zionist training farms in Yugoslavia, to see them safely across the border.) According to the story told by Jon and David Kimche, with 'the full and generous cooperation of all the chief actors' (*The Secret Roads: The 'Illegal' Migration of a People, 1938-1948*, London, 1954), these Jews from Palestine spoke a language not totally different from that of Eichmann. They had been sent to Europe by the communal settlements in Palestine, and they were not interested in rescue operations: 'That was not their job.' They wanted to select 'suitable material,' and their chief enemy, prior to the extermination program, was not those who made

life impossible for Jews in the old countries, Germany or Austria, but those who barred access to the new homeland; that enemy was definitely Britain, not Germany. Indeed, they were in a position to deal with the Nazi authorities on a footing amounting to equality, which native Jews were not, since they enjoyed the protection of the mandatory power; they were probably among the first Jews to talk openly about mutual interests and were certainly the first to be given permission 'to pick young Jewish pioneers' from among the Jews in the concentration camps. Of course, they were unaware of the sinister implications of this deal, which still lay in the future; but they too somehow believed that if it was a question of selecting Jews for survival, the Jews should do the selecting themselves. It was this fundamental error in judgment that eventually led to a situation in which the non-selected majority of Jews inevitably found themselves confronted with two enemies—the Nazi authorities and the Jewish authorities. As far as the Viennese episode is concerned, Eichmann's preposterous claim to have saved hundreds of thousands of Jewish lives, which was laughed out of court, finds strange support in the considered judgment of the Jewish historians, the Kimches: 'Thus what must have been one of the most paradoxical episodes of the entire period of the Nazi regime began: the man who was to go down in history as one of the arch-murderers of the Jewish people entered the lists as an active worker in the rescue of Jews from Europe.'"[5]

Nazi Zionist Adolf Eichmann stated,

"[H]ad I been a Jew, I would have been a fanatical Zionist. I could not imagine being anything else. In fact, I would have been the most ardent Zionist imaginable."[6]

Eichmann was a Jew and a Zionist—indeed, as he stated, the most ardent Zionist imaginable, one who fulfilled Jewish prophecy and mass murdered assimilatory Jews in his staged rôle as a crypto-Jewish Nazi leader. Eichmann was an ardent Zionist who selected out genetically superior Jews for survival, one who fulfilled the desire of Orthodox Jews to live a segregated life in a Ghetto. Adolf Eichmann was an ardent Zionist who helped found the "Jewish State". Eichmann likened himself to Paul, a Jew who persecuted Jews and who had converted to Christianity in an effort to preserve the Jewish nation. Adolf Eichmann stated,

"I issued the cloth [yellow cloth for the badges Jews were forced to wear] to my Jewish functionaries and they trotted off with them. [***] There was

---

[5]. H. Arendt, *Eichmann in Jerusalem: A Report on the Banality of Evil*, Viking, New York, (1963), pp. 55-56; in the revised 1964 edition at pp. 60-61.
[6]. A. Eichmann, "Eichmann Tells His Own Damning Story", *Life Magazine*, Volume 49, Number 22, (28 November 1960), pp. 19-25, 101-112; at 22.

a Jewish lawyer in Vienna who said to me, 'Sir, I wear this star with pride.' This man impressed me. He was an idealist. So I let him emigrate soon afterward. [***] We even had some Jewish SS men who had taken part in the early struggles of the Nazis—about 50 of them in Germany and Austria. I remember giving my attention to a Jewish SS sergeant, a good man, who wanted to leave for Switzerland. I had instructed the border control to let him pass [***] He was a 100% Jew, a man of the most honorable outlook. [***] I am no anti-Semite. I was just politically opposed to Jews because they were stealing the breath of life from us. [***] Certainly I too had been aiming at a solution of the Jewish problem, but not like this. [***] I would not say I originated the ghetto system. That would be to claim too great a distinction. The father of the ghetto system was the orthodox Jew, who wanted to remain by himself. In 1939, when we marched into Poland, we had found a system of ghettos already in existence, begun and maintained by the Jews. We merely regulated those, sealed them off with walls and barbed wire and included even more Jews than were already dwelling in them. The assimilated Jew was of course very unhappy about being moved to a ghetto. But the Orthodox were pleased with the arrangement, as were the Zionists. The latter found ghettos a wonderful device for accustoming Jews to community living. Dr. Epstein from Berlin once said to me that Jewry was grateful for the chance I gave it to learn community life at the ghetto I founded at Theresienstadt, 40 miles from Prague. He said it made an excellent school for the future in Israel. The assimilated Jews found ghetto life degrading, and non-Jews may have seen an unpleasant element of force in it. But basically most Jews feel well and happy in their ghetto life, which cultivates their peculiar sense of unity. [***] [W]e did not want to punish individual Jews. We wanted to work toward a political solution. [***] Himmler would not stand for that kind of thing. That is sadism. [***] 'I will gladly jump into my grave in the knowledge that five million enemies of the Reich have already died like animals.' ('Enemies of the Reich,' I said, not 'Jews.') [***] Long before the end, any of the Jews I dealt with would have set up foreign exchange for me in any country I had named, if I had promised any special privileges for them. [***] It would be too easy to pretend that I had turned suddenly from a Saul to a Paul. No, I must say truthfully that if we had killed all the 10 million Jews that Himmler's statisticians originally listed in 1933, I would say, 'Good, we have destroyed an enemy.' But here I do not mean wiping them out entirely. That would not be proper—and we carried on a proper war. Now, however, when through the malice of fate a large part of these Jews whom we fought against are alive, I must concede that fate must have wanted it so. I always claimed that we were fighting against a foe who through thousands of years of learning and development had become superior to us. I no longer remember exactly when, but it was even before Rome itself had been founded that the Jews could already write. It is very depressing for me to think of that people writing laws over 6,000 years of written history. But it tells me that they must be a people of the first magnitude, for law-givers have always been great."[7]

Bryan Mark Rigg estimates the total number of Jewish soldiers and sailors in the Nazi military perhaps ranges upwards to 150,000.[8]

At his trial, Session Number 90, 26 Tammuz 5721, 10 July 1961,[9] Eichmann confirmed that he twice requested permission to learn Hebrew from a Rabbi. He also stated that the annihilation (*Vernichtung*) of the Jews to him meant deportation and Zionism, however, he further stated that Hitler later changed course in the middle of the war and sought the physical annihilation of the Jews. Yet again Eichmann stated that he was a convinced Zionist, who wanted to put segregated soil under the feet of the Jewish populace, and that it was Adolf Böhm's book *Die Zionistische Bewegung*, which convinced him that the root of all evil was the fact that the Jews did not have a homeland.

Julius Streicher was another crypto-Jewish poseur, who pretended to be the most anti-Semitic man alive and archenemy of the Jews, so that he could forward the Jewish Zionists' "final solution of the Jewish question", which solution was Zionism, not extermination—*see:* Ernst Heimer's *Der Giftpilz: Ein Stürmerbuch für Jung u. Alt*, Der Stürmer, Nürnberg, (1938)—a children's book designed to lure kids into believing that the Nazis would protect them from the Jewish bankers, who in fact used the Nazis to destroy the Germans. Just as the crypto-Jewish *Dönmeh* Turks appeared to be the most zealous Moslems in their communities, while keeping to the Jewish faith in private and subverting the Turkish Empire; and just as the Marrano Jews of Spain and South America pretended to be the most pious Catholics in all of the Spanish Empire, while forwarding the interests of Jews around the world; crypto-Jews including Julius Streicher, Adolf Hitler, Adolf Eichmann, Reinhard Heydrich and Joseph Goebbels pretended to be the most ardent anti-Semites in the world, the most feared foes of the Jewish bankers, and they thereby gained the trust of the German People by pretending to fight the Jewish bankers who were bent on

---

[7]. A. Eichmann, "Eichmann Tells His Own Damning Story", *Life Magazine*, Volume 49, Number 22, (28 November 1960), pp. 19-25, 101-112; at 23, 24, 102, 104, 106, 110; and "Eichmann's Own Story: Part II", *Life Magazine*, (5 December 1960), pp. 146-161; at 150, 158, 161.

[8]. "Who Were Hitler's Jewish Soldiers", *The Jewish Chronicle*, (6 December 1996), p. 1. *See also:* W. Hoge, "Rare Look Uncovers Wartime Anguish of Many Part-Jewish Germans", *The New York Times*, (6 April 1997), p. 16. *See also:* B. M. Rigg, *Hitler's Jewish Soldiers: The Untold Story of Nazi Racial Laws and Men of Jewish Descent in the German Military*, University Press of Kansas, Lawrence, Kansas, (2002); **and** *Rescued from the Reich: How One of Hitler's Soldiers Saved the Lubavitcher Rebbe*, Yale University Press, New Haven, (2004).

[9]. A. Eichmann, Israel: Bet ha-mishpat ha-mehozi (Jerusalem), Israel: Misrad ha-mishpatim, Yisra'el: Misrad ham-Mispatîm, *The Trial of Adolf Eichmann: Record of Proceedings in the District Court of Jerusalem*, Volume 4, Israel State Archives, Trust for the Publication of the Proceedings of the Eichmann Trial, in co-operation with the Israel State Archives and Yad Vashem, the Holocaust Martyrs' and Heroes' Remembrance Authority, Jerusalem, (1993).

destroying Germany. They did this in order to subvert German interests and fulfill Judaic Messianic prophecy and the evil designs of the Jewish bankers to ruin Germany. In so doing, the Jewish bankers put the foxes in charge of the hen house and the Jewish bankers used their crypto-Jewish Zionist Nazis to ruin Europe and chase the Jews to Palestine.

Streicher was fond of the old Zionist maxim,

> "Without a solution of the Jewish question
> there will be no salvation of humanity!"
>
> "Ohne Lösung der Judenfrage
> keine Erlösung der Menschheit!"[10]

The following article appeared in *The Jewish Chronicle* on 22 September 1922 on page 31, which states that there would be no salvation of humanity without a solution of the Jewish question,

"5682.

THE YEAR'S RETROSPECT.

THE year just closing will be for ever memorable in Jewish annals as the year which saw the confirmation of the Mandate, with its formal and solemn establishment of the Jewish claim to Palestine as the National Home of the race. That one great central, irrevocable fact, however it be construed or whittled down by individual statesmen, stamps 5682 as *annus mirabilis* in Jewish history. It calls a halt to two thousand years of aimless drifting, and sets a definite direction in which the Jew may march with confidence. It comes at a moment of immense opportuneness to lift, if ever so little, an almost intolerable burden of suffering, confusion, and despair. It represents a movement which, whatever deductions may legitimately be made from its value upon this or that ground, is, at all events in essence, constructive. It embodies the recognition by the nation that it has a second problem of 'reparations' to solve—reparation to the Jew for two thousand years of martyrdom; and that the solution of the Jewish question is indispensable to world peace. Whether the Jewish Palestine, as the politicians are at the moment fashioning it, be a great bright light, illuminating the darkness of the Diaspora, or a will-o'-the-wisp full with fatality for the hopes of our people, the world-approved Mandate we cannot away with. Hold destiny what it may, the future of the Jewish People after the Mandate's confirmation can never be like the past. It is that which makes the year now ending a year of years in our people's chequered career, and its story a tale to linger over in the depressing procession of tragedies called Jewish

---

10. E. Heimer, *Der Giftpilz: Ein Stürmerbuch für Jung u. Alt*, Der Stürmer, Nürnberg, (1938), pp. 62, 63.

history."

It is interesting to note that the prosecutor at Eichmann's trial stated that Eichmann's accusation that Chaim Weizmann had declared a Jewish war against Germany was a "lie", when in fact it was true and was reported in *The London Times*, on 6 September 1939 on page 8 under the title, "Jews Fight for Democracies".[11] This was but one of the countless Jewish declarations of war against Germany, including repeated provocations from Weizmann, as proven in Hartmut Stern's book, *Jüdische Kriegserklärungen an Deutschland: Wortlaut, Vorgeschichte, Folgen*, FZ-Verlag, München, Second Edition, (2000), ISBN: 3924309507; *see also:* Stern's response to Goldhagen,[12] *KZ-Lügen: Antwort auf Goldhagen*, FZ-Verlag, München, Second Edition, (1998), ISBN: 3924309361; *see also:* Rabbi Moshe Shonfeld, *The Holocaust Victims Accuse: Documents and Testimony on Jewish War Criminals*, Neturei Karta of U.S.A., Brooklyn, (1977). Many believe that these Jewish declarations of war against Germany were deliberate provocations meant to worsen the situation of Jews in Germany so as to force them towards embracing Zionism and into emigrating to Palestine, against their own wishes.

The Editors of *The World's Work*, presumably French Strother and Burton J. Hendrick, revealed that the Zionists had established a governing body at least as early as 1921, and that Chaim Weizmann was at the head of this government, which means that he had the power to declare war against Germany as the leading official of the Zionist organization, and bear in mind that leading Zionists openly declared that Jews were a foreign and hostile nation within Germany,

> "*The situation which provoked the controversy at Cleveland arose from the arrival in this country of Dr. Chaim Weizmann from London to share in its deliberations. Dr. Weizmann is the head of the world organization of Zionists. This world organization has a highly centralized form of government. This consists of an international committee, including representatives from all countries that have a local organization. But the real control is vested in what is known as the 'Inner Actions Council.' This is a compact body of only seven men; and it is dominated by the Jews of Europe.*"[13]

In 1921, Jewish anti-Zionist Henry Morgenthau saw the writing on the wall and sought to distinguish between Jews in general and the nationalistic Zionists, who would provoke a war that would alienate Jews,

---

11. Chaim Weizmann's letter of 29 August 1939 to Prime Minister Chamberlain, that the Jews had declared war on Germany, "Jews Fight for Democracies" *The London Times*, (6 September 1939), p. 8. ***See also:*** *The Jewish Chronicle*, (8 September 1939).
12. D. J. Goldhagen, *Hitler's Willing Executioners: Ordinary Germans and the Holocaust*, Knopf, New York, (1996).
13. H. Morgenthau, "Zionism a Surrender, Not a Solution", *The World's Work*, Volume 42, Number 3, (July, 1921), pp. i-viii, at i.

"I for one, will not forego this vision of the destiny of the Jews. I do not presume to say to my co-religionists of Europe that the shall accept my programme. But neither do I intend to allow them to impose their programme upon me. They may continue, if they will, a practice of our common faith which invites martyrdom, and which makes the continuance of oppression a certainty. I have found a better way (and when I say *I*, it is to speak collectively as one of a great body of American Jews of like mind). I resent the activities of Dr. Weizmann and his followers in this country. In the foregoing pages I have given my reasons for opposing Zionism. They make plain why I asserted in its first paragraphs that Zionism is not a solution; that it is a surrender. It looks backward, and not forward. It would practically place in the hands of seven men, steeped in a foreign tradition, the power to turn back the hands of time upon all which I and my predecessors of the same convictions have won for ourselves here in America. We have fought our way through to liberty, equality, and fraternity. We have found rest for our souls. No one shall rob us of these gains. We enjoy in America exactly the spiritual liberty, the financial success, and the social position which we have earned. Any Jew in America who wishes to be a saint of Zion has only to practice the cultivation of his spiritual gifts—there is none to hinder him. Any Jew in America who seeks material reward has only to cultivate the powers of his mind and character—there are no barriers between him and achievement. Any Jew in America who yearns for social position has only to cultivate his manners—there are no insurmountable discriminations here against true gentlemen. The Jews of France have found France to be their Zion. The Jews of England have found England to be their Zion. We Jews of America have found America to be our Zion. Therefore, I refuse to allow myself to be called a Zionist. I am an American."[14]

Adolf Eichmann stated that he had sought a deal with the Western Allies to exchange one million Jews for 10,000 trucks to be used on the Eastern front. Jewish Communist turned Zionist, Joel Brand had established a relationship with the Nazis and tried to arrange the deal with the Western Allies.[15] The offer was declined. This story was first publicly exposed in 1956.[16] Eichmann told another

---

14. H. Morgenthau, "Zionism a Surrender, Not a Solution", *The World's Work*, Volume 42, Number 3, (July, 1921), pp. i-viii, at viii.
15. J. Brand, *Desperate Mission: Joel Brand's Story*, Criterion Books, New York, (1958); **and** *Advocate for the Dead: The Story of Joel Brand*, A. Deutsch, London, (1958, ©1956); **and** *ha-Satan veha-nefesh*, Ladori, Tel Aviv, (1960); **and** *Adolf Eichmann: Fakten gegen Fablen*, Ner-Tamid-Verlag, München, (1961).
16. A. Weissberg-Cybulski, *Bi-shelihut nidonim la-mavet*, Tel-Aviv, `Ayanot, (1956); **German** *Die Geschichte von Joel Brand*, Kiepenheuer & Witsch, Köln, (1956); **English** *Advocate for the Dead: The Story of Joel Brand*, A. Deutsch, London, (1958); **Danish** *Joel Brands Historie*, Gyldendal, København, (1957); **Italian** *La Storia di Joel Brand*,

story of his dealings with the Zionist Dr. Rudolf Kastner, which ultimately resulted in the deaths of countless assimilated Hungarian Jews, and the survival of the fittest Zionists for Israel, who were Kastner's friends. Eichmann stated, *inter alia*,

> "As a matter of fact, there was a very strong similarity between our attitudes in the SS and the viewpoint of these immensely idealistic Zionist leaders who were fighting what might by their last battle. As I told Kastner: 'We, too, are idealists and we, too, had to sacrifice our own blood before we came to power.' I believe that Kastner would have sacrificed a thousand or a hundred thousand of his blood to achieve his political goal. He was not interested in old Jews or those who had become assimilated into Hungarian society. But he was incredibly persistent in trying to save biologically valuable Jewish blood—that is, human material that was capable of reproduction and hard work. 'You can have the others,' he would say, 'but let me have this group here.' And because Kastner rendered us a great service by helping keep the deportation camps peaceful, I would let his groups escape. After all, I was not concerned with small groups of a thousand or so Jews."[17]

In an article by Chris Johnston and Nassim Khadem, "War Crime Suspect Admits to His Leading Fascist Role", *The Age*, (15 February 2006); Lajos Polgar, a leader of the "Arrow Cross Party" in Hungary, was quoted as stating, among other things (for a similar view of Jews, and in particular Hungarian Jews, *see:* Douglas Reed, "How Odd of God", *Disgrace Abounding*, Chapter 23, Jonathan Cape, London, (1939), pp. 228-262. *See also:* Rebecca Dana and Peter Carlson's article on the diary of Harry "S" Truman in *The Washington Post*, on 11 July 2003 on page A1),

> "'The Jews were not wanted in Hungary. They were taking over. When they come into power and money they are terrible; they don't know anything. The thing is, you can't help but want to get rid of them.
>
> 'The party wanted to be free from the Jews, and there was only one way that was possible and that was by getting rid of them, by sending them out, but the biggest problem actually was that the Jews have no real home to send them to.'
>
> Mr Polgar said Arrow Cross was not anti-Semitic but Zionist, or pro-Jewish. 'The idea was to put them into ghettoes... where they would be protected. Then after the war they would be sent back to settle peacefully in Palestine. So in a noble sense, I am a Zionist. Zionism wants a home for the Jews.'"[18]

Feltrinelli, Milano, (1958).
17. A. Eichmann, "Eichmann Tells His Own Damning Story", *Life Magazine*, Volume 49, Number 22, (28 November 1960), pp. 19-25, 101-112; and "Eichmann's Own Story: Part II", *Life Magazine*, (5 December 1960), pp. 146-161; at 146.
18.<http://www.theage.com.au/news/national/war-crime-suspect-admits-to-fascist-

Jewish Zionists again sought to terrorize Hungarian Jews in 1956, in an attempt to scare them into fleeing to Israel.[19]

The Zionist "Rescue Committee" published a memorandum "for Zionist eyes only" written by Apolinari Hartglass entitled *Comments on Aid and Rescue*, in 1943, in which it predicted that 7 million European Jews would be murdered. The Zionists' concern was not to save Jews, especially not the German assimilationist Jews they so hated, but to let them die off and to ensure that the remnant of the Jews had no option but to live in Palestine, even if against their wishes. The Zionists wanted to cut off all other nations to Jewish emigration even though they asserted that it meant certain death for many millions of Jews. They then planned to exploit the tragedy to promote Zionism and to hand pick those Jews who would survive in Europe and to condemn those Jews whom they resented to death. This memorandum is contained in the Central Zionist Archives in Jerusalem, S/26 1232, and parts of it are quoted in Tom Segev's book, *The Seventh Million: The Israelis and the Holocaust*, Hill and Wang, New York, (1993), pp. 99-101. I quote from the *Comments*, as translated to English on page 100 of *The Seventh Million*:

> "Whom to save: ... Should we help everyone in need, without regard to the quality of the people? Should we not give this activity a Zionist-national character and try foremost to save those who can be of use to the Land of Israel and to Jewry? I understand that it seems cruel to put the question in this form, but unfortunately we must state that if we are able to save only 10,000 people from among 50,000 who can contribute to building the country and to the national revival of the people, as against saving a million Jews who will be a burden, or at best an apathetic element, we must restrain ourselves and save the 10,000 that can be saved from among the 50,000—despite the accusations and pleas of the million. I take comfort from the fact that it will be impossible to apply this harsh principle 100 percent and that the million will get something also. But let us see that it does not get too much."

The "something" that the million assimilating Jews the Zionists hated "got" was humiliation and death. There is a troubling contradiction in the statements of the Rescue Committee; in that they claimed to have little influence, and yet spoke as if they had absolute control over the fate of European Jews. Did the Zionists control Nazi policy? Were they at the head of it?

The Zionists, steeped in the same racist and eugenic ideology which permeated Nazism, only wanted the very best genetic stock of the European Jews to emigrate to Palestine, and only a very limited number of those. The Zionists

---

role/2006/02/15/1139890808344.html>
19. R. H. Williams, *The Ultimate World Order—As Pictured in "The Jewish Utopia"*, CPA Book Publisher, Boring, Oregon, (1957?), pp. 58-60.

calculated that Palestine simply could not house a large number of Jews.[20] Among very early references to "soap which they make of human fats"—which is today known to have been a myth, and "gas chambers and blood-poisoning stations", Sholem Asch wrote in his article, "In the Valley of Death", *The New York Times*, (7 February 1943), pages 16 and 36, at 16,

> "The population of the Warsaw ghetto, into which 500,000 Jews were driven, was reduced last September to 120,000, and in October to only 40,000, as proved by the number of food cards issued. Those remaining are the strongest; they have not been killed yet because they are being used as slave labor."

Judaism is replete with stories and prophecies that filter the Jewish people primarily based on three criteria: "racial" purity, craftiness and deceitfulness, and obedience to God. The stories of Cain and Abel, Isaac and Ishmael, Jacob and Esau, etc. teach the Jews that God has elected some to be his people and some to be weeded off. Numerous prophecies tell that only the "remnant", the "elect" of God among the Jews, will survive. The rest of the Jews will be killed off (*Isaiah* 1:9; 6:9-13; 10:20-22; 11:11-12; 17:6; 37:31-33; 41:9; 42; 43; 44; 65; 66. *Ezekiel* 20:38; 25:14. *Daniel* 12:1, 10. *Amos* 9:8-10. *Joel* 2:32. *Obadiah* 1:18. *Micah* 5:8. *Romans* 9:27-28; 11:1-5. *See also: Enoch*)

Since Palestine could not house all of the Jews of Europe; and since the Jews of Europe would not go to Palestine until terrified in Europe; and, further, since the Jews of Europe had suffered from inbreeding as a result of Jewish self-segregation; and, still further, since the Jews had an ancient history of martyrdom and the ritual sacrifice of their own children; the Zionists instituted the Holocaust as a means to artificially select Jews for emigration to Palestine, and those slated for immediate death, and they believed that natural selection would improve the "tribe" through death by disease, starvation, exposure and overwork. The Zionists felt they had the Lord on their side. As but one example among many, *Isaiah* 6:9-13 states:

> "9 And he said, Go, and tell this people, Hear ye indeed, but understand not; and see ye indeed, but perceive not. 10 Make the heart of this people fat, and make their ears heavy, and shut their eyes; lest they see with their eyes, and hear with their ears, and understand *with* their heart, and convert, and be healed. 11 Then said I, Lord, how long? And he answered, Until the cities be wasted without inhabitant, and the houses without man, and the land be utterly desolate, 12 And the LORD have removed men far away, and *there be* a great forsaking in the midst of the land. 13 But yet in it *shall be* a tenth, and *it* shall return, and shall be eaten: as a teil tree, and as an oak, whose substance *is* in them, when they cast *their leaves: so* the holy seed *shall be*

---

20. A. T. Clay, "Political Zionism", *The Atlantic Monthly*, Volume 127, Number 2, (February, 1921), pp. 268-279, at 269. H. Morgenthau, "Zionism a Surrender, Not a Solution", *The World's Work*, Volume 42, Number 3, (July, 1921), pp. i-viii, at iv.

the substance thereof."

Benjamin Disraeli, who was to become Prime Minister of England, wrote of the allegedly destructive effects of racial integration in his *Coningsby; or, The New Generation*, H. Colburn, London, (1844), pp. 249-254 (other Jews, like Nicolai and Boas, strongly contested these unproven racial theories),

> "The party broke up. Coningsby, who had heard Lord Eskdale announce Sidonia's departure, lingered to express his regret, and say farewell.
> 'I cannot sleep,' said Sidonia, 'and I never smoke in Europe. If you are not stiff with your wounds, come to my rooms.'
> This invitation was willingly accepted.
> 'I am going to Cambridge in a week,' said Coningsby. 'I was almost in hopes you might have remained as long.'
> 'I also; but my letters of this morning demand me. If it had not been for our chase, I should have quitted immediately. The minister cannot pay the interest on the national debt; not an unprecedented circumstance, and has applied to us. I never permit any business of State to be transacted without my personal interposition; and so I must go up to town immediately.'
> 'Suppose you don't pay it,' said Coningsby, smiling.
> 'If I followed my own impulse, I would remain here,' said Sidonia. 'Can anything be more absurd than that a nation should apply to an individual to maintain its credit, and, with its credit, its existence as an empire, and its comfort as a people; and that individual one to whom its laws deny the proudest rights of citizenship, the privilege of sitting in its senate and of holding land? for though I have been rash enough to buy several estates, my own opinion is, that, by the existing law of England, an Englishman of Hebrew faith cannot possess the soil.'
> 'But surely it would be easy to repeal a law so illiberal—'
> 'Oh! as for illiberality, I have no objection to it if it be an element of power. Eschew political sentimentalism. What I contend is, that if you permit men to accumulate property, and they use that permission to a great extent, power is inseparable from that property, and it is in the last degree impolitic to make it the interest of any powerful class to oppose the institutions under which they live. The Jews, for example, independently of the capital qualities for citizenship which they possess in their industry, temperance, and energy and vivacity of mind, are a race essentially monarchical, deeply religious, and shrinking themselves from converts as from a calamity, are ever anxious to see the religious systems of the countries in which they live flourish; yet, since your society has become agitated in England, and powerful combinations menace your institutions, you find the once loyal Hebrew invariably arrayed in the same ranks as the leveller and the latitudinarian, and prepared to support the policy which may even endanger his life and property, rather than tamely continue under a system which seeks to degrade him. The Tories lose an important election

at a critical moment; 'tis the Jews come forward to vote against them. The Church is alarmed at the scheme of a latitudinarian university, and learns with relief that funds are not forthcoming for its establishment; a Jew immediately advances and endows it. Yet the Jews, Coningsby, are essentially Tories. Toryism, indeed, is but copied from the mighty prototype which has fashioned Europe. And every generation they must become more powerful and more dangerous to the society which is hostile to them. Do you think that the quiet humdrum persecution of a decorous representative of an English university can crush those who have successively baffled the Pharaohs, Nebuchadnezzar, Rome, and the Feudal ages? The fact is, you cannot destroy a pure race of the Caucasian organization. It is a physiological fact; a simple law of nature, which has baffled Egyptian and Assyrian Kings, Roman Emperors, and Christian Inquisitors. No penal laws, no physical tortures, can effect that a superior race should be absorbed in an inferior, or be destroyed by it. The mixed persecuting races disappear; the pure persecuted race remains. And at this moment, in spite of centuries, or tens of centuries, of degradation, the Jewish mind exercises a vast influence on the affairs of Europe. I speak not of their laws, which you still obey; of their literature, with which your minds are saturated; but of the living Hebrew intellect.

'You never observe a great intellectual movement in Europe in which the Jews do not greatly participate. The first Jesuits were Jews; that mysterious Russian Diplomacy which so alarms Western Europe is organized and principally carried on by Jews; that mighty revolution which is at this moment preparing in Germany, and which will be, in fact, a second and greater Reformation, and of which so little is as yet known in England, is entirely developing under the auspices of Jews, who almost monopolize the professorial chairs of Germany. Neander the founder of Spiritual Christianity, and who is Regius Professor of Divinity in the University of Berlin, is a Jew. Benary, equally famous, and in the same University, is a Jew. Wehl, the Arabic Professor of Heidelberg, is a Jew. Years ago, when I was in Palestine, I met a German student who was accumulating materials for the History of Christianity, and studying the genius of the place; a modest and learned man. It was Wehl; then unknown, since become the first Arabic scholar of the day, and the author of the life of Mahomet. But for the German professors of this race, their name is Legion. I think there are more than ten at Berlin alone.

'I told you just now that I was going up to town tomorrow, because I always made it a rule to interpose when affairs of State were on the carpet. Otherwise, I never interfere. I hear of peace and war in the newspapers, but I am never alarmed, except when I am informed that the Sovereigns want treasure; then I know that monarchs are serious.

'A few years back we were applied to by Russia. Now, there has been no friendship between the Court of St. Petersburg and my family. It has Dutch connections, which have generally supplied it; and our representations in favour of the Polish Hebrews, a numerous race, but the most suffering and degraded of all the tribes, have not been very agreeable to the Czar. However, circumstances drew to an approximation between the Romanoffs and the Sidonias. I resolved to go myself to St. Petersburg. I had, on my arrival, an interview with the Russian Minister of Finance, Count Cancrin; I beheld the son of a Lithuanian Jew. The loan was connected with the affairs of Spain; I resolved on repairing to Spain from Russia. I travelled without intermission. I had an audience immediately on my arrival with the Spanish Minister Senor Mendizabel; I beheld one like myself, the son of Nuevo Christiano, a Jew of Arragon. In consequence of what transpired at Madrid, I went straight to Paris to consult the President of the French Council; I beheld the son of a French Jew, a hero, an imperial marshal, and very properly so, for who should be military heroes if not those who worship the Lord of Hosts?'

'And is Soult a Hebrew?'

'Yes, and others of the French marshals, and the most famous; Massena, for example; his real name was Manasseh: but to my anecdote. The consequence of our consultations was, that some Northern power should be applied to in a friendly and mediative capacity. We fixed on Prussia; and the President of the Council made an application to the Prussian Minister, who attended a few days after our conference. Count Arnim entered the cabinet, and I beheld a Prussian Jew. So you see, my dear Coningsby, that the world is governed by very different personages from what is imagined by those who are not behind the scenes.'

'You startle, and deeply interest me.'

'You must study physiology, my dear child. Pure races of Caucasus may be persecuted, but they cannot be despised, except by the brutal ignorance of some mongrel breed, that brandishes fagots and howls extermination, but is itself exterminated without persecution, by that irresistible law of Nature which is fatal to curs.'

'But I come also from Caucasus,' said Coningsby.

'Verily; and thank your Creator for such a destiny: and your race is sufficiently pure. You come from the shores of the Northern Sea, land of the blue eye, and the golden hair, and the frank brow: 'tis a famous breed, with whom we Arabs have contended long; from whom we have suffered much: but these Goths, and Saxons, and Normans were doubtless great men.'

'But so favoured by Nature, why has not your race produced great poets, great orators, great writers?'

'Favoured by Nature and by Nature's God, we produced the lyre of David; we gave you Isaiah and Ezekiel; they are our Olynthians, our Philippics. Favoured by Nature we still remain: but in exact proportion as we have been favoured by Nature we have been persecuted by Man. After a thousand struggles; after acts of heroic courage that Rome has never equalled; deeds of divine patriotism that Athens, and Sparta, and Carthage have never excelled; we have endured fifteen hundred years of supernatural slavery, during which, every device that can degrade or destroy man has been the destiny that we have sustained and baffled. The Hebrew child has entered adolescence only to learn that he was the Pariah of that ungrateful Europe that owes to him the best part of its laws, a fine portion of its literature, all its religion. Great poets require a public; we have been content with the immortal melodies that we sung more than two thousand years ago by the waters of Babylon and wept. They record our triumphs; they solace our affliction. Great orators are the creatures of popular assemblies; we were permitted only by stealth to meet even in our temples. And as for great writers, the catalogue is not blank. What are all the schoolmen, Aquinas himself, to Maimonides? And as for modern philosophy, all springs from Spinoza.

'But the passionate and creative genius, that is the nearest link to Divinity, and which no human tyranny can destroy, though it can divert it; that should have stirred the hearts of nations by its inspired sympathy, or governed senates by its burning eloquence; has found a medium for its expression, to which, in spite of your prejudices and your evil passions, you have been obliged to bow. The ear, the voice, the fancy teeming with combinations, the imagination fervent with picture and emotion, that came from Caucasus, and which we have preserved unpolluted, have endowed us with almost the exclusive privilege of MUSIC; that science of harmonious sounds, which the ancients recognised as most divine, and deified in the person of their most beautiful creation. I speak not of the past; though, were I to enter into the history of the lords of melody, you would find it the annals of Hebrew genius. But at this moment even, musical Europe is ours. There is not a company of singers, not an orchestra in a single capital, that is not crowded with our children under the feigned names which they adopt to conciliate the dark aversion which your posterity will some day disclaim with shame and disgust. Almost every great composer, skilled musician, almost every voice that ravishes you with its transporting strains, springs from our tribes. The catalogue is too vast to enumerate; too illustrious to dwell for a moment on secondary names, however eminent. Enough for us that the three great creative minds to whose exquisite inventions all nations at this moment yield, Rossini, Meyerbeer, Mendelssohn, are of Hebrew race; and little do your men of fashion, your muscadins of Paris, and your dandies of London, as they thrill into raptures at the notes of a Pasta or a Grisi, little do they suspect that they are offering their homage to 'the sweet singers of Israel!'"[21]

Disraeli wrote in 1852 in his *Lord George Bentinck: A Political Biography*, Chapter 24, Third Revised Edition, Colburn, (1852), pp. 482-507, at 491-497,

"But having made this full admission of the partial degradation of the Jewish race, we are not prepared to agree that this limited degeneracy is any

---

21. B. Disraeli, *Coningsby; or, The New Generation*, The Century Co., New York, (1904), pp. 230-235.

justification of the prejudices and persecution which originated in barbarous or mediæval superstitions. On the contrary, viewing the influence of the Jewish race upon the modern communities, without any reference to the past history or the future promises of Israel, dismissing from our minds and memories, if indeed that be possible, all that the Hebrews have done in the olden time for man, and all which it may be their destiny yet to fulfil, we hold that instead of being an object of aversion, they should receive all that honour and favour from the northern and western races which, in civilised and refined nations, should be the lot of those who charm the public taste and elevate the public feeling. We hesitate not to say that there is no race at this present, and following in this only the example of a long period, that so much delights, and fascinates, and elevates, and ennobles Europe, as the Jewish.

We dwell not on the fact, that the most admirable artists of the drama have been and still are of the Hebrew race: or, that the most entrancing singers, graceful dancers, and exquisite musicians, are sons and daughters of Israel: though this were much. But these brilliant accessories are forgotten in the sublimer claim.

It seems that the only means by which in these modern times we are permitted to develop the beautiful is music. It would appear definitively settled that excellence in the plastic arts is the privilege of the earlier ages of the world. All that is now produced in this respect is mimetic, and, at the best, the skilful adaptation of traditional methods. The creative faculty of modern man seems by an irresistible law at work on the virgin soil of science, daily increasing by its inventions our command over nature, and multiplying the material happiness of man. But the happiness of man is not merely material. Were it not for music, we might in these days say, the beautiful is dead. Music seems to be the only means of creating the beautiful in which we not only equal but in all probability greatly excel the ancients. The music of modern Europe ranks with the transcendent creations of human genius; the poetry, the statues, the temples of Greece. It produces and represents as they did whatever is most beautiful in the spirit of man, and often expresses what is most profound. And who are the great composers, who hereafter will rank with Homer, with Sophocles, with Praxiteles, or with Phidias? They are the descendants of those Arabian tribes who conquered Canaan, and who by the favour of the Most High have done more with less means even than the Athenians.

Forty years ago—not a longer period than the children of Israel were wandering in the desert—the two most dishonoured races in Europe were the Attic and the Hebrew, and they were the two races that had done most for mankind. Their fortunes had some similarity: their countries were the two smallest in the world, equally barren and equally famous; they both divided themselves into tribes; they both built a famous temple on an acropolis; and both produced a literature which all European nations have accepted with reverence and admiration. Athens has been sacked oftener than Jerusalem, and oftener rased to the ground; but the Athenians have escaped expatriation, which is purely an oriental custom. The sufferings of the Jews however have been infinitely more prolonged and varied than those of the Athenians. The Greek nevertheless appears exhausted. The creative genius of Israel on the contrary never shone so bright; and when the Russian, the Frenchman, and the Anglo-Saxon, amid applauding theatres or the choral voices of solemn temples yield themselves to the full spell of a Mozart or a Mendelssohn, it seems difficult to comprehend how these races can reconcile it to their hearts to persecute a Jew.

We have shown that the theological prejudice against the Jews has no foundation, historical or doctrinal; we have shown that the social prejudice, originating in the theological but sustained by superficial observations irrespective of religious prejudice, is still more unjust, and that no existing race is so much entitled to the esteem and gratitude of society as the Hebrew. It remains for us to notice the injurious consequences to European society of the course pursued by the communities of this race, and this view of the subject leads us to considerations which it would become existing statesmen to ponder.

The world has by this time discovered that it is impossible to destroy the Jews. The attempt to extirpate them has been made under the most favourable auspices and on the largest scale; the most considerable means that man could command have been pertinaciously applied to this object for the longest period of recorded time. Egyptian pharaohs, Assyrian kings, Roman emperors, Scandinavian crusaders, Gothic princes, and holy inquisitors, have alike devoted their energies to the fulfilment of this common purpose. Expatriation, exile, captivity, confiscation, torture on the most ingenious and massacre on the most extensive scale, a curious system of degrading customs and debasing laws which would have broken the heart of any other people, have been tried, and in vain. The Jews, after all this havoc, are probably more numerous at this date than they were during the reign of Solomon the wise, are found in all lands, and unfortunately prospering in most. All which proves, that it is in vain for man to attempt to baffle the inexorable law of nature which has decreed that a superior race shall never be destroyed or absorbed by an inferior.

But the influence of a great race will be felt; its greatness does not depend upon its numbers, otherwise the English would not have vanquished the Chinese, nor would the Aztecs have been overthrown by Cortez and a handful of Goths. That greatness results from its organisation, the consequences of which are shown in its energy and enterprise, in the strength of its will and the fertility of its brain. Let us observe what should be the influence of the Jews, and then ascertain how it is exercised. The Jewish race connects the modern populations with the early ages of the world, when the relations of the Creator with the created were more intimate than in these days, when angels visited the earth, and God himself even spoke with man. The Jews represent the Semitic principle; all that is spiritual in our nature. They are the trustees of tradition, and the conservators of the religious element. They are a living and the most striking evidence of the falsity of that pernicious doctrine of modern times, the natural equality of man. The particular equality of a particular race is a matter of municipal arrangement, and depends entirely on political considerations and circumstances; but the natural equality of man now in vogue, and taking the form of cosmopolitan fraternity, is a principle which, were it possible to act on it, would deteriorate the great races and destroy all the genius of the world. What would be the consequences on the great Anglo-Saxon republic, for example, were its citizens to secede from their sound principle of reserve, and mingle with their negro and coloured populations? In the course of time they would become so deteriorated that their states would probably be reconquered and regained by the aborigines whom they have expelled, and who would then be their superiors. But though nature will never ultimately permit this theory of natural equality to be practised, the preaching of this dogma has already caused much mischief, and may occasion much more. The native tendency of the Jewish race, who are justly proud of their blood, is against the doctrine of the equality of man. They have also another characteristic, the faculty of acquisition. Although the European laws have endeavoured to prevent their obtaining property, they have nevertheless become remarkable for their accumulated wealth. Thus it will be seen that all the tendencies of the Jewish race are conservative. Their bias is to religion, property, and natural aristocracy; and it should be the interest of statesmen that this bias of a great race should be encouraged, and their energies and creative powers enlisted in the cause of existing society."

Disraeli wrote in his *Endymion*, D. Appleton and Company, New York, (1880), pp. 251-252,

"Baron Sergius and Endymion were sitting together rather apart from the rest. The baron said, 'You have heard to-day a great deal about the Latin race, their wondrous qualities, their peculiar destiny, their possible danger. It is a new idea, or rather a new phrase, that I observe is now getting into the political world, and is probably destined to produce consequences. No man will treat with indifference the principle of race. It is the key of history, and why history is often so confused is that it has been written by men who were ignorant of this principle and all the knowledge it involves. As one who may become a statesman and assist in governing mankind, it is necessary that you should not be insensible to it; whether you encounter its influence in communities or in individuals, its qualities must ever be taken into account. But there is no subject which more requires discriminating knowledge, or where your illustrating principle, if you are not deeply founded, may not chance to turn out a will-o'-the-wisp. Now this great question of the Latin race, by which M. de Vallombrosa may succeed in disturbing the world—it might be well to inquire where the Latin race is to be found. In the North of Italy, peopled by Germans and named after Germans, or in the South of Italy, swarming with the descendants of Normans and Arabs? Shall we find the Latin race in Spain, stocked by Goths, and Moors, and Jews? Or in

France, where there is a great Celtic nation, occasionally mingled with Franks? Now I do not want to go into the origin of man and nations—I am essentially practical, and only endeavor to comprehend that with which I have personally to deal, and that is sufficiently difficult. In Europe I find three great races with distinct qualities—the Teutons, the Sclaves, and the Celts; and their conduct will be influenced by those distinctive qualities. There is another great race which influences the world, the Semites. Certainly, when I was at the Congress of Vienna, I did not believe that the Arabs were more likely to become a conquering race again than the Tartars, and yet it is a question at this moment whether Mehemet Ali, at their head, may not found a new empire in the Mediterranean. The Semites are unquestionably a great race, for among the few things in this world which appear to be certain, nothing is more sure than that they invented our alphabet. But the Semites now exercise a vast influence over affairs by their smallest though most peculiar family, the Jews. There is no race gifted with so much tenacity, and such skill in organization. These qualities have given them an unprecedented hold over property and illimitable credit. As you advance in life, and get experience in affairs, the Jews will cross you everywhere. They have long been stealing into our secret diplomacy, which they have almost appropriated; in another quarter of a century they will claim their share of open government. Well, these are races; men and bodies of men influenced in their conduct by their particular organization, and which must enter into all the calculations of a statesman. But what do they mean by the Latin race? Language and religion do not make a race—there is only one thing which makes a race, and that is blood.'"

5.11 Zionist Lies

The First World War emancipated the Jews of Russia. Turkey and Germany were greatly weakened and were further crippled by unjust debts placed on them through treacherous treaties. Jews in Eastern Europe were segregated and seemed ready for emigration to Palestine—though most did not wish to go. In 1916, France and Britian divided up the Mid-East amongst themselves in the Sykes-Picot Agreement. At the San Remo conference in 1920, the British granted themselves the right to rule of Palestine and the French granted themselves the right to rule Syria. The Jews pushed for the ratification of the Palestine Mandate in the League of Nations so that they could enforce their bogus interpretation of the Balfour Declaration of 1917.

The political Zionists remained a fanatical minority group among Jews, and though many Eastern European Jews would have been happy to have moved to Palestine, if it had meant a good job and a stable life, very few Western Jews desired to leave their comfortable homes and head for the desert. Most Jews knew that the political Zionists were totalitarian zealots, and dangerous terrorists. An article appeared in 1921, which, while naïve and inaccurate on some points, made several important arguments against the utterly selfish, undemocratic, totalitarian political Zionist movement, which are valid to this day. It was published in: *The Atlantic Monthly*, Volume 127, Number 2, (February, 1921), pp. 268-279 (note that the racist political Zionists dominated and censored the mass media at the time when they made the racist political Zionist Albert Einstein an international celebrity and censored his critics):

"POLITICAL ZIONISM
BY ALBERT T. CLAY
I

A TRAVELER returning from the Near East is at once struck by the utter ignorance of Europeans and Americans concerning the true situation in Palestine—an ignorance due largely to the fact that in London there is, practically, only one of the important daily papers that will print anything detrimental to the schemes of the Political Zionists. Besides the English press, the other sources of information upon which America has been dependent for its news of Palestine have been the Jewish Telegraphic Agency and the Zionist propaganda. The latter, with its harrowing stories of pogroms in Europe, and its misrepresentations of the situation in the Near East, has been able to awaken not a little sympathy for the Zionist programme. But there certainly are reasons why Americans should endeavor to realize fully what is happening in Syria, and this quite promptly.

In discussing the existing conditions in Palestine, and the serious problem that the League of Nations will very probably have to face, it is necessary to differentiate briefly between what have been called the three aspects of Zionism, namely, the religious, economic, and political aspects.

Religious Zionism is an expression used to represent the belief of orthodox Judaism that the Jews are the chosen people of the one and only God; that a Messiah will be sent to redeem Israel; and that Jehovah will gather his people, restore the Temple and its service, and reëstablish the priesthood and the Jewish kingdom. For the restoration of their kingdom and the fulfillment of prophecy, they look to God in his own time and way, and not to Jewish financiers and politicians, or to peace conferences. Only a small group of orthodox Jews, 'the Eastern,' take an active part in the political movement to establish a Jewish state. Tolerance for the religious ideals of different faiths precludes any criticism or lack of respect for Religious Zionism. The Christian faith, it might be added, is, in certain respects at least, inseparably identified with some of its ideals.

Economic Zionism, so-called, has as its object the amelioration of the deplorable conditions in which Jews have lived in certain lands, where they have been outrageously persecuted, and many instances foully murdered. Since the governments concerned could no be induced to alleviate their sufferings, the Jews, in recent years, have been urged to emancipate themselves by seeking a new home, where they might live in security, and carry on their activities as free citizens. About fifty years ago organizations sprang up which encouraged colonization in Palestine. However, most Jews preferred to go to South and North America, with the result that some thousands went to Palestine and two millions moved westward. About forty colonies, some large and others containing only a few houses have been established in Palestine, numbering about 13,000 souls. The entire Jewish population, including those who are indigenous, numbers 65,300. For comparison, it may be stated that there are also about 62,500 Christians and over a half million Moslems in the land. Economic Zionism is not a theory, nor is it an experiment. The Balfour declaration sanctions the movement; it reads: 'His Majesty's government view with favor the establishment in Palestine of a national home for the Jewish people, and will use their best endeavors to facilitate the achievement of this object, it being clearly understood that nothing shall be done which may prejudice the civil and religious rights of existing non-Jewish communities in Palestine.' The San Remo Conference has interpreted the Peace Treaty as implying this, and there is no alternative; moreover, the movement is already a substantial reality.

A visit to some of the better established Jewish colonies will not fail to awaken sympathy for Economic Zionism. No unbiased observer of past events could think of throwing obstacles in the way of those Jews who, being persecuted in certain lands, prefer to live in a community solely Jewish; or who, through historical sentiment, long reside in a purely Jewish cultural community in the land of their ancestors. Only an extremist would deny the gratification of their desires to as many of these people as can be accommodated; yet it must be borne in mind that, as estimated by experts, the tiny country can support only about a million and a half additional inhabitants; which number, if all were Jews, would represent only one tenth of the fifteen millions in the world.

II

Political Zionism was launched by Herzl, in 1896, in a monograph on 'The Jewish State'; and since that time this has become the dominant note in the whole movement. He and others have claimed that the establishment of a Jewish commonwealth would become an active force, by bringing diplomatic pressure to bear upon the nations, to secure protection for Jews in all lands. A clannish sense of pride in the Jewish race, however, seems to be uppermost in their minds. They apparently think that their status in society will be enhanced everywhere if a Jewish nation exists in Palestine. This phase of Zionism is the crux of the whole Palestine problem.

Political Zionism is strongly opposed by many orthodox Jews in Palestine; especially because they recognize that, through the fanaticism of the Zionist leaders, it has become most difficult for them to maintain their former amicable relations with the other natives. It is opposed also by many of the leading Jews throughout the world, especially, as the Political Zionists themselves admit, by the upper circles of Jewish society. The Central Conference of American Rabbis, which has a membership of about three hundred, representing many of the largest and most important synagogues in America, has year after year discussed the problem; and while favoring the idea of the country's being open to Jews who, because of religious persecution, desire to reside there, it denies that the Jews are 'a people without a country'; and even refuses to 'subscribe to the phrase in the [Balfour] declaration which says, 'Palestine is to be the national home-land for the Jewish people.'

When we consider the feelings of the Jews who desire to spend their lives in study and meditation in Palestine and be buried there, we must not lose sight of the fact that the same impulse also draws, and has drawn, the Christian and the Moslem. It is the Holy Land for the three great religions. It is not the birthplace of Islam; yet Mohammed, who claimed to be the successor of a line of prophets from Abraham to Christ, would have made Jerusalem the centre of his religion if the Jews and the Christians had recognized him as a prophet. As it is, Jerusalem is one of three most revered cities in Islam; moreover, the sites identified with Abraham, Jacob, Rachel, Joseph, Moses, Samuel, David, Solomon, and other Old Testament characters, are regarded with as much veneration by the Moslem as by the Jew.

One need only recall the immense and magnificent hospices built by the Eastern and the Western branches of the Christian Church, as well as the many monasteries, hospitals, homes, and schools, throughout the land, to reach some conception of what the country is to the Christian. The inhabitants of Bethlehem and Nazareth, as well as of some other cities, are largely Christian. Moreover, practically every country in Christian Europe is represented among the inhabitants of Palestine by colonies, settlements, or communities.

The Political Zionists, through their propaganda, systematically endeavor to give the world a false conception of the Palestinians. They would have us believe, to quote the words of Zangwill, that 'Palestine is not so much occupied by the Arabs as over-run by them. They are nomads... . And therefore we must gently persuade them to 'trek.'' Examine the literature of the leaders of Zionism, and it will be found that this false position is reiterated again and again. True, nomads are found in Palestine, as everywhere throughout the Orient; but to foist upon the intelligent public the idea that the population of this land is made up of Bedouins, or even of Arabs, is a deliberate attempt to deceive it.

The inhabitants of the land should be called Syrians—or Palestinians, if Palestine is to be separated from Syria. True, there are many Arabs living there, more, for example, than Greeks, Germans, or Latins, because of the proximity of Arabia; but these are not the real Palestinians, nor do they represent the bulk of the substantial part of the nation. The people whom the Jews conquered when they entered Palestine were called by the general name of Amorites or Canaanites. While many were massacred by the Jews in certain cities, still only a portion of the country was conquered. Even after David took Jerusalem, Amorites continued to live in that city; besides, many foreign peoples, as the Hittites and Philistines, also lived in the land. There can be no question that the blood of the present Palestinian, or Syrian, includes that of the Jew as well as of the Amorite, Hittite, Phœnician, Philistine, Persian, Greek, Latin, and Arab. Such a fusion is not unlike that found in the veins of many Americans whose ancestors have lived here for several generations. When the whole population of Palestine became Mohammedan, there is little doubt that a large percentage of the Jews were also forced to accept this faith; their descendants are now classed by the Political Zionists as 'Arabs.' The Yemenites, who we know migrated from Arabia, and who in every respect resemble the Arab in physique, appearance, and bearing, they, none the less, call Jews, because of their faith. Then, also, in such Christian cities as Bethlehem and Ramallah a type is seen that is distinctively European, and doubtless largely represents remnants or descendants of the Crusaders, or of Christians who migrated to the Holy Land in the past centuries. Moreover, the Palestinian or Syrian is a composite race, largely Semitic, which has developed from the association of the different racial elements inhabiting the land for at least five thousand years past. And while the Arabs have in all periods filtered in from Arabia, and the language, as in Egypt and Mesopotamia, is Arabic, it is a deliberate misrepresentation to classify the inhabitants as 'Arabs.'

These are the people whose status the Political Zionist proposes to reduce by securing the control of the country; and who—what is still worse—must be persuaded to 'trek.' As Zangwill says, 'After all, they have all Arabia with its million square miles, and Israel has not a square inch. There is no particular reason for the Arabs to cling to these few kilometres. To fold their tents and silently steal away is their proverbial habit; let them exemplify it now.' *Palestine*, the organ of the British Palestine Committee, for July 10, 1920, says: 'For the Arab nation there are vast areas outside of Palestine in which to develop its national life, and Arabs of Palestine will be free to develop there, also'

III

Much has been written upon the historic claims of the Jews to this territory, which they held for less than five hundred years, prior to two thousand five hundred years ago. But how about the claims of the Palestinian, who possessed the land before the Jew, and who is still in possession, having lived there for over five thousand years? The Aramæans, who came from Aram, whom we call Hebrews, under Joshua conquered, and even ruthlessly exterminated, the people of a portion of Palestine; and later on, under David and Solomon, extended their rule over the whole country. But, if we are to decide the question of actual ownership of r the territory, the Palestinian who has continuously lived there surely has a clearer title than the Jew. Moreover, this decision is based upon the records handed down by the Jew himself. Even the Hebrew language, which the Jews are attempting to revive as their spoken tongue, originally belonged to the people they are trying to oust. The language in Aram—Abraham's ancestral home—was Aramæan; when the Aramieans came to Palestine, they adopted the Canaanite language, now called Hebrew.

The Palestine News, the official journal of the Egyptian Expeditionary Force under Allenby, published, on November 14, 1918, a declaration, which had been agreed to by the British and French Governments, and communicated to the President of the United States, informing the people that their aim in waging the war in the East was 'to ensure the complete and final emancipation of all those people so long oppressed by the Turks, to establish national governments and administrations which shall derive their authority from the initiative and free will of the peoples themselves,' and 'to assure, by their support and practical aid, the normal workings of such governments and administrations as the people themselves have adopted.'

In the twelfth of the fourteen points enumerated by President Wilson to Congress, January 8, 1918, he demanded that the nationalities then under Turkish rule should be assured of 'an absolutely unmolested opportunity of autonomous development.' His second principle, stated in his address at Mount Vernon, July 4, 1918, reads: 'The settlement of every question, whether of territory, of sovereignty, of economic arrangement, or of political relationship shall be upon the basis of the free acceptance of that settlement by the people immediately concerned, and not upon the basis of the material interest or advantage of any other nation or people which may desire a different settlement for the sake of its exterior influence or mastery.'

The edict of England and France, which was published in every town and village in the land about the time the Armistice was signed, has been violated in every essential particular; nor have the principles and demands of Mr. Wilson been observed. 'An unmolested opportunity of autonomous development' has been denied the inhabitants. The questions 'of territory, of economic arrangement, or political relationship' have been settled contrary to the will of 'the people immediately concerned'; and it has been done 'upon the basis of the material interest or advantage' of another people 'for the sake of its exterior interest or mastery.'

Not only have these principles and demands been ignored, but the twenty-second article of the League of Nations Covenant, in which they were incorporated, has been grossly violated. The middle section of this article reads: 'Certain communities formerly belonging to the Turkish Empire have reached a stage of development where their existence as independent nations can be provisionally recognized, subject to the rendering of administrative advice and assistance by a Mandatory Power until such time as they are able to stand alone. The wishes of these communities must be a principal consideration in the selection of the Mandatory Power.' It is needless to point out that their existence as independent nations has not been provisionally recognized, nor have the wishes of the people been a principal consideration in the selection of the Mandatory Power.

The circulation of the self-determination edict by England and France in November, 1918, which the people accepted placidly, calmed the popular feeling for a time; but after a few months the people saw clearly that the Political Zionists were favored by the British authorities, to their disadvantage; and they began to appreciate that they were being dealt with falsely. National anti-foreign sentiment grew apace, and in the spring of 1919 conditions had reached such a point that General Money had difficulty in quieting the people. He continually represented the necessity for his government to make a clear declaration of its policy—either one of repression of the people in favor of the Jews, or one of equality of treatment, which would have been acceptable to all, including the Palestinian Jews, but not, of course, to the Political Zionists. The Peace Conference, as a result of the dissatisfaction, appointed an inter-Allied commission to ascertain the wishes of the people. France, who claimed the whole of Syria, which included Palestine, declined send out her representatives; and her example was followed by England. The work of the Commission, therefore, devolved upon the two American representatives, Ambassador Crane and President King. This Commission held a most impartial and exhaustive inquiry, hearing delegates from almost every town and village. In order to be ready to give useful information before the Commission, branches of the Moslem and Christian League were formed at Jaffa, Gaza, Hebron, Djenin, Nablus, Acre, Haifa, Safed, and other places. All branches worked under a constitution approved by the Military Governor of Jerusalem. It was decided to up three resolutions to be presented the Commission:—

1. The independence of Syria, from the Taurus Mountains to Rafeh, the frontier of Egypt.

2. Palestine not to be separated from Syria, but to form one whole country.

3. Jewish immigration to be restricted.

The entire Christian and Moslem population agreed to these resolutions.

IV

It should be said here that there is no justification, from an ethnological or geographical point of view, for dividing Syria into the northern part under the French and the southern part, namely Palestine, under the British. This has already been pointed out by the greatest authority on the history and geography of Palestine, Sir George Adam Smith. One race, the Syrian, or Palestinian, is dominant throughout the territory, from Aleppo to Beersheba; and there is no natural frontier that an divide the two halves of this land. France for decades had regarded herself as the protector of the country. Although, being occupied with the enemy, she had done practically nothing toward driving out the Turks, the situation was such that, when the British army entered Jerusalem, in deference to the French a company of French soldiers was invited to be present. The question arises, then, why should the land and people be separated, and two separate administrations be established, with all the expense that this implies? For the entire territory, from Aleppo to Beersheba, is only about 400 miles and 100 miles wide—about the length of Pennsylvania, and one third its width? Why divide this small land and its people? Let us ask another question at the same time: why was the Balfour pronouncement made in 1917?

The Turkish government, when approached during the war on the problem of a Jewish state, said that it would continue to maintain, as it always had done, a favorable attitude toward the Jews in their efforts to promote flourishing settlements, within the limits of the capacity of the country, and toward the free development of their civilization and their economic enterprises; but it looked with disfavor upon Zionists who have political ambitions for Palestine, and it regards them as enemies to the government. But what the Turks refused to grant the Jews, Britain promised them, even before she had captured the country. The Political Zionists inform us that the text of the Balfour declaration was revised in the Zionist offices in America as well as in England, and that it was put into the form in which they desired it. Moreover, they intimate that this stroke of British policy had the desired effect upon the Zionists in Germany during the war. The financial assistance rendered by the Jewish plutocrats during the war, it is said, was a matter of no small consideration. But besides this, and the bid for Jewish favor everywhere, there can be little doubt that uppermost in the minds of the Cabinet, because of France's interest in the land, was the idea of creating a buffer state between the portion they would let the French retain and the Suez Canal. The Canal, according to English opinion, is the chief asset of the Empire. The strategic value of this territory to England has been referred to recently by Lord Curzon in the House of Lords. Hence, the reason that the Balfour declaration was made, and that Syria has been divided. It might be added, that this division is yet to be ratified by the League of Nations.

When the first body of representatives appeared before the Commission sent out by the Peace Conference, Aref Pasha el Dajani, the President of the Moslem and Christian League, was asked what mandatory government the League preferred. He replied that at one time they would unanimously have asked for Great Britain, but the Balfour declaration had so shocked them that they now requested that America should have the mandate for Palestine and Syria. The Commission interviewed all the communities separately, getting in each instance the reply that their requests had been made through the Moslem and Christian League, except in the case of the Zionists, who asked for a British mandate and a separate rule for Palestine. The Commission then traveled throughout the country, making an impartial and exhaustive inquiry, hearing deputations from almost every town. Everywhere they found the same unanimity for the three resolutions.

The report of the Commissioners has never been published. The Conference, apparently under the influence of the Political Zionists, took no notice of it except in so far as to announce that no political privilege would be granted to the Jews, who were in the minority in the land; but that they would be given economic privileges in connection with its development. As a result, not a few natives who had returned from America and elsewhere with their gains, for this very purpose, were naturally disappointed. Some British firms were ready to invest capital in the development of the country, particularly for the improvement of the ports of Haifa and Jaffa; but they were turned down under instructions from the Foreign Office, so that the Zionist could have the first option in such undertakings.

Relying upon the decision they had given the Americans on the

Commission, as well as upon the fact that they had made their views perfectly clear to the British authorities, the Moslems and Christians did not send a deputation to the Conference held at San Remo, which, as is well known, gave the mandate over Palestine to Great Britain. Through the efforts of the Zionist Commission, which had powerful representatives present, a clause was interpolated in the mandate, establishing a 'Jewish homeland' in accordance with the Balfour declaration.

The Grand Mufti, who is the ecclesiastical head of the Moslems in Jerusalem, on hearing the news concerning the mandate, still refused to believe that the British, who had pledged themselves to protect small powers, and who had promised that their rights should not be violated, would allow the Christians and Moslems of Palestine to be ruled by Political Zionists. The Moslems, he said, looked upon Great Britain as their best friend; they had welcomed the arrival of the British armies and in spite of all appearance to the contrary, he still believed that Great Britain would treat them fairly. The Grand Mufti was anxious that it should be understood that he and his followers were not anti-Jews, but that they objected to their country's being exploited by Jewish foreigners, and to their efforts to make both Christians and Moslems their vassals. While the Zionists during the past years had collected through propaganda immense sums from all parts of the world, he said, the Moslem and Christian natives of Palestine, by reason of the Turkish oppression and the war, were without funds. All that they asked for was a number of years in which to get on their feet economically. The Moslems, the Grand Mufti told the writer, had objections to the same quiet development of Jewish colonies going on as in the past. What they did strenuously object to was the plan of the British government to turn over their land to the Political Zionists, for the purpose of establishing a Jewish state.

The highly respected Aref Pasha, President of the Moslem and Christian League, which had been formed to stem the tide of Jewish immigration, said that the Moslems, understanding Great Britain's love for justice, decided to fight their coreligionists and to throw in their lot with her. Not less than 130,000 Moslems, many of them deserters from the Turkish army, fought with the British. The Moslems of India figured prominently in the same cause. Now, however, they find that the British victory means for them vassalage under the Jews; the people, he said, preferred the tyranny of the Turk to being ruled by the Jew.

The Christian inhabitants of the land hold the same view. Last spring no less than 20,000 people held a demonstration in Jerusalem, in order to show the administration and the foreign consuls their bitter opposition to this Jewish movement. Following this demonstration, many of the Christians proceeded to the Church of the Holy Sepulchre, and took a solemn oath that they would resist with their lives the Jews' efforts to rule them. So far as is known to me, not a single representative of any of the religious communities in Palestine favors the project. The views of the Christians are summed up in the following message, which a highly honored citizen of the country dictated to the writer as he was leaving the port of Jaffa, requesting that it should be made public. 'The Moslems and Christians welcomed the British occupation because they did not know that their country had been sold to the Jews. The honor of England is in jeopardy. The Christians of the whole world do not know of this treachery, nor did the three hundred millions of Moslems know of it. But some day it will be known, because it will surely mean another war. Had the people known what was to happen, they would have worn crape when the British entered.'

To show the consideration with which the Political Zionists are treated by the British government, the following is offered. The conflict between the British troops and the Turco-Germans left many cities and villages of Palestine in a condition not unlike that of those in Northern France and Belgium. Few people in Europe and America appreciate what the Syrian inhabitants of the land have suffered because of the conflict. The herds and farm-stock of the people had been carried away by the Turks, and they were naturally sorely pressed in their efforts to secure plough animals and grain for the cultivation of their fields. The Anglo-Palestine Bank, a Zionist concern, lent money to these people at a very exorbitant rate. The Chief Administrator, appreciating the embarrassment of the natives, and in order to ensure that the economic restoration of the country should speedily be effected, revived the Turkish system of making loans to the farmers, and made arrangements with a British bank, the Anglo-Egyptian, to lend them the money at six and a half per cent, payable over a period which could be extended to five years. In the event of failure of payment, the land would become the property of the government, not of the Zionist bank.

The Zionist Commissioner, realizing that this defeated their plans to secure titles to lands, set their forces at work in London, with the result that orders were actually sent from the Foreign Office to suspend this arrangement, which had been such a boon to the war-ridden inhabitants. It was not long afterward that General Money resigned, and Colonel Vivian Gabriel, his chief financial adviser, was relieved of his post, because it was stated that he had adopted 'an attitude inconsistent with the Zionist policy of the Government.' The injustice of the interference, however, on the part of the Zionists, became so clear to everyone that, after several months, even Dr. Weizmann, the President of the Zionists, thought it necessary to withdraw the embargo; and the British government again permitted the loans to be made.

The departure of General Money, a thoroughly sound and upright governor of the best British type, was a great loss to the people, and it was the signal for a recrudescence of the Zionist claims. The Zionist Commission claimed the right to a previous scrutiny and veto of all the acts of the administration; they asked the British government for the lands and farms of the interned German colonists; they asked for the possession of the magnificent German Hospice on the Mount of Olives (then occupied by the Administration), for their projected Jewish University. They offended the Moslems by trying to acquire lands adjoining the Mosque of Omar, for which they offered £150,000. There seemed to be no limit to their arrogance; moreover, the aggressiveness of individuals, on the street and everywhere, was most marked.

The old resident Jews of Palestine certainly have other than religious grounds for their indifference toward the efforts of the Political Zionists. Last winter the Council of Jerusalem Jews appointed a commission of representative men holding leading positions, to visit parents who were sending their children to proscribed schools, in order to secure their withdrawal. Among these schools, which included those conducted by the convents and churches, some of which have existed in Jerusalem for a long time, are the British High School for Girls, the English College for Boys, and the Jewish School for Girls. In the latter, conducted by Miss Landau, an educated English Jewess, all the teachers are Jewish; most of the teaching is in the English language. This school, which is financed by enlightened Jews of England, was denounced more severely than the others, because, not being in sympathy with the programme of the Political Zionists, Miss Landau refused to teach the Zionist curriculum. She was even informed that her school would be closed.

In a series of articles that appeared in *Doar Hayom*, the Hebrew daily paper, last December, it was stated that the parents who refused to comply with the requests of the Commission were to be boycotted, cast out from all intercourse with Jews, denied share in Zionist funds, and deprived of all custom for their shops and hotels. 'Anyone who refused, let him know that it is forbidden for him to be called by the name of Jew; and there is to be for him no portion or inheritance with his brethren.' They were given notice that they would 'be fought by all lawful means.' Their names were to be put 'upon a monument of shame, as a reproach forever, and their deeds writte unto the last generation.' 'If they are supported, their support will cease; if they are merchants, the finger of scorn will be pointed at them; if they are rabbis, they will be moved far from their office; they shall be put under the ban and persecuted, and all the people of the world shall know that there is no mercy in justice.'

A month later the results of this 'warfare' were reviewed. We were informed that some Jews had been influenced, 'but others—and the greater number, and those of the Orthodox,—those who fear God—having read the letters [signed by the head of its delegates and the Zionist Commission] became angry at the 'audacity' of the Council of Jerusalem Jews 'which mix themselves up in private affairs,' have torn the letter up, and that finished it.'

Then followed a long diatribe against these parents, boys, and girls, in which it was demanded that the blacklist of traitors to the people be sent to 'those who perform circumcision, who control the cemeteries and hospitals'; that an order go forth so that 'doctors will not visit their sick, that assistance when in need, if they are on the list of the American Relief Fund, will not be given to them.' 'Men will cry to them, 'Out of the way, unclean, unclean.' ... They are in no sense Israelites.'

It is to be regretted that only these few paraphrases and quotations from the series of articles published can be presented here.

The work of the Councils Committee met with not a little success; pupils left schools, and teachers gave up their positions. Two instructors in the English College, whose fathers were rabbis, and a third, whose brother was a teacher in a Zionist school, resigned. Another refused to do so, and declared himself ready, in the interests of the Orthodox Jews, who were suffering under this tyranny, which they deplored, to give the fullest testimony to the authorities concerning this persecution. The administration, under Governor Bols, finally intervened, and at least no further public efforts to carry out their programme were made.

If, in this early stage of the development of Political Zionism, even the Palestinian Religious Jews already find themselves under such a tyranny, what will happen if these men are allowed to have full control of the government? And what kind of treatment can the Christian and th Moslem expect in their efforts to educate their children, if the Political Zionists are allowed to develop their Jewish state to such a point that they can dispense with their mandatory and tell the British to clear out? When such things happen under British administration, what will take place if the Jewish State is ever realized, and such men are in full control?

V

The appointment of a Jew and Political Zionist, Sir Herbert Samuel, as the High Commissioner of Palestine, although he is considered to be an impartial and fair-minded man, was regarded as a serious mistake by practically every non-Jew in Palestine, because of the powerful, and even fanatical, forces that would be brought to bear upon him. The question arises, what was done on his advent in July with regard to the civil rights of the people, which were guaranteed by England's edict, by the Balfour declaration, the League of Nations, and the San Remo Conference? In his inaugural address, Samuel informed the people of Palestine that he would nominate an advisory council,—which would be composed mostly of British officials, with ten unofficial members, whom he would choose from the various sections of the people,—to meet under his presidency at frequent intervals; to this council matters of importance would be submitted for advice; and the unofficial members would be free also to raise questions to which they desired the attention of the government to be directed.

Palestine and Syria have, perhaps, more intelligent men in proportion to the inhabitants than any other country in the Near East, for which fact, of course, there are abundant reasons. Despite all that has been said with regard to the self-determination of small nations, and all that has been promised these people, by official statements and edicts, concerning their civil rights and their wishes, we learn that they are to be represented by ten unofficial members, appointed by the leader of the Political Zionists, who, when called by him, shall have the privilege of meeting, to hear reports, to give advice, and to ask questions. Certainly, this is a remarkable realization of the much heralded doctrine of self-determination of the small nation, and a remarkable fulfillment of all the promises that have been made to these unhappy people.

It is also deemed most unfortunate that the British government has placed the judicial department of the country in the hands of a Jew and Political Zionist, who even has the appointment of the judges of Palestine, about twenty of whom are Moslems. The demoralizing effect of this is fully appreciated by non-Jews. Protests against his occupying this position have been made, but without avail. The case, however, is different when the Jews endeavor to oust a Christian judge who is not favorable to their programme. Even a man of the highest type and standing, credited with a long career of faithful judicial service, has been disposed of through their influence.

Those who are familiar with life in Palestine, where the feeling between Moslem and Christian and Jew is perhaps more intense than in any other land, are fully cognizant that this scheme for a Jewish state not only accentuates and increases the animosities that have always existed, but invites another tragic chapter in the history of the Hebrews. The Political Zionists are simply intensifying this feeling, as well as the bigotry and fanaticism of the masses, by their efforts to force themselves into a sovereign position. And there can be no question that anti-Semitism, not only in Palestine but throughout the world, will increase more and more as the world, Christian and Moslem, becomes familiar with the situation.

The British politicians in London seem to have little comprehension of the difficulties they are helping to create for their Empire. The Political Zionists will never be satisfied with the country west of the Jordan, and only as far north as the Litany. All kinds of intrigues on the part of their politicians, to secure the territory that will be held by the French and Arabs, can be looked for. They have already claimed that the boundaries of the Solomonic kingdom, which extended to the Euphrates, should be those of their state. Already an outlet on the Gulf of Akaba has been demanded. Since there are 50,000 Jews in Bagdad, what is to prevent their plutocrats, when Great Britain is again hard pressed, from exacting another declaration from the government, which will embrace this territory?

In *Palestine*, for September 25 the statement is made that the boundary-line set by France would make it impossible to get water for electric power. This would rob them, they claim, of all hope of economic prosperity. There can be no other result but that Britain's difficulties with France and Arabia will be increased, and that the estrangement between these countries will be accentuated.

It is the opinion of nearly every non-Jewish British official in Palestine, not only that Britain's reputation for justice and fair dealing is at stake, and that a great wrong is being done the inhabitants of the land, but that there are serious dangers ahead for the Empire. They believe that, if immigration from Russia, Roumania, and Poland is to be allowed to any great extent, so that the Jews will be in the majority,—will have, as they say, at least fifty-one per cent,—not only racial riots and massacres will result, but there will be a continual menace to the Empire, especially because of the interest of the Moslems of other lands in Jerusalem and in their coreligionists. Moreover, these officials feel keenly the change in the attitude toward the British that has come over the inhabitants since they entered, for they know that they are now hated and despised.

The propagandists endeavor to have the world believe that, since Sir. H. Samuel's appointment, the opposition of the inhabitants is disappearing; and we are told that many have signed petitions asking for Jewish rule. To one familiar with the actual situation, this, to say the least, is ludicrous. Thousands of signatures could easily be ohtamed at the cost of three or four for a shilling. Order has been maintained the last few months in this little land with the assistance of 24,000 soldiers. But we are informed that anti-Zionist sentiment has increased since the arrival of Sir H. Samuel, to whom quite recently national associations at Jaffa, Hebron, and Gaza sent the following resolution:—

'With all due respect to His Britannic Majesty and to your person, we beg to protest against the decision taken at San Remo [that is, the granting of the mandate to Great Britain], and against your appointment.'

The Palestine problem can be easily and effectively disposed of by the British government with dignity and honor, to the satisfaction of the Christians and Moslems in Palestine and throughout the world, as well as of the many Jews who are opposed to this political movement. This can be accomplished by simply carrying out the provisions of the League of Nations and all the pronouncements that Great Britain has made. The loosely worded and ambiguous Balfour declaration does not prevent this; for if the non-Jewish inhabitants are granted their civic rights, which can mean only that they will have a voice in the government in proportion to their population, then justice will be rendered them, and the problem will be solved. Unless this is done, governing by a mandate, as many British maintain, is simply another phrase for a power's taking possession of a country, and ruling it as it desires. And unless this is done now, before the status of the Christian and the Moslem is compromised, and before the country becomes full of Russian, Roumanian, and Polish Jews, so that they will be in a majority, a grave injustice will be committed, which will be resented more and more by the Christians and Moslems of the world as they become familiar with the situation in their Holy Land."

Lord Islington, Lord Sydenham, and others, repeatedly reminded the House of Lords that the British had promised Palestine to the Palestinians, and prohibited the formation of a Jewish Government in that territory, in the Balfour Declaration of 1917 itself; as well as in the correspondence between Sir Henry McMahon, His Majesty's High Commissioner at Cairo and the Sherif Husayn (Hussein) of Mecca of July 14th, August 30th, September 9th, October 24th, November 5th, and December 14th of 1915—most especially the letter from McMahon to Husayn of 24 October 1915; and in General Allenby's Proclamation of 14 November 1918. The Allies had sought the help of the Arabs in defeating the Turkish Empire and promised them sovereignty in their own lands. They then stabbed the Arabs in the back with the Sykes-Picot Agreement of 1916, and yet worse after the war. *The New York Times* reported on 20 July 1922 on page 19,

> "JERUSALEM, June 22 (Correspondence of the Associated Press).— The inhabitants of Palestine, both Moslem and Christian, are immeasurably pleased that the British House of Lords yesterday passed the Islington motion disapproving the Balfour declaration of 1917. The native press is jubilant; pan-Arab demonstrations are being held and the local cable office is swamped with congratulatory messages from Arabs to the House of Lords.
>
> The Balfour declaration pledged the erection of a Jewish homeland in Palestine. The resolution passed yesterday by a vote of 60 to 29 set forth that 'the mandate for Palestine in its present form is unacceptable to this House, because it directly violates the pledges made by his Majesty's Government to the people of Palestine in the declaration of October, 1915, and again in the declaration of November, 1918 (pledges given to the Arabs), and is as at present framed opposed to the sentiments and wishes of the great majority of the people of Palestine. That, therefore, its acceptance by the Council of the League of Nations should be postponed until such modifications have therein been effected as will comply with pledges given by his Majesty's Government.'
>
> The Arabs regard this incident as a great victory. 'It is the bounden duty,' says an Arab call to a demonstration of celebration, 'of all of us to set forth our gratitude to the House of Lords for having proved to the world that God and justice still live in Great Britain.'
>
> Miraat el Shark, a Jerusalem newspaper, says: 'We will win our fight for freedom; we have God and right on our side.' Beit el Makdes, another local paper, says: 'Our victory in the House of Lords is the beginning of the end of political Zionism.'
>
> The Zionists are correspondingly disappointed at the news. They have not failed to cable strong protests to London. The Chairman of the Zionist organization here said to the Associated Press:
>
> 'All our hopes have been shattered on the rocks of political expediency. If the House of Commons follows the lead of the House of Lords, then Jews of the world will have been dealt a more staggering blow than that administered by the Emperor Hadrian 1,800 years ago, when his persecutions brought about the last dispersion of the Jewish race.'"

Jewish prophecy had long held that Gentiles should soldier and slave for Israel. In other words, Israel is a leech on the Gentile nations, which has no right to exist, and which forever throws the world into turmoil. *The London Times* published a Letter to the Editor from Lord Sydenham of Combe, "British Policy in Palestine. Divergence from Balfour Declaration." on 4 April 1923, on page 6, which stated, *inter alia*,

"I do not think any useful purpose can be served by further discussion of the terms of the correspondence between Sir H. McMahon and the Sherif of Mecca. There can be no doubt that Palestine was included in the area in which 'Great Britain is prepared to recognize and support the independence of the Arabs.' [***] Into Palestine we are dumping successive shiploads of impecunious aliens, we are imposing a loan equal to the whole annual revenue, and we have ordained a third official language perfectly useless to the people. All this, together with minor inflictions, we are doing in opposition to the strongly expressed wishes of a huge majority of Palestinians. It would be interesting if the 'Zionist Organization' would explain what 'civil rights' are left to a little people so circumstanced, and how the declaration, 'revised in the Zionist offices in America as well as in England,' can be reconciled with this use of British military forces."

Lord George Sydenham Clarke Sydenham of Combe, author of *The Jewish World Problem*,[22] told the House of Lords of the,

"mad policy of protecting the Jews against the Arabs in Palestine with the help of English bayonets, which cost the British taxpayer five hundred thousand pounds a month."[23]

On 7 April 1922, on page 8, *The London Times* published a Letter to the Editor from Lord Sydenham,

"JEWISH 'NATIONAL HOME.'

LORD SYDENHAM URGES INQUIRY.
TO THE EDITOR OF THE TIMES.

Sir,—I have read the important articles of your Correspondent in the Near East on Palestine with great interest. We have established what you justly call 'a powerful irritant' in the Near East, and the entire responsibility for the consequences must fall upon us.

Next year the taxpayers will have to provide another £4,000,000, which might be largely increased by events, and, as you point out, 'the extent of our financial commitments is very imperfectly understood.' I hold strongly that some solution of our difficulty must be found before it becomes obviously dangerous.

It has already been proved that the two parts of the Declaration are incompatible. You cannot make Palestine into a 'national home' in the sense which the Zionists proclaim, and at the same time insist that 'nothing shall

---

22. Lord George Sydenham Clarke Sydenham of Combe, *The Jewish World Problem*, reprinted from *The Nineteenth Century and After*, (November, 1921), pp. 888-901.
23. K. G. W. Ludecke, *I Knew Hitler: The Story of a Nazi A Who Escaped the Blood Purge*, Charles Scribner's Sons, New York, (1937), p. 213.

be done which may prejudice the civil and religious rights' of the owners of the soil. The civil rights of the Palestinians are being violated in many ways, and before their eyes, every day, and the natural result is growing exasperation.

If this contention is correct, why should we not say plainly that the 'national home' must be conditioned by inexorable facts? It is now clear that the Declaration was made without any inquiry into the economic possibilities of the country. I cordially agree with Lord Northcliffe's proposal that 'an impartial Commission should be appointed to inquire into the results of the experiment'; but I suggest that the inquiry should be extended to ascertain what additional population beyond the natural increase can be economically supported, by what means, and in what time. We have officers trained in India who are well able to conduct such an inquiry, and the long-established Jewish colonies would provide valuable data. Are these colonies or any of them being worked on an economic basis to-day? Palestine does not lend itself to cheap irrigation; but that aspect of the question needs investigation.

My own strong opinion is that the national home must eventually break down on economic grounds, because you cannot indefinitely maintain colonies unable to pay their way. This is also the view of some leading American Jews besides Mr. Morgenthau. If, then, as Dr. Weizmann proposes, 'between 50,000 and 60,000 Jews per annum' are deposited in the Holy Land, we shall soon be confronted with appalling difficulties—partly economic and partly arising from the hostility of the rightful owners of the land, who would find themselves displaced by the growing horde of immigrants.

My conclusion is that, in the interests of the Jews as well as the Arabs, immigration must be stopped until a full inquiry has taken place, if serious troubles are to be averted. For moral as well as economic reasons, the 'powerful irritant' must be removed.

I am, Sir, yours obediently,
April 3. SYDENHAM."

In a Letter to the Editor published in *The London Times* on 24 August 1922 on page 11, Lord Sydenham accused the crypto-Jewish Zionist spokesman Winston Churchill of being a Zionist dictator, and one might add a typical Zionist liar and sophist seeking to stifle debate (*see also:The Jewish Chronicle* issues from about 15 June 1922 to 17 June 1922, which republish portions of the debates in the House of Commons and in the House of Lords),

"THE RESPONSIBILITIES OF CRITICISM.

TO THE EDITOR OF THE TIMES.

Sir,—In his remarkable letter to Lord Islington, Mr. Churchill propounds a doctrin which is new and disturbing. Stated baldly, that doctrine appears to be that the critics of any policy which any Government may adopt are responsible for any disasters which that policy entails because they stated their opinions 'without having the power of altering the policy.'

Does Mr. Churchill really wish us to believe that the opponents of Mr. Montagu's policy, whose only thought was the welfare of the masses of India, are responsible for the heavy loss of life—unparalleled since the Mutiny—which that policy inevitably entailed? Everything which we foretold has happened or is happening, and if the Prime Minister's recent speech has any meaning it is that he intends to reverse the main principles of that disastrous policy.

Again, are those who consistently opposed the total change of policy in Ireland, which the Government suddenly adopted last summer, really responsible for the appalling destruction of life and property which they foresaw?

In Palestine the policy of forcing by British bayonets a horde of aliens, some of them eminently undesirable, upon the rightful owners of the country, in violation of Lord Balfour's promise that the 'civil rights' of the Arabs should not be prejudiced, led to risings before the delegation came to London. Are we, who opposed the policy because we knew that its injustice must lead to loss of life, responsible for anything that may now happen?

I humbly venture to suggest that Mr. Churchill's new doctrine can apply only under a dictatorship, that it is wholly unsuited to this country, and that even the Coalition Government may benefit from honest criticism. 'No people,' it has been well said, 'can deserve freedom except there is a healthy criticism of public men and of national policy.'

I am, Sir, yours obediently,
SYDENHAM.
The Priory, Lamberhurst, Kent, Aug. 19."

On 8 September 1922 on page 9, *The London Times* published correspondence which had taken place between Lord Sydenham and Winston Churchill (this correspondence also appeared under the heading "British Policy in Palestine. Mr. Churchill and Lord Sydenham. Amusing Correspondence." in *The Jewish Chronicle* on 15 September 1922 on page 17),

"OUR PALESTINE POLICY.

LORD SYDENHAM'S CHARGES.

CORRESPONDENCE WITH MR. CHURCHILL.

The following correspondence has passed between Mr. Churchill and Lord Sydenham:—

FROM MR. CHURCHILL TO LORD SYDENHAM.

26th August, 1022.

Dear Lord Sydenham,—I observe in your letter to *The Times* of August

19, in reference to my correspondence with Lord Islington, you write as follows:—

'In Palestine the policy of forcing by British bayonets a horde of aliens, some of them eminently undesirable, upon the rightful owners of the country, in violation of Lord Balfour's promise that the 'civil rights' of the Arabs should not be prejudiced, led to risings before the delegation came to London. Are we, who opposed the policy because we knew that its injustice must lead to loss of life, responsible for anything that may now happen?'

I observe also that at the time of Mr. Balfour's declaration in 1917 you are reported to have expressed yourself as follows:

'I earnestly hope that one result of the war will be to free Palestine from the withering blight of Turkish rule, and to render it available as the national home of the Jewish people, who can restore its ancient prosperity.'

It seems to me that before you take further part in this particular controversy you owe at to the public, and, I may add, to yourself to offer some explanation of the apparent discrepancy between these positions. In particular it would be interesting to know what has occurred in the interval to convert 'the Jewish people' for whom you hoped to make Palestine 'the national home' into 'a horde of aliens.' Your opinions as to the expediency of the policy of Zionism may no doubt quite naturally have turned a complete somersault in the last five years, but the relation of the Jewish race to Palestine has not altered in that period. Either, therefore, you were mistaken then in thinking that the Jews were entitled to regard Palestine as 'the national home' or you are mistaken now in describing them as 'a horde of aliens.'

It is to this point that it would be specially interesting to see you address yourself.

From Lord Sydenham to Mr. Churchill.

Aug. 29, 1922.

Dear Mr. Churchill,—It is my strong impression that I have already sent an explanation of my change of view to the Jewish paper which asked me for a message by telegram in 1917. This explanation, however, seems not to have been supplied to you with the text of the message which you read in the House of Commons. I was in the country away from books and papers, and I, too, hurriedly sent the reply which you again quote.

I was grievously mistaken, and for three reasons:—

1. I had no knowledge of the economic conditions of Palestine, which can never support a large population, and can only receive carefully selected immigrants gradually without grave injury to the inhabitants.

2. I was quite unaware that the Balfour Declaration was obtained by the prolonged underhand methods, which are, in part, described in the Zionist Political Report. This remarkable document came to me as a revelation.

3. A 'Jewish National Home' can be interpreted in several ways, and Mr. Balfour's undertaking—that the 'civil rights' of the Palestinians would not be prejudiced naturally reassured me. I never dreamed that a Jewish Government would be set up, and I imagined only a slow immigration of

desirable Jews under a purely British Government. In 1917, it was not yet clear that there would be a rush of Russian and Central European Jews to other countries, and that a portion of them would be dumped down in Palestine. I was further reassured in 1918 by General Allenby's Proclamation, which appeared to render impossible what is now happening, while the text of the Treaty with the Hedjaz, which is disputed, was unknown to me at the time.

Since 1917 I have given much thought and study to the Jewish problems, and I have been forced to change my opinions. I was, as you suggest, 'mistaken in thinking that the Jews were entitled to regard Palestine as the 'National Home.' I consider that they have no more claim to Palestine than the modern Italians to Britain, or the Moors to Southern Spain. I also think that 'a horde of aliens' correctly describes the immigrants.

I am sure you will agree that, when a man finds himself obliged to change his opinions, he is not only justified in pressing what be has come to believe just and right, but he is actually bound to do so. When the Government changed their minds in regard to the 'murder gang' in Ireland, they were not only right, but bound to make a complete change in their policy.

I have tried to answer your questions quite frankly, and I have only to add that I greatly regret my mistake, due mainly to my absence from London, and to the fact that I was then absorbed in studying the course of the war, which engrossed my thoughts at the time.

FROM MR. CHURCHILL TO LORD SYDENHAM.

August 31, 1922.

Dear Lord Sydenham,—I am obliged to you for your letter of the 29th instant, in which you admit that you were grievously mistaken when you promised to support the Zionist policy, and have entirety changed your view on the question of establishing a Jewish national home. In the face of so complete an admission, expressed, as it is, in language of the utmost courtesy, I do not wish to press my point unduly. If, however, the only reasons which have changed you from an ardent advocate into an active opponent are those set out in your letter, I cannot but feel that they are inadequate, even where they are not based on misconceptions.

(1) The policy of his Majesty's Government has always been to bring in only 'carefully selected immigrants gradually, without grave injury to the inhabitants,' or, I may add, any kind of injury to the inhabitants.

(2) Lord Balfour's declaration did not arise from underhand methods of any kind, but from wide and deep arguments which have been clearly explained.

(3) No Jewish Government has been set up in Palestine, but only a British . Government, in which Jews as well as Arabs participate. A reference to the White Paper recently published should reassure you in this respect.

There is, however, one reason for a change of view, which I am glad to see you do not give—namely, that it was an easy and popular thing to advocate a Zionist policy in the days of the Balfour Declaration, and that it is a laborious and much-criticized task to try to give honourable effect at the present time to the pledges which were given then. Still, it seems to me that if a public man, like yourself, has mistakenly supported the giving of the pledge, he should, even if he has changed his mind, show a little forbearance, and even consideration, to those who are endeavouring to make it good. Might you not well have left to others the task of inflicting censure and creating difficulties, and reserved your distinguished controversial gifts for some topic upon which you have an unimpeachable record? To change your mind is one thing; to turn on those who have followed your previous advice another.

P.S.—I am sending this correspondence to the Press.

FROM LORD SYDENHAM TO MR. CHURCHILL.

4th September, 1922.

Dear Mr. Churchill,—I beg to thank you for your letter of the 31st ultimo, which I received this morning.

We are all of us liable to 'misconceptions'; but I regret that I cannot admit as such the three points you refer to, for the following reasons:—

(1) I was glad to learn that latterly some care has been observed on the selection of immigrants; but I have abundant evidence that for some time most unsuitable persons were freely admitted, and this is proved by the official inquiry into the Jaffa riots. I am still not satisfied that persons who do not fulfil the economic requirements of Palestine and whose importation may adversely affect the interests of the Palestinians are excluded.

(2) I cannot, of course, tell why the Government, at a time when the Empire was fighting for its life and the conquest of Palestine had not been accomplished, adopted the policy of Lord Balfour's declaration. The Zionists, however, who do not represent all Jews, have explained some of the elaborate steps they took to bring pressure on the Government, and I have a good deal of information on this subject. They have further hinted not obscurely that the first High Commissioner was their selection. I must assume that the Government, in yielding to this pressure, envisaged some great advantage to the Empire, though I can see only danger.

(3) As Government in Palestine is an autocracy under an Administrator whom you have described as an 'ardent Zionist,' and as important posts are increasingly being conferred upon Jews, I must adhere to my contention that it is, in actual fact, a 'Jewish Government.'

I am sorry that I cannot accept your proposition that a man who has once expressed a mistaken opinion is thereafter debarred from opposing a policy which he has been forced to believe unjust and dangerous. If your principle had held the field in the past, much of our political history would have read differently.

The moral I draw is that it is unwise to he beguiled into any expression of opinion failing time to make a careful study of the question raised. To this unwisdom I have pleaded guilty with extenuating circumstances."

Henry Morgenthau pointed out that leading Jews misrepresented the precise language of the Balfour Declaration, which did not offer to give Palestine to the Jews, but merely expressed support for the idea that Jews might wish to live there under the rule of the indigenous population,

> "It is worth while at this point to digress for a moment from my main argument, to point out that the Balfour Declaration is itself not even a compromise. It is a shrewd and cunning delusion. I have been astonished to find that such an intelligent body of American Jews as the Central Conference of American Rabbis should have fallen into a grievous misunderstanding of the purport of the Balfour Declaration. In a resolution adopted by them, they assert that the declaration says: 'Palestine is to be a national home-land for the Jewish people.' Not at all! The actual words of the declaration (I quote from the official text) are: 'His Majesty's Government views with favor the establishment *in Palestine of a national home for the Jewish people.*' These two phrases sound alike, but they are really very different. I can make this obvious by an analogy. When I first read the Balfour Declaration I was temporarily making my home in the Plaza Hotel. Therefore I could say with truth: 'My home is in the Plaza Hotel.' I could not say with truth: 'The Plaza Hotel is my home.' If it were 'my home,' I would have the freedom of the whole premises, and could occupy any room in the house with impunity. Quite obviously, however, I would not venture to trespass in the rooms of my friend, Mr. John B. Stanchfield, who happened at the same time also to have found 'a home-land in the Plaza,' nor in the private quarters of any other resident of that hostelry, whose right to his share in it was as good as mine, and in many cases of much longer standing."[24]

5.12 Zionists Declare that Anti-Semitism is the Salvation of the "Jewish Race"

Why would any Jew sponsor Adolf Hitler, or found an anti-Jewish society? After Jewish emancipation, the vast majority of European Jews wanted to assimilate into Western society. Racist Zionist Jews, a small minority in the Jewish community, feared that the "Jewish race" would disappear through the "final solution to the Jewish question" of "assimilation", or so they stated in their writings and speeches of the Nineteenth Century. The Nazis did not coin the phrase "final solution to the Jewish question", nor did the Nazis intend it to mean the extermination of the Jews. The Zionists used the expression to refer to the integration of Jews, which process the Zionists loathed. The political Zionists were and are racist segregationists. Both the political Zionists and the Nazis, who were in fact political Zionists, offered an alternative "final solution to the Jewish

---

[24]. H. Morgenthau, "Zionism a Surrender, Not a Solution", *The World's Work*, Volume 42, Number 3, (July, 1921), pp. i-viii, at iii-iv.

question" to that of assimilation, one of Jewish segregation in a "World Ghetto"—which is another Zionist phrase.[25] Before the Nazis even came into existence, the political Zionists called for the segregation of Jews into a ghetto.

The "final solution" of extermination was not proposed by a German Nazi, but rather by an American Jew; and it was not the extermination of Jews which he proposed, but the genocide of German Gentiles. Theodor Newman Kaufman advocated the genocidal sterilization of all Germans as a "final solution" in 1941 in his book *Germany Must Perish!*, Argyle Press, Newark, New Jersey, (1941), before the Wannsee-Konferenz occurred.

Kaufman's book was merely a more modern manifestation of the ancient racist Jewish divine commandment that Jews must exterminate the seed of Esau/Edom/Amalek/Haman (*Deuteronomy* 25:19), lest God exterminate the Jews, themselves; which "race" of Esau came to signify Gentiles in general. I do not touch upon the question of whether and which Nazis did in fact attempt to exterminate European Jews under their control. There certainly was an ancient Jewish tradition that assimilatory Jews must be exterminated. Numerous Jewish prophets called for the genocide of Jews and Gentiles in their pursuit of Jewish world domination and a Messianic Age, when all religions other than Judaism would be suppressed, when all the nations would be destroyed, and when Gentile cattle would serve the Jews as slaves or face certain death.

The political Zionists, Albert Einstein chief racist among them, embraced the myth that anti-Semitism is the salvation of the "Jewish race", in that it forces Jews to segregate against their will and better natures. Einstein hated non-racist Jews, though he himself had married a non-Jew. At least since Spinoza, prominent Jewish racists have openly stated that anti-Semitism is the only means to preserve the divine race.

Jewish terror organizations have long tried to alienate and terrify Jews, and to promote anti-Semitism as a means to force Jews to flee to Palestine. Jews have often posed as anti-Semites and committed terrorist acts against other Jews in order to frighten them into segregation and emigration. In its article "Zionism", the *Great Soviet Encyclopedia: A Translation of the Third Edition*, Volume 23, Macmillan, New York, (1979), pp. 745-746, at 745, wrote,

> "After the state of Israel was formed in 1948 on part of Palestine's territory by a resolution of the United Nations, Zionism became Israel's official ideology. Its main goals are to secure the unconditional support of Israel by the world's Jews, to gather the world's Jews in Israel, and to inculcate a Zionist spirit among Jews in various countries. Zionism seeks to expand Israel to the boundaries of the 'Greater Land of Israel.' To this end, Zionists evoke the thesis of 'eternal anti-Semitism,' a situation which they often deliberately instigate."

---

25. T. Herzl, English translation by H. Zohn, R. Patai, Editor, *The Complete Diaries of Theodor Herzl*, Volume 1, Herzl Press, New York, (1960), p. 172.

See also: N. S. Alent'eva, Editor, *Tseli i metody voinstvuiushchego sionizma*, Izd-vo polit. lit-ry, Moskva, (1971). Н. С. Алентьева, Редактор, Цели и методы воинствующего сионизма, Издательство Политической Литературы, Москва, (1971).

Jewish terrorist organizations do not care about the negative repercussions of their vile actions for other Jews should they be found out, because they feel that anti-Semitism benefits their cause of forcing Jews to Israel against their will. They recklessly promote Jewish disloyalty and deceit around the world in the belief that if their deceit is unearthed innocent as well as guilty Jews might be forced to flee to Israel. Victor Ostrovsky wrote in his book *By Way of Deception*,

> "The one problem with the system [of sayanim] is that the Mossad does not seem to care how devastating it could be to the status of the Jewish people in the diaspora if it was known. The answer you get if you ask is: 'So what's the worst that could happen to those Jews? They'd all come to Israel? Great.'"[26]

Jewish racists helped to put Hitler into power in order to herd up the Jews of Europe and force them into segregation. Jewish racists collaborated with the Nazis to kill off the weakest Jews and preserve the best genetic stock for deportation to Palestine, which could not possibly house the numerous Jews of Europe. Western Jews in general hated Eastern Jews. Political Zionists encouraged the Nazis to force assimilatory Jews, especially Eastern Jews, into segregation. They also encouraged the Soviets towards anti-Semitism in order to leave "red assimilationist" Jews no option but to create a Jewish state in formerly Russian territory, in China, or in Palestine, or face annihilation.

The worst enemy of persons of Jewish descent has always been the Zionist, especially the Zionist in anti-Semite's clothing. Too many Zionists have carried on, and carried out, the bloodthirsty and treacherous tradition of ancient Jewish racism, which they see as the product of "superior Jewish racial instincts", and which admonishes Jews to exterminate other Jews who would otherwise assimilate.

Einstein claimed that anti-Semites were correct to be believe that Jews exercised undue influence in Germany. Einstein wrote in the *Jüdische Rundschau*, on 21 June 1921, on pages 351-352,

> "This phenomenon [*i. e.* Anti-Semitism] in Germany is due to several causes. Partly it originates in the fact that the Jews there exercise an influence over the intellectual life of the German people altogether out of proportion to their number. While, in my opinion, the economic position of the German Jews is very much overrated, the influence of Jews on the Press, in literature, and in science in Germany is very marked, as must be apparent

---

26. V. Ostrovsky and C. Hoy, *By Way of Deception*, St. Martin's Paperbacks, New York, (1990), pp. 87-88.

to even the most superficial observer. This accounts for the fact that there are many anti-Semites there who are not really anti-Semitic in the sense of being Jew-haters, and who are honest in their arguments. They regard Jews as of a nationality different from the German, and therefore are alarmed at the increasing Jewish influence on their national entity. [***] But in Germany the judgement of my theory depended on the party politics of the Press[.][27]

Einstein also stated,

"The way I see it, the fact of the Jews' racial peculiarity will necessarily influence their social relations with non-Jews. The conclusions which—in my opinion—the Jews should draw is to become more aware of their peculiarity in their social way of life and to recognize their own cultural contributions. First of all, they would have to show a certain noble reservedness and not be so eager to mix socially—of which others want little or nothing. On the other hand, anti-Semitism in Germany also has consequences that, from a Jewish point of view, should be welcomed. I believe German Jewry owes its continued existence to anti-Semitism."[28]

Nazi Zionist Joseph Goebbels, sounding very much like political Zionist Albert Einstein, was quoted in *The New York Times*, on 29 September 1933, on page 10,

"It must be remembered the Jews of Germany were exercising at that time a decisive influence on the whole intellectual life; that they were absolute and unlimited masters of the press, literature, the theatre and the motion pictures, and in large cities such as Berlin, 75 percent of the members of the medical and legal professions were Jews; that they made public opinion, exercised a decisive influence on the Stock Exchange and were the rulers of Parliament and its parties."

Einstein had a reputation as a rabid anti-assimilationist, which is to say that Einstein was a rabid racist segregationist. On 15 March 1921, Kurt Blumenfeld wrote to Chaim Weizmann,
"Einstein [***] is interested in our cause most strongly because of his revulsion from assimilatory Jewry."[29]

Einstein expressed his virulently segregationist viewpoint in 1921,

---

27. A. Einstein, "Jewish Nationalism and Anti-Semitism", *The Jewish Chronicle*, (17 June 1921), p. 16.
28. A. Einstein, A. Engel translator, "How I became a Zionist", *The Collected Papers of Albert Einstein*, Volume 7, Document 57, Princeton University Press, (2002), pp. 234-235, at 235.
29. J. Stachel, *Einstein from 'B' to 'Z'*, Birkhäuser, Boston, (2002), p. 79, note 41.

"To deny the Jew's nationality in the Diaspora is, indeed, deplorable. If one adopts the point of view of confining Jewish ethnical nationalism to Palestine, then one, to all intents and purposes, denies the existence of a Jewish people. In that case one should have the courage to carry through, in the quickest and most complete manner, entire assimilation. We live in a time of intense and perhaps exaggerated nationalism. But my Zionism does not exclude in me cosmopolitan views. I believe in the actuality of Jewish nationality, and I believe that every Jew has duties towards his coreligionists. [***] [T]he principal point is that Zionism must tend to strengthen the dignity and self-respect of the Jews in the Diaspora. I have always been annoyed by the undignified assimilationist cravings and strivings which I have observed in so many of my friends."[30]

In 1921, Einstein declared, referring to Eastern European Jews,

"These men and women retain a healthy national feeling; it has not yet been destroyed by the process of atomisation and dispersion."[31]

On 1 July 1921, Einstein was quoted in the *Jüdische Rundshau* on page 371,

"Let us take brief look at the *development of German Jews* over the last hundred years. With few exceptions, one hundred years ago our forefathers still lived in the Ghetto. They were poor and separated from the Gentiles by a wall of religious tradition, secular lifestyles and statutory confinement and were confined in their spiritual development to their own literature, only relatively weakly influenced by the forceful progress which intellectual life in Europe had undergone in the Renaissance. However, these little noticed, modestly living people had one thing over us: *Every one of them belonged with all his heart to a community*, into which he was incorporated, in which he felt a worthwhile member, in which nothing was asked of him which conflicted with his normal processes of thought. Our forefathers of that era were pretty pathetic both bodily and spiritually, but—in social relations— in an enviable state of mental equilibrium. Then came emancipation. It offered undreamt of opportunities for advancement. The isolated individual quickly found their way into the upper financial and social circles of society. They eagerly absorbed the great achievements of art and science which the

---

30. A. Einstein, "Jewish Nationalism and Anti-Semitism", *The Jewish Chronicle*, (17 June 1921), p. 16.
31. A. Einstein quoted in: J. Stachel, "Einstein's Jewish Identity", *Einstein from 'B' to 'Z'*, Birkhäuser, Boston, (2002), p. 65. Stachel cites, *About Zionism: Speeches and Letters*, Macmillan, New York, (1931), pp. 48-49. For Zionist Ha-Am's use of the image of atomisation and dispersion, see: A. Hertzberg, *The Zionist Idea*, Harper Torchbooks, New York, (1959), p. 276.

Occidentals[32] had created. They contributed to the development with passionate affection, and themselves made contributions of lasting value. They thereby took on the lifestyle of the Gentile world, turning away from their religious and social traditions in growing masses—took on Gentile customs, manners and mentality. It appeared as if they were being completely dissolved into the numerically superior, politically and culturally better organized host peoples, such that no trace of them would be left after a few generations. The complete eradication of the Jewish nationality in Middle and Western Europe appeared to be inevitable. However, it didn't turn out that way. It appears that racially distinct nations have instincts which work against interbreeding. The adaptation of the Jews to the European peoples among whom they have lived in language, customs and indeed even partially in religious practices *was unable to eliminate all feelings of foreignness* which exist between Jews and their European host peoples. In short, this spontaneous feeling of foreignness is ultimately due to a loss of energy.[33] For this reason, *not even well-meant arguments can eradicate it*. Nationalities do not want to be mixed together, rather they want to go their own separate ways. A state of peace can only be achieved by mutual tolerance and respect."

Einstein stated that Jews should not participate in the German Government,

"I regretted the fact that [Rathenau] became a Minister. In view of the attitude which large numbers of the educated classes in Germany assume towards the Jews, I have always thought that their natural conduct in public should be one of proud reserve."[34]

Einstein merely parroted the Zionist Party line. Werner E. Mosse wrote,

"While the leaders of the CV saw it as their special duty to represent the interests of the German Jews in the active political struggle, Zionism stood for... systematic Jewish non-participation in German public life. It rejected as a matter of principle any participation in the struggle led by the CV."[35]

---

32. At the time Einstein made his statement, Jews and Gentiles often referred to Jews as "Orientals".
33. Einstein repeatedly spoke of the Germans as "greedy" to acquire territory and of the "loss of energy" when different "races" attempted to live together. He have been speaking literally. Georg Friedrich Nicolai wrote of the struggle of life to aquire the energy of the sun and he applied this struggle to humanity. G. Nicolai, *Die Biologie des Krieges, Betrachtungen eines deutschen Naturforschers*, O. Füssli, Zürich, (1917); English translation: *The Biology of War*, Century Co., New York, (1918), pp. 36-39, 44-53.
34. A. Einstein quoted in: R. W. Clarck, *Einstein, the Life and Times*, World Publishing Company, USA, (1971), p. 292. Clarck refers to: *Neue Rundschau*, Volume 33, Part 2, pp. 815-816.
35. W. E. Mosse, "Die Niedergang der deutschen Republik und die Juden", *The Crucial Year 1932*, p. 38; English translation in: K. Polkehn, "The Secret Contacts: Zionism and

The Jewish Central-Verein fought against Nazi racism, while many Zionists embraced it. In 1925, Einstein wrote in the official Zionist Party organ *Jüdische Rundschau*,

> "By study of their past, by a better understanding of the spirit [Geist] that accords with their race, they must learn to know anew the mission that they are capable of fulfilling. [***] What one must be thankful to Zionism for is the fact that it is the only movement that has given many Jews a justified pride, that it has once again given a despairing race the necessary faith, if I may so express myself, given new flesh to an exhausted people."[36]

On 12 October 1929, Albert Einstein wrote to the *Manchester Guardian*,

> "In the re-establishment of the Jewish nation in the ancient home of the race, where Jewish spiritual values could again be developed in a Jewish atmosphere, the most enlightened representatives of Jewish individuality see the essential preliminary to the regeneration of the race and the setting free of its spiritual creativeness."[37]

Einstein's public racism eventually waned, but he continued to publicly express his segregationist philosophy in the same terms as anti-Semites, as well as his belief that Jews "thrived on" and owed their "continued existence" to anti-Semitism.

Einstein stated in December of 1930 to an American audience,

> "There is something indefinable which holds the Jews together. Race does not make much for solidarity. Here in America you have many races, and yet you have the solidarity. Race is not the cause of the Jews' solidarity, nor is their religion. It is something else—which is indefinable."[38]

Einstein's confusing public statement perhaps resulted from his desire to promote multi-culturalism in America, which had the benefit of freeing up Jewish immigration to the United States.[39] Einstein was also likely parroting, or

---

Nazi Germany, 1933-1941", *Journal of Palestine Studies*, Volume 5, Number 3/4, (Spring-Summer, 1976), pp. 54-82, at 56-57.

36. English translation by John Stachel in J. Stachel, "Einstein's Jewish Identity", *Einstein from 'B' to 'Z'*, Birkhäuser, Boston, (2002), p. 67. Stachel cites, "Botschaft", *Jüdische Rundschau*, Volume 30, (1925), p. 129; French translation, *La Revue Juive*, Volume 1, (1925), pp. 14-16.

37. J. Stachel, "Einstein's Jewish Identity", *Einstein from 'B' to 'Z'*, Birkhäuser, Boston, (2002), p. 65. Stachel cites, *About Zionism: Speeches and Letters*, Macmillan, New York, (1931), pp. 78-79.

38. A. Einstein quoted in "Einstein on Arrival Braves Limelight for Only 15 Minutes", *The New York Times*, (12 December 1930), pp. 1, 16, at 16.

39. E. A. Ross, *The Old World in the New: The Significance of past and Present Immigration to the American People*, Century Company, New York, (1914), p. 144.

trying to parrot, a fellow anti-assimilationist political Zionist whose pamphlet was well known in America, Solomon Schechter and his *Zionism: A Statement*, Federation of American Zionists, New York, (1906), in which Schechter states, among other things, "Zionism is an ideal, and as such is indefinable."[40]

Einstein stated in 1938,

"JUST WHAT IS A JEW?

The formation of groups has an invigorating effect in all spheres of human striving, perhaps mostly due to the struggle between the convictions and aims represented by the different groups. The Jews, too, form such a group with a definite character of its own, and anti-Semitism is nothing but the antagonistic attitude produced in the non-Jews by the Jewish group. This is a normal social reaction. But for the political abuse resulting from it, it might never have been designated by a special name.

What are the characteristics of the Jewish group? What, in the first place, is a Jew? There are no quick answers to this question. The most obvious answer would be the following: A Jew is a person professing the Jewish faith. The superficial character of this answer is easily recognized by means of a simple parallel. Let us ask the question: What is a snail? An answer similar in kind to the one given above might be: A snail is an animal inhabiting a snail shell. This answer is not altogether incorrect; nor, to be sure, is it exhaustive; for the snail shell happens to be but one of the material products of the snail. Similarly, the Jewish faith is but one of the characteristic products of the Jewish community. It is, furthermore, known that a snail can shed its shell without thereby ceasing to be a snail. The Jew who abandons his faith (in the formal sense of the word) is in a similar position. He remains a Jew.

[***]

WHERE OPPRESSION IS A STIMULUS

[***]

Perhaps even more than on its own tradition, the Jewish group has thrived on oppression and on the antagonism it has forever met in the world. Here undoubtedly lies one of the main reasons for its continued existence through so many thousands of years."[41]

Albert Einstein was parroting racist political Zionist leader Theodor Herzl, who wrote in his book *The Jewish State*,

---

[40]. Reprinted in the relevant part in A. Hertzberg, *The Zionist Idea*, Harper Torchbooks, New York, (1959), p. 505.

[41]. A. Einstein, "Why do They Hate the Jews?", *Collier's*, Volume 102, (26 November 1938); reprinted in *Ideas and Opinions*, Crown, New York, (1954), pp. 191-198, at 194, 196. Einstein expressed himself in a similar way to Peter A. Bucky, P. A. Bucky, Einstein, and A. G. Weakland, *The Private Albert Einstein*, Andrews and McMeel, Kansas City, (1992), p. 87.

"Oppression and persecution cannot exterminate us. No nation on earth has survived such struggles and sufferings as we have gone through. Jew-baiting has merely stripped off our weaklings; the strong among us were invariably true to their race when persecution broke out against them. This attitude was most clearly apparent in the period immediately following the emancipation of the Jews. Later on, those who rose to a higher degree of intelligence and to a better worldly position lost their communal feeling to a very great extent. Wherever our political well-being has lasted for any length of time, we have assimilated with our surroundings. I think this is not discreditable. Hence, the statesman who would wish to see a Jewish strain in his nation would have to provide for the duration of our political well-being; and even Bismarck could not do that. [***] The Governments of all countries scourged by Anti-Semitism will serve their own interests in assisting us to obtain the sovereignty we want. [***] Great exertions will not be necessary to spur on the movement. Anti-Semites provide the requisite impetus. They need only do what they did before, and then they will create a love of emigration where it did not previously exist, and strengthen it where it existed before. [***] I imagine that Governments will, either voluntarily or under pressure from the Anti-Semites, pay certain attention to this scheme; and they may perhaps actually receive it here and there with a sympathy which they will also show to the Society of Jews."[42]

Einstein went along with the crowd of prominent political Zionists who openly stated that anti-Semitism is welcomed, encouraged and useful to the Zionists. They based their belief on Spinoza's declaration that emancipation leads to assimilation and that the Jews only exist in modern times because of anti-Semitism. Prominent Zionist and author of the *Encyclopaedia Judaica; das Judentum in Geschichte und Gegenwart*, Jakob Klatzkin stated in 1925,

"The national viewpoint taught us to understand the true nature of antisemitism, and this understanding widens the horizons of our national outlook. [***] In the age of enlightenment antisemitism was included among the phenomena that are likely to disappear along with other forms of prejudice and iniquity. The antisemites, so the rule stated, were the laggard elements in the march of progress. Hence, our fate is dependent on the advance of human culture, and its victory is our victory. [***] In the period of Zionism, we learned that antisemitism was a psychic-social phenomenon that derives from our existence as a nation within a nation. Hence, it cannot change, until we attain our national end. But if Zionism had fully understood its own implications, it would have arrived, not merely as a psycho-sociological explanation of this phenomenon, but also as a justification of it. It is right to protest against its crude expressions, but we are unjust to it and distort its nature so long as we do not recognize that essentially it is a

---

[42]. T. Herzl, *A Jewish State: An Attempt at a Modern Solution of the Jewish Question*, The Maccabæan Publishing Co., New York, (1904), pp. 5-6, 25, 68, 93.

defense of the integrity of a nation, in whose throat we are stuck, neither to be swallowed nor to be expelled. [***] And when we are unjust to this phenomenon, we are unfair to our own people. If we do not admit the rightfulness of antisemitism, we deny the rightfulness of our own nationalism. If our people is deserving and willing to live its own national life, then it is an alien body thrust into the nations among whom it lives, an alien body that insists on its own distinctive identity, reducing the domain of their life. It is right, therefore, that they should fight against us for their national integrity. [***] Know this, that it is a good sign for us that the nations of the world combat us. It is proof that our national image is not yet utterly blurred, our alienism is still felt. If the war against us should cease or be weakened, it would indicate that our image has become indistinct and our alienism softened. We shall not obtain equality of rights anywhere save at the price of an explicit or implied declaration that we are no longer a national body, but part of the body of the host-nation; or that we are willing to assimilate and become part of it. [***] Instead of establishing societies for defense against the antisemites, who want to reduce our rights, we should establish societies for defense against our friends who desire to defend our rights. [***] When Moses came to redeem the children of Israel, their leaders said to him, 'You have made our odor evil in the eyes of Pharaoh and in the eyes of his servants, giving them a sword with which to kill us.' Nevertheless, Moses persisted in worsening the situation of the people, and he saved them."[43]

Karl Kautsky predicted that the Jews would disappear due to their assimilation following World War I. The First World War, which the Zionists planned would fulfill their dream of a Jewish state, instead rendered it obsolete, and they were the only group that had a vested interest in promoting discord in Europe, anti-Semitism and the segregation and expulsion of Jews. Others had learned that the emigration of large numbers of Jews from their country resulted in economic hardship. The Zionists unwisely promised profits for all from racism directed against Jews.

Albert Einstein's anti-assimilationist beliefs hailed from an ancient tradition. Simon Dubnow wrote in 1905,

"Assimilation is common treason against the banner and ideals of the Jewish people. [***] But one can never 'become' a member of a natural group, such as a family, a tribe, or a nation. One may attain the rights or privileges of citizenship with a foreign nation, but one cannot appropriate for himself its nationality too. To be sure, the emancipated Jew in France calls himself a Frenchman of Jewish faith. Would that mean, however, that he became a part of the French nation, confessing to the Jewish faith? Not at all. Because,

---

[43]. J. Klatzkin, *Tehumim: Ma'amarim*, Devir, Berlin, (1925). English translation in J. B. Agus, *The Meaning of Jewish History*, Volume 2, Abelard-Schuman, New York, (1963), pp. 425-426.

in order to be a member of the French nation one must be a Frenchman by birth, one must be able to trace his genealogy back to the Gauls, or to another race in close kinship with them, and finally one must also possess those characteristics which are the result of the historic evolution of the French nation. A Jew, on the other hand, even if he happened to be born in France and still lives there, in spite of all this, he remains a member of the Jewish nation, and whether he likes it or not, whether he is aware or unaware of it, he bears the seal of the historic evolution of the Jewish nation."[44]

Long before the First World War, Voltaire stated in the end of Chapter 104 of his *Essai sur les Moeurs et l'Esprit des Nations, et sur les Principaux faits de l'Histoire Depuis Charlemagne Jusqu'à Louis XIII*, (1769); that should Gentiles—in Voltaire's view—become wise to the ways of Jews and prevent Jews from exploiting them, then rich Jews would abandon their religious superstitions and assimilate and the poor Jews would become thieves like Gypsies. According to Voltaire, whose work was well known, Jews would disappear through assimilation.[45]

The emancipation of Jews in Bolshevik lands, and the assimilation of affluent Jews in capitalistic societies, greatly concerned the Zionists, who feared it would be the end of all Jews. Before Voltaire, Spinoza noted that assimilation was causing the Jewish ethnicity to disappear. After Voltaire, Wellhausen, relying on Spinoza's observations, noted that emancipation was leading the Jews to assimilate and therefore to disappear—a fact that terrified the Zionists, many of whom were hypocrites who had themselves married Gentiles. Julius Wellhausen wrote in 1881,

> "The Jews, through their having on the one hand separated themselves, and on the other hand been excluded on religious grounds from the Gentiles, gained an internal solidarity and solidity which has hitherto enabled them to survive all the attacks of time. The hostility of the Middle Ages involved them in no danger; the greatest peril has been brought upon them by modern times, along with permission and increasing inducements to abandon their separate position. It is worth while to recall on this point the opinion of Spinoza,[*Footnote: Tract. Theol. Polit. 0. 4, ad fin.*] who was well able to form a competent judgment :—'That the Jews have maintained themselves

---

44. S. Dubnow, *Die Grundlagen des Nationaljudentums*, Jüdischer Verlag, Berlin, (1905). English translation from K. A. Strom, Editor, *The Best of Attack! and National Vanguard Tabloid*, National Alliance, Arlington, Virginia, (1984), p. 60.

45. F. M. Arouet de Voltaire, *Histoire de Charles XII, Roi de Suède*, (1731); **and** *Dictionnaire Philosophique*, Multiple Editions; multiple English translations, including: W. F. Flemming, *A Philosophical Dictionary*, Volume 6, Dingwall-Rock, New York, (1901), pp. 266-313; **and** *Essai sur les Moeurs et l'Esprit des Nations, et sur les Principaux faits de l'Histoire Depuis Charlemagne Jusqu'à Louis XIII*, Chapter 104, (1769); **and** *Philosophie Génerale: Métaphysique, Morale et Théologie*, Chez Sanson et Compagnie, Aux Deux-Ponts, (1792).

so long in spite of their dispersed and disorganised condition is not at all to be wondered at, when it is considered how they separated themselves from all other nationalities in such a way as to bring upon themselves the hatred of all, and that not only by external rites contrary to those of other nations, but also by the sign of circumcision, which they maintain most religiously. Experience shows that their conservation is due in a great degree to the very hatred which they have incurred. When the king of Spain compelled the Jews either to accept the national religion or to go into banishment, very many of them accepted the Roman Catholic faith, and in virtue of this received all the privileges of Spanish subjects, and were declared eligible for every honour; the consequence was that a process of absorption began immediately, and in a short time neither trace nor memory of them survived. Quite different was the history of those whom the king of Portugal compelled to accept the creed of his nation; although converted, they continued to live apart from the rest of their fellow-subjects, having been declared unfit for any dignity. So great importance do I attach to the sign of circumcision also in this connection, that I am persuaded that it is sufficient by itself to maintain the separate existence of the nation for ever.' The persistency of the race may, of course, prove a harder thing to overcome than Spinoza has supposed; but nevertheless he will be found to have spoken truly in declaring that the so-called emancipation of the Jews must inevitably lead to the extinction of Judaism wherever the process is extended beyond the political to the social sphere. For the accomplishment of this centuries may be required."[46]

Spinoza's observations are antedated by Biblical writings, which tell that God will punish assimilated Jews and pious Jews to remind all of Israel that God is a Jew. God punishes them with the sword and with fire and renders them ash. The punishment of assimilatory Jews through murderous anti-Semitism in order to drive them back to God is perhaps most strongly advocated in the books of *Deuteronomy* and *Ezekiel*, and in *Malachi* 4:1-6 it states,

> "1 For, behold, the day cometh, that *shall* burn as an oven; and all the proud, yea, and all that do wickedly, shall be stubble: and the day that cometh shall burn them up, saith the LORD of hosts, that it shall leave them neither root nor branch. 2 But unto you that fear my name shall the Sun of righteousness arise with healing in his wings; and ye shall go forth, and grow up as calves of the stall. 3 And ye shall tread down the wicked; for they shall be ashes under the soles of your feet in the day that I *shall* do *this*, saith the LORD of hosts. 4 Remember ye the law of Moses my servant, which I commanded unto him in Horeb for all Israel, *with* the statutes and judgments. 5 Behold, I *will* send you Elijah the prophet before the coming of the great and dreadful

---

[46]. J. Wellhausen, *Sketch of the History of Israel and Judah*, Third Edition, Adam and Charles Black, London, (1891), pp. 201-203.

day of the LORD: 6 And he shall turn the heart of the fathers to the children, and the heart of the children to their fathers, lest I come and smite the earth *with* a curse."

Long before the Holocaust, some authors[47] cited *Exodus* 1:11-12 and *Exodus* 3:2 as instances where persecution benefitted the Jews and increased their numbers,

> "1:11 Therefore they did set over them taskmasters to afflict them with their burdens. And they built for Pharaoh treasure cities, Pithom and Raamses. 1:12 But the more they afflicted them, the more they multiplied and grew. And they were grieved because of the children of Israel. [***] 3:2 And the angel of the LORD appeared unto him in a flame of fire out of the midst of a bush: and he looked, and, behold, the bush burned with fire, and the bush was not consumed."

Jewish Zionists Theodor Herzl and Albert Einstein concluded that anti-Semitism was a good and useful thing, in that it forced Jews towards Zionism and segregation. Spiritual Zionist Ahad Ha-Am noted that Western Zionists thrived on anti-Semitism, because their racist political Zionism is "a product of anti-Semitism, and is dependent on anti-Semitism for its existence[.]"[48]

Zionist leader Chaim Weizmann wrote to Ha-Am in mid-December, 1914,

> "I pointed out to [Balfour] that we too are in agreement with the cultural antisemites, in so far as we believe that Germans of the Mosaic faith are an undesirable, demoralizing phenomenon, but that we totally disagree with [Cosima] Wagner and [Houston Stewart] Chamberlain as to the diagnosis and the prognosis; and I also said that, after all, all these Jews have taken part in building Germany, contributing much to her greatness, as other Jews have to the greatness of France and England, at the expense of the whole Jewish people, whose sufferings increase in proportion to 'the withdrawal' from that people of the creative elements which are absorbed into the surrounding communities—those same communities later reproaching us for this *absorption*, and reacting with antisemitism."[49]

---

47. *See:* John Gill's (1697-1771) *Exposition* works on: *Exodus* 3:2. *Psalm* 80:8; 129:2. II *Corinthians* 4:16. ***See also:*** "The Modern Jews", *The North American Review*, Volume 60, Number 127, (April, 1845), pp. 329-368, at 330.

48. A. Ha-Am, "The Jewish State and the Jewish Problem", in A. Hertzberg, *The Zionist Idea*, Harper Torchbooks, New York, (1959), pp. 262-269, at 266. *Contrast Ha-Am's comment with:* Dr. Birnbaum, "Dr BIRNBAUM mentioned that it had often been contended that Zionism was but a reaction against anti-Semitism... .", *The Jewish Chronicle*, (3 September 1897), p. 12.

49. Letter from C. Weizmann to A. Ha-Am of December 14 and 15, 1914, in: C. Weizmann, Edited by L. Stein, *et al.*, *The Letters and Papers of Chaim Weizmann*, Volume 7, Israel Universities Press, Jerusalem, (1975), pp. 81-83, at 81-82. Reprinted in:

Even after obtaining the Balfour Declaration in exchange for bringing America into the war on the side of Great Britain, the Zionists faced a seemingly insurmountable challenge after the First World War. The vast majority of Jews did not want to segregate, much less steal the Palestinian's land and move to the desert. The question prompts itself, to what extent did the Zionists promote the anti-Semitism of the Holocaust, which ultimately led to formation of the State of Israel?

Israel Zangwill, in consort with many other Zionists—including Einstein, stated in 1911,

> "But if from the Gentile point of view the Jewish problem is an artificial creation, there is a very real Jewish problem from the Jewish point of view—a problem which grows in exact proportion to the diminution of the artificial problem. Orthodox Judaism in the diaspora cannot exist except in a Ghetto, whether imposed from without or evolved from within. Rigidly professing Jews cannot enter the general social life and the professions. Jews *qua* Jews were better off in the Dark Ages, living as chattels of the king under his personal protection and to his private profit, or in the ages when they were confined in Ghettos. Even in the Russian Pale a certain measure of autonomy still exists. It is emancipation that brings the 'Jewish Problem.' It is precisely in Italy with its Jewish Prime Minister and its Jewish Syndic of Rome that this problem is most acute. The Saturday Sabbath imposes economic limitations even when the State has abolished them. As Shylock pointed out, his race cannot eat or drink with the Gentile. Indeed, social intercourse would lead to intermarriage. Unless Judaism is reformed it is, in the language of Heine, a misfortune, and if it is reformed, it cannot logically confine its teachings to the Hebrew race, which, lacking the normal protection of a territory, must be swallowed up by its proselytes. [***] Nor is there anywhere in the Jewish world of to-day any centripetal force to counteract these universal tendencies to dissipation. The religion is shattered into as many fragments as the race. After the fall of Jerusalem the Academy of Jabneh carried on the authoritative tradition of the *Sanhedrin*. In the Middle Ages there was the *Asefah* or Synod to unify Jews under Judaism. From the middle of the sixteenth to the middle of the eighteenth century, the *Waad* or Council of Four Lands legislated almost autonomously in those Central European regions where the mass of the Jews of the world was then congregated. To-day there is no center of authority, whether religious or political. Reform itself is infinitely individual, and nothing remains outside a few centers of congestion but a chaos of dissolving views and dissolving communities, saved from utter disappearance by persecution and racial sympathy. The notion that Jewish interests are Jesuitically federated or that Jewish financiers use their power for Jewish ends is one of the most ironic

---

L. Brenner, Editor, *51 Documents: Zionist Collaboration with the Nazis*, Barricade Books Inc., Fort Lee, New Jersey, (2002), pp. 21-22, at 22.

of myths. No Jewish people or nation now exists, no Jews even as sectarians of a specific faith with a specific center of authority such as Catholics or Wesleyans possess; nothing but a multitude of individuals, a mob hopelessly amorphous, divided alike in religion and political destiny. There is no common platform from which the Jews can be addressed, no common council to which any appeal can be made. Their only unity is negative—that unity imposed by the hostile hereditary vision of the ubiquitous Haman. [***] The labors of Hercules sink into child's play beside the task the late Dr. Herzl set himself in offering to this flotsam and jetsam of history the project of political reorganization on a single soil. But even had this dauntless idealist secured co-operation instead of bitter hostility from the denaturalized leaders of all these Jewries, the attempt to acquire Palestine would have had the opposition of Turkey and of the 600,000 Arabs in possession. It is little wonder that since the great leader's lamentable death, Zionism—again with that idealization of impotence—has sunk back into a cultural movement which instead of ending the Exile is to unify it through the Hebrew tongue and nationalist sentiment. But for such unification, a religious revival would have been infinitely more efficacious: race alone cannot survive the pressure of so many hostile milieux—or still more parlous—so many friendly. [***] In the diaspora anti-Semitism will always be the shadow of Semitism. The law of dislike for the unlike will always prevail. And whereas the unlike is normally situated at a safe distance, the Jews bring the unlike into the heart of every milieu and must thus defend a frontier-line as large as the world. The fortunes of war vary in every country, but there is a perpetual tension and friction even at the most peaceful points, which tend to throw back the race on itself. The drastic method of love—the only human dissolvent—has never been tried upon the Jew as a whole, and Russia carefully conserves—even by a ring fence—the breed she designs to destroy. But whether persecution extirpates or brotherhood melts, hate or love can never be simultaneous throughout the diaspora, and so there will probably always be a nucleus from which to restock this eternal type. But what a melancholy immortality! 'To be *and* not to be'—that is a question beside which Hamlet's alternative is crude. [***] But abolition of the Pale and the introduction of Jewish equality will be the deadliest blow ever aimed at Jewish nationality. Very soon a fervid Russian patriotism will reign in every Ghetto and the melting-up of the race will begin. But this absorption of the five million Jews into the other hundred and fifty millions of Russia constitutes the Jewish half of the problem. It is the affair of the Jews. [***] Moreover, while as already pointed out the Jewish upper classes are, if anything, inferior to the classes into which they are absorbed, the marked superiority of the Jewish masses to their environment, especially in Russia, would render *their* absorption a tragic degeneration."[50]

---

[50]. I. Zangwill, *The Problem of the Jewish Race*, Judaen Publishing Company, New York, (1914), pp. 4-5, 10-11, 17-20; which was first published as an article, "The Jewish Race", *The Independent*, Volume 71, Number 3271, (10 August 1911), pp. 288-295, at 289-291,

As early as 1903, Zangwill wrote,

> "At present, though orthodox rabbis are working amicably with ultra-modern thinkers, the movement is political, and more indebted to the pressure of the external forces of persecution than to internal energy and enkindlement. [***] Apart from its political working, Zionism forces upon the Jew a question the Jew hates to face. Without a rallying centre, geographical or spiritual; without a Synhedrion; without any principle of unity or of political action; without any common standpoint about the old Book; without the old cement of dictory laws and traditional ceremonies; without even ghetto walls built by his friend the enemy, it is impossible for Israel to persist further, except by a miracle—of stupidity. It is a wretched thing for a people to be saved only by its persecutors or its fools. As a religion, Judaism has still magnificent possibilities, but the time has come when it must be denationalized or renationalized."[51]

In 1914, Zionist Joseph Chaim Brenner stated that Jews owed their survival to anti-Semitism, a thought echoed by Albert Einstein,

> "Then they come and tell us: All praise to our history of martyrdom! All praise to the martyr-people who suffered everything and yet survived despite all persecution, all oppression by authorities, and all hatred of the people. But here, too, who can tell us what might have happened if not for the oppression and the hatred? Who can tell us whether, had there been no universal and understandable hatred of such a strange being, the Jew, that strange being would have survived at all? But the hatred was inevitable, and hence survival was equally inevitable! A form of survival such as befits that kind of being, survival with no struggle for worldly things (apart from those familiar livelihoods by which we live a dog's or a loan-shark's life) but, of course, full of martyrdom for the sake of the world-to-come, yes, certainly, in the name of the Kingdom of Heaven. [***] The expulsions and the ghettos—these assured our survival. [***] History! History! But what has history to tell? It can tell that wherever the majority population, by some fluke, did not hate the Jews among them, the Jews immediately started aping them in everything, gave in on everything, and mustered the last of their meager strength to be like everyone else. Even when the yoke of ghetto weighed most heavily upon them—how many broke through the walls? How many lost all self-respect in the face of the culture and beautiful way of life of the others! How many envied the others! How many yearned to approach them!"[52]

293-295.
51. I. Zangwill, "Zionism and the Future of the Jews", *The World's Work*, Volume 6, Number 5, (September, 1903), pp. 3895-3898, at 3897-3898.
52. J. H. Brenner, "Self-Criticism", in A. Hertzberg, *The Zionist Idea*, Harper

Before the Holocaust, political Zionists warned assimilatory Jewry that the Holocaust was coming, then political Zionists encouraged it. While the Holocaust was occurring, political Zionists rejoiced in the fact that the prophecies were being fulfilled and gloated over their warnings, which were made good by their own actions. It is some magician who holds up a cup of blood, predicts that it will spill, and then deliberately pours it onto the ground. After the Holocaust, Jewish and Christian Zionists poured blame on assimilatory Jewry for the demise of the Jews in Europe the Zionists had deliberately caused.[53] The Zionists had a road map to Jerusalem in the book of *Ezekiel*, and the road was paved by Hitler. *Ezekiel* 20:30-49:

"30 Wherefore say unto the house of Israel, Thus saith the Lord GOD; Are ye polluted after the manner of your fathers? and commit ye whoredom after their abominations? 31 For when ye offer your gifts, when ye make your sons to pass through the fire, ye pollute yourselves with all your idols, even unto this day: and shall I be enquired of by you, O house of Israel? As I live, saith the Lord GOD, I will not be enquired of by you. 32 And that which cometh into your mind shall not be at all, that ye say, We will be as the heathen, as the families of the countries, to serve wood and stone. 33 As I live, saith the Lord GOD, surely with a mighty hand, and with a stretched out arm, and with fury poured out, will I rule over you: 34 And I will bring you out from the people, and will gather you out of the countries wherein ye are scattered, with a mighty hand, and with a stretched out arm, and with fury poured out. 35 And I will bring you into the wilderness of the people, and there will I plead with you face to face. 36 Like as I pleaded with your fathers in the wilderness of the land of Egypt, so will I plead with you, saith the Lord GOD. 37 And I will cause you to pass under the rod, and I will bring you into the bond of the covenant: 38 And I will purge out from among you the rebels, and them that transgress against me: I will bring them forth out of the country where they sojourn, and they shall not enter into the land of Israel: and ye shall know that I *am* the LORD. 39 As for you, O house of Israel, thus saith the Lord GOD; Go ye, serve ye every one his idols, and hereafter *also*, if ye will not hearken unto me: but pollute ye my holy name no more with your gifts, and with your idols. 40 For in mine holy mountain, in the mountain of the height of Israel, saith the Lord GOD, there shall all the house of Israel, all of them in the land, serve me: there will I accept them, and there will I require your offerings, and the firstfruits of your oblations, with all your holy things. 41 I will accept you with your sweet savour, when I bring you out from the people, and gather you out of the countries wherein ye have been scattered; and I will be sanctified in you before the heathen. 42 And ye shall know that I *am* the LORD, when I shall bring you into the

---

Torchbooks, New York, (1959), pp. 307-312, at 308, 310.
53. An interesting dialog on this issue takes place in the 1991 film *The Quarrel* directed by Eli Cohen.

land of Israel, into the country *for* the which I lifted up mine hand to give it to your fathers. 43 And there shall ye remember your ways, and all your doings, wherein ye have been defiled; and ye shall lothe yourselves in your own sight for all your evils that ye have committed. 44 And ye shall know that I *am* the LORD when I have wrought with you for my name's sake, not according to your wicked ways, nor according to your corrupt doings, O ye house of Israel, saith the Lord GOD. 45 Moreover the word of the LORD came unto me, saying, 46 Son of man, set thy face toward the south, and *drop thy* word toward the south, and prophesy against the forest of the south field; 47 And say to the forest of the south, Hear the word of the LORD; Thus saith the Lord GOD; Behold, I will kindle a fire in thee, and it shall devour every green tree in thee, and every dry tree: the flaming flame shall not be quenched, and all faces from the south to the north shall be burned therein. 48 And all flesh shall see that I the LORD have kindled it: it shall not be quenched. 49 Then said I, Ah Lord GOD! they say of me, Doth he not speak parables?"

Ezekiel 21:31-32,

"31 And I will pour out mine indignation upon thee, I will blow against thee in the fire of my wrath, and deliver thee into the hand of brutish men, *and* skilful to destroy. 32 Thou shalt be for fuel to the fire; thy blood shall be in the midst of the land; thou shalt be no *more* remembered: for I the LORD have spoken it."

Ezekiel 28:18, 25,

"18 Thou hast defiled thy sanctuaries by the multitude of thine iniquities, by the iniquity of thy traffick; therefore will I bring forth a fire from the midst of thee, it shall devour thee, and I will bring thee to ashes upon the earth in the sight of all them that behold thee. [***] 25 Thus saith the Lord GOD; When I shall have gathered the house of Israel from the people among whom they are scattered, and shall be sanctified in them in the sight of the heathen, then shall they dwell in their land that I have given to my servant Jacob."

The political Zionists relied upon the hope that anti-Semitism would tend to force Jews into unity and segregation, and away from assimilation. Even after the nation of Israel was founded, the Israelis have been fighting a demographic battle for existence, which they believe compels them to propagandize for immigration.[54] Even today, the demographics of the Moslem versus Jewish populations in the region of Israel cause some to provoke international anti-Semitism, or to exaggerate the appearance of anti-Semitism, or to stage anti-

---

54. J. B. Agus, *The Meaning of Jewish History*, Volume 2, Abelard-Schuman, New York, (1963), pp. 446-447.

Semitic incidents in order to persuade more Jews to emigrate to Israel. It was only after the horrors of the Holocaust—shortly after—that the Jewish-State became a reality—after two-thousand years of failed attempts.

Again, the question prompts itself, to what extent did the Zionists promote the anti-Semitism of Fascism and Communism, which ultimately led to formation of the State of Israel? It was already clear to Jewish leaders in 1901, that the Zionists were threatening fellow Jews with a holocaust and were working with anti-Semites to make it happen,

> "Now behold Satan has come and confused the world. There are threats from the leaders of the Zionists that a powerful danger is lurking behind our walls and that the power of the enemies of Israel is prevailing—Heaven forbid. It is therefore all the more incumbent upon us to protect ourselves from confusing the masses of the people. Everyone who has a brain in his skull will realize that the Zionists, through their nonsensical writings, will only increase hostility; if they continue in their brazenness to spread the libel that we are in revolt against the peoples and that we are a danger to the lands in which we reside, then their evil prophecy will be fulfilled—Heaven forbid. [***] A thick cloak rests over the eyes of the leaders of the Zionists. Only owing to their lack of faith and absence of belief in God do they fail to realize the extent of the danger involved in their promises to the masses of the peoples among whom we live, of all the delights of the world provided they give aid to the Zionists. They even urge them to expel Jews from their midst and every sensible person will realize the help which they are giving to the enemies of Israel."[55]

In 1896, Theodor Herzl wrote his book *The Jewish State*,

> "Great exertions will not be necessary to spur on the movement. Anti-Semites provide the requisite impetus. They need only do what they did before, and then they will create a love of emigration where it did not previously exist, and strengthen it where it existed before. [***] I imagine that Governments will, either voluntarily or under pressure from the Anti-Semites, pay certain attention to this scheme; and they may perhaps actually receive it here and there with a sympathy which they will also show to the Society of Jews."[56]

Most Jews had no desire to colonize Palestine until after the Holocaust, and even then only very few of the Jews who had themselves suffered the Holocaust elected to move to Palestine after the Second World War and most of that few

---

[55]. M. Selzer, Editor, "Statement by the Holy Gerer Rebbe, the Sfas Emes, on Zionism (1901)", *Zionism Reconsidered: The Rejection of Jewish Normalcy*, Macmillan, New York, (1970), pp. 19-22, at 19-20.

[56]. T. Herzl, *A Jewish State: An Attempt at a Modern Solution of the Jewish Question*, The Maccabæan Publishing Co., New York, (1904), pp. 68, 93.

were forced, in one way or another, to do so.

"Christian" Zionists who were hoping for the Apocalypse also saw anti-Semitism as a good and useful thing, in that it forced Jews towards Zionism and segregation. Christian Zionist William Blackstone, who was praying for the end times when the anti-Christ would come and when Jews would be destroyed, wrote in a very popular book *Jesus Is Coming* in 1908,

> "The anti-semitic agitations in Germany, Austria and France, and the fierce persecutions in Russia and Roumania, have stirred up the Jews of the world as the eagle doth her nest. Deut. 32:11.
> [***]
> The Universelle Israelite Alliance was organized in Paris in 1860, and later the Anglo-Jewish Association in England. Through these powerful organizations the Jews can make themselves felt throughout the world. And now, within a few years, there have been organized Chovevi (lovers of) Zion and Shova (colonizers of) Zion societies, mostly among the orthodox Jews of Russia, Roumania, Germany, and even in England and the United States. This is really the first practical effort they have made to regain their home in Palestine.
>
> In a few words, followers of the status quo are striving to reconcile the genius of Judaism with the requirements of modern times, and in Western Europe are in a great majority.
>
> The Reformed Jews or Neologists have rapidly thrown away their faith in the inspiration of the Scriptures. They have flung to the wind all national and Messianic hopes. Their Rabbis preach rapturously about the mission of Judaism, while joining with the most radical higher critics in the destruction of its very basis, the inspiration of the Word of God. Some have gone clear over into agnosticism.
>
> Strange to say, from these agnostics now comes the other wing of the Zionist party. And not only have they joined this party, but they furnished the leaders, viz.: Dr. Max Nordau of Paris, and Dr. Theodore Herzl of Vienna.
>
> The orthodox Jews who have enlisted under the Zionist banner, are animated by the most devout religious motives. But the agnostics aver that this is not a religious movement at all. It is purely economic and nationalistic. Dr. Herzl, its founder and principal leader, espoused it as a *dernier resort*, to escape the persecutions of anti-semitism, which has taken such a firm hold of the masses of the Austrian people. He conceived the idea that if the Jews could regain Palestine and establish a government, even under the suzerainty of the Sultan, it would give them a national standing which would expunge anti-semitism from the other nations of the world, and make it possible for all Jews to live comfortably in any nation they may desire.
>
> Not all the orthodox Jews have joined this movement. Indeed, the leaders of the Chovevi Zion Societies hold aloof.
>
> The call, issued by Dr. Herzl, for the Zionist Congress, held in Basle,

Switzerland in 1897 met with severe opposition from the German Rabbis and also a large portion of the Jewish press, as well as the mass of rich reformed Jews. Nevertheless, over 200 delegates, from all over Europe and the Orient and some from the United States, met and carried through the program of the congress with tremendous enthusiasm.

Memorials, approving the object of the congress, came in from all sections, signed by tens of thousands of Jews.

The congress elected a central committee and authorized the raising of $50,000,000 capital.

It has certainly marked a wonderful innovation in the attitude of the Jews and a closer gathering of the dry bones of Ezekiel.

And now, after ten years of wonderful growth and progress it remains to be seen what the providential openings in the Ottoman Empire may be that shall give opportunity to realize its object.

Zionism is now the subject of the most acrimonious debate among the Jews. Many of the orthodox criticise it as an attempt to seize the prerogatives of their God.

While others say that God will not work miracles to accomplish that which they can do themselves.

Most of the reformed Jews, now that they can no longer ridicule the movement, decry it, as an egregious blunder that will increase instead of diminishing anti-semitism.

They have no desire to return to Palestine. They are like the man in Kansas, who, in a revival meeting said he did not want to go to heaven, nor did he wish to go to hell but he said he wanted to stay right there in Kansas.

Just so these reformed Jews are content to renounce all the prophesied glory of a Messianic kingdom in the land of their ancestors, preferring the palatial homes and gathered riches which they have acquired in Western Europe and the United States. They coolly advise their persecuted brethren, in Russia, Roumania, Persia and North Africa, to patiently endure their grievous persecutions until anti-semitism shall die out.

But these brethren retort that their prudent advisers would think very differently if they lived in Morocco or Russia, and that even in Western Europe anti-semitism instead of dying out, is rather on the increase.

In the midst of these disputes, the Zionists have seized the reins and eschewing the help of Abraham's God they have accepted agnostics as leaders and are plunging madly into this scheme for the erection of a Godless state.

But the Bible student will surely say, this godless national gathering of Israel is not the fulfilment of the glorious divine restoration, so glowingly described by the prophets.

No, indeed! Let it be carefully noted that while God has repeatedly promised to gather Israel, with such a magnificent display of *His* miraculous power, that it shall no more be said, 'The Lord liveth that brought up the children of Israel out of the land of Egypt; but the Lord liveth, that brought up the children of Israel from the land of the north and from all the lands whither he had driven them,' Jer. 16:14; yet has He also said, 'Gather yourselves together, yea, gather together, O nation that hath no longing, before the decree bring forth, before the day pass as the chaff, before the fierce anger of the Lord come upon you.' Zeph. 2:1, 2. Could this prophecy be more literally fulfilled than by this present Zionist movement?

One of the speakers at the first congress said of the Sultan, 'If His majesty will now receive us, we will accept Him as our Messiah.'

God says, 'Ye have sold yourselves for nought and ye shall be redeemed without money.' Isa. 52:3.

But Dr. Herzl is reported to have said, 'We must buy our way back to Palestine, salvation is to be by money.'

What a sign is this that the end of this dispensation is near.

If it stood alone we might well give heed to it. But when we find it supported by all these other signs, set forth in the Word, how can we refuse to believe it?

Shall we Christians condemn the Jews for not accepting the cumulative evidence that Jesus is the Messiah; and ourselves refuse this other cumulative evidence that His second coming is near?

It is significant that this first Zionist congress assembled just 1,260 years after the capture of Jerusalem by the Mohammedans in A. D. 637. Dan. 12:7.

It is probable that 'the times of the Gentiles' are nearing their end, and that the nations are soon to plunge into the mighty whirl of events connected with Israel's godless gathering, 'Jacob's trouble' (Jer. 30:6, 7), that awful time of tribulation, like which there has been none in the past, nor shall be in the future. Mat. 24:21.

But we, brethren, are not of the night. We are to watch and pray always that we may escape all these things that shall come to pass and stand before the Son of Man. Lu. 21:36.

Oh! glorious Hope. No wonder the Spirit and the Bride say come. No wonder the Bridegroom saith, 'Surely I come quickly,' and shall not we all join with the enraptured apostle,

'Even so come, Lord Jesus'?"[57]

The belief among some Jews that anti-Semitism has had beneficial consequences is not dead. In a work which is yet to be released, but which has been reviewed, *The Paradox of Anti-Semitism*, Continuum International Publishing Group, (2006), Rabbi Dan Cohn-Sherbock apparently asserts that anti-Semitism has had positive, as well as negative, consequences for the Jewish People. Jay Lefkowitz, director of Cabinet affairs in President George Herbert Walker Bush's administration, reiterated an old refrain,

---

[57]. W. E. Blackstone, *Jesus Is Coming*, Third Revision, The Bible house, Los Angeles, (1908), pp. 211, 238-241.

"Deep down, I believe that a little anti-Semitism is a good thing for the Jews—it reminds us who we are."[58]

5.13 Communist Jews in America

It was very persuasive to argue to anyone ignorant of the facts that the *Protocols* were fictions on their face and that there were no Zionist or financial groups operating behind the scenes to influence governments and the outcome of wars, as Louis Marshall did argue—just as it was persuasive to argue to anyone ignorant of the facts that the charges of an Italian organized crime syndicate operating at the same time were fictions. Joe Valachi has since bolstered the allegations that these secret, or not so secret, societies exist and that their corrupt actions and intentions pose a real threat to humanity. In fact, the Italian mafia was overseen by the Jewish mafia.

Benjamin Harrison Freedman, a man with firsthand knowledge of Zionist and Communist inner circles, came forward with allegations that Zionists and Communists had corrupted the Government of the United States of America and were responsible for America's involvement in World War I, and deliberately contributed to the tensions of post-World War I Germany.[59] It was also alleged

---

58. *The New York Times*, Magazine Section, (12 February 1995), pp. 38, 65.
59. B. H. Freedman, "Palestine", *Economic Council Letter*, (15 October 1947); **and** *Destiny Magazine*, (29 January 1948); "A Jewish Defector Warns America", Address at the Willard Hotel in Washington, D. C. in 1961, a transcription of which is widely available on the internet; **and** *The Hidden Tyranny Revealed*, New Christian Crusade Church, Metairie, Louisiana, (1970); **and** *Why Congress is Crooked or Crazy or Both*, New York, (1975). *See also:* C. B. Dall and B. Freedman, *Israel's Five Trillion Dollar Secret*, Liberty Bell Publications, Reedy, West Virginia, (1977). *See also:* D. Reed, *Somewhere South of Suez*, Devin-Adir, New York, (1951), pp. 330-332. *See also:* J. Rorty, "Storm Over the Investigating Committees", *Commentary*, Volume 19, Number 2, (February, 1955), pp. 128-136, at 131. *See Also:* A. Forster, *Square One*, D. I. Fine, New York, (1988), p. 121. *See also:* "Ready to Meet Suit, Jewish Group Says", *The New York Times*, (8 July 1946), p. 3. *See also:* "Mufti Mentioned at Libel Hearing" *The New York Times*, (4 May 1948), p. 20. *See also:* "Anti-Zionist Lists Policy 'Dictators'", *The New York Times*, (5 May 1948), p. 35. *See also:* "Witness Admits Aiding Arab Cause", *The New York Times*, (7 May 1948), p. 7. *See also:* "Anti-Zionist Tells of Dinner in Capital", *The New York Times*, (8 May 1948), p. 4. *See also:* "Anti-Zionist Tells of Threats on Life", *The New York Times*, (14 May 1948), p. 6. *See also:* "Denies Link to Smith", *The New York Times*, (18 May 1948), p. 48. *See also:* "Richardson Admits Link to M'Williams", *The New York Times*, (19 May 1948), p. 22. *See also:* "Anti-Nazi Leader Cleared of Libel", *The New York Times*, (27 May 1948), p. 23. *See also:* "Accuser Disputed by Mrs. Rosenberg", *The New York Times*, (9 December 1950), p. 18. *See also:* "Night Session Held on Rosenberg Case", *The New York Times*, (12 December 1950), p. 41. *See also:* "Two Deny Charges in Rosenberg Case", *The New York Times*, (13 December 1950), p. 25. *See also:* "Rankin Role Told in Rosenberg Case", *The New York Times*, (14 December 1950), p. 41. *See also:* "Committee Clears Anna M. Rosenberg", *The New York Times*, (15 December 1950), p. 24. *See also:* "Rosenberg Accusers Face Perjury Inquiry", *The New York Times*, (16 January 1951), p. 26. *See also:* "Sues for

that the Communists Ethel Greenglass Rosenberg and her husband Julius Rosenberg had treasonously provided Communists with American nuclear secrets for building atomic bombs. Ethel and Julius were convicted and executed.

Communist leaders like Jacob Abraham Stachel, a. k. a. "Jack" Stachel (deceased), were prosecuted by the United States Government. *The New York Times* stated in Stachel's obituary on 2 January 1966, *inter alia*,

> "Less well known than such party leaders as Eugene Dennis and Gus Hall, Jacob A. Stachel was one of the first 11 Communists convicted under the Smith Act in 1949 for conspiring to overthrow the United States Government and served five years in prison."[60]

Jacob Abraham Stachel (deceased), foreign born of Galician-Jewish origin, was a follower of the "self-hating Jew" Karl Marx.[61] Galician Jews had an especially bad reputation and were criticized by Gentiles and Jews alike, from Herzl to Hitler. A typical characterization is found in: E. A. Ross, *The Old World in the New: The Significance of Past and Present Immigration to the American People*, The Century Company, (1914), p. 146,

> "Besides the Russian Jews we are receiving large numbers from Galicia, Hungary, and Roumania. The last are said to be of a high type, whereas the Galician Jews are the lowest. It is these whom Joseph Pennell, the illustrator, found to be 'people who, despite their poverty, never work with their hands; whose town... is but a hideous nightmare of dirt, disease and poverty' and its misery and ugliness 'the outcome of their own habits and way of life and not, as is usually supposed, forced upon them by Christian persecutors.'"

There was a high concentration of Frankist Hasidic Jews in Galicia and one wonders how many of those Jewish Communist subversives who emigrated to America from Galicia were Frankists. Frankists often promoted anti-Semitism as means to promote themselves and as a means to take over Gentile governments. Communist Jews used this tactic in America.

Nathaniel Weyl wrote in his book *The Jew in American Politics*,

> "Although Communist leaders were normally taciturn about the extent to which Party membership was Jewish, Jack Stachel complained in *The Communist* for April 1929 that in Los Angeles 'practically 90 per cent of the membership is Jewish.' In 1945, John Williamson, another national leader of the American Communist Party, observed that, while a seventh of the Party membership was concentrated in Brooklyn, it was not the working-

---

$12,000,000", *The New York Times*, (22 March 1952), p. 30. *See also:* "Goldberg Urged to Reverse Pro-Israeli Policies of U. S.", *The New York Times*, (20 August 1965), p. 8.
60. "Jack Stachel, U. S. Communist And Party Official, Dies at 65", *The New York Times*, (2 January 1966), p. 73.
61. J. Prinz, *Wir Juden*, Erich Reiss, Berlin, (1934), p. 44.

class districts, but in Brownsville, Williamsburg, Coney Island and Bensonhurst, which he characterized as 'primarily Jewish American communities.' [***] The extent to which American Communism remained an organization of the foreign-born was revealed by a boast in *The Communist* for July 1936 that 45% of Party section organizers were now native-born as against none native-born in 1934. [***] In 1929, massacres of Jews by Palestine Arabs were described by the *Freiheit*, New York's Communist Party Yiddish organ, as a 'pogrom'. The Party promptly reprimanded the *Freiheit* for having failed to realize that these murders were a 'class war... against British imperialism and their Zionist agents.' The *Freiheit* proceeded to report the Palestine struggle in a Nazi fashion. 'Indeed,' comments Glazer, 'the cartoons it ran of hook-nosed and bloated Jews sadistically attacking Arabs could have appeared in any German anti-Semitic newspaper.'"[62]

Associate Justice of the United States Supreme Court Felix Frankfurter was suspected of being the power behind the throne of the Franklin Delano Roosevelt administration and was suspected of having been a Communist. It was alleged in 1950, that Frankfurter together with Henry Morgenthau, Jr. and Herbert H. Lehman corrupted the Government of the United States in the interests of Communism and Zionism. These three Jews were called, "A GOVERNMENT IN THEMSELVES".[63] Albert Einstein had an ongoing affair with a Soviet spy, Margarita Konenkova, and had other connections to Communism.[64] Max Born wrote, "Einstein was well known to be politically left-wing, if not 'red'."[65] In 1919, Einstein denied being a Bolshevist, but acknowledged that he was universally considered to be one. Albert Einstein wrote to Heinrich Zangger in mid-December, 1919, "Another comical thing is that I myself count everywhere as a Bolshevist[.]"[66] However, on 27 January 1920, Einstein informed Born that he was reading communist material and found the Bolsheviks appealing and believed that they would succeed in Germany.[67] Einstein defended Pacifist Georg Nicolai against an alleged conspiracy of the "pan-German press". Both

---

62. N. Weyl, *The Jew in American Politics*, Arlington House, New Rochelle, New York, (1968), pp. 118-120. Weyl cites: M. Epstein, *The Jew and Communism*, Trade Union Sponsoring Committee, New York, (1959); **and** N. Glazer, *The Social Basis of American Communism*, Harcourt, Brace, New York, (1961), pp. 151-152.
63. W. Trohan, "3 Men Called a Government in Themselves", *The Chicago Daily Tribune*, (29 May 1950, Final Edition), pp. 1-2.
64. F. S. Litten, "Note: Einstein and the Noulens Affair", *British Journal for the History of Science*, Volume 24, (1991), pp. 465-467.
65. M. Born, *My Life: Recollections of a Nobel Laureate*, Charles Scribner's Sons, New York, (1975), p. 185.
66. Letter from A. Einstein to H. Zangger of 15 or 22 December 1919, English translation by A. Hentschel, *The Collected Papers of ALbert Einstein*, Volume 9, Document 217, Princeton University Press, (2004), pp. 185-186, at 185.
67. Letter from A. Einstein to H. and M. Born of 27 January 1920, *The Collected Papers of Albert Einstein*, Volume 9, Document 284, Princeton University Press, (2004).

Einstein and Nicolai were signatories to the "Manifesto to the Europeans" and a protest against the murder of the Communist leaders Karl Liebknecht and Rosa Luxemburg.

Benjamin Harrison Freedman was active in the prosecutions of alleged Communist traitors. Freedman also made it his mission to expose the undue influence of Zionists on the American Government and over American public opinion. *The New York Times* reported (among other things) on 5 May 1948 on page 35,

> "Benjamin H. Freedman, who says he has spent more than $100,000 of his own money fighting Zionism, charged yesterday that outstanding Americans of the Jewish faith were the 'dictators' of our policy on Palestine."

Freedman made an address at the Biltmore Hotel and *The New York Times* reported on 20 August 1965 on page 8, in an article entitled "Goldberg Urged to Reverse Pro-Israeli Policies of U. S.":

> "Mr. Freedman declared that the presence of Israel in the Middle East was due to a world Zionist plot involving the British. The existence of a Jewish state in the Middle East, he said, could provoke a world nuclear war."

Today, there are plans in the ready to attack Iran with nuclear weapons in order to secure Israel's hegemony in the Middle East.

No one doubts the existence of the Mossad, nor their corrupt use of disloyal citizens of various nations throughout the world to infiltrate the mass media, financial markets and the governments of many nations. The Mossad is sponsored by a nation born out of Theodor Herzl's racist vision. The fact that the Mossad is a state sponsored institution renders it no less secretive and no less deadly than the Cosa Nostra. Of course, as with the Italian mafia, no generalization to all persons of Jewish descent can fairly be made based on the activities of those who are aggressively disloyal. To do so would be a gross injustice to millions of very fine people. Numerous Israeli agents, many, if not most of whom were American Jews, have been investigated and prosecuted by the Government of the United States of America for espionage. Israel has proven itself again and again to be an aggressive enemy of the United States.

In assessing the rôle some Jews played in the politics of the late Nineteenth and Early Twentieth Centuries, it must be recognized that there was a definite and urgent need for social change in the Nineteenth and early Twentieth Centuries and many of the Jews who participated in entirely reasonable efforts to bring about that much needed change are to be commended, admired and emulated. Their efforts to bring about social justice were in conflict with the perceived interests of monarchies and oligarchies around the world, making them the targets of smear campaigns by very powerful forces, who stood to lose much from equitable wealth distribution. Furthermore, lower level Jewish political Zionists and Communists have often been bitter enemies of each other. But the lower level games of these pawns ought not to distract attention from the genocidal Jewish financiers who oversaw and regulated both the Zionists and the Communists, and the Zionist Communists. The real goal, and it was one many Jews even on the lowest levels sensed, was to fulfill the Judaic Messianic prophecies.

The *Times* articles meant to refute the *Protocols* were in turn refuted by Paquita de Shishmareff who argued that Maurice Joly's book was itself derived from other sources, *i. e.* Karl Marx's good friend Jacob Venedey's *Macchiavel, Montesquieu, Rousseau*, Berlin, (1850); Niccolo Machiavelli's *The Prince*; and Charles de Secondat, Baron de Montesquieu's, *De l'esprit des lois, ou Du rapport que les loix doivent avoir avec la constitution de chaque gouvernement, les moeurs, le climat, la religion, le commerce, &c., à quoi d'auteu a ajouteé des recherches nouvelles, sur les loix romaines touchant les successions, sur les loix françoises, & sur les loix féodales*, Barrillot & Fils, Geneve, 1748; and Joly would likely have been introduced to these works by Adolphe Isaac Crémieux. Shishmareff argues that a prayer book which quotes the Bible is not rendered a forgery merely because it makes use of an earlier source.[68] In addition, there is a distinction between a forgery and a fabrication, and to call the book a forgery is to assert that the content of it is authentic.

Racist Zionist blackmailer Louis Dembitz Brandeis asserted in 1918 (therefore three years before the *Times* article appeared) that the *Protocols* were a forgery and asked that no response be published to refute them.[69] Brandeis intimated that he had evidence that they were a forgery. The first such evidence to come to the fore was a copy of Joly's book. Perhaps Brandeis had an original copy of the authentic *Protocols* and therefore had reason to believe that the Russian copy was a forgery.

The London Times published a letter from Zionist Israel Zangwill,[70] who alleged that Count A. M. du Chayla had seen the original handwritten *Protocols* in French, though others claim no such original ever existed. Chayla later testified at a trial meant to ban the publication of the *Protocols*. This trial took place in Bern in 1934, after having been instigated in 1933. A verdict was rendered in 1935. The outcome of the corrupt trial, which found that the *Protocols* must be suppressed, and the defendants must pay 28,000 francs, was overturned on appeal in 1937. The results of the original trial and of the appeal were miscast by some elements of the press to give the false impression that the *Protocols* had been proven a forgery, when in fact the defendants, and the right to free speech, had been vindicated.[71] Chayla smeared Nilus in a Russian

---

68. L. Fry, *Waters Flowing Eastward: The War Against the Kingship of Christ*, TBR Books, Washington, D. C., (2000), pp. 98-105.
69. L. D. Brandeis, M. I. Urofsky and D. W. Levy, Editors, *Letters of Louis D. Brandeis* Volume 4, State University of New York Press, Albany, New York, (1975), pp. 365, 452-453, 507-508.
70. I. Zangwill, Letter to the Editor, *The London Times*, (20 August 1921), p. 7.
71. *Confer:* H. de Vreis de Heekelingen, *Les Protocols des Sages de Sion Constituent-ils un Faux*, Lausanne, (1938); *partial English translation in:* L. Fry, "The Berne Trials", *Waters Flowing Eastward: The War Against the Kingship of Christ*, Appendix 2, TBR

language newspaper published in Paris, *Posledniya Novosty*, in 1921.[72] Nilus was persecuted by the Bolshevists in Russia, who made it a capital offense to possess copies of the *Protocols*. Chayla claimed that the *Protocols*, in their original French, were written by more than one person, and were in poor French. Tatiana Fermor claimed that Chayla was an *agent provocateur*, who was arrested for espionage, defiled Catholic churches, called for pogroms, etc., and cannot be considered a credible source.[73]

5.14 The Attempted Assassination of Henry Ford

Though the American Jewish leader Louis Marshall, president of the American Jewish Committee from 1912-1929, spoke out against the *Protocols* and pressured Putnam to not publishing them,[74] racist American Zionist leader and blackmailer, U. S. Supreme Court Justice Louis Brandeis refused to sign

---

Books, Washington, D. C., (2000), pp. 263-267; *see also* pp. 106-111. *See also:* W. Creutz, *Les Protocoles des Sages de Sion: Leur Authenticité*, Les Nouvelles Éditions Nationales, Brunoy, (1934); English translation, *New Light on the Protocols: Latest Evidence on the Veracity of this Remarkable Document*, The Right Cause, Chicago, (1935). *See also: The Jewish Victory at Berne: Side Lights on the Verdict*, Christian Aryan Protection League, London, (1935). *See also:* U. Fleischhauer, *Die echten Protokolle der Weisen von Zion; Sachverständigengutachten, erstattet im Auftrage des Richteramtes V in Bern*, U. Bodung, Erfurt, (1935). *See also: The Berne Trail— Concerning the "Protocols of the Elders of Zion." Fourteenth Session; Tuesday, May 14, 1935—P.M.*, (1935). *See also:* H. Jonak von Freyenwald, *Der Berner Prozess um die Protokolle der Weisen von Zion. Akten und Gutachten*, U. Bodung, Bund Nationalsozialistischer Eidgenossen, Erfurt, (1939). *See also:* G. Schwartz-Bostunitsch, *Jüdischer Imperialismus*, O. Ebersberger, Landsberg am Lech., (1935). *See also:* R. E. Edmondson, *The Damning Parallels of the Protocol "Forgeries" as Adopted and Fulfilled in the United States by Jewish-Radical Leadership: A Diabolical Capitalist-Communist Alliance Unmasked*, New York, (1935). *See also:* S. Vász, *Das Berner Fehlurteil über die Protokolle der Weisen von Zion: eine kritische Betrachtung über das Prozessverfahren*, U. Bodung, Erfurt, (1935). *See also: Berner Bilderbuch vom Zionisten-Prozess um die "Protokolle der Weisen von Zion."*, U. Bodung-Verlag, Erfurt, (1936). *See also:* G. Barroso, *Os Protocolos dos Sábios de Sião; o Imperialismo de Israel, o Plano dos Judeus para o Conquista do Mundo, o Código do Anti-Cristo, Provas de Autenticidade, Documentos*, Agencia Minerva, São Paulo, (1936). *See also:* K. Bergmeister, *Der jüdische Weltverschwörungsplan; die Protokolle der Weisen von Zion vor dem Strafgerichte in Bern*, U. Bodung, Erfurt, (1937); English translation, *The Jewish World Conspiracy; The Protocols of the Elders of Zion before the Court in Berne*, Sons of Liberty, Liberty Bell Publications, Hollywood, California, (1938).

72. H. Bernstein, *The Truth about 'The Protocols of Zion'*, Exhibit G, Ktav Publishing House, New York, (1971), pp. 36-369.

73. L. Fry, *Waters Flowing Eastward: The War Against the Kingship of Christ*, TBR Books, Washington, D. C., (2000), pp. 92-97.

74. L. Fry, *Waters Flowing Eastward: The War Against the Kingship of Christ*, TBR Books, Washington, D. C., (2000), pp. 79-90.

Marshall's protest[75] and defended Henry Ford, whose newspaper published articles which endorsed the *Protocols* and aggressively and personally attacked Louis Marshall.[76] Zionists placed enormous pressure on Marshall, Jacob Schiff and the American Jewish Committee to submit to their will, and Marshall feared them. THE DEARBORN INDEPENDENT attacked Marshall on 26 November 1921 (*see also:* "Hylan in Attack upon Untermyer", *The New York Times*, (2 November 1921), p. 3):

"America's' Jewish Enigma—Louis Marshall

SOMETHING of an enigma is Louis Marshall, whose name heads the list of organized Jewry in America, and who is known as the arch-protester against most things non-Jewish. He is head of nearly every Jewish movement that amounts to anything, and he is chief opponent of practically every non-Jewish movement that promises to amount to something. Yet he is known mostly as a name—and not a very Jewish name at that.

It would be interesting to know how the name of 'Marshall' found its way to this Jewish gentleman. It is not a common name, even among Jews who change their names. Louis Marshall is the only 'Marshall' listed in the Jewish Encyclopedia, and the only Jewish 'Marshall' in the index of the publications of the American Jewish Historical Society. In the list of the annual contributors to the American Jewish Committee are to be found such names as Marshutz, Mayer, Massal, Maremort, Mannheimer, Marx, Morse, Mackler, Marcus, Morris, Moskowitz, Marks, Margolis, Mareck—but only one 'Marshall,' and that is Louis. Of any other prominent Jew it may be asked, 'Which Straus?' 'Which Untermeyer?' 'Which Kahn?' 'Which Schiff?'—but never, 'Which Marshall?' for there is only one.

This in itself would indicate that Marshall is not a Jewish name. It is an American, or an Anglo-Saxon name transplanted into a Jewish family. But how and why are questions to which the public as yet have no answer.

Louis Marshall is head of the American Jewish Committee, and the American Jewish Committee is head of all official Jewish activity in the United States.

As head of the committee, he is also head of the executive committee of the New York Kehillah, an organization which is the active front of organized Jewry in New York, and the center of Jewish propaganda for the United States. The nominal head of the Kehillah is Rabbi Judah L. Magnes, a brother-in-law of Louis Marshall. Not only are the American Jewish Committee and the Kehillah linked officially (see chapter 33, Volume II, reprint of this series), but they are linked domestically as well.

Louis Marshall was president of all the Jewish Committees of the world

---

[75]. *The "Protocols," Bolshevism and the Jews: An Address to Their Fellow-Citizens*, American Jewish Committee, New York, (1921).
[76]. L. D. Brandeis, M. I. Urofsky and D. W. Levy, Editors, *Letters of Louis D. Brandeis* Volume 4, State University of New York Press, Albany, New York, (1975), p. 507.

at the Versailles Peace Conference, and it is charged now, as it has been charged before, that the Jewish Program is the only program that went through the Versailles conference as it was drawn, and the so-called League of Nations is busily carrying out its terms today. A determined effort is being made by Jews to have the Washington Conference take up the same matter. Colonel House was Louis Marshall's chief aid at Paris in forcing the Jewish Program on an unwilling world.

Louis Marshall has appeared in all the great Jewish cases. The impeachment of Governor Sulzer was a piece of Jewish revenge, but Louis Marshall was Sulzer's attorney. Sulzer was removed from the office of governor. The case of Leo Frank, a Jew, charged with the peculiarly vicious murder of a Georgia factory girl, was defended by Mr. Marshall. It was one of those cases where the whole world is whipped into excitement because a Jew is in trouble. It is almost an indication of the racial character of a culprit these days to note how much money is spent for him and how much fuss is raised concerning him. It seems to be a part of Jewish loyalty to prevent if possible the Gentile law being enforced against Jews. The Dreyfus case and the Frank case are examples of the endless publicity the Jews secure in behalf of their own people. Frank was reprieved from the death sentence, and sent to prison, after which he was killed. That horrible act can be traced directly to the state of public opinion which was caused by raucous Jewish publicity which stopped at nothing to attain its ends. To this day the state of Georgia is, in the average mind, part of an association of ideas directly traceable to this Jewish propaganda. Jewish publicity did to Georgia what it did to Russia—grossly misrepresented it, and so ceaselessly as to create a false impression generally. It is not without reason that the Ku Klux Klan was revived in Georgia and that Jews were excluded from membership.

Louis Marshall is chairman of the board and of the executive committee of the Jewish Theological Seminary of America, whose principal theologian, Mordecai M. Kaplan, is the leading exponent of an educational plan by which Judaism can be made to supercede Christianity in the United States. Under cover of synagogal activities, which he knows that the well known tolerance of the American people will never suspect, Rabbi Kaplan has thought out and systemized and launched a program to that end, certainly not without the approval of Mr. Marshall.

Louis Marshall is not the world leader of Jewry, but he is well advanced in Jewry's world counsel, as is seen by the fact that international Jewry reports to him, and also by the fact that he headed the Jews at the 'kosher conference'—as the Versailles assemblage was known among those on the inside. Strange things happened in Paris. Mr. Marshall and 'Colonel' House had affairs very well in hand between them. President Wilson sent a delegation to Syria to find out just what the contention of the Syrians was against the Jews, but that report has never seen the light of day. But it was the easiest thing imaginable to keep the President informed as to what the Jews of New York thought (that is, the few who had not taken up their residence in Paris). For example, this prominent dispatch in the New York

Times of May 27, 1919:
> 'Wilson gets Full Report of Jewish Protest Here.
> 'Copyright, 1919, by the New York Times Co.
> 'By Wireless to *The New York Times.*
> 'Paris, May 26.—Louis Marshall, who has succeeded Judge Mack as head of the Jewish Committee in Paris, was received by President Wilson this afternoon, and gave him a long cabled account of the Jewish mass meeting recently held in Madison Square Garden, including the full text of the resolutions adopted at the meeting .... and editorial comment in *The Times* and other papers ....'

When Russia fell, Louis Marshall hailed it with delight. The New York *Times* begins its story on March 19, 1917:

'Hailing the Russian upheaval as the greatest world event since the French Revolution, Louis Marshall in an interview for the New York *Times* last night said'—a number of things, among which was the statement that the events in Russia were no surprise. Of course they were not, the events being of Jewish origin, and Mr. Marshall being the recipient of the most intimate international news.

Even the new Russian revolutionary government made reports to Louis Marshall, as is shown by the dispatch printed in the New York *Times* of April 3, 1917, in which Baron Gunzburg reports what had been done to assure to the Jews the full advantage of the Russian upheaval.

This glorification of the Jewish overthrow of Russia, it must be remembered, occurred before the world knew what Bolshevism was, and before it realized that the revolution meant the withdrawal of the whole eastern front from the war. Russia was simply taken out of the war and the Central Powers left free to devote their whole attention to the western front. One of the resulting necessities was the immediate entrance of America into the conflict, and the prolongation of the hostilities for nearly two more years.

As the truth became known, Louis Marshall first defended, then explained, then denied—his latest position being that the Jews are against Bolshevism. He was brought to this position by the necessity of meeting the testimony of eye-witnesses as given to congressional investigation committees. This testimony came from responsible men whom even Mr. Marshall could not dispose of with a wave of his hand, and as time has gone on the testimony has increased to mountainous proportions that *Bolshevism is Jewish in its origin, its method, its personnel and its purpose.* Herman Bernstein, a member of Mr. Marshall's American Jewish Committee, has lately been preparing American public opinion for a great anti-Semitic movement in Russia. Certainly, it will be an anti-Semitic movement, because it will be anti-Bolshevist, and the Russian people, having lived with the hybrid for five years, are not mistaken as to its identity.

During the war, Mr. Marshall was the arch-protestor. While Mr. Baruch was running the war from the business end ('I probably had more power than perhaps any other man did in the war; doubtless that is true'), Mr. Marshall was running another side. We find him protesting because an army officer gave him instructions as to his duties as a registration official. It was Mr. Marshall who complained to the Secretary of War that a certain camp contractor, after trying out carpenters, had advertised for Christian carpenters only. It was to the discrimination in print that Mr. Marshall chiefly objected, it may be surmised, since it is the policy of his committee to make it impossible, or at least unhealthy, to use print to call attention to the Jew.

It was Mr. Marshall who compelled a change in the instructions sent out by the Provost Marshal General of the United States Army to the effect that 'the foreign-born, especially Jews, are more apt to malinger than the native-born.' It is said that a Jewish medical officer afterward confirmed this part of the instruction, saying that experience proved it. Nevertheless, President Wilson ordered that the paragraph be cut out.

It was Mr. Marshall who compelled the revision of the Plattsburg Officers' Training Manual. That valuable book rightly said that 'the ideal officer is a Christian gentleman.' Mr. Marshall wrote, wired, demanded, and the edition was changed. It now reads that 'the ideal officer is a courteous gentleman,' a big drop in idealism.

There was nothing too unimportant to draw forth Mr. Marshall's protest. To take care of protests alone, he must have a large organization.

And yet with all this high-tension pro-Jewish activity, Mr. Marshall is not a self-advertising man, as is his law partner, Samuel Untermyer, who has been referred to as the arch-inquisitor against the Gentiles. Marshall is a name, a power, not so much a public figure.

As an informed Jew said about the two men:

'No, Marshall doesn't advertise himself like Sam, and he has never tried to feature himself in the newspapers for personal reasons. Outside his professional life he devotes himself exclusively to religious affairs.' That is the way the American Jew like to describe the activities referred to above—'religious affairs.' We shall soon see that they are political affairs.

Mr. Marshall is short, stocky, and aggressive. Like his brother-in-law, Rabbi Magnes, he works on the principle that 'the Jew can do no wrong.' For many years Mr. Marshall has lived in a four-story brownstone house, of the old-fashioned type, with a grilled door, in East Seventy-second street. This is an old-time 'swell' neighborhood, once almost wholly occupied by wealthy Jews. It was as close as they could crowd to the choice Fifth Avenue corners, which had been pre-empted by the Vanderbilts, the Astors, and other rich families.

That Mr. Marshall regards the whole Jewish program in which he is engaged, not in its religious aspect alone, but in its world-wide political aspect, may be judged from his attitude on Zionism. Mr. Marshall wrote in 1918 as follows:

'I have never been identified and am not now in any way connected with the Zionist organization. I have never favored the creation of a sovereign Jewish state.'

BUT—

Mr. Marshall says, 'Let the Zionists go on. Don't interfere with them.' Why? He writes:

'Zionism is but an incident of a far-reaching plan. It is merely a convenient peg on which to hang a powerful weapon. All the protests that non-Zionists may make would be futile to affect that policy.'

He says that opposition to Zionism at that time would be dangerous. 'I could give concrete examples of a most impressive nature in support of what I have said. I am not an alarmist, and even my enemies will give me credit for not being a coward, but my love for our people is such that even if I were disposed to combat Zionism, I would shrink from the responsibilities that might be entailed were I to do so.'

And in concluding this strange pronouncement, he says:

'Give me the credit of believing that I am speaking advisedly.'

Of course, there is more to Zionism than appears on the surface, but this is as close as anyone can come to finding a Jewish admission on the subject.

If in this country there is apprehension over the Jewish Problem, the activities of Louis Marshall have been the most powerful agents to evoke it. His propagandas have occasioned great resentment in many sections of the United States. His opposition to salutary immigration laws, his dictation to book and periodical publishers, as in the recent case of G. P. Putnam's Sons, who modified their publishing program on his order; his campaign against the use of 'Christological expressions' by Federal, State and municipal officers; all have resulted in alarming the native population and harming the very cause he so indiscreetly advocates.

That this defender of 'Jewish rights,' and restless advocate of the Jewish religious propaganda, should make himself the leader in attacking the religion of the dominant race in this country, in ridiculing Sunday laws and heading an anti-Christianity campaign, seems, to say the least, inconsistent.

Mr. Marshall, who is regarded by the Jews as their greatest 'constitutional' lawyer, since the decline of Edward Lauterbach (and that is a tale!) originated, in a series of legal arguments, the contention that 'this is not a Christian country nor a Christian government.' This argument he has expounded in many writings. He has built up a large host of followers among contentious Jews, who have elaborated on this theme in a variety of ways. It is one of the main arguments of those who are endeavoring to build up a 'United Israel' in the United States.

Mr. Marshall maintains that the opening of deliberative assemblies and conventions with prayer is a 'hollow mockery'; he ridicules 'the absurd phrase 'In the name of God, Amen,'' as used in the beginning of wills. He opposes Sunday observance legislation as being 'the cloak of hypocrisy.' He advocates 'crushing out every agitation which tends to introduce into the body politic the virus of religious controversy.'

But Mr. Marshall himself has spent the last twenty years of his life in the 'virus of religious controversy.' A few of his more impertinent interferences have been noted above. These are, in the Jewish phrase, 'religious activities' with a decidedly political tinge.

The following extracts are quoted from the contentions of Mr. Marshall, published in the *Menorah Journal,* the official organ of the Jewish Chautauqua, that the United States is not a Christian country:

IS OURS A CHRISTIAN GOVERNMENT?
BY LOUIS MARSHALL

When, in 1892, Mr. Justice Brewer, in rendering the decision of the Supreme Court of the United States in the case of the Church of the Holy Trinity against the United States (144 U.S. 457), which involved an interpretation of the Alien Labor Law, indulged in the *obiter* remark that 'this is a Christian nation,' a subject was presented for the consideration of thoughtful minds which is of no ordinary importance.

The dictum of Mr. Justice Story in Vidal against Girard's Executors (2 How. U.S., 198), to the effect that Christianity was a part of the common law of Pennsylvania, is also relied upon, but is not an authoritative judicial determination of that proposition. The remark was not necessary to the decision.

The remarks of Mr. Justice Brewer, to which reference has already been made, were also unnecessary to the decision rendered by the court.

The fact that oaths are administered to witnesses, that the hollow mockery is pursued of opening deliberative assemblies and conventions with prayer, that wills begin with the absurd phrase 'In the name of God, Amen,' that gigantic missionary associations are in operation to establish Christian missions in every quarter of the globe, were all instanced. But none of these illustrations affords any valid proof in support of the assertion that 'this is a Christian nation.'

Our legislation relative to the observance of Sunday is such a mass of absurdities and inconsistencies that almost anything can be predicated thereon except the idea that our legislators are impressed with the notion that there is anything sacred in the day. According to the views of any section of the Christian church, the acts which I have enumerated as permitted would be regarded as sinful. Their legality in the eye of the law is a demonstration that the prohibitory enactments relating to Sunday are simply police regulations, and it should be the effort of every good American citizen to liberalize our Sunday legislation still more, so that it shall cease to be the cloak of hypocrisy.

As a final resort, we are told by our opponents that this is a Christian government because the majority of our citizens are adherents of the Christian faith; that this is a government of majorities, because government means force and majorities represent the preponderance of strength. This is a most dangerous doctrine ... .

If the Christianity of the United States is to be questioned, the last person to initiate the inquiry should be a member of that race which had no hand in creating the Constitution or in the upbuilding of the country. If Christian prayers in public are a hollow mockery, and Sunday laws unreasonable, the last person in the world to oppose them should be a Jew.

Mr. Marshall has the advantage of being an American by birth. He was born in Syracuse, New York, in 1856, the son of Jacob and Zilli Marshall. After practicing law in Syracuse, he established himself in New York, became a Wall Street corporation lawyer, and his native country has afforded him generous means to win a large fortune.

The question arises whether it is patriotic for Mr. Marshall to implant into the minds of his foreign-born co-religionists the idea that this is not a Christian country, that Sunday laws should be opposed, and that the manners and customs of the native-born should be scorned and ridiculed. The effect has been that thousands of immigrant Jews from Eastern Europe are persistently violating Sunday laws in the large industrial centers of the country, that they are haled to court, lectured by judges, and fined. American Jews who are carrying into practice the teachings of Mr. Marshall and his followers are reaping the whirlwind of a natural resentment.

Mr. Marshall was the leader of the movement which led to the abrogation of the treaty between the United States and Russia. Whenever government boards or committees are appointed to investigate the actions, conduct or conditions of foreign-born Jews, great influences are immediately exerted to have Mr. Marshall made a member of such bodies to 'protect' the Jewish interests.

As head of millions of organized Jews in the United States, Mr. Marshall has invariably wielded this influence by means of a campaign of 'protests,' to silence criticisms of Jewish wrongdoing. He thus protested when testimony was made before the Senate Sub-Committee in Washington, in 1919, that the Jewish East Side of New York was the hotbed of Bolshevism. Again he protested to Norman Hapgood against the editorial in *Harper's Weekly*, criticising the activities of Jewish lobbyists in Washington.

Mr. Marshall describes himself in 'Who's Who' as a leader in the fight for the abrogation of the treaty with Russia. That was a distinct interference in America's political affairs and was not a 'religious activity' connected with the preservation of 'Jewish rights' in the United States. The limiting expression 'in the United States' is, of course, our own assumption. It is doubtful if Mr. Marshall limits anything to the United States. He is a Jew and therefore an internationalist. He is ambassador of the 'international nation of Jewry' to the Gentile world.

The pro-Jewish fights in which Mr. Marshall has been engaged in this country make a considerable list:

He fought the proposal of the Census Bureau to enumerate Jews as a race. As a result, there are no official figures, except those prepared by the American Jewish Committee, as to the Jewish population of the United States. The Census has them listed under a score of different nationalities, which is not only a non-descriptive method, but a deceptive one as well. At a pinch the Jewish authorities will admit of 3,500,000 Jews in the United States. The increase in the amount of Passover Bread required would indicate that there are 6,000,000 in the United States now! But the

Government of the United States is entirely at sea, officially, as to the Jewish population of this country, except as the Jewish government in this country, as an act of courtesy, passes over certain figures to the government. The Jews have a 'foreign office' through which they deal with the Government of the United States.

Mr. Marshall also fought the proposed naturalization laws that would deprive 'Asiatics' of the privilege of becoming naturalized citizens. This was something of a confession!

Wherever there were extradition cases to be fought, preventing Jewish offenders from being extradited, Mr. Marshall was frequently one who assisted. This also was part of his 'religious activities,' perhaps.

He fought the right of the United States Government to restrict immigration. He has appeared oftener in Washington than any other Jew on this question.

In connection with this, it may be suggested to Mr. Marshall that if he is really interested in upholding the law of the land and restraining his own people from lawless acts, he could busy himself with profitable results if he would look into the smuggling of Jews across the Mexican and Canadian borders. And when that service is finished, he might look into the national Jewish system of bootlegging which, as a Jew of 'religious activities,' he should be concerned to break up.

Louis Marshall is leader of that movement which will force the Jew by law into places where he is not wanted. The law, compelling hotel keepers to permit Jews to make their hotels a place of resort if they want to, has been steadily pushed. Such a law is practically a Bolshevik order to destroy property, for it is commonly known what Jewish patronage does for public places. Where a few respectable Jews are permitted, the others flock. And when one day they discover that the place they 'patronize' is becoming known as 'a Jew hotel' or a 'Jew club,' then all the Jews abandon it—but they cannot take the stigma with them. The place is known as 'a Jew place,' but lacks both Jew and Gentile patronage as a result.

When Louis Marshall succeeded in compelling by Jewish pressure and Jewish threats the Congress of the United States to break the treaty with Russia, he was laying a train of causes which resulted in a prolongation of the war and the utter subjugation of Russia. Russia serves the world today as a living illustration of the ruthlessness, the stupidity and the reality of Jewish power—endless power, fanatically mobilized for a vengeful end, but most stupidly administered. Does Mr. Marshall ever reflect on the grotesque stupidity of Jewish leadership?

It is regretted that space does not permit the publication here of the correspondence between Mr. Marshall and Major G. H. Putnam, the publisher, as set forth in the annual report of the American Jewish Committee. It illustrates quite vividly the methods by which Mr. Marshall secures the suppression of books and other publications which he does not like. Mr. Marshall, assisted by factors which are not mentioned in his letter, procured the suppression of the Protocols, after the house of Putnam had them ready to publish, and procured later the withdrawal of a book on the Jewish Question which had attracted wide attention both here and in England.

Mr. Marshall apparently has no confidence in 'absurdities' appearing absurd to the reader, nor of 'lies' appearing false; but he would constitute himself a censor and a guide of public reading, as well as of international legislation. If one might hazard a guess—Mr. Marshall's kind of leadership is on the wane."

The correspondence between Marshall and Putnam appeared in the *American Jewish Year Book 5682* (1921-22), pp. 327ff. It is also reproduced, together with editorial comment, in L. Fry, *Waters Flowing Eastward: The War Against the Kingship of Christ*, TBR Books, Washington, D. C., (2000), pp. 79-90; and, with a very different editorial comment, in: *Louis Marshall: Champion of Liberty; Selected Papers and Addresses*, Volume 1, The Jewish Publication Society of America, Philadelphia, (1957), pp. xxxix, 320-389. Marshall attempted to explain his comments when writing to Max Senior (that letter which THE DEARBORN INDEPENDENT called, what "Mr. Marshall wrote in 1918") in a letter from Louis Marshall to John Spargo of 11 December 1920, and the context of the remainder of the letter is indeed important—as is the broader context of the Zionists' known intimidation of the American Jewish Committee.[77] It is interesting, though, that Marshall himself feared the consequences of a Congressional investigation of the charges made in THE DEARBORN INDEPENDENT and implicit in the *Protocols*. Was he worried about what might turn up?[78] Any investigation may turn up evidence of wrongdoing or embarrassing facts, which does not necessarily mean that the wrongdoing sought exits. Witch hunts may scare up goblins, instead. They might also turn up witches.

Aaron Sapiro sued Henry Ford for libel for attacking him and Jews in general, in 1927. The suit did not go well for Sapiro; but, mysteriously, Ford settled the suit and retracted the articles published in THE DEARBORN INDEPENDENT—after a mistrial had already been declared making it likely Ford would eventually win the case. Eventually, after many strange events and allegations, and the attempted assassination of Henry Ford, Louis Marshall and Aaron Sapiro forced Ford to retract his anti-Semitic campaign in 1927 in a written apology allegedly signed by Ford, which was widely published and

---

[77]. L. Marshall to J. Henkin, J. Spargo and M. Senior (and the Editors' notes to same), *Louis Marshall: Champion of Liberty; Selected Papers and Addresses*, Volume 1, The Jewish Publication Society of America, Philadelphia, (1957), pp. 350-355; Volume 2, pp. 721-723.

[78]. L. Marshall to J. Henkin, J. Spargo and M. Senior (and the Editors' notes to same), *Louis Marshall: Champion of Liberty; Selected Papers and Addresses*, Volume 1, The Jewish Publication Society of America, Philadelphia, (1957), pp. 350-355; Volume 2, pp. 721-723.

which was written by Marshall and others.[79]

Brandeis often wrote of his admiration for Ford. Marshall was confused by Ford and wrote to him,

> "What seemed most mysterious was the fact that you whom we had never wronged and whom we had looked upon as a kindly man, should have lent yourself to such a campaign of vilification apparently carried on with your sanction."[80]

Henry Ford's apology was not written by Ford nor by his lawyers,[81] but was instead written by Arthur Brisbane, Samuel Untermyer and Louis Marshall; and was signed by Ford's employee Harry Herbert Bennett with Ford's name. Marshall then wrote a letter to Ford graciously accepting the apology Marshall himself had written. Marshall had a well deserved reputation as a liar and a crooked lawyer. Like many Jewish leaders of his era, Marshall was immensely wealthy. Jewish corruption was one of the leading causes of economic inequality and wealth condensation in America. It was especially pernicious, because it tended to result from vice, theft and political corruption, rather than production.

Marshall wanted Ford to halt all publication of *The International Jew* around the world. On 7 December 1927, Adolf Hitler published an article in the *Völkischer Beobachter* which published Henry Ford's letter (written by Samuel Untermyer and Louis Marshall and published in *The New York Times*) to Theodor Fritsch, who published Ford's *The International Jew* in Germany, demanding that Fritsch cease publication of the German translation.[82] Fritsch wrote back to Ford and claimed that Ford's retraction and apology were

---

79. *The New York Times*, (9 July 1927), pp. 9, 12; *The New York Times*, (10 July 1927), p. 12. ***See also:*** H. Ford and L. Marshall, *Statement by Henry Ford Regarding Charges Against Jews, Made in His Publications: The Dearborn independent, and a Series of Pamphlets Entitled "The International Jew," Together with an Explanatory Statement by Louis Marshall, President of the American Jewish Committee, and His Reply to Mr. Ford*, American Jewish Committee, New York, (1927). ***See also:*** *Louis Marshall: Champion of Liberty; Selected Papers and Addresses*, Volume 1, The Jewish Publication Society of America, Philadelphia, (1957), pp. 321-389. ***See also:*** M. Rosenstock, *Louis Marshall, Defender of Jewish Rights*, Chapters 5-7, Wayne State University Press, Detroit, (1965). ***See also:*** H. Bennett and P. Marcus, *We Never Called Him Henry*, Chapter 7, Gold Medal Books, Fawcett Publications, (1951), pp. 46-56. ***See also:*** H. Bennett, *True* ("Man's Magazine"), (October, 1951), p. 125.

80. Marshall to Ford, 5 July 1927, *Louis Marshall: Champion of Liberty; Selected Papers and Addresses*, Volume 1, The Jewish Publication Society of America, Philadelphia, (1957), pp. 379-380 at 380.

81. *The New York Times*, (9 July 1927), pp. 1, 9; *The New York Times*, (10 July 1927), p. 1.

82. "Ford Stops Book on Jews in Europe", *The New York Times*, (14 November 1927), p.23. "Asks Ford to Press Ban on Sale of Book", *The New York Times*, (26 November 1927), p. 17. "Gets Ford's Ban on Book", *The New York Times*, (10 December 1927), p. 35.

insincere, and that Jewish bankers forced Ford to sign it, which was true. Since Hitler's article published only excerpts of Fritsch's letter to Ford, Marshall wrote to Ford requesting the entire letter so that he could tell Ford what to say in response to it. Therefore, we know that Ford was controlled by Marshall on these issues and was willing to put his signature on statements he had not written—probably nothing new for Ford. Marshall wrote to Ford,

> "This will enable me to indicate what I believe would be a desirable answer to [Fritsch's] unwarrantable remarks."[83]

The articles which were highly critical of Jews that were published in THE DEARBORN INDEPENDENT and republished in *The International Jew* were likely written by William J. Cameron, who replaced E.G. Pipp as editor of THE DEARBORN INDEPENDENT in April of 1920, just before the paper kicked off its anti-Jewish campaign on 22 May 1920. In turn, Cameron received his information from Boris Brasol[84] and Paquita de Shishmareff, who wrote *Waters Flowing Eastward: The War Against the Kingship of Christ* under the *nom de plume* Leslie Fry, which book attempts to prove the authenticity of the *Protocols*.[85] Paquita de Shishmareff was later named, then cleared, in President Roosevelt's Sedition Trials. Cameron believed in the myth that the British were a lost tribe of Israel—the so-called "British-Israel" movement.[86] This movement had a long association with Zionism and many of its founders and members were crypto-Jews and Zionists.

Henry Ford gave an interesting interview, which was published in *The New York Times* on 29 October 1922 on page 5. Ford had knowledge of Moloch, or Baal worship. Ford equated war to human sacrifice. Ford also stated that the beauty of the automobile was that it would promote "mixing". He asserted that wars would soon end. He asserted that he was not religious. It is strange that Ford had a difficult time identifying Benedict Arnold, but knew of such obscure beliefs as Moloch worship. Alex Jones has videotaped events at the "Bohemian Grove" where views not unlike Ford's were expressed. Ford's philosophy mirrors Cabalistic Judaism. Ford may well have instigated his campaign at the behest of, or in collusion with, very powerful forces, who wanted to fulfill Judaic

---

83. Marshall to Ford, 21 December 1927, *Louis Marshall: Champion of Liberty; Selected Papers and Addresses*, Volume 1, The Jewish Publication Society of America, Philadelphia, (1957), pp. 386-387 at 387.
84. B. Brasol, *The Protocols and World Revolution, Including a Translation and Analysis of the "Protocols of the Meetings of the Zionist Men of Wisdom"*, Small, Maynard & Company, Boston, (1920); **and** *Socialism Vs. Civilization*, C. Scribner's Sons, New York, (1920); **and** *The World at the Cross Roads*, Small, Mayhard & Co., Boston, (1921); **and** *The Balance Sheet of Sovietism*, Duffield, New York, (1922).
85. L. Fry, "About the Author", *Waters Flowing Eastward: The War Against the Kingship of Christ*, TBR Books, Washington, D. C., (2000), front matter.
86. M. Barkun, *Religion and the Racist Right: The Origin of the Christian Identity Movement*, Revised Edition, University of North Carolina Press, (1997).

prophecies. Ford's campaign against the Jews came at a time when powerful American Jews wanted to accomplish two goals: One, to stop, or at least slow, the influx of Eastern Jews into the United States who had been "freed" from the Pale of Settlement; and, two, to populate Palestine with Jews—American Jews, even American Zionists, clearly had no interest in trekking to the desert and they wanted to redirect Russian Jews to head towards Palestine.

It might also have been that Ford had heard of Nachum Sokolow's pronouncement that the First World War was an act of human sacrifice to Moloch. Sokolow's statement was published under the heading "Begrüssung für Sokolow. Zionistische Massendemonstration in Berlin", *Jüdische Rundschau*, Number 82/83, (14 October 1921), pp. 595-596 (front page and second page of the issue), at 595.

The question prompts itself, was Henry Ford a "useful idiot" for the Zionists and Bolshevists. The interview in *The New York Times* on 29 October 1922 on page 5:

"FORD, DENYING HATE, LAYS WAR TO JEWS"

Asserts They Are the Greatest
Victims of a Money System
That Is All Wrong.

HE ADMIRES THEIR POWER

Sees Education as Great Need and
Thinks Automobile Is Con-
tributing a Large Part.
Special to The New York Times.

BOSTON, Mass., Oct. 28.—'I curry favor with no man,' snapped Henry Ford, the automobile king, in answer to my question as he let his chair, which had been tilted back against the wall in his apartment at the Copley Plaza, fall forward with an abrupt jerk.

'But when I do say that I have no hatred in my heart for the Jew I mean it. In fact, I do not blame the Jew money-lender for bunking humanity just as long as humanity lets him get away with it. As a matter of fact, I admire the Jew because when things get stuck he is the only one who seems to have the power to start it up again and pull it over.' Tilting back the chair again, he resumed more quietly. 'However, that does not wipe out the fact that the Jew, who is a victim of a false money system, is the very foundation of the world's greatest curse today—war.

'He is the cause of all the abnormality in our daily life because he is the money maniac. One cannot blame him as long as he is able to play his game. Our money system is all wrong, and the Jew, who is the money specialist, is its greatest victim. There is the fact.

'No, I have no hatred for the Jew, and those Jews who play hardest at the money game are very much in the minority.'

As he paused and stroked his iron gray hair, I said:
'The money system—how would you change that?'
He came back quickly.

Would abolish Interest.

'I believe the whole world would benefit tremendously if all interest on money were abolished.' It was a startling statement, and I attempted to follow it up, whereupon the Detroit manufacturer dismissed the subject as quickly as he had broached it by answering:

'I cannot go into that further at this time because I am now writing something on that subject out in Detroit.

'To get back to the Jew again,' he continued voluntarily, 'the only reason that the Jew money lender doesn't take the pocketbook of the everyman is because the everyman won't let him. Through education the everyman will one day refuse to let the Jew bunk him with this institution called war, because it is these same money lenders who create war today. War is purely a financial institution. I learned that through my peace ship expedition. That expedition was a college of experience.'

'Where does patriotism fit here?' I asked.

'Patriotism,' he retorted, is as Johnson said, 'a last refuge for the scoundrel.' It is worked up by these money lenders who are playing their money game. Poverty, misery and the slaughtering of the flower of young manhood mean nothing to them as long as their money game goes on successfully.' The chair had come forward again and his thin hand was jerking back and forth.

'And the mob, true to its emotion,' he went on, 'swallows the stuff, hook, line and sinker, whereupon bands play and even mothers in the hysteria of it all place their own offspring upon the altar of murder, just as the ignorant mothers of years ago fed their babies into the flaming bowls of the god Moloch.'

'And how near are we to the end of it all?' he was asked.

'We will have more wars,' he answered, 'but we are nearer the end than most people think.'

Motor as an Educator.

'You spoke of education as the remedy. Just what kind?'

'Do you know,' he replied, 'the automobile is contributing a great part. It has opened new roads. It allows people to mix as never before. It is the mixing of people that will on some far day turn the trick. This idea that money is all there is to business is all wrong. The present system of business is simply an inheritance handed down through the ages. Doing something for humanity through business should be the dominating feature. This idea is the warp and woof of my Detroit industry.

'We are on the threshold of remarkable advances in industry. The main reason why I am here in the East at present is to inspect one of my new plants at Green Island, Troy, N. Y.—the only plant of its kind in the world where the heat, light and manufacturing power are all to be furnished by electricity.

'Coué? Oh, yes, I have read his philosophy. He has the right idea. People do not dream hard enough. I absolutely believe that if a person dreams his dreams intensely enough those dreams cannot help but come true. There is a reason for everything in this world, no matter how terrible it may seem. We are always going on for the better. Oh, no, I am not orthodox in my religion. Doing for your fellow-men is religion enough for me.'

'God? Why, God is in everything, always working for perfection. My motto is, one world at a time. Make this as fine a world as possible and don't worry about the next. That will take care of itself. Three worlds from now the Ford will be a better car than ever before, because of the experience gained. Life is experience. The whole process of reaching the ultimate perfection is naught but experience. Facts are facts, and we should not be afraid of them.'

'You are more or less of a fatalist, Mr. Ford?'

'Perhaps so, in the proper sense. Surely there is an inevitable law of action and reaction. Selfishness has little or nothing to do with it. If humanity suddenly discovers that by doing something for somebody else it itself will accrue greater benefit, the brotherhood of man idea will quickly prevail, purely from selfish motives. So, you see, selfishness has not so much to do with it as people think.'

Manner Suggests the Motor.

Henry Ford, tall and lithe, with his steel-gray eyes and quick motions and speech, suggest the motor. For the most part his quick answers have an air of finality to them, while at other times he turns questions that he does not care to answer aside with a kind of impatience.

'I know nothing about this President talk,' he almost snapped at one time. He could not be inveigled into discussing party politics in any way which was significant in itself.

However, his 'go through' spirit is an inspiring thing to see.

I first met him in the lobby of the hotel yesterday morning. He had not time to talk just then, but said if I cared to see him at 8 o'clock tonight he would be glad to do so.

'Where will I meet you?' I asked.

'Right here where we are now,' he answered, and I left him.

At 8 o'clock tonight, ten hours later, I stood in the lobby making a bet with myself that Henry Ford, who was being covered by the hotel authorities, who was not even listed on the register, would forget the appointment.

The theatre throngs had left the lobby, only a few people remained. The hands of the clock read 8:02. Suddenly an elevator door off to my left opened and Henry Ford stepped out. 'Ah, there you are,' he said.

'No wonder you turn out so many Fords a day with great precision,' I remarked. 'Why do you say that?' he questioned, his eyes wide.

'Why, ten hours have intervened since my seeing you for a brief moment this morning and you did not forget your appointment.'

'I never forget appointments. It is one of the first principles of business,' he answered in a matter-of-fact way."

Harry Herbert Bennett claimed,

"In the early 1920's Mr. Ford was getting an average of five threatening letters a week. When he rode down the street, his driver had a gun under each arm. Mr. Ford had two loaded Magnum revolvers in holsters that were built into the car, and if I rode with him, I carried a gun, too. [Many of Bennett's statements must be taken with a grain of salt. For example, the first "Magnum revolver", the .357 Magnum Smith & Wesson Model 27, did not appear until 1935.]"[87]

At 8:30 PM, on 27 March 1927, two men tried to murder Henry Ford. The attempted assassination occurred shortly before Ford was scheduled to testify in the Sapiro libel suit against him. Harry Herbert Bennett, an employee of Ford's and Ford's spokesman to the press when the attempt was made on Ford's life during the Sapiro trial, stated in his book *We Never Called Him Henry*,[88] and in *True* ("Man's Magazine"), October, 1951, page 125, that Arthur Brisbane, Samuel Untermyer and Louis Marshall had drawn up the apology which they wrongfully attributed to Ford; and that he, Harry Herbert Bennett, signed Ford's name on it—all of which was done with Ford's knowledge and consent. Ford did not read the "apology" and wanted it to be as "bad" as possible.[89] Why did Henry Ford allow himself to be controlled by Louis Marshall? *THE DEARBORN INDEPENDENT* had railed against Marshall on 26 November 1921 in an article entitled "America's Jewish Enigma—Louis Marshall".

After Aaron Sapiro filed his libel suit against Ford, Ford had investigators try to determine if there was any corruption involved in the prosecution of the case against him. Ford's investigation turned up evidence of jury tampering by Sapiro in the form of bribes. Due to the exposure of this scandal in the press, a mistrial was declared, even though Sapiro was cleared of the charges. Ford had essentially won the case.

However, shortly before Ford was scheduled to testify in the trial, two men in a large Studebaker sedan attempted to murder Henry Ford by forcing his Ford coupé off of a road and down a steep embankment immediately after Ford had crossed the bridge spanning the Rouge River on his way home. On 31 March 1927, *The New York Times* reported that there was suspicion that there had been a plot to murder Ford. The front page headline read, "Henry Ford Hurt in Crash as Other Car Upsets His; Plot to Kill Him Suspected".

On 2 April 1927, Harry Bennett, temporarily Ford's spokesman, told the

---

[87]. H. Bennett and P. Marcus, *We Never Called Him Henry*, Chapter 7, Gold Medal Books, Fawcett Publications, New York (1951), p. 42.

[88]. H. Bennett and P. Marcus, *We Never Called Him Henry*, Chapter 7, Gold Medal Books, Fawcett Publications, New York (1951), pp. 46-56 at 56.

[89]. H. Bennett and P. Marcus, *We Never Called Him Henry*, Chapter 7, Gold Medal Books, Fawcett Publications, (1951), p. 56.

press that the crash was an accident and that those who had run Ford off of the road were known and would not be prosecuted. However, many years later, after Henry Ford had died, Bennett published a polemic against Ford in 1951, which changed the alleged facts, as documented in the press of the 1920's. Bennett, in his later story, gave no indication that those who had chased Ford off of the road were known, but instead implied that Ford had staged the event, though Bennett offered up no proof and had made no such statements in 1927. Bennett claimed in 1951 that Ford's car had been run off of the bridge into the river and that he had investigated this accident scene. However, press accounts from the 1920's state that Ford's car was chased down an embankment after crossing the river and had missed the water. Bennett's conflicting accounts cannot be accurate, and in any event he had not witnessed what had occurred, though others had and they confirmed Ford's initial story.

It appears that Ford was frightened by the experience and, in spite of the fact that his lawyers had essentially won the case for him, Ford settled with Sapiro and Marshall. The American crime syndicate was run by Jews and Marshall had easy access to their services. After the attempted murder, Louis Marshall told Henry Ford, one of the most powerful men of industry in the world, what to do and what not to do. Douglas Reed states that murder and violent intimidation were common practice for political Zionists before and after Ford was attacked and that Zionists often murdered with impunity, especially in Palestine, due to their corrupt influence over the courts.[90] After the attack, Louis Marshall again and again stated that Ford would sign anything Marshall prepared for him. Louis Marshall wrote to Robert Marshall on 11 January 1928:

"[Henry Ford] expressed his readiness to do anything that I might at any time suggest to enable him to minimize the evil that had been done. In fact, for several months past, I have prepared letters for him in order to bring about the withdrawal and destruction of the re-published articles from the Dearborn Independent under the title 'The International Jew,' which have been circulated in various European countries in half a dozen languages. Ford is ready to sign anything that I prepare for him and has made a 'a holy show' of Fritsch—the most bitter of German anti-Semites who has now shown himself to be a low blackmailer."[91]

In a letter to Herman Bernstein of 21 February 1928, Louis Marshall wrote:

"I was very much amused at what Henry Ford told me when he called on me some weeks ago. He said that Cameron is out of a job and had indicated his willingness to write on the Jewish side of the subject. I replied that we

---

90. D. Reed, *Somewhere South of Suez*, Devin-Adair Company, U. S. A., (1951), pp. 324-327.
91. L. Marshall to R. Marshall, 11 January 1928, *Louis Marshall: Champion of Liberty; Selected Papers and Addresses*, Volume 1, The Jewish Publication Society of America, Philadelphia, (1957), pp. 387-389 at 388.

did not need his help."[92]

5.15 How the Zionists Blackmailed President Wilson

The Zionists asserted their influence in the uppermost positions in the United States Government through corrupt means. It is widely known that while serving as president at Princeton University, Woodrow Wilson, who was to become President of the United States of America, had an affair with a married woman known as "Mrs. Peck" (Mary Allen Peck a. k. a. Mary Allen Hulbert). Mrs. Peck divorced her husband and remarried, which second marriage also failed. Mrs. Peck retained Louis Marshall's law partner Samuel Untermyer (Zionist patron,[93] together with Brandeis a Rothschild partisan in the banking investigations,[94] corrupt war profiteer, coauthor of "Ford's" apology and later one of the chief organizers of the international boycott against German goods in 1933[95]) to bring suit against President Wilson for breach of promise. She alleged that Wilson had promised to marry her when his wife died.

Mrs. Peck offered up Wilson's love letters as proof of her allegation; but Wilson did not marry Mrs. Peck when his first wife died and instead married Mrs. Edith Bolling Galt. Mrs. Peck demanded $75,000.00USD from the

---

[92]. L. Marshall to H. Bernstein, 11 January 1928, *Louis Marshall: Champion of Liberty; Selected Papers and Addresses*, Volume 1, The Jewish Publication Society of America, Philadelphia, (1957), p. 388.

[93]. "Zionists in a Row at Big Convention", *The New York Times*, (29 June 1915), p. 8. "$100,000 Raised by the Zionists", *The New York Times*, (30 June 1915), p. 8.

[94]. **On Untermyer, see:** Corp Author: United States., Congress., House., Committee on rules., *Investigation of the Money Trust. No. 1-[2] Hearings Before the Committee on Rules of the House of Representatives, on House Resolutions 314 and 356. Friday, January 26, 1912.*, Washington, D. C., U. S. Govt. Print. Off., (1912). *See also:* Corp Author: United States., Congress., House., Committee on Banking and Currency., *Money Trust Investigation... Statistical and Other Information Compiled under Direction of the Committee.*, Washington, D. C., U. S. Govt. Print. Off., (1912). *See also:* A. P. Pujo and Arsène Paulin and E. A. Hayes. Corp Author: United States., Congress., House., Committee on Banking and Currency, *Money Trust Investigation. Investigation of Financial and Monetary Conditions in the United States under House Resolutions Nos.429 and 504, Before a Subcommittee of the Committee on Banking and Currency.*, Washington, D. C., U. S. Govt. Print. Off., (1913). *See also:* J. G. Milburn, W. F. Taylor, *Money Trust Investigation: Brief on Behalf of the New York Stock Exchange*, New York, C.G. Burgoyne, (1913). *See also:* J. P. Morgan, *Testimony of Mr. J. Pierpont Morgan and Mr. Henry P. Davison Before the Money Trust Investigation*, J.P. Morgan & Co., New York, (1913). **On Brandeis, see:** L. D. Brandeis, *Other People's Money and How the Bankers Use It*, F.A. Stokes, New York, (1914). *See also:* L. D. Brandeis, M. I. Urofsky and D. W. Levy, Editors, *Letters of Louis D. Brandeis*, In Five Volumes, State University of New York Press, Albany, New York, (1975). This set has numerous letters as well as editorial comment related to Brandeis and Untermyer's campaigns against some bankers.

[95]. "Text of Untermyer's Address", *The New York Times*, (7 August 1933), p. 4. *See also:* "Untermyer Back, Greeted in Harbor", *The New York Times*, (7 August 1933), p. 4.

President for breach of promise. Wilson did not have the money. If made public, these letters could have destroyed Wilson.

Samuel Untermyer and Louis Brandeis blackmailed President Wilson with Wilson's love letters from the affair with Mrs. Peck, forcing Wilson to nominate the outspoken and unpopular racist Zionist Louis Dembitz Brandeis for the United States Supreme Court. Brandeis was the least respected lawyer in the United States. In return, Untermyer paid Mrs. Peck[96] $65,000.00USD through the Zionist banker and multi-millionaire Bernard Baruch, who became Chairman of the War Industries Board under Wilson, and was a notorious war profiteer—Baruch proclaimed that he had more power during the war than any other person.[97] The Jewish leadership in America profiteered immensely from the First World War and cared not about the American lives lost to generate their profits. *The New York Times* reported on 25 August 1917 on the front page,

**"AMERICAN BOARD
TO BUY FOR ALLIES**
Baruch, Lovett, and Brookings
Named to Make All
Purchases Here.

BIG ECONOMIES EXPECTED

European Allies Have Been
Boosting Prices by Competitive
Dealings—More Loans.
Special to The New York Times.
 WASHINGTON, Aug. 24.—Official announcement was made tonight that an agreement had been reached between the Governments of the United States, Great Britain, France, and Russia, by which all purchases in this country for these allied Governments would be made by an American commission composed of Bernard M. Beruch, Robert S. Lovett, and Robert S. Brookings.

The announcement followed conferences today between the Secretary of the Treasury, Lord Northcliffe, special representative of Great Britain;

---

96. L. D. Brandeis, M. I. Urofsky and D. W. Levy, Editors, *Letters of Louis D. Brandeis* Volume 4, State University of New York Press, Albany, New York, (1975), pp. 264-265. *See also:* H. R. Miller, *Scandals in the Highest Office; Facts and Fictions in the Private Lives of our Presidents*, Random House, New York, (1973), pp. 172-199, *especially* 182-183, 196. Mrs. Peck published a book: M A. Hulbert, *The Story of Mrs. Peck, an Autobiography*, Minton, Balch & Company, New York, (1933).
97. "Hears Baruch Sold on Peace Note Tip", *The New York Times*, (4 January 1917), p.1. "American Board to Buy for Allies", *The New York Times*, (25 August 1917), p. 1. "May Sieze Oil Rich Lands", *The New York Times*, (28 April 1918), p. 11. J. A. Schwarz, *The Speculator: Bernard Baruch in Washington, 1917-1965*, University of North Carolina Press, Chapel Hill, North Carolina, (1981).

Ambassador Jusserand of France, and Ambassador Bakhmeteff of Russia. The agreement provides that hereafter all purchases of supplies of every description shall be made for account of this Government and the allied Governments concerned.

It is understood that Italy will assent to the agreement.

The official announcement, issued by Secretary McAdoo, was as follows:

'Formal agreements were signed today by the Secretary of the Treasury, with the approval of the President, on behalf of the United States, and by the representatives of Great Britain, France, and Russia for the creation of a commission with headquarters at Washington, through which all purchases made by those Governments in the United States shall proceed. It is expected that similar agreements will be signed with representatives of other allied Governments within the next few days.

'The agreements name Bernard M. Baruch, Robert S. Lovett and Robert S. Brookings as the commission. These gentlemen are also members of the recently created War Industries Board of the Council of National Defense, and will thereby be able thoroughly to coordinate the purchases of the United States Government with the purchases of the allied powers.

'It is believed that these arrangements will result in a more effective use of the combined resources of the United States and foreign Governments in the prosecution of the war.'

As rapidly as practicable other countries engaged in the war against the Central Powers will be brought into the arrangement. The purchasing commission will have headquarters in Washington and will avail itself of all the organized facilities already in operation for the prosecution of the war. The War Industries Board has had charge of enormous buying projects in the short time it has been in existence. Its members are intimately acquainted with every phase of the many business conditions involved in the supply of munitions and war supplies. They have acted with the constant co-operation and direction of President Wilson.

The action taken in forming the purchasing commission to take charge of the buying for all the Allies has been rendered necessary because of the continual disadvantages in the markets for various supplies resulting from the competitive buying of the many representatives of the different belligerent countries in the United States.

One of the most distinct difficulties occurring in this line became known within the past ten days, when it was found that France was buying copper in very large amounts in this country at a price far in excess of the likely to be paid by the United States under existing agreements with the copper syndicate. Similar instances were also found in the matter of buying wheat and meat supplies. In some cases it was found that agents of the allied countries had combed the Western markets for grain months in advance of any efforts of American buyers and had large quantities of materials stored awaiting favorable conditions of shipment, while prices went upward in consequence of the steadily increasing scarcity of certain staples.

The commission will begin its work at once. All programs for the purchase of war supplies will be laid before it and will receive its consideration and be carried out under its direction.

In the conferences today it was developed that the monthly program of advances of money by this Government to the Allies would be subject to a material increase in totals. The Italian campaign will require a larger credit, and other allowances will be larger hereafter. The ttotal of $500,000,000 a month heretofore loaned will be increased to $600,000,000. This money will be for the greater part expended in this country in the purchase of war supplies for the Allies and under the direction of the new Purchasing Commission."

Brandeis became the first Jewish Supreme Court Justice appointed to the United States Supreme Court, though not the first nominated. Untermyer was very active in Brandeis' nomination and subsequent appointment. It should be noted that Brandeis and Untermyer were men of ill repute and Brandeis' nomination was scandalous and was strongly opposed by many newspapers, the bar association, senators, President Taft, etc.[98] Brandeis and Untermyer worked together to secure the banking interests of the United States for the Rothschild family. Both Brandeis and Untermyer (and Untermyer's law partner Louis Marshall) were notorious "shysters".[99] Many former government officials and numerous active officials in the government sought to prevent Brandeis' appointment to the Supreme Court and a massive campaign was waged against him in fear that he might be appointed, which story was well covered in *The New York Times* over the period of several months.

If Untermyer and Brandeis did not blackmail Wilson, Brandeis, who was so widely hated and of such poor reputation, never would have been nominated or appointed to the Supreme Court. Nicholas Murray Butler wrote in 1936,

"When on January 28, 1916, President Wilson nominated Louis D. Brandeis of Boston to be an Associate Justice of the Supreme Court of the United States, there was furious criticism and opposition to the confirmation of this appointment from many members of the bench and bar."[100]

Brandeis had been recruited into racist Zionism by Theodor Herzl's honorary secretary in London, Jacob Judah Aaron de Haas,[101] and Brandeis was

---

98. M. I. Urofsky, "Attorney for the People—The 'Outrageous' Brandeis Nomination", *Supreme Court Historical Society 1979 Yearbook*, Volume 4, Supreme Court Historical Society, Washington, D. C., (1979).
99. "Hylan in Attack upon Untermyer", *The New York Times*, (2 November 1921), p. 3.
100. N. C. Butler, "Across the Busy Years: Part VI. Things Seen and Heard in Politics", *Scribner's Magazine*, Volume 100, Number 3, (September, 1936), p. 161.
101. L. D. Brandeis, M. I. Urofsky and D. W. Levy, Editors, *Letters of Louis D. Brandeis* Volume 2, State University of New York Press, Albany, New York, (1975), p. 659. A. Bein, *Theodore Herzl: A Biography of the Founder of Modern Zionism*, Meridian Books,

privy to Zionist secrets and, being a United States Supreme Court Justice, was a powerful and very well-connected mouthpiece for, and instrument of, Zionist policy in America. De Haas maintained a strong influence over Brandeis, and Brandeis controlled Wilson. The Zionists had an American dictator in their pocket. The Zionists used their influence over Woodrow Wilson to bring America into the First World War on the side of British, in exchange for the Balfour Declaration.

5.15.1 Before the War, the Zionists Plan a Peace Conference After the War—to be Led by a Zionist Like Woodrow Wilson

The Zionists orchestrated the First World War to disrupt the world, knowing that there would eventually be a need for a peace conference where the fate of small nations would be discussed, which would provide the Zionists with an opportunity to petition for a nation-state.[102] Political Zionists gave speeches before and during the First World War, which likened the situation of the Zionists, in terms of the war, to the efforts of Mazzini, Garibaldi and Cavour to solve the "Italian Question" through means of a peace conference.

The Sardinians had entered the Crimean War (1854-1856) on the side of the Turks, English and French, against the Russians, solely for the purpose of raising the "Italian Question" at the inevitable peace conference which would of necessity follow the Crimean War.[103] Giuseppe Mazzini, who was Jewish,[104] used Masonic lodges and secret societies to forward the agenda of World Jewry. He joined the "Carbonari". His crypto-Jewish revolutionary "Young Italy" movement served as the model for the crypto-Jewish "Young Turk" movement, which destroyed the Turkish Empire for the Zionists in the Balkan Wars and in the First World War. The crypto-Jewish Zionist Mustafa Kemal "Atatürk" then completely dissolved the Empire and promoted Turkish nationalism and anti-Imperialism.

The Masonic Lodges of Italy spread to Salonika through the Grand Master of the Macedonia Risorta Masonic Lodge, the Jewish Zionist Emmanuel Carasso, who headed the Committee of Union and Progress of the Jewish and crypto-Jewish "Young Turks", which committed genocide against the Armenian Christians, and which destroyed the Turkish Empire on behalf of the Zionists. Carasso worked with the Jewish Zionist Revolutionary terrorists Vladimir Jabotinsky and Israel Lazarevich Helphand, a. k. a. "Alexander Parvus"—who inspired the crypto-Jewish Zionist terrorist Lev Bronstein, a. k. a. "Leon

---

New York, (1962).
102. S. E. Weltmann, "Germany, Turkey, and the Zionist Movement, 1914-1918", *The Review of Politics*, Volume 23, Number 2, (April, 1961), pp. 246-269.
103. J. G. Sperling, "CRIMEAN WAR", *Encyclopedia International*, Grolier Incorporated, New York, (1966), pp. 320-321, at 320.
104. Maj.-Gen. Count Cherep-Spiridovich, *The Secret World Government or "The Hidden Hand": The Unrevealed in History 100 Historical Mysteries Explained*, (1926/2000), The Book Tree, Escondido, California, p. 139.

Trotsky", to adopt a policy of "permanent revolution" against the human race—and who brought Lenin from Switzerland to Russia.[105]

Following Mazzini's example, the Zionists sought a peace conference. In order to have a peace conference, there must first be war. The Zionists required a world war, in order to destroy the Turkish Empire and free up Palestine for the Jews; and in order to break apart the Russian Empire and free up the Jews for deportation to Palestine. The Zionists promoted a racist vision of small "racially" homogenous nations, and planned to make a plea for a Jewish State after they had ruined the Turkish Empire and brought the World into war.

President Wilson gave the Warburgs and other Jewish financiers great powers in the United States Government. During the war, Wilson appointed Bernard Baruch as Chairman of the War Industries Board. Before America had entered the war, President Wilson's advisor "Colonel" Edward Mandell House, who had close connections with the New York financiers, had begun work on President Wilson's "Fourteen Points". Before the war had even begun, House essentially defined the League of Nations in his book *Philip Dru: Administrator* published in 1912, which League of Nations—as defined in the Covenants House drafted in 1918—paved the way for the Zionists' Mandate for Palestine of 1922. America declared war against Germany in April of 1917, and in the same month, "Colonel" House met with Balfour to discuss the terms of peace. Later in 1917, Balfour issued the famous Balfour Declaration to the most famous financier of them all, "Lord" Rothschild. House also organized "The Inquiry" in 1917, which was a board that planned peace negotiations. President Wilson issued the "Fourteen Points" in 1918, which misled Germany into surrendering; and in 1919, "Colonel" House betrayed President Wilson, America and Germany to British, French and Zionist interests at the Paris Peace Conference. At that point, Wilson had finally had enough—though his health suddenly began to fail him.

Zionist Louis Brandeis stated in 1915,

"The war is developing opportunities which make possible the solution of the Jewish problem. [***] While every other people is striving for development by asserting its nationality, and a great war is making clear the value of small nations, shall we voluntarily yield to anti-Semitism, and instead of solving our 'problem' end it by noble suicide? Surely this is no time for Jews to despair. Let us make clear to the world that we too are a nationality striving for equal rights to life and to self-expression. That this should be our course has been recently expressed by high non-Jewish authority. Thus Seton-Watson; speaking of the probable results of the war, said:

---

105. D. Fahey, *The Mystical Body of Christ in the Modern World*, Browne and Nolan Limited, London, (1935); **and** *The Rulers of Russia*, Third Revised and Enlarged American Edition, Condon Printing Co., Detroit, (1940). *See also:* J. M. Landau, *Pan-Turkism, from Irredentism to Cooperation*, Indiana University Press, Bloomington, (1995).

'There are good grounds for hoping that it [the war] will also give a new and healthy impetus to Jewish national policy, grant freer play to their splendid qualities, and enable them to shake off the false shame which has led men who ought to be proud of their Jewish race to assume so many alien disguises and to accuse of anti-Semitism those who refuse to be deceived by mere appearances. It is high time that the Jews should realize that few things do more to foster anti-Semitic feeling than this very tendency to sail under false colors and conceal their true identity. The Zionists and the orthodox Jewish Nationalists have long ago won the respect and admiration of the world. No race has ever defied assimilation so stubbornly and so successfully; and the modern tendency of individual Jews to repudiate what is one of their chief glories suggests an almost comic resolve to fight against the course of nature.' [***] The Zionist movement is idealistic, but it is also essentially practical. It seeks to realize that hope; to make the dream of a Jewish life in a Jewish land come true as other great dreams of the world have been realized, by men working with devotion, intelligence, and self-sacrifice. It was thus that the dream of Italian independence and unity, after centuries of vain hope, came true through the efforts of Mazzini, Garibaldi and Cavour; that the dream of Greek, of Bulgarian and of Serbian independence became facts."

Zionists had been planning for an international peace conference following a devastating world war at least since the Congress of Vienna of 1814-1815 failed to achieve the results the Rothschilds sought. They thought to use the arguments of small nations for independence, based on the historic unity of the peoples of those territories, as a basis to argue for a Jewish state. In 1923, racist Zionist Israel Zangwill lamented that the League of Nations and the First World War had failed to achieve the Zionist's goals. In an article entitled, "Mr. Zangwill on Zionism", *The London Times* wrote, on 16 October 1923, on page 11,

"The only hope for the Jewish Diaspora lay in the clause of the Treaty of Versailles providing for the protection of minorities. But the League of Nations had only moral power, and was as yet only spurious institution."

Racist Zionist Theodor Herzl spoke at the first Zionist Congress of 1897 and disclosed the machinations of the Zionists and their centuries' old desire to destroy the Turkish Empire and bankrupt the Sultan. Herzl had a covert plan to have Turks mass murder Armenians, which would cause an outrage around the world, so as to leave the Turkish Empire at the mercy of the Jewish controlled media, which Herzl pledged would cover up the atrocities if the Sultan would agree to give the Zionists Palestine.[106] *The New York Times* reported on 31

---

[106]. "The Turkish Situation by One Born in Turkey", *The American Monthly Review of Reviews*, Volume 25, Number 2, (February, 1902), pp. 182-191, at 186-188. "Zionism", *Encyclopædia Britannica*, Eleventh Edition, (1911).

August 1897 on page 7,

"ZIONIST CONGRESS IN BASEL.

The Delegates Adopt Dr. Herzl's Programme
for Re-establishing the Jews in Palestine.

BASEL, Switzerland, Aug., 30.—At to-day's session of the Zionist Congress the delegates present unanimously adopted, with great enthusiasm, the programme for re-establishing the Hebrews in Palestine, with publicly recognized rights.

A dispatch was sent to the Sultan of Turkey, thanking his Majesty for the privileges enjoyed by the Hebrews in his empire.

The Zionist Congress opened at Basel yesterday with 200 delegates in attendance from various parts of Europe. Dr. Theodor Herzl, the so-called 'New Moses' and originator of the scheme to purchase Palestine and resettle the Hebrews there, was elected President and Dr. Max Nordau was elected Vice President of the Congress.

Dr. Herzl has only recently come into prominence. He seeks to float a limited-liability company in London for the purpose of acquiring Palestine from the Sultan of Turkey and thoroughly organizing it for resettlement by the Hebrews. He has, it is said, already won converts to the Zionistic movement in all parts of the world.

When asked to outline his plans, Dr. Herzl said:

'We shall first send out an exploring expedition, equipped with all the modern resources of science, which will thoroughly overhaul the land from one end to the other before it is colonized, and establish telephonic and telegraphic communication with the base as it advances. The old methods of colonization will not do here.

'See here,' continued Dr. Herzl, showing a good-sized book, 'this is one of the four books which contain the records of the movement—the logbooks of the Mayflower,' he added, with a smile. That one watchword, the 'Jewish State,' has been sufficient to rouse the Jews to a state of enthusiasm in the remotest corners of the earth, though there are those forming the so-called philanthropic party who predict that the watchword will provoke reprisals from Turkey. Inquiries in Constantinople and Palestine show that nothing is further from the truth.

'My plan is simple enough. We must obtain the sovereignty over Palestine—our never-to-be-forgotten, historical home. At the head of the movement will be two great and powerful agents—the Society of Jews and the Jewish Company. The first named will be a political organization, and spread the Jewish propaganda. The latter will be a limited-liability company, under English laws, having its headquarters In London and a capital of, say, a milliard of marks. Its task will be to discharge all the financial obligations of the retiring Jews and regulate the economic conditions in the new country. At first we shall send only unskilled labor—that is, the very poorest, who will make the land arable. They will lay out streets, build bridges and railroads, regulate rivers, and lay down telegraphs according to plans prepared at headquarters. Their work will bring trade, their trade the market, and the markets will cause new settlers to flock to the country. Every one will go there voluntarily, at his or her own risk, but ever under the watchful eye and protection of the organization.

'I think we shall find Palestine at our disposal sooner than we expected. Last year I went to Constantinople and had two long conferences with the Grand Vizier, to whom I pointed out that the key to the preservation of Turkey lay in the solution of the Jewish question.

'The Jews, in exchange for Palestine, would regulate the Sultan's finances and prevent disintegration, while for Europe we should form a new outpost against Asiatic barbarism and a guard of honor to hold intact the sacred shrines of the Christians.

'We can afford to play a waiting game, and either take over Palestine from the European Congress called together to divide the spoils of disintegrated Turkey, or look out for another land, such as Argentina, and say: 'Your Zion Is there.'

'It is to confer over this point that the congress was arranged for at Basel.

'I am sure that the Jews are even better colonists than Englishmen. There are already colonies of Jews in Palestine, and I have on my table excellent Bordeaux, Sauterne, and cognac grown in that country. It is well known that in Galicia and the Balkans the Jews perform the roughest kind of manual labor. There the wealth they bring is not their money, but themselves.'"

When Herzl's designs failed to achieve their ends, the First World War and the Jewish-led revolution of the "Young Turks"[107] achieved the same objectives. The crypto-Jewish Young Turks committed genocide against the Armenian Christians. In exchange for the Zionists having brought America into the war against America's own best interests and without the consent of American People, the Allies destroyed the Turkish Empire and took Palestine by force in the First World War, which had been the Zionists' goal for centuries. The Zionists created the war in order to achieve these ends, and had been planning and fomenting the war for many generations.

When the First World War had only just begun, Chaim Weizmann wrote to Shmarya Levin, in New York, on 23 September 1914, that the war provided a means to establish a Jewish state in Palestine,

"But will it be possible to raise a Jewish voice also when there is talk of

---

107. I. Zangwill, *The Problem of the Jewish Race*, Judaen Publishing Company, New York, (1914), pp. 9, 11; which was first published as an article, "The Jewish Race", *The Independent*, Volume 71, Number 3271, (10 August 1911), pp. 288-295, at 290-291. J. Prinz, *The Secret Jews*, Random House, New York, (1973), pp. 111-112.

peace, when the interests of small nations are to be safeguarded? This, my dear friends, is what will fall, in part at least, to your lot, for America will play an enormous role in the clarification of all these questions. We in Europe can, and should, prepare for that time, and I'd very much like to know your views about it."[108]

The *Encyclopedia International* wrote in its article on Weizmann,

"As director of the Admiralty laboratories (1916-19), [Chaim Weizmann] contrived a process for extracting acetone, a solvent used in making cordite [an explosive propellent used as a smokeless replacement for black powder], from cereal and horse chestnuts. This significant discovery gave Weizmann diplomatic leverage in negotiating with the British wartime government on the future of Zionism, a cause he had adopted in 1898. A product of these negotiations was the Balfour Declaration, a promissory statement of support for 'the establishment in Palestine of a national home for the Jewish people,' issued on Nov. 2, 1917, by the Foreign Secretary."[109]

British Prime Minister Lloyd George stated in his *War Memoirs*,

"When our difficulties were solved through Dr. Weizmann's genius, I said to him: 'You have rendered great service to the State, and I should like to ask the Prime Minister to recommend you to His Majesty for some honour.' He said: 'There is nothing I want for myself.' 'But is there nothing we can do as a recognition of your valuable assistance to the country?' I asked. He replied: ' Yes, I would like you to do something for my people.' He then explained his aspirations as to the repatriation of the Jews to the sacred land on Palestine they had made famous. That was the fount and origin of the famous declaration about the National Home for Jews in Palestine.

As soon as I became Prime Minister I talked the whole matter over with Mr. Balfour, who was then Foreign Secretary. As a scientist he was immensely interested when I told him of Dr. Weizmann's achievement. We were anxious at that time to enlist Jewish support in neutral countries, notably in America. Dr. Weizmann was brought into direct contact with the Foreign Secretary. This was the beginning of an association, the outcome of which after long examination, was the famous Balfour Declaration which

---

108. C. Weizmann to S. Levin, *The Letters and Papers of Chaim Weizmann*, Volume 7, Series A, Israeli University Press, Jerusalem, (1975), pp. 10-13, at 11.
109. "Weizmann, Chaim", *Encyclopedia International*, Volume 19, Grolier, New York, (1966), pp. 288-289, at 288. Weizmann's autobiography and his letters and papers provide a wealth of information about his involvement in Zionism during World War I, *see:* C. Weizmann, *The Letters and Papers of Chaim Weizmann*, Volumes 7 and 8, Series A, Israeli University Press, Jerusalem, (1975); **and** *Trial and Error: The Autobiography of Chaim Weizmann*, Harper & Brothers, New York, (1949).

became the charter of the Zionist movement. So that Dr. Weizmann with his discovery not only helped us to win the War, but made a permanent mark upon the map of the world."[110]

Harry Elmer Barnes wrote several books which detailed the propaganda the Allies and Americans used to draw America into the First World War.[111] He records that President Wilson desired to enter the war in the Spring of 1917, in order to give America a voice in the planned Peace Conference—one of the chief aims of the Zionists,

> "Having been converted to intervention by these various influences, Mr. Wilson rationalized his change of mind in terms of noble moral purpose. As he told Jane Addams in the spring of 1917, he felt that, if there was to be any hope of a just and constructive peace, the United States must be represented at the peace conference following the war. Mr. Wilson could only be at the peace conference if the United States had previously entered the conflict."[112]

Barnes again stated in 1940,

> "When, as an outcome of these various influences, Wilson had been converted to intervention, he rationalized his change of attitude on the basis of a noble moral purpose. As he told Jane Addams in the spring of 1917, he felt that the United States must be represented at the peace conference which would end the World War if there was to be any hope of a just and constructive peace. But Wilson could be at the peace conference only if the United States had previously entered the World War."[113]

Louis Marshall, President of the American Jewish Committee, wrote to John Spargo on 11 December 1920,

> "I was strongly pro-Ally from the day that Germany declared war, and I labored constantly to see to it that the Jews of the United States, so far as

---

110. D. Lloyd George, *War Memoirs of David Lloyd George*, Volume 2, I. Nicholson & Watson, London, (1933-1936), pp. 586-587.
111. H. E. Barnes, *The Genesis of the World War: An Introduction to the Problem of War Guilt*, A.A. Knopf, New York, London, (1927); **and** *In Quest of Truth and Justice: Debunking the War Guilt Myth*, National Historical Society, Chicago, (1928); **and** *War in the Twentieth Century*, Dryden Press, New York, 1940.
112. H. E. Barnes, "When Last We Were Neutral", *The American Mercury*, (November, 1939), pp. 276-284, at 279. Entered in the Congressional Record of the United States of America, *Congressional Record: Proceedings and Debates of the 76th Congress: Second Session*, Volume 85, Part 2, United States Government Printing Office, Washington, D.C., (1939), pp. 578-580, at 579.
113. H. E. Barnes, *War in the Twentieth Century*, Random House, Dryden Press, New York, (1940), pp. 75-76.

my influence could accomplish that result, would say nothing and do nothing that would in any way militate against the Entente. I can say, with all becoming modesty, that I was most successful in that endeavor. When the Balfour Declaration was made, I looked upon it as only incidentally of interest to the Jews. I interpreted it as an important political move, undoubtedly inspired by altruism, but at the same time intended to strengthen the Entente, and especially England, in the Near East, to protect the Suez Canal and the road to India."[114]

Some asserted after the war that England had been duped into Palestine and the Balfour Declaration by the Zionists, who led the British to believe that it would be in their best interests for the Jews to control the land around the Suez and secure the British route to India and to oil. Some claim that this arrangement instead cost the British dearly.[115] In May of 1916, France and England divided Palestine in half in the Sykes-Picot Pact.[116] After the war, pursuant to the San Remo Conference, France sought to control all of Syria, including much of Palestine. Henry Morgenthau pointed out that to give the Jews Palestine on the premise that it would secure the Suez for the British was a false notion. Instead, it would have inflamed the Moslem world against England and would have caused unrest among the millions of Moslems in India. This might have cost the British India and thereby made the Suez of next to no value to the British— except perhaps as an escape route on their way out of India. Moslem support of the British was crucial to their interests. Arousing Moslem wrath by placing Jews in charge of Palestine and its Holy places was against British interests, despite the Zionists propaganda. Morgenthau, himself a Jew, wrote in 1921,

"POLITICAL IMPOSSIBILITY OF A JEWISH STATE

I HAVE just said that it may be politic for the British Government to coddle the aspirations of the Jews. There are, however, profound reasons why this coddling will not take the form of granting to them even the name and surface appearance of a sovereign government ruling Palestine. In the first place, Britain's hold upon India is by no means so secure that the Imperial Government at London can afford to trifle with the fanatical sensibilities of the millions of Mohammedans in its Indian possessions. Remember that Palestine is as much the Holy Land of the Mohammedan as it is the Holy Land of the Jew, or the Holy Land of the Christian. His shrines cluster there as thickly. They are to him as sacredly endeared. In 1914 I visited the famous caves of Macpelah, twenty miles from Jerusalem; and I shall never forget the mutterings of discontent that murmured

---

114. L. Marshall to J. Spargo, *Louis Marshall: Champion of Liberty; Selected Papers and Addresses*, Volume 1, The Jewish Publication Society of America, Philadelphia, (1957), pp. 351-353, at 351.
115. L. Fry, *Waters Flowing Eastward: The War Against the Kingship of Christ*, TBR Books, Washington, D. C., (2000), p. 68.
116. F. Owen, *Tempestuous Journey: Lloyd George: His Life and Times*, McGraw-Hill Company, Inc. New York, (1955), p. 426.

in my ears, nor the threatening looks that confronted my eyes, from the lips and faces of the devout Mohammedans whom I there encountered. For these authentic tombs of Abraham, Isaac, and Jacob are as sacred to them, because they are saints of Islam, as they are to the most orthodox of my fellow Jews, whose direct ancestors they are, not only in the spiritual, but in the actual physical sense. To these Mohammedans, my presence at the tombs of my ancestors was as much a profanation of a Mohammedan Holy Place as if I had laid sacrilegious hands upon the sacred relics in the mosque at Mecca. To imagine that the British Government will sanction a scheme for a political control of Palestine which would place in the hands of the Jews the physical guardianship of these shrines of Islam, is to imagine something very foreign to the practical political sense of the most politically practical race on earth. They know too well how deeply they would offend their myriad Mohammedan subjects to the East.

Exactly the same political issue of religious fanaticism applies to the question of Christian sensibilities. Any one who has seen, as in 1914 I saw at Easter-tide, the tens of thousands of devout Roman Catholics from Poland, Italy, and Spain, and the other tens of thousands of devout Greek Catholics from Russia and the East, who yearly frequent the shrines of Christianity in Palestine, and who thus consummate a lifetime of devotion by a pilgrimage undertaken at, to them, staggering expense and physical privation; and who has observed, as I have observed, the suppressed hatred of them all for both the Jew and the Mussulman; and who has noted, further, the bitter jealousies between even Protestant and Catholic, between Greek Catholic and Roman—such an observer, I say, can entertain no illusions that the placing of these sacred shrines of Christian tradition in the hands of the Jews would be tolerated. The most enlightened Christians might endure it, but the great mass of Christian worshippers of Europe would rebel. They regard the Jew not merely as a member of a rival faith, but as the man whose ancestors rejected their fellow Jew, the Christ, and crucified Him. Their fanaticism is a political fact of gigantic proportions. A Jewish State in Palestine would inevitably arouse their passion. Instead of such a State adding new dignity and consideration to the position of the Jew the world over (as the Zionists claim it would do), I am convinced that it would concentrate, multiply, and give new venom to the hatred which he already endures in Poland and Russia, the very lands in which most of the Jews now dwell, and where their oppressions are the worst.

The political pretensions of Zionism are fantastic. I think the foregoing paragraphs have demonstrated this."[117]

In 1922 and 1923, Lord George Sydenham Clarke Sydenham of Combe published several Letters to the Editor in *The London Times*, in which he

---

[117]. H. Morgenthau, "Zionism a Surrender, Not a Solution", *The World's Work*, Volume 42, Number 3, (July, 1921), pp. i-viii, at v-vi.

demonstrated that Jewish colonies in Palestine were a terrible financial drain on Great Britain. Lord Sydenham proved what an irritant it was to the Moslem world to have a large influx of Jews into Palestine. He pointed out the injustice and provocation which arose from the appointment of ardent Jewish Zionists to rule over Palestine in a *de facto* Jewish Government and how these irritations served to undermine British interests in India and throughout the Middle East.

It would take another world war, the Holocaust, the independence of India from Great Britain and the creation of Pakistan, as well as pervasive corruption both clerical and profane to overcome these political and religious obstacles. The Jews used the French under Napoleon, and then the British in the First World War, to chase the Turks out of Palestine and Greater Syria. The Jews lured the French and the British into the region by leading them to believe that a route to their colonies was vitally important to their national interests.

The Jews created the illusion that only Jews could be their friends in the Middle East to secure this route, while Moslems could not. The opposite was true as both the French and the British soon learned after the First World War. When the Turks were finally forced out of Palestine and Greater Syria, the French and British went to war over who would control this region, into which they had been led by the Jews. The Jews then felt a need to destroy the French and the British Imperial interests in Asia. The Jews accomplished this goal in the Second World War with their Zionist National Socialists, with the Nazis; and with their old friends, the Imperial Japanese—Jewish monies and political influence deliberately caused the deaths of hundred of thousands of Americans in the Second World War alone. Zionist Jews murdered one hundred million people in two world wars in order to create a racist "Jewish State" in Palestine which would house one to five million Jews in a place where they did not want to live. They murdered another hundred million through Communism.[118] Boris Brasol told of the Zionists' plan in 1920 to create a Socialist German army that would crush British Imperialism and secure Palestine for the Jews, and note that the army was the Nazi army, an army Walther Rathenau began to build in cooperation with the Bolsheviks in 1922 with the Rappallo Treaty (Poale-Zion were Russian Communist Jewish Zionists),

> "Mr. Eberlin, a Jew himself, and one of the foremost leaders of the Poale-Zionist movement, in a book recently published in Berlin, entitled 'On the Eve of Regeneration,' stated:
>
> 'The foreign policy of England in Asia Minor is determined by its interests in India. There was a saying about Prussia that she represents the army with an admixture of the people. About England it could be said that she represents a colonial empire with a supplement of the metropolis.... It is obvious that England desires to use Palestine as a shield against India.

---

118. S. Courtois, *et al.*, *The Black Book of Communism : Crimes, Terror*, Repression, Harvard University Press, Cambridge, Massachusetts, (1999).

This is the reason why she is feverishly engaged in the construction of strategic railroad lines, uniting Egypt to Palestine, Cairo to Haifa, where work is started for the construction of a huge port. In the near future Palestine will be in a position to compete with the Isthmus of Suez, which is the main artery of the great sea route from the Mediterranean to the Indian Ocean.' [*Footnote:* Translation from Russian, 'On the Eve of Regeneration,' by I. Eberlin, pp. 129, 130, Berlin, 1920.]

But this Poale-Zionist goes a step farther when he asserts that:

'It is only Socialism attainted in Europe which will prove capable of giving honestly and without hypocrisy Palestine to the Jews, thus assuring them unhampered development... . The Jewish people will have Palestine only when British Imperialism is broken.'"[119]

The Second World War unhitched England from the East and largely destroyed British Imperialism. The Zionists deliberately caused those events and created those circumstances. The lost lives and misery were a deliberate human sacrifice the Zionists made to their Jewish God.

Article 22 of the Covenant of the League of Nations gave the winning powers of World War I the arbitrary authority to divide the spoils of war amongst themselves under the guise of acting as the benefactor of small nations. This product of the war was anticipated by the Zionists before the war began, when they correctly guessed that at the closure of the war, which had not yet happened, negotiations over the fate of small nations would occur where they could make a bid for a Jewish State. The mandate power the League of Nations fit the purpose of creating a Jewish State so well as to leave little doubt that it was custom tailored to suit the purpose of the creation of a Jewish State in Palestine, which territory had previously been held by Turkey:

"Article 22 of the Covenant of the League of Nations, 28 June 1919
Article 22. To those colonies and territories which as a consequence of the late war have ceased to be under the sovereignty of the States which formerly governed them and which are inhabited by peoples not yet able to stand by themselves under the strenuous conditions of the modern world, there should be applied the principle that the well-being and development of such peoples form a sacred trust of civilization and that securities for the formance of this trust should be embodied in this Covenant.

The best method of giving practical effect to this principle is that the tutelage of such peoples should be entrusted to advanced nations who by reason of their resources, their experience or their geographical position can best undertake this responsibility, and who are willing to accept it, and that

---

[119]. B. L. Brasol, *The World at the Cross Roads*, Small, Mayhard & Co., Boston, (1921), pp. 371-379.

this tutelage should be exercised by them as Mandatories on behalf of the League.

The character of the mandate must differ according to the stage of the development of the people, the geographical situation of the territory, its economic conditions and other similar circumstances.

Certain communities formerly belonging to the Turkish empire have reached a stage of development where their existence as independent nations can be provisionally recognized subject to the rendering of administrative advice and assistance by a Mandatory until such time as they are able to stand alone. The wishes of these communities must be a principal consideration in the selection of the Mandatory.

Other peoples, especially those of Central Africa, are at such a stage that the Mandatory must be responsible for the administration of the territory under conditions which will guarantee freedom of conscience and religion, subject only to the maintenance of public order and morals, the prohibition of abuses such as the slave trade, the arms traffic and the liquor traffic, and the prevention of the establishment of fortifications or military and naval bases and of military training of the natives for other than police purposes and the defence of territory, and will also secure equal opportunities for the trade and commerce of other Members of the League.

There are territories, such as South-West Africa and certain of the South Pacific Islands, which, owing to the sparseness of their population, or their small size, or their remoteness from the centres of civilization, or their geographical contiguity to the territory of the Mandatory, and other circumstances, can be best administered under the laws of the Mandatory as integral portions of its territory, subject to the safeguards above-mentioned in the interests of the indigenous population.

In every case of Mandate, the Mandatory shall render to the Council an annual report in reference to the territory committed to its charge.

The degree of authority, control or administration to be exercised by the Mandatory shall, if not previously agreed upon by the Members of the League, be explicitly defined in each case by the Council.

A permanent Commission shall be constituted to receive and examine the annual reports of the Mandatories and to advise the Council on all matters relating to the observance of the mandates."

Lord Northcliffe, principal owner of *The Times* of London, opposed Zionism and called for an inquiry into the results of the Zionist experiment. He planned to personally report on his findings. He was prevented from doing so in his own newspaper. Douglas Reed, who worked for *The London Times*, alleged in his book *The Controversy of Zion*[120] that Lord Northcliffe, principal owner of the *Times* and an anti-Zionist, believed that he was being poisoned after he

---

[120]. D. Reed, "The End of Lord Northcliffe", *The Controversy of Zion*, Chapter 34, Bloomfield, Sudbury, (1978).

openly opposed Zionism, which was at the critical time the Palestine Mandate came under consideration in the League of Nations. Northcliffe suffered from some of the same symptoms as President Wilson.

An editor at *The Times*, Wickham Steed, wished to suppress Northcliffe's anti-Zionist views. Northcliffe sought to fire Steed. Steed hired Northcliffe's own lawyer to defend him—Steed. Northcliffe wanted to take over as editor of *The Times*, and would have spoken out against the League of Nations Mandate for Palestine. An unnamed doctor, at Steed's instigation, declared Northcliffe insane and committed him to an asylum. Northcliffe died soon thereafter on 14 August 1922. Reed presents the history of events that led to Northcliffe's demise, but comes to no conclusions as to the ultimate cause of his death.

Douglas Reed wrote, *inter alia*,

> "Lord Northcliffe was removed from control of his newspapers and put under constraint on June 18, 1922; on July 24, 1922 the Council of the League of Nations met in London, secure from any possibility of loud public protest by Lord Northcliffe, to bestow on Britain a 'mandate' to remain in Palestine and by arms to install the Zionists there (I describe what events have shown to be the fact; the matter was not so depicted to the public, of course).
>
> This act of 'ratifying' the 'mandate' was in such circumstances a formality. The real work, of drawing up the document and of ensuring that it received approval, had been done in advance, in the first matter by drafters inspired by Dr. Weizmann and in the second by Dr. Weizmann himself in the ante-chambers of many capitals. The members of Mr House's 'Inquiry' had drafted the Covenant of the League of Nations: Dr. Weizmann, Mr. Brandeis, Rabbi Stephen Wise and their associates had drafted the Balfour Declaration; now the third essential document had to he drafted, one of a kind that history never knew before. Dr. Weizmann pays Lord Curzon (then British Foreign Secretary) the formal compliment of saying that he was 'in charge of the actual drafting of the mandate' but adds, 'on our side we had the valuable assistance of Mr. Ben V. Cohen... one of the ablest draughtsmen *in America*'. Thus a Zionist in America (Mr. Cohen was to play an important part in a much later stage of this process) in fact drafted a document under which 'the new world order' was to dictate British policy, the use of British troops and the future of Palestine."[121]

The League of Nations followed from the "New World Order" proposed by the "progressive" U. S. President Woodrow Wilson, who had been blackmailed by the Zionists and was under the control of an enigmatic man, who was sort of a mixture of Svengali, Karl Marx, Huey Long and Karl Rove—one "Colonel" Edward Mandell House (House never actually was a colonel). The League, created by Wilson and "Colonel" House, organized the distribution of Third

---

121. D. Reed, *The Controversy of Zion*, Omnia Veritas Ltd.

World colonies among the major powers after World War I.

The League was a first step towards world government of the type envisioned in Jewish Messianic prophecy, though it was very weak compared to the absolute tyranny proposed by the ancient Jews. A more absolute world government was envisioned by H. G. Wells in 1913 in his book on world war and atomic bombs, *The World Set Free: A Story of Mankind*, Macmillan, London, (1914); also published in Leipzig, Germany, by B. Tauchnitz; and carried still further in Well's *The Open Conspiracy; Blue Prints for a World Revolution*, V. Gollancz Ltd., London, (1928); which was itself derivative of Ivan Stanislavovich Bloch's *The Future of War in Its Technical, Economic, and Political Relations; Is War Now Impossible?*, Doubleday & McClure Co., New York, (1899); preceded by William Winwood Reade's *The Martyrdom of Man*, Trübner & Co., London, (1872); and Baron Edward Bulwer-Lytton's *The Coming Race: Or the New Utopia*, (1848).[122]

The Zionists learned early on that Liberal and Socialist revolution led to assimilation. In 1898, Nachman Syrkin,[123] who despised assimilation, combined Zionism with Marxist internationalism in a way that would prevent the assimilation of Jews and would conform to Jewish Messianic supremacism. In the hands of the Zionists, Communism was an intermediary means to achieve Jewish nationalism, as well as a means to subjugate Gentile peoples and place them under absolute autocratic government led overtly, or in some instances covertly, by Jews. As is clear in Syrkin's writings, the Zionists tended to label every other group of human beings as their enemy, which allowed them justify their inhumanity by blaming their victims.

Syrkin deduced Jewish Nationalism from Communist Internationalism by presuming that Internationalism is merely partisan international cooperation; and that individual liberty, equality and fraternity depend upon national status and ethnic segregation. In order for there to be an international understanding, there must first be dignified segregated and ethnically based nations, which mutually respected one another, and which compete on a level playing field. In Syrkin's eyes, a Jew had no right to choose his or her own individuality in an international community of humanity. He or she must first be a nationalistic Jew and place Jewish interests ahead of all others, before acquiring the free will to become a dignified representative in the international community as a Jewish member of the community of nations. This *Blut und Boden* belief system, this volatile blend of Zionist Nationalism and Communist Internationalism later became known as Nazism and mirrors the Nazi Party's original platform as iterated in "The 25

---

122. E. Bulwer-Lytton, *Rienzi: The Pilgrims of the Rhine; The Coming Race*, Brainard, New York, Continental Press, New York, (1848); and *Rienzi, Two Volumes in One; The Pilgrims of the Rhine; the Coming Race*, Boston, Dana Estes & Co., 1848; and *The Pilgrims of the Rhine. The Coming Race*, Dana Estes & Co., Boston, (1849); and *The Coming Race: Or the New Utopia*, Francis B. Felt & Co., New York, (1871).

123. N. Syrkin, under the nom de plume "Ben Elieser", *Die Judenfrage und der socialistische Judenstaat*, Steiger, Bern, (1898); English translation in A. Hertzberg, *The Zionist Idea*, Harper Torchbooks, New York, (1959), pp. 333-350.

Points" of Nazism in 1920.[124] Bernard Lazare made similar Zionist arguments at about the same time as Syrkin.[125] Einstein later parroted their thoughts.[126]

The Zionists wanted to establish the precedent of separating out small, ethnically segregated nations from international unions and empires in order to justify the creation of the small Jewish nation they sought to create—and in order to put an end to the assimilation of Jews occurring in the Turkish, pan-Slavic and pan-Germanic Empires, which were very cosmopolitan and tolerant communities into which Jews easily and happily dissolved. At the same time, the Zionists required strong international organizations which would have the authority and the power needed to establish this territorial Jewish State, while protecting the right of Jews to live wherever they chose and to have full rights and privileges in all nations.

The Zionists hoped that a ruined Europe could be led by an American controlled movement calling itself "international", that would use its collective force to destroy international unions and establish tiny impotent nations in the place of the empires which had existed before World War I, while concurrently weakening the sovereignty of European states in favor of the dominance of America, which was itself dominated by Jews. They would do this through the League of Nations. This American-led "international" institution could then insist upon the creation of the State of Israel. The Balfour Declaration, Wilson's Fourteen Points, the League of Nations British-Palestine Mandate, etc. tended toward the destruction of cosmopolitan assimilated international societies for the sake of ethnically segregated small nations. Even with this international support for ethnically segregated Nationalism (not to be confused with a truly internationalist and cosmopolitan spirit), the Zionists failed to persuade the majority of Jews to follow them, and so lacked the large numbers of decent citizens needed to make a nation-state viable. Communism, which was meant to ruin the Gentiles and liberate the Jews, failed them. Pseudo-Internationalism for the sake of Jewish Nationalism, *viz.* the League of Nations, failed them. Most significantly, the Jewish People refused to oblige the Zionists, but the Zionists never gave up their struggle to force the Jews to move the Palestine.

The Zionists determined that they needed a rapid rise in anti-Semitism to force Jews to move. They knew that bad economic conditions were the best

---

124. *Cf.* E. T. S. Dugdale, "The 25 Points", *The Programme of the N.S.D.A.P. and its General Conceptions*, Franz Eher Nachf., Munich (1932), pp. 18-20; reprinted as an Appendix "The Program of the National Socialist German Workers Party" in K. G. W. Ludecke, *I Knew Hitler: The Story of a Nazi AWho Escaped the Blood Purge*, Charles Scribner's Sons, New York, (1937), pp. 793-796; reprinted as "Appendix A" in A. C. Sutton, *Wall Street and the Rise of Hitler*, GSG & Associates, San Pedro, California, (2002), pp. 179-181.

125. B. Lazare, *Job's Dungheap*, Schocken Books, New York, (1948); quoted in A. Hertzberg, *The Zionist Idea*, Harper Torchbooks, New York, (1959), p. 476.

126. *See, for example:* Einstein's postscript in his letter to P. Epstein of 5 October 1919, *The Collected Papers of Albert Einstein*, Volume 9, Document 122, Princeton University Press, (2004).

conditions for anti-Semitism to grow and for a dictator to come to power.

Albert Einstein wrote to Adriaan Fokker on 30 July 1919 that the German political mentality led Germans to follow an unscrupulous minority in blind obedience, and that the German people were fools to be outraged at the dictated peace and the Treaty of Versailles. The Germans had laid down their arms based on the false promises of a just peace iterated in Woodrow Wilson's Fourteen Points. Instead of demanding that those promises be kept, Jewish traitors of Einstein's ilk forced Germany into accepting the Treaty of Versailles, which destroyed the Turkish Empire, the pan-German Empire, the German nation and the German economy. The Germans never would have laid down their arms if they had known the treachery that awaited them. There was a large delegation of Jews at the peace talks, who decided Germany's sorry fate.

Some Germans planned to continue the war. Einstein held out hope for the future of the League of Nations, because,

> "It is especially encouraging that America, which has not retained the fatal traditions of Europe, is in charge."

> "Insbesondere ist erfreulich, dass Amerika, welches nicht mit den fatalen Traditionen Europas belastet ist, die Führung hat."[127]

European "traditions" and resultant Nationalism were common topics of the era,[128] and the derogatory commonplaces that emerged often vilified Germans. The Zionists had planned that America would lead the League, because they led America.

Einstein wrote to Hedwig Born on 31 August 1919,

> "The greatest danger for future developments is, in my opinion, the potential withdrawal of the Americans; it is to hoped that Wilson can prevent it. I don't believe that humanity as such can change in essence, but I do believe that it is possible and even *necessary* to put an end to anarchy in international relations, even though the sacrifice of autonomy will be significant for individual states."[129]

Wilson was not so spiteful towards the German People as Einstein was. Though Wilson tried to prevent the Zionists from corrupting his intentions to the point where even he could no longer tolerate their unfairness, Wilson could not

---

[127]. Letter from A. Einstein to A. D. Fokker of 30 July 1919, *The Collected Papers of Albert Einstein*, Volume 9, Document 78, Princeton University Press, (2004), pp. 117-119, at 117.

[128]. *See, for example:* R. Tagore, *Nationalism*, Macmillan, New York, (1917); German translation: *Nationalismus*, Kurt Wolff, München, (1918).

[129]. Letter from A. Einstein to H. Born of 31 August 1919, English translation by A. Hentschel, *The Collected Papers of Albert Einstein*, Volume 9, Document 97, Princeton University Press, (2004), pp. 79-80, at 80.

prevent the injustices done to Germany after the First World War, which injustices Einstein and his hateful ilk sought. Wilson's Zionist partner as President, "Colonel" Edward Mandell House, betrayed him and the United States to the British, French and Zionists in the League of Nations. They instituted the punitive measures against Germany Einstein had long espoused, which measures ultimately led to Hitler's rise to power and to the Second World War, which ultimately led large numbers of Jews to Zionism making it possible to create the State of Israel in Palestine.

All along Zionists encouraged anti-Semitism in order to leave assimilated and assimilating Jews no option but to join them. When even medieval-style anti-Semitism failed to inspire large numbers of Jews to become Zionists in the 1930's, the worst of the horrors began at the behest of the Zionists. Syrkin knew in 1898 that the Jewish masses could be united by anti-Semitic criminals, even by crypto-Jews posing as anti-Semitic criminals. He probably did not realize that even Zionist sponsored criminals could not persuade patriotic assimilated Jews to leave their homes in their various nations.

Einstein was quoted in *The Literary Digest* during his visit to America in 1921, and made clear his inconsistent support of nationalistic Zionism (nationalism and segregation for Jews) and concomitant Internationalism and anti-Nationalism (no freedom of sovereignty and "racial" integration for Gentiles). Einstein lacked the wit of Syrkin, though not his willingness to employ sophistry. Einstein failed to speak out against the injustices done to Germany, which, if corrected, would promote the "Internationalism" he allegedly espoused. Einstein asserts the positivist dogma that science ought to play a fundamental rôle in politics, which inevitably leads to politics playing a fundamental rôle in science through censorship, destructive partisanship, etc.:

"EINSTEIN FINDS THE WORLD NARROW

PROFESSOR ALBERT EINSTEIN, whose theories on space, light, and infinity have made his name familiar throughout the world, thinks that this small particular planet on which we live is suffering from narrowness in its point of view. Too much nationalism—that is Professor Einstein's definition for the 'disease from which mankind is suffering to-day.' Even before the war sectional prejudices were bad enough, but the 'prewar internationalism' was infinitely preferable to the present state of mind of most of humanity, he says, and he urges that the people of this sphere return to charity and mutual understanding. The great German scientist arrived in this country early in April, to lecture upon Zionism as well as upon his revolutionary theory of relativity. A New York *Times* reporter, who was among the many newspaper men assembled to greet him at the pier, gives this picture of the thinker whom the nations have decided to honor:

> A man in a faded gray raincoat and a flopping black felt hat that nearly concealed the gray hair that straggled over his ears stood on the boat deck of the steamship *Rotterdam* yesterday, timidly facing a battery of camera

men. In one hand he clutched a shiny briar pipe and the other clung to a precious violin. He looked like an artist—a musician. He was.

But underneath his shaggy locks was a scientific mind whose deductions have staggered the ablest intellects of Europe—a mind whose speculative imagination was so vast that its great scientific theories puzzled and appalled the reasoning faculty.

The man was Dr. Albert Einstein, propounder of the much-debated theory of relativity that has given the world a new conception of space, and time, and the size of the universe.

Dr. Einstein comes to this country as one of a group of prominent Jews who are advocating the Zionist movement and hope to get financial aid and encouragement for the rebuilding of Palestine and the founding of a Jewish university. He is of medium height, with strongly built shoulders, but an air of fragility and self-effacement. Under a high, broad forehead are large and luminous eyes, almost childlike in their simplicity and unworldliness.

Professor Einstein does not like to be interviewed, and the questions of the reporters bothered him a great deal. One of the few real interviews he has ever given was forwarded from Berlin to the New York *Evening Post*, shortly before Einstein's departure for this country. 'I had come to Professor Einstein to hear what he had to say about the plight of German science,' writes Mr. Tobinkin. The subject was just then occupying much space in the newspapers of Berlin. Professor Einstein, however, spoke not of science, but of humanity:

'Of course,' he said, 'science is suffering from the terrible effects of the war, but it is humanity that should be given first consideration. Humanity is suffering in Germany, everywhere in eastern Europe, as it has not suffered in centuries.

'Humanity,' he continued, 'is suffering from too much and too narrow a conception of nationalism. The present wave of nationalism, which at the slightest provocation or without provocation passes over into chauvinism, is a sickness.

'The internationalism that existed before the war, before 1914, the internationalism of culture, the cosmopolitanism of commerce and industry, the broad tolerance of ideas—this internationalism was essentially right. There will be no peace on earth, the wounds inflicted by the war will not heal, until this internationalism is restored.'

'Does this imply you oppose the formation of small nations?' the interviewer asked.

'Not in the least,' he replied. 'Internationalism as I conceive it implies a rational relationship between countries, a sane union and understanding between nations, mutual cooperation, mutual advancement without interference with a country's customs or inner life.'

'And how would you proceed to bring back this internationalism that existed prior to 1914?'

'Here,' he said, 'is where science, scientists, and especially the scientists of America, can be of great service to humanity. Scientists, and the scientists of America in the first place, must be pioneers in this work of restoring internationalism.'

'America is already in advance of all other nations in the matter of internationalism. It has what might be called an international 'psyche.' The extent of America's leaning to internationalism was shown by the initial success of Wilson's ideas of internationalism, the popular acclaim with which they met from the American people.

'That Wilson failed to carry out his ideas is beside the point. The enthusiasm with which the preaching of these ideas by Wilson was received shows the state of mind of the American public. It shows it to be internationally inclined. American scientists should be among the first to attempt to develop these ideas of internationalism and to help carry them forward. For the world, and that means America, also, needs a return to international friendship. The work of peace can not go forward in your own country, in any country, so long as your Government or any Government is uneasy about its international relations. Suspicion and bitterness are not a good soil for progress. They should vanish. The intellectuals should be among the first to cast them off.'

There are two men in Germany to-day who are traditionally inaccessible to newspaper men, Mr. Tobinkin notes. One is the financier, Hugo Stinnes. The other is Einstein. We are told:

Einstein has been greatly abused by a section of the German press, and he therefore shuns publicity. He lives in a quiet section of Berlin on the top floor of a fairly up-to-date apartment-house. His study consists of a reception-room, or rather a conference-room, and of his private workroom. The walls of the conference-room are lined with books of a general character. The large number of English books is especially noticeable. There is an *édition de luxe* of Dickens in English and a costly Shakespeare edition in German. Alongside of Shakespeare stands Goethe in a similarly luxurious edition. Einstein is an admirer of both Goethe and Schiller, and has the busts of the two poets prominently displayed.

Adjoining the conference-room is a large music-room. When he is not in his study, Mrs. Einstein told me, her husband is in the music-room. Music and cigars are the scientist's only relaxations. The number of cigars he smokes is controlled by Mrs. Einstein for his health's sake, but there is no control over the amount of time he chooses to spend at the piano or with his violin, for he plays both instruments well.

His workroom is exceedingly simple. There is a telescope in it. The windows give an exceptionally good view of the sky. There are also a number of globes and various metal representations of the solar system. There are two engravings of Newton on the walls. They are the only pictures in the room. The table he works at is simple and rather small. There is a small typewriter, which is used by his secretary. Einstein has a large correspondence, receiving on an average sixty letters a day.

He was pacing up and down the room as I entered his study. He was drest in a pair of worn-out trousers and a sweater-coat. If he had a collar on, the collar was very unobtrusive, for I can not recall having seen it. He was at work. His hair was disheveled and his eye had a roving look. His wife told me that when the professor is seized by a problem the fact becomes known to her by this peculiar wandering look which comes into his eyes and by his feverish pacing up and down the room. At such times, she said, the professor is never disturbed. His food is brought to him in his workroom. Sometimes this mode of living lasts for three or four days at a time. It is when the professor rejoins his family at the table that they know that his period of intense concentration, and abstraction, is over.

After such a period of concentration, Einstein often rests himself by reading fiction. He is fond of reading Dostoyefsky. He walks a lot through the parks, and in the summer often goes out with his family in the fields. But he is never asked by his wife or children to go for a walk. It is he who has to do the asking, and when he asks them for a walk they know that his mind is relieved of work. His hours of work are indefinite. He sometimes struggles through a whole night with a problem and goes to bed only late in the morning.

Dr. Einstein asked whether he could not see a copy of my interview with him before it was printed. I told him that I would not write the interview until after my return to America.

'In that event,' he said, 'when you write it, be sure not to omit to state that I am a convinced pacifist, that I believe that the world has had enough of war. Some sort of an international agreement must be reached among nations preventing the recurrence of another war, as another war will ruin our civilization completely. Continental civilization, European civilization, has been badly damaged and set back by this war, but the loss is not irreparable. Another war may prove fatal to Europe.'

The New York *World* extends a welcome, and a hearty congratulations, in the following editorial:

It is not invidious to say that of the many distinguished visitors from abroad recently arriving in New York the one inspiring the most spontaneous popular demonstration at the pier is not a great general or statesman but a plain man of science—Dr. Albert Einstein, who comes with prominent Jews in aid of the Zionist movement.

Plain, that is, as respects his unaffected personality, but a scientific investigator who has progressed into a higher sphere of speculative thought unfathomable to the ordinary intelligence. What he has to exhibit is not a new play or a new theory of life but a new hypothesis of the celestial mechanism, involving a radically altered conception of time and space and the size of the universe.

It is something when New York turns out to honor a stranger bringing gifts of this recondite character. Perhaps by the time he is ready to return the public will be glibly discussing the Einstein theory of relativity, whether or not it proves capable of understanding it. But behind the outward demonstration there is discernible a sincere tribute of admiration to the physicist who, amid the turmoil of war and the distractions of material interests, has kept his mind fixt on the star of pure science and has mounted to the heights with Newton and the other great leaders of scientific thought."[130]

In promoting the League of Nations, Einstein was not so concerned about the fate of Europe, as he was the creation of a Jewish state in Palestine. Einstein and friends wanted to achieve the Messianic Jewish goals of the destruction of all Gentile governments and the creation of a Jewish State. Therein lies the resolution of the apparent contradiction between Einstein's Zionism and his anti-nationalistic Internationalism. The Old Testament tells the Jews that they will ruin all other nations and forever keep their own. The contradictory, simplistic, absolute and arbitrary nature of Einstein's pronouncements are the result of his mediocre intellect and his reliance upon others to craft his speeches and beliefs. Einstein's request to read the interview before it was published and his insistence that it contain his scripted political messages is further evidence that much of the man's public persona was a fraud.

After the First World War, the Zionists had their Peace Conference and their League of Nations and their Palestine Mandate, but they lacked the broad support of the Jewish People. They decided to bring on a Second World War, which would result in another Peace Conference; and, the second time around, they would torture the Jewish People into embracing Zionism.

Lenni Brenner wrote in his exposé *Zionism in the Age of the Dictators*, "The Wartime Failure to Rescue", Chapter 24, Lawrence Hill Books, Chicago, (1983), pp. 235-238 [Brenner cites in his notes: "22. Michael Dov-Ber Weissmandel, *Min HaMaitzer* (unpublished English translation). 23. Ibid. 24. Ibid. (Hebrew edn), p. 92. 25. Ibid., p. 93."],

"'For only with Blood Shall We Get the land'

The Nazis began taking the Jews of Slovakia captive in March 1942. Rabbi Michael Dov-Ber Weissmandel, an Agudist, thought to employ the traditional weapon against anti-Semitism: bribes. He contacted Dieter Wisliceny, Eichmann's representative, and told him that he was in touch with the leaders of world Jewry. Would Wisliceny take their money for the lives of Slovakian Jewry? Wisliceny agreed for 50,000 in dollars so long as it came from outside the country. The money was paid, but it was actually

---

[130]. "Personal-Glimpses: Einstein Finds the World Narrow", *The Literary Digest*, (16 April 1921), pp. 33-34.

raised locally, and the surviving 30,000 Jews were spared until 1944 when they were captured in the aftermath of the furious but unsuccessful Slovak partisan revolt.

Weissmandel, who was a philosophy student at Oxford University, had Volunteered on 1 September 1939 to return to Slovakia as the agent of the world Aguda. He became one of the outstanding Jewish figures during the Holocaust, for it was he who was the first to demand that the Allies bomb Auschwitz. Eventually he was captured, but he managed to saw his way out of a moving train with an emery wire; he jumped, broke his leg, survived and continued his work of rescuing Jews. Weissmandel's powerful post-war book, *Min HaMaitzer* (From the Depths), written in Talmudic Hebrew, has unfortunately not been translated into English as yet. It is one of the most powerful indictments of Zionism and the Jewish establishment. It helps put Gruenbaum's unwillingness to send money into occupied Europe into its proper perspective. Weissmandel realised: 'the money is needed here – by us and not by them. For with money here, new ideas can be formulated.'[22] Weissmandel was thinking beyond just bribery. He realised immediately that with money it was possible to mobilise the Slovak partisans. However, the key question for him was whether any of the senior ranks in the SS or the Nazi regime could be bribed. Only if they were willing to deal with either Western Jewry or the Allies, could bribery have any serious impact. He saw the balance of the war shifting, with some Nazis still thinking they could win and hoping to use the Jews to put pressure on the Allies, but others beginning to fear future Allied retribution. His concern was simply that the Nazis should start to appreciate that live Jews were more useful than dead ones. His thinking is not to be confused with that of the Judenrat collaborators. He was not trying to save some Jews. He thought strictly in terms of negotiations on a Europe-wide basis for all the Jews. He warned Hungarian Jewry in its turn: do not let them ghettoise you! Rebel, hide, make them drag the survivors there in chains! You go peacefully into a ghetto and you will go to Auschwitz! Weissmandel was careful never to allow himself to be manoeuvred by the Germans into demanding concessions from the Allies. Money from world Jewry was the only bait he dangled before them.

In November 1942, Wisliceny was approached again. How much money would be needed for all the European Jews to be saved? He went to Berlin, and in early 1943 word came down to Bratislava. For $2 million they could have all the Jews in Western Europe and the Balkans. Weissmandel sent a courier to Switzerland to try to get the money from the Jewish charities. Saly Mayer, a Zionist industrialist and the Joint Distribution Committee representative in Zurich, refused to give the Bratislavan 'working group' any money, even as an initial payment to test the proposition, because the 'Joint' would not break the American laws which prohibited sending money into enemy countries. Instead Mayer sent Weissmandel a calculated insult: 'the letters that you have gathered from the Slovakian refugees in Poland are exaggerated tales for this is the way of the '*Ost-Juden*' who are always demanding money'.[23]

The courier who brought Mayer's reply had another letter with him from Nathan Schwalb, the HeChalutz representative in Switzerland Weissmandel described the document:

> There was another letter in the envelope, written in a strange foreign language and at first I could not decipher at all which language it was until I realised that this was Hebrew written in Roman letters, and written to Schwalb's friends in Pressburg [Bratislava] ... It is still before my eyes, as if I had reviewed it a hundred and one times. This was the content of the letter:
>
> 'Since we have the opportunity of this courier, we are writing to the group that they must constantly have before them that in the end the Allies will win. After their victory they will divide the world again between the nations, as they did at the end of the first world war. Then they unveiled the plan for the first step and now, at the war's end, we must do everything so that Eretz Yisroel will become the state of Israel, and important steps have already been taken in this direction. About the cries coming from your country, we should know that all the Allied nations are spilling much of their blood, and if we do not sacrifice any blood, by what right shall we merit coming before the bargaining table when they divide nations and lands at the war's end? Therefore it is silly, even impudent, on our part to ask these nations who are spilling their blood to permit their money into enemy countries in order to protect our blood—for only with blood shall we get the land. But in respect to you, my friends, *atem taylu*, and for this purpose I am sending you money illegally with this messenger.'[24]

Rabbi Weissmandel pondered over the startling letter:

> After I had accustomed myself to this strange writing, I trembled, understanding the meaning of the first words which were 'only with blood shall we attain land'. But days and weeks went by, and I did not know the meaning of the last two words. Until I saw from something that happened that the words '*atem taylu*' were from '*tiyul*' [to walk] which was their special term for 'rescue'. In other words: you, my fellow members, my 19 or 20 close friends, get out of Slovakia and save your lives and with the blood of the remainder—the blood of all the men, women, old and young and the sucklings—the land will belong to us. Therefore, in order to save their lives it is a crime to allow money into enemy territory—but to save you beloved friends, here is money obtained illegally.
>
> It is understood that I do not have these letters, for they remained there and were destroyed with everything else that was lost.[25]

Weissmandel assures us that Gisi Fleischman and the other dedicated

Zionist rescue workers inside the working group were appalled by Schwalb's letter, but it expressed the morbid thoughts of the worst elements of the WZO leadership. Zionism had come full turn: instead of Zionism being the hope of the Jews, their blood was to be the political salvation of Zionism."

5.15.2 "Colonel" Edward Mandell House

Einstein was not alone in this regard, "Colonel" Edward Mandell House was President Wilson's intimate friend—then betrayer—during his presidency. House initially prompted Wilson to run for the presidency, and then dominated Wilson's presidency as the true "power behind the throne". In 1911, "Colonel" Edward Mandell House, who was perhaps a crypto-Jew and who was certainly a Zionist, anonymously authored the fictional novel *Philip Dru: Administrator*, B. W. Huebsch, New York, (1912); which pitted a corrupt conservative Senator against a progressive Socialist hero in a second civil war in America. The novel was propaganda for a Socialist revolution.

After publishing his novel, House, with the assistance of a few of the large banking houses in New York, recruited Woodrow Wilson to run for the Presidency and guided and funded Wilson's campaign. In his novel, House vilifies a financier named "Thor". This campaign against specific bankers matched the real campaign of the Zionists Louis Brandeis and Samuel Untermyer.[131] Brandeis and Untermyer pretended to fight banker corruption, but really only attacked the Rothschilds' competition and secured control over American finances for the Rothschild family.

---

131. **On Untermyer, see:** Corp Author: United States., Congress., House., Committee on rules., *Investigation of the Money Trust. No. 1-[2] Hearings Before the Committee on Rules of the House of Representatives, on House Resolutions 314 and 356. Friday, January 26, 1912.*, Washington, D. C., U. S. Govt. Print. Off., (1912). *See also:* Corp Author: United States., Congress., House., Committee on Banking and Currency., *Money Trust Investigation... Statistical and Other Information Compiled under Direction of the Committee.*, Washington, D. C., U. S. Govt. Print. Off., (1912). *See also:* A. P. Pujo and Arsène Paulin and E. A. Hayes. Corp Author: United States., Congress., House., Committee on Banking and Currency, *Money Trust Investigation. Investigation of Financial and Monetary Conditions in the United States under House Resolutions Nos.429 and 504, Before a Subcommittee of the Committee on Banking and Currency.*, Washington, D. C., U. S. Govt. Print. Off., (1913). *See also:* J. G. Milburn, W. F. Taylor, *Money Trust Investigation: Brief on Behalf of the New York Stock Exchange*, New York, C.G. Burgoyne, (1913). *See also:* J. P. Morgan, *Testimony of Mr. J. Pierpont Morgan and Mr. Henry P. Davison Before the Money Trust Investigation*, J.P. Morgan & Co., New York, (1913). **On Brandeis, see:** L. D. Brandeis, *Other People's Money and How the Bankers Use It*, F.A. Stokes, New York, (1914). *See also:* L. D. Brandeis, M. I. Urofsky and D. W. Levy, Editors, *Letters of Louis D. Brandeis*, In Five Volumes, State University of New York Press, Albany, New York, (1975). This set has numerous letters as well as editorial comment related to Brandeis and Untermyer's campaigns against some bankers.

House's book was written in 1911. In 1912-1913, the Congressional House of Representatives investigated bankers in the "Money Trust Investigation" which explored some of the corruption which was rampant at the time.[132] The scandal made it obvious that many reforms were needed. The bankers initiated the investigation so that it would point out the need for reforms, and then they instituted "reforms" which would give them absolute control and shield them from further investigations. President Wilson and "Colonel" House took this manufactured opportunity to place financiers at the reins of government. They enacted several laws which gave the banks control over the money supply through the creation of the Federal Reserve System.[133] This corruption eventually led to the Great Depression, as pools of rich financiers artificially ran up stock prices and then sold off their interests, to then profit a second time by short selling the stocks that they had at first collusively inflated.[134]

"Colonel" House came from Texas. He expresses a sympathy for the South in his novel and an antagonism towards the blacks President Wilson later betrayed while in office. Wilson also came from the South and the Civil War greatly affected him.

House advocated the use of propaganda in his propaganda novel. He also advocated Illuminati and Freemasonry methods of subtle manipulation. Many of the things he promoted were very important social reforms. He was a strong advocate of women's suffrage, equal pay for women and the rights of women to dignity and independence. He held out a helpful hand so that he could let go at just the right moment and let the nation fall. House wrote in Chapter 43 of his novel *Philip Dru: Administrator*,

---

132. Corp Author: United States., Congress., House., Committee on rules., *Investigation of the Money Trust. No. 1-[2] Hearings Before the Committee on Rules of the House of Representatives, on House Resolutions 314 and 356. Friday, January 26, 1912.*, Washington, D. C., U. S. Govt. Print. Off., (1912). *See also:* Corp Author: United States., Congress., House., Committee on Banking and Currency., *Money Trust Investigation... Statistical and Other Information Compiled under Direction of the Committee.*, Washington, D. C., U. S. Govt. Print. Off., (1912). *See also:* A. P. Pujo and Arsène Paulin and E. A. Hayes. Corp Author: United States., Congress., House., Committee on Banking and Currency, *Money Trust Investigation. Investigation of Financial and Monetary Conditions in the United States under House Resolutions Nos.429 and 504, Before a Subcommittee of the Committee on Banking and Currency.*, Washington, D. C., U. S. Govt. Print. Off., (1913). *See also:* J. G. Milburn, W. F. Taylor, *Money Trust Investigation: Brief on Behalf of the New York Stock Exchange*, New York, C.G. Burgoyne, (1913). *See also:* J. P. Morgan, *Testimony of Mr. J. Pierpont Morgan and Mr. Henry P. Davison Before the Money Trust Investigation*, J.P. Morgan & Co., New York, (1913).

133. E. C. Mullins, *Secrets of the Federal Reserve: The London Connection*, Bankers Research Institute, Staunton, Virginia, (1983); **and** *The Federal Reserve Conspiracy*, Common Sense, Union, New Jersey, (1954); **and** *A Study of the Federal Reserve*, Kasper and Horton, New York, (1952).

134. "The Crash of 1929", *The American Experience*, PBS Video Documentary, (1990).

'In many ways,' said Dru. 'Have clubs for them, where they may sing, dance, read, exercise and have their friends visit them. Have good women in charge so that the influence will be of the best. Have occasional plays and entertainments for them, to which they may each invite a friend, and make such places pleasanter than others where they might go. And all the time protect them, and preferably in a way they are not conscious of. By careful attention to the reading matter, interesting stories should be selected each of which would bear its own moral. Quiet and informal talks by the matron and others at opportune times, would give them an insight into the pitfalls around them, and make it more difficult for the human vultures to accomplish their undoing. There is no greater stain upon our vaunted civilization,' continued Dru, 'than our failure to protect the weak, the unhappy and the abjectly poor of womankind.

'Philosophers still treat of it in the abstract, moralists speak of it now and then in an academic way, but it is a subject generally shunned and thought hopelessly impossible.

'It is only here and there that a big noble-hearted woman can be found to approach it, and then a Hull House is started, and under its sheltering roof unreckoned numbers of innocent hearted girls are saved to bless, at a later day, its patron saint.

'Start Hull Houses, Senator Selwyn, along with your other plan, for it is all of a kind, and works to the betterment of woman. The vicious, the evil minded and the mature sensualist, we will always have with us, but stretch out your mighty arm, buttressed as it is by fabulous wealth, and save from the lair of the libertines, the innocent, whose only crime is poverty and a hopeless despair.

'In your propaganda for good,' continued Dru, 'do not overlook the education of mothers to the importance of sex hygiene, so that they may impart to their daughters the truth, and not let them gather their knowledge from the streets."

The use of reading material for propaganda purposes was a tactic Schiff had used to propagandize bored Russian prisoners of war in Japan in 1905 with revolutionary propaganda. Before Schiff, the Illuminati and Freemasons used reading rooms as a platform from which to propagandize a populace. *Philip Dru: Administrator* was typical of the sentimental Bolshevist propaganda that evolved from the literature of Victor Hugo's *Les Misérables* and Charles Dickens' *A Christmas Carol*, among many other like works.

In Bolshevik propaganda, the emotional presentation of the suffering of the poor is used to justify violent revolution, mass murder and absolute dictatorship in a totalitarian state; as an allegedly necessary step towards a democracy—which never comes. For example, House's propaganda exploited the suffering of the poor to justify dictatorship, revolutionary war resulting in countless unnecessary deaths, and the militaristic Imperialism of the United States over Mexico and Latin America, again resulting in countless unnecessary deaths. It seems the Zionists hold fast to these objectives even today.

As the Russian Jewish immigrants to America began to impose their influence, the American news media began to fill with communist propagandists, and many Hollywood script writers, film makers, producers, directors and actors owed their allegiance to Joseph Stalin and the Soviet Union. When the United States Government investigated their activities, they relentlessly lied to the American public in order to protect one another. They produced motion pictures which exploited the good natures of most Americans and which appealed to the liberal sentiments of most Americans.

However, though some were innocent dupes, many of these Communists had no loyalty to America or to the principles of Liberalism. They sought to subjugate America under Soviet-Stalinist-style despotism. Had they been successful, it would have meant the utter destruction of the American People. Hardcore Communists had little or no respect for human life—the movement took the form of a pernicious cult bent on destroying society. Members were sheepishly loyal to each other, to their leaders and to their cause. They blindly obeyed orders, had no regard at all for the rights of others to self-determination or even to life. They were quick to betray American interests to the Soviets. Hardcore Communists were perfectly willing to commit any and all acts, no matter how heinous or deceitful, to further the advancement of Communism and destroy the lives of their fellow human beings—all in the name of their false Liberalism.

Preaching false Liberalism and appealing to the good nature of human beings in order to exploit a gullible population is an old Biblical tradition, not only in Christianity, but also in the Old Testament. Christianity, itself, is based on the human sacrifice of Jesus[135] and countless Jews and converts followed his example in the first centuries of the movement, offering their lives to God. According to the Old Testament, before the Jews fled Egypt, their God committed atrocities against the Egyptians and miraculously made the Egyptians gullible. The Jews then borrowed their Egyptian neighbors' jewels of gold and silver. The Jews stole this treasure as they left, betraying the trust and generosity of the Egyptian People—or so goes the story. However, there is no archeological evidence to support the Exodus myth.

The story is evidently an allegory, where the firstborn of Egypt are sacrificed to Baal in the pursuit of Zionism, of Greater Israel. The possibility exists that the Jews absorbed an Egyptian sub-population and that Moses was a secessionist Egyptian leader, perhaps the Egyptian Pharaoh Akhenaton IV, who brought monotheism, circumcision, and other ancient Egyptian beliefs and practices that ended up with the Jews. It could be that the Jews demanded that the Egyptian monotheist exiles give up their gold and their firstborn children as ritual sacrifices to Baal.

It was the Canaanites, the Judeans, not the Egyptians, who worshiped Baal. Baalism demanded as a sacrifice the child that opens the womb—the firstborn. This child would be burned as a "holocaust", a burnt offering to Baal. The Jews

---

135. W. W. Reade, *The Martyrdom of Man*, Trübner & Co., London, (1872).

never fully surrendered this practice in the Old Testament, nor in the history of the ancient world. Although it allegedly inspired God's wrath—angered the Egyptian monotheists, many if not most Jews continued the practice of sacrificing their children in a holocaust. *Exodus* 10; 11:

> "10:1 And the LORD said unto Moses, Go in unto Pharaoh: for I have hardened his heart, and the heart of his servants, that I might shew these my signs before him: 2 And that thou mayest tell in the ears of thy son, and of thy son's son, what things I have wrought in Egypt, and my signs which I have done among them; that ye may know how that I *am* the LORD. 3 And Moses and Aaron came in unto Pharaoh, and said unto him, Thus saith the LORD God of the Hebrews, How long wilt thou refuse to humble thyself before me? let my people go, that they may serve me. 4 Else, if thou refuse to let my people go, behold, to morrow will I bring the locusts into thy coast: 5 And they shall cover the face of the earth, that one cannot be able to see the earth: and they shall eat the residue of that which is escaped, which remaineth unto you from the hail, and shall eat every tree which groweth for you out of the field: 6 And they shall fill thy houses, and the houses of all thy servants, and the houses of all the Egyptians; which neither thy fathers, nor thy fathers' fathers have seen, since the day that they were upon the earth unto this day. And he turned himself, and went out from Pharaoh. 7 And Pharaoh's servants said unto him, How long shall this man be a snare unto us? let the men go, that they may serve the LORD their God: knowest thou not yet that Egypt is destroyed? 8 And Moses and Aaron were brought again unto Pharaoh: and he said unto them, Go, serve the LORD your God: *but* who *are* they that shall go? 9 And Moses said, We will go with our young and with our old, with our sons and with our daughters, with our flocks and with our herds will we go; for we *must hold* a feast unto the LORD. 10 And he said unto them, Let the LORD be so with you, as I will let you go, and your little ones: look *to it*; for evil *is* before you. 11 Not so: go now ye *that are* men, and serve the LORD; for that ye did desire. And they were driven out from Pharaoh's presence. 12 And the LORD said unto Moses, Stretch out thine hand over the land of Egypt for the locusts, that they may come up upon the land of Egypt, and eat every herb of the land, *even* all that the hail hath left. 13 And Moses stretched forth his rod over the land of Egypt, and the LORD brought an east wind upon the land all that day, and all *that* night; *and* when it was morning, the east wind brought the locusts. 14 And the locusts went up over all the land of Egypt, and rested in all the coasts of Egypt: very grievous *were they*; before them there were no such locusts as they, neither after them shall be such. 15 For they covered the face of the whole earth, so that the land was darkened; and they did eat every herb of the land, and all the fruit of the trees which the hail had left: and there remained not any green thing in the trees, or in the herbs of the field, through all the land of Egypt. 16 Then Pharaoh called for Moses and Aaron in haste; and he said, I have sinned against the LORD your God, and against you. 17 Now therefore forgive, I pray thee, my sin only this once,

and intreat the LORD your God, that he may take away from me this death only. 18 And he went out from Pharaoh, and intreated the LORD. 19 And the LORD turned a mighty strong west wind, which took away the locusts, and cast them into the Red sea; there remained not one locust in all the coasts of Egypt. 20 But the LORD hardened Pharaoh's heart, so that he would not let the children of Israel go. 21 And the LORD said unto Moses, Stretch out thine hand toward heaven, that there may be darkness over the land of Egypt, even darkness *which* may be felt. 22 And Moses stretched forth his hand toward heaven; and there was a thick darkness in all the land of Egypt three days: 23 They saw not one another, neither rose any from his place for three days: but all the children of Israel had light in their dwellings. 24 And Pharaoh called unto Moses, and said, Go ye, serve the LORD; only let your flocks and your herds be stayed: let your little ones also go with you. 25 And Moses said, Thou must give us also sacrifices and burnt offerings, that we may sacrifice unto the LORD our God. 26 Our cattle also shall go with us; there shall not an hoof be left behind; for thereof must we take to serve the LORD our God; and we know not with what we must serve the LORD, until we come thither. 27 But the LORD hardened Pharaoh's heart, and he would not let them go. 28 And Pharaoh said unto him, Get thee from me, take heed to thyself, see my face no more; for in *that* day thou seest my face thou shalt die. 29 And Moses said, Thou hast spoken well, I will see thy face again no more. 11:1 And the LORD said unto Moses, Yet will I bring one plague *more* upon Pharaoh, and upon Egypt; afterwards he will let you go hence: when he shall let *you* go, he shall surely thrust you out hence altogether. 2 Speak now in the ears of the people, and let every man borrow of his neighbour, and every woman of her neighbour, jewels of silver, and jewels of gold. 3 And the LORD gave the people favour in the sight of the Egyptians. Moreover the man Moses *was* very great in the land of Egypt, in the sight of Pharaoh's servants, and in the sight of the people. 4 And Moses said, Thus saith the LORD, About midnight will I go out into the midst of Egypt: 5 And all the firstborn in the land of Egypt shall die, from the firstborn of Pharaoh that sitteth upon his throne, even unto the firstborn of the maidservant that *is* behind the mill; and all the firstborn of beasts. 6 And there shall be a great cry throughout all the land of Egypt, such as there was none like it, nor shall be like it any more. 7 But against any of the children of Israel shall not a dog move his tongue, against man or beast: that ye may know how that the LORD doth put a difference between the Egyptians and Israel. 8 And all these thy servants shall come down unto me, and bow down themselves unto me, saying, Get thee out, and all the people that follow thee: and after that I will go out. And he went out from Pharaoh in a great anger. 9 And the LORD said unto Moses, Pharaoh shall not hearken unto you; that my wonders may be multiplied in the land of Egypt. 10 And Moses and Aaron did all these wonders before Pharaoh: and the LORD hardened Pharaoh's heart, so that he would not let the children of Israel go out of his land."

In 1887, Communist Frederick Engels knew that the First World War was coming and that it would destroy the Empires of Europe and leave them ripe for revolution. He also knew that it would murder millions of people, and he welcomed the holocaust as a necessary sacrifice to Communism,

> "No other war is now possible for Prussia-Germany than a world war, and indeed a world war of hitherto unimagined sweep and violence. Eight to ten million soldiers will mutually kill each other off, and in the process devour Europe barer than any swarm of locusts ever did. The desolation of the Thirty Years' War compressed into three or four years and spread over the entire continent: famine, plague, general savagery, taking possession both of the armies and of the masses of the people, as a result of universal want; hopeless demoralization of our complex institutions of trade, industry and credit, ending in universal bankruptcy; collapse of the old states and their traditional statecraft, so that crowns will roll over the pavements by the dozens and no one be found to pick them up; absolute impossibility of foreseeing where this will end, or who will emerge victor from the general struggle. Only *one* result is absolutely sure: general exhaustion and the creation of the conditions for the final victory of the working class."[136]

Like other religious cults, Communists recruited initiates by telling them tales of Utopia, filling their days and thoughts with comradeship and eventually demanding that they become obedient servants to the cause. They were masters of propaganda and had the means to disseminate it. They had no scruples whatsoever and eventually succeeded in manipulating public opinion to the point where those who accused them of what they were doing were themselves treated like criminals by society.[137] The only way they could offer competition to better reasoned and far more effective systems of government was to weaken those systems through corruption, and concurrently blame the destruction they deliberately caused on those who had tried to prevent it. Communists have perpetrated tens of millions, if not hundreds of millions, of murders—which they view as human sacrifices to the cause, the dogmas and dictatorship of Communism—ultimately human sacrifices to Judaism.

The truth behind "Colonel" House's feigned Liberalism was that Mexico had oil fields, gold mines, silver mines and rubber plantations, which House's financier friends wanted to exploit. Jewish financiers had been working toward a "race war" between Latin Catholics and Anglo-Saxon Protestants centered in Mexico and spreading to the United States, France, Austria and North Germany,

---

136. B. D. Wolfe, *Marxism: One Hundred Years in the Life of a Doctrine*, Dial Press, New York, (1965), p. 67. Wolfe cites: "From Engels's introduction to the reissue of a pamphlet by Sigismund Borkheim. Borkheim's pamphlet, *Zur Errinnerung fuer die deutschen Mordspatrioten 1806-07* [***] The introduction is reproduced in *Werke*, Vol. XXI, pp. 350-351."

137. R. Radosh and A. Radosh, *Red Star over Hollywood: The Film Colony's Long Romance with the Left*, Encounter Books, San Francisco, (2005).

at least since the time of the Civil War. The Rothschilds desired to divide America up between France and Great Britain.[138] The North would join with Canada and return to the British Empire. The South would go to Mexico, which would in turn serve as a colony of France. The Rothschilds would then have a profitable division between Latin and French Catholics in the South, and Anglo-Saxon Protestants in the North. The Rothschilds could then use the model they had so successfully employed in Europe to create perpetual wars[139] between the North and South which would earn the Rothschilds immense profits, place both Empires further in the Rothschilds' debt, and destroy the competitive threat that American finance posed. Bismarck, who had close contacts with Jewish finance, stated,

> "The division of the United States into federations of equal force was decided long before the Civil War by the high financial powers of Europe. These bankers were afraid that the United States, if they remained in one block and was one nation, would attain economic and financial independence, which would upset their financial domination over Europe and the world. Of course, in the 'inner circle' of Finance, the voice of the Rothschilds prevailed. They saw an opportunity for prodigious booty if they could substitute two feeble democracies burdened with debt to the financiers, ... in place of a vigorous Republic sufficient unto herself. Therefore, they sent their emissaries into the field to exploit the question of slavery and to drive a wedge between the two parts of the Union... . The rupture between the North and the South became inevitable; the masters of European finance employed all their forces to bring it about and to turn it to their advantage."[140]

On 10 June 1862, on page 3, *The Chicago Tribune* reported,

"FRANCE AND MEXICO.

THE SECRET HISTORY OF THE EXPEDITION.

---

[138]. G. E. Griffin, *The Creature from Jekyll Island: A Second Look at the Federal Reserve*, Fourth Edition, American Media, Westlake Village, California, (2002), pp. 222-224. R. McNair Wilson, *Monarchy or Money Power*, Eyre and Spottiswoode Ltd., London, (1933), pp. 81-83.

[139]. G. E. Griffin, "The Rothschild Formula", *The Creature from Jekyll Island: A Second Look at the Federal Reserve*, Chapter 11, Fourth Edition, American Media, Westlake Village, California, (2002), pp. 217-234.

[140]. G. E. Griffin, *The Creature from Jekyll Island: A Second Look at the Federal Reserve*, Fourth Edition, American Media, Westlake Village, California, (2002), p. 374. Griffin cites C. Siem, *La Vieille France*, Number 216, (17-24 March 1921), pp. 13-16.

THE ACTUAL ATTITUDE OF THE FRENCH GOVERNMENT.

New Mutterings of Intervention.

[New York Times Correspondent.]

PARIS, May 23, 1862.

The Mexican affair has assumed all at once at Paris a most serious aspect. Never before has the Emperor been attacked by the liberal press with such violence, or rather, with such an outspoken energy, as within the last few days, on this unfortunate Mexican expedition. It is the all-absorbing topic of the moment, and I cannot do better than to give you an *apercu* of the situation, as we understand it here.

It so happens that, so far as regards the Press, the three papers which have thus far defended the cause of the rebellion in the United States, are exactly those which sustain the Almonte-Maximilian programme for Mexico; while the rest of the journals, with the exception of the Catholics, defend the cause of the Union in the United States, and combat the monarchical programme in Mexico. This striking concurrence in the division of views on the two subjects, indicates, beyond any question, that for the French there is an important connection between the two. It is this connection which gives the question its gravity.

For a long time the Emperor has dreamed of two things:

First—The acquisition of Sonora, with its gold and silver mines.

Second—The reconstruction of the Latin race, and the pitting of this race and Catholicism against the Anglo Saxon race and Protestantism.

The two governments of France and England, and no doubt of Spain also, did not believe till lately that there was any possibility of the suppression of the rebellion in the United States and the reconstruction of the Union. When, therefore, the treaty of London, of last year, in regard to the expedition to Mexico, was drawn up, it was drawn up with an almost complete indifference as to what the United States might think or do about it, and there is now every reason to believe that each of the contracting parties had ulterior views, which were not only concealed from the world, but from each other. The treaty was therefore drawn up in a loose and vague manner, so as to admit of deviations at will, so that each might seize upon whatever advantages offered themselves. And here I ought to recall, for its historical value, an observation made by Mr. Dayton nine months ago, and put upon record at the time in this correspondence, to the effect that, although the French government was full of kind and frank expressions towards the United States in connection with this Mexican expedition, yet that there seemed to be a vagueness and a confusion in their own understanding of the objects and the details of the expedition which foreboded no good to the future relations between France and the United States.

At the time of the arrival of the Soledad Convention at Paris there had been nothing done toward changing the belief of the French Government that a final dissolution of the Union was inevitable, and Napoleon is known at that time to have given Gen. Lorencez hasty and imperative orders to hurry on to the City of Mexico, without regard to consequences. Why? Because, the Government papers here now say, it was recognized as impossible to gain the objects of the expedition without displacing Jaurez from power and establishing in his stead a stable government, capable of offering, besides indemnity for the present, security for the future. And here is where the English and Spaniards deserted Napoleon, and where the great majority of Napoleon's own subjects also deserted him. They divided on the question of an interference in the internal affairs of Mexico, after having obtained satisfaction for the first objects of the expedition. It came out all at once that Napoleon had been serious in his secret transactions with Almonte at Paris, and that the plan of erecting a throne for an Austrian Prince was not an illusion. Knowing the mind of the Mexican people, the Allies and the Liberals of Paris naturally and legitimately jumped to the conclusion that the Emperor was bent on a conquest of the country, for that was the only condition on which he could maintain a foreign Prince in power, and that sooner or later it would terminate with an acquisition of territory and a war with the United States.

The news of the breaking up of the alliance at Orizaba arrived in Europe with that of the capture of New Orleans, and it is hard to tell which event caused most consternation at the Palace. For the first time the fact that the Southern Confederacy might possible prove a failure, penetrated the short vision of the French Government; and now we believe that under the influence of these two events, the French Government has modified its intentions, and that it has sent to Mexico orders not to push matters to the extreme point at first designed.

The opposition press here has said to the Emperor: Your Mexican expedition, under the present aspect of the case, (that is to say, as an agent of the monarchial party,) is either an aberration or a scheme for the ransom of Venetia. If it be the first, comment is unnecessary—there is but one course to follow: withdraw as quickly as possible after securing what Mexico owes us; if it be the ransom of Venetia that is intended, permit us to suggest that a war with Austria in the quadrilateral will cost us infinitely less in time; men, money, and especially in honor, than a war with the United States.

The opposition press also points out with telling effect on the public mind the analogy which exists between the entrance of the allies into France in 1815, bringing with them the exiles who were selling their country in order to gain power for a minority. For whatever may be the faults of Juarez, he is fighting for his native country against the foreigner, which constitutes his patriotism—quite another thing to that of Almonte, Miramon and company.

As we understand the question then, to-day, Napoleon, at the moment he heard of the treaty of Soledad, gave to Gen. Lorencez instructions which conveyed with them the perspective of a monarchy, a more or less permanent occupation, an acquisition of territory, and a strengthening of the Latin race in America. But the late Union victories have changed the programme, and by this time we have every reason to believe Gen. Lorencez has received a modification to his previous orders. But how far this modification extends no one knows or pretends even to conjecture. That the Emperor will renounce the monarchical programme is, however, generally believed, but whether, when his troops arrive at the capital, they will treat with Juarez or insist on putting Almonte into the Presidential chair before treating, is all in doubt. If Almonte is put into the chair provisionally, every one can see that then the reign of anarchy will only have commenced, and that the French will be obliged to remain to carry out their unfortunate programme by force. And yet, up to the present moment, the Ministerial papers here declare that it will be degrading to the dignity of France to treat with such a man as Juarez, and that such a thing cannot be thought of for a moment. But who can see the end if they go beyond Juarez? One step beyond him and everything is darkness and confusion. Every one in France seems to understand that, if the power of the Federal Government is again consolidated by the suppression of the rebellion, Mexico will at once occupy the attention of the United States, and that France cannot afford, for the benefit of an Austrian Duke and a score of Mexican exiles, to bring upon herself a war with the United States.

The Republicans in France, in view of this war with the United States, declare that it will bring with it the downfall of the Bonaparte dynasty, and they are quite elated at the prospect.

Among the persons who have been indicated as having used their influence with the Emperor since the commencement of the rebellion, in urging on the Sonora programme, are Messrs. Michel-Chevalier, Fould, Rouher, and De Rothschild. These gentlemen do not see why France should not make an acquisition of valuable gold mines—which, by the way, she much needs—as well as the United States.

As regards the more utopian scheme of reconstructing and strengthening the Latin and Catholic elements in America, some of the most influential imperialist writers of France have long been urging it. To these must be added a demented party not far removed from the Emperor's person, who dream of nothing less than setting up in America what has been repudiated in Europe—a nobility system, based upon the divine right, and which shall give an asylum and an occupation to the castoff kings and princes of Europe. They would have the Grand Duke Maxamilian or Ferdinand II., of Naples, placed on the throne of Mexico, surrounded by the European rejected princes, and this try to gain a new foothold for a system which is here growing weaker every day.

But the Emperor has generally shown great judgment in seizing the right side of questions as they pass before him, and great wisdom in retreating from mistaken positions, into which, like the ablest of men, he has sometimes fallen; and we have great confidence that he will yet, with the new light which has broken in upon him from the United States, retire from Mexico before he has become so far entangled in the meshes that await him.

A new secession pamphlet is also just out, to which M. Marc de Haut, advocate at the Imperial Court, has put his name. It is entitled: *The American Crisis: its causes, probable results, and connection with France and Europe.* The pamphlet is but a repetition of several of those which have preceded it, and appears to prove that the secessionists think it necessary to keep certain arguments continually, in one form or another, before the public. The following are the stereotyped heads of arguments found in this book: Republics, when the grow too large, must divide. The Americans of the North are ancient English Puritans, sombre, intolerant, taciturn and commercial. The Southerners are descendants of the Cavaliers, grand, historical *seigneurs*, who love a large and free existence, who don't build workshops or counters, but furnish orators, statesmen and presidents. The sole cause of the dissolution of the Union is the tariff—slavery was only the pretext. The Yankees abandoned slavery in the Northern States, not from principle, but because free labor was more profitable in their climate. The proof of this is found in their well known antipathy to the person of the negro. The present struggle is one of free trade against protection. A reunion can never take place. And then the writer terminates with that funny appeal for the sympathy of the French—that the South is French. 'Does not,' he exclaims, 'the General-in-Chief of the Southern forces bear a French name—Beauregard? And what souvenirs do the following names of *Southern* towns recall to the French hear—Louisburg, Montmorency, St. Louis, Vincennes, Duquesne, New Orleans?'

Thus you will see that the French secessionists demand sympathy for the South because it is French, while, the other day, the London *Times* demanded the sympathy of the English for the South because it is English! We hope they will settle the question between them.

MALAKOFF."

Oil magnates wanted to steal Mexican oil and the American Government readied to invade Mexico in order to seize their oil fields during World War I, but President Wilson did not approve the plan. Bernard Baruch tells this story and according to him, the financiers asked the President to invade Mexico without a declaration of war by the Congress.[141] House, while exorcizing his real power over the United States Government, used banker corruption to justify "reforms" which resulted in greater banker corruption and eventually in the Great Depression.

At the instigation of the Jewish bankers, House and Wilson led America into bloody world war allegedly for the sake of peace and to "make the world safe for democracy"—democracies like Bolshevik Russia. They were unjust to

---

[141]. B. M. Baruch, *Baruch: My Own Story*, Henry Holt and Company, New York, (1957), pp. 213-217.

Germany in the name of justice, and oppressed and exploited the Third World in the name of freedom and equality for small nations. The First World War yielded them immense profits, which the Jewish bankers then used to corrupt the stock markets, which then led to the Depression, which then enabled them to buy whatever they wanted to buy at deflated prices, which then led to the Second World War, which yielded them immense profits. Smedley D. Butler's book *War Is a Racket*, Round Table Press, New York, (1935), tells of the ungodly profits the warmongers made under the Wilson Administration at the expense of the American People they were duty bound to protect. Wilson was the perfect puppet dictator House had envisioned in his book.

The Zionists knew a great deal about dictators and revolutions. George Orwell warned in 1945 that revolutions most often result in a mere shift of power, and ultimately return to the same, or even worse, unfair conditions,[142] Zionist Max Simon Nordau explored this common wisdom in 1909 in a book translated into English in 1910,

> "Revolutions do not, as a rule, transform anything, with the exception of the hierarchy of rank. Generally they leave everything essentially as it is: the weak continue to be exploited, and the strong exploit. New modes of adaptation to what is disagreeable prolong the endurance of what is endurable. Only, other individuals and classes take the place of those individuals and classes hitherto privileged to exploit. Revolution gives to some what it takes from others. It is a practical test of the symbols and prestige of power, which are tried and found wanting. It gives the strong the position inherited by the weak man, who maintained it simply because his strength was a tradition which had never been tested. It destroys an appearance which corresponded to no reality. But its effect does not last. 'Red men are white men on the way; white men are red men arrived,' as Alphonse Karr has said. A new order soon becomes petrified to a new routine; the new real strength soon dissipates itself in new symbols; new weakly heirs begin to live on the prestige of new strong ancestors. A long period of time presents the aspect of a succession of waves of more or less equal size. The noisiest revolutions are very limited in their effect, and do not go very deep. Tocqueville [*Footnote:* Quoted by Robert Flint, 'The Philosophy of History in France and Germany,' Edinburgh and London, 1874, p. 313.] declares that 'even the great French Revolution has had far less influence upon the course of development of French history than is believed.' Lotze [*Footnote:* Hermann Lotze, 'Microcosm: Idea of a History and Natural History of Mankind—an Attempted Anthropology,' vol. iii, Leipzig, 1864, p. 49.] lets fall a stimulating remark: 'The unrest and variety manifest in constant revolutions and reconstructions, for which a connected meaning is sought, simply represents the history of the male sex: women make their way through the storm and stress, hardly affected by its changing

---

142. G. Orwell, *Animal Farm: A Fairy Story*, Secker & Warburg, London, (1945).

aspects, renewing with perpetual uniformity the grand, simple forms of the life of the human soul.' This needs one limitation, however. History is not that of the male sex, but of a small section of it; what Lotze says of women is true of the great majority of men.

We have been speaking of revolutions. It might be objected that historical advance is not always, perhaps not even mainly, due to revolution, but to at least an equal extent to slow, tentative, and peaceful innovations, limited in extent, directed by authority. The objection would be invalid. From a psychological point of view there is no difference between the revolution and the cautious, official reform. Every innovation breaks in upon habit, and compels new adaptations. Even the picture on a postage-stamp cannot be altered without disturbing someone and overcoming some opposition. The difference between revolution and reform or evolution is not a difference of essential, but of mass, extent, energy, rhythm. Revolution requires greater strength on the part of those who rouse it than reform does, because it has against it the weight of habit, the whole routine of life, the interests of the powerful, the symbols connected in the minds of the multitude with the ideas of power, legality, order, and respectability: on its side, only the superior will-power of its leaders, the sense of discontent of their followers, and the adaptability of the young, whose habits are not yet stereotyped, and whose discontent is less patient than that of the older generation. The advantage of reform is that it can be undertaken with smaller powers. It is set going with the aid of the whole machinery of society and the State, which embodies the habits of the multitude. It therefore departs less from routine, offends fewer people, and demands less new adaptation than revolution does. But the same cause operates in both—the discontent that is felt and understood as the need for change."[143]

The horrors of the Civil War and the destruction of the South still haunted Americans, who were not eager for revolution nor war. Americans had to be shocked and propagandized into the war. House had to create his revolution and dictatorship by operating behind the scenes through a puppet President. He had to find someone with charisma—someone he could control.

House maintained an almost surreal relationship with President Wilson. Wilson thought of House as his soul mate or "alter ego", until House betrayed him for the sake of Great Britain and Zionism at the Paris Peace Conference in 1919, where Wilson's (originally House's) Fourteen Point plan for a just peace with Germany (and the colonial exploitation of the Third World) was abandoned for punitive measures that crushed Germany.

Much has been written by and about Edward Mandell House.[144] Sigmund

---

[143]. M. Nordau, *The Interpretation of History*, Willey Book Company, New York, (1910), pp. 311-313; which is an English translation by M. A. Hamilton of *Der Sinn der Geschichte*, C. Duncker, Berlin, (1909).
[144]. E. M. House, *Philip Dru: Administrator*, B. W. Huebsch, New York, (1912). **See also:** A. D. H. Smith, *The Real Colonel House*, George H. Doran Company, New York,

Freud coauthored a book with William Bullitt, *Thomas Woodrow Wilson, Twenty-Eighth President of the United States: A Psychological Study*, Houghton Mifflin, Boston, (1966/1967); which famously employed the use of psychology as a political weapon to smear Wilson, and which expresses the authors' opinions about the formation and nature of Wilson's personality and its relationship to, and impact on, world events. Many of the disastrous actions Woodrow Wilson took as President of the United States were forced on him by the ardent Zionists Louis Dembitz Brandeis and "Colonel" Edward Mandell House.

House's intentions were revealed in his book *Philip Dru: Administrator* of 1911. He planned to corrupt the Senate and install a puppet President of the United States, who would do his bidding and that of the financiers House favored, and who favored him. With a puppet President in power, House planned to stack the Supreme Court with appointees of his choosing and House planned to name all of the President's other appointees. "Colonel" Edward Mandell House succeeded in his plans. In his book he makes a Socialist dictator the hero. House was the corrupt "Selwyn". House wrote, *inter alia:*

"Chapter XI
Selwyn Plots with Thor
For five years Gloria and Philip worked in their separate fields, but, nevertheless, coming in frequent touch with one another. Gloria proselyting the rich by showing them their selfishness, and turning them to a larger purpose in life, and Philip leading the forces of those who had consecrated themselves to the uplifting of the unfortunate. It did not take Philip long to

---

(1918); **and** *Mr. House of Texas*, Funk & Wagnalls Co., New York, London, (1940). *See also:* E. J. Dillon, *The Inside Story of the Peace Conference*, Harper & Brothers, New York, (1920). *See also:* E. H. House and C. Seymour, *What Really Happened at Paris; the Story of the Peace Conference, 1918-1919*, C. Scribner's Sons, New York, (1921). *See also:* B. J. Hendrick, *The Life and Letters of Walter H. Page*, Volume 1, Doubleday, Page & Co., Garden City, New York, (1922), p. 270ff. *See also:* E. M. House and S. Charles, *The Intimate Papers of Colonel House*, Boston, New York, Houghton Mifflin Company, (1926-1928). *See also:* A. Macphail, *Three Persons*, L. Carrier & Co., J. Murray, New York, Montreal, London, (1929). *See also:* G. S. Viereck, *The Strangest Friendship in History: Woodrow Wilson and Colonel House*, Liveright, Inc., New York, (1932). *See also:* W. C. Caton, *Die Rolle des obersten House im Rahmen der friedensaktion Wilsons im Jahre 1916/17 ...*, Buchdr. B. Müller, Heidelberg, (1937). *See also:* A. H. Kober, *Wilson und der Weltkrieg: Rätsel einer Freundschaft*, Societäts-Verlag, Frankfurt a.M., (1938). *See also:* A. L. George and J. L. George, *Woodrow Wilson and Colonel House: A Personality Study*, The John Day Company, New York, (1956). *See also:* R. N. Richardson, *Colonel Edward M. House: The Texas years, 1858-1912*, Volume 1, Publications in History, Hardin-Simmons University, Abilene, Texas, (1964). *See also:* S. Freud and W. Bullitt, *Thomas Woodrow Wilson, Twenty-Eighth President of the United States: A Psychological Study*, Houghton Mifflin, Boston, (1966/1967). *See also:* J. Reinertson, *Colonel House, Woodrow Wilson and European Socialism, 1917-1919*, Madison, Wisconsin, (1971). *See also:* R. W. Wall, *Edward M. House's influence on United States foreign policy 1913-1917*, Hardin-Simmons University, Abilene, Texas, (1972).

discern that in the last analysis it would be necessary for himself and co-workers to reach the results aimed at through politics. Masterful and arrogant wealth, created largely by Government protection of its profits, not content with its domination and influence within a single party, had sought to corrupt them both, and to that end had insinuated itself into the primaries, in order that no candidates might be nominated whose views were not in accord with theirs.

By the use of all the money that could be spent, by a complete and compact organization and by the most infamous sort of deception regarding his real opinions and intentions, plutocracy had succeeded in electing its creature to the Presidency. There had been formed a league, the membership of which was composed of one thousand multi-millionaires, each one contributing ten thousand dollars. This gave a fund of ten million dollars with which to mislead those that could be misled, and to debauch the weak and uncertain.

This nefarious plan was conceived by a senator whose swollen fortune had been augmented year after year through the tributes paid him by the interests he represented. He had a marvelous aptitude for political manipulation and organization, and he forged a subtle chain with which to hold in subjection the natural impulses of the people. His plan was simple, but behind it was the cunning of a mind that had never known defeat. There was no man in either of the great political parties that was big enough to cope with him or to unmask his methods.

Up to the advent of Senator Selwyn, the interests had not successfully concealed their hands. Sometimes the public had been mistaken as to the true character of their officials, but sooner or later the truth had developed, for in most instances, wealth was openly for or against certain men and measures. But the adroit Selwyn moved differently.

His first move was to confer with John Thor, the high priest of finance, and unfold his plan to him, explaining how essential was secrecy. It was agreed between them that it should be known to the two of them only.

Thor's influence throughout commercial America was absolute. His wealth, his ability and even more the sum of the capital he could control through the banks, trust companies and industrial organizations, which he dominated, made his word as potent as that of a monarch.

He and Selwyn together went over the roll and selected the thousand that were to give each ten thousand dollars. Some they omitted for one reason or another, but when they had finished they had named those who could make or break within a day any man or corporation within their sphere of influence. Thor was to send for each of the thousand and compliment him by telling him that there was a matter, appertaining to the general welfare of the business fraternity, which needed twenty thousand dollars, that he, Thor, would put up ten, and wanted him to put up as much, that sometime in the future, or never, as the circumstances might require, would he make a report as to the expenditure and purpose therefor.

There were but few men of business between the Atlantic and Pacific, or between Canada and Mexico, who did not consider themselves fortunate in being called to New York by Thor, and in being asked to join him in a blind pool looking to the safe-guarding of wealth. Consequently, the amassing of this great corruption fund in secret was simple. If necessity had demanded it twice the sum could have been raised. The money when collected was placed in Thor's name in different banks controlled by him, and Thor, from time to time, as requested by Selwyn, placed in banks designated by him whatever sums were needed. Selwyn then transferred these amounts to the private bank of his son-in-law, who became final paymaster. The result was that the public had no chance of obtaining any knowledge of the fund or how it was spent.

The plan was simple, the result effective. Selwyn had no one to interfere with him. The members of the pool had contributed blindly to Thor, and Thor preferred not to know what Selwyn was doing nor how he did it. It was a one man power which in the hands of one possessing ability of the first class, is always potent for good or evil.

Not only did Selwyn plan to win the Presidency, but he also planned to bring under his control both the Senate and the Supreme Court. He selected one man in each of thirty of the States, some of them belonging to his party and some to the opposition, whom he intended to have run for the Senate.

If he succeeded in getting twenty of them elected, he counted upon having a good majority of the Senate, because there were already thirty-eight Senators upon whom he could rely in any serious attack upon corporate wealth.

As to the Supreme Court, of the nine justices there were three that were what he termed 'safe and sane,' and another that could be counted upon in a serious crisis.

Three of them, upon whom he could not rely, were of advanced age, and it was practically certain that the next President would have that many vacancies to fill. Then there would be an easy working majority.

His plan contemplated nothing further than this. His intention was to block all legislation adverse to the interests. He would have no new laws to fear, and of the old, the Supreme Court would properly interpret them.

He did not intend that his Senators should all vote alike, speak alike, or act from apparently similar motives. Where they came from States dominated by corporate wealth, he would have them frankly vote in the open, and according to their conviction.

When they came from agricultural States, where the sentiment was known as 'progressive,' they could cover their intentions in many ways. One method was by urging an amendment so radical that no honest progressive would consent to it, and then refusing to support the more moderate measure because it did not go far enough. Another was to inject some clause that was clearly unconstitutional, and insist upon its adoption, and refusing to vote for the bill without its insertion.

Selwyn had no intention of letting any one Senator know that he controlled any other senator. There were to be no caucuses, no conferences of his making, or anything that looked like an organization. He was the center, and from him radiated everything appertaining to measures affecting 'the interests.'

Chapter XII
Selwyn Seeks a Candidate

Selwyn then began carefully scrutinizing such public men in the States known as Presidential cradles, as seemed to him eligible. By a process of elimination he centered upon two that appeared desirable.

One was James R. Rockland, recently elected Governor of a State of the Middle West. The man had many of the earmarks of a demagogue, which Selwyn readily recognized, and he therefore concluded to try him first.

Accordingly he went to the capital of the State ostensibly upon private business, and dropped in upon the Governor in the most casual way. Rockland was distinctly flattered by the attention, for Selwyn was, perhaps, the best known figure in American politics, while he, himself, had only begun to attract attention. They had met at conventions and elsewhere, but they were practically unacquainted, for Rockland had never been permitted to enter the charmed circle which gathered around Selwyn.

'Good morning, Governor,' said Selwyn, when he had been admitted to Rockland's private room. 'I was passing through the capital and I thought I would look in on you and see how your official cares were using you.'

'I am glad to see you, Senator,' said Rockland effusively, 'very glad, for there are some party questions coming up at the next session of the Legislature about which I particularly desire your advice.'

'I have but a moment now, Rockland,' answered the Senator, 'but if you will dine with me in my rooms at the Mandell House to-night it will be a pleasure to talk over such matters with you.'

'Thank you, Senator, at what hour?'

'You had better come at seven for if I finish my business here to-day, I shall leave on the 10 o'clock for Washington,' said Selwyn.

Thus in the most casual way the meeting was arranged. As a matter of fact, Rockland had no party matters to discuss, and Selwyn knew it. He also knew that Rockland was ambitious to become a leader, and to get within the little group that controlled the party and the Nation.

Rockland was a man of much ability, but he fell far short of measuring up with Selwyn, who was in a class by himself. The Governor was a good orator, at times even brilliant, and while not a forceful man, yet he had magnetism which served him still better in furthering his political fortunes. He was not one that could be grossly corrupted, yet he was willing to play to the galleries in order to serve his ambition, and he was willing to forecast his political acts in order to obtain potential support.

When he reached the Mandell House, he was at once shown to the Senator's rooms. Selwyn received him cordially enough to be polite, and asked him if he would not look over the afternoon paper for a moment while he finished a note he was writing. He wrote leisurely, then rang for a boy and ordered dinner to be served.

Selwyn merely tasted the wine (he seldom did more) but Rockland drank freely though not to excess. After they had talked over the local matters which were supposed to be the purpose of the conference, much to Rockland's delight, the Senator began to discuss national politics.

'Rockland,' began Selwyn, 'can you hold this state in line at next year's election?'

'I feel sure that I can, Senator, why do you ask?'

'Since we have been talking here,' he replied, 'it has occurred to me that if you could be nominated and elected again, the party might do worse than to consider you for the presidential nomination the year following.

'No, my dear fellow, don't interrupt me,' continued Selwyn mellifluously.

'It is strange how fate or chance enters into the life of man and even of nations. A business matter calls me here, I pass your office and think to pay my respects to the Governor of the State. Some political questions are perplexing you, and my presence suggests that I may aid in their solution. This dinner follows, your personality appeals to me, and the thought flits through my mind, why should not Rockland, rather than some other man, lead the party two years from now?

'And the result, my dear Rockland, may be, probably will be, your becoming chief magistrate of the greatest republic the sun has ever shone on.'

Rockland by this time was fairly hypnotized by Selwyn's words, and by their tremendous import. For a moment he dared not trust himself to speak.

'Senator Selwyn,' he said at last, 'it would be idle for me to deny that you have excited within me an ambition that a moment ago would have seemed worse than folly. Your influence within the party and your ability to conduct a campaign, gives to your suggestion almost the tender of the presidency. To tell you that I am deeply moved does scant justice to my feelings. If, after further consideration, you think me worthy of the honor, I shall feel under lasting obligations to you which I shall endeavor to repay in every way consistent with honor and with a sacred regard for my oath of office.'

'I want to tell you frankly, Rockland,' answered Selwyn, 'that up to now I have had someone else in mind, but I am in no sense committed, and we might as well discuss the matter to as near a conclusion as is possible at this time.'

Selwyn's voice hardened a little as he went on. 'You would not want a nomination that could not carry with a reasonable certainty of election, therefore I would like to go over with you your record, both public and private, in the most open yet confidential way. It is better that you and I, in the privacy of these rooms, should lay bare your past than that it should be done in a bitter campaign and by your enemies. What we say to one another here is to be as if never spoken, and the grave itself must not be more silent. Your private life not only needs to be clean, but there must be no public act at which any one can point an accusing finger.'

'Of course, of course,' said Rockland, with a gesture meant to convey the complete openness of his record.

'Then comes the question of party regularity,' continued Selwyn, without noticing. 'Be candid with me, for, if you are not, the recoil will be upon your own head.'

'I am sure that I can satisfy you on every point, Senator. I have never scratched a party ticket nor have I ever voted against any measure endorsed by a party caucus,' said Governor Rockland.

'That is well,' smiled the Senator. 'I assume that in making your important appointments you will consult those of us who have stood sponsor for you, not only to the party but to the country. It would be very humiliating to me if I should insist upon your nomination and election and then should for four years have to apologize for what I had done.'

Musingly, as if contemplating the divine presence in the works of man, Selwyn went on, while he closely watched Rockland from behind his half-closed eyelids.

'Our scheme of Government contemplates, I think, a diffuse responsibility, my dear Rockland. While a president has a constitutional right to act alone, he has no moral right to act contrary to the tenets and traditions of his party, or to the advice of the party leaders, for the country accepts the candidate, the party and the party advisers as a whole and not severally.

'It is a natural check, which by custom the country has endorsed as wise, and which must be followed in order to obtain a proper organization. Do you follow me, Governor, and do you endorse this unwritten law?'

If Rockland had heard this at second hand, if he had read it, or if it had related to someone other than himself, he would have detected the sophistry of it. But, exhilarated by wine and intoxicated by ambition, he saw nothing but a pledge to deal squarely by the organization.

'Senator,' he replied fulsomely, 'gratitude is one of the tenets of my religion, and therefore inversely ingratitude is unknown to me. You and the organization can count on my loyalty from the beginning to the end, for I shall never fail you.

'I know you will not ask me to do anything at which my conscience will rebel, nor to make an appointment that is not entirely fit.'

'That, Rockland, goes without saying,' answered the Senator with dignity. 'I have all the wealth and all the position that I desire. I want nothing now except to do my share towards making my native land grow in prosperity, and to make the individual citizen more contented. To do this we must cease this eternal agitation, this constant proposal of half-baked measures, which the demagogues are offering as a panacea to all the ills that flesh is heir to.

'We need peace, legislative and political peace, so that our people may turn to their industries and work them to success, in the wholesome knowledge that the laws governing commerce and trade conditions will not be disturbed over night.'

'I agree with you there, Senator,' said Rockland eagerly.

'We have more new laws now than we can digest in a decade,' continued Selwyn, 'so let us have rest until we do digest them. In Europe the business world works under stable conditions. There we find no proposal to change the money system between moons, there we find no uncertainty from month to month regarding the laws under which manufacturers are to make their products, but with us, it is a wise man who knows when he can afford to enlarge his output.

'A high tariff threatens to-day, a low one to-morrow, and a large part of the time the business world lies in helpless perplexity.

'I take it, Rockland, that you are in favor of stability, that you will join me in my endeavors to give the country a chance to develop itself and its marvelous natural resources.'

As a matter of fact, Rockland's career had given no evidence of such views. He had practically committed his political fortunes on the side of the progressives, but the world had turned around since then, and he viewed things differently.

'Senator,' he said, his voice tense in his anxiety to prove his reliability, 'I find that in the past I have taken only a cursory view of conditions. I see clearly that what you have outlined is a high order of statesmanship. You are constructive: I have been on the side of those who would tear down. I will gladly join hands with you and build up, so that the wealth and power of this country shall come to equal that of any two nations in existence.'

Selwyn settled back in his chair, nodding his approval and telling himself that he would not need to seek further for his candidate.

At Rockland's earnest solicitation he remained over another day. The Governor gave him copies of his speeches and messages, so that he could assure himself that there was no serious flaw in his public record.

Selwyn cautioned him about changing his attitude too suddenly. 'Go on, Rockland, as you have done in the past. It will not do to see the light too quickly. You have the progressives with you now, keep them, and I will let the conservatives know that you think straight and may be trusted.

'We must consult frequently together,' he continued, 'but cautiously. There is no need for any one to know that we are working together harmoniously. I may even get some of the conservative papers to attack you judiciously. It will not harm you. But, above all, do nothing of importance without consulting me.

'I am committing the party and the Nation to you, and my responsibility is a heavy one, and I owe it to them that no mistakes are made.'

'You may trust me, Senator,' said Rockland. 'I understand perfectly.'

[***]

Chapter XIV
The Making of a President

Selwyn now devoted himself to the making of enough conservative senators to control comfortably that body. The task was not difficult to a man of his sagacity with all the money he could spend.

Newspapers were subsidized in ways they scarcely recognized themselves. Honest officials who were in the way were removed by offering them places vastly more remunerative, and in this manner he built up a strong, intelligent and well constructed machine. It was done so sanely and so quietly that no one suspected the master mind behind it all. Selwyn was responsible to no one, took no one into his confidence, and was therefore in no danger of betrayal.

It was a fascinating game to Selwyn. It appealed to his intellectual side far more than it did to his avarice. He wanted to govern the Nation with an absolute hand, and yet not be known as the directing power. He arranged to have his name appear less frequently in the press and he never submitted to interviews, laughingly ridding himself of reporters by asserting that he knew nothing of importance. He had a supreme contempt for the blatant self-advertised politician, and he removed himself as far as possible from that type.

In the meantime his senators were being elected, the Rockland sentiment was steadily growing and his nomination was finally brought about by the progressives fighting vigorously for him and the conservatives yielding a reluctant consent. It was done so adroitly that Rockland would have been fooled himself, had not Selwyn informed him in advance of each move as it was made.

After the nomination, Selwyn had trusted men put in charge of the campaign, which he organized himself, though largely under cover. The opposition party had every reason to believe that they would be successful, and it was a great intellectual treat to Selwyn to overcome their natural advantages by the sheer force of ability, plus what money he needed to carry out his plans. He put out the cry of lack of funds, and indeed it seemed to be true, for he was too wise to make a display of his resources. To ward heelers, to the daily press, and to professional stump speakers, he gave scant comfort. It was not to such sources that he looked for success.

He began by eliminating all states he knew the opposition party would certainly carry, but he told the party leaders there to claim that a revolution was brewing, and that a landslide would follow at the election. This would keep his antagonists busy and make them less effective elsewhere.

He also ignored the states where his side was sure to win. In this way he was free to give his entire thoughts to the twelve states that were debatable, and upon whose votes the election would turn. He divided each of these states into units containing five thousand voters, and, at the national headquarters, he placed one man in charge of each unit. Of the five thousand, he roughly calculated there would be two thousand voters that no kind of persuasion could turn from his party and two thousand that could not be changed from the opposition. This would leave one thousand doubtful ones to win over. So he had a careful poll made in each unit, and eliminated the strictly unpersuadable party men, and got down to a complete analysis of the debatable one thousand. Information was obtained as to their race, religion, occupation and former political predilection. It was easy then to know how to reach each individual by literature, by persuasion or perhaps by some more subtle argument. No mistake was made by sending the wrong letter or the wrong man to any of the desired one thousand.

In the states so divided, there was, at the local headquarters, one man for each unit just as at the national headquarters. So these two had only each other to consider, and their duty was to bring to Rockland a majority of the one thousand votes within their charge. The local men gave the conditions, the national men gave the proper literature and advice, and the local man then applied it. The money that it cost to maintain such an organization was more than saved from the waste that would have occurred under the old method.

The opposition management was sending out tons of printed matter, but they sent it to state headquarters that, in turn, distributed it to the county organizations, where it was dumped into a corner and given to visitors when asked for. Selwyn's committee used one-fourth as much printed matter, but it went in a sealed envelope, along with a cordial letter, direct to a voter that had as yet not decided how he would vote.

The opposition was sending speakers at great expense from one end of the country to the other, and the sound of their voices rarely fell on any but friendly and sympathetic ears. Selwyn sent men into his units to personally persuade each of the one thousand hesitating voters to support the Rockland ticket.

The opposition was spending large sums upon the daily press. Selwyn used the weekly press so that he could reach the fireside of every farmer and the dweller in the small country towns. These were the ones that would read every line in their local papers and ponder over it.

The opposition had its candidates going by special train to every part of the Union, making many speeches every day, and mostly to voters that could not be driven from him either by force or persuasion. The leaders in cities, both large and small, would secure a date and, having in mind for themselves a postmastership or collectorship, would tell their followers to turn out in great force and give the candidate a big ovation. They wanted the candidate to remember the enthusiasm of these places, and to leave greatly pleased and under the belief that he was making untold converts. As a matter of fact his voice would seldom reach any but a staunch partisan.

Selwyn kept Rockland at home, and arranged to have him meet by special appointment the important citizens of the twelve uncertain states. He would have the most prominent party leader, in a particular state, go to a rich brewer or large manufacturer, whose views had not yet been crystallized, and say, 'Governor Rockland has expressed a desire to know you, and I would like to arrange a meeting.' The man approached would be flattered to think he was of such importance that a candidate for the presidency had expressed a desire to meet him. He would know it was his influence that was wanted but, even so, there was a subtle flattery in that. An appointment would be arranged. Just before he came into Rockland's presence, his name and a short epitome of his career would be handed to Rockland to read. When he reached Rockland's home he would at first be denied admittance. His sponsor would say,—'this is Mr. Munting of Muntingville.' 'Oh, pardon me, Mr. Munting, Governor Rockland expects you.'

And in this way he is ushered into the presence of the great. His fame, up to a moment ago, was unknown to Rockland, but he now grasps his hand cordially and says,—'I am delighted to know you, Mr. Munting. I recall the address you made a few years ago when you gave a library to Muntingville. It is men of your type that have made America what it is to-day, and, whether you support me or not, if I am elected President it is such as you that I hope will help sustain my hands in my effort to give to our people a clean, sane and conservative government.'

When Munting leaves he is stepping on air. He sees visions of visits to Washington to consult the President upon matters of state, and perhaps he sees an ambassadorship in the misty future. He becomes Rockland's ardent supporter, and his purse is open and his influence is used to the fullest extent.

And this was Selwyn's way. It was all so simple. The opposition was groaning under the thought of having one hundred millions of people to reach, and of having to persuade a majority of twenty millions of voters to take their view.

Selwyn had only one thousand doubtful voters in each of a few units on his mind, and he knew the very day when a majority of them had decided to vote for Rockland, and that his fight was won. The pay-roll of the opposition was filled with incompetent political hacks, that had been fastened upon the management by men of influence. Selwyn's force, from end to end, was composed of able men who did a full day's work under the eye of their watchful taskmaster.

And Selwyn won and Rockland became the keystone of the arch he had set out to build.

There followed in orderly succession the inauguration, the selection of cabinet officers and the new administration was launched.

Drunk with power and the adulation of sycophants, once or twice Rockland asserted himself, and acted upon important matters without having first conferred with Selwyn. But, after he had been bitterly assailed by Selwyn's papers and by his senators, he made no further attempts at independence. He felt that he was utterly helpless in that strong man's hands, and so, indeed, he was.

One of the Supreme Court justices died, two retired because of age, and all were replaced by men suggested by Selwyn.

He now had the Senate, the Executive and a majority of the Court of last resort. The government was in his hands. He had reached the summit of his ambition, and the joy of it made all his work seem worth while.

But Selwyn, great man that he was, did not know, could not know, that when his power was greatest it was most insecure. He did not know, could not know, what force was working to his ruin and to the ruin of his system.

Take heart, therefore, you who had lost faith in the ultimate destiny of the Republic, for a greater than Selwyn is here to espouse your cause. He comes panoplied in justice and with the light of reason in his eyes. He comes as the advocate of equal opportunity and he comes with the power to enforce his will.

[***]

Chapter XVII
Selwyn and Thor Defend Themselves

In the meantime Selwyn and Thor had issued an address, defending their course as warranted by both the facts and the law.

They said that the Government had been honeycombed by irresponsible demagogues, that were fattening upon the credulity of the people to the great injury of our commerce and prosperity, that no laws unfriendly to the best interests had been planned, and no act had been contemplated inconsistent with the dignity and honor of the Nation. They contended that in protecting capital against vicious assaults, they were serving the cause of labor and advancing the welfare of all.

Thor's whereabouts was a mystery, but Selwyn, brave and defiant, pursued his usual way.

President Rockland also made a statement defending his appointments of Justices of the Supreme Court, and challenged anyone to prove them unfit. He said that, from the foundation of the Government, it had become customary for a President to make such appointments from amongst those whose views were in harmony with his own, that in this case he had selected men of well known integrity, and of profound legal ability, and, because they were such, they were brave enough to stand for the right without regard to the clamor of ill-advised and ignorant people. He stated that he would continue to do his duty, and that he would uphold the constitutional rights of all the people without distinction to race, color or previous condition.

Acting under Selwyn's advice, Rockland began to concentrate quietly troops in the large centers of population. He also ordered the fleets into home waters. A careful inquiry was made regarding the views of the several Governors within easy reach of Washington, and, finding most of them favorable to the Government, he told them that in case of disorder he would honor their requisition for federal troops. He advised a thorough overlooking of the militia, and the weeding out of those likely to sympathize with the 'mob.' If trouble came, he promised to act promptly and forcefully, and not to let mawkish sentiment encourage further violence.

He recalled to them that the French Revolution was caused, and continued, by the weakness and inertia of Louis Fifteenth and his ministers and that the moment the Directorate placed Bonaparte in command of a handful of troops, and gave him power to act, by the use of grape and ball he brought order in a day. It only needed a quick and decisive use of force, he thought, and untold suffering and bloodshed would be averted.

President Rockland believed what he said. He seemed not to know that Bonaparte dealt with a ragged, ignorant mob, and had back of him a nation that had been in a drunken and bloody orgy for a period of years and wanted to sober up. He seemed not to know that in this contest, the clear-brained, sturdy American patriot was enlisted against him and what he represented, and had determined to come once more into his own.

[***]

Chapter XXXVI
Selwyn's Story, Continued

Flushed though I was with victory, and with the flattery of friends, time servers and sycophants in my ears, I felt a deep sympathy for the boss. He was as a sinking ship and as such deserted. Yesterday a thing for envy, today an object of pity.

I wondered how long it would be before I, too, would be stranded.

The interests, were, of course, among the first to congratulate me and to assure me of their support. During that session of the legislature, I did not change the character of the legislation, or do anything very different from the usual. I wanted to feel my seat more firmly under me before attempting the many things I had in mind.

I took over into my camp all those that I could reasonably trust, and strengthened my forces everywhere as expeditiously as possible. I weeded out the incompetents, of whom there were many, and replaced them by big-hearted, loyal and energetic men, who had easy consciences when it came to dealing with the public affairs of either municipalities, counties or the State.

Of necessity, I had to use some who were vicious and dishonest, and who would betray me in a moment if their interests led that way. But of these there were few in my personal organization, though from experience, I knew their kind permeated the municipal machines to a large degree.

The lessons learned from Hardy were of value to me now. I was liberal to my following at the expense of myself, and I played the game fair as they knew it.

I declined re-election to the next legislature, because the office was not commensurate with the dignity of the position I held as party leader, and again, because the holding of state office was now a perilous undertaking.

In taking over the machine from the late boss, and in molding it into an almost personal following I found it not only loosely put together, but inefficient for my more ambitious purposes.

After giving it four or five years of close attention, I was satisfied with it, and I had no fear of dislodgment.

I had found that the interests were not paying anything like a commensurate amount for the special privileges they were getting, and I more than doubled the revenue obtained by the deposed boss.

This, of course, delighted my henchmen, and bound them more closely to me.

I also demanded and received information in advance of any extensions of railroads, standard or interurban, of contemplated improvements of whatsoever character, and I doled out this information to those of my followers in whose jurisdiction lay such territory.

My own fortune I augmented by advance information regarding the appreciation of stocks. If an amalgamation of two important institutions was to occur, or if they were to be put upon a dividend basis, or if the dividend rate was to be increased, I was told, not only in advance of the public, but in advance of the stockholders themselves.

All such information I held in confidence even from my own followers, for it was given me with such understanding.

My next move was to get into national politics. I became something of a factor at the national convention, by swinging Pennsylvania's vote at a critical time; the result being the nomination of the now President, consequently my relations with him were most cordial.

The term of the senior Senator from our State was about to expire, and, although he was well advanced in years, he desired re-election.

I decided to take his seat for myself, so I asked the President to offer him an ambassadorship. He did not wish to make the change, but when he understood that it was that or nothing, he gracefully acquiesced in order that he might be saved the humiliation of defeat.

When he resigned, the Governor offered me the appointment for the unexpired term. It had only three months to run before the legislature met to elect his successor.

I told him that I could not accept until I had conferred with my friends. I had no intention of refusing, but I wanted to seem to defer to the judgment of my lieutenants.

I called them to the capital singly, and explained that I could be of vastly more service to the organization were I at Washington, and I arranged with them to convert the rank and file to this view.

Each felt that the weight of my decision rested upon himself, and their vanity was greatly pleased. I was begged not to renounce the leadership, and after persuasion, this I promised not to do.

As a matter of fact, it was never my intention to release my hold upon the State, thus placing myself in another's power.

So I accepted the tender of the Senatorship, and soon after, when the legislature met, I was elected for the full term.

I was in as close touch with my State at Washington as I was before, for I spent a large part of my time there.

I was not in Washington long before I found that the Government was run by a few men; that outside of this little circle no one was of much importance.

It was my intention to break into it if possible, and my ambition now leaped so far as to want, not only to be of it, but later, to be *it*.

I began my crusade by getting upon confidential terms with the President.

One night, when we were alone in his private study, I told him of the manner and completeness of my organization in Pennsylvania. I could see he was deeply impressed. He had been elected by an uncomfortably small vote, and he was, I knew, looking for someone to manage the next campaign, provided he again received the nomination.

The man who had done this work in the last election was broken in health, and had gone to Europe for an indefinite stay.

The President questioned me closely, and ended by asking me to undertake the direction of his campaign for re-nomination, and later to manage the campaign for his election in the event he was again the party's candidate.

I was flattered by the proffer, and told him so, but I was guarded in its acceptance. I wanted him to see more of me, hear more of my methods and to become, as it were, the supplicant.

This condition was soon brought about, and I entered into my new relations with him under the most favorable circumstances.

If I had readily acquiesced he would have assumed the air of favoring me, as it was, the rule was reversed.

He was overwhelmingly nominated and re-elected, and for the result he generously gave me full credit.

I was now well within the charmed circle, and within easy reach of my further desire to have no rivals. This came about naturally and without friction.

The interests, of course, were soon groveling at my feet, and, heavy as my demands were, I sometimes wondered like Clive at my own moderation.

The rest of my story is known to you. I had tightened a nearly invisible coil around the people, which held them fast, while the interests despoiled them. We overdid it, and you came with the conscience of the great majority of the American people back of you, and swung the Nation again into the moorings intended by the Fathers of the Republic.

When Selwyn had finished, the fire had burned low, and it was only now and then that his face was lighted by the flickering flames revealing a sadness that few had ever seen there before.

Perhaps he saw in the dying embers something typical of his life as it now was. Perhaps he longed to recall his youth and with it the strength, the nervous force and the tireless thought that he had used to make himself what he was.

When life is so nearly spilled as his, things are measured differently, and what looms large in the beginning becomes but the merest shadow when the race has been run.

As he contemplated the silent figure, Philip Dru felt something of regret himself, for he now knew the groundwork of the man, and he was sure that under other conditions, a career could have been wrought more splendid than that of any of his fellows.

Chapter XXXVII
The Cotton Corner

In modeling the laws, Dru called to the attention of those boards that were doing that work, the so-called 'loan sharks,' and told them to deal with them with a heavy hand. By no sort of subterfuge were they to be permitted to be usurious. By their nefarious methods of charging the maximum legal rate of interest and then exacting a commission for monthly renewals of loans, the poor and the dependent were oftentimes made to pay several hundred per cent, interest per annum. The criminal code was to be invoked and protracted terms in prison, in addition to fines, were to be used against them.

He also called attention to a lesser, though serious, evil, of the practice of farmers, mine-owners, lumbermen and other employers of ignorant labor, of making advances of food, clothing and similar necessities to their tenants or workmen, and charging them extortionate prices therefor, thus securing the use of their labor at a cost entirely incommensurate with its value.

Stock, cotton and produce exchanges as then conducted came under the ban of the Administrator's displeasure, and he indicated his intention of reforming them to the extent of prohibiting, under penalty of fine and imprisonment, the selling either short or long, stocks, bonds, commodities of whatsoever character, or anything of value. Banks, corporations or individuals lending money to any corporation or individual whose purpose it was known to be to violate this law, should be deemed as guilty as the actual offender and should be as heavily punished.

An immediate enforcement of this law was made because, just before the Revolution, there was carried to a successful conclusion a gigantic but iniquitous cotton corner. Some twenty or more adventurous millionaires, led by one of the boldest speculators of those times, named Hawkins, planned and succeeded in cornering cotton.

It seemed that the world needed a crop of 16,000,000 bales, and while the yield for the year was uncertain it appeared that the crop would run to that figure and perhaps over. Therefore, prices were low and spot-cotton was selling around eight cents, and futures for the distant months were not much higher.

By using all the markets and exchanges and by exercising much skill and secrecy, Hawkins succeeded in buying two million bales of actual cotton, and ten million bales of futures at an approximate average of nine and a half cents. He had the actual cotton stored in relatively small quantities throughout the South, much of it being on the farms and at the gins where it was bought. Then, in order to hide his identity, he had incorporated a company called 'The Farmers' Protective Association.'

Through one of his agents he succeeded in officering it with well-known Southerners, who knew only that part of the plan which contemplated an increase in prices, and were in sympathy with it. He transferred his spot-cotton to this company, the stock of which he himself held through his dummies, *and then had his agents burn the entire two million bales.* The burning was done quickly and with spectacular effect, and the entire commercial world, both in America and abroad, were astounded by the act.

Once before in isolated instances the cotton planter had done this, and once the farmers of the West, discouraged by low prices, had used corn for fuel. That, however, was done on a small scale. But to deliberately burn one hundred million dollars worth of property was almost beyond the scope of the imagination.

The result was a cotton panic, and Hawkins succeeded in closing out his futures at an average price of fifteen cents, thereby netting twenty-five dollars a bale, and making for himself and fellow buccaneers one hundred and fifty million dollars.

After amazement came indignation at such frightful abuse of concentrated wealth. Those of Wall Street that were not caught, were open in their expressions of admiration for Hawkins, for of such material are their heroes made.

[***]

Chapter XLIII
The Rule of the Bosses

General Dru was ever fond of talking to Senator Selwyn. He found his virile mind a never-failing source of information. Busy as they both were they often met and exchanged opinions. In answer to a question from Dru, Selwyn said that while Pennsylvania and a few other States had been more completely under the domination of bosses than others, still the system permeated everywhere.

In some States a railroad held the power, but exercised it through an individual or individuals.

In another State, a single corporation held it, and yet again, it was often held by a corporate group acting together. In many States one individual dominated public affairs and more often for good than for evil.

The people simply would not take enough interest in their Government to exercise the right of control.

Those who took an active interest were used as a part of the boss' tools, be he a benevolent one or otherwise.

'The delegates go to the conventions,' said Selwyn, 'and think they have something to do with the naming of the nominees, and the making of the platforms. But the astute boss has planned all that far in advance, the candidates are selected and the platform written and both are 'forced' upon the unsuspecting delegate, much as the card shark forced his cards upon his victim. It is all seemingly in the open and above the boards, but as a matter of fact quite the reverse is true.

'At conventions it is usual to select some man who has always been honored and respected, and elect him chairman of the platform committee. He is pleased with the honor and is ready to do the bidding of the man to whom he owes it.

'The platform has been read to him and he has been committed to it before his appointment as chairman. Then a careful selection is made of delegates from the different senatorial districts and a good working majority of trusted followers is obtained for places on the committee. Someone nominates for chairman the 'honored and respected' and he is promptly elected.

'Another member suggests that the committee, as it stands, is too unwieldy to draft a platform, and makes a motion that the chairman be empowered to appoint a sub-committee of five to outline one and submit it to the committee as a whole.

'The motion is carried and the chairman appoints five of the 'tried and true.' There is then an adjournment until the sub-committee is ready to report.

'The five betake themselves to a room in some hotel and smoke, drink and swap stories until enough time has elapsed for a proper platform to be written.

'They then report to the committee as a whole and, after some wrangling by the uninitiated, the platform is passed as the boss has written it without the addition of a single word.

'Sometimes it is necessary to place upon the sub-committee a recalcitrant or two. Then the method is somewhat different. The boss' platform is cut into separate planks and first one and then another of the faithful offers a plank, and after some discussion a majority of the committee adopt it. So when the sub-committee reports back there stands the boss' handiwork just as he has constructed it.

'Oftentimes there is no subterfuge, but the convention, as a whole, recognizes the pre-eminent ability of one man amongst them, and by common consent he is assigned the task.'

Selwyn also told Dru that it was often the practice among corporations not to bother themselves about state politics further than to control the Senate.

This smaller body was seldom more than one-fourth as large as the House, and usually contained not more than twenty-five or thirty members.

Their method was to control a majority of the Senate and let the House pass such measures as it pleased, and the Governor recommend such laws as he thought proper. Then the Senate would promptly kill all legislation that in any way touched corporate interests.

Still another method which was used to advantage by the interests where they had not been vigilant in the protection of their 'rights,' and when they had no sure majority either in the House or Senate and no influence with the Governor, was to throw what strength they had to the stronger side in the factional fights that were always going on in every State and in every legislature.

Actual money, Selwyn said, was now seldom given in the relentless warfare which the selfish interests were ever waging against the people, but it was intrigue, the promise of place and power, and the ever effectual appeal to human vanity.

That part of the press which was under corporate control was often able to make or destroy a man's legislative and political career, and the weak and the vain and the men with shifty consciences, that the people in their fatuous indifference elect to make their laws, seldom fail to succumb to this subtle influence."

House's 1911 novel appeared as if a stage play scripted to give life to the *Übermensch* discussed in Max Simon Nordau's *The Interpretation of History* of 1909, which was translated into English in 1910, and (in chapters not here reprinted) to make the ambitious "superior man"—House's "dictator"—appear benevolent and necessary to reform. Nordau, who was not in this particular instance writing sentimental propaganda, had a more pessimistic view of the man who aspired to lead,

> "The superior man reckons with the organized habits of the average crowd. His egoism employs different means for its satisfaction in an old, compact, and firmly established State from those applicable to the simple conditions of primitive barbarism. He no longer waves his axe above the head of the individual whom he wishes to subdue; he does not even permit armed servants to spread terror before them; instead he masters the machinery of State, and thus acquires at a single blow the power that in an unorganized crowd could only have been won by a series of acts of violence directed against individuals. He disturbs the habits of the multitude as little as possible; he makes them useful.
>
> The parasitic egoism of the strong man assumes the most different forms, and passes, according to the degree of energy it possesses, through every stage, from the lowest desire for pleasure, through greed, vanity, and ambition, to the hunger for power and that inability to endure the thought of resistance, any limitation of personal omnipotence, which is allied to the hypertrophy of self that develops into megalomania. One is content with small satisfactions: he seeks to win his way to political power by his pliancy and observation of the idiosyncrasies of the men who are its guardians. He is the typical opportunist. At school he acquires the good graces of his teacher by flattery and obsequiousness; at the examination he studies the little preferences of the examiners; when an official, he pays court to those above him; by means of invitations, intrigues, and the influence of women, he becomes an academician, obtains titles and orders, and ends by dying as a pillar of society and the State, respectable and influential, surrounded by toadies, and envied by people in general. Another looks higher: he would not receive but distribute honours. In an absolute monarchy he attaches himself to the person of the ruler, studies him, and tries to make himself indispensable to him—in other words, he tries to master him and use him for the accomplishment of his own will. Under a modern democracy he comes forward at popular meetings; is at pains to acquire an influence over the crowd and to win their votes by appealing to their emotions and prejudices, by making promises and juggling with illusions; at the same time he tries to force himself into the inner circles of the leading people. Once in office, he continues his activity until he has become a minister, party leader, or, in a republic, President. Others, though these are more rare, will not stop short of supreme power. They do not employ, or not to any great extent, the method of subservience, but rather that of force, much after the fashion of primitive man—that of mutiny, rising, military revolt, dictatorship, *coup*

d'état. They are represented on a small scale by such men as Nicola di Rienzi, Jack Cade, Masaniello; on a big scale, and on the biggest, by Oliver Cromwell, Washington, Napoleon I. and III., and Louis Kossuth.

The instinct of exploitation that the man of will and deeds retains enables him to display his organic superiority in another sphere, in other fields of action, when it is directed to the amassing of wealth by speculations on the Stock Exchange, company promoting, the formation of trusts, cartels, and monopoly undertakings. Mighty financiers manage average men in the same way as do politicians, courtiers, and military despots. They begin by conjuring up illusions and intoxicating weak heads with their delights; then, as their power grows, they intimidate some and rouse the cupidity of the others by rewards and promises, purchase useful allies by a cleverly graduated system of shares, and so build up a human pyramid, on to the top of which they climb over backs, shoulders, and heads. The amassers of gold belong to the same family as the demagogue, the party leader, and the king-maker; this is not the place to enter into the psychic differences between them. Member of the same family, but a poor relation, an unsuccessful cousin, is the professional criminal, who has to content himself with the poorest and least remunerative form of exploitation, because he only possesses the parasitic instinct, without the intellectual equipment in himself, or the social forces behind him, to enable him to satisfy it on a large scale or in the grand style.

All these activities and careers conform to a single type. A man who is richly endowed by nature in any direction employs or misuses his superiority in order to subjugate others to his will, obtain possession of the fruits of their labour, or use them simply and solely for his own profit or pleasure. According to the degree and quality of his superiority, he makes them serviceable to himself by compulsion, fascination, illusion, or gross deception. To take a few examples. The politician uses the parliamentary system as a ladder up which he may climb from being a secretary to a member, parliamentary reporter, or honorary secretary to some political club, to member of a parliamentary committee, member of Parliament itself, party leader, and finally minister. The scholar can use the organization of the University or academy as a means to obtaining a position and reputation independent of the worth of his scientific attainments. The financier employs the mechanism of the Stock Exchange and the limited liability company to draw the small competences of the many into his net and combine them into a vast fortune. Even the criminal has arrangements at his disposal which render his evil-doing less arduous, such as the Mafia, the Camorra, the Mano Negra, and the unions of thieves and burglars, with a far-reaching system of division of labour, that exist in large towns and are also international in their scope.

From the psychological point of view all institutions represent organized habits. They have been materialized by the human brain, and have no existence apart from man. The superior man must therefore approach men through habit, and try to turn it to his advantage. He may either adapt himself to it or try to alter it. The lower order of aspirant adapts himself. Rabagas acquired reputation and influence as a revolutionary, but became reactionary when he attained the ministry. The powerful personality alters it: Robespierre found a loyal people, and taught it to convey its king and queen to execution on a tumbril. Yet there are some habits so deeply rooted and so strongly organized that no individual can stand against them. Cromwell failed to destroy the habit of loyalty in the English people; which made the Restoration possible immediately after his death. Napoleon could not overcome the habit of religion in the French people, or avoid a concordat with Rome. Were a negro of the highest genius to arise in the United States, a Napoleon in generalship, a Cavour in diplomacy, a Gladstone in eloquence, and a Bismarck in strength of will, he could never attain the highest position there, because the habit of race hatred would ever be more powerful than his genius. In Russia today it would be impossible for a Jew, whether he had been baptized or no, to rouse a mass movement like that led by Lasalle in Germany in the fifties and sixties; or to rise to the premiership, as Disraeli did in England. Each time that a personality endeavours to subdue others to its will there is a clash between this will and the habits opposed to it: the more deeply rooted, general, and essential are their habits, the more powerful must be the will that is to overcome them, until it reaches a limit beyond which the power of a single will cannot go. Napoleon was one of the most powerful personalities the species has hitherto produced. Yet he was overcome by weak contemporaries like Alexander I., Francis II., Frederick William III., and George III., because they were supported by the habits of the whole of Europe, with the exception of France, and could demand and obtain from their peoples exertions which even Napoleon's mighty intellect could not call forth.

It is necessary to guard against the possibility of misunderstanding. All the preceding examples show the exploiter rising above his fellows in order to satisfy his desires at their expense. Nothing has been said of the nobler type of ambition, which strives for power and influence for the sake of serving mankind, and is impelled only by the desire of making the world better, more beautiful, and happier. The reason for this apparent omission is that the expression 'superior man' is used in a purely biological, not in an ethical, sense. It merely represents the individual who is equipped with organic energy above the average, especially in the sphere of judgment and will. The superior man in this sense uses his superiority selfishly for his own advantage, not selflessly for the good of others. That this is so is painful to anyone who seeks to see history as governed by a moral ideal; but it is an observed fact which admits of no exception. The selfless friends of man are not opportunists. They have no ambition. They are incapable of making incessant efforts to subdue the many to their will. Their influence is confined to their words and example. They spend their lives as settlers penitents, or teachers, like Buddha Cakya-Muni; they are crucified like Jesus, or, to take smaller instances, burned like Savonarola, or hanged like John Brown, the enemy of negro slavery. The influence of men who wish to save their fellows is felt, as I have already shown, through others—disciples, perhaps, of developed will-power, who work for some reward, real or imagined, earthly or hereafter; or rulers and politicians, who find something in the doctrine of salvation which they can use for their own selfish ends. Elaborate psychological analysis would be necessary before the rare instances of the use of power by those in authority for the good of their subjects could be ascribed to pure altruism. Titus, 'the delight of the human race,' did not seem so benevolent to all the people under his sway as he did to the Romans. Alfred the Great was certainly a benefactor to his realm, but, in giving peace, order, well-being, and education to his disordered State, he was in the first instance working for himself. Joseph II. is probably the best and most indubitable example of a philanthropist on the throne. But it is very doubtful whether his qualities were such as to have raised him, by his own strength, above his fellow-men. He was Emperor because born in the purple. He was the inheritor, not the founder, of a dynasty. It is on a materially lower plane that the altruists who combine strength of will with love for their fellows are to be found—St. Francis of Assisi, St. Vincent de Paul, Peabody, Dr. Barnardo, Dunant, perhaps General Booth. But the men who scale the heights of power and make their mark on history have been spurred on by selfishness, and delayed by no backward glances at their fellow-men.

At the lowest stage of civilization there is probably little difference between the individuals composing any race or horde. No one rises high above the others: exploitation is confined to the family, the wife, and growing children. The arrangements of life are determined by custom—that is, by habit; such institutions as there are exist, not to afford privilege to anyone, but to economize effort by sparing the need for fresh decisions; there are no leaders or rulers, or they possess small dignity or power. Another case where mutual exploitation within the race or people is impossible is that of a body composed of individuals of remarkable judgment and will-power, who are, to use the phrase a match for one another. Such a community is superficially denominated a democracy; as a matter of fact it is a loose confederation of aristocrats who, impatient of any overlordship, live side by side in proud and jealous independence, remaining poor because each is dependent on his own labour, and this in a primitive State, under natural conditions, can provide the bare necessities of life, but allow no one to become rich. Such, according to Vico, was the condition of the Quirites in the early days of Rome. History teaches that this condition of things did not last long. The gifted people overflowed its boundaries, first to plunder, then to conquer; it made itself master of foreign peoples of less force, among whom it formed a ruling nobility, and then carried out the exploitation made possible by its organic superiority, first in the countries it had subdued, then in colonies; finally, with the help of the power and riches thus acquired, in its own land upon compatriots who had been slower and less adaptable, and had remained at home in poverty.

The limited extent to which the multitude are able to free themselves from their habits, and direct their thought and will along lines outside their organized associations, not only makes it easier for the superior man to master and exploit them with the aid of existing institutions which they occupy and utilize; it also renders it possible for power to be retained by individuals who are not themselves in any sense superior men, and never could have risen above the crowd by their own strength."[145]

The German Government was very much aware of "Colonel" House's influence over President Wilson. The German Ambassador to America, Count von Bernstorff, wrote to Count von Montgelas of the German Foreign Office on 5 May 1914,

"Colonel and Mrs. House will soon be arriving in Berlin and, as far as I know, will be staying with the American Ambassador. Gerard will certainly receive him, for Colonel House is President Wilson's most intimate friend. He is one of the few people with whom the hermit-like President lives at all on terms of friendship. He sees other people only on business. Here Colonel House is thought to be 'the power behind the throne'. If this may be one of those common American exaggerations, yet it is so far true that Colonel House possesses great influence. He has interests in Texas and was able therefore often to advise the President regarding the Mexican question, mostly in the direction of energetic action, in opposition to Bryan.

If an opportunity occurs of treating Colonel House in a friendly fashion it would be to our interests. If you get to know him, you will find him an agreeable member of society. He knows a great deal about Wall Street. I met him at the houses of Speyer and Warburg."[146]

"Demnächst werden Colonel und Mrs. House nach Berlin kommen und, soviel ich weiß, bei dem amerikanischen Botschafter wohnen. Jedenfalls wird sich Gerard ihrer annehmen, da Colonel House der intimste Freund des Präsidenten Wilson ist. Er gehört zu den wenigen Leuten, mit welchen der einsiedlerische Präsident überhaupt freundschaftlich verkehrt. Sonst sieht Herr Wilson die Menschen nur zu geschäftlichen Besprechungen. Colonel House gilt daher hier als „the power behind the throne". Wenn hierin auch eine der üblichen amerikarlischen Übertreibungen liegen mag, so ist jedenfalls so viel wahr, daß Colonel House großen Einfluß besitzt. Er hat Interessen in Texas und konnte daher auch oft den Präsidenten in der mexikanischen Frage beraten, meistens in der energischen Richtung im

---

145. M. Nordau, *The Interpretation of History*, Willey Book Company, New York, (1910), pp. 290-297; which is an English translation by M. A. Hamilton of *Der Sinn der Geschichte*, C. Duncker, Berlin, (1909).
146. Letter from Count von Bernstorff to Count von Montgelas of 5 May 1914, English translation by E. T. S. Dugdale, *German Diplomatic Documents 1871-1914*, Volume 4, Harper & Brothers, New York, London, (1931), p. 348.

Gegensatze zu Bryan.

Wenn sich Gelegenheit bieten sollte, Colonel House freundlich zu behandeln, so würde dies in unserem Interesse liegen. Sie werden, falls Sie ihn kennen lernen, in Colonel House einen angenehmen Gesellschafter finden. Er weiß auch in Wall Street gut Bescheid. Ich traf ihn bei Speyers und Warburgs."[147]

The German Ambassador to America, Count von Bernstorff, wrote to the German Foreign Office on 6 May 1914,

"A letter from myself to Count Montgelas is on the way begging that House be treated as well as possible; he may be described as the only personal friend Wilson has. Being a Texan, he exercised special influence in the question of Mexico. He lives now in New York, where he knows the great bankers well. I have often met him with Speyer and Warburg. I recommend his being received by His Majesty, if that is possible."[148]

"Ich schrieb bereits Privatbrief, der unterwegs an Graf Montgelas mit der Bitte um möglichst freundliche Behandlung von House, den man vielleicht als den einzigen persönlichen Freund Wilsons bezeichnen kann. Als Texaner hat er besonders in Mexikofrage Einfluß ausgeübt. Er lebt jetzt in New York, wo er mit den großen Bankiers gut bekannt. Ich traf ihn öfters bei Speyer und Warburg. Ich befürworte Empfang durch Seine Majestät, falls angängig."[149]

Boris L. Brasol wrote in 1921,

"Because of America's tremendous natural resources and her unlimited financial wealth, because of her great man-power and immense technical assets, also on account of Russia's withdrawal from the Entente combination, America's entry into the war gave her instantaneously the advantage of becoming the leading power among the belligerents. But there were two angles to America's leadership in the trend of world events — the

---

[147]. Letter from Count von Bernstorff to Count von Montgelas of 5 May 1914, *Die Grosse Politik der Europäischen Kabinette 1871-1914: Sammlung der diplomatischen Akten des Auswärtigen Amtes. Im Auftrage des Auswärtigen Amtes herausgegeben von Johannes Lepsius, Albrecht Mendelssohn Bartholdy, Friedrich Thimme.*, Deutsche Verlagsgesellschaft für Politik und Geschichte, Berlin, (1926), pp. 110-111.

[148]. Letter from Count von Bernstorff to German Foreign Office of 6 May 1914, English translation by E. T. S. Dugdale, *German Diplomatic Documents 1871-1914*, Volume 4, Harper & Brothers, New York, London, (1931), pp. 347-348.

[149]. Letter from Count von Bernstorff to German Foreign Office of 6 May 1914, *Die Grosse Politik der Europäischen Kabinette 1871-1914: Sammlung der diplomatischen Akten des Auswärtigen Amtes. Im Auftrage des Auswärtigen Amtes herausgegeben von Johannes Lepsius, Albrecht Mendelssohn Bartholdy, Friedrich Thimme.*, Deutsche Verlagsgesellschaft für Politik und Geschichte, Berlin, (1926), p. 110.

purely practical influence which she was able to exert upon the financial resources of the military conflict; and second, the political phase pertaining to the terms of the peace settlement. The first element was negative and destructive, for its aim was to accelerate the defeat of Germany and the victory of the Allies. The second element was positive and constructive; it sought to build up a new political and social order along the lines of the Wilsonian doctrine. However, the political credit given by Europe to America was by no means an unconditional surrender of Europe to the New World. Europe was prepared to follow America so long as she retained the hope that her prescriptions would bring an immediate solution of the European troubles. The failure to fulfill this hope was bound to produce a radical change in the attitude of European Nations toward the Wilsonian ideology, and eventually toward America herself.

It was obviously impossible to solve European problems by merely proclaiming a series of moral commandments or scholastic principles, however commendable they may have been. Above all, in order to present tangible schemes for the reconstruction of European States, it was absolutely necessary to acquire a deep knowledge of the political and social history of Europe. But a comprehensive knowledge of political phenomena does not spring up like a deus ex machina; it is rather attained by constant participation in the everyday political life of the different national bodies, evolving a firm historical tradition in foreign policy. America, however, has never had such a tradition and, therefore, she could not have had the experience which was indispensable for the maintenance of her political leadership in European affairs.

As to the controversy between the Senate and the President, it will be recalled that Article II, Section 2, of the Constitution of the United States, vests the President with the power to make treaties '*by and with the advice of the Senate ... providing two-thirds of the Senators present concur.*' Although the making of treaties forms part of the executive prerogative, and in spite of the fact that the President is the Chief Executive, nevertheless, his right to enter into treaties is limited by the above provision. European statesmen were cognizant of this limitation, but Europe at large was unconcerned about such 'technicalities' of the American Constitution. Mr. Wilson appeared on the European Continent not only as the Chief Executive of, but also as the sole spokesman for, America. The peoples of Europe were inclined to believe that whatever he said, proclaimed, admitted or agreed upon was absolutely binding upon the United States. It was a matter of great disappointment to the outside world when gradually the controversy between the President and the Senate divulged the fact that President Wilson, no more than the Senate, had the authority to enter into alliances with European Nations, and that both the President and the Senate, with regard to the framing of treaties, had equal rights, neither of them having authority to act independently of the other. The executive power of the United States was represented at Paris by the person of Mr. Wilson himself, while 'His Majesty's opposition' was kept arrested in Washington, D. C. It so happened, however, that while Mr. Wilson's administration was Democratic, the majority of the Senate was Republican. This was precisely why Mr. Wilson should have secured a strong Republican representation at the Peace Conference, thus avoiding any possible surprises in the future. But Mr. Wilson was nevered considered an able psychologist. In all his political doings the human touch was distinctly lacking. Senator Lodge may have been wrong in some points of his criticism of the Peace Treaty, but that did not alter the nature of the case itself. In a matter of such vital importance as the framing of the Covenant, a Republican Senate certainly was entitled to have its voice heard in Europe before the treaty was actually completed.

The struggle which arose between the Senate and the President of the United States did not add to the prestige of the latter. On the contrary, it tended to make Mr. Wilson's position in Europe all the more difficult. The statesmen assembled at Versailles were put face to face with an undeniable fact, that America had two foreign policies — one advocated by the President, and the other maintained by the Senate. For European diplomacy such a condition would have been impossible. Messrs. Orlando, Lloyd George and Clemenceau had free hands with regard to their own countries, while Mr. Wilson was handicapped in each of his enterprises regardless of their particular merits. For a short period Mr. Wilson was regarded in Europe as almighty; very soon, however, he proved his impotency on the soil of his own country.

The American delegation to the Peace Conference, headed by President Wilson himself, was composed of men of varied abilities, but above all scarcely familiar with the basic facts of European history and the underlying psychological factors of European relationship. Although cunning politicians, most of these men were pronounced amateurs in State affairs, sometimes without even elementary administrative experience, as was the case with Colonel House. It is true that during the two years preceding the armistice there was in Washington a commission at work engaged in gathering data for the future Peace Conference. This body succeeded in accumulating tons of memoranda pertaining to the different national problems, but much of the information thus obtained was distinctly erroneous and hopelessly misleading. Persons who themselves were quite ignorant of international affairs were requested to present their views and render their 'expert' opinions on problems of the utmost complexity. The work of the commission was purely mechanical and, therefore, absolutely discoördinated. Besides, with regard to the Eastern problem, which proved to be the heel of Achilles in the European situation, the information collected by the commission came mostly from Semitic sources.

No sooner had Mr. Wilson proclaimed his motto of self-determination than Washington became a meeting place for innumerable promoters of different mushroom States, all of whom claimed their allegiance to the Wilsonian dogma. None of these ten-days-old republics was absent from the American capital: Ukrainians, Lithuanians, Czechs, Slovaks, Letts, Finns, Georgians, Esthonians, Armenians, White Russians, Zionists and what not;

all of them offered evidence in support of their claims for independence. Their respective representatives enjoyed free entry to the State Department. They were attentively listened to, while their contradictory statements were scrupulously added to the files of Colonel House s commission. Indeed, it was an orgy of self-determination.

Referring to the personnel of the American delegation, it is noteworthy that their very names, with the exception of Mr. Lansing and Colonel House, have remained almost unknown to the general public. [*Footnote:* Hon. Henry White and General Tasker H. Bliss were the other two delegates.] The delegates were simply absorbed by the personality of Mr. Wilson. From time to time the papers alluded to a new name in the American delegation, but it meant nothing either to the hearts or to the minds of the American people.

Colonel House was next to President Wilson to attract public attention. Notrndy knew who he was, from whence he came, nor what he stood for, and his prestige was largely due to his mysteriousness. It was understood at Paris that he exerted a tremendous, almost boundless, influence upon the President. In fact, one of Colonel House's intimate friends, Mr. Arthur D. Howden Smith, in his volume 'The Real Colonel House,' frankly admitted:

'He holds a power never wielded before in this country by any man out of office, a power greater than that of any political boss or Cabinet member. He occupies a place in connection with the Administration which is anomalous, because no such place ever existed before Woodrow Wilson became President of the United States.'[*Footnote:* 'The Real Colonel House, An Intimate Biography,' by Arthur D. Howden Smith, p. 14, George H. Doran Company, New York, 1918.]

It was rumored that Colonel House was very radical in his political views, that he shared Mr. Wilson's admiration for the 'chosen people' and was bitterly anti-Russian. In addition it was positively known that he was sent to Germany by President Wilson prior to America's entry into the war, but until now the object of his mission was never discovered.

Mr. Keynes, in his able characterization of the personnel of the Peace Conference, referring to the American Peace Delegation and Mr. Wilson personally, stated that:

'His fellow-plenipotentiaries were dummies; and even the trusted Colonel House, with vastly more knowledge of men and of Europe than the President, from whose sensitiveness the President's dullness had gained so much, fell into the background as time went on.... Thus day after day and week after week, he (Mr. Wilson) allowed himself to be closeted, unsupported, unadvised, and alone, with men much sharper than himself, in situations of supreme difficulty, where he needed for success every description of resource, fertility, and knowledge.'[*Footnote:* John Maynard Keynes, C.B., 'The Economic Consequences of the Peace,' p. 45, Harcourt,

Brace, and Howe, New York, 1920.]

That the members of the American delegation were dummies is a generally recognized fact. One has only to recall the manner in which the Shantung settlement was brought about. In his testimony before the Foreign Relations Committee of the Senate Mr. Lansing frankly admitted that:

'President Wilson alone approved the Shantung decision, that the other members of the American Delegation made no protest against it, and that President Wilson alone understood whether Japan has guaranteed to return Shantung to China.'

The same applies to the delicate question of Fiume. Mr. Wilson disagreed on all points with Signor Orlando. It was a personal altercation between the President and the Italian plenipotentiary, no other members of the American delegation having participated in the controversy. Mr. Wilson's sudden decision to appeal to the Italian people 'over the heads of the Italian Government,' unwise as it may have been, was taken quite independently, while the other members of the delegation, when they read this proclamation in *Le Temps*, were probably as much surprised as Signor Orlando himself.

On every question of international importance the President acted autocratically, without advice from his colleagues. Had he consulted them beforehand, he probably would have avoided many false steps as well as his erroneous move concerning the Fiume settlement. The whole affair was caused by groundless rumors accusing the Italian Government of the intention to incorporate Fiume in the territory of the Italian Kingdom in spite of Mr. Wilson's determination to cede the city to Jugoslavia. Had Mr. Lansing been consulted he would certainly have drawn the President's attention to the fact that the decision of converting the Fiume problem into an international scandal was all the more detrimental to the general cause of peace, since it came on the eve of the arrival of the German Peace Mission.

When Mr. Baruch arrived in Paris he became very active with regard to the framing of the financial policy of the Allies, and especially that of America; and because he was not only a member of the American delegation, but also a prominent figure in the Jewish delegation, it was not impossible that he had much to do with the President's peculiar stand with regard to the notorious 'Jewish Minority Rights.'

Mr. Dillon, whose knowledge of the inside story of the intrigue at the Peace Conference is so profound, did not hesitate to state that the Allied policy toward the Zionist claims was:

'Looked upon as anything but disinterested.' Elucidating this point, Mr. Dillon added:

'Unhappily this conviction was subsequently strengthened by certain of the measures decreed by the Supreme Council between April and the close of the Conference. The misgivings of other delegates turned upon a matter which at first sight may appear so far removed from any of the pressing issues of the twentieth century as to seem wholly imaginary. They feared that a religious — some would call it racial — bias lay at the root of Mr. Wilson's policy. It may seem amazing to some readers, but it is none the less a fact that a considerable number of delegates believed that the real influences behind the Anglo-Saxon peoples were Semitic.'[*Footnote:* E. J. Dillon, 'The Inside Story of the Peace Conference,' p. 496, Harper & Brothers, New York, 1920.]

This observation is quite correct, but scarcely can it be confined to the Anglo-Saxon peoples only. It is true that Mr. Wilson's policy at all times was distinctly pro-Jewish and that Mr. Lloyd George's affiliations with Sir Philip Sassoon aroused much comment among the general public. Nor can the fact be denied that the British policy, ever since Mr. Balfour's declaration on the Zionist claims of November 2, 1917, has been developing under the coordinated pressure of Messrs. Rufus Isaacs, Louis Namier, Mond and Montagu, all of whom are Jews, manifesting a spirit of deep loyalty to the cause of Israel. But almost every plenipotentiary at the Peace Conference had his own Jew to guide him in matters of international importance. Mr. Clemenceau himself, whose reputation of a French 'tiger' was so exaggerated, had Mr. Mendel as private secretary, acting as intermediary between the Quai d'Orsay and the Stock Exchange. In the same way the Italian policy was largely controlled by Baron Sonnino, Minister of Foreign Affairs. The German Peace Delegation, in turn, was so obviously dominated by Jewish banking interests that it became known as 'The Warburg Delegation,' while the Spa Conference was labeled as the 'Hugo Stinnes Conference.' Thus, Mr. Dillon's remark being correct in itself, is to be interpreted in a larger sense, namely, that the Jews as a united nation brought upon the Peace Conference a twofold pressure: First, that of the international finance whose fundamental aim it was to save Germany from economic ruin; and, second, the influence of international Bolshevism, which, as *The Jewish Chronicle* justly remarked, is:

'At many points consonant with the finest ideals of Judaism.'[*Footnote:* See *The Jewish Chronicle*, No. 2609, April 4, 1919, p. 7, article entitled 'Peace, War, and Bolshevism.']

The effect of this double pressure was most disastrous. On one hand it left the German problem unsolved, while on the other hand it gave tremendous impetus to the revolutionary movement throughout the world.

Many excellent articles and books have been written on the proceedings of the Peace Conference, giving a detailed account of the happenings at Paris. Therefore, it would scarcely be advis able here to repeat all that has been said about the diplomatic achievements and of the Peace Treaty itself. The object of this volume is to depict the world crisis so far as it reflects upon the international situation.

It was a correct assertion on the part of Mr. Sarolea when he stated that:

'To us the present social convulsion is but an untoward incident and an aftermath of the war. To posterity the war itself will only appear as the preliminary to the revolutionary catastrophe which has just begun, and which is spreading with such inexorable directness in the two hemispheres. We are still totally in the dark as to its meaning and as to its future possibilities. In the meantime we can only see that until it has spent its force it is futile to talk about concluding peace. For a peace settlement means an agreement between the Allied Governments and the Governments of Germany, Austria, and Russia. And there are no sovereign German, Austrian, or Russian Governments left with whom we can conclude peace. There will be no such settled governments for years to come. No agreements made to-day can bind the future, or can have either reality or finality.'[*Footnote:* Charles Sarolea, 'Europe and the League of Nations,' pp. 8 and 9.]

The Peace Treaty itself is neither real nor final. The series of conferences which were held by the Allied and German statesmen, after the signing of the Treaty, have considerably amended the provisions of the Covenant, especially with regard to its economic clauses. Therefore, a final analysis of the treaty, whether it be considered from a narrow legal viewpoint or treated in the light of a broad political event, would have to be considered as premature. In a preceding chapter it was pointed out that the World War and subsequent events were but links in an endless chain of causes and consequences, extending as far back as the middle of the Nineteenth Century.

However, out of the turmoil of political babbling which accompanied the work of the Peace Conference, two factors of international significance have arisen, both of which will bear a lasting influence upon the future destinies of humanity. They are: *The League of Nations* and *International Bolshevism*. Both factors express the modern tendency of internationalism as opposed to the principle of national existence of the state. But while the idea of an association of nations is the moderate ramification of the principle of internationalism, Bolshevism is its revolutionary manifestation. Nevertheless, both phenomena work in the same direction, tending to undermine the fundamental basis of national development.

The Peace Conference was not the originator of either of these two factors but it promoted both, and the future historian will always associate their perpetuation with the policies of the Peace Conference."[150]

---

[150]. B. L. Brasol, *The World at the Cross Roads*, Small, Mayhard & Co., Boston, (1921), pp. 196-210.

Like many of the Jewish critics of the day, notably Alfred Rosenberg, Brasol sought to prove that "Jews" as a general group promoted internationalism and Bolshevism, controlled world affairs, and that the only solution was to promote the common interest of the anti-Semites with the Zionists in the formation of an absolutely independent Jewish State in Palestine. Brasol sought to establish that British imperialism had subverted the Balfour Declaration. Brasol was a Zionist, and like Zionist Winston Churchill, and the Zionist Chaim Weizmann, Brasol offered up the carrot and the stick of Zionism versus Bolshevism:

> "If Lord Milner was instrumental in forcing upon the English people a disastrous policy in Egypt, his Majesty's Government as a body is to be blamed for the shortsighted, and also extremely harmful, attitude towards Palestine. At present it cannot be doubted that Mr. Balfour's declaration of November 2, 1917, with regard to British support of the Zionist claim, was a clever move to keep France out of the Promised Land. The ambition of the Jews to establish a homeland of their own in Palestine was used by British as a pretext to include that part of Asia in the orbit of British influence. Mr. Herbert Adams Gibbons was right when as far back as in January, 1919, he asserted that the Britishers 'have planned, through using Zionism, to prevent codominium with France and other nations in Palestine, to establish an all-rail British route from Haifa to Bassorah.' [*Footnote:* See Mr. Gibbons's article 'Zionism and the World Peace,' published in the *Century Magazine*, January, 1919, pp. 368-3 78.]
>
> So far, so good, or at least, so long as political Zionism, advocated by British diplomats, had a definite political object to serve, criticism was confined to the question of whether England or France, or both, ought to control Palestine and Mesopotamia. It is not impossible that Messrs. Weizmann and Sokolow intended to doublecross British diplomacy, while the British intended to double-cross their Zionist friends, and it was difficult to forecast who, in the long run, would prove to be the user and who the used. Still there was logic in the declaration of November 2, 1917, because there was a chance for Britain to expand her influence in Asia Minor through the wise realization of the Palestine scheme. Moreover, in a way, Palestine could have been used as a new stronghold for British rule in the East, thus strengthening England's position with regard to India. Instead, England appointed Sir Herbert Samuel High Commissioner of Palestine, which renders the whole Palestine scheme hopeless.
>
> It is important to remember that according to Jewish sources the population of Palestine is divided thus: Mohammedans, six hundred and fifty thousand; Christians, one hundred and fifty thousand; Jews, ninety thousand. The bulk of the population is composed of Arabs, part of whom profess the Koran, while others have been converted to Christianity. The latter group, which is but a minor section of the total Arabian populace, is ravaged by internal strife, belonging to different denominations of the Christian Church: Roman Catholic, Protestant, Russian Greek Orthodox, etc. Nevertheless, the Arabs, whether Christians or Mohammedans, are

united in their hatred of the Jew. As everywhere, the Jew in Palestine is an urban element, while the Arabs are mostly farmers. The Jew in Palestine, as all over the world, is a middleman and not a producer. He is engaged in small trade. Only few Jews have settled as farmers.

The antagonism between the Arabs and the Jews is so accentuated that often the country has been on the brink of an open anti-Semitic revolt. The Ottoman Empire had great trouble in suppressing the anti-Semitic feeling among both its Christian and Mohammedan subjects.

The appointment of Sir Herbert Samuel, which was so much applauded by the Zionist group in England, is a direct challenge to the Arabs. To appoint a Jew to a post which requires holding the balance between the Jews and the Arabs, is a measure which is apt to ruin the very idea of British prestige. What England gained through the gallant efforts of General Allenby is now nullified by Samuel's appointment. It is immaterial whether Sir Herbert Samuel is good or bad, whether he is able or inefficient, the point is that he is a Jew, and as such, he cannot maintain an equilibrium between the two parts of the Palestine population, so bitterly hostile to each other. Nor does it add to British prestige when orders are given, as they were given by Sir Herbert Samuel, to British governmental employees to stand up when the Zionist anthem, Atikva, is played.

When the Zionist claim was first established, and Theodore Hertzl, in 1897, came out with his specific program of a Jewish State, the world at large gave a sigh of relief as it was trusted that henceforth the Jews would have a country of their own where they would be able to develop freely and unhampered their racial peculiarities, their cultural traditions and their religious thought. Christian countries have been so accustomed to innumerable complaints made by the Jews of their oppression, of anti-Semitism breeding throughout the world, of pogroms ravaging the Jewish masses, that there was every reason to hope that the Jews would dash to Palestine, leaving those cruel Christians to their own destinies. What better scheme for a fair solution of the Jewish problem could be hoped for by both Gentiles and Jews? The enormous wealth of Jewish bankers could be easily used for the reconstruction of Palestine, which could thus be made a model state. There is a place for everybody under the sun, and there is no reason whatsoever why the Jews should not have their place in Asia Minor, with Jerusalem once more becoming their metropolis, with the Rothschilds and Warburgs conferring the blessings of their benevolent rule on the hitherto downtrodden people.

With this understanding, the greatest statesmen of Europe, long before Mr. Balfour's declaration, promised Theodore Hertzl their utmost support to the Zionist scheme. Kaiser Wilhelm II was the first to migrate to Palestine, thus setting the example for the Jews to follow. The Turkish Sultan assured Mr. Hertzl that he would favorably look upon the Zionist efforts in the Ottoman Empire. The Russian Minister of the Interior, Mr. V. K. Plehve, promised his help to facilitate Jewish emigration from Russia. Another reason why so many Gentiles were willing to give their enthusiastic support to the Zionist movement was because it was justly argued that should the Jews build up a state of their own, they would be relieved of the necessity of bearing the burden of double-citizenship and double-allegiance on one hand to their own nation, and on the other hand to the countries of their adoption. This would also enable them to abandon their traditional policy of intermeddling in foreign matters, giving them a chance to enjoy genuine independence and civic freedom. From a legal point of view, then, the Jews would be considered, outside of Palestine, as aliens, just as Americans are considered in Japan, or the Japanese in America. While, of course, as Jewish citizens, they would not enjoy the rights of citizenship in any other country outside of their own Jewish State, they would also be relieved of all duties to Gentile countries. Consequently, they would be relieved of the hardship of serving simultaneously God and Mammon.

But when the time came, and the restoration of Palestine was announced by the Great Powers, many people, including some of the Jews themselves, became bitterly disappointed. Palestine has been restored not as a *Jewish State*, but merely as a *Homeland* for those restless spirits who, while residing in New York, London or Paris, would use Palestine as their summer resort, or perhaps as an additional base for their Third Internationale.

The British protectorate over Palestine converted that country into a British colony, with the British administration ruling over the population.

The most representative Zionists, themselves, came out with bitter criticism against such a solution. Thus, Israel Zangwill, in the London *Jewish Chronicle*, violently denounced the Judo-British pact proposing to make Palestine a purely Jewish State, with the expulsion of all Arabs to Arabia.

The *Jewish Guardian*, referring to this situation, remarked:

'Zionists were aiming for a Jewish Palestine but the Jews received a British Palestine.'

Mr. Eberlin, a Jew himself, and one of the foremost leaders of the Poale-Zionist movement, in a book recently published in Berlin, entitled 'On the Eve of Regeneration,' stated:

'The foreign policy of England in Asia Minor is determined by its interests in India. There was a saying about Prussia that she represents the army with an admixture of the people. About England it could be said that she represents a colonial empire with a supplement of the metropolis... . It is obvious that England desires to use Palestine as a shield against India. This is the reason why she is feverishly engaged in the construction of strategic railroad lines, uniting Egypt to Palestine, Cairo to Haifa, where work is started for the construction of a huge port. In the near future Palestine will be in a position to compete with the Isthmus of Suez, which is the main artery of the great sea route from the Mediterranean to the Indian

Ocean.'[*Footnote:* Translation from Russian, 'On the Eve of Regeneration,' by I. Eberlin, pp. 129, 130, Berlin, 1920.]

But this Poale-Zionist goes a step farther when he asserts that:

'It is only Socialism attainted in Europe which will prove capable of giving honestly and without hypocrisy Palestine to the Jews, thus assuring them unhampered development... . The Jewish people will have Palestine only when British Imperialism is broken.'

That the present policy towards Palestine is hopelessly erroneous can scarcely be denied. The Jews blame England for making it a British colony, while the Arabs are outraged by the appointment of Sir Herbert Samuel, because he is a Jew. The British public itself is at the cross roads — whether to consider Palestine as the Promised Land for the Jews, or for the English — and so, everybody on the Thames is waiting for Mr. Lloyd George and his parliamentary secretary, Mr. Sassoon, to solve the mystery of the Sphinx with regard to their Asia Minor policy.

However, there is nothing humorous in the whole situation because Lenin, the Argus of international dissension, is closely watching the developments in Syria, Mesopotamia and Palestine, and his agents are hard at work inciting the Jews against the British and the Arabs against the Jews. Moscow Soviet propagandists are always headed for political mischief; wherever there is natural cause for unrest, they stimulate it, converting it into an international scandal. All the more serious is the situation because Palestine is literally the shield for British rule in India."[151]

Another surreal part of the very odd "Colonel" House story occurred posthumously in October of 1939. England and France had just declared war on Germany on 3 September 1939. Many Americans worried that history would repeat itself and that America would be dragged into another bloody world war. The German Government had an incentive to undermine the relationship between the United States and Great Britain. It was not unusual for faked documents to be used as war propaganda.

Congressman Jacob Thorkelson of Montana submitted a letter into the Congressional Record, which was allegedly written by "Colonel" House and was addressed to David Lloyd George, Prime Minister of England. The letter was dated 10 June 1919 and was allegedly written on stationary from the British Embassy in New York, though Thorkelson did not have the supposed original. Therefore, there was no original signature, which could be checked for authenticity. The letter made it appear that Great Britain sought to recapture America as a colony, through the League of Nations. If the letter were authentic,

---

[151]. B. L. Brasol, *The World at the Cross Roads*, Small, Mayhard & Co., Boston, (1921), pp. 371-379.

one would have to believe that House and J. P. Morgan & Company were corrupting agents of the British Government and were undermining American sovereignty—and there is evidence that they were, especially if one considers the fact that Rothschilds essentially ruled England, Morgan and House.[152] While this may sound preposterous, there were published calls for the melding of Great Britain and the United States, such as those of Clarence K. Streit, who was an aide with the American mission at the Versailles Peace Conference in the First World War, and later a *New York Times* reporter who covered the League of Nations. In 1938, Streit openly called for the two nations to unite, and more broadly for a world government.[153]

House's secretary, Francis B. Denton, immediately stated that the alleged House letter was spurious. *The New York Times* published several articles denouncing the letter as a fake.[154] Fabulously wealthy banker and oil man, George Washington Armstrong,[155] tried to explain the discrepancies between the

---

152. E. C. Mullins, *Secrets of the Federal Reserve: The London Connection*, Bankers Research Institute, Staunton, Virginia, (1983); **and** *The Federal Reserve Conspiracy*, Common Sense, Union, New Jersey, (1954); **and** *A Study of the Federal Reserve*, Kasper and Horton, New York, (1952).

153. C. K. Streit, *"Where Iron Is, There Is the Fatherland!" A Note on the Relation of Privilege and Monopoly to War*, B. W. Huebsch, Inc., New York, (1920); **and** *Union Now: A Proposal for a Federal Union of the Democracies of the North Atlantic*, Harper, New York, (1938); **and** *World Government or Anarchy? Our Urgent Need for World Order*, World Citizens Association, Chicago, (1939); **and** *A Constructive Proposal: a Federal Union Now of Democracies...*, New York, (1939); **and** *Union Now with Britain*, Harper & Bros., New York, London, (1941); **and** *Union Now: a Proposal for Atlantic Federal Union of the Free*, Harper & Bros., New York, London, (1949); **and** *Freedom's Frontier: Atlantic Union Now, the Vast Opportunity the Two American Revolutions Offer Sovereign Citizens Today*, Freedom & Union Press, Washington, (1961).

154. "Doubt Col. House Wrote Letter Ascribed to Him", *The New York Times*, (13 October 1939), p. 10. "A Historical Report", *The New York Times*, (14 October 1939), p. 18. "Defends Colonel House", *The New York Times*, (18 October 1939), p. 22. "Asks Thorkelson Reprimand", *The New York Times*, (21 October 1939), p. 6.

155. G. W. Armstrong, *The Crime of '20: the Unpardonable Sin of "Frenzied Finance"*, Press of the Venney Co., E.G. Senter, Dallas, Texas, (1922); **and** *Truth: The Story of the Dynasty of the Money Trust in America, Part 1*, Fort Worth, Texas, Stafford-Lowdon Co., (1923); **and** *The Calamity of '30, Its Cause and the Remedy*, Fort Worth, Texas, (1931); **and** *A State Currency System: To Hell with Wall Street*, Stafford-Lowdon Co., Fort Worth, Texas, (1932); **and** *The Rothschild Money Trust*, Sons of Liberty, Metairie, Louisiana, (1940); **and** *The March of Bolshevism*, Babcock Co., Fort Worth, Texas, (1945); **and** *World Empire*, Judge Armstrong Foundation, Babcock, Fort Worth, Texas, (1945/1947); **and** *Traitors*, Judge Armstrong Foundation, Fort Worth, Texas, (1948); **and** *Zionist Wall Street*, Judge Armstrong Foundation, Fort Worth, Texas, (1949); **and** *Money and the Tariff*, Judge Armstrong Foundation, Fort Worth, Texas, (1949); **and** *The Zionists*, Judge Armstrong Foundation, Fort Worth, Texas, (1950); **and** *The Truth about My Alleged $50,000,000.00 Donation*, Judge Armstrong Foundation, Fort Worth, Texas, (1950); **and** *Our Constitution*, Judge Armstrong Foundation, Fort Worth, Texas, (1950); **and** *Third Zionist War*, Judge Armstrong Foundation, Fort Worth, Texas, (1951); **and** *Memoirs of George W. Armstrong*, Steck, Austin, Texas, (1958).

"House Report" and contrary facts, in 1950-1951,[156] by claiming that Great Britain was controlled by the Zionists and the N. M. Rothschild & Son Bank, which controlled the Bank of England, the railroads and the press. Armstrong attributed the letter to Lord Northcliffe.

It is interesting to note that a "secret society" had been formed by Cecil John Rhodes in the Nineteenth Century with the expressed purpose of reunifying the British Empire.[157] Rhodes attempted to unite English-speaking financiers to pool their wealth and rule the world, and one of his main goals was to bring America back under British control. Rhodes was long-term associate of Nathaniel M. Rothschild and Alfred Beit, and was a *de facto* Rothschild agent.[158] Rhodes' not so secret society was founded on racist principles and promoted the Jewish Messianic ambitions of one world government, one world language, etc.

President Wilson's "progressive" movement in America, like the "progressive" movements of Bolshevism and Zionism, was in practice a repressive movement which included segregationist laws and punitive government censorship. Wilson betrayed the American blacks who had voted him into office, by promoting segregation; and Wilson made it a Federal offense in the United States to speak out against the war, or on behalf of Germany, and imprisoned those who had dared to do either, or both. The Congress passed the "Espionage Act" on 15 June 1917 and amended it on 16 May 1918 to make it even more oppressive. Many anti-war protestors were beaten, arrested and imprisoned as a result.

These "progressive" movements for "international peace", instead proved to be fronts for centralized racist international tyranny, exploitive colonization, and the promotion of Entente European and of American interests, at the expense of the rest of the world—including Germany. Einstein had a long relationship with the League of Nations, which was advocated by Wilson. Zionist spokesman Samuel Landman wrote in 1936,

---

156. G. W. Armstrong, *The Zionists*, Judge Armstrong Foundation, Fort Worth, Texas, (1950). *Third Zionist War*, Judge Armstrong Foundation, Fort Worth, Texas, (1951).

157. "Mr. Rhodes's Ideal of Anglo-Saxon Greatness", *The New York Times*, (9 April 1902), p. 1. *See also:* W. T. Stead, "Cecil John Rhodes", *The American Monthly Review of Reviews*, Volume 25, Number 5, (May, 1902), pp. 548-560, at 556-557. *See also:* "The Progress of the World", *The American Monthly Review of Reviews*, Volume 25, Number 5, (May, 1902), pp. 515-598. *See also:* S. G. L. Millin, *Cecil Rhodes*, Harper & Brothers, New York, (1933); **and** *Rhodes*, Chatto & Windus, London, (1952). *See also:* C. Quigley, *Tragedy and Hope: A History of the World in Our Time*, Macmillan, New York, (1966), pp.130-133, 144-153, 950-956, 1247-1278; **and** *The Anglo-American Establishment: From Rhodes to Cliveden*, Books in Focus, New York, (1981), p. ix; **and** "The Round Table Groups in Canada, 1908-38", *Canadian Historical Review*, Volume 43, Number 3, (September, 1962), pp. 204-224. *See also:* J. E. Flint, *Cecil Rhodes*, Little, Brown, Boston, (1974). *See also:* R. I. Rotberg and M. F. Shore, *The Founder: Cecil Rhodes and the Pursuit of Power*, Oxford University Press, New York, (1988). *See also:* A. Thomas, *Rhodes*, St. Martin's Press, New York, (1997).

158. G. E. Griffin, *The Creature from Jekyll Island: A Second Look at the Federal Reserve*, Fourth Edition, American Media, Westlake Village, California, (2002), p. 208.

> "Moreover, the fact that the very existence of the future of Jewish Palestine depends, from the point of view of international law, on a Mandate of the League of Nations has powerfully contributed towards making the Jews everywhere into strong supporters of the League of Nations. In France, for instance, it is well known that the Jews are among the leaders of the pro-League policy. In other lands it is equally true, though less well known. For instance, the views of such a man as Dr. Einstein—a convinced Zionist believer in the League—count heavily in the land where he now dwells—the U.S.A. [***] In the opinion of Lord Cecil and General Smuts, the League of Nations and a Jewish Palestine are the two greatest positive results of the Great War. The two things are interdependent to a large extent. A Government that has let the world understand clearly that Great Britain stands unshakably by the League cannot logically do otherwise with regard to Zionism and Palestine."[159]

The formation of the League of Nations after the apocalypse of the First World War, and the attempted formation of the State of Israel, were the fulfillment of Jewish prophecy.

5.15.3 The Balfour Declaration—*QUID PRO QUO*

Zionist Jews betrayed Germany in the middle of the First World War by bringing America into the war on the side of the British. The Zionists controlled President Wilson through blackmail. They struck a deal with the British and agreed to use their influence over Woodrow Wilson and the American Press to bring America into the war on England's side. For their part, the British agreed to issue the Balfour Declaration and conquer Palestine. The entire world suffered as a consequence.

Albert Einstein's anti-German rhetoric in the post-war period especially irked many Germans, because they knew that Zionist traitors like Einstein had betrayed Germany to England and Russia in exchange for a deal with the British to take Palestine from Turkey and make it available to the Jews for a homeland. This stab in the back came after Germany had done so much for Jews and it betrayed the generally very positive relationship between Jews and Germany. Albert Einstein stated in 1938,

> "When the Germans had lost the World War hatched by their ruling class, immediate attempts were made to blame the Jews, first for instigating the war and then for losing it. In the course of time, success attended these efforts. The hatred engendered against the Jews not only protected the privileged classes, but enabled a small, unscrupulous, and insolent group to

---

[159]. S. Landman, *Great Britain, the Jews and Palestine*, New Zionist Press (New Zionist Publication Number 1), London, (1936), pp. 7, 10.

place the German people in a state of complete bondage."[160]

Albert Einstein told Peter A. Bucky,

> "For instance, after the First World War, many Germans accused the Jews first of starting the war and then of losing it. This is nothing new, of course. Throughout history, Jews have been accused of all sorts of treachery, such as poisoning water wells or murdering children as religious sacrifices. Much of this can be attributed to jealousy, because, despite the fact that Jewish people have always been thinly populated in various countries, they have always had a disproportionate number of outstanding public figures."[161]

Einstein's opinion that many Germans blamed Jews for the First World War, and for Germany's defeat in that war, is correct. Hitler wrote in his unpublished sequel to *Mein Kampf*,

> "The war against Germany was fought by an overpowering world coalition in which only a part of the states could have a direct interest in Germany's destruction. In not a few countries the shift to war was brought by influences which in no way sprang from the real domestic interests of these nations or even which could also be to their benefit. A monstrous war propaganda began to befog public opinion of these peoples and to stir it into enthusiasm for a war which for these very peoples in part could not bring any gain at all and indeed sometimes ran downright counter to their real interests.
>
> International world Jewry was the power which instigated this enormous war propaganda. For as senseless as the participation in the war by many of these nations may have been, seen from the viewpoint of their own interests, it was just as meaningful and logically correct seen from the viewpoint of the interests of world Jewry."[162]

Einstein does not tell us how or why the Germans came to this conclusion, how this message was spread, or why it was widely believed. A factual analysis based on primary source material answers these questions.

Marxist and Secretary of State in the German Foreign Office,[163] Karl Kautsky wrote in 1921,

---

160. A. Einstein, "Why do They Hate the Jews?", *Collier's*, Volume 102, (26 November 1938); reprinted in *Ideas and Opinions*, Crown, New York, (1954), p. 192.
161. P. A. Bucky, Einstein, and A. G. Weakland, *The Private Albert Einstein*, Andrews and McMeel, Kansas City, (1992), p. 87.
162. A. Hitler, translated by S. Attanasio, *Hitler's Secret Book*, Bramhall House, New York, (1962), p. 211.
163. D. K. Buchwald, *et al.*, Editors, *The Collected Papers of Albert Einstein*, Volume 9, Document 37, Princeton University Press, (2004), p. 60, note 4.

"Neither of the two belligerent groups [in the First World War] had the upper hand from the outset. Each was obliged to utilise every resource at its disposal. On both sides of the trenches, each government sought to obtain the full support of its proletarians, and also of its Jews. The cheapest concession that could be made to the latter was in the form of promises to support Zionism. For these promises were all to be realised at the expense of Turkey. The Central Powers, as well as the Entente, permitted the Jews to believe that their victory would result in a Jewish homeland in Palestine."[164]

Influential Jews in America, and those Jews in the press throughout the Western World, were often of German-Jewish descent, and were perceived as being quite pro-German prior to the middle of the First World War. Jews had strongly defended German Protestants in the *Kulturkampf*. There were also millions of Russian Jews in America at the time, and they hated the Czar and were pro-German because Germany was the enemy of Russia. German-Americans, many of whom were of Jewish descent, were an influential group in the 1916 Presidential campaigns.[165] President Wilson, in part, won his campaign on the slogan, "He kept us out of the war!"

Republican Theodore Roosevelt was forced out of the race for the Republican nomination because he had alienated the German-American vote—the "hyphenates", which included many Jews. American Jews who had emigrated from Germany and Austria were very concerned by the rhetoric of the advocates for "preparedness", *i. e.* war against Germany. The statements of the advocates of "preparedness" attacked pacifists as if disloyal to America and claimed that immigrants from Entente countries were loyal Americans, but immigrants from Germany and Austria were traitors. This so affected the Jewish community, that some of the advocates for "preparedness" made exceptions to their ethnic attacks for German Jews.[166]

Strangely, some political Zionists claimed that all Jewish newspapers around the world, outside of Germany itself, became anti-German in 1914. Germany tried very hard to help Jews fulfill their dreams of emancipation in Russia, and to achieve a homeland in Palestine. In search of an explanation for the fact that some of the leadership of the pro-German Jews of the world suddenly became anti-German, many Germans concluded that they were rewarded for helping their Jewish neighbors by an international Jewish betrayal. Though many Jews took bold actions to distance themselves from the anti-German activities of a prominent few, leading Jews in the German press, in the

---

164. K. Kautsky, *Are the Jews a Race?*, International Publishers, New York, (1926), pp. 193-194.
165. "The German Favorite Son", *The New York Times*, (23 May 1916), p. 10. "The Hyphen Vote, the Silent Vote, and the American Vote", *The New York Times*, (2 October 1916), p. 10.
166. S. Joseph, "In Dispute at America's Door", *The New York Times*, (10 June 1916), p. 10.

German Government, and in the German-Jewish financial community, subverted German interests during and after the war. These were often the same Jews who had beat the drums of war most loudly when the war began.

Lisa Endlich tells a revealing story of the conflicts among German-Jewish financiers, who split along Zionist lines, in her book *Goldman Sachs: The Culture of Success*,

> "World War I divided Europe and Goldman Sachs. Henry Goldman, highly conscious and fiercely proud of his German-Jewish heritage, was a staunch and vocal supporter of Germany and its war efforts. An intense, high-strung, and didactic man, his outspoken support and deep admiration for everything German did untold damage to the firm's reputation. When Sam Sachs returned from Europe shortly after the outbreak of the war, after assuring the Kleinwort partners of the firm's pro-British stance, he was horrified by his brother-in-law's open support for the enemy. The Sachs's German origins were just as recent and just as strong, but their allegiance was to England and France.
>
> In 1915, Goldman, despite pressure from his partners and sisters, rejected Goldman Sachs's participation in the $500 million J. P. Morgan sponsored Anglo-French loan to fund the war effort, to which virtually all the leading Wall Street firms of the day were subscribing. The firm had a longstanding policy requiring unanimous agreement of the partnership for participation in any piece of business. Out of their own strong beliefs and to save face for their firm, Sam and Harry Sachs marched down to the offices of J. P. Morgan and personally subscribed $125,000 toward the loan.
>
> As the war continued, the ill will between Goldman and the Sachses grew. One can only imagine the uncomfortable atmosphere that must have prevailed in the firm's small offices. Even after the United States entered the war Goldman continued to speak out publicly in support of Germany, despite the fact that two of his partners and one of his partner's sons were on duty in Europe. The episode was a painful one for the Goldman and Sachs families both personally and professionally. Finally, Kleinwort cabled Goldman Sachs that it was in danger of being blacklisted in London. The British merchant bank had been embarrassed when called before the Ministry of Blockade and shown a large number of cables between Goldman Sachs, its partner of two decades, and German banks. It was clear to the Kleinwort partners that the firm was doing an active exchange business with the Germans. They wrote to Goldman Sachs in 1916: 'We were frankly astonished at the evident importance of these operations, and we are therefore not surprised to find the authorities skeptical as to the possibility of entirely avoiding any indirect connection between such business and your sterling account with us.' The Bank of England eventually prevented Kleinwort from doing exchange business with Goldman Sachs, cutting off much of Goldman Sachs's London business until after the war.
>
> The firm's business had come to an almost complete standstill, despite its growing stature in the financial community. [***] Shortly after this,

Henry Goldman announced his departure on Goldman Sachs letterhead with the words 'Save & Serve. Buy Liberty Bonds!' emblazoned in red at the top of the page. He wrote, 'I am not in sympathy with many trends which are now stirring the world and which are now shaping public opinion. I retire with the best of feeling towards the firm (and all of its members) with which I have been associated for thirty-five years and to which I have given all there is in me.'"[167]

The Zionists, who had President Wilson under their control through blackmail, struck a deal with the British. The Zionists brought America into the war on England's side and Britain issued the Balfour Declaration promising Palestine to the Zionists as a potential homeland for Jews—both before and after securing Palestine for the Palestinians. This greatly changed the face of international Jewish propaganda.

Jewish interests in the media in France, England and America had long agitated against Russia in the hopes that these nations would pressure Russia to free Jews from the Pale of Settlement. Russia and the Czar were regularly ridiculed in the press in the West and story after story appeared in the newspapers telling of atrocities allegedly committed against Jews by Russians. Many prominent and influential Jews actively agitated against Russia with governmental leaders in Italy, France and England—Russia's allies—the Allies.[168] However, when the Zionists decided to turn against Germany, the press suddenly began to laud Russia in the middle of World War I and urged Russian Jews to fight for the Allies for the sake of taking Palestine from the Turks; while Jewish financiers conspired with the German Government to destroy the Russian State and its people.

Formerly openly anti-Russian Jews suddenly became pro-Russian[169] and urged all Russian Jews to fight to capture Palestine—a move that cost the Turks and Germans, who were the enemies of Russia in the First World War. The Russian Revolution freed Russian Jews and the entrance of America into the war on the side of the Allies secured Germany's eventual defeat. This was part of a Zionist strategy to elicit the Balfour Declaration. As a result, many Germans came to stereotype all Jews as if duplicitous and believed that Jews had caused them to lose the war and had caused the terrible hardships the Germans faced in the post-war period. Many prominent Jews published works claiming that Germans are inherently evil and that Germany must be divided and made

---

167. L. Endlich, *Goldman Sachs: The Culture of Success*, Alfred A. Knopf, New York, (1999), pp. 41-43.

168. V. Jabotinsky, "Evidence Submitted to the Palestine Royal Commission (11 February 1937)", in A. Hertzberg, *The Zionist Idea*, Harper Torchbooks, New York, (1959), pp. 559-570, at 562-563.

169. As but one example of many articles in *The New York Times*, see: "Russian Jews Hope for Allied Victory: See Their Chances of Future Reform in Triumph of Democracy: Germany Their Worst Foe: Most Reactionary Elements in Muscovite Officialdom Said to be of Teutonic Origin.", *The New York Times*, (16 October 1916), p. 8.

agrarian and primitive.

After the Zionists made their deal with the British, a wave of anti-German propaganda appeared in American and British journals, newspapers and books linking Germans with the persecution of Americans in Germany and of Jews in Russia. Zionist headquarters moved from Berlin to London. As opposed to the double dealings of the Zionists, some German Jews revealed that many of the Zionists in Palestine were savages and that Germany represented the best hope of "World Jewry".

The Germans were about to win the war in 1916. The Zionists in England interceded with the British government, who were largely resigned to defeat, and promised them that they could bring America into the war on the side of the British, with their influence over the press, their financial power, and their power in the American government.[170] It was well-known that the Zionists had President Wilson in their pocket. The Zionists Louis Brandeis and Samuel Untermyer blackmailed him President Wilson with love letters he had written to Mrs. Peck.

Jews in Germany were enjoying unprecedented power and equal rights in Germany and many fought valiantly to defend the "Fatherland" in the First World War. They, and the Germans in general, felt that the Zionists had stabbed them in the back in the pursuit of racist Zionism. German Jews tried to obstruct the immigration of Eastern European Jews, whom they considered to be primitive and decadent.[171] In America, German Jews and their decedents had tended toward assimilation. Some exploited immigrant Eastern European Jews—the impoverished *Ostjuden* who emigrated to America and soon greatly outnumbered the German Jews—in garment factories owned by German Jews; which resulted in the formation of some of the earliest labor unions in America.

German Jews encouraged Eastern European Jews to assimilate, and feared that the massive influx of orthodox Eastern European Jews to America would result in increased anti-Semitism. Sephardic and German Jews were quite successful in American, and looked down upon the less sophisticated *Ostjuden*, the Jews of Eastern Europe. Burton J. Hendrick wrote in 1923 in a pro-Jewish article meant to refute the accusations of THE DEARBORN INDEPENDENT,

> "In all that has been said of the economic progress of the Jews in America one fact should not escape observation. The Jewish names in this list are especially significant; Lewisohn, Kahn, Wolf, Guggenheim, Warburg, Schiff; they are all names of German Jews. The same statement is true of the great Jewish department store proprietors: Straus, Stern, Gimbel, Altman. An examination of the occasional Jewish name that appears as a director of banks would bring out the same fact. The important Jewish banking houses—Kuhn, Loeb & Co., Speyer & Co., Goldman, Sachs & Co.,

---

[170]. B. Freedman, *The Hidden Tyranny*, New Christian Crusade Church, Metairie, Louisiana, (1970).

[171]. See Einstein's articles and the editorial comment in: *The Collected Papers of Albert Einstein*, Volume 7, Documents 34, 35, 37, Princeton University Press, (2002).

Hallgarten—are almost exclusively Germanic. In the financial advertisements of this magazine a few Jewish names figure; they are invariably the names of German Jews. The big Jewish lawyers of New York—Untermyer, Marshall—and of Chicago—Levy Meyer, Samuel Alshuler—also belong to the German branch of the race. Most of the Jews who have reached important public position—Henry Morgenthau, Oscar Straus, Eugene Meyer, Louis Brandeis, Abraham Elkus—are likewise German Jews; a few others, Bernard Baruch, Benjamin Cardozo, belong to that Spanish-Portuguese element which has been established in this country for nearly three hundred years. Yet these German and Spanish branches represent only a small minority of the Jewish population of America. Of the three million Jews in this country, probably not far from 2,500,000 are Russian Jews. Of New York City's 1,500,000 Jews not far from 1,300,000 have come from the East of Europe. What progress have these Jews made? How do they earn their living? What fields of business do they 'dominate'? This phase of the subject will be treated in the next article."[172]

Burton J. Hendrick iterated a typical pro-Sephardic Jew and pro-German Jew attitude common among the Jewish elite in the West, those Jews who prevented the exodus of Jews seeking refuge from the pogroms and from Nazism. Hendrick wrote in 1923 in an anti-Communist—anti-Polish-Jew— article entitled "Radicalism among the Polish Jews"—in contrast to a series of otherwise philo-Semitic articles he wrote on "The Jews in America" in *The World's Work*,[173]

"There is only one way in which the United States can be protected from the anti-Semitism which so grievously afflicts the eastern sections of Europe. That is by putting up bars against these immigrants until the day comes when those already here are absorbed."[174]

Racist Marxist Zionist Ber Borochov stated,

"Anti-Semitism menaces both the poor helpless Jews and the all-powerful

---

[172]. B. J. Hendrick, "The Jews in America: II Do the Jews Dominate American Finance?", *The World's Work*, Volume 44, Number 3, (January, 1923), pp. 266-286, at 286. The next article in the series is: "The Jews in America: III The Menace of the Polish Jew", *The World's Work*, Volume 44, Number 4, (February, 1923), pp. 366-377.

[173]. B. J. Hendrick, "The Jews in America: I How They Came to This Country", *The World's Work*, Volume 44, Number 2, (December, 1922), pp. 144-161; **and** "The Jews in America: II Do the Jews Dominate American Finance?", *The World's Work*, Volume 44, Number 3, (January, 1923), pp. 266-286; **and** "The Jews in America: III The Menace of the Polish Jew", *The World's Work*, Volume 44, Number 4, (February, 1923), pp. 366-377.

[174]. B. J. Hendrick, "Radicalism among the Polish Jews", *The World's Work*, Volume 44, Number 6, (April, 1923), pp. 591-601, at 601.

Rothschilds. The latter, however, understand very well where the source of trouble lies; the poverty-ridden Jewish masses are at fault. The Jewish plutocracy abhors these masses, but anti-Semitism reminds it of its kinship to them. Two souls reside within the breast of the Jewish upper bourgeoisie—the soul of a proud European and the soul of an unwilling guardian of his eastern coreligionists. Were there no anti-Semitism, the misery and poverty of the Jewish emigrants would be of little concern to the Jewish upper bourgeoisie. It is impossible, however, to leave them in some west European city (on their way to a place of refuge) in the care of the local governments, for that would arouse anti-Semitic ire. Therefore, in spite of themselves and despite their efforts to ignore the Jewish problem, the Jewish aristocrats must turn philanthropists. They must provide shelter for the Jewish emigrants and must make collections for pogrom-ridden Jewry. Everywhere the Jewish upper bourgeoisie is engaged in the search for a Jewish solution to the Jewish problem and a means of being delivered of the Jewish masses. This is the sole form in which the Jewish problem presents itself to the Jewish upper bourgeoisie."[175]

Many American Jews sought to prevent public awareness of the discord between German Jews and Eastern European Jews. They tried to prevent the press from covering the strikes by Eastern European Jews against factories owned by German Jews.[176] Some believe that American Jewish financiers funded Hitler in order to block the flow of *Ostjuden* to the West, to provide a buffer against the spread of Bolshevism, to profit from the wars Hitler was liable to provoke, and to promote Zionism.[177] Some of these reasons might also have been behind the failure of Great Britain to act against the Nazi regime until they were forced into war—and Hitler was allegedly somewhat surprised that England actually declared war against Germany when Germany invaded Poland.

Speaking in general terms, Eastern European Jews resented the assimilationist attitudes of the German Jews. Even before Herzl, in the 1880's when the Pogroms heated up in Russia, Russian Jews like Peretz Smolenskin railed against rich assimilated Jews in the West, Jews who had allegedly disowned their "Volk".[178] By choosing England over Germany, the Zionists were able to create discord between German Jews, who were the most ardent anti-Zionist—pro-assimilationists among Jews, and German Gentiles; thereby

---

175. B. Borochov, *Nationalism and Class Struggle*, New York, (1935), pp. 135-136, 183-205; quoted in A. Hertzberg, *The Zionist Idea*, Harper Torchbooks, New York, (1959), pp. 360-366, at 361.

176. Video documentary, "Jewish Chicago 1833-1948", *Chicago Stories*, Network Chicago WTTW 11, Chicago, (2003).

177. H. Kardel, *Adolf Hitler, Begründer Israels*, Verlag Marva, Genf, (1974); English translation *Adolf Hitler: Founder of Israel*, Modjeskis' Society Dedicated to Preservation of Cultures, San Diego, (1997).

178. Some of Smolenskin's writings are found in A. Hertzberg, *The Zionist Idea*, Harper Torchbooks, New York, (1959), pp. 142-157.

forcing German Jews towards Zionism and weakening Germany in preparation for Marxist revolution—revolution which came at war's end. By siding with the British, the Zionists were also siding against the Turkish Empire, which ruled Palestine and Greater Syria.

Anti-Semites, many of whom worked for the Zionists, exploited this opportunity to stereotype all Jews based on the actions of a few. They wanted to create an animus against all Jews for the mere fact of being Jews, so as to obstruct assimilation. Hitler's friend, Dietrich Eckart, wrote in his *Bolshevism from Moses to Lenin: A Dialogue Between Adolf Hitler and Me*,

> "'Completely aside from that, it's clear that they have had America by the throat for quite a while,' I continued. 'No country, writes Sombart, displays more of a Jewish character than the United States. [*Notation:* Werner Sombart, *Die Juden und das Wirtschaftsleben* (Leipzig, 1911), p. 39.] We have already seen a consequence of this in the World War. In 1915, at a time when the true Americans hadn't the slightest thought of a war against us and, in fact, were so disposed toward us that any indication of a possible conflict of interest could have been smoothly and amicably settled, a secret advisory committee met with President Wilson for the sole purpose of preparing the country for war against Germany. [*Notation by English translator deleted—its evidentiary content demonstrated below*.] And who was the chief wire-puller in these nefarious activities, which were set into motion a full two years before the engagement of the United States in the war? The previously unknown Jew, Bernard Baruch. 'I believed that the war would come, long before it came,' he later calmly explained to the special committee of Congress which confirmed all this. And no one got up and beat the crafty scoundrel to a pulp.'"[179]

The Germans knew of the deal struck between the Zionists, President Wilson and the British, as it happened. *The New York Times* reported on 12 November 1917 on page 13,

"ZIONISTS HERE SEE TEUTON PLAN HALTED

British Victories in the Holy Land
Thwart Germany's Ambition
to Control Palestine.

---

[179]. D. Eckart and A. Hitler, *Der Bolschewismus von Moses bis Lenin: Zwiegespräch zwischen Adolf Hitler und mir*, Hoheneichen-Verlag, München, (1924); English translation by W. L. Pierce, "Bolshevism from Moses to Lenin", *National Socialist World*, (1966). URL: <http://www.jrbooksonline.com/DOCs/Eckart.doc> p. 9. Sombart's book was translated into English by M. Epstein, *The Jews and Modern Capitalism*, The Free Press, Glencoe, Illinois, (1951).

HER PRESS CAMPAIGN BARED

Its Aim Was to Save Enough Eastern
Territory to Menace
the Suez Canal.

American Zionists who have been watching with interest the various military operations near the Holy Land have been tremendously relieved by the events of the last few days. The British victories at Beersheba and Gaza, forecasting the eventual occupation of Jerusalem, and the promise given last week by Mr. Balfour, in the name of the British Government, that they would 'use their best endeavors to facilitate the establishment of Palestine as a national home for the Jewish people,' have apparently spiked a German scheme for setting up in Palestine a Jewish State, nominally autonomous, but really under German control.

A statement issued yesterday by the Provisional Executive Committee for General Zionist Affairs gave a detailed account of a press campaign supporting this scheme which has been going on in Germany and Austria for some time. This is held to indicate that the German military leaders foresaw the collapse of the Berlin-to-Bagdad plan and were preparing another arrangement by which it was hoped that Germany might save from the wreck of its plans in the Near East enough to form a constant menace to the Suez Canal, Egypt, and India.

'To accomplish this purpose,' says the committee, 'Germany was evidently preparing to ride roughshod, if need be, over its present ally, should Turkey refuse to recognize that it was to her 'best interests' to fall in with the new project. To give 'punch' to its publicity campaign, Germany unearthed a conspiracy between America and the Zionist Organization, including United States Supreme Court Justice Louis D. Brandeis, Judge Julian W. Mack, head of the American Military Insurance Department; Felix Frankfurter of the War Department, as well as Lord Walter Rothschild, leader of the English Zionists, and former Ambassador Henry W. Morgenthau to seize Palestine for exploitation by the Jews, Christian missionaries, and capitalists.

'In the end, if General Allenby hadn't gotten the jump on her by striking hard and quickly, Germany would one day soon have blandly announced the establishment of a Jewish republic under its auspices and suzerainty, and in response to Turkey's protests would have pointed to the overwhelming demand of the German people, and quoted for the benefit of its ravished ally, 'Vox populi, vox Dei.'

'If it had carried out its new plan, the establishment of an autonomous Jewish State in Palestine under its overlordship, whether with the consent of the Ottoman Government or in utter disregard of Turkey's wishes, Germany would have had, in addition to the strategical advantage that this would mean for the next war,' also the satisfaction of 'beating the Allies to it.' England, France, Italy, and Russia have already made it clear that the establishment of a Jewish State in Palestine is one of their aims in this war, and in Jewish circles in America it is held that Washington's view as to the desirability of this coincides with that of the Allies.

'Some echoes of these whisperings must have reached Germany, and several of its leading publications speak harshly of these 'infamous American Zionist proposals.' Thus Die Kölnische Zeitung, published in Cologne, publishes a long screed impugning the honesty of President Wilson, and ending with these complimentary allusions to Americans in general:

The Americans belong to that class of ?????? that have been for the last sixty years undermining the proud edifice of the Turkish Empire, and haven't stopped it yet. The Palestine action fully reveals Wilson's intentions. America has dropped its mask and shown itself in its true colors—a power that has the greatest interest from the capitalistic and religious point of view to bring Turkey under the influence of missionaries and capitalists. This is the true American humanity, which is based on the alliance of the religious men with the king of trusts. Turkey has watched this campaign with the utmost patience, and now it has received the cruelest reward. It can see now that America is not far behind the other Entente Powers in their enmity to Turkey and their plans for its destruction.

Kaiser Visits Palestine.

'For Germany to give its consent to the establishment of the Jewish nationality on its historic soil, requires a reversal of its previous attitude toward Palestine. Attempts have been made to establish German colonies in the Holy Land, and Kaiser Wilhelm has paid several visits to Palestine in order to win favor with the peoples of that country, and to encourage his subjects in their vain attempts to gain a strong footing there.

'The way was being prepared by a rather obvious campaign which began with the publication of apparently innocent scientific articles, by experts, on the near East, which discussed at great length, and with much detail, the accomplishments of the Jewish colonists and the vast possibilities of Palestine from an economic standpoint. A remarkable array of such articles, studying Palestine from every conceivable angle, has been published in over a hundred periodicals in Germany and Austria. These were followed by 'letters to the editor' and now the propaganda has attained the editorial stage.'

Among the first of these articles was one by Major Carl Frank Enders to make clear to the German people that it had better give up all hope of colonization in the Holy Land, and at the same time warn Turkey not to put any obstacles in the way of the Jewish operations there. Major Enders wrote:

The realization of the Zionists idea means infinitely more to our economic life than those fantasies and dreams of the German people that the Near East will create for us the lost world markets. * * * It will not be politically wise for Turkey to hinder the Jewish immigration into

Palestine * * * German colonization in Palestine is nothing but a dream, beyond the realm of realization, which I would advise the German people to forego.

'The Munich Neueste Nachrichten makes the frank statement that 'Zionism has become a question of the first magnitude, and Germany and Turkey have no choice but to give it serious consideration.' Gustave von Dobeller said: 'For many years the object which our Kaiser tried to accomplish by arduous political effort has been the making of a strong Turkey. A method not to be despised would be the establishment of a strong Jewish State, under Turkish suzerainty. As the Jewish people favor republics, let them, therefore, establish a republic, which must, however, be under the protection of the Ottoman Empire. It is always a question of importance whether you or your opponent has the key of the door. The idea of establishing a Jewish State is good for that power which effects it.'

Sees No Gain to Jews.

'The Vice President of the Austrian Parliament, Professor Paul Rohrbach, whose job was that of persuading the Jews of Germany and Austria-Hungary that the political schemes of the Allies are not to be trusted, wrote: 'The national aspirations of the Jews will be listened to with more sympathy by the allies of Middle Europe than by the Entente, even though certain papers and politicians on that side have lately been promising great things to the Jews. I do not believe that, even if the Entente were victorious and Turkey dismembered so that Palestine came under the suzerainty of either England of France, the Jews would benefit by this. Jews will have nothing to gain by the imperialistic schemes of England.'

'The Frankfurter Zeitung said:

'Pan Turkish ideas have no meaning in Palestine, where practically no Turks dwell.'

'Die Reichsbote, the mouthpiece of the Junkers, is calling upon the German Government to act promptly for the establishment of a Jewish State to 'offset the American Zionist proposals.' This must be done, it insists, to counteract the Wilson intrigue and 'to prevent England from making use of these American Zionist proposals as a backdoor which will enable her to pass freely from Egypt to India. For this purpose,' it says, 'the German-Austrian Zionist plans for a Jewish settlement must be strengthened. This is the opportune moment for the Zionist movement to attain its ideal.'

'These 'American Zionist proposals' are creating a real panic in the minds of Germany. The indications are that the German Press is alluding to the Palestine Commission appointed by President Wilson last Summer, consisting of Former Ambassador Morgenthau and Felix Frankfurter of War Secretary Baker's Advisory Council. At any rate, the Deutsche Worte speaks of them as a 'graver calamity than a declaration of war by a small or even medium-sized nation would be,' and charges the enemies of Germany with 'trying to enlist in their service the Zionist movement.' But it sees through the game of the Allies. 'We know very well what Mr. Morgenthau and Lord Rothschild are doing in this behalf for America and England,' it declares, the while it admits that if 'this plan of our enemies succeeds, it will go very badly with us.'

'These editorials will suffice to indicate how Germany was making ready to 'beat the Allies to it' in Palestine. General Allenby had not beaten Germany by taking Beersheba and capturing the highway to Jerusalem. The unfurling of the Union Jack over the hills of the Holy City will signalize the end of the 'Berlin to Bagdad' dream.'"

An article entitled, "Christians and Jews Rejoice: How the British Occupation of Jerusalem Was Received in Different Circles", *The New York Times Current History of the European War*, Volume 14, (January-March, 1918), pp. 315-316, quoted from the *Kölnische Volkszeitung*,

"The associations of the word Jerusalem are so deeply rooted that the conquest of the city gives considerable kudos to the conqueror. Especially in the case of the Anglo-Saxon world stimulation of war spirit has been attained which, owing to the lack of successes in the main war theatres, would otherwise have been difficult to effect. The interests of the Jews in the Entente countries, especially of the supporters of Zionism, in the Palestine campaign has shown itself in unambiguous form.

In view of the tremendous influence which Jewish capital possesses in warfare, Entente financiers and politicians will welcome the favorable effects of the capture of Jerusalem on these powerful Israelite circles. From the military standpoint it cannot be denied that the battles which led to the capture were well prepared and cleverly planned, but regarding the war situation in the Orient as a whole there is no reason to overestimate the event. Jerusalem can, at the most, serve as a valuable base on the line of communications, but it lies too far from the really important aims of the British to give ground for anxiety. It may with good reason be expected that on a line more to the rear, more easy to defend, the Turks will call a halt to the British advance."

Political Zionist leader Samuel Landman repeatedly confirmed the Germans' and Austrians' belief that Zionist Jews had used President Woodrow Wilson to bring America into the war on the side of the Allies in exchange for the Zionist Balfour Declaration. If Germany should win the war, the Zionists would obtain Palestine as a concession to the Jewish bankers for financing the war; and should England win the war, the Zionists still would obtain Palestine as a concession for bringing America into the war on the side of the British. Zionist Jews had no loyalty to Turkey, Russia, England, Germany *or America*. Many Jews delighted in the destruction of the war, which many Jews hoped would leave Europe ripe for Jewish Bolshevik revolution. Landman wrote in 1936,

"During the critical days of 1916 and of the impending defection of Russia, Jewry, as a whole, was against the Czarist regime and had hopes that Germany, if victorious, would in certain circumstances give them Palestine. Several attempts to bring America into the War on the side of the Allies by influencing influential Jewish opinion were made and had failed. Mr. James A. Malcolm, who was already aware of German pre-war efforts to secure a foothold in Palestine through the Zionist Jews and of the abortive Anglo-French démarches at Washington and New York; and knew that Mr. Woodrow Wilson, for good and sufficient reasons, always attached the greatest possible importance to the advice of a very prominent Zionist (Mr. Justice Brandeis, of the US Supreme Court); and was in close touch with Mr. Greenberg, Editor of the *Jewish Chronicle* (London); and knew that several important Zionist Jewish leaders had already gravitated to London from the Continent on the *qui vive* awaiting events; and appreciated and realised the depth and strength of Jewish national aspirations; spontaneously took the initiative, to convince first of all Sir Mark Sykes, Under-Secretary to the War Cabinet, and afterwards Monsieur Georges Picot, of the French Embassy in London, and Monsieur Goût of the Quai d'Orsay (Eastern Section), that the best and perhaps the only way (which proved so to be) to induce the American President to come into the War was to secure the co-operation of Zionist Jews by promising them Palestine, and thus enlist and mobilise the hitherto unsuspectedly powerful forces of Zionist Jews in America and elsewhere in favour of the Allies on a *quid pro quo* contract basis. Thus, as will be seen, the Zionists, having carried out their part, and greatly helped to bring America in, the Balfour Declaration of 1917 was but the public confirmation of the necessarily secret 'gentleman's' agreement of 1916 made with the previous knowledge, acquiescence and/or approval of the Arabs and of the British, American, French and other Allied Governments, and not merely a voluntary altruistic and romantic gesture on the part of Great Britain as certain people either through pardonable ignorance assume or unpardonable ill-will would represent or misrepresent.

Sir Mark Sykes was Under-Secretary to the War Cabinet specially concerned with Near Eastern affairs, and, although at the time scarcely acquainted with the Zionist movement, and unaware of the existence of its leaders, he had the flair to respond to the arguments advanced by Mr. Malcolm as to the strength and importance of this movement in Jewry, in spite of the fact that many wealthy and prominent international or semi-assimilated Jews in Europe and America were openly or tacitly opposed to it (Zionist movement), or timidly indifferent. MM. Picot and Goût were likewise receptive.

An interesting account of the negotiations carried on in London and Paris, and subsequent developments, has already appeared in the Jewish press and need not be repeated here in detail, except to recall that immediately after the 'gentleman's' agreement between Sir Mark Sykes, authorized by the War Cabinet, and the Zionist leaders, cable facilities through the War Office, the Foreign Office and British Embassies, Legations, etc., were given to the latter to communicate the glad tidings to their friends and organizations in America and elsewhere, and the change in official and public opinion as reflected in the American press in favour of joining the Allies in the War, was as gratifying as it was surprisingly rapid. [***] In Germany, the value of the bargain to the Allies, apparently, was duly and carefully noted. In his 'Through Thirty Years' Mr. Wickham Steed, in a chapter appreciative of the value of Zionist support in America and elsewhere to the Allied cause, says General Ludendorff is alleged to have said after the War, that: 'The Balfour Declaration was the cleverest thing done by the Allies in the way of propaganda, and that he wished Germany had thought of it first.' [*Footnote:* Volume 2, page 392.] As a matter of fact, this was said by Ludendorff to Sir Alfred Mond (afterwards Lord Melchett), soon after the War. The fact that it was Jewish help that brought U.S.A. into the War on the side of the Allies has rankled ever since in German—especially Nazi—minds, and has contributed in no small measure to the prominence which anti-Semitism occupies in the Nazi programme."[180]

W. J. M. Childs wrote in 1924, and note that Jews stabbed the Armenians, Germany, Turkey, Russia and America in the back—truly backstabbed the entire world by manufacturing the First World War, destroying the Russian Nation and by bringing America into the war at a point when Germany was about to win the war and bring about an equitable peace—all this death and destruction at the behest of treacherous Jews so that a few hundred thousand Jews could be forced against their will and better judgement to move to Palestine,

> "Much is heard of the [Balfour] Declaration as an instrument conferring upon the Jewish race unwarrantable privileges in a land from which that race had been effectively dispersed. There has been remarkably little said as to the reasons of high policy which impelled the Allies to adopt the purpose of the Declaration as one of their war aims.
> To some extent altruistic motives influenced certain Gentile protagonists of the Zionism expressed in the Declaration. At a time when justice for oppressed races and small peoples had become an Allied slogan it was at least consistent to include the Jews among those whose wrongs might be righted as an outcome of the War. But we well may doubt how far such considerations, standing alone, would have carried the Allied Governments towards accepting the restoration of the Jewish people to

---

180. S. Landman, *Great Britain, the Jews and Palestine*, New Zionist Press (New Zionist Publication Number 1), London, (1936), pp. 4-6. ***See also:*** B. Freedman, *The Hidden Tyranny*, New Christian Crusade Church, Metairie, Louisiana, (1970). Samuel Landman repeated his story in: S. Landman, "Origins of the Balfour Declaration: Dr. Hertz's Contribution", in I. Epstein, J. H. Hertz, E. Levine, and C. Roth, Editors, *Essays in Honour of the Very Rev. Dr. J. H. Hertz, Chief Rabbi of the United Hebrew Congregations of the British Empire, on the Occasion of His Seventieth Birthday, September 25, 1942 (5703)*, E. Goldston, London, (1942); and in: S. Landman, "Balfour Declaration: Secret Facts Revealed", *World Jewry: Independent Weekly Journal*, Volume 2, Number 43, J. H. Castel, London, (22 February 1935).

Palestine as a war aim. The truth is, of course, that for Great Britain and her Allies the policy indicated in the Declaration was most definitely a war measure, well calculated to yield results of immense importance to the Allied cause. And, further, that for Great Britain special reasons existed why she should adopt and support the policy of the Declaration.

These may be found in the obvious advantages of covering the Suez Canal by an outpost territory, in which important elements of the population would not only be bound to her by every interest, but would command the support of world Jewry. That was the long view of British Imperial interests, taken in 1916 and 1917; it counted for much then, but for even more after the war.

But apart from exclusive British interests, the Declaration may be described as essentially a war measure adopted by the Powers of the Entente in the furtherance of their own vital interests. Defined in greater detail, it was a bold, imaginative, and statesmanlike effort to prevent the incalculable and universal influence of Jewry being exerted on the side of the Central Powers—as, indeed, it was, to a serious extent, then being exerted—and to transfer this highly important influence to the cause of the Entente. Nor was it a project of sudden origin, or hastily embraced. The advantages to be gained if the policy of the Declaration were adopted had long been urged; opposition to that policy had long been active. Before the British Government gave the Declaration to the world it had been closely examined in all its bearings and implications, weighed word by word, and subjected to repeated change and amendment. Unless full weight be given to these antecedent facts, no correct judgment upon the Declaration and its policy in operation can be formed.

2. *The Zionists and the Declaration.* Zionism had been a living and ambitious force in the Jewish world long before 1914. While awaiting its real opportunity it had, in 1905, rejected the tempting offer of territory for the creation of a Zionist State in Uganda, under the British flag. It had steadily looked to Palestine as the one land which could provide the historical connexion essential to Zionist aims. The entry of Turkey into the war brought the hitherto impracticable dreams of Zionism within the bounds of possible attainment. If the goodwill of the Allies, particularly of Great Britain, could be secured, and provided that ultimate success should attend the Allied arms, much might be done to realize the dearest ambitions of Zionism. It lay with Zionist leaders to bring their ideal before the British Government as a scheme likely to be of advantage to the Entente.

Suffice to say that at this crisis of its fortunes Zionism was fortunate, that in Dr. C. Weizmann and Mr. N. Sokolov it found two leaders equal to the great occasion, that British Statesmen, including Mr. (now Lord) Balfour, Lord Milner, Mr. Lloyd George, Lord Robert Cecil, immediately recognized the political importance and value of the Zionist suggestions, and that in the subsequent long negotiations and discussions by which the aims of Zionism were harmonized with the political realities of the situation, the British negotiators were Mr. Balfour and the late Sir Mark Sykes, both of them convinced and ardent supporters of Zionist aspirations. These British representatives and the Zionist leaders just named must be credited with the chief part in framing the policy of the Declaration.

Support of Zionist ambitions, indeed, promised much for the cause of the Entente. Quite naturally Jewish sympathies were to a great extent anti-Russian, and therefore in favour of the Central Powers. No ally of Russia, in fact, could escape sharing that immediate and inevitable penalty for long and savage Russian persecution of the Jewish race. But the German General Staff desired to attach Jewish support yet more closely to the German side. With their wide outlook on possibilities they seem to have urged, early in 1916, the advantages of promising Jewish restoration to Palestine under an arrangement to be made between Zionists and Turkey, backed by a German guarantee. The practical difficulties were considerable; the subject perhaps dangerous to German relations with Turkey; and the German Government acted cautiously. But the scheme was by no means rejected or even shelved, and at any moment the Allies might have been forestalled in offering this supreme bid. In fact in September 1917 the German Government were making the most serious efforts to capture the Zionist movement.

Another most cogent reason why the policy of the Declaration should be adopted by the Allies lay in the state of Russia herself. Russian Jews had been secretly active on behalf of the Central Powers from the first; they had become the chief agents of German pacifist propaganda; by 1917 they had done much in preparation for that general disintegration of Russian national life, later recognized as the revolution. It was believed that if Great Britain declared for the fulfilment of Zionist aspirations in Palestine under her own pledge, one effect would be to bring Russian Jewry to the cause of the Entente.

It was believed, also, that such a declaration would have a potent influence upon world Jewry in the same way, and secure for the Entente the aid of Jewish financial interests. It was believed, further, that it would greatly influence American opinion in favour of the Allies. Such were the chief considerations which, during the later part of 1916 and the next ten months of 1917, impelled the British Government towards making a contract with Jewry.

But when the matter came before the Cabinet for decision delays occurred. Amongst influential English Jews Zionism had few supporters, at all events for a Zion in Palestine. It had still fewer in France. Jewish influence both within and without the Cabinet is understood to have exerted itself strenuously and pertinaciously against the policy of the proposed Declaration.

Under the pressure of Allied needs the objections of the anti-Zionists were either over-ruled or the causes of objection removed, and the Balfour Declaration, as we have seen, was published to the world on 2nd November 1917. That it is in purpose a definite contract with Jewry is beyond question. Subsequently the Declaration was accepted and endorsed by the Governments of France, Italy, and Japan.

That it is in purpose a definite contract between the British Government and Jewry represented by the Zionists is beyond question. In spirit it is a pledge that in return for services to be rendered by Jewry the British Government would 'use their best endeavours' to secure the execution of a certain definite policy in Palestine. No time limit is set for performance; completion alone appears to have been intended as the conclusion of the contract. It would thus seem to be an agreement incapable of being greatly varied except by consent.

How far the implied services of Jewry have been or may yet be rendered cannot be estimated, and must always remain a matter of opinion. The Declaration certainly rallied world Jewry, as a whole, to the side of the Entente. The war was won by the Entente; and to the Declaration as a measure to that end may be attributed a share in achieving the great result. And it is possible to understand from many sources that directly, and indirectly, the services expected of Jewry were not expected in vain, and were, from the point of view of British interests alone, well worth the price which had to be paid. Nor is it to be supposed that the services already rendered are the last—it well may be that in time to come Jewish support will much exceed in importance any thought possible in the past. That, however, is a possibility for Palestine of the future to demonstrate."[181]

Bernard Shaw wrote in 1930,

"The controversy proved superfluous after all; for the foreign trade department at the Admiralty, in the sensible hands of Sir Richard Webb, consented to pay for the confiscated cargoes; the support of the American Jews was purchased by Lord Balfour at the price of Jerusalem (Zion); and the sinking of the Lusitania by a German submarine not only removed the danger of America coming into the war on the German side, but practically forced her in on our side."[182]

Concerned that Chaim Weizmann had not recognized James A. Malcolm's[183] leading role in drawing America into the war through the influence of American Jews like Supreme Court Justice Louis D. Brandeis through British support of the Zionist cause, Malcolm Thomson[184] wrote in a Letter to the Editor

---

181. W. J. M. Childs, in H. W. V. Temperley, Editor, *A History of the Peace Conference of Paris*, Volume 6, Published under the auspices of the British Institute of International Affairs, Henry Frowde and Hodder & Stoughton, London, (1924), pp. 171-174.
182. B. Shaw, *What I Really Wrote about the War*, Constable, London, (1930), p. 118; *reprinted:* Brentano's, New York, (1931), p. 99.
183. J. A. Malcolm, *Origins of the Balfour Declaration: Dr. Weizmann's Contribution*, British Museum, London, (1944).
184. R. John, *Behind the Balfour Declaration: The Hidden Origins of Today's Mideast Crisis*, Institute for Historical Review, Costa Mesa, California, (1988); **and** "Behind the Balfour Declaration: Britain's Great War Pledge To Lord Rothschild", *The Journal for*

published as "Origin of the Balfour Declaration" in *The [London] Times Literary Supplement* of 22 July 1949 on page 473, in response to their review of Chaim Weizmann's *Trial and Error*,[185] quoting from Adolf Böhm's *Die Zionistische Bewegung*,

> "'Mr. Malcolm, President of the Armenian National Committee in London, advised Sir Mark Sykes to influence Wilson through Brandeis, and to guarantee Palestine forthwith to the Jews, in order to gain their support. After discussion with Lord Milner, Sykes begged Mr. Malcolm to put him into touch with the Zionist leaders, because Sir Edward Grey and Mr. Balfour were convinced of the justice of the Zionist demand for Palestine. Through Greenburg, Malcolm made contact with Weizmann.' [***] [T]he Foreign Office had sent word to Brandeis and through him had worked on Wilson, in Washington."

> "Mr. *Malcolm*, Präsident des Armenischen National-Komitees in London, riet Sir *Mark Sykes*, *Wilson* durch *Brandeis* zu beeinflussen und den Juden, um sie günstig zu stimmen, gleichzeitig Palästina zu sichern. Nach Rücksprache mit Lord *Milner* bat *Sykes* Mr. *Malcolm*, ihn mit den zionistischen Führern in Verbindung zu setzen, da Sir *Edward Grey* und Mr. *Balfour* von der Gerechtigkeit der zionistischen Forderung auf Palästina überzeugt seien. Durch *Greenberg* trat *Malcolm* auch mit *Weizmann* in Verbindung. [*Footnote:* Über die hier dargestellten Vorgänge siehe den Bericht über die „Balfour-Declaration" von *S. Landmann*, der von 1917-1922 Sekretär der zionistischen Exekutive war, in „World Jewry", London, 1935, Nr. 42 und 43.]"[186]

Malcolm Thomson wrote in a Letter to the Editor "The Balfour Declaration" in *The London Times* on 2 November 1949 on page 5,

> "A change of attitude was, however, brought about through the initiative of Mr. James A. Malcolm, who pressed on Sir Mark Sykes, then Under-Secretary to the War Cabinet, the thesis that an allied offer to restore Palestine to the Jews would swing over from the German to the allied side the very powerful influence of American Jews, including Judge Brandeis, the friend and adviser of President Wilson."[187]

Historical Review, Volume 6, Number 4, (Winter, 1985-1986), pp. 389ff.:
<http://www.ihr.org/jhr/v06/v06p389_John.html>
185. C. Weizmann, *Trial and Error: The Autobiography of Chaim Weizmann*, H. Hamilton, London, (1949).
186. A. Böhm, *Die Zionistische Bewegung*, Volume 1, Jüdischer Verlag, Berlin, Hozaah Ivrith Co., Ltd., Tel Aviv, (1935), p. 656.
187. Response from V. I. Gaster was published "The Balfour Declaration", *The London Times*, (10 November 1949), p. 5.

See also: The Secret History of the Balfour Declaration and the Mandate, Pamphlets on Arab Affairs, Number 6, Arab Office, London, (1947).

Lloyd George stated before the House of Commons on 19 June 1936,

> "The obligations of the Mandate are specific and definite. They are to encourage the establishment of a national home for the Jews without detriment to any of the rights of the Arab population. I agree that it is a dual undertaking, and we must see that both parts of the Mandate are thoroughly enforced. But look at the conditions under which we entered into it. It was one of the darkest periods of the War when Mr. Balfour prepared his Declaration. Let me recall the circumstances to the House. At the time the French army had mutinied, the Italian army was on the eve of collapse and America had hardly started preparing in earnest. There was nothing left but Britain confronting the most powerful military combination the world has ever seen. It was important for us to seek every legitimate help we could get. We came to the conclusion, from information we received from every part of the world, that it was vital we should have the sympathies of the Jewish community. I can assure the Committee that we did not come to that conclusion from any predilections or prejudices, certainly we had no prejudices against the Arabs, because at that moment we had hundreds and thousands of troops fighting for Arab emancipation from the Turk.
>
> In these circumstances and on the advice which we received we decided that it was desirable to secure the sympathy and co-operation of that most remarkable community, the Jews throughout the world. They were helpful in America and in Russia, which at that moment was just walking out and leaving us alone. In these conditions we proposed this to our Allies. France accepted it, Italy accepted it, and the United States accepted it, all the other Allies accepted it, and all the nations which constitute the League of Nations accepted it. And the Jews—I am here to bear testimony to the fact—with all the influence they possess responded nobly to the appeal which was made. I do not know whether the House realises how much we owe to Dr. Weizmann with his marvellous scientific brain. He absolutely saved the British army at a critical moment when a particular ingredient which was essential we should have for our great guns was completely exhausted. His great chemical genius enabled us to solve that problem. But he is only one out of many who rendered great services to the Allies. It is an obligation of honour which we undertook, to which the Jews responded. We cannot get out of it without dishonour."[188]

Frank Owen wrote in his book *Tempestuous Journey: Lloyd George: His Life and Times,*

---

[188]. D. Lloyd George, *The Parliamentary Debates. Official Report. House of Commons*, Series 5, Volume 313, (19 June 1936), cols. 1339-1345, at 1341-1342.

"Enough for a day? No. There was trouble in the House of Lords about Honours. And there was always Ireland. But something—or rather, somebody—else was about to cause still more division in the War Cabinet.

There was another persistent people knocking at the door—and one with a still older history of oppression and exile. The Jews.

For nearly 2,000 years, the Jews had been wanting and waiting to return to the Land of their Fathers. ('Next Year in Jerusalem' they toasted at their Passover.) But it was not until about the dawn of the present century that the powerful Zionist Movement had been born, a world-wide organization pledged to restore Palestine as the national homeland of the Jewish people. They were not likely to overlook the possibilities of action opened up by a world war, and when the contemporary tyrant occupier of their ancient country (the Turk) took the side of the Central Powers, the Zionists naturally sought succour from the Allies. One of their leading members was a Russian Jew named Dr. Weizmann.

The reader has met him already, with Lloyd George one day in 1915 at the Ministry of Munitions, when the brilliant scientist set to work to produce the then vitally-needed acetone. In declining any honour or award to himself for his services, he had told Lloyd George of the national aspirations of his own people. Dr. Weizmann already knew Balfour, and had worked under him at the Admiralty. To him, too, the ardent Zionist confided his dreams, and Balfour had been perhaps more impressed.

Asquith, who was still Prime Minister in those days, had not been so encouraging. He had his good reasons. One was that secret Sykes-Picot Pact of May, 1916, whereby the Allies had agreed to carve up the Turkish Empire in the Middle East into Russian, French and British zones; the proposed Anglo-French dividing line cut right through Palestine. By the autumn of that year, however, a still stronger reason had arisen for revising this arrangement. This was the urgent necessity of winning over the goodwill of American Jewry to the Allied cause. For the Germans had not been idle in courting Zionism, either, notably addressing themselves to the Russian Jews.

So, under a new War Cabinet which included Lloyd George, Balfour and Smuts (another strong sympathizer with the ideas of Zionism), there had gone forth secret assurances to the Zionist leaders that Britain would support their claims, if she could carry her Allies with her. One thus addressed was Justice Brandeis, an outstanding figure of the Movement in the United States, and a close personal friend of President Wilson. A Zionist delegation, which included Dr. Weizmann, Sir Herbert Samuel and Mr. James de Rothschild, M.P., had journeyed to Paris, and there secured the agreement of the French Government.

Throughout the summer of 1917, Balfour kept up his talks with the Zionists, and on 3 September, he laid before the War Cabinet the draft of a public statement to be made by the British Government endorsing and proclaiming all that had been promised in private.

But not everybody was pro-Zionist, and perhaps the least unanimous (in fact, they were about equally divided) were the people most concerned. Within the War Cabinet itself two more meetings were required before a bridge could be built to span the differences, and in public life, outside, the rifts long remained. Fiercest opposition of all came from wealthy Jews, who feared that if a Jewish National State were established they might lose their own status as citizens of the countries where they and their forbears had long dwelt and prospered. Lloyd George's own old friend, Sir Charles Henry, M.P., was foremost among these Anti-Zionists, and he did not delay any longer to found an anti-Zionist newspaper, *The Jewish Guardian*, to express his views.

In the War Cabinet, the new Secretary of State for India, Edwin Montagu, led the Anti-Zionist party. In a stormy meeting on 4 October, 1917, Balfour warned of a new German drive to capture the Zionist forces for the enemy side, and he claimed that though some rich Jews in Britain might oppose the idea of Zionism, it was enthusiastically backed by those in America and Russia. On whose side were those influential people to be ranged? There was no inconsistency whatever in having a Jewish National Home and Jews being members of other States. The French Government were sympathetic to the idea, and so, as he personally knew, was President Wilson.

Edwin Montagu rose. He most strongly objected to a 'National Home' for Jews, insisting that the Jews were really only a religious community and that he was himself a 'Jewish Englishman'. He turned to Lloyd George. 'All my life,' he said, 'I have been trying to get out of the Ghetto. You want to force me back there!'

Curzon was opposed to the proposal on other grounds. Ah! well did he recollect a journey he had made through the Promised Land, many years ago now. Alas! It was a barren land, with little cultivation even on the terraced slopes, and watered by all too few streams. How could this place of stone and sand become a home for millions more Jews? Moreover, what about the Moslems already living there?

Milner interposed to declare himself in favour of the National Home far Jews—provided nothing was done to prejudice the civil and religious rights of the non-Jews in Palestine, or the political status of Jews elsewhere.

The Prime Minister ruled that the War Cabinet had heard enough for one day. There was still a war on. Resolved: to hear the further views of Zionists, Anti-Zionists, Non-Zionists, and President Wilson.

The days passed. A week. Three weeks.

The Jews (at any rate, the pro-Zionist Jews) were getting restive. In particular, Lord Rothschild, the Head of his House. He had been in correspondence with Balfour since mid-July, and was beginning to wonder if anything was going to happen in the War Cabinet or not? Because, decidedly, something was happening in Palestine.

The British Army was marching in.

After three years' hold-up, 80 per cent of it by Turkish bluff (the considerable contribution of British Army Intelligence in accepting it must not be entirely overlooked), our far more powerful forces in Egypt had begun to take the offensive against a war-weary enemy, who now counted as many deserters as troops remaining on his battle strength.

'Jerusalem by Christmas!' Lloyd George had demanded of General Allenby, in appointing him to the Egypt Command in the summer of 1917. Now Allenby had crossed the desert from Egypt, turned the weak Turkish line at Gaza by a brilliant manœuvre and was moving on the Holy City. This he would take, entering humbly on foot a fortnight before Christmas Day.

At a third War Cabinet, 31 October, 1917, Balfour once more brought up the question of the National Home. How could its establishment possibly prejudice Jews elsewhere? Surely, on the analogy of a European immigrant in the United States, it would help that they had a recognized land of origin? As for the present poverty of Palestine, the scientific development of her resources might yet make it a land flowing with milk and honey.

Curzon followed. He delivered another reminiscent address on his travels in the Middle East, which the Prime Minister this time interrupted to ask if he agreed with some expression of sympathy? Resolved:

'His Majesty's Government view with favour the establishment in Palestine of a National Home for the Jewish People, and will use their best endeavours to facilitate the achievement of this object, it being clearly understood that nothing shall be done which may prejudice the civil and religious rights of existing non-Jewish communities in Palestine or the rights and political status enjoyed by Jews in any other country.'

Next day, Lloyd George presented this draft to the leaders of British Jewry. Of eight of them, four accepted it, including the Chief Rabbi, Dr. Hertz, one was neutral and three were hostile. Thus, the famous Balfour Declaration was delivered to the world. Next year, France, Italy and the United States all declared their accord with this policy.

But what *was* the policy? Lloyd George himself, in later years, insisted that what he had meant was that Jews should be free to go to Palestine and settle there in such strength as the land could support—or be made to support. Then, in due course, they should set up their own autonomous Jewish Administration. By no means all Jews would go there, any more than all the Irish-born return to Ireland.

It did not work out that way. The Jewish Question, like the Irish Question, had been too long part of History to be dismissed from it overnight. But the troubles this generation has known were far ahead in October, 1917. [***] There was also a new row raging between the Zionist and the anti-Zionist Jews. His Foreign Secretary, Balfour, was no Jew, but he was the foremost and certainly the most famous Christian Zionist."[189]

---

[189]. F. Owen, *Tempestuous Journey: Lloyd George: His Life and Times*, McGraw-Hill Company, Inc. New York, (1955), pp. 426-428, 492.

William D. Rubinstein argues that one of the drafts of the Balfour Declaration was written by a crypto-Jew named Leopold Charles Moritz Stennett Amery.[190] Amery's family feigned conversion to Protestantism. His mother was perhaps the child of Frankist Jews who fled Hungary after the revolution of 1848, who eventually settled in England by way of Constantinople—many Jews and crypto-Jews emerged from Turkish *Dönmeh* training grounds to become prominent Zionist spokesmen and leaders, as well as revolutionaries who sought to subvert the societies into which they moved.[191] Perhaps beginning with Poland, Salonika and Paris, these crypto-Jewish *Dönmeh* have established subversive groups around the world. Amery was a leading force in unseating Chamberlain's government and installing longtime Zionist Winston Churchill as Prime Minister. Leopold Amery's son John, outwardly an anti-Semite and a Fascist—like so many Jewish Zionists of the period, betrayed England and helped the Zionist Nazis. He was hanged for treason after the war. A typical Zionist leader of his time, Leopold Amery, together with Chaim Weizmann, also helped betray a million, by his own account, Hungarian Jews to death.

Benjamin Harrison Freedman wrote (Bear in mind the ill will between Armenians and the Turks who controlled Palestine. The Zionists —Jewish bankers and the Young Turks under Jewish leadership,[192] *Dönmeh* Turks who had long feigned Moslem conversion while undermining Turkish society and eventually succeeded in overthrowing the Sultan and destroying much of Turkish culture— the Zionists secretly and artificially created this ill-will to bring about the ruin of the Turkish Empire during the First World War. Jewish bankers and other Jewish Zionists, forever destroyed the Turkish Empire and mass murdered the Armenians.),

> "Mr. James A. Malcolm was an Oxford-educated Armenian who had been appointed to take charge of Armenian interests during and after the War. In his official capacity as advisor to the British Government on Eastern affairs... he had frequent contact with the Cabinet Office, the Foreign Office, the War Office and the French and other Allied embassies in London and made visits to Paris for consultation with his colleagues and leading French officials.
>
> He was passionately devoted to an Allied victory. While his home in London was being bombed by the Germans in 1944, he prepared the following account which speaks for itself. Mr. Malcolm feared he would not survive, and prepared the following which he deposited in the British

---

190. W. D. Rubinstein, "The Secret of Leopold Amery", *History Today*, Volume 49, Number 2, (February, 1999), pp. 17-23. D. Davis, "Balfour Declaration's Author was a Secret Jew", *The Jerusalem Post*, (12 January 1999), p. 1.
191. Y. Küçük, *Şebeke = Network*, YGS Yayinlari, Kadiköy, Istanbul, (2002).
192. I. Zangwill, *The Problem of the Jewish Race*, Judaen Publishing Company, New York, (1914), pp. 9, 11; which was first published as an article, "The Jewish Race", *The Independent*, Volume 71, Number 3271, (10 August 1911), pp. 288-295, at 290-291. J. Prinz, *The Secret Jews*, Random House, New York, (1973), pp. 111-112.

Museum for the benefit of posterity. It has become one of the most important documents explaining how the United States was railroaded into World War I, and follows here:

> During one of my visits to the War Cabinet Office in Whitehall Gardens in the late summer of 1916 I found Sir Mark Sykes less buoyant than usual... I enquired what was troubling him... [H]e spoke of military deadlock in France, the growing menace of submarine warfare, the unsatisfactory situation which was developing in Russia and the general bleak outlook... [T]he Cabinet was looking anxiously for United States intervention...
>
> [H]e had thought of enlisting the substantial Jewish influence in the United States but had been unable to do so...
>
> [R]eports from America revealed a very pro-German tendency among the wealthy American-Jewish bankers and bond houses, nearly all of German origin, and among Jewish journalists who took their cue from them... I inquired what special argument or consideration had the Allies put forward to win over American Jewry... Sir Mark replied that he made use of the same argument as used elsewhere, viz., that we shall eventually win and it was better to be on the winning side...
>
> I informed him that there was a way to make American Jewry thoroughly pro-Ally, and make them conscious that only an Allied victory could be of permanent benefit to Jewry all over the world... I said to him, 'You are going the wrong way about it... do you know of the Zionist Movement?'... Sir Mark admitted ignorance of this movement and I told him something about it and concluded by saying, 'You can win the sympathy of the Jews everywhere in one way only, and that way is by offering to try and secure Palestine for them'... Sir Mark was taken aback. He confessed that what I had told him was something quite new and most impressive...
>
> He told me that Lord Milner was greatly interested to learn of the Jewish Nationalist movement but could not see any possibility of promising Palestine to the Jews... I replied that it seemed to me the only way to achieve the desired result, and mentioned that one of President Wilson's most intimate friends, for whose humanitarian views he has the greatest respect, was Justice Brandeis of the Supreme Court, who was a convinced Zionist...
>
> [I]f he could obtain from the War Cabinet an assurance that help would be given towards securing Palestine for the Jews, it was certain that Jews in all neutral countries would become pro-British and pro-Ally... I said I thought it would be sufficient if I were personally convinced of the sincerity of the Cabinet's intentions so that I could go to the Zionists and say, 'If you help the Allies, you will have the support of the British in securing Palestine for the Jews'...
>
> [A] day or two later, he informed me that the Cabinet had agreed to my suggestion and authorized me to open negotiations with the

Zionists... the messages which were sent to the Zionist leaders in Russia were intended to hearten them and obtain their support for the Allied cause... other messages were sent to Jewish leaders in neutral countries and the result was to strengthen the pro-Allied sympathies of Jews everywhere...

[A] wealthy and influential anti-Zionist Jewish banker there was shown the telegram announcing the provisional promise of Palestine to the Jews... he was very much moved and said, 'How can a Jew refuse such a gift?'...

[A]ll these steps were taken with the full knowledge and approval of Justice Brandeis, between whom and [Zionist leader] Dr. Weizmann there was an active interchange of cables... [A]fter many anxious weeks and months, my seed had borne fruit and the Government had become an ally of Zionism... the Declaration is dated 2nd November, 1917, and is known to history as the Balfour Declaration... its obligation to promise British help for the Jews to obtain Palestine."[193]

The *Jewish Daily Bulletin* allegedly wrote on 30 October 1934, on page 3,

"The New Germany persists toward the complete extermination of the Jew because it was Jews who instigated the United States to enter the World War, accomplishing the defeat of Germany, and who later caused the inflation in Germany, Herr Richard Kunze, a leading Nazi Parliament figure, declared at a mass meeting in Magdeburg yesterday."[194]

Winston Churchill told William Griffin in August of 1936 in an interview published in the *New York Enquirer*,

"America should have minded her own business and stayed out of the World War. If you hadn't entered the war, the Allies would have made peace with Germany in the spring of 1917. Had we made peace then there would have been no collapse in Russia followed by Communism, no breakdown in Italy followed by Fascism, and Germany would not have signed the Versailles Treaty, which has enthroned Nazism in Germany. If America had stayed out of the war, all these 'isms' wouldn't today be sweeping the continent of Europe and breaking down parliamentary government, and if England had made peace early in 1917, it would have saved over one million British, French, American and other lives."[195]

---

193. B. H. Freedman, *Why Congress is Crooked or Crazy or Both*, New York, (1975). <http://www.natvan.com/pdf/12-04-04.pdf>

194. As quoted in: B. Freedman, *The Hidden Tyranny*, New Christian Crusade Church, Metairie, Louisiana, (1970). I have not verified that the primary source quotation is accurate.

195. As quoted in: B. Freedman, *The Hidden Tyranny*, New Christian Crusade Church, Metairie, Louisiana, (1970); who cites "Scribner's Commentator in 1936", which perhaps

Zionist[196] British Prime Minister David Lloyd George wrote in 1939, "The Germans were equally alive to the fact that the Jews of Russia wielded considerable influence in Bolshevik circles. The Zionist Movement was exceptionally strong in Russia and America. The Germans were, therefore, engaged actively in courting favour with that Movement all over the world. A friendly Russia would mean not only more food and raw material for Germany and Austria, but fewer German and Austrian troops on the Eastern front and, therefore, more available for the West. These considerations were brought to our notice by the Foreign Office, and reported to the War Cabinet.

The support of the Zionists for the cause of the Entente would mean a great deal as a war measure. Quite naturally Jewish sympathies were to a great extent anti-Russian, and therefore in favour of the Central Powers. No ally of Russia, in fact, could escape sharing that immediate and inevitable penalty for the long and savage Russian persecution of the Jewish race. In addition to this, the German General Staff, with their wide outlook on possibilities, urged, early in 1916, the advantages of promising Jewish restoration to Palestine under an arrangement to be made between Zionists and Turkey, backed by a German guarantee. The practical difficulties were considerable; the subject was perhaps dangerous to German relations with Turkey; and the German Government acted cautiously. But the scheme was by no means rejected or even shelved, and at any moment the Allies might have been forestalled in offering this supreme bid. In fact in September, 1917, the German Government were making very serious efforts to capture the Zionist Movement.

Another most cogent reason for the adoption by the Allies of the policy of the declaration lay in the state of Russia herself. Russian Jews had been secretly active on behalf of the Central Powers from the first; they had become the chief agents of German pacifist propaganda in Russia; by 1917 they had done much in preparing for that general disintegration of Russian society, later recognised as the Revolution. It was believed that if Great Britain declared for the fulfilment of Zionist aspirations in Palestine under her own pledge, one effect would be to bring Russian Jewry to the cause of the Entente.

It was believed, also, that such a declaration would have a potent influence upon world Jewry outside Russia, and secure for the Entente the

---

refers to *Scribner's Magazine*, which later merged with *Commentator*. I have not verified that the primary source quotation is accurate. A quite similar quotation appears in G. Allen and L. Abraham, *None Dare Call It Conspiracy*, (1971); which cites *Social Justice: Father Coughlin's Weekly Review*, (3 July 1939), p. 4.

<u>196</u>. "Eine große Rede Weizmanns in Jerusalem Vor der Abreise aus Palästina", *Jüdische Rundschau*, Volume 25, Number 4, (16 January 1920), p. 4. R. Sharif, "Christians for Zion, 1600-1919", *Journal for Palestine Studies*, Volume 5, Number 3/4, (Spring-Summer, 1976), pp. 123-141, at 136-139. A. Hertzberg, *The Zionist Idea*, Harper Torchbooks, New York, (1959), pp. 528, 594.

aid of Jewish financial interests. In America, their aid in this respect would have a special value when the Allies had almost exhausted the gold and marketable securities available for American purchases. Such were the chief considerations which, in 1917, impelled the British Government towards making a contract with Jewry."[197]

Sigmund Freud and William C. Bullitt wrote in 1932,

> "Balfour had replaced Grey as British Foreign Secretary. He came to America in April 1917 to inform Wilson that the condition of the Allies was desperate, that Russia was more than likely to withdraw from the war, that the morale of France was collapsing, that the financial condition of England threatened calamity and that the United States would have to carry a war burden enormously greater than either Wilson or anyone else in America had anticipated. He was prepared to reveal to Wilson some at least of the secret treaties of the Allies and to discuss war aims, assuming naturally that Wilson would insist on defining the precise aims for which he must ask the people of the United States to pour out a flood of blood and wealth.
> Wilson wished to settle the question of war aims with Balfour definitely and at once. At that moment he might have written his own peace terms and might possibly have turned the war into the crusade for peace which he had proclaimed. The Allies were completely at his mercy. But House persuaded him not to demand a definition of war aims from Balfour by the argument that the discussion which would ensue would interfere with the prosecution of the war. Both Wilson and House overlooked the fact that all the warring powers had discussed their peace terms in detail while prosecuting the war with notable efficiency. House also inserted in Wilson's mind the picture of a Peace Conference at which England would loyally cooperate with the United States in establishing a just and lasting peace. And Wilson, always anxious to 'dodge trouble,' let slip this opportunity to avoid the terms of the Treaty of Versailles and secure the just peace of which he dreamed. Both the President and House seem to have misunderstood totally the sort of respect that the governments of Europe had for Wilson. For the President as wielder of the physical strength of America, they had the greatest respect; for Woodrow Wilson as a moral leader, they had no respect. So long as the physical assistance of the United States was vital to the Allies they had to defer to the President of the United States; but Woodrow Wilson was never able to make any European statesman 'drunk with this spirit of self-sacrifice.'
> Balfour mentioned the existence of some of the secret treaties to Wilson and promised to send them to Wilson; but he never sent them and, having arranged for the utmost physical assistance from the United States, went

---

197. D. Lloyd George, *Memoirs of the Peace Conference*, Volume 2, Howard Fertig, New York, (1939/1972), pp. 725-726. *See also:* Volume 2, pp. 586-587; and Volume 4, pp. 1836-1843.

home happy."[198]

Many revisionists have argued that the great debts the Allies had accrued caused Wilson to enter the war in order to ensure that America could recover its loans.[199] This argument does not seem plausible for the simple reason that America incurred more expenses by going to war and making additional loans to the Allies, than the total monies it stood to lose if England and France were to default on their initial loans. America could not recover these internal expenses and America itself was financed by its own citizens, who invested large sums in bonds.

Prior to the close of World War I, Germany had provided Jews with more opportunities than any other nation on Earth. In return, Germany benefitted from Jewish contributions in Mathematics, the Arts and Sciences, the professions, high finance, and from Jewish educators. Many of the most prosperous of the Americans of Jewish descent had emigrated to America from Germany and promoted German businesses and culture in America—until the political Zionists began to smear the Germans, who had done so much to help Jews throughout the world. Then, Germany became a pariah nation in the American press. Germans and those of German descent, including German-Jewish immigrants, were resented and persecuted in America, and America entered the war on England's side. Many Germans knew that the British then issued the Balfour declaration (actually drafted by Zionists) to Rothschild in fulfilment of a contract with Zionists to win the war for England in exchange for Palestine by bringing in America on the Allies' side:

"Foreign Office.
November 2nd, 1917.

Dear Lord Rothschild,
I have much pleasure in conveying to you, on behalf of His Majesty's Government, the following declaration of sympathy with Jewish Zionist aspirations which has been submitted to, and approved by, the Cabinet

'His Majesty's Government view with favour the establishment in Palestine of a national home for the Jewish people, and will use their best endeavours to facilitate the achievement of this object, it being clearly understood that nothing shall be done which may prejudice the civil and religious rights of existing non-Jewish communities in Palestine, or the rights and political status enjoyed by Jews in any other country.'

I should be grateful if you would bring this declaration to the knowledge of the Zionist Federation."

---

198. S. Freud and W. C. Bullitt, *Thomas Woodrow Wilson, A Psychological Study*, Discus Books, New York, (1968), pp. 232-233.
199. P. Birdsall, "Neutrality and Economic Pressures, 1914-1917", *Science and Society*, (Spring, 1939).

The British had no lawful authority to make this declaration. The British did not control Palestine, and even if they had, they would have had no right to offer it up to the Jews for settlements. Henry Morgenthau pointed out that leading Jews misrepresented the precise language of the Balfour Declaration, which did not offer to give Palestine to the Jews, but merely expressed support for the idea that Jews might wish to live there under the rule of the indigenous population,

> "It is worth while at this point to digress for a moment from my main argument, to point out that the Balfour Declaration is itself not even a compromise. It is a shrewd and cunning delusion. I have been astonished to find that such an intelligent body of American Jews as the Central Conference of American Rabbis should have fallen into a grievous misunderstanding of the purport of the Balfour Declaration. In a resolution adopted by them, they assert that the declaration says: 'Palestine is to be a national home-land for the Jewish people.' Not at all! The actual words of the declaration (I quote from the official text) are: 'His Majesty's Government views with favor the establishment *in Palestine of a national home for the Jewish people*.' These two phrases sound alike, but they are really very different. I can make this obvious by an analogy. When I first read the Balfour Declaration I was temporarily making my home in the Plaza Hotel. Therefore I could say with truth: 'My home is in the Plaza Hotel.' I could not say with truth: 'The Plaza Hotel is my home.' If it were 'my home,' I would have the freedom of the whole premises, and could occupy any room in the house with impunity. Quite obviously, however, I would not venture to trespass in the rooms of my friend, Mr. John B. Stanchfield, who happened at the same time also to have found 'a home-land in the Plaza,' nor in the private quarters of any other resident of that hostelry, whose right to his share in it was as good as mine, and in many cases of much longer standing."[200]

5.16 A Newspaper History of Zionist Intrigues During the First World War, which Proves that Jewish Bankers Betrayed Germany

The London Times reported on 17 August 1914, on page 7,

"AMERICAN SYMPATHY INCREASING.

CHANCELLOR'S 'FUTILE PLEA.'

---

[200]. H. Morgenthau, "Zionism a Surrender, Not a Solution", *The World's Work*, Volume 42, Number 3, (July, 1921), pp. i-viii, at iii-iv.

FEARS OF JAPAN'S INTENTIONS.

(FROM OUR OWN CORRESPONDENT.)
WASHINGTON, AUG. 16.

'A futile plea' is the *New York World's* comment on the German Chancellor's latest effort to justify Germany in American eyes. Like the *New York Times*, the *Tribune*, and various other organs reflecting respectable opinion, the *New York World* resents German efforts to cloak the scandal of the violation of Belgian neutrality under vague references to a life and death struggle between Teuton and Slav. The *New York Times* is particularly indignant at the attempt to make out that England entered upon the war in order to further her commercial ambitions at the expense of Germany.

There are also signs of indignation at the clumsy propaganda of Pro-Germans in the United States. The responsible American Press is doing its best to be fair. Its Readers are constantly reminded that the news which comes is mainly from Anglo-French sources. Some newspapers are publishing daily extracts from German-American organs side by side with extracts from Franco-American contemporaries. Accusations of ignorance and prejudice are therefore annoying.

Unmistakable evidence is reaching Washington that South American sympathies are equally with us. The only discordant note is an agitation in the Japanophobe Press over the reported determinations of Japan to make war.[201] In spite of a reassuring statement by Count Okuma, the opinion is widely expressed that Japan espies an opportunity of expansion into China. There is reason to believe that the State Department is not immune from such fears, though there is no basis for reports that it has already taken a hand in current Far Eastern diplomacy. Should Japan take up arms, the State Department's policy will be one of cautious championship of the integrity of China outside foreign zones."

The *London Times* reported on 18 August 1914, on page 5,

"THROUGH GERMAN EYES.

THE BRITISH FLEET'S
MOVEMENTS.

BID FOR AMERICAN FAVOUR.

GREAT NUMBERS OF PRISONERS.

A party of Americans who left Berlin on August 13 were each presented

---

201. *See:* Letter from A. D. Fokker to A. Einstein of 18 November 1919, *The Collected Papers of Albert Einstein*, Volume 9, Document 168, Princeton University Press, (2004).

at the station of departure with a packet of 12 *Lokalanzeiger*. On the outside of the packet, one of which, by the kindness of one of the tourists, has come into our possession, is fixed a handbill addressed to 'The returning Citizens of the-to-us-friendly United States.'

The enclosed newspapers, it is stated, must 'serve to destroy the web of lies which the hostile Press has spread over us, and give truth its place of honour.' Then, in still larger type: 'Redistribution for publication in American papers is solicited.'

The newspapers in question seem chiefly anxious to convey two impressions—that Germany is everywhere victorious, and that American public opinion is favourable to Germany's cause. The ultimatum of Japan to Germany followed hard upon the gift which the Japanese Colony in Berlin are said to have given to their 'dear, brave friends.'

The Russians have, according to these papers, been beaten back all along the line. The French have been thoroughly beaten in Alsace, and the event is published in the following *communiqué*:—

At Mulhausen German troops have taken prisoner 10 French officers and 513 men. In addition four guns and a great number of rifles were taken. German soil is cleansed of the enemy.

At Lagarde 'more than 1,000 unwounded prisoners of war have fallen into our hands, more than a sixth of the two French regiments which were in the fight.'

According to a telegram from Hannover, 500 Belgian prisoners have been brought into the province, and 700 French prisoners of war are announced from Worms to be on their way to internment in Germany.

In the paper of August 13 is a notice to the effect that German submarines 'in the course of the last few days' have run along the East Coast of England and Scotland as far as the Shetlands. As to the results of this expedition—so runs the notice—nothing can, for obvious reasons, be published.

A telegram from Copenhagen purports to give the movements of the English Fleet. A great number of English men-of-war are said to have been sighted off Grimsby, going in a south-easterly direction, but the main British fleet is assembled to the east of Pentland Firth.

The news of victories generally seemed to be given out by the Kaiser himself. Liége is said to have fallen, with all its forts, into German hands (August 9). In spite of the demand of the *Lokalanzeiger* that the German losses should be published, no such list is given, on the ground that the number has not yet been ascertained."

The London Times reported on 19 August 1914, on page 5,

"PRESIDENT WILSON CRITICIZED.
(FROM OUR CORRESPONDENT.)
NEW YORK, Aug. 17.

The recent announcement of the State Department as to the attitude of this Government stating that 'loans by Americans bankers to any foreign nation which is at war are inconsistent with the true spirit of neutrality' is the subject of much comment here. Assuming that this is intended to apply to such arrangements as the Morgan French loan proposal, which was not a war loan in the ordinary sense, but merely a proposition to buy foodstuffs for France on a credit to be established here, the leading newspapers sharply criticize and condemn the Government's policy.

The *New York Sun* inquires whether, 'if Dr. Wilson and Mr. Bryan hold that it is a violation of the true spirit of neutrality to lend a belligerent funds to buy foodstuffs, it is not equally a violation of that spirit to sell a belligerent foodstuffs'; the *Sun* thinks the position of the Administration inconsistent with the modern theory of international law.

The *New York Times* feels that Dr. Wilson and Mr. Bryan 'are betrayed by their natural benevolent idealism into taking a somewhat extreme attitude against loaning American credit in time of war.' Food, it remarks, is needed for non-combatants as well as for the armies.

The *World* says:—

A national loan would be inadmissible, but to discourage loans by individuals while exerting the Government's utmost power to encourage the sale of our surplus products in belligerent markets is neither sound business, correct sentiment, nor true neutrality. It is statesmanship at cross purposes.

It is feared that the attitude of the Government may delay the resumption of shipments of grain and cotton commodities. Such shipments, it is argued in many quarters, will soon exhaust European balances here, and it will be almost impossible for Europe to purchase grain, &c., here unless credit in some way is arranged. We need the proceeds from our surplus grain and cotton quite as much as Europe will need those products.

KAISER'S PROTEST TO AMERICA.

WASHINGTON, AUG. 18.

The Kaiser has made a protest to President Wilson stating that Germany has been maligned and her motives misunderstood, misconstrued, and misrepresented in a campaign organized to foster anti-German sentiment.

The United States Government is protesting against these allegations through Mr. Gerard, the American Ambassador at Berlin, and Mr. Bryan.— *Exchange Telegraph Company*."

The London Times published the following letter on 19 August 1914, on page 7,

"GERMAN SOCIALISTS AND THE WAR.
TO THE EDITOR OF THE TIMES.

Sir,—If the German Social Democratic Party was as wholeheartedly against this war as Mr. H. M. Hyndman would have us believe, would he kindly explain how it is that the official organ of that party, the *Vorwärts*, which had hitherto seldom shown any tenderness for the Kaiser, broke out, just as the most acute stage of this crisis, into a sudden outburst of praise for Germany's War Lord as a great prince of peace?

About 20 years ago, I was watching, with Herr Bebel, a Prussian regiment of Foot Guards marching out of the Brandenburg Gate at Berlin. The Socialist leader told me, with some pride, that more than half of them probably were Social Democrats. I asked him whether, in the event of war, that would make the slightest difference, and he replied to me quite frankly, 'No, I am afraid, not the slightest. Nothing will happen until Germany has been sobered by a great military catastrophe. *Das Volk ist noch immer siegestrunken*' (The people are still drunk with victory).

It is folly to attempt to disguise from ourselves that this war is at present a popular war, and probably more popular against England than against any other of the allied Powers. Do not let us forget that no movement has received more enthusiastic support throughout Germany than the German big Navy movement. In this island country of ours no Navy League has ever secured, in all these years, a tithe of the popular support which the German Navy League has received in Continental Germany. Founded under exalted patronage, it could boast within a few years a membership of over one million, recruited all over the country, and largely through University professors and school teachers, who were the most active instruments of this essentially anti-British propaganda.

Yours obediently,
VALENTINE CHIROL.

August 18."

In English and American newspapers, the Zionist cause was said to be championed by the Czar, by the Germans, by the Turks, by the British, by the Armenians, etc., depending on the complexion of the world at the time and which nation/side appeared to be winning once war broke out. There are too many relevant articles to reproduce all of them here, but I will reprint a few.

The New York Times reported on 1 July 1914,

> "Prof. Levin of Berlin told the convention that European countries, including Turkey, were friendly to the Zionists, and that there was a great need of a university at Jerusalem."[202]

Early in the war in 1915, more than two years before the Balfour Declaration of the British pledged Palestine to the Zionists for a homeland, the Russians stated that one of the reasons for their war against the Turks was to capture Palestine for the Zionists. *The New York Times* wrote on 15 July 1915, page 3,

"SENT JEWS TO CAUCASUS.

---

[202]. "Zionists Plan Campaign", *The New York Times*, (1 July 1914), p. 11.

**Grand Duke Told Them to Retake
Palestine, German Paper Says.**

Special Cable to THE NEW YORK TIMES.

ZURICH, July 14, (Dispatch to The London Daily Graphic.)—The Munich journal Neueste Nachrichten publishes a dispatch from Lemberg stating that before the fall of that town Grand Duke Nicholas issued an order of the day to the Jewish soldiers in his army, stating that he had decided to give them a special opportunity of showing courage and patriotism. One of the aims of the struggle with Turkey was said to be in order to reconquer Palestine for the Jews so they could live there united and independent. The order of the day concluded as follows:

'We will therefore pave the way for you to join the Army of the Caucasus. It now depends on you what treatment your race and co-religionists will receive during the war and after. Reconquer Palestine for yourselves and a new day of glory will dawn for Jewry.'

Jewish soldiers in the Galacian army were then transferred to the Army of the Caucasus."

Later, the Russian Revolution was said to favor the Zionists. Bolsheviks were said to have freed the Zionists, then banned them. Two themes emerged at war's end, and they were not lost upon the Germans—the Zionists were loyal only to themselves, and the combatant nations' loyalty to Zionism came not from love, but desperation—and the need for money and to bring America into the war as an ally. Such illusions were created by the enormous wealth and influence of Jewish high finance.

Maurice Paléologue recorded the cruelly conducted concentration of Jews by the Russians and the rôle of the alleged influence of American Jews on America's war policy, as well as the use of the Jewish question to promote Jews and alternatively to condemn Jews as allies of the Germans or of the Russians, throughout Paléologue's *An Ambassador's Memoirs*. For example, we find his entry of 28 October 1914,

"*Wednesday, October* 28, 1914.

For the Jews of Poland and Lithuania the war is one of the greatest disasters they have ever known. Hundreds of thousands of them have had to leave their homes in Lodz, Kielce, Petrokov, Ivangorod, Skiernewice, Suvalki, Grodno, Bielostock, etc. Almost everywhere the prelude to their lamentable exodus has been the looting of their shops, synagogues, and houses. Thousands of families have taken refuge in Warsaw and Vilna; the majority are wandering aimlessly like a flock of sheep. It's a miracle that there have been no *pogroms* — organized massacres. But not a day passes in the zone of the armies without a number of Jews being hanged on a trumped-up charge of spying.

Incidentally, Sazonov and I have been talking of the Jewish question and all the religious, political, social and economic problems it raises. He informed me that the Government was considering what modifications could be made in the far too arbitrary and vexatious regulations to which the Russian Jews are subjected. A new law is about to be issued in favour of the Jews of Galicia who will become subjects of the Tsar. I have encouraged him to be as tolerant and liberal as possible:

'I'm speaking to you as an ally. In the United States there is a very large, influential and wealthy Jewish community who are very indignant at your treatment of their co-religionists. Germany is very skillfully exploiting this quarrel with you—which means a quarrel with us. It's a matter of importance for us to win the sympathy of Americans.'"[203]

Political Zionist leader Israel Zangwill published a letter in *The London Times* on 19 August 1914 on page 7, which precipitated his Zionist campaign to draw America into World War I on the side of the Allies and against Germany, and to convince German Jews around the world to side with Zionists against Germany,

"EQUALITY FOR JEWS IN RUSSIA.
TO THE EDITOR OF THE TIMES.

Sir,—The rumour in your issue of to-day that the Tsar is about to give civil and political rights to his Jews will, if confirmed, do much to relieve the feelings of those who, like myself, believe that the Entente with Russia was too high a price to pay even for safety against the German peril. Not that the Russians are not a fine people; it is only with the Russian Government that civilization has a quarrel, and the quarrel is as much on behalf of her Russian as her Jewish subjects. The offer of autonomy to Poland—even if it is only a good stroke of business—shows that that Government is entering upon an era of greater intelligence, and learning at last from her British ally that minorities and dependencies are attached more closely by love than by fear. The emancipation of the Russian Jews would be felt as an immense relief in many countries, not only among Jews, who have felt bitterly that the old land of freedom was helping involuntarily to perpetuate the Pale, but among Christians also, for all civilization suffers under this medieval survival with its sequelæ in massacre and emigration. In Russia there is a colossal field—half of Europe and half of Asia—for the energies of the six million Jews now cooped up in a province of which they are forbidden even the villages.

Their enfranchisement would, indeed, be a logical consequence of the redemption of Poland, for how could Russia permit the Jews in her Polish dominion to be freer than in Russia proper? But there is no logic in Russia, and it is, alas! far from improbable that the Poles, now engaged in a barbarous boycott of their Jews, would be stupid enough to imitate Russia

---

[203]. M. Paléologue, *An Ambassador's Memoirs*, Volume 1, Chapter 4, George H. Doran, New York, (1923), p. 173.

and deny them equality. In that case the Jews now in Austrian and German Poland would lose their hard-won rights just as the Jews of Khiva and Bokhara lost theirs when these regions were assigned to Russia. And Russian Jews would only assuredly count as human beings if Russia, instead of conquering German and Austrian Poland, herself loses to Germany her German Balkan-speaking provinces. In these—and they include the bulk of the Jewish Pale—the Jews would be seised at a stroke of the rights they have so long vainly demanded from Russia. Is it not tragic that in this instance civilization should have more to gain from German militarism than from our Eastern ally? I hope that in the final issue of this cosmic cataclysm England will not be found the catspaw of Powers opposed to her noblest traditions, but that by her insistence on justice and freedom all round she will retrospectively justify her Entente, show a glorious profit on her outlay in armaments, resume her moral hegemony of the world, and her old place in the affections of mankind."

To which J. E. C. Bodley replied in *The London Times* on 21 August 1914, on page 4,

"MR. ZANGWILL'S ANTI-BRITISH THEORIES.
TO THE EDITOR OF THE TIMES.

Sir,—Mr. Israel Zangwill informs us in *The Times* of yesterday that, because of Jewish disabilities in Russia, 'the Entente with Russia was too high a price to pay, even for safety against the German peril.' Mr. Zangwill is welcome to consider that the interests of his Russian compatriots are more important than those of the land of his adoption and of the British Empire. But before trying to convert us to his inopportune theory he should have a word to say to the proceedings of his fellow Hebrews in the United States, as recorded in the instructive dispatches from *The Times* Washington Correspondent on August 15, &c. [*Refer to the Endnote.*[204]] showing that the powerful Jewish Press of America is German in sympathy and bitterly anti-English in its unscrupulous propaganda.

Most of us are willing to believe that the majority of British Jews are (unlike Mr. Zangwill) first Englishmen and then Hebrews. But utterances such as his make it necessary to recall the unpleasant fact that, in the Press of Europe and America, Jewish influence means German influence. French anti-Semitism in its origin was entirely an anti-German movement, roused by the undue influence of German Jews in the Press and politics of France; and at that time the long-settled Jewish communities of Bordeaux and Baynonne excited no animosity.

---

204. "America and the War", *The London Times*, (15 August 1914), p. 5. However, ***See also:*** "Jew and German", *The London Times*, (15 August 1914), p. 3. "American Sympathy Increasing", *The London Times*, (17 August 1914), p. 7. "Through German Eyes", *The London Times*, (18 August 1914), p. 5. "Kaiser's Protest to America", *The London Times*, (19 August 1914), p. 5.

What right has Mr. Zangwill to lecture us and to talk lightly of 'the German peril'—which is no peril to him or to his people—when England alone in the world has given, at the expense of her working classes and of her ratepayers, a reckless hospitality to the Russian Jews, whose interests he puts above those of the British race?

I deplore anti-Semitism, especially at a crisis which has united British subjects of all races. But to propagate that *doctrine de haine* in England seems to be the object of Mr. Israel Zangwill."

Bodley referred to the fact that most Jews of German Jewish descent sided with Germany and expressed their pro-German stance in their newspapers. His charge that French anti-Semitism arose from the belief that Jewish liberalism was a stalking horse for German militarism found examples in the Dreyfus Affair and in anti-Semitic propaganda of the period (*see, for example, the period cartoon reproduced in:* R. M. Seltzer, *Jewish People, Jewish Thought: The Jewish Experience In History*, Macmillan, New York, (1980), p. 631).

The machinations of Jewish financiers in the anti-Catholic French Revolution as well as Rothschild's theft of the wealth of France in Napoleon's anti-Catholic campaigns, left many French suspicious of Jewish bankers. Jews had been accused of robbing nations of their gold from the times of the Roman Empire, when Flaccus charged the Jews with stealing the gold of Rome and sending it to Jerusalem.[205] Before Flaccus, the Jews accused themselves of stealing the Egyptians' gold by asking to borrow it from their trusting Egyptian neighbors, then emigrating without giving it back (*Exodus* 3:22; 11:2; 12:35-36). Joseph stole the gold of Egypt and ruined its money, then enslaved the Egyptian People (*Genesis* 47). Many have interpreted the Old Testament to predict that the when the Messiah arrives, the Jews will horde all the gold, silver and jewels of the world and keep this treasure in Jerusalem (*Isaiah* 23:17-18). Michael Higger wrote in his book published in 1932, *The Jewish Utopia*, divulging the intentions of Jews who wish to fulfill Judaic Messianic prophecy,

"All the treasures and natural resources of the world will eventually come in possession of the righteous. This would be in keeping with the prophecy of Isaiah: 'And her gain and her hire shall be holiness to the Lord; it shall not be treasured nor laid up; for her gain shall be for them that dwell before the Lord, to eat their fill and for stately clothing.[*Isaiah* 23:18]'[20] Similarly, the treasures of gold, silver, precious stones, pearls, and valuable vessels that have been lost in the seas and oceans in the course of centuries will be raised up and turned over to the righteous.[21] Joseph hid three treasuries in Egypt: One was discovered by Korah, one by Antoninus, and one is reserved for the righteous in the ideal world.[22] [***] Gold will be of secondary importance in the new social and economic order. Eventually, all the

---

205. M. T. Cicero, *Pro Flaccus*, Chapter 28; translated by C. D. Yonge, *The Orations of Marcus Tullius Cicero*, Volume 2, George Bell & Sons, London, (1880), pp. 454-455.

friction, jealousy, quarrels, and misunderstandings that exist under the present system, will not be known in the ideal Messianic era.[319] The city of Jerusalem will possess most of the gold and precious stones of the world. That ideal city will be practically full of those metals and stones, so that the people of the world will realize the vanity and absurdity of wasting their lives in accumulating those imaginary valuables.[320]"[206]

The Messianic prophecy found in *Haggai* 2:7-8 states,

"7 And I will shake all nations, and the desire of all nations shall come: and I will fill this house *with* glory, saith the LORD of hosts. 8 The silver *is* mine, and the gold *is* mine, saith the LORD of hosts."

An American Zionist Jew, Mordecai Manuel Noah, wrote in 1818,

"In *Russia*, the Jews are to be found in immense numbers, but greatly scattered over this extensive empire; they receive every protection and kindness from the mild and humane sovereign, and if he studies his own interest, he will go yet further in advancing their condition. They have it in their power to make a princely return. In 1745, the czarina, by an ukase, expelled the Jews on a charge of sending coin into foreign countries, but they were subsequently recalled."[207]

By robbing a nation of its gold and silver coinage, Jews were able to destroy that nation's economy by depriving it of its money supply. Jews were also able to increase the value of gold and silver as they accrued a monopoly over it.
The Chicago Tribune, reported on 17 August 1870 on page 4,

"FRANCE.
Special Despatch to The Chicago Tribune.

NEW YORK, Aug. 16,—A Paris letter to the *World* remarks that Messrs. de Rothschild are said to lose several thousand dollars a day on the money they keep idle in their safe, or, rather, vault. One of their most lucrative branches of business is dealing in bullion, and melting and refining gold. The government has ordered them to discontinue this business. The Messrs. de Rothschild are not looked upon with a favorable eye by the government. It is notorious that their sympathies are all German. They have not contributed a son to any of the war funds. You know nine-tenths of the banking business of Paris is in the hands of German bankers. The police

---

206. M. Higger, *The Jewish Utopia*, Lord Baltimore Press, Baltimore, (1932), pp. 12-13, 57.
207. M. M. Noah, *Discourse Delivered at the Consecration of the Synagogue of [K. K. She`erit Yisra`el] in the City of New-York on Friday, the 10th of Nisan, 5578, Corresponding with the 17th of April, 1818*, Printed by C.S. Van Winkle, New-York, (1818), p. 42.

watch them very closely. It is even rumored that one of the wealthiest of them was arrested yesterday for sending large amounts of money out of France. The Bank of France refuses to touch the paper of men suspected of extorting bullion. At the last discount day one of the firms under this suspicion sent in 600,000 francs worth of paper to be discounted. Every cent of it was returned, refused.

In Switzerland matters are still worse. The banks have suspended specie payments, and have refused to discount any notes except those of manufacturers in the neighborhood, and these only in sums sufficient to keep the manufactories running. The interdiction to export gold from France presses with a heavy weight upon Switzerland. It and the banks' refusal to discount have forced all commercial firms in Switzerland to suspend payments. Men whose books show them to be worth millions are compelled to suspend payments, because none of their assets are available. It is impossible to get a bill on Paris cashed anywhere, and all but impossible to get a bill on London cashed. Travellers are advised by bankers here to take with them gold enough to pay their expenses.

The outflow of gold from France continues to be enormous, despite all the measures taken. This necessarily so. Last week the French Government was obliged to send $15,000 in gold to Spain to pay for the wheat, wine, oil, brandy, etc., bought there by the government agents."

In 1890, the Marquis de Mores and other Frenchman alleged that Jews had taken over France. There allegations were met with the threat that the Jews controlled the money markets of Europe, and had enormous influence in America, and that those who stood against the Jews, especially noblemen, would face the Guillotine—as they had in the French Revolution. *The Chicago Daily Tribune* reported on 4 February 1890 on page 5,

"DE MORES ON THE JEWS.

AIMS OF THE LEADERS OF THE ANTI-SEMITIC MOVEMENT IN FRANCE.

It Is Claimed That the Country Is Really Governed by the Jews, Who Find No Difficulty in Getting the Department Officials and the Legislators in Their Power—The London 'Times' Pays Parnell $25,000 and His Libel Suit Is Withdrawn—General Foreign News.
SPECIAL CABLE DISPATCH TO THE TRIBUNE.
(Copyright, 1890, by Jaqmes Gordon Bennett.)
PARIS, Feb. 3.—The Marquis de Mores, who fought a duel with Camille Dreyfus yesterday, is one of the recognized leaders of the anti-Semitic party in France, which is actively working against the Jews. The Marquis, when interviewed by a *Herald* correspondent, said: 'Foolish rumors are being circulated to the effect that we are attempting to drive the Jews from France. This is the most utter nonsense. We have no objection to Jews because they

are Jews—in fact, we regard them as useful and necessary members of society so long as they remain in their proper place—but we object most decidedly to their monopolizing the entire country. We object to a state of things which permits a sect only a few thousand strong to govern a nation which numbers millions. We are not stirring up an agitation with a view to depriving the Jews of any of their rights, but of securing French people in rights which the Jews have succeeded in swindling them out of. When I say that this nation is governed by Jews I speak advisedly. It is true that we are living under a régime which we call republican, but unfortunately we are a republic in name only. In all its machinery the administration of today has retained the policy of centralization and redtapeism just as it existed when we were ruled by Emperors and Kings. The only difference is that the country has lost all the advantages of stability and responsibility which she used to enjoy, and has in exchange gained none of the benefits of a real, enlightened democracy. Local self-government, as understood in America, is unknown to us. We write the word 'liberty' in large letters on our public buildings, and then in our private lives continue to submit to oppression just as if the great revolution had never occurred. The Anglo-Saxon, it has often been said, will never fight except for something tangible, but we Frenchmen will tear down the heavens for an idea only. The trouble is that, having once established the external truth of our ideas, we never dream of putting them into practice; we are content to lay down to the world great principles of action which must lead to prosperity. The world acts upon them quickly, while we, vain theorists, accept the empty shadow for the reality. That is the way it has been with our Republic, and that is how the Jews who are not at all theorists, but shrewd, farseeing schemers, have by getting hold of all centralized power been able to wield an influence in the management of national affairs none the less absolute for being exercised in secret.

A GOVERNMENT BY CLERKS.

'Theoretically the French people govern themselves; practically they are governed by a certain number of clerks and under secretaries in the bureaus of the Paris Ministers. These men are paid only a few hundred francs per month, and can consequently be tempted by a few hundred francs over and above their meager salaries. The Ministers and nominal heads of the departments may or may not be honest men, they may or may not be ignorant of what is going on among their subordinates, but even if they are so disposed they can do little to remedy the evil. With the present kaleidoscope system Ministers succeed Ministers so rapidly that they have neither time nor inducement to learn to discharge the duties of the office. The result is that clerks and secretaries who have held their positions long enough to understand the work are left to transact the business of the country, and these young gentlemen or old gentlemen, as the case may be, hardly able to live on their official salaries, manage to live comfortably on the supplemental salaries paid them by the Jews.

'It is needless to add that the Jews do not pay these salaries for nothing. Not only is bribery carried on throughout the various executive departments of the State, but far from uncommonly in the Chamber itself. Last year the salaries of not less than 180 deputies were attached for debt. I mention this to show how welcome a few 1,000-franc notes would be to a debtor thus embarrassed. In such cases 1,000-franc notes are not always forthcoming from the Jews, but always for a consideration. The result is that all serious legislation for the real interests of the people is impossible. No great reform can pass the Chamber, although the people are clamoring for reform. No great abuse can be done away with, although the people are groaning under numberless abuses, for it must be borne in mind that whenever there is a popular abuse there is money to be wrung from the people; hence the Jews believe in popular abuses and fight against reform. And such is the insidious influence of the Rothschilds and their followers and such the perfection of their organization throughout France that the real voice of the people is not heard even at the general elections.

JEWS KEPT BOULANGER OUT.

'The pressure of immense sums of money, used as the Jews know how to use it, is simply incalculable. There is not the slightest doubt that but for this hostile influence last September the dissatisfied elements led by Boulanger would have swept the country at these elections. It would be hard to say how many millions of francs have been furnished M. Constans, Minister of the Interior, from the Jewish coffers. Thus for the time they have stifled the voice of universal discontent, but it will not be stifled forever. A storm is brewing and will burst ere long. Boulangism, even without Boulanger, is today stronger than ever, for Boulangism never meant anything but discontent, and every week and every day gives France new causes for being discontented.

'And not only have the Jews been able to prevent all legislation tending toward reform and toward bettering the people's lot, but they have paralyzed the industrial activity of the country by a long series of financial swindles, which in the end always result in taking the people's savings in exchange for more of less valueless bonds and shares. They have succeeded in obtaining the absolute control of the Bank of France and all our great institutions of credit, and they can at will refuse or grant a needed loan. If a man wants to raise 2,000,000 francs for any enterprise he is absolutely at the mercy of the Jews, and if the enterprise is not big enough to suit them they refuse to bother about it.

'Now, what we demand is that this financial tyranny shall cease; that the workingman shall be able to get in his purchases something like the value of his money; that the consumer shall be allowed to deal directly with the producer, thus saving the middleman's or Jew's enormous profits; that the Government shall grant credit to workingmen's societies, organized on a socialistic and coöperative basis; in short, that the Jew be forced to attend to his own business and allow other people to attend to theirs.

'THE PAST, PRESENT AND THE FUTURE.'

'It is curious to study the causes that will bring a crisis to a head. The Jews, after taking the past in the shape of savings and the future in the shape of loans, in their insatiable greed have laid their hands on the present. They have now touched the daily life of the people, and this will bring the crisis. Speaking of the meat question in Paris, German dressed mutton, under cover of existing treaties and tariffs, is flooding Paris, and all the men who used to live from work in the slaughter-houses, tanneries, etc., are idle and hungry, and, under existing circumstances, nothing can be done for them. From that savage quarter, when hungry, will start the bolt, as these people have a right to live and will not allow men rolling in easily-gotten millions to regulate their appetites. Out object is to execute social reforms, and we begin our social experiments in Paris. Our ideas are that when individual enterprise has created a national monopoly the duty of society is to step in, indemnify creative genius, and give the benefit of the instrument to all and not to one. If in Paris we arrive at a majority and execute some reforms the country, now sick of talk, will follow us farther and we will be able to force other reforms. But remember we are not fanatics; we only want to pull off the masks and give every man his due.'

AN EDITORIAL OPINION.

The Paris *Herald* prints the following editorial on the above interview:

We are agreeably surprised with our interview with the Marquis de Mores on the anti-Jewish question. It has made a deep impression in political circles and is an important contribution to a grand question which is of current interest. The Marquis was moderate, conservative, with sound ideas on the evils of bureaucracy in France—evils which we fear are almost inseparable from a shifting and evanescent republican government everywhere. It has had a rank growth in the United States for generations.

As to the crusade against the Jews, the Marquis should remember that he who sows the wind shall reap the whirlwind. Among the facts of modern civilization which must be accepted is that the Jews control the money markets of Europe and have a vast influence in America. If they were not possessed of great ability it would be otherwise. In the struggle for wealth more, perhaps, than in any other the fittest survives. Is it wise for the Marquis and noblemen like him, under pretense of an anti-Semitic crusade, to excite the mob against the rich, to teach poverty and crime to war upon property? A crusade against 100,000,000 francs means very soon a crusade against 1,000 francs. It is Belleville against the rentes, crime against thrift.

The Marquis should not forget the sinister lessons of the revolution. It was noblemen of his class—Mirabeau, Talleyrand, Lafayette, Egalite d'Orleans—who sharpened the pike which the mob drove home to the heart of France. Even Princes of the blood dallied with the fashionable movement until it was too late, as, in the Arabian tale, the spirit they summoned from the bottle became a demon that swept the earth.

Noblemen like the Marquis were to blame for the revolution of 1789, as many of the highest nobles in France are to blame for the Boulanger revolution of 1889, which, but for the steadier nerve of the French people, might have come to issues as grave as those of the 'Terror.'

The cry against the Jews is a cry against the rich, the outcast against

Dives. The Marquis and his noble friends are playing with fire as their ancestors did before them. Fight bureaucracy to the end—that is all right and it will be a good campaign—but let the Jews alone. Avoid all mad, eccentric politics, like Knownothingism and anti-Masonry. Especially let France remember that the nineteenth century came in under the shadows of the guillotine, and not invoke that appalling specter for its closing years."

The Roman Catholic Church was suspicious that Jewish liberals; who trumpeted the ideals of the French Revolution of liberty, equality and fraternity; tended toward the atheism, or the paganism, that attended the French Revolution. Reformed Judaism and reformed Catholicism in the form of Protestantism were merging. There was obvious collusion between Jewish liberalism in the press and the *Kulturkampf* against Catholicism.[208] A similar set of circumstances occurred in Vienna, where Karl Lueger eventually became Mayor of the city—Vienna had suffered a stock market crash in 1873 on "Black Friday", which had been caused by corrupt Jews, and Jewish firms such as the House of Rothschild openly profiteered from the calamity.[209] *The Catholic Encyclopedia*, Volume 9, Robert Appleton, New York, (1910), wrote,

> "Lueger, KARL, burgomaster of Vienna, Austrian political leader and municipal reformer, b. at Vienna, 24 October, 1844; d. there, 10 March, 1910. His father, a custodian in the Institute of Technology in Vienna, was of a peasant family of Neustadtl in Lower Austria, his mother, the daughter of a Viennese cabinet maker. After completing the elementary schools, in 1854 he entered the Theresianum,Vienna, from which he passed in 1862 to the University of Vienna, enrolling in the faculty of law, taking his degree four years later. After serving his legal apprenticeship from 1866 to 1874, he opened an office of his own and soon attained high rank in his profession by his sure and quick judgment, his exceptionally thorough legal knowledge, and his cleverness and eloquence in handling cases before the court. His generosity in giving his services gratuitously to poor clients, who flocked to him in great numbers, was remarkable, and may account largely for the fact that, although he practised law until 1896, he never became a wealthy man.
>
> In 1872, having decided upon a political career, he joined an independent Liberal political organization, the Citizens' Club of the Landstrasse, one of the districts, or wards, of Vienna. Liberalism, which had guided Austria from aristocracy to democracy in government, was at this period the one political creed the profession of which offered any prospect of success in practical politics. But Liberalism had come to mean economic advancement for the capitalist at the cost of the small tradesman, the capitalist being usually a Jew. The result was an appalling material moral

---

208. P. W. Massing, *Rehearsal for Destruction: A Study of Political Anti-Semitism in Imperial Germany*, Howard Fertig, New York, (1967), pp. 278-279, 296, 304, 314-315.
209. "Vienna's 'Black Friday'", *The Chicago Daily Tribune*, (2 June 1873), front page.

degradation and a regime of political corruption focussed at Vienna, which city in the seventies of the last century was the most backward capital in Europe, enormously overtaxed, and with a population sunk in a lazy indifference, political, economic, and religious. The Jewish Liberalism ruled supreme in city and country; public opinion was moulded by a press almost entirely Jewish and anti-clerical; Catholic dogmas and practices were ridiculed; priests and religious insulted in the streets. In 1875 Lueger was elected to the Vienna city council for one year. Re-elected in 1876 for a full term of three years, he resigned his seat in consequence of the exposure of corruption in the city administration. Having now become the leader of the anti-corruptionist movement, he was again elected councillor in 1878 as an independent candidate, and threw himself heart and soul into the battle for purity in the municipal government.

In 1882 Lueger's party, called the Democratic was joined by the Reform and by the German National organizations, the three uniting under the name Anti-Semitic party. In 1885 Lueger associated himself with Baron Vogelsang, the eminent social-political worker, whose influence and principles had great weight in the formation of the future Christian Socialists. The year 1885 witnessed, too, Lueger's election to the Reichsrat, where, although the only member of his party in the house, he quickly assumed a leading position. He made a memorable attack on the dual settlement between Austria and Hungary, and against what he bitterly called 'Judeo-Magyarism' on the occasion of the Ausgleich between Austria and Hungary in 1886. A renewal of this attack in 1891 almost caused him to be hounded from the house. At his death there were few members of the Austrian Reichsrat who did not share his views. In 1890 Lueger had been elected to the Lower Austrian Landtag; here again he became the guiding spirit in the struggle against Liberalism and corruption. In municipal, state, and national politics he was now the leader of the Anti-Semitic and Anti-Liberal party, the back-bone of which was the union of Christians called variously the Christian Socialist Union and, in Vienna especially, the United Christians, This union developed later into the present (1910) dominant party in Austria, the Christian Socialists. In 1895 the United Christians were strong enough to elect Lueger burgomaster of Vienna, but his majority in the council was too small to be effective and he would not accept. His party returning after the September elections with an increased majority, Lueger was once more elected burgomaster, but Liberal influence prevented his confirmation by the emperor. The council stubbornly reelected him and was dissolved. In 1896 he was again chosen. Not, however, until the brilliant victory of his party, now definitely called the Christian Socialist party, in the Reichsrat elections in 1897, when he was for the fifth time chosen burgomaster, did the emperor confirm the choice.

Lueger's subsequent activity was devoted to moulding and guiding the policy of the Christian Socialist party and to the re-creation of Vienna, of which he remained burgomaster until his death, his re-election occurring in 1903 and 1909. The political ideal of the Christian Socialists is a German-

Slav-Magyar state under the Habsburg dynasty, federal in plan, Catholic in religion but justly tolerant of other beliefs, with the industrial and economic advancement of all the people as an enduring political basis. The triumph of the party has conditioned an ever-increasing revival of Catholic religious life and organization of every kind. Under Lueger's administration Vienna was transformed. Nearly trebled in size, it became, in perfection of municipal organization and in success of municipal ownership, a model to the world, in beauty it is now unsurpassed by any European capital. A born leader of the people, Lueger joined to a captivating exterior a fiery eloquence tempered by a real Viennese wit, great organizing power, unsullied loyalty to the Habsburg dynasty, and unimpeachable integrity. Among all classes his influence and popularity were unbounded. A beautiful characteristic was his tender love of his mother; he was himself in turn idolized by children, He was anti-Semitic only because Semitism in Austria was politically synonymous with political corruption and oppressive capitalism. Lueger never married. A fearless outspoken Catholic, the defence of Catholic rights was ever in the forefront of his programme. His cheerfulness, resignation, and piety throughout his last illness edified the nation. His funeral was the most imposing ever accorded in Vienna to anyone not a royal personage."

Hermann Bielohlawek vented his rage about the alleged defamations of the "Viennese Jewish press beasts" against Lueger, and the alleged "muzzling and terrorism" of the Social Democrats who prevented fair and open debate, before the Vienna City Council in 1902.[210]

Germans who were opposed to Jewish tribalism sometimes called on the Jews to assimilate. Martin Luther resented Jews for not accepting Christ as their Messiah and initially strove to convert them. In more modern times, Heinrich von Treitschke demanded that Jews assimilate. Assimilation was often coerced by laws which allowed only converted Jews to become university professors, etc.

The London Times reported on 15 August 1914 on page 5,

"AMERICA AND THE WAR.

GERMAN BAIT REFUSED.

ENTIRE SYMPATHY WITH THE ALLIES.

BRITISH UNITY ADMIRED.

---

[210]. H. Bielohlawek, "Yes, We Want to Annihilate the Jews!" in R. S. Levy, *Antisemitism in the Modern World: An Anthology of Texts*, D. C. Heath and Company, Toronto, (1991), pp. 115-120.

(FROM OUR OWN CORRESPONDENT.)

WASHINGTON, AUG. 14.

The outcome of the impending battle is awaited here with intense anxiety. As days pass the realization grows that the conflict is unlike the Balkan war, which was regarded primarily as a spectacle. This concerns the United States almost as closely as it does the belligerents, and people are learning that there is no place in the twentieth century world for the isolated United States of Washington or Monroe. That undoubtedly is one reason for what a German apologist calls the 'amazing volume of anti-Teutonic prejudice' displayed by the American Press. A perusal of the majority of the leading newspapers of the United States fails to reveal anything except sympathy in varying degrees with us and our Allies.

Disapprobation was registered at the initial Austrian attack upon Servia, and still more at the way in which Germany took up arms. The treatment of Belgium especially seems to have awakened Americans to the real significance of German policy.

Regarding England's course there is only one view, and that was weightily expressed in Admiral Mahan's recent statement. Honour and expediency alike are deemed to have demanded our participation.

Thus in a few days the obvious effects of the Kaiser's sedulous missionary work from Prince Henry's visit downwards have been obliterated. His manifestations of friendship are forgotten, and only the sabre-rattling and epigrams of the 'War Lord' are remembered. Hence there is a marked tendency to saddle on the Kaiser the responsibility for the cataclysm.

BRITAIN UNITED.

Germany, too, has suffered—unjustly to a great extent, or so level-headed people here are inclined to believe—by stories that have been published of her treatment of defenceless belligerents and stranded Americans suspected of espionage, and even of outrages upon officials of other countries. An impression of hysterical ruthlessness has been spread.

And if Germany has suffered, we have scored. The evident effectiveness of our military preparations, the wholehearted cooperation of the Government and Opposition, Mr. Redmond's great speech, the reconciliation of Lord Charles Beresford and Mr. Churchill, the suffragist truce, and the general coolness of the public have all been reported with a wealth of approving detail. There is no longer talk of the decay of British statesmanship and nerve. The crisis, it is proclaimed, has been met in a way whereof every Anglo-Saxon should be proud. The war has brought the American people closer to us than any amount of exhibitions or 'hands-across-the-sea' celebrations could have done.

The question prompts itself: Will there be reaction? Barring accident, it seems impossible that there should be. Yet there are factors which cannot be overlooked. German-Americans, especially Jewish-German-Americans, are active, and their influence is not to be despised. Of this I am convinced by recent investigations. There are signs of pro-German activity in high financial circles. The newspapers of New York, Boston, and even Chicago, are by no means immune from that kind of suasion which business sometimes tries to apply to journalism. While it is doubtful whether any independent newspaper will yield more to such influences that to emphasize obvious facts—such as that the bulk of such war news as we get comes from Anglo-French sources, and is, therefore, not uncoloured—the existence of propaganda should not be overlooked.

GERMAN JEWISH PRESS.

The German-American-Jewish Press is also active. The *Wahrheit* and the *Tageblatt*, the two chief German-Jewish organs, inveigh against our helping Russia and Slavs. The *Wahrheit* even says that Germany, Austria, and Italy are the only European countries not openly antagonistic to Jews. The German-American Press and German-American societies, led by the excellent New York *Staats-Zeitung*, similarly hammer away in defence of the Fatherland, helped by a widely-scattered band of German or Germanophil professors, of whom Professor Hugo Munsterberg,[211] of Harvard, is the most important, and by a new English weekly just started in New York in the interests of a true understanding of Germany's position. It is doubtful, however, whether even German-Americans are solid. The *Staats-Zeitung* to-day proclaims that they are, but the statement is contradicted by my experience of the big Eastern cities. A good many thoughtful and influential German-Americans seem to make no secret of their disapproval of what one of them called the 'militaristic madness of the Kaiser.' I even heard talk of the probability of a German Republic should Germany be beaten.

Among Irish-Americans there is the same division of opinion. While Mr. Redmond has many followers, there are some extremists, represented in New York, for instance, by the Irish National Volunteer Organization, who deem him a traitor. But a discussion of this subject is premature until this week's Irish-American newspapers are available."

Zionist leader Israel Zangwill responded to J. E. C. Bodley in *The London Times* of 28 August 1914 on page 5,

"JEWS AND THE WAR.
TO THE EDITOR OF THE TIMES.

Sir,—Mr. Bodley labels as 'anti-British theories' views which I hold in common with many distinguished Englishmen, and by a further jugglery suggests that 'anti-British' and 'Jewish' are synonymous. These dialectical methods only need pointing out.

But I welcome his letter, for it enables me to correct a slip of the pen in my own. 'German Balkan-speaking provinces' should, of course, have been 'Baltic German-speaking provinces.' Because I drew attention to the fate of the Jews in those provinces, Mr. Bodley accuses me of putting Russo-Jewish

---

[211]. H. Münsterberg, *The War and America*, D. Appleton, New York, London, (1914).

interests before those of my own native land. But since the Russian Jews are England's allies in this war—some 200,000 of them fighting on our side—why should a mention of their interests expose me to Mr. Bodley's labels? Rather does his indifference to the interests of an oppressed race seem to me 'anti-British.' If England has lost the Palmerston tradition, it has been because of 'the German peril.' Once relieved from that nightmare, England would indeed cease to be 'the England of our dreams' if she continued callous to those great civilized ideals which she has so often served and not infrequently initiated.

As to the argument about newspapers into which the Chief Rabbi has been betrayed, a newspaper is not Jewish because it is owned by a frequently anti-Semitic Jew, and there is no real Jewish newspaper in the world (except naturally the German) which is not wholeheartedly on the side of England and against Germany. There is, indeed, no country so beloved by the Jews as England (has not even Zionism placed its legalized centre in London?). And for Mr. Bodley to say I talk lightly of 'the German peril' comes as 'the most unkindest cut of all' to the author of *The War God*, which, through the mouth of Sir Herbert Tree at His Majesty's Theatre, gave German *Militarismus* the warning which I hope will yet prove prophetic:—

> Why squat here spinning crafty labyrinths,
> Getting your filthy network o'er the globe?
> You think to bind the future? Poor grey spinner!
> Fate, the blind housewife, with her busy broom
> Shall shrivel at one sweep your giant web
> And leave a little naked scuttling spider!"

The venom Zangwill directed at Germany, and the Zionists' move from Berlin, where they were well treated and were more prosperous than any other Jews on earth, to London, prompted many Germans to suspect that the Zionists had cut, or sought to cut, a deal with the British and the Russians to bring America into the war on the Allies' side and against the Germans and Turks, in exchange for a planned Zionist takeover of Palestine, which would be free from German, or Turkish, oversight. The Jewish population represented a scant percentage of the total population of Palestine, and the Turks would have been more sympathetic to the rights of Moslems than would the British. Many Germans believed that the sudden shift among German-Jewish newspapers to an anti-German stance from their decidedly pro-German posture demonstrated the collusion of the Zionists and the Allies against the Germans. In *The London Times* of 28 September 1914 on page 9, Zangwill wrote,

"THE KAISER'S AMERICAN AGENTS.

TO THE EDITOR OF THE TIMES.

Sir,—Your American Correspondent's article on the failure of the German Press campaign will give pleasure to English Jews, not only as patriots, but because the suggestion that this campaign was largely one of Jewish journals seems to have vanished. Indeed, the *Wahrheit*, the German-Jewish paper with the largest circulation, which has hitherto been represented as playing a peculiarly malign part, astonished me by sending me a lengthy editorial entitled 'Zangwill and the War,' declaring:—

Although we know the majority of our readers are German or pro-German, we are convinced, exactly as Zangwill is convinced, that there could be no greater misfortune for humanity than a victory for the German arms. [It goes on] And even were we convinced that the momentary interest of the Jews is with Germany and not with the Allies we would—and should—be ready, exactly as Zangwill teaches, to sacrifice the momentary interest of the Jewish people in the name of the general culture and civilization of all humanity.

I should add that, since receiving Sir Edward Grey's assurance that England's sympathies lay with the emancipation of the Russian Jews, I have had a number of applications from Jews—Rumanian and English, as well as Russian Jews outside Russia—anxious to enlist in the Jewish Territorial Organization under the idea it is a branch of the British Army! It would certainly be easy to form a foreign legion of Jews grateful for Britain's sympathy—apart from the thousands in our Regular Forces, whose names are being published in the *Jewish Chronicle*. The only pity is that the Tsar does not at once remove Jewish disabilities as a concession to his British Ally, not to mention the strengthening of his own position. But in justice to his Imperial Majesty it must be said that he has as yet made no promise whatever, and that therefore the doubts thrown upon his honour by the entire Jewish Press of America are without foundation."

Before the war began, a Jewish leader in Berlin, Dr. Paul Nathan, had warned the world of the inhumanity of political Zionists in Palestine—who employed terrorist tactics some Germans later came to associate with Zionists and Bolsheviks in general and which later led some Germans to believe that the Zionists had instigated the First World War through terrorist and propagandist tactics, and had made it impossible for Germany to win that war. *The New York Times* reported on 18 January 1914 in Section 3 on page 3:

"SAYS THE ZIONISTS DISTURB PALESTINE

Dr. Paul Nathan of Berlin
Asserts Jewish Cause Is
Imperilled.

TERRORISM, RUSSIAN STYLE

Statements Exaggerated, New York
Jews Say—Dispute Over

Question of Language.

Special Cable to THE NEW YORK TIMES.

BERLIN, Jan. 17.—Grave charges against what he terms the 'arrogant Zionist activity' in Palestine, with which prominent men in New York are identified, are preferred by Dr. Paul Nathan, a well-known Jewish leader in Berlin. Dr. Nathan has just returned from the Holy Land, where he went on a trip of investigation on behalf of the German Jewish National Relief Association, through which the Jewish philanthropists in America and Europe operate.

In a pamphlet issued to-day Dr. Nathan accuses the Zionist elements in Palestine of stirring up discord, even among the Mohammedan and Christian populations, to such an extent that the entire cause is imperiled. Allegations based on documentary evidence are made that alleged Zionists are carrying on a campaign of terrorism modeled almost on Russian pogrom lines. Their hostility is directed mainly under the auspices of the German Association and is said to spring from failure to realize their desire to establish a great technical college at Haifa, where wealthy American Jews already have endowed various institutions.

Dr. Nathan declares that attempts have been made to blow up some of the German schools, and that the Zionists have not shrunk even from organizing riots. Matters recently reached such a pass that the Mohammedan Governor of Jerusalem was compelled to issue a public warning against further disturbances of the peace. Only strong resistance on the part of the religious elements in the country resulted in effecting a partial restoration of order.

Dr. Nathan says he desires to raise his voice against the 'overwrought Jewish nationalist chauvinism.' As a friend of Zionist works he appeals to their supporters throughout the world to suppress the 'officious intriguing elements at work in Palestine,' which threaten Judaism's interests with incalculable and irreparable harm. [...]"

The article continues and Louis Marshall and others denied Nathan's charges by shouting them down and tried to change the subject to the issue of which language should be spoken in Palestinian schools, which issues were discussed in later articles—the inhumanity of the Zionists having conveniently disappeared from the debate.[212] Marshall said, "No responsible Zionists there have been guilty of the acts charged in the cablegram[.] It's all nonsense." But Marshall did not say that the acts were not committed nor did he deny the fact that they were committed by political Zionists, who were, by definition, irresponsible for having committed them.

Political Zionist Israel Zangwill tried to turn all Jews against Germany. *The*

---

[212]. "Zionists Answer Nathan", *The New York Times*, (20 January 1914), p. 3. "Nathan Answeres Haifa Criticism", *The New York Times*, (1 February 1914), Section 3, p. 3.

New York Times reported on 10 September 1914 on the front page,

"ZANGWILL URGES JEWS TO SUPPORT ALLIES

Has Sir Edward Grey's Assurance That He Will Seek Emancipation of Russian Jews.

Special Cable to THE NEW YORK TIMES.

LONDON, Sept. 9.—Israel Zangwill has sent to The Standard an appeal to the Jews of neutral countries to support the Allies against Germany. Mr. Zangwill appeals especially to the Jews of America. He says:

'Although the most monstrous war in human history was 'made in Germany,' and although Germany's behavior in the war is as barbarous as her temper in peace, I note with regret that a certain section of Jewry in America and other neutral countries seems to withhold sympathy from Great Britain and her Allies.

'In so far as these Jews are German born their feeling for Germany is as intelligible as is mine for England, but in so far as they are swayed by consideration for the interests of the Russian Jews, to whom Germany and Austria are offering equal rights, let me tell them that it is better for the Jewish minority to continue to suffer, and that I would far sooner lose my own rights as an English citizen than that the great interests of civilization should be submerged by the triumph of Prussian militarism.

'And in saying this I speak not as a British patriot, but as a world patriot, dismayed and disgusted by the inhuman ideal of the Gothic superman. I am well aware that Germany's press agent paints Germany as the guardian of civilization and as an angel fighting desperately against hordes of savages imported from Africa and Asia, but if we are using black forces it is for white purposes, while she is using white forces for black purposes.

'But it is not even certain that the Jews of Russia will continue to suffer once England is relieved from this Teutonic nightmare. Assurances I have been privileged to obtain from Sir Edward Grey that he would neglect no opportunity of encouraging the emancipation of the Russian Jews mark a turning point in their history, replacing, as it does, windy Russia rumors by a solid political basis of hope. Nor is this a mere utterance of a politician in a crisis.

'I am in a position to state that it represents the attitude of all that is best in English thought. It is with confidence therefore that I appeal to American and other 'neutral' Jews not to let the shadow of Russia alienate their sympathies from this indomitable island, which now, as not seldom before, is fighting for mankind and which may yet civilize Russia and Germany.'"

The American Hebrew countered Zangwill's British intrusion into American internal politics in the name of the Jews. *The New York Times* reported on 15 September 1914,

"OPPOSE ZANGWILL'S APPEAL

It Lacks a Jewish Point of View,
American Hebrew Says.

The appeal of Israel Zangwill for Jewish support of the cause is criticised adversely in The American Hebrew, which says, editorially,

'We regret the intrusion of Israel Zangwill's appeal to American Jews at this time as an advocate of the cause of England. As an Englishman, Mr. Zangwill is within his rights—and it would be his patriotic duty—to support the cause of his Government. But his advice to American Jews lacks a Jewish point of view, notwithstanding his assurance that the English Prime Minister has experienced a change of heart regarding the Jewish question in Russia.

'We remember too well Sir Edward Grey's attitude when approached by English Jews on the Russian passport question. He was reluctant then to interfere in Russia's internal affairs. With Russia victorious on the field, will his British prudence allow him to overcome this reluctance? Mr. Zangwill as a Jewish leader is in the wrong galley as a press agent of the British Empire.'"

The New York Times wrote on 20 February 1916, on page 10,

"HOPE FOR JEWS IN RUSSIA.

Prof. Basch Quotes Letter from
Alfred Dreyfus to Show Trend.

All the Jews of the world, and, indeed, all men who are interested in the cause of humanity, says Victor Basch, Professor at the University of Paris, who is now at Columbia University, should make every effort to secure emancipation and the rights of citizenship for the Russian and Polish Jews. There are six millions of Russian Jews, asserts Professor Basch, who have entered the war with the greatest enthusiasm, and who are rewarded with nothing but renewed persecution. It is for this reason, he points out, that there is a Jewish aspect to the present war.

In France today there is, he says, no trace of antisemitism left. Whatever conditions may have been in the past, Professor Basch points out that today in France there is nothing but admiration for the Jews who have been fighting so bravely and dying for their country. A personal letter from Colonel Alfred Dreyfus, the famous prisoner of the Isle du Diable, to Professor Basch voices these sentiments:

* * * France is the country which first proclaimed equality for all men and put the Jews on the same plane with other citizens. It is the country which accepts and sustains all the persecuted peoples, all the martyrs. But it is also the country where justice triumphs always in spite of iniquity and the sophistry of the 'raison d'Etat.'

Need I quote my own case to you who so generously felt the same indignation as did all noble and generous minds? Accused, then condemned for an infamous crime of which I was innocent, I received finally a brilliant reparation which was the triumph of truth and right. The victory of France in the present war means the victory of right and humanity and the liberation of all the oppressed peoples. We are the champions of liberty and of civilization. The defeat of France would be the defeat of civilization. The martyrdom of Belgium, the crushing of the Serbs, and the extermination of the Armenians are but a foretaste of what a Germanized world would be. France, on the contrary, would realize a Europe where would reign greater justice, greater kindness, and greater humanity.

Jan. 29, 1916. ALFRED DREYFUS.

Professor Basch showed in the course of his remarks that the Russian Jews should find in the present war a favorable opportunity to achieve emancipation. The entire liberal party in Russia, he said, was widely proclaiming that the first step toward the regeneration of their country must be the emancipation of the Jews, that only the bureaucracy which is of German origin was their oppressor and that the Russian peasant had few racial or religious prejudices.

Antisemitism, he declared, was born in Germany, and came from Germany to Russia. Consequently he thinks a bill of complete Jewish emancipation, social as well as legal, is possible in Russia, more possible than in Germany, and more possible today than ever before."

This article published the claim that anti-Semitism had completely disappeared in France, which claim was not only false, it was absurd. The wartime propagandists brought forth Alfred Dreyfus to play upon the emotions of the Jewish community in their efforts to vilify the Germans, but it was the French, not the Germans, who had persecuted Dreyfus believing him to be an agent of the Germans, because the Jews had so often betrayed France to Germany and Catholicism to Protestantism; and it was the Jews, not the Turks, who were behind the genocide of the Armenians. Basch strangely claimed that the Russian *muzhiks* "had few racial or religious prejudices[,]" and sought to place the blame for all the hardships of the Jews in Russian controlled lands on the Germans—who were fighting against the Russians and who had made great advances in emancipating the Jews. Conventional wisdom molded by Jewish propagandists held that Germans in and around the Russian Royal family had brought anti-Semitism to Russia. The fact is that Jews had deliberately segregated themselves for centuries and had encouraged the Czars to mild persecutions so as to keep the Jews segregated and promote Jewish emigration to Palestine, Germany, England and America. By far the largest concentration of Jews in the world was to be found in "Russia", though Sephardic Jews considered these people to be converted Khazars and not real Jews, not the "chosen people".

Anti-Semitism was not created in Germany and Germany had done far more for the interests of the Jews than had Russia or France, which is to say Germany provided Jews with an environment in which they could thrive and do more for themselves and for humanity. If the true goal of the Zionists were the emancipation of Russian Jews, a most noble and necessary pursuit that promised to spare millions their misery, the logical choice would have been to have sided with Germany against Russia in the First World War, though that might not have achieved the political Zionists' goals of ensuring that Zionism would succeed in the creation of an autonomous state free from German or Turkish oversight whichever side won the war. The words "civilized" and "civilization" were, understandably, code words for states in which Jews enjoyed equal rights with the rest of humanity, and the French Revolution had emancipated Jews in France. The false messages Basch and Dreyfus expressed above were that German victory meant Jewish oppression and French victory meant complete Jewish emancipation.

Germany was working hard to secure the liberty of Russian Jews and was at war with the Russian Czarist regime that allegedly oppressed Jews. The article itself points out that the Jews fighting for Russia were rewarded only with renewed persecution—perhaps at the instigation of the Zionists who feared that the emancipation of the Jews without a national homeland would lead to assimilation and the death of the race. The ardent political Zionist Israel Zangwill voiced this concern even before the First World War had begun,

> "But the abolition of the Pale [of Settlement] and the introduction of Jewish equality will be the deadliest blow ever aimed at Jewish nationality."[213]

The political Zionist Theodor Herzl conspired with the Turks to cover up the persecution of the Armenians caused by the Jews. France was not just, nor kind, nor humane, nor did it free all oppressed people, at war's end. In fact, tragically, France's injustice and inhumanity to Germany created an environment where Nazism could flourish. France was also the nation which most strongly opposed the British takeover of all of Palestine at war's end and thus placed an obstacle in the way of the Zionists.

Yet more alarming sories than the involvement of Zionists in bringing America into the war emerged after the German loss. Adolf Hitler claimed in 1923 that the Bolshevists, with their alleged control of the press, instigated World War I so that the German and Russian autocracies would weaken each other in their fight against one another, which would provide an opportunity for revolutionary Jews (there was a Jewish Bolshevik revolution in Bavaria, Germany, in 1918, and a series of Jewish revolutionary attempts took place in Russia, finally succeeding in the Jewish Bolshevik Revolution in 1917) to overthrow the monarchies and then fully emancipate the Jews of Russia and

---

[213]. I. Zangwill, *The Problem of the Jewish Race*, Judaen Publishing Company, New York, (1914), p. 19; which was first published as an article, "The Jewish Race", *The Independent*, Volume 71, Number 3271, (10 August 1911), pp. 288-295, at 294.

Germany, as the Jews had been emancipated by the French Revolution.[214] Some philo-Semites had come to similar conclusions. Karl Kautsky wrote in 1921,

> "It is not in Palestine, but in Eastern Europe, that the destinies of the suffering and oppressed portion of Jewry are being fought out. Not for a few thousand Jews, or at most a few hundred thousand, but for a population of between eight and ten million. Emigration abroad cannot help them, no matter whither it may be turned. Their destiny is intimately connected with that of the *revolution,* in *their own country.*
>
> The methods of the Bolsheviks are not those of the Western European Social-Democracy. The Bolsheviks will not be able to found a modern socialist state. What they are really establishing is a bourgeois revolution, which will assume forms corresponding to the social condition of present-day Russia, resembling in many ways the forms of the great French Revolution toward the end of the Eighteenth Century. Among its other effects, the French Revolution liberated the Jews in France, giving them full rights of citizenship. The same accomplishment will be included among the permanent achievements of the Russian Revolution for all of Eastern Europe, unless the Revolution succumbs to the most savage counter-revolution. But the struggle in Eastern Europe now is not only a struggle for political freedom and for the rights of the Jews to change their domicile. The conditions are also being prepared for an enhancement of their economic situation. In addition to the emancipation of the Jews, the emancipation of the peasants also will be one of the achievements of the revolution in Eastern Europe. A more prosperous peasantry will take the place of the present impoverished peasantry, thus creating a greater internal market for urban industry. Once peace has been reestablished in Eastern Europe, industry, and with it transportation, will necessarily develop with giant strides; the urban population will find abundant employment and food, and the great mass of the Jewish population will find it possible to rise from conditions of life in which they have hardly emerged from the *lumpenproletariat,* to the conditions of the proletariat in large-scale industry, as a portion of which class they may then take part in the upward struggle of the entire class.
>
> Herein only is there a possibility for the Jewish masses to achieve a truly human status. Zionism cannot strengthen them in this effort. Zionism will weaken them at the historically decisive moment by promulgating an ambition which amounts practically to a desertion of the colours.
> [***]
> The only force capable of a thorough overturning of the present order and of a complete destruction of all oppression, of all legal and social inequality, now remains the proletariat, which must achieve this end in order to achieve its own liberation. Only a victorious proletariat can bring

---

214. A. Hitler, "The International Jew and the International Stock Exchange: Guilty of World War [1923]", in R. S. Levy, *Antisemitism in the Modern World: An Anthology of Texts,* D. C. Heath, Lexington, Massachussetts, Toronto, (1991).

complete emancipation for the Jews; all of Jewry, except in so far as it is already fettered to capitalism, is interested in a proletarian victory.
[***]
'The 'Yiddish' daily press, after having been in existence for ten years, exceeds the Polish press in circulation and in Russia is second in this respect only to the Russian press proper.' [*Footnote:* Hersch, *Le Juif*, p. 9]"[215]

Many countered such claims by pointing out that the war resulted in great suffering for Jews and that the Bolsheviks eventually persecuted Jews and specifically targeted Zionists. The Bolsheviks were in fact very good the Jews, and Bolshevik "anti-Semitism" was simply a Jewish means to preserve the "Jewish race".

The New York Times reported on 22 February 1916 on page 7,

"SEES CHANCE FOR ZIONISTS.

War Will Open Palestine to Them,
Dr. Mossinsohn Says.

The University Zionist Society held a meeting last night at 347 Amsterdam Avenue. Eugene Meyer, Jr., President of the club, presided, and the speakers were Dr. B. Z. Mossinsohn, director of the Hebrew Gymnasium at Jaffa; Dr. Leo Motzkin, head of the Larger Action Committee on Zionism and organizer of the International Bureau at Copenhagen, and Z. W. Gluskin, who was one of the pioneers in the educational and industrial development of Palestine.

Dr. Mossinsohn discussed the war as it affected affairs in Palestine, and told of the possible political combinations at the end of the conflict. In the readjustment that is coming in the Near East, he sees great opportunity for the permanent establishment of the Jews in Palestine. It is going to be desirable to develop that country, and he believes that a share in this task will fall to the Jews."

The New York Times reported on 13 November 1916 on page 13,

"TO GET RIGHTS FOR JEWS.

International Committee Suggested
to Solve Problem After War.

An International Committee of Correspondence to facilitate a world-wide demand for the settlement of the Jewish problem was proposed by Oscar S. Straus, Chairman of the Public Service Commission, at the tenth annual convention of the American Jewish Committee, held at the Hotel

---

[215]. K. Kautsky, *Are the Jews a Race?*, International Publishers, New York, (1926), pp. 213-215, 240-241, 243.

Astor yesterday. It was voted to submit Mr. Straus's proposal to the American Jewish Congress, which will be held some time before the end of the war.

In offering his suggestion Mr. Straus said that such a committee would be able when peace was discussed at the war's close to present a strong case for the Jews in countries where they are oppressed. He called attention to the good work done in this country before the Revolutionary War by the Colonial Committee of Correspondence, which was formed in Boston in 1722 and soon had branches which kept each informed of sentiment and action in the different colonies.

'There is need,' Mr. Straus said, 'of some instrumentality through which the Jews in all countries may address themselves to our common object, which, shorn of all details, is this—the securing of equal rights for Jews in countries where they are oppressed. I believe we should name such a committee here and now.'

Jacob H. Schiff opposed immediate action, and Henry M. Goldfogle moved that the proposal of Mr. Straus be referred to the Executive Committee of the American Jewish Committee for consideration of its submission to the Congress Committee, and after Mr. Straus said this would be satisfactory to him, the motion was carried. [...]"

While the political Zionists were promoting rabid nationalism and continued war, most Jews opposed the political Zionists. *The New York Times* reported on 16 January 1917 on page 3,

"Har Sinai Temple was crowded tonight at the opening religious service, the feature of which was the sermon by the Rev. Dr. David Philipson of Cincinnati. He protested against the Zionistic movement, holding that internationalism alone would enable the Jews to retain their place among the nations. This important question will be discussed in the convention and action will be taken.

'We protagonists of universalism,' said Dr. Philipson, 'are being laughed to scorn. Our claim that Israel is an international religious community is being held up to ridicule. We are told that Israel can only survive by stressing its separatistic nationalism; that only by drawing ourselves off from our fellow inhabitants in the lands in which we live as a separate nationalistic group can we perpetuate Jewish life.

'But that we will not do. We internationalists, basing our claim on what has been Israel's task in the world, taking our stand on the religious idealistic interpretation of history whereof we believe Israel presents the most striking symbol, as over against the materialistic interpretation whereof the present war, the apotheosis of nationalism, is the climax—we internationalists, despite all the frightfully distressing days through which we are passing, must hold our rudder true, feeling that the mists will disappear before a rearising sun.'"

The Russian Revolution was funded by German-Jewish financiers, who intended to free the Jews from the oppressive Pale of Settlement and pogroms and to further the cause of Jews in Palestine. They also wanted to take over the Russian Government and steal the wealth of Russia. They further sought murderous revenge and committed genocide against the Russians.

Revolution in Russia was also promoted by the German Government, in particular by Ludendorff, especially after the Balkan Wars lead to the First World War. An unstable government in Russia, or a friendly government in Russia, would profit the Germans immensely as America entered the war on the side of the Allies. Ludendorff admitted after the war that he had been duped by the Jews. After the war, Walter Rathenau secured the Rappollo Treaty, in anticipation of the Second World War. *The New York Times* reported on 28 March 1917 on page 13,

"SEES NEW LIFE FOR ZIONISM.

Leo Motzkin Says the Russian Revolution
Will Aid the Movement.

Leo Motzkin of Kieff, Russia, one of the leading Zionist publicists and the head of the international press bureau which had much to do with the acquittal of Mendel Beilis of the charge of ritual murder, is now in New York, and no one has followed recent events in Russia with greater interest than he, especially in their relation to possibilities for the Jews. Mr. Motzkin said yesterday that he was confident that the Russian revolution would mean the ultimate liberation of the Jews and unprecedented progress for the Zionist movement. But he saw many things to be done and admitted that there were still difficulties and uncertainties to be encountered.

'The Russian revolution,' said Mr. Motzkin, 'will ultimately lead to the full emancipation of the Jews in Russia, both social and national. But we cannot base too much on what we are hearing now about Jewish rights, because these rights can be established only by law, and laws cannot be made until the Constituent Assembly meets. There is no doubt, however, that the condition of the Jews in Russia was materially ameliorated in an administrative way when the temporary authorities came into power, and there is no doubt that the Constituent Assembly will grant equality to the Jews.

'There are naturally various parties among the Jews in Russia, but all agree that the present régime will give all of them equal rights. The Zionists, especially, expect the establishment of the new Government to advance their cause, for two main reasons:

'First—because the persecution of the Zionists will cease. Under the old régime the Zionist party, with other progressive parties, was persecuted and hindered. Zionism was illegal, as was evidenced by the fact that when the war began 100 Zionist cases were awaiting trial in courts. Of course, Zionism will now become legal, as will other progressive movements, and the hindrances will be removed.

'Second—With the growth of democracy and the removal of restrictions from speech and the press Zionists will be permitted to extend their propaganda and educated persons will be able to learn something of Zionism and to understand its ideal. They will learn to respect its purpose, which is simply the creation of a national cultural home for Jewish people in their ancient country. This view is based upon the fact that the present Foreign Minister of Russia has recently expressed his sympathy with the Zionist aim, and the same sentiments have been heard from other progressive statesmen in all democratic countries.'

Mr. Motzkin added that big commercial organizations in Petrograd had attempted to establish relations with similar organizations in England and America, but had been handicapped by the old régime. The fact that many members of these organizations were among the revolutionists, he said, made it certain that international business would be developed with other democratic countries."

The New York Times reported on 23 July 1917 on page 9,

"JEWISH SOCIALISTS FOR FREE PALESTINE

Appeal to Brethren Here and in Russia to Oppose Anything That Hinders Allies, Who Aid It.

A notable appeal from a Jewish Socialistic labor association exiled from Palestine to Socialist brethren in the United States and Russia to oppose any movement 'having the effect of putting in question the liberation of Palestine by the allied armies,' has reached this country through official sources. The appeal seems to align the Jewish Socialists of Asia Minor firmly on the side of the Allies and against the Turks and Germans evidently with the idea that through allied victory alone can the dream of Zionism for an independent Palestine come true.

The appeal comes from the Poale Zion, a Socialist labor organization consisting of sixty to eighty members, most of them prominent in the more advanced thought of the sections from which Turkish oppression has exiled them. They are now refugees in Egypt. They belong to the artisan class, for the most part, and are now connected with Mospruds Jewish Relief Committee in Cairo.

The text of their resolution, in which they adopt for the first time a nationalistic point of view, is as follows:

'We, the Poale Zion, who are refugee Palestinians in Egypt, beg you to communicate with our Socialistic companions in America and Russia, putting the following appeal before them:

'Considering that we find ourselves at an epoch of history in which it is our duty to put events to the best possible purpose, and considering that the allied powers have openly claimed that they are fighting for the liberation of small nationalities, and considering that the advance of the British armies toward Palestine signifies for us and for our country the inauguration of an era of independence and liberty and justice, we address you, comrades, with the appeal to redouble your vigilance in proclaiming among all of those who take part in the International Socialistic Conference that for safeguarding the interests of the Jewish masses of Palestine, oppressed in the home of its ancestors by the Turkish regime, they should with all their forces oppose any resolution having the effect of putting in question the liberation of Palestine by the allied armies.'"

The New York Times reported on 9 November 1917 on page 3,

"BRITAIN FAVORS ZIONISM.

Balfour Gives Cabinet View In a
Letter to Rothschild.

LONDON, Nov. 8.—Arthur J. Balfour, Secretary of State for Foreign Affairs, has written the following letter to Lord Rothschild expressing the Government's sympathy with the Zionist movement:

'The Government view with favor the establishment of Palestine as a national home for the Jewish people, and will use their best endeavors to facilitate the achievement of this object, it being clearly understood that nothing will be done that may prejudice the civil or religious rights of existing non-Jewish communities in Palestine.'

Mr. Balfour adds that this declaration of sympathy with the Jewish Zionist aspirations has been submitted to and approved by the Cabinet.

The Jewish Chronicle, commenting on Mr. Balfour's letter, says:

'With one step the Jewish cause has made a great bound forward. It is the perceptible lifting of the cloud of centuries; a palpable sign that the Jew—condemned for two thousand years by unparalleled wrong—is at last coming to his right. He is to be given the opportunity and means by which in place of being a hyphenation he can become a nation, in place of being a wanderer in every clime there is to be a home for him in his ancient land. The day of his exile is to be ended.'"

The New York Times reported on 12 November 1917 on page 13,

"ZIONISTS HERE SEE
TEUTON PLAN HALTED

British Victories in the Holy Land
Thwart Germany's Ambition
to Control Palestine.

HER PRESS CAMPAIGN BARED

Its Aim Was to Save Enough Eastern
Territory to Menace
the Suez Canal.

American Zionists who have been watching with interest the various military operations near the Holy Land have been tremendously relieved by the events of the last few days. The British victories at Beersheba and Gaza, forecasting the eventual occupation of Jerusalem, and the promise given last week by Mr. Balfour, in the name of the British Government, that they would 'use their best endeavors to facilitate the establishment of Palestine as a national home for the Jewish people,' have apparently spiked a German scheme for setting up in Palestine a Jewish State, nominally autonomous, but really under German control.

A statement issued yesterday by the Provisional Executive Committee for General Zionist Affairs gave a detailed account of a press campaign supporting this scheme which has been going on in Germany and Austria for some time. This is held to indicate that the German military leaders foresaw the collapse of the Berlin-to-Bagdad plan and were preparing another arrangement by which it was hoped that Germany might save from the wreck of its plans in the Near East enough to form a constant menace to the Suez Canal, Egypt, and India.

'To accomplish this purpose,' says the committee, 'Germany was evidently preparing to ride roughshod, if need be, over its present ally, should Turkey refuse to recognize that it was to her 'best interests' to fall in with the new project. To give 'punch' to its publicity campaign, Germany unearthed a conspiracy between America and the Zionist Organization, including United States Supreme Court Justice Louis D. Brandeis, Judge Julian W. Mack, head of the American Military Insurance Department; Felix Frankfurter of the War Department, as well as Lord Walter Rothschild, leader of the English Zionists, and former Ambassador Henry W. Morgenthau to seize Palestine for exploitation by the Jews, Christian missionaries, and capitalists.

'In the end, if General Allenby hadn't gotten the jump on her by striking hard and quickly, Germany would one day soon have blandly announced the establishment of a Jewish republic under its auspices and suzerainty, and in response to Turkey's protests would have pointed to the overwhelming demand of the German people, and quoted for the benefit of its ravished ally, 'Vox populi, vox Dei.'

'If it had carried out its new plan, the establishment of an autonomous Jewish State in Palestine under its overlordship, whether with the consent of the Ottoman Government or in utter disregard of Turkey's wishes, Germany would have had, in addition to the strategical advantage that this would mean for the next war,' also the satisfaction of 'beating the Allies to it.' England, France, Italy, and Russia have already made it clear that the establishment of a Jewish State in Palestine is one of their aims in this war, and in Jewish circles in America it is held that Washington's view as to the desirability of this coincides with that of the Allies.

'Some echoes of these whisperings must have reached Germany, and several of its leading publications speak harshly of these 'infamous American Zionist proposals.' Thus Die Kölnische Zeitung, published in Cologne, publishes a long screed impugning the honesty of President Wilson, and ending with these complimentary allusions to Americans in general:

> The Americans belong to that class of ?????? that have been for the last sixty years undermining the proud edifice of the Turkish Empire, and haven't stopped it yet. The Palestine action fully reveals Wilson's intentions. America has dropped its mask and shown itself in its true colors—a power that has the greatest interest from the capitalistic and religious point of view to bring Turkey under the influence of missionaries and capitalists. This is the true American humanity, which is based on the alliance of the religious men with the king of trusts. Turkey has watched this campaign with the utmost patience, and now it has received the cruelest reward. It can see now that America is not far behind the other Entente Powers in their enmity to Turkey and their plans for its destruction.

Kaiser Visits Palestine.

'For Germany to give its consent to the establishment of the Jewish nationality on its historic soil, requires a reversal of its previous attitude toward Palestine. Attempts have been made to establish German colonies in the Holy Land, and Kaiser Wilhelm has paid several visits to Palestine in order to win favor with the peoples of that country, and to encourage his subjects in their vain attempts to gain a strong footing there.

'The way was being prepared by a rather obvious campaign which began with the publication of apparently innocent scientific articles, by experts, on the near East, which discussed at great length, and with much detail, the accomplishments of the Jewish colonists and the vast possibilities of Palestine from an economic standpoint. A remarkable array of such articles, studying Palestine from every conceivable angle, has been published in over a hundred periodicals in Germany and Austria. These were followed by 'letters to the editor' and now the propaganda has attained the editorial stage.'

Among the first of these articles was one by Major Carl Frank Enders to make clear to the German people that it had better give up all hope of colonization in the Holy Land, and at the same time warn Turkey not to put any obstacles in the way of the Jewish operations there. Major Enders wrote:

> The realization of the Zionists idea means infinitely more to our economic life than those fantasies and dreams of the German people that the Near East will create for us the lost world markets. * * * It will not be politically wise for Turkey to hinder the Jewish immigration into Palestine * * * German colonization in Palestine is nothing but a dream, beyond the realm of realization, which I would advise the German

people to forego.

'The Munich Neueste Nachrichten makes the frank statement that 'Zionism has become a question of the first magnitude, and Germany and Turkey have no choice but to give it serious consideration.' Gustave von Dobeller said: 'For many years the object which our Kaiser tried to accomplish by arduous political effort has been the making of a strong Turkey. A method not to be despised would be the establishment of a strong Jewish State, under Turkish suzerainty. As the Jewish people favor republics, let them, therefore, establish a republic, which must, however, be under the protection of the Ottoman Empire. It is always a question of importance whether you or your opponent has the key of the door. The idea of establishing a Jewish State is good for that power which effects it.'

Sees No Gain to Jews.

'The Vice President of the Austrian Parliament, Professor Paul Rohrbach, whose job was that of persuading the Jews of Germany and Austria-Hungary that the political schemes of the Allies are not to be trusted, wrote: 'The national aspirations of the Jews will be listened to with more sympathy by the allies of Middle Europe than by the Entente, even though certain papers and politicians on that side have lately been promising great things to the Jews. I do not believe that, even if the Entente were victorious and Turkey dismembered so that Palestine came under the suzerainty of either England of France, the Jews would benefit by this. Jews will have nothing to gain by the imperialistic schemes of England.'

'The Frankfurter Zeitung said:

'Pan Turkish ideas have no meaning in Palestine, where practically no Turks dwell.'

'Die Reichsbote, the mouthpiece of the Junkers, is calling upon the German Government to act promptly for the establishment of a Jewish State to 'offset the American Zionist proposals.' This must be done, it insists, to counteract the Wilson intrigue and 'to prevent England from making use of these American Zionist proposals as a backdoor which will enable her to pass freely from Egypt to India. For this purpose,' it says, 'the German-Austrian Zionist plans for a Jewish settlement must be strengthened. This is the opportune moment for the Zionist movement to attain its ideal.'

'These 'American Zionist proposals' are creating a real panic in the minds of Germany. The indications are that the German Press is alluding to the Palestine Commission appointed by President Wilson last Summer, consisting of Former Ambassador Morgenthau and Felix Frankfurter of War Secretary Baker's Advisory Council. At any rate, the Deutsche Worte speaks of them as a 'graver calamity than a declaration of war by a small or even medium-sized nation would be,' and charges the enemies of Germany with 'trying to enlist in their service the Zionist movement.' But it sees through the game of the Allies. 'We know very well what Mr. Morgenthau and Lord Rothschild are doing in this behalf for America and England,' it declares, the while it admits that if 'this plan of our enemies succeeds, it will go very badly with us.'

'These editorials will suffice to indicate how Germany was making ready to 'beat the Allies to it' in Palestine. General Allenby had not beaten Germany by taking Beersheba and capturing the highway to Jerusalem. The unfurling of the Union Jack over the hills of the Holy City will signalize the end of the 'Berlin to Bagdad' dream.'"

Morgenthau later published a Zionist appeal which is consistent with the accusation: "The Future of Palestine", *The New York Times*, (12 December 1917), p. 14; and he published a racist polemic against the Germans and the Kaiser, *Ambassador Morgenthau's Story*, Doubleday, Page, Garden City, New York, (1918). He later came to oppose the Zionists. His son, Henry Junior, became an arch political Zionist. However, Morgenthau Senior published an anti-Zionist article "Zionism a Surrender, Not a Solution", *The World's Work*, Volume 42, Number 3, (July, 1921), pp. i-viii; when Chaim Weizmann and the Eastern European Jews took over the Zionist movement in America at the infamous Cleveland Convention of American Zionists in the summer of 1921.

An article entitled, "Christians and Jews Rejoice: How the British Occupation of Jerusalem Was Received in Different Circles", *The New York Times Current History of the European War*, Volume 14, (January-March, 1918), pp. 315-316, quoted from the *Kölnische Volkszeitung*,

"The associations of the word Jerusalem are so deeply rooted that the conquest of the city gives considerable kudos to the conqueror. Especially in the case of the Anglo-Saxon world stimulation of war spirit has been attained which, owing to the lack of successes in the main war theatres, would otherwise have been difficult to effect. The interests of the Jews in the Entente countries, especially of the supporters of Zionism, in the Palestine campaign has shown itself in unambiguous form.

In view of the tremendous influence which Jewish capital possesses in warfare, Entente financiers and politicians will welcome the favorable effects of the capture of Jerusalem on these powerful Israelite circles. From the military standpoint it cannot be denied that the battles which led to the capture were well prepared and cleverly planned, but regarding the war situation in the Orient as a whole there is no reason to overestimate the event. Jerusalem can, at the most, serve as a valuable base on the line of communications, but it lies too far from the really important aims of the British to give ground for anxiety. It may with good reason be expected that on a line more to the rear, more easy to defend, the Turks will call a halt to the British advance."

The New York Times reported on 14 November 1917 on page 3,

"ZIONISTS GET TEXT OF BRITAIN'S PLEDGE

Balfour's Declaration Promises
Defense of Jews' Rights in
Palestine and Elsewhere.

The declaration by Great Britain of its purpose to facilitate the effort of the Zionists to establish a national home for the Jewish people in Palestine, which was formally announced by Arthur J. Balfour, Secretary of State of Foreign Affairs, in a letter to Baron Rothschild, Vice President of the British Zionist Federation, on Nov. 3, carries with it a proviso that the establishment of a Jewish State in the Holy Land shall not in any way conflict with the rights of non-Jewish communities now existing in Palestine. It also carries pledges of Great Britain to oppose any project offered at the peace conference which might in any way impair the rights and political status enjoyed by Jews in any other country.

The Provisional Zionist Committee in this city has received from Dr. Chaim Weitzman, President of the British Zionist Committee, and Dr. Nachum Sokolow of the Inner Actions Committee a cable giving the complete text of the British proposal, which differs somewhat from the first reports published in this country. The full text of the British declaration is:

'His Majesty's Government views with favor the establishment in Palestine of a national home for the Jewish people, and will use its best endeavors to facilitate the achievement of this object, it being clearly understood that nothing shall be done which may prejudice the civil and religious rites of existing non-Jewish communities in Palestine or the rights and political status enjoyed by Jews in any other country.'"

The Armenian Christians had for a long time been persecuted by the Jews through the Turks. The Young Turks, led by crypto-Jews[216] who carried out a revolution against the Sultan which had been planned for centuries by the *Dönmeh* Jews, and who pretended to be Moslem, slaughtered the Armenians. The Jews committed the Armenian genocide. The Armenian people were largely blind to the fact that it was the Zionists who had caused the persecutions. Their well paid leaders, who worked for the Zionists, betrayed them. *The New York Times* reported on 19 November 1917 on page 5, giving evidence of the cooperation of the Armenian leadership with the Zionists (Freedman stated that, "James A. Malcolm was an Oxford-educated Armenian"), in spite of the fact that Zionist Theodor Herzl had secretly conspired with the Sultan of Turkey to cover up the persecution of Armenians, and the Young Turks under crypto-Jewish leadership mass murdered them,

---

216. I. Zangwill, *The Problem of the Jewish Race*, Judaen Publishing Company, New York, (1914), pp. 9, 11; which was first published as an article, "The Jewish Race", *The Independent*, Volume 71, Number 3271, (10 August 1911), pp. 288-295, at 290-291. J. Prinz, *The Secret Jews*, Random House, New York, (1973), pp. 111-112.

"JOIN ZIONIST MOVEMENT.

Enlistment of Two Rothschilds
Reported in London Dispatch.

The Jewish Morning Journal published the following yesterday as a special dispatch from London:

'At a reception held in Princess Hall, Piccadilly, London, given by Lord Rothschild, the head of the Rothschild family in England, in celebration of the official declaration by the British Government in favor of a Jewish home land in Palestine, Lord Rothschild announced that his younger brother, Charles, and Baron Edmund De Rothschild of Paris, head of the French branch of the Rothschild family, had joined the Zionist movement.

'The reception was attended by all the Zionist leaders in England as well as by prominent Jews and gentiles. One of the latter, a priest, presented Lord Rothschild with a handsome volume of suitable texts relating to the return of the Jews to Palestine.

'The prevailing opinion in well-informed Zionist circles in London is that Russia will urge the interallied conference, to be held soon in Paris, to give its approval to Zionism. The Armenian Consul in London congratulated the Zionist leaders on their excellent prospect of getting Palestine, and expressed a hope that the Jews would prove good neighbors.

'Lord Swaythling, Lucien Wolf, the publicist, who is the foreign editor of the London Daily Graphic, and Sir Philip Magnus, a Member of Parliament, formed a league of British Jews to combat the view that the Jews form a nation, as manifest by the Palestine declaration of the British Government. This league, however, expresses the readiness to facilitate the settlement of the Jews in Palestine.

'The German newspaper, Germania, organ of the German Catholic Party, urges the German Government to take steps against the alliance of Great Britain and the Zionists.'"

The Armenians were Christians. The "Young Turks", led by Jewish positivists, slaughtered the Armenians, and accomplished, in part, the ancient Judaic goals of ruining Christendom and secularizing the Turks. Dönmeh Jews pretended to convert Islam, changed their names to escape detection and undermined Turkish society, much like the Frankists, who came from this movement in Turkey, pretended to convert to Catholicism, became Polish aristocracy and destroyed Poland, which never recovered after having been one of the most advanced societies on Earth. The Jews hoped the ruined Poles would venture into the world spreading modern culture and monotheism to prepare the way for Jews to migrate to the ends of the Earth and dominate all cultures—just as the Jews had spread culture and monotheism when they were chased out of Palestine and traveled to the ends of the Earth in the Diaspora (*Genesis* 12:3; 28:14. *Deuteronomy* 28:64-66. *Isaiah* 27:6; 49:6. *Jeremiah* 24:9)—note that the Jews were promised all lands upon which they had slapped the soles of their feet, and thus believed the ends of the Earth and all points in between were theirs

(*Deuteronomy* 6:10-11; 11:24-25. *Joshua* 1:2-5. *Isaiah* 2:1-4; 40:15-17, 22-24; 54:1-4; 60:5, 8-12; 61:5-6). The Jews used Roman Christians to condition the world to accept eventual Jewish domination and the destruction of the Gentiles themselves.

The *Encyclopaedia Judaica* writes in its article "Messianic Movements":

> "Even Josephus—who tried to conceal the messianic motives of the great revolt—once had to reveal that 'what more than all else incited them to the war was an ambiguous oracle, likewise found in the sacred Scriptures, to the effect that at that time one from their country would become ruler of the world' [***] One trend of Jewish messianism which left the national fold was destined 'to conquer the conquerers'—by the gradual Christianization of the masses throughout the Roman Empire. Through Christianity, Jewish messianism became an institution and an article of faith of many nations. Within the Jewish fold, the memory of glorious resistance, of the fight for freedom, of martyred messiahs, prophets, and miracle workers remained to nourish future messianic movements."[217]

Many Spanish Cabalist Jews had emigrated to Turkey when Ferdinand, a Jew,[218] and Isabella, expelled many of the Jews from Spain in 1492. Turkey became a center for Jewish mysticism and the production of Cabalist revolutionaries, crypto-Jewish leaders, and Jewish heads of state. The Spanish aristocracy had perhaps expelled the Jews in order to "save" the Sephardic Jewish "race" from extinction through assimilation—the Sephardics were considered to be the true Judeans by most Jews of the age, though some later argued that they were merely religious Jews descended from Phoenician sailors who had settled in Spain. Another myth, which Spanish Jews initiated during the Inquisition, was that they had migrated to Spain long before the crucifixion of Christ, and therefore could not be held to account for killing Christ. German and other Jews fabricated similar fictions. The Jews of Worms told that their ancestors' Sanhedrin had written to the King of Judea and asked that Christ not be put to death.[219] The question naturally arises, was the entire British-Israel movement, which was so vital to Zionist interests, initiated by Jews who sought to distance themselves from the crucifixion of Christ?

Note that 1492 was the year that Columbus sailed to the Americas. Some argue that he was a crypto-Jew in search of a homeland for the Jews, where they would not assimilate. He was financed by Jews and Jews accompanied him on

---

217. "Messianic Movements", *Encyclopaedia Judaica*, Volume 11 LEK-MIL, Encyclopaedia Judaica, Jerusalem, The Macmillan Company, New York, (1971), cols. 1417-1427, at 1421.

218. B. J. Hendrick, "The Jews in America: I How They Came to This Country", *The World's Work*, Volume 44, Number 2, (December, 1922), pp. 144-161, at 149.

219. "The Modern Jews", *The North American Review*, Volume 60, Number 127, (April, 1845), pp. 329-368, at 333-334, 355.

his voyage.[220] The Jews of this age welcomed and perhaps intentionally caused their own suffering as an artificial means to hasten the arrival of the Messiah—which is to say the unimaginably rich Jews intentionally caused the less wealthy Jews to suffer, in collusion with Ferdinand, himself a Jew, and Isabella. The genocidal Zionists believed in the Messianic myth of "hevlei Mashiah", or "the birth pangs of the Messiah".[221] The *Encyclopaedia Judaica* wrote in its article on "Messianic Movements",

> "Even on the eve of the expulsion of the Jews from Spain, both Jews and *anusim* actively harbored these hopes. About 1481 a Converso told a Jew, when at his request the latter read the messianic prophecies to him: 'Have no fear! Until the appearance of the Messiah, whom all of us wait for, you must disperse in the mountains. And I—I swear it by my life—when I hear that you are banished to separate quarters or endure some other hardship, I rejoice; for as soon as the measure of your torments and oppression is full, the Messiah, whom we all await, will speedily appear. Happy the man who will see him!' One Marrano was certain that the Messiah would possess the philosopher's stone and be able to turn iron into silver. He also hoped that 'in 1489 there will be only one religion' in the world. Even after the expulsion many Marranos expressed these hopes and were punished for them by the Inquisition (*ibid.*, 350ff.)"[222]

In order to restrain the Christians from reacting to a Jewish Messiah as the anti-Christ, the Spanish Jews may have sought to destroy the Catholic Church with the "Spanish Popes", who were likely of Jewish descent, and who would perhaps have permitted the ascendency of the Jewish anointed King, and who perhaps sought to turn God's eye from the Christians to the Jews, by making the Christians decadent.

The New York Times reported on 30 November 1917,

> "Those of the Zionist movement here believe that after the war even Germany will not place obstacles in the way of the realization of Jewish

---

220. A. Leroy-Beaulieu, *Israel chez les nations: Les Juifs et l'antisémitisme*, C. Lévy, Paris, (1893); English translation by F. Hellman, *Israel among the Nations: A Study of the Jews and Antisemitism*, G. P. Putnam's Sons, New York, W. Heinemann, London, (1895), p. 356. H. N. Casson, "The Jew in America", *Munsey's Magazine*, Volume 34, Number 4, (January, 1906), pp. 381-395. B. J. Hendrick, "The Jews in America: I How They Came to This Country", *The World's Work*, Volume 44, Number 2, (December, 1922), pp. 144-161.
221. "Messianic Movements", *Encyclopaedia Judaica*, Volume 11 LEK-MIL, Encyclopaedia Judaica, Jerusalem, The Macmillan Company, New York, (1971), cols. 1417-1427, at 1418. G. Scholem, *Kabbalah*, New American Library, New York, (1974), p. 284. Compare to: *Matthew* 24:7-8. *Mark* 13:7-8.
222. "Messianic Movements", *Encyclopaedia Judaica*, Volume 11 LEK-MIL, Encyclopaedia Judaica, Jerusalem, The Macmillan Company, New York, (1971), cols. 1417-1427, at 1425.

hopes."[223]

The New York Times reported on 3 December 1917 on page 4,

"ZIONISTS PLAN BIG LOAN.

$101,000,000 to Create and Maintain
Proposed Palestinian Government
Special to The New York Times.

BALTIMORE, Dec. 2.—At two great meetings held tonight in the Hippodrome and Palace Theatres under the auspices of the Baltimore Conference for Jewish National Restoration in Palestine the declaration of the British Government, promulgated by Mr. Balfour, favoring the establishment of a Jewish national home in Palestine, was unanimously and enthusiastically approved.

Prior to the submission of the resolution, Jacob De Haas, at one time the secretary to Dr. Herzl, the founder of the Zionist movement, said in the course of an address that in the near future subscriptions would be asked to a $1,000,000 fund to be used in the creation of the Government in Palestine, and subsequently a $100,000,000 liberty loan would be issued to provide for its maintenance.

While all the principal speakers dwelt upon the benefit to be derived from nationalization, Mr. De Haas devoted himself more particularly to the political significance of the movement. He made the assertion that not only were the European Allies back of the declaration, but that this Government would in the very near future announce its endorsement and concur in the establishment of a national Jewish home."

The New York Times reported on 7 December 1917 on page 4, after recalling the tyranny of the Bolsheviks,

"Jews Turn to Palestine.

Then there are the Jews. Besides their manifold efforts in general Russian politics, they are swelling the tide of national movements. The Zionists now are the strongest party among Russian Jews, and they are overjoyed at the British promise of Palestine. At Odessa last Friday there was a huge Zionist demonstration, with a procession twenty blocks long. Grusenberg, the newly elected member for Odessa, made a speech of triumph and gratitude, to which the British Council, Picton Bage, replied. Toward the close of the demonstration members of the Bund, or Jewish Socialist Party, began agitating against the Zionists and England. There was a scuffle, and a shot was fired, but no harm was done."

---

[223]. "Jewish Socialists Acclaim Zionism", *The New York Times*, (30 November 1917), p. 6.

The New York Times reported on 7 December 1917 on page 4,

"VOTING FAVORS BOLSHEVIKI.

But Constitutional Democrats Make Strong Showing Also.

PETROGRAD, Dec. 6.—According to the preliminary returns from the provinces the Bolsheviki in the elections obtained 2,704,000 votes; the Constitutional Democrats, 2,230,500, and the Social Revolutionaries, who form the majority of the Left, 221,260.

The Central Executive Committee has given its consent to a decree granting to the Councils of Electoral Districts the right to proceed with re-elections for all elective bodies, including the Constituent Assembly, in accordance with the demands of the electors. Thus it will be possible for the electors to revoke their choice in the case of those representatives whose politics no longer correspond with their own.

The project provoked great opposition on the part of the moderate element of the committee, who termed it an attempt to curtail the rights of members of the Constituent Assembly. In defending the measure Leon Trotzky, the Bolshevist Foreign Minister, said:

'Should there be a majority of the Constitutional Democrats, members of the Right and Social Revolutionists, the people would forcibly dissolve the Constituent Assembly. This measure is meant to avoid the possibility of dissolution.'

Since the system of representation is proportional, an objection to one member of the Constituent Assembly would necessitate the recall of all the members of a given election district."

The New York Times reported on 10 December 1917, on page 4,

"ARMENIANS FAVOR ZION.

London Association Sends Resolutions to Justice Brandeis.

The Provisional Zionist Committee yesterday announced that Justice Louis D. Brandeis of the United States Supreme Court has received a letter of congratulation from the Armenian United Association of London on the British declaration in favor of the establishment of a national Jewish home in Palestine, to which the Cabinet promises that 'his Majesty's Government will exert its best endeavors.'

The resolution accompanying the letter follows:

The council of the Armenian United Association of London, having read in the press that the British Government had now formally expressed its sympathy with the project for the reconstruction of Palestine as the national home of the Jewish people, at their meeting held on Nov. 10, 1917, at the offices of the association,

Resolved, To record their unalloyed gratification and to convey their cordial congratulations and sincere and neighborly greetings to the President, Dr. C. F. Weitzman, committee and members of the Zionist Federation of Great Britain, and through them to all other Zionist leaders and Zionist organizations, and especially those in the United States, Russia, France, Italy, Poland, and Rumania, upon the recognition of Jewish nationality and their righteous, inalienable claim to the historic soil and country of their ancestry.

Resolved, further, to request the Honorary Secretary to send copies of this resolution to Chief Rabbi, Dr. Weitzman, to Lord Rothschild, to Baron Edmond de Rothschild, to Mr. Nahoum Sokolow, to Dr. Tschlenow of Moscow, to Judge Louis D. Brandeis of the United States Supreme Court, and to the press."

The New York Times reported on 14 December 1917,

"The Jews of Russia, he predicts, will have an important influence. The capture of Jerusalem by the British, he says, will be a weighty factor in the situation."[224]

The New York Times reported on 21 December 1917 on page 6 that German Zionists had betrayed Germany,

"ENGLAND'S RECOGNITION.

Appreciative Comment of a German
Jewish Paper on Britain's Attitude.

Judische Rundschau, the official organ of the German Zionists, commenting on the British Government's declaration of its attitude toward Zionism, says that this is the first occasion on which a great power has officially declared itself in relation to Zionism. For the first time the claim ah the Jewish Nation to a renewal of its national existence in Palestine has been lifted by a European Government into the circle of the weighty political problems of the present time, and it must be admitted that the recognition of this claim by the British Government is an event of world-wide historic importance."

The New York Times reported on 24 December 1917 on page 9,

"SEES ZIONISTS' HOPE
IN ALLIED VICTORY

---

[224]. "Russian Battle is Still Raging", *The New York Times*, (14 December 1917), pp. 1-2, at 2.

Britain's Pledge to Restore Jerusalem
Urged Upon Jews as
Reason for War Effort.

GREAT MEMORIAL MEETING

Aged Men Declare Themselves
Young Again and Anxious to
Start Anew in the Holy Land.

In celebration of the British promise to restore Jerusalem and the Holy Land to the Jewish people, thousands of New York Zionists packed Carnegie Hall last night in a commemoration meeting. Thousands more crowded the streets around the building, unable to get in, until long after the beginning of the meeting. Inside American, British and Zionist flags were intertwined, and with songs in the Hebrew language interspersed between the speeches, a group of leaders of Zionism in New York and the Old World told of the significance of the British promise.

The last and most enthusiastically welcomed speaker was Dr. Schmarya Levin, who spoke in Yiddish, declaring that the act of Great Britain was not an act of politics or diplomacy, but something far deeper, a stage in the development of history which in effect added another chapter to the Bible a modern chapter by which the Jews of today could link something of their own time to the story of the old Jewish kingdom.

Dr. Levin spoke as a representative of the international Zionist organization, but the speaker who stirred most enthusiasm, next to him, was a Christian, the Rev. Otis A. Glazebrook, late American Consul at Jerusalem, who had charge of the distribution of Jewish relief funds in the Holy City.

Hope Centred in the Allies.

And one of the most enthusiastic outbursts of the evening occurred when Dr. Glazebrook declared: 'It is the duty of every Jew who loves Palestine, who fosters the hope of the restoration of Israel, to use his influence, his material wealth, and his life to see that England and the Allies win this war.

'We have seen a vision of the restoration of the Jewish people,' he said, 'and we pray that this vision may not be spoiled by war, but may be crowned by a war, ending gloriously in victory for the Entente Powers. If Palestine is to be restored to Israel, remember that Palestine and Syria must remain in the hands of the Allies. And the one most important lesson just now, more important than the immediate working out of the details of the Zionistic state, is that you see and do your whole and complete duty in this war for the success of Great Britain, France, Italy, and America.'

Dr. Stephen S. Wise, Chairman of the meeting, said that what Zionists were rejoicing over was only a scrap of paper, 'but that scrap of paper is written in English. It is signed by the British Government, and therefore is sacred and inviolable. It represents not an unconsidered policy of a temporary Government, but all the great political parties of England have united in giving their adherence to this declaration. It is true to the finest traditions of the British people, and is a symbol of the will of the Allies to right wrongs, however ancient, to undo injustice, however hoary, to supplant the Prussian ideal of rule by might with the changelessly true principles of justice and right.

'Liberation, Justice, Peace.'

'This meeting is a challenge to every American Jew to unite with us. We offer our hands in welcome to those who up to this time have not worked with us. Let them come to us.

'More than all else, this meeting has been called in order to reaffirm the faith of every living American Jew not only in the certainty of the triumph of our arms, but in the righteousness of our aims. The American Jew by this assembly tonight reaffirms his faith that there shall be no faltering until victory shall crown our arms, and such a triumph be granted to our aims and the aims of our allies as shall bring the boon of liberation, justice and peace to all the nations.'

Nathan Straus, who was repeatedly interrupted by applause, spoke as a man who was seeing the realization of the dream of a lifetime. 'There are only a few things that can enthuse a man of my years,' he said. 'I have come to the place where I am skeptical and hard to be impressed, for I have seen so many things go wrong, but now they are going right. The moment of realization has come.

'I stand before you in appearance and somewhat in fact an old man. Many of these gray hairs have come through years of striving for the national cause of our people. My eyes have grown weak watching, my heart heavy with praying; but all this time, as the soldiers say, I carried on. And this moment is my reward.

'All we who have worked for Zionism are rejuvenated now. But the support which is most necessary is that of the masses of Jews, and the masses of Jews are Zionists. If they are not I'm sorry for them. In Zionism the Jew and the non-Jew have found a bond of brotherhood.

'This promise of England has made me young again. All Jews are young now. I feel that this appearance of mine is camouflage: I want to buy a horse and plow, a cow—for I can't be separated from the milk business—and begin a new life in the old land. All Jews are young now and we shall make our old country flow with milk and honey.

Abram I. Elkus, former Ambassador to Turkey, praised the work of the various American consular officials in that empire, 'who spent their time and energy without stint to alleviate the suffering of those of all races and creeds.'

Other speakers were Dr. Aaron Aaronson, director of the Zionist agricultural experiment station in Palestine; Morris Rothenberg, Chairman of the Zionist Council of Greater New York, and Jacob de Haas, Secretary of the Provisional Zionist Executive Committee.

'The Star Spangled Banner' and 'The Hatikvah,' the Jewish national anthem, were sung at the beginning and end of the meeting. Palestinian songs were sung by the Hadassah Choral Union, directed by A. W. Binder.

Declare for a Jewish State.

PHILADELPHIA, Dec. 23.—Resolutions in favor of making Palestine a Jewish State, to be populated by Jews from all parts of the earth, were adopted here today at a conference of Jewish labor organizations held under the auspices of the workmen's wing of the Zionist movement. Speakers explained that this State should be a Jewish nation in fact and a centre of Judaic literature, art and law."

The New York Times reported on 30 December 1917 on page 5, that German Zionist financiers had betrayed Germany,

"JEWS IN GERMANY FIRM.

Won't Support War Loan Until
Palestine Independence Is
Sanctioned.
Special Cable to THE NEW YORK TIMES.

THE HAGUE, Dec. 29.—It is reported here that the leading Jewish financiers of Germany refused to support the German war loan unless the German Government undertook to refrain from all opposition to the establishment of a Jewish State in Palestine, independent of any Turkish suzerainty or control.

By Associated Press.

THE HAGUE, Dec. 29.—The Jewish Correspondence Bureau here has received a telegram from Berlin stating that at a Zionist conference in Germany a resolution was adopted in which satisfaction was expressed that Great Britain had recognized the right of the Jewish people to a national existence in Palestine."

Eduard Bernstein wrote after the war,

"To many Social Democrats the war really seemed to be one for national existence; and to many passionate natures the opposition of so many Jews to the war credits might have seemed to betray un-German or anti-German thinking. How little such feeling had to do with anti-Semitism can be seen from the fact that those Jews who voted for the war loans were more highly esteemed and sought after than ever."[225]

---

225. P. W. Massing, *Rehearsal for Destruction: A Study of Political Anti-Semitism in*

The Manufacture and Sale of Saint Einstein

The New York Times reported on 2 April 1918 on page 10,

"ZIONISTS CELEBRATE NEW JEWISH FUTURE

2,500 in Carnegie Hall Pledge
Loyalty to America and
the Allies.

CHEER PALESTINE SOLDIERS

Dr. Wise Says Jewish Freedom Is
Secure Because It Is Written in
the English Language.

In a tremendous demonstration in Carnegie Hall last night the Zionists of New York attested their patriotism to America, their loyalty to the cause of the Allies, and their joy over the prospect of a land for the Jews in Palestine. The meeting was arranged by the Zionist Council of this city and it was preceded by a parade in which 2,500 Zionists marched.

As the marchers filed into Carnegie Hall the banners they carried were ranged along the wall and their flags hung out from platform and galleries. The meeting was full of enthusiasm from the start, and there were three periods when it reached the greatest pitch. One of these was when the blue and white flag of the House of David, the flag of the new Jewish home land, was raised, and again when Louis Lipsky, Chairman of the Executive Committee of the Federation of American Zionists, mentioned the name of President Wilson. Then again when the Rev. Dr. Stephen S. Wise declared that the charter of Jewish freedom was secure and sacred because it was written in the English language by the English people.

Over the stage there was a great American banner and stacked to one side of the stage were the flags of the Allies. Hung from one side of the stage was the Jewish flag. This is a white field upon which are two broad blue stripes. In the centre is the six-pointed star of the House of David. When this flag was put up the entire audience from boxes to topmost gallery, arose and cheered. Among those on the platform were young men in the khaki of the Jewish Legion. There were about 250 of them and they were honored by the speakers.

Morris Rothenberg, Chairman of the meeting and President of the Zionist Council of New York, said in opening the meeting that the arrival of the Zionist Commission in Palestine to lay the foundation of the new Jewish national freedom was worthy of celebration. Nathan Straus spoke

---

Imperial Germany, Howard Fertig, New York, (1967), p. 325.

briefly, but he called upon Dr. Wise to deliver to the Zionists the message of patriotism and devotion to the cause of America and her allies.

The storm of applause and the cheering broke out again when Dr. Wise declared that Germany would never win the war. It came again when, lifting his hands above his head, he said: 'England, France, and America have said to Germany, 'Thus far shall thou go, and no further.' '

'If Germany could win the war,' he said, 'as she cannot, she would give Palestine back again to those hands to which our Holy Land under God shall never be restored—to the Turks. And, gentlemen, there was a time when some of you would have felt differently, but I speak for myself tonight, not for you. I speak not as Chairman of the Provisional Zionist Committee. I speak as a Jew: I speak as an American. I say to you, the charter of Jewish freedom is secure, is sacred, because it is written in the English language by the English people, and if men say to you: 'How do you know but if Germany could win the war Germany might give Palestine to the Jews?' I answer, 'We want never to be the receivers of stolen property, we want never, never; never will we accept any gift from foul and murderous hands. We are going into Palestine with heart directly facing the world, as self-revering free men. We will go to Palestine as one of the victorious Allies, or else shall stay out until another and better day dawns.'

Mr. Straus was applauded when he said: 'We are going to Palestine this year and we will stay there.' In referring to the arrival of the Jewish Administrative Commission in Palestine, Mr. Lipsky said:

'While jubilant over the change in our national status, the Zionist organization desires to express its feelings with regard to the Governments and peoples that have made this change possible. The magnanimity of the British Government in making its historic declaration on Nov. 2, 1917, will never be forgotten by the Jewish people. Relations have been established that will forever link our destiny with the interests of the great empire. In the days to come Nov. 2 will be a day of Jewish rejoicing, and our traditions will be enriched by the memory of the act of reparation achieved by a great Government in the midst of a gigantic struggle, in which its own future had to be defended by its heroic sons. As a token of that relationship the sons of Israel, under their own banner, will soon stand shoulder to shoulder on the Palestinian frontier with the gallant and heroic Englishmen. The blood there shed will be an everlasting covenant between the two peoples, which nothing shall ever erase.'"

The New York Times reported on 24 December 1915 on page 3,

"SEMITIC ISSUE IN GERMANY.

Some Berlin Newspapers Accused of
Reviving Anti-Jewish Feeling.

BERLIN, Dec. 23.—Anti-Semitism, an issue which has been almost dead since the beginning of the war, has been revived this week by the Tageszeitung and other newspapers. In consequence a controversy which may be described as almost bitter has broken out between papers of the Tageszeitung stamp on the one hand and those like the Tageblatt, which adopt a liberal attitude in regard to the Jewish question, on the other.

The more liberal papers resent intensely every anti-Jewish movement, particularly as it is asserted that German Jews have borne their share of the war's burdens liberally and are doing their utmost for the Fatherland in both a military and an economic sense.

The present revival of the anti-Semitic movement began with a savage attack in the Tageszeitung against Eugen Dietrich of Jena, who had accused 'a Berlin morning newspaper'—inferentially, the Tageszeitung—of being anti-Jewish. The Tageszeitung denied it was the newspaper attacked and further roused Jewish feeling by putting the blame on still another Berlin journal.

The Tagesblatt entered the controversy, calling the Tageszeitung utterances 'base defamation of German Jews, many of whom died for the Fatherland after voluntarily joining the army—in which they were notably different from certain anti-Semitic Nationalist typewriter heroes, who have not lived up to the war propaganda they preached for years.'

Theodor Wolff of the Tageblatt, who is perhaps the most prominent editor of Germany, declares that notwithstanding the recent revival of anti-Semitism the feeling against Jews in Germany is gradually on the wane, existing nowhere to a great extent except possibly among the minor nobility.

'I am glad to be able to say there is absolutely no anti-Jewish movement in Government circles or in the high nobility,' Mr. Wolff said. 'The Jew now has equal rights in the army and may become an officer along with a Christian. In virtually all strata the Jew is found intermingled with all others.

'It is only among the minor nobility that the Jew is still unwelcome, on account of the fact that he is able to outstrip his competitors, who are jealous of him. But do not forget there are notable exceptions in this class—fine examples of Germans who are too broad to be anti-Jewish or anti-anything. A few German newspapers which represent this class of the minor nobility, such as the Tageszeitung and the Taegliche Rundschau, are naturally anti-Jewish, but their agitation is becoming less effective each month. I look for eventual liberty for Jews in Germany, such as exists in America today.'"

Early in the war, it was alleged that Jews avoided military service in Germany by working for Jewish war profiteers under the direction of Walter Rathenau. Jews have often been accused of cowardice in war, allegedly preferring to shuffle goods in the Quartermasters Corps to the front lines.[226] Jews were also accused of supplying substandard arms at inflated prices. After the

---

[226]. R. Earley, *War, Money and American Memory: Myths of Virtue, Valor and Patriotism*, Diane Publications, Collingdale, Pennsylvania, (2000).

war, it was frequently alleged that Jews had reaped their alleged war profits in hopes of using the money to achieve their Zionist aims—the implication being that Zionists started the war in order to found and to fund their new state.[227]

The German Ministry of War ordered a census taken in October of 1916 to determine the percentage of Jews serving in the military. The results showed that Jews represented a lower percentage in the military than in the general population. Some claimed that Jews were, in part, deliberately excluded from the census. The results of the census were not published by the German Government, which feared they might cause conflict between Gentile and Jewish soldiers. However, the results were leaked and published in pamphlet form.

Walther Rathenau was widely accused of profiteering from the war, as was Bernard Baruch, an American who was Chairman of the American War Industries Board. Rathenau was also accused of making statements which indicated that he had hoped that Germany would lose the war.[228] Rathenau was further accused of profiteering from the reparations he encouraged Germany to pay after the war, and from the profits to be made through the Rapallo Treaty.

Jews had long been accused of war profiteering. Schopenhauer and Wagner were among the many pacifists who have made the same accusation against the Jews. Schopenhauer wrote; in terms Einstein would later, in part, copy;

> "War is a word as heavy as lead. It is the scourge of humanity and of nations, the antithesis of all reason, although not seldom a harvest for the great, for ministers, generals, contractors, and Jews. War is mankind's obscene picture, and war first begot despotism. War begot the feudal system. War made of free men the first slaves."[229]

In December of 1915, Theodor Wolff, Chief Editor of the *Berliner Tageblatt*, stated that there was no anti-Semitic movement in the German government or higher nobility. Anti-Semitism, as basic bigotry, and as a complex political, racist and religious belief system, doubtlessly continued on many levels, conscious and unconscious, as did Wolff's somewhat juvenile and provocative approach to confronting it. Einstein criticized the *Berliner Tageblatt*,[230] in spite of the fact that he used it as an organ to unfairly denigrate his critics. The *Berliner Tageblatt*'s approach to redressing anti-Semitism was counterproductive. Willi Buch (Wilhelm Buchow) wrote in 1937,

---

227. L. Fry, *Waters Flowing Eastward: The War Against the Kingship of Christ*, TBR Books, Washington, D. C., (2000), p. 43.
228. T. Fritsch, "The Desperate Act of a Desperate People", (1922), in R. S. Levy, *Antisemitism in the Modern World: An Anthology of Texts*, D. C. Heath and Company, Toronto, (1991), pp. 192-199.
229. G. Nicolai, *Die Biologie des Krieges, Betrachtungen eines deutschen Naturforschers*, O. Füssli, Zürich, (1917); English translation: *The Biology of War*, Century Co., New York, (1918), p. 493.
230. *The Collected Papers of Albert Einstein*, Volume 9, Documents 16 and 217, Princeton University Press, (2004).

"Besides, other Jewish newspapers like the *Berliner Tageblatt* and the *Freisinnige Zeitung* worked in the same direction as the philo-Semitic defense publications. The defense against anti-Semitism was so reckless, the attacks against its representatives so full of hate and obvious lies that their effect upon the sober and realistic German was mostly contrary to the intended one."[231]

It was only after America entered the war on the Allies' side when Germany was about to win it and bring peace to the world; and after the Zionists moved their headquarters from Berlin to London and then attempted to blackmail Germany in 1917 and made very public their allegiance, including the allegiance of Zionist financiers, to the Allies; that anti-Semitism began to rise as a political movement in Germany in 1918—especially after the short-lived Bolshevist revolution in Bavaria. The political Zionists believed that the strife between Gentile and Jew benefitted their cause. Failed Communist takeovers of Germany in January and March of 1919 and March, 1920, further resulted in concerns that Jewish Bolshevists had Germany forever in their sights. Indeed, the Communist finally took Eastern Germany after the Second World War, and the Nazi Party was a Communist organization.

The unfortunate Jews in Poland, Lithuania, Ukraine and Russia suffered terribly as pawns and scapegoats caught between all rival forces as the First World War progressed—though not nearly so badly as they later would in the Holocaust to come in the Second World War. They had the Zionists to blame for their suffering.

The Bolsheviks also played no small part in the misery the Jews of Eastern Europe endured. The policy was often to segregate Jews into concentrated masses meant for expulsion; which was done at the behest of the political Zionists. The Jews did not wish to leave Europe. The Zionists took it upon themselves to insist that the Jews of Eastern Europe migrate to Palestine in order to provide the Zionists with soldiers and slaves. When the First World War could not accomplish this end, the Zionists took it upon themselves to promote anti-Semitism in order force the expulsion of the Jews from Europe to Palestine. Just as the Zionists ignored the desires of the majority of Jews, the American people were never asked if they wanted to fight war after war to found a racist "Jewish State" in Palestine and maintain it. Zionists have absolutely no respect for the principle of self-determination, be it on a national or a personal level.

5.17 The Germans' Side of the First World War

---

[231]. W. Buch, *50 Jahre antisemitische Bewegung. Beiträge zu ihrer Geschichte*, Deutscher Volksverlag G.m.b.H., H. Bachmann, München, (1937); English translation in: P. W. Massing, *Rehearsal for Destruction: A Study of Political Anti-Semitism in Imperial Germany*, Howard Fertig, New York, (1967), p. 310.

Prior to the *quid pro quo* arrangement between the British and the Zionist Jews to bring America into the war on the side of Great Britain in exchange for the Balfour Declaration, a great many books and pamphlets were published in America defending Germany,[232] and the financiers backed both sides in the war until the time of the Balfour Declaration. After the Jewish deal to bring America into the war was struck, a great many books were published in America attacking Germany—many of which adopted the vilification of all Germans propagandized by Émile Durkheim in 1915.[233] The Jewish anti-German propaganda campaign, and their efforts to bring the German People into world wars, have been very successful. A "Suppressed Speech by Company Sergeant-Major" made during the First World War stated,

> "What is the use of a wounded German anyway? He goes into hospital and the next thing that happens is that you meet him again in some other part of the line. That's no good to us, is it? So when you see a German laid out, just finish him off... . Kill them, every mother's son of them. Remember that your job is to kill them... exterminate the vile creatures. Murder that vile animal called a German."[234]

At least as early as the 1860's, Zionist racist and National Socialist Moses

---

[232]. H. Münsterberg, *The War and America*, D. Appleton, New York, London, (1914). *See also: Truth about Germany: Facts about the War*, Trow Press, New York, (1914). *See also:* E. v. Mach, *What Germany Wants*, Little, Brown and Company, Boston, (1914). *See also:* R. G. Usher, *Pan-Germanism*, Second Revised and Enlarged Edition, Grosset & Dunlap, New York, (1914). *See also: Deutschland und der Weltkrieg*, B. G. Teubner, Berlin, Leipzig, (1915); English translation by W. W. Whitelock, *Modern Germany in Relation to the Great War*, Mitchell Kennerley, New York, (1916). *See also:* C. A. Lindbergh, *Why Is Your Country at War and What Happens to You after the War, and Related Subjects*, National Capital Press, Inc., Washington, D.C., (1917); *reprinted as: Your Country at War and What Happens to You after a War*, Dorrance & Co., Philadelphia, (1934).

[233]. É. Durkheim, *"Germany above All" The German Mental Attitude and the War*, Librairie Armand Colin, Paris, (1915). *See also:* "By a German", *I Accuse! (J'Accuse!)*, Grosset & Dunlap, New York, (1915). *See also:* W. F. Barry, *The World's Debate: An Historical Defence of the Allies*, George H. Doran, New York, (1917). *See also:* W. T. Hornaday, *A Searchlight on Germany: Germany's Blunders, Crimes and Punishment*, American Defense Society, New York, (1917). *See also:* D. W. Johnson, *Plain Words from America: A Letter to a German Professor*, London, New York, Toronto, Hodder & Stoughton, (1917). **More recently:** D. J. Goldhagen, *Hitler's Willing Executioners: Ordinary Germans and the Holocaust*, Knopf, New York, (1996). **See Hartmut Stern's response to Goldhagen:** *KZ-Lügen: Antwort auf Goldhagen*, FZ-Verlag, München, Second Edition, (1998), ISBN: 3924309361; **and** *Jüdische Kriegserklärungen an Deutschland: Wortlaut, Vorgeschichte, Folgen*, FZ-Verlag, München, Second Edition, (2000), ISBN: 3924309507.

[234]. "This is War!", *The American Mercury*, Volume 42, Number 188, (August, 1939), pp. 437-443, at 443; which cites: "Suppressed Speech by Company Sergeant-Major (Published in *No More War*, London)."

Hess argued that the "German race" had a genetically programmed antagonism towards the "Jewish race"—the implication being that one must destroy the other in order to survive. Two World Wars did nearly accomplish the destruction of Germany, and ended their prominence in world affairs. Hess wrote in 1862,

> "It seems that German education is not compatible with our Jewish national aspirations. Had I not once lived in France, it would never have entered my mind to interest myself with the revival of Jewish nationality. Our views and strivings are determined by the social environment which surrounds us. Every Living, acting people, like every active individual, has its special field. .Indeed, every man, every member of the historical nations, is a political, or as we say at present, a social animal; yet within this sphere of the common social world, there are special places reserved by Nature for individuals according to their particular calling. The specialty of the German of the higher class, of course, is his interest in abstract thought; and because he is too much of a universal philosopher, it is difficult for him to be inspired by national tendencies. 'Its whole tendency,' my former publisher, Otto Wigand, once wrote to me, when I showed him an outline of a work on Jewish national aspirations, 'is contrary to my pure human nature.'
>
> The 'pure human nature' of the Germans is, in reality, the character of the pure German race, which rises to the conception of humanity in theory only, but in practice it has not succeeded in overcoming the natural sympathies and antipathies of the race. German antagonism to Jewish national aspiration has a double origin, though the motives are really contrary to each other. The duplicity and contrariety of the human personality, such as we can see in the union of the spiritual and the natural, the theoretical and the practical sides, are in no other nation so sharply marked in their points of opposition as in the German. Jewish national aspirations are antagonistic to the theoretical cosmopolitan tendencies of the German. But in addition to this, the German opposes Jewish national aspirations because of his racial antipathy, from which even the noblest Germans have not as yet emancipated themselves. The publisher, whose 'pure human' conscience revolted against publishing a book advocating the revival of Jewish nationality, published books preaching hatred to Jews and Judaism without the slightest remorse, in spite of the fact that the motive of such works is essentially opposed to the 'pure human conscience.' This contradictory action was due to inborn racial antagonism to the Jews. But the German, it seems, has no clear conception of his racial prejudices; he sees in his egoistic as well as in his spiritual endeavors, not German or Teutonic, but 'humanitarian tendencies'; and he does not know that he follows the latter only in theory, while in practice he clings to his egoistic ideas.

[***]

In 1858, there appeared, at Leipzig, a work written by Otto Wigand under the title *Two discourses concerning the desertion from Judaism*, being an analysis of the views on this question expressed in the recently published correspondence of Dr. Abraham Geiger. The author endeavors to prove that the conclusions of Dr. Geiger are untenable both from a philosophic and from a social standpoint. Here are his social arguments:

'My friend,' says the author, 'there are certain conclusions which you cannot escape. The stamp of slavery, if we may use this expression, which centuries of oppression have deeply impressed upon the Jewish features, might have been obliterated by the blessed hand of regained civil liberty. The gait of the Jews, buoyed up by the happy reminiscences of the victory won in the struggle for the noble possession of liberty, might have been straighter and prouder. The Jewish face may certainly beam with pride, as it views the tremendous progress made by the Jews in a brief time, their mighty flight to the spiritual height upon which they now stand, which is especially notable considering the fact that their poets and writers at whose greatness the nation is astonished, and of whose talents the entire people takes account, have sprung from those who, a generation ago, could hardly converse correctly in the language of the land. Such a state of affairs should undoubtedly call forth admiration in the hearts of the present German generation, and yet, in spite of these achievements, the wall separating Jew and Christian still stands unshattered, for the watchman that guards them is one who will not be caught napping. It is the race difference between the Jewish and Christian populations. If this assertion of mine surprises or astonishes you, I ask you to consider whether it is not almost a rule with the Germans that race differences generate prejudices which cannot be overcome by any manifestation of good-will on the part of the other race. The relations existing between the German and the Slavic populations in Bohemia, in Hungary and Transylvania, between the Germans and the Danes in Schleswig, or between the Irish and the Anglo-Saxon settlers in Ireland, illustrates well the power of race antagonism in the German world. In all these countries the different elements of the population have lived side by side for centuries, sharing equally all political rights, and yet, so strong are the national or racial differences, that a social amalgamation of the various elements of the population is even at the present day quite unthinkable. And what comparison is there between the race differences of a German and Slav, a Celt and Anglo-Saxon, or a German and Dane, and the race antagonism between the children of the Sons of Jacob, who are of Asiatic descent, and the descendants of Teut and Herman, the ancestors of whom have inhabited Europe from time immemorial; between the proud and the tall blond German and the small of figure, black-haired and black-eyed Jew? Races which differ in such a degree oppose each other instinctively and against such opposition reason and good sense are powerless.'

These expressions are certainly frank and sincere in their meaning, though they by no means prove the conclusions to which the author wishes to arrive, namely, the desirability of conversion; for conversion will not turn a Jew into a German. But they at least contain the confession, that an instinctive race antagonism triumphs in Germany above all humanitarian sentiments. The 'pure human nature' resolves itself, according to the Germans, in the nature of pure Germanism. The 'high-born blond race' looks with contempt upon the regeneration of the 'black-haired, quick-moving mannikins,' without regard to whether they are descendants of the Biblical patriarchs, or of the ancient Romans and Gauls.

While other civilized western nations mention the shameful oppression to which the Jews were formerly subjected, only as an act of theirs of which they are ashamed, the German remembers only the 'stamp of slavery' which he impressed upon 'the Jewish physiognomy.'

In a *feuilleton* which appeared recently in the *Bonnerzeitung*, entitled 'Bonn Eighty Years Ago,' the author speaks of the Jews in mocking terms and describes them as people who lived in separate quarters and supported themselves by petty trades. I believe that we should wonder less at the fact that the Jews, who were forbidden to participate in the important branches of industry and commerce, lived on petty trade, than at the fact that they were able to live at all in those centuries of oppression. As a matter of fact, almost every means of existence, including the right of domicile, was denied them. It was only by means of bribes that every Jewish generation could procure anew the 'privilege' not to be driven out of their homes in Bonn, and they felt happy indeed if, in spite of the contract, they were not robbed of their property and exiled, or attacked by a fanatical mob in the bargain. I, also, can tell a story of 'eighty years ago.' A Jew won the high favor of the Kurfuerst of Bonn, that he and his descendants were granted the 'privilege' to settle in Ebendich.

[***]

Gabriel Riesser, the editor of the magazine, *The Jew*, as far as I can recollect, never fell into the error, common to all modern German Jews, that the emancipation of the Jews is irreconcilable with the development of Jewish Nationalism. He demanded emancipation for the Jews on the one condition only, that of their receiving all civil and political rights in return for their assuming all civil and political burdens."[235]

Racist Zionist Moses Hess stated that emancipation ended Jewish nationalism in Germany, making Jewish liberty and Germany the enemies of Zionism. Racist Zionist Adolf Hitler put an end to both Jewish freedom and Germany. Hess, in the express terms Hitler would later adopt, relied upon racist mythologies and National Socialism to solve the "dilemma" of Jewish nationalism.

The racist hatred against Germans by some Jews reached its climax in the proposed genocide of Germans by Theodor Newman Kaufman, who claimed to have connections to Franklin Delano Roosevelt and Winston Churchill, in 1941

---

235. M. Hess, *Rom und Jerusalem: die letzte Nationalitätsfrage*, Eduard Wengler, Leipzig, (1862); English: "Fourth Letter", "Note III" and "Note IV", *Rome and Jerusalem: A Study in Jewish Nationalism*, Bloch, New York, (1918), pp. 56-57, 240-244.

in Kaufman's genocidal book *Germany Must Perish!*[236] After the Balfour Declaration, German Zionist financiers attempted to blackmail Germany into unconditionally securing Palestine as a Jewish State without any Turkish or German oversight. Since Turkey was Germany's ally, this was an unreasonable request, though Germany did attempt to gain Palestine as a land of settlement for Jews with almost complete independence.

President Wilson won his declaration of war against Germany in the United States Congress based on false reports of the sinking of the S. S. Sussex and through the arranged attack on the Lusitania. Wilson was elected with Jewish financier's money, twice, and surrounded himself with appointees, who were themselves Jewish financiers, or who were selected by Jewish financiers.

Francis Neilson wrote in his book *The Makers of War*,

"In America, Woodrow Wilson, desperate to find a pretext to enter the war, found it at last in the 'sinking' of the *Sussex*, in mid-channel. Someone invented the yarn that American lives had been lost. With this excuse he went to Congress for a declaration of war. Afterwards, the Navy found that the *Sussex* had not been sunk, and no American lives were lost."[237]

Though much was initially published exonerating Germany,[238] the German

---

236. T. N. Kaufman, *Germany Must Perish!*, Argyle Press, Newark, New Jersey, (1941).
237. F. Neilson, *The Makers of War*, C. C. Nelson, Appleton, Wisconsin, (1950), pp. 149-150.
238. R. E. Martin, *Stehen wir vor einem Weltkrieg?*, F. Engelmann, Leipzig, (30 June 1908); **and** *Der Weltkrieg und sein Ende*, Rudolf Martin, Berlin, (1915). *See also:* E. v. Mach, *What Germany Wants*, Little, Brown and Company, Boston, (1914). *See also: Truth about Germany: Facts about the War*, Trow Press, New York, (1914). *See also: Deutschland und der Weltkrieg*, Berlin, Leipzig, B. G. Teubner, (1915); English translation by W. W. Whitelock, *Modern Germany in Relation to the Great War*, Mitchell Kennerley, New York, (1916). *See also:* C. A. Lindbergh, *Why Is Your Country at War and What Happens to You after the War, and Related Subjects*, National Capital Press, Inc., Washington, D.C., (1917); *reprinted as: Your Country at War and What Happens to You after a War*, Dorrance & Co., Philadelphia, (1934). *See also:* G. v. Jagow, *Ursachen und Ausbruch des Weltkrieges*, Reimar Hobbing, Berlin, (1919). *See also:* L. Brentano, *Der Weltkrieg und E. D. Morel: Ein Beitrag sur englishcen Vorgeschichte des Krieges*, Drei Masken Verlag, München, (1921). *See also:* C. E. Playne, *The Neuroses of the Nations*, G. Allen & Unwin, Ltd., London, (1925). *See also:* I. C. Willis, *England's Holy War; a Study of English Liberal Idealism During the Great War*, Knopf, New York, (1928). *See also: Deshalb haben wir den Krieg nicht verloren!*, Norddeutsches Druck- und Verlagshaus, Hannover, (1926). *See also:* H. Bauer, *Sarajewo: Die Frage der Verantwortlichkeit der serbischen Regierung an dem Attentat von 1914*, W. Kohlhammer, Stuttgart, (1930). *See also:* M. H. Cochran, *Germany not Guilty in 1914 (Examining a Much Prized Book)*, Stratford, Boston, (1931). *See also:* H. C. Peterson, *Propaganda for War; the Campaign Against American Neutrality, 1914-1917*, University of Oklahoma Press, (1939); *See also:* J. R. Mock and C. Larson, *Words that Won the War; the Story of the Committee on Public Information, 1917-1919*, Princeton University Press, (1939). *See also:* F. Neilson, *The Makers of War*, C. C. Nelson, Appleton, Wisconsin, (1950). B.

side of the story as to how England and America entered into the First World War is not often told today, but is essential to an understanding of the political climate in Germany in the post-World War I period. Theobald von Bethmann-Hollweg, Chancellor of Germany, spoke to the Reichstag on 4 August 1914, and stated,

"THE CHANCELLOR'S SPEECH IN THE REICHSTAG,
AUGUST 4, 1914

A TERRIBLE fate is breaking over Europe. Since we won in war the respect of the world for our German Empire we have lived in peace forty-four years, and have guarded the peace of Europe. In peaceful labor we have grown strong and mighty; and people have envied us. In nervy patience we have suffered hostilities to be fanned in the east and the west, and fetters to be forged against us. The wind was sown there, and now we have the whirlwind. We wanted to go on living and working in peace, and like a silent vow, from the Emperor down to the youngest recruit, this was the will: Our sword shall not be drawn except in a just cause. Now the day has come when we must draw it. Russia has put the torch to our house. We have been forced into a war with Russia and France.

Gentlemen, a number of papers penned in the stress of hurrying events have been distributed to you. [*Footnote:* These papers the New York Times printed as 'The German White Paper,' perhaps a misnomer. While the Times deserves thanks for having published this information, the comparison of this hurried compilation with the well arranged British White Paper has been unfavorable to the cause of Germany.] Let me single out the facts which characterize our action.

From the first moment of the Austrian conflict we strove and labored that this conflict might be confined to Austria-Hungary and Servia. All the cabinets, notably the English cabinet, took the same ground, only Russia insisted that she would have to say a word. This was the beginning of the danger threatening Europe. As soon as the first definite news of military preparations in Russia reached us, we declared in St. Petersburg, kindly but firmly, that military preparations against us would force us to take similar steps, and that mobilization and war are not far apart. Russia assured us in the most friendly way that she was taking no measures against us. England in the meanwhile was trying to mediate between Austria and Servia, and was receiving our hearty support. On July 28 the Emperor telegraphed to the Czar asking him to consider that Austria had the right and the duty to protect herself against the Greater-Servian plots which threatened to undermine her existence. The Emperor called the Czar's attention to their common monarchical interest against the crime of Serajevo, and asked the

---

Freedman, *The Hidden Tyranny*, New Christian Crusade Church, Metairie, Louisiana, (1970).

Czar to help him personally to smooth away the difficulties between Vienna and St. Petersburg. At about the same time, and before he had received this telegram, the Czar asked the Emperor to help him and to counsel moderation in Vienna. The Emperor accepted the part of mediator, but he has hardly begun to act, when Russia mobilizes all her troops against Austria-Hungary. Austria-Hungary on the other hand had mobilized only her army corps on the Servian frontier, and two other corps in the north, but far removed from Russia. The Emperor at once points out to the Czar that the Russian mobilization makes his mediation, undertaken at the Czar's request, very difficult if not impossible. We nevertheless continue our mediation even to the extreme limit permitted by our alliance. During this time Russia of her own accord repeats her assurance that she is taking no military preparations against us.

Then there arrives the 31st of July. In Vienna a decision is due. We have already succeeded so far that Vienna has renewed a personal exchange of opinion with St. Petersburg, which had stopped for some time, but even before a decision is made in Vienna, we receive the news that Russia is mobilizing her entire army — that is, she is mobilizing also against us. The Russian Government, which from our repeated representations knows what a mobilization on our frontier means, does not notify us, and gives us no explanatory reply. Not until July 31st in the afternoon a telegram is received from the Czar in which he says that his army is taking no provocative attitude towards us. But — the Russian mobilization on our frontier was vigorously begun as early as during the night of July 30th. While we are still trying to mediate in Vienna at Russia's request, the whole Russian military force rises on our long, almost open frontier; and France, while she is not yet mobilizing, confesses that she is making military preparations. And we? We had intentionally refrained, up to that moment, from calling a single reservist to the colors — for the sake of the peace of Europe. Should we now be waiting any longer, until the powers between whom we are wedged in would choose their own moment of attack? To expose Germany to this danger would have been a crime! For this reason we demanded at once, on July 31st, that Russia demobilize, which action alone could still have preserved the peace of Europe. The Imperial Ambassador in St. Petersburg was simultaneously instructed to declare that we should have to consider ourselves at war with Russia, if she declined. The Imperial Ambassador has followed his instructions.

Even today we do not yet know Russia's reply to our demand that she demobilize. No telegraphic news has reached us, although the telegraph went on for a while communicating many less important matters. So it came that when the time limit was long past the Emperor was obliged to mobilize our military forces at five o'clock in the afternoon of August 1st. At the same time we had to ask for assurances as to the attitude of France. She replied to our definite inquiry whether she would be neutral in a Russian-German war by saying that she would do what her interests demanded. This was an evasion of our question if not a negative reply. The Emperor nevertheless ordered that the French frontier be respected in its entirety. This order has been rigorously obeyed with one single exception. France, who mobilized at the same hour that we did, declared that she would respect a zone of ten kilometers on our frontier. And what did really happen? Bomb throwing, flyers, cavalry scouts, and companies invading Alsace-Lorraine. Thus France attacked us before war had been declared.

As regards the one exception I mentioned, I have received this report from the General Staff: 'As regards the French complaints concerning our transgressing her frontier, only one case is to be acknowledged. Contrary to definite orders a patrol of the 14th Army Corps, led it would seem by an officer, crossed the frontier on August 2d. It appears that all were shot except one man, who returned. But long before this one act of crossing the frontier took place, French flyers dropped bombs as far from France as South Germany, and near the Schluchtpass French troops made an attack on our frontier guards. Thus far our troops have confined themselves to the protection of our frontier.' This is the report of the General Staff.

We have been forced into a state of self-defence, and the necessity of self-defence knows no other law. Our troops have occupied Luxemburg, and have perhaps already been obliged to enter Belgian territory. That is against the rules of international law. It is true that the French Government announced in Brussels that it would respect Belgian neutrality as long as its opponents would do so. But we knew that France was ready for an invasion of Belgium. France could afford to wait. We could not wait. An attack on our flank on the lower Rhine might have been fatal. We were therefore obliged to disregard the protest of the Luxemburg and Belgian governments. For the wrong we have done thereby we shall try to atone, as soon as our military end is obtained. People who like ourselves are fighting for their lives and homes must think of naught but how they may survive.

Gentlemen, we are standing shoulder to shoulder with Austria-Hungary. As regards the attitude of England, Sir Edward Grey's remarks yesterday in the lower house of Parliament have shown what her stand will be. We have assured the English Government that we shall not attack the north coast of France as long as England remains neutral, and that we shall not infringe the territorial integrity and independence of Belgium. This assurance I here repeat before the whole world; and I may add, as long as England remains neutral, we shall not even take any hostile measures against the French merchant marine, provided France will treat our merchantmen in the same way.

Gentlemen, this was the course of events. Germany enters this war with a clear conscience. We are fighting to protect the fruits of our peaceful labor, and our heritage of the great past. We are fighting for our future. The fifty years are not yet past during which Moltke used to say we should have to remain armed if we were to protect our heritage and our achievements of 1870.

Now the supreme hour has come which will test our people. But it finds us ready and full of confidence. Our army is in the field, our fleet is well prepared, and back of them stands the whole German people — *The Whole German People.*"²³⁹

The telegraphic correspondence referred to in the above speech is reproduced in *Truth about Germany: Facts about the War*, Throw Press, New York, (1914).Theobald von Bethmann-Hollweg, Chancellor of Germany, again spoke to the Reichstag in 1914 and stated, *inter alia*,

> "Where the responsibility in this greatest of all wars lies is quite evident to us. Outwardly responsible are the men in Russia who planned and carried into effect the general mobilization of the Russian army. But in reality and truth the British Government is responsible. The London Cabinet could have made war impossible if they had unequivocally told Petersburg that England was not willing to let a continental war of the Great Powers result from the Austro-Hungarian conflict with Serbia. Such words would have compelled France to use all her energy to keep Russia away from every warlike measure. Then our good offices and mediation between Vienna and Petersburg would have been successful, and there would have been no war! But England has chosen to act otherwise. She knew that the clique of powerful and partly irresponsible men surrounding the Czar were spoiling for war and intriguing to bring it about. England saw that the wheel was set a-rolling, but she did not think of stopping it. While openly professing sentiments of peace, London secretly gave St. Petersburg to understand that England stood by France and therefore by Russia too. This has been clearly and irrefutably shown by the official publications which in the meantime have come out, more particularly by the Blue Book edited by the British Government. Then St. Petersburg could no longer be restrained. In proof of this we possess the testimony of the Belgian Chargé d'Affaires at St. Petersburg, a witness who is surely beyond every suspicion. He reported (you know his words, but I will repeat them now), he reported to his Government on July 30th that 'England commenced by making it understood that she would not let herself be drawn into a conflict. Sir George Buchanan said this openly. To-day, however, everybody in St. Petersburg is quite convinced,—one has actually received the assurance—that England will stand by France. This support is of enormous weight and has contributed largely toward giving the war-party the upper hand. Up to this summer English statesmen have assured their Parliament that no treaty or agreement existed influencing England's independence of action, should a war break out, England was free to decide whether she would participate in a European war or not. Hence, there was no treaty obligation, no compulsion, no menace of the homeland which induced the English statesmen to originate the war and then at once to take part in it. The only

---

239. E. v. Mach, "Appendix A", *What Germany Wants*, Little, Brown and Company, Boston, (1914), pp. 146-153.

conclusion left is that the London Cabinet allowed this European war, this monstrous world war, because they thought it was an opportune moment with the aid of England's political confederates, to destroy the vital nerve of her greatest European competitors in the markets of the world. Therefore, England, together with Russia (I have spoken about Russia on the 4th of August), is answerable before God and man for this catastrophe which has come over Europe and over mankind."[240]

At least as early as 1908, even before the Balkan Wars, German writers were anticipating the events which would result in the "World War" with England, France and Russia; and revealed the existence of a British alliance with France to attack Germany, whether or not Germany had invaded Belgium in an act of self-defense—the pretext for the British and French declarations of war against Germany. The English-French Entente had created Belgium. The defensive German invasion of Belgium was the excuse the British and French gave for their entrance into the war—a completely unnecessary war made most horrible by the entrance of the Entente, and then made to last by the entrance of the Americans. Rudolf Emil Martin, Regierungsrat im Reichsamt des Innern in Berlin, wrote in his book *Stehen wir vor einem Weltkrieg?* F. Engelmann, Leipzig, pp. 142-145, on 30 June 1908, as quoted in his *Der Weltkrieg und sein Ende*, Rudolf Martin, Berlin, (1915), pp. 62-64:

"Eine Voraussage des Weltkrieges aus dem Jahre 1908.

In meinem am 30. Juni 1908 erschienenen Buche „Stehen wir vor einem Weltkrieg?" finden sich auf Seite 142 bis 145 folgende Ausführungen:

„Seit der Zusammenkunft der englischen Königsfamilie mit der russischen Zarenfamilie am 9. und 10. Juni auf der Rhede von Reval ist die politische Lage um vieles ernster geworden. Die eifrigen Versicherungen der russischen, englischen und französischen Blätter, daß die Zusammenkunft von Reval sich gegen Deutschland richte, bilden den besten Beweis für die hochgradige Gespanntheit der internationalen Lage. Nachdem in Paris, London und Petersburg alle Vorbereitungen zum Kampfe gegen Deutschland getroffen worden sind, pocht den verantwortlichen Leitern der Politik das Herz vor Aufregung, denn niemand weiß, wie dieser Weltkrieg enden wird. Man diskutiert in den politischen Zirkeln in Paris und London die Frage, ob Deutschland sich diese beispiellose und vollkommene Einkesselung wohl gefallen lassen werde. (Seite 142.)

„Schon heute ist sicher, daß König Eduard jede direkte

---

[240]. C. F. Horne, Editor, "Speech of the Chancellor Before the Reichstag", *Source Records of the Great War*, Volume 1, The American Legion, Indianapolis, (1931), pp. 409-415, at 409-410.

Auseinandersetzung zwischen Deutschland und Frankreich über Marokko verhindern wird. Weit hinter uns liegen die Zeiten vor und während des Burenkrieges, als in den Jahren 1899 bis 1901 Chamberlain nicht abgeneigt war, dem Deutschen Reiche einige Häfen an der atlantischen Küste Marokkos einzuräumen und sich mit Deutschland allein über Marokko zu verständigen. König Eduard ist heute entschlossen, dem eingekesselten Deutschland keinerlei Zugeständnisse zu machen. Diesen ruhigen, besonnenen Herrscher, dem jede Leidenschaft für das Militärwesen abgeht, schreckt die Möglichkeit eines Krieges gegen Deutschland nicht mehr zurück. Diese seine Stellung zu Krieg und Frieden hat König Eduard am 9. und 10. Juni 1908 vor aller Welt dargetan. Aber nur die Eingeweihten verstanden international jede Nüance des Schauspiels von Reval. Ostentativ stellte König Eduard den General French und den Admiral Fisher dem Zaren vor.

„General French ist der Generalinspektor des englischen Landheeres und Admiral Fisher ist der Höchstkommandierende der englischen Flotte. General French befehligt in dem kommenden Kriege die englische Landarmee auf dem Kontinent. Ihn mußte der Zar kennen lernen.

„Als der französische Ministerpräsident Rouvier in der zweiten Woche des Mai 1905 die letzte Hoffnung aufgab, daß es zwischen Deutschland und Frankreich über Marokko zur Verständigung kommen werde, schloß er mit England die geheime englisch-französische Militärkonvention ab, die in viel höherem Maße die Bezeichnung eines Schutz- und Trutzbündnisses verdient, als etwa das Bündnis zwischen Deutschland und Österreich-Ungarn. (Seite 143.) Deutschland hat Österreich-Ungarn nur beizustehen, wenn Österreich-Ungarn von Rußland angegriffen wird. England aber will Frankreich Beistand leisten auch in dem Falle, wenn Frankreich den Krieg gegen Deutschland eröffnet. So ist der Sinn dieser englisch-französischen Militärkonvention. Und so will es König Eduard.

„Unmittelbar nach dem Abschluß dieses wichtigsten aller gegenwärtig bestehenden Bündnisse, welches aber öffentlich noch heut in sehr geschickter Weise abgeleugnet wird, reiste General French mit zwei englischen Generalstabsoffizieren nach Frankreich, um längs der Meuse in Nordfrankreich das Terrain zu inspizieren, welches die englische Armee von 100 000 Mann unter seinem Oberbefehl zu besetzen hatte und noch hat. General French denkt gar nicht daran, diese Feststellung zu dementieren. Die Zeiten sind eben vorbei, wo man in England auf strenge Geheimhaltung des englisch-französischen Kriegsplanes Wert legte. Während General French mit seinen Generalstabsoffizieren in der Gegend von Sedan unter Führung der französischen Generalstäbler Tag für Tag studierte, besuchte der englische Botschafter in Berlin das Auswärtige Amt, um im Laufe der Unterhaltung anzudeuten, daß England im Falle eines deutsch-französischen Krieges and der Seite Frankreichs kämpfen werde.

„König Eduard weiß ganz genau, daß man in Berlin die Aufgabe des General French im Kriegsfalle kennt. Wenn König Eduard dessenungeachtet den General French und den Admiral Fisher zu dem

Familienfest in Reval zuzog, so wollte er Deutschland dadurch zu verstehen geben, daß zwischen England und Rußland eine Militärkonvention gegen Deutschland geschlossen werde. Aus dem Briefwechsel zwischen dem Deutschen Kaiser und Lord Tweedmouth ist bekannt, daß Admiral Fischer die Seele einer unternehmungslustigen Flottenpolitik ist. König Eduard will den Krieg nicht. Er will uns nur in wohlwollender Weise gewarnt haben. Wenn wir uns absolut fügen, geschieht uns nicht. (Seite 144.)

„Überdies will König Eduard den bewundernswerten Bau der diplomatischen Einkesselung Deutschlands im Frieden noch vollständig beenden. Erst in den nächsten Monaten beginnt der wichtige Schlußakt des gewaltigen Baununternehmens. Österreich-Ungarn soll uns abspenstig gemacht werden. König Eduard wird diesen schwierigsten Teil der Aufgabe persönlich übernehmen. Wahrscheinlich wird man Österreich-Ungarn die künftige Erwerbung der ganzen europäischen Türkei mit Ausnahme von Konstantinopel versprechen.

„Jetzt ist der letzte Augenblick, wo Deutschland seine Kriegsrüstung mit äußerster Energie vermehren muß, wenn es nicht schweren Schaden erleiden will. Große Bewilligungen für die Vermehrung unserer Luft- und Seemacht werden aber bei dem zerrütteten Zustand unserer Finanzen von dem Reichstag nur zu erreichen sein, wenn ihm ein großes nationales Ziel vor Augen geführt wird.

„Eine Nation, die sich derartig einkesseln läßt, gibt freiwillig ihren Rang auf. Die einzig würdige Antwort auf diese Einkesselung ist eine riesenhafte Verstärkung unserer Kriegsrüstung." (S. 145.)

Diese von mir am 30. Juni 1908 veröffentlichten Details des englisch-französischen Abkommens sind ein historischer Beweis dafür, daß England auch dann Frankreich im Weltkrieg beigestanden haben würde, wenn wir nicht durch Belgien marschiert wären."

In the 1880's Friedrich Engels anticipated the events of the First World War. Eduard Bernstein recounted that,

"Friedrich Engels had predicted something like this during the eighties when he warned me not to think lightly of a war with Russia. A war between Germany and Russia, he wrote, would automatically draw in France on the side of Russia."[241]

In 1887, Frederick Engels knew that the First World War was coming and that it would destroy the Empires of Europe and leave them ripe for revolution,

"No other war is now possible for Prussia-Germany than a world war, and indeed a world war of hitherto unimagined sweep and violence. Eight to ten

---

241. P. W. Massing, *Rehearsal for Destruction: A Study of Political Anti-Semitism in Imperial Germany*, Howard Fertig, New York, (1967), p. 325.

million soldiers will mutually kill each other off, and in the process devour Europe barer than any swarm of locusts ever did. The desolation of the Thirty Years' War compressed into three or four years and spread over the entire continent: famine, plague, general savagery, taking possession both of the armies and of the masses of the people, as a result of universal want; hopeless demoralization of our complex institutions of trade, industry and credit, ending in universal bankruptcy; collapse of the old states and their traditional statecraft, so that crowns will roll over the pavements by the dozens and no one be found to pick them up; absolute impossibility of foreseeing where this will end, or who will emerge victor from the general struggle. Only *one* result is absolutely sure: general exhaustion and the creation of the conditions for the final victory of the working class."[242]

Before America entered the war, Germany was close to winning it. They would have settled it with a comparatively large degree of restraint and justice (compared to the punitive Treaty of Versailles, orchestrated by a large cabal of Jews, which destroyed Germany), had not America interceded on behalf of England. As it happened, the Germans knew that Zionists made a deal with England to bring America into the war on England's side in exchange for the Balfour Declaration, but even before that declaration was made public and even before German Zionist financiers attempted to blackmail the German Government, the Germans knew that Wilson was maneuvering for war and sought a pretext. Wilson wanted a League of Nations and a Palestine Mandate, which would fulfill Jewish Messianic prophecy. Shortly before America declared war on Germany, *The New York Times* published the following article of 24 March 1917, on page 2,

"ACCUSES WILSON OF 'CRIMINAL ERRORS'

Berlin Paper Says 'Monstrous Guilt of War' Would Fall On His Administration.

BERLIN, March 22, (via London, March 23.)—The Lokal-Anzeiger accuses President Wilson of criminal carelessness in his conduct of American-German relations. The paper says:

'Dispatches from America and other neutral countries repeatedly play with the idea of the possibility of Germany according American ships

---

[242]. B. D. Wolfe, *Marxism: One Hundred Years in the Life of a Doctrine*, Dial Press, New York, (1965), p. 67. Wolfe cites: "From Engels's introduction to the reissue of a pamphlet by Sigismund Borkheim. Borkheim's pamphlet, *Zur Errinnerung fuer die deutschen Mordspatrioten 1806-07* [***] The introduction is reproduced in *Werke*, Vol. XXI, pp. 350-351."

different treatment from that given other neutral steamers on the ground that Germany must have an interest in avoiding a conflict with America. It seems a fact that America also is keeping alive the hope that at the last moment we may find a way to compromise with the American standpoint. After the Chancellor, as well as the other officials involved, has repeatedly emphasized that there can be no going back for us, it is only necessary now to lay stress upon the following:

'The policy of President Wilson, since the breaking off of diplomatic relations, has been characterized by careless and criminal errors. He has played with the destinies of great peoples. He desires to make his further course depend upon whether Germany commits an overt act, that is, an openly hostile action against an armed American merchantman. At the same time he lets it be known that he has commanded these armed merchantmen to open fire on their part on all submarines immediately.

'In the face of the reasons we have given the whole world as a basis for unrestricted submarine warfare, it is unparalleled rashness if the President risks the lives of American citizens in the careless belief that we will not dare to injure them. Even apart from the fact that our naval authorities declare that it is practically impossible to distinguish American from non-American merchantmen, the German Government must emphatically decline to consider any discrimination. If President Wilson rashly wants war, he should start it and he will have it. On our side it only remains to assure him that we have put an end to negotiations about submarine warfare once and for all. The monstrous guilt for a German-American war, should it come, would fall alone upon President Wilson and his Government.'"

On 2 April 1917 (Lenin left Switzerland and entered Petrograd on 3 April 1917), President Woodrow Wilson, in a speech grounded in hypocrisy, without provocation and with no vital American national interest at stake, called for the Congress of the United States of America to declare war on the German Nation,

"Gentlemen of the Congress:

I have called the Congress into extraordinary session because there are serious, very serious, choices of policy to be made, and made immediately, which it was neither right nor constitutionally permissible that I should assume the responsibility of making.

On the 3d of February last I officially laid before you the extraordinary announcement of the Imperial German Government that on and after the 1st day of February it was its purpose to put aside all restraints of law or of humanity and use its submarines to sink every vessel that sought to approach either the ports of Great Britain and Ireland or the western coasts of Europe or any of the ports controlled by the enemies of Germany within the Mediterranean. That had seemed to be the object of the German submarine warfare earlier in the war, but since April of last year the Imperial Government had somewhat restrained the commanders of its undersea craft in conformity with its promise then given to us that passenger boats should not be sunk and that due warning would be given to all other vessels which its submarines might seek to destroy, when no resistance was offered or escape attempted, and care taken that their crews were given at least a fair chance to save their lives in their open boats. The precautions taken were meagre and haphazard enough, as was proved in distressing instance after instance in the progress of the cruel and unmanly business, but a certain degree of restraint was observed The new policy has swept every restriction aside. Vessels of every kind, whatever their flag, their character, their cargo, their destination, their errand, have been ruthlessly sent to the bottom without warning and without thought of help or mercy for those on board, the vessels of friendly neutrals along with those of belligerents. Even hospital ships and ships carrying relief to the sorely bereaved and stricken people of Belgium, though the latter were provided with safe-conduct through the proscribed areas by the German Government itself and were distinguished by unmistakable marks of identity, have been sunk with the same reckless lack of compassion or of principle.

I was for a little while unable to believe that such things would in fact be done by any government that had hitherto subscribed to the humane practices of civilized nations. International law had its origin in the attempt to set up some law which would be respected and observed upon the seas, where no nation had right of dominion and where lay the free highways of the world. By painful stage after stage has that law been built up, with meagre enough results, indeed, after all was accomplished that could be accomplished, but always with a clear view, at least, of what the heart and conscience of mankind demanded. This minimum of right the German Government has swept aside under the plea of retaliation and necessity and because it had no weapons which it could use at sea except these which it is impossible to employ as it is employing them without throwing to the winds all scruples of humanity or of respect for the understandings that were supposed to underlie the intercourse of the world. I am not now thinking of the loss of property involved, immense and serious as that is, but only of the wanton and wholesale destruction of the lives of noncombatants, men, women, and children, engaged in pursuits which have always, even in the darkest periods of modern history, been deemed innocent and legitimate. Property can be paid for; the lives of peaceful and innocent people can not be. The present German submarine warfare against commerce is a warfare against mankind.

It is a war against all nations. American ships have been sunk, American lives taken, in ways which it has stirred us very deeply to learn of, but the ships and people of other neutral and friendly nations have been sunk and overwhelmed in the waters in the same way. There has been no discrimination. The challenge is to all mankind. Each nation must decide for itself how it will meet it. The choice we make for ourselves must be made with a moderation of counsel and a temperateness of judgment befitting our character and our motives as a nation. We must put excited feeling away.

Our motive will not be revenge or the victorious assertion of the physical might of the nation, but only the vindication of right, of human right, of which we are only a single champion.

When I addressed the Congress on the 26th of February last, I thought that it would suffice to assert our neutral rights with arms, our right to use the seas against unlawful interference, our right to keep our people safe against unlawful violence. But armed neutrality, it now appears, is impracticable. Because submarines are in effect outlaws when used as the German submarines have been used against merchant shipping, it is impossible to defend ships against their attacks as the law of nations has assumed that merchantmen would defend themselves against privateers or cruisers, visible craft giving chase upon the open sea. It is common prudence in such circumstances, grim necessity indeed, to endeavour to destroy them before they have shown their own intention. They must be dealt with upon sight, if dealt with at all. The German Government denies the right of neutrals to use arms at all within the areas of the sea which it has proscribed, even in the defense of rights which no modern publicist has ever before questioned their right to defend. The intimation is conveyed that the armed guards which we have placed on our merchant ships will be treated as beyond the pale of law and subject to be dealt with as pirates would be. Armed neutrality is ineffectual enough at best; in such circumstances and in the face of such pretensions it is worse than ineffectual; it is likely only to produce what it was meant to prevent; it is practically certain to draw us into the war without either the rights or the effectiveness of belligerents. There is one choice we can not make, we are incapable of making: we will not choose the path of submission and suffer the most sacred rights of our nation and our people to be ignored or violated. The wrongs against which we now array ourselves are no common wrongs; they cut to the very roots of human life.

With a profound sense of the solemn and even tragical character of the step I am taking and of the grave responsibilities which it involves, but in unhesitating obedience to what I deem my constitutional duty, I advise that the Congress declare the recent course of the Imperial German Government to be in fact nothing less than war against the Government and people of the United States; that it formally accept the status of belligerent which has thus been thrust upon it, and that it take immediate steps not only to put the country in a more thorough state of defense but also to exert all its power and employ all its resources to bring the Government of the German Empire to terms and end the war.

What this will involve is clear. It will involve the utmost practicable cooperation in counsel and action with the governments now at war with Germany, and, as incident to that, the extension to those governments of the most liberal financial credits, in order that our resources may so far as possible be added to theirs. It will involve the organization and mobilization of all the material resources of the country to supply the materials of war and serve the incidental needs of the nation in the most abundant and yet the most economical and efficient way possible. It will involve the immediate full equipment of the Navy in all respects but particularly in supplying it with the best means of dealing with the enemy's submarines. It will involve the immediate addition to the armed forces of the United States already provided for by law in case of war at least 500,000 men, who should, in my opinion, be chosen upon the principle of universal liability to service, and also the authorization of subsequent additional increments of equal force so soon as they may be needed and can be handled in training. It will involve also, of course, the granting of adequate credits to the Government, sustained, I hope, so far as they can equitably be sustained by the present generation, by well conceived taxation... .

While we do these things, these deeply momentous things, let us be very clear, and make very clear to all the world what our motives and our objects are. My own thought has not been driven from its habitual and normal course by the unhappy events of the last two months, and I do not believe that the thought of the nation has been altered or clouded by them I have exactly the same things in mind now that I had in mind when I addressed the Senate on the 22d of January last; the same that I had in mind when I addressed the Congress on the 3d of February and on the 26th of February. Our object now, as then, is to vindicate the principles of peace and justice in the life of the world as against selfish and autocratic power and to set up amongst the really free and self-governed peoples of the world such a concert of purpose and of action as will henceforth ensure the observance of those principles. Neutrality is no longer feasible or desirable where the peace of the world is involved and the freedom of its peoples, and the menace to that peace and freedom lies in the existence of autocratic governments backed by organized force which is controlled wholly by their will, not by the will of their people. We have seen the last of neutrality in such circumstances. We are at the beginning of an age in which it will be insisted that the same standards of conduct and of responsibility for wrong done shall be observed among nations and their governments that are observed among the individual citizens of civilized states.

We have no quarrel with the German people. We have no feeling towards them but one of sympathy and friendship. It was not upon their impulse that their Government acted in entering this war. It was not with their previous knowledge or approval. It was a war determined upon as wars used to be determined upon in the old, unhappy days when peoples were nowhere consulted by their rulers and wars were provoked and waged in the interest of dynasties or of little groups of ambitious men who were accustomed to use their fellow men as pawns and tools. Self-governed nations do not fill their neighbour states with spies or set the course of intrigue to bring about some critical posture of affairs which will give them an opportunity to strike and make conquest. Such designs can be successfully worked out only under cover and where no one has the right to ask questions. Cunningly contrived plans of deception or aggression, carried, it may be, from generation to generation, can be worked out and kept from the light only within the privacy of courts or behind the carefully guarded confidences of a narrow and privileged class. They are happily impossible where public opinion commands and insists upon full information concerning all the nation's affairs.

A steadfast concert for peace can never be maintained except by a partnership of democratic nations. No autocratic government could be trusted to keep faith within it or observe its covenants. It must be a league of honour, a partnership of opinion. Intrigue would eat its vitals away; the plottings of inner circles who could plan what they would and render account to no one would be a corruption seated at its very heart. Only free peoples can hold their purpose and their honour steady to a common end and prefer the interests of mankind to any narrow interest of their own.

Does not every American feel that assurance has been added to our hope for the future peace of the world by the wonderful and heartening things that have been happening within the last few weeks in Russia? Russia was known by those who knew it best to have been always in fact democratic at heart, in all the vital habits of her thought, in all the intimate relationships of her people that spoke their natural instinct, their habitual attitude towards life. The autocracy that crowned the summit of her political structure, long as it had stood and terrible as was the reality of its power, was not in fact Russian in origin, character, or purpose; and now it has been shaken off and the great, generous Russian people have been added in all their naive majesty and might to the forces that are fighting for freedom in the world, for justice, and for peace. Here is a fit partner for a league of honour.

One of the things that has served to convince us that the Prussian autocracy was not and could never be our friend is that from the very outset of the present war it has filled our unsuspecting communities and even our offices of government with spies and set criminal intrigues everywhere afoot against our national unity of counsel, our peace within and without our industries and our commerce. Indeed it is now evident that its spies were here even before the war began; and it is unhappily not a matter of conjecture but a fact proved in our courts of justice that the intrigues which have more than once come perilously near to disturbing the peace and dislocating the industries of the country have been carried on at the instigation, with the support, and even under the personal direction of official agents of the Imperial Government accredited to the Government of the United States. Even in checking these things and trying to extirpate them we have sought to put the most generous interpretation possible upon them because we knew that their source lay, not in any hostile feeling or purpose of the German people towards us (who were, no doubt, as ignorant of them as we ourselves were), but only in the selfish designs of a Government that did what it pleased and told its people nothing. But they have played their part in serving to convince us at last that that Government entertains no real friendship for us and means to act against our peace and security at its convenience. That it means to stir up enemies against us at our very doors the intercepted note to the German Minister at Mexico City is eloquent evidence.

We are accepting this challenge of hostile purpose because we know that in such a government, following such methods, we can never have a friend; and that in the presence of its organized power, always lying in wait to accomplish we know not what purpose, there can be no assured security for the democratic governments of the world. We are now about to accept gage of battle with this natural foe to liberty and shall, if necessary, spend the whole force of the nation to check and nullify its pretensions and its power. We are glad, now that we see the facts with no veil of false pretence about them, to fight thus for the ultimate peace of the world and for the liberation of its peoples, the German peoples included: for the rights of nations great and small and the privilege of men everywhere to choose their way of life and of obedience. The world must be made safe for democracy. Its peace must be planted upon the tested foundations of political liberty. We have no selfish ends to serve. We desire no conquest, no dominion. We seek no indemnities for ourselves, no material compensation for the sacrifices we shall freely make. We are but one of the champions of the rights of mankind. We shall be satisfied when those rights have been made as secure as the faith and the freedom of nations can make them.

Just because we fight without rancour and without selfish object, seeking nothing for ourselves but what we shall wish to share with all free peoples, we shall, I feel confident, conduct our operations as belligerents without passion and ourselves observe with proud punctilio the principles of right and of fair play we profess to be fighting for.

I have said nothing of the governments allied with the Imperial Government of Germany because they have not made war upon us or challenged us to defend our right and our honour. The Austro-Hungarian Government has, indeed, avowed its unqualified endorsement and acceptance of the reckless and lawless submarine warfare adopted now without disguise by the Imperial German Government, and it has therefore not been possible for this Government to receive Count Tarnowski, the Ambassador recently accredited to this Government by the Imperial and Royal Government of Austria-Hungary; but that Government has not actually engaged in warfare against citizens of the United States on the seas, and I take the liberty, for the present at least, of postponing a discussion of our relations with the authorities at Vienna. We enter this war only where we are clearly forced into it because there are no other means of defending our rights.

It will be all the easier for us to conduct ourselves as belligerents in a high spirit of right and fairness because we act without animus, not in enmity towards a people or with the desire to bring any injury or disadvantage upon them, but only in armed opposition to an irresponsible government which has thrown aside all considerations of humanity and of right and is running amuck. We are, let me say again, the sincere friends of the German people, and shall desire nothing so much as the early reestablishment of intimate relations of mutual advantage between us — however hard it may be for them, for the time being, to believe that this is spoken from our hearts. We have borne with their present government through all these bitter months because of that friendship — exercising a patience and forbearance which would otherwise have been impossible. We shall, happily, still have an opportunity to prove that friendship in our daily attitude and actions towards the millions of men and women of German birth and native sympathy, who live amongst us and share our life, and we shall be proud to prove it towards all who are in fact loyal to their neighbours and to the Government in the hour of test. They are, most of them, as true and loyal Americans as if they had never known any other fealty or allegiance. They will be prompt to stand with us in rebuking and restraining the few who may be of a different mind and purpose. If there should be disloyalty, it will be dealt with with a firm hand of stern repression; but, if it lifts its head at all, it will lift it only here and there and without countenance except from a lawless and malignant few.

It is a distressing and oppressive duty, gentlemen of the Congress, which I have performed in thus addressing you. There are, it may be, many months of fiery trial and sacrifice ahead of us. It is a fearful thing to lead this great peaceful people into war, into the most terrible and disastrous of all wars, civilization itself seeming to be in the balance. But the right is more precious than peace, and we shall fight for the things which we have always carried nearest our hearts — for democracy, for the right of those who submit to authority to have a voice in their own governments, for the rights and liberties of small nations, for a universal dominion of right by such a concert of free peoples as shall bring peace and safety to all nations and make the world itself at last free. To such a task we can dedicate our lives and our fortunes, everything that we are and everything that we have, with the pride of those who know that the day has come when America is privileged to spend her blood and her might for the principles that gave her birth and happiness and the peace which she has treasured. God helping her, she can do no other."[243]

---

[243]. W. Wilson, "War Message", Sixty-Fifth Congress, First Session, Senate Document Number 5, Serial Number 7264, Washington, D.C., (1917) pp. 3-8.

6 ZIONISM IS RACISM

Jews have always been tribalistic and racist. Ancient Jews dubbed themselves the "chosen people" of a racist and genocidal God, and in so doing justified their racism and bloodlust with religion. Institutionalizing their racism as a religion guaranteed them that their progeny would remain forever segregated from the outside world of sub-human "cattle". The racism must have come before the religious mythology, because Jewish religious mythology is based upon supremacist racism.

> "The General Assembly [***] Determines that Zionism is a form of racism and racial discrimination."—UNITED NATIONS GENERAL ASSEMBLY RESOLUTION NUMBER 3379[244]

> "For thou *art* an holy people unto the LORD thy God: the LORD thy God hath chosen thee to be a special people unto himself, above all people that *are* upon the face of the earth."—DEUTERONOMY 7:6

6.1 Introduction

Deuteronomy, Chapter 7, states,

> "When the LORD thy God shall bring thee into the land whither thou goest to possess it, and hath cast out many nations before thee, the Hittites, and the Girgashites, and the Amorites, and the Canaanites, and the Perizzites, and the Hivites, and the Jebusites, seven nations greater and mightier than thou; 2 And when the LORD thy God shall deliver them before thee; thou shalt smite them, *and* utterly destroy them; thou shalt make no covenant with them, nor show mercy unto them: 3 Neither shalt thou make marriages with them; thy daughter thou shalt not give unto his son, nor his daughter

---

[244]. "3379 (XXX). Elimination of All Forms of Racial Discrimination", General Assembly—Thirtieth Session, Resolutions adopted on the reports of the Third Committee, 2400th Plenary Meeting, (10 November 1975), pp. 83-84. URL:

http://www.un.org/documents/ga/res/30/ares30.htm

Confer: Zionism & Racism: Proceedings of an International Symposium, International Organization for the Elimination of All Forms of Racial Discrimination, Tripoli, (1977), pp. 249-250. *Cf.* F. A. Sayegh, *Zionism: A Form of Racism And Racial Discrimination" Four Statements Made at the U.N. General Assembly*, Office of the Permanent Observer of the Palestine Liberation Organization to the United Nations, (1976), pp. 40-41. URL:

http://www.ameu.org/uploads/sayegh_march1_03.pdf

After the fall of the Soviet Union, which had long sponsored racial integration (*see:* "Circus" a motion picture released in 1936 directed by Grigori Alexandrov starring Lyubov Orlova), the U. N. withdrew this resolution under great pressure from Zionists.

shalt thou take unto thy son. 4 For they will turn away thy son from following me, that they may serve other gods: so will the anger of the LORD be kindled against you, and destroy thee suddenly. 5 But thus shall ye deal with them; ye shall destroy their altars, and break down their images, and cut down their groves, and burn their graven images with fire. 6 For thou *art* an holy people unto the LORD thy God: the LORD thy God hath chosen thee to be a special people unto himself, above all people that *are* upon the face of the earth. 7 The LORD did not set his love upon you, nor choose you, because ye were more in number than any people; for ye *were* the fewest of all people: 8 But because the LORD loved you, and because he would keep the oath which he had sworn unto your fathers, hath the LORD brought you out with a mighty hand, and redeemed you out of the house of bondmen, from the hand of Pharaoh king of Egypt. 9 Know therefore that the LORD thy God, he *is* God, the faithful God, which keepeth covenant and mercy with them that love him and keep his commandments to a thousand generations; 10 And repayeth them that hate him to their face, to destroy them: he will not be slack to him that hateth him, he will repay him to his face. 11 Thou shalt therefore keep the commandments, and the statutes, and the judgments, which I command thee *this* day, to do them. 12 Wherefore it shall come to pass, if ye hearken to these judgments, and keep, and do them, that the LORD thy God shall keep unto thee the covenant and the mercy which he sware unto thy fathers: 13 And he will love thee, and bless thee, and multiply thee: he will also bless the fruit of thy womb, and the fruit of thy land, thy corn, and thy wine, and thine oil, the increase of thy kine, and the flocks of thy sheep, in the land which he sware unto thy fathers to give thee. 14 Thou shalt be blessed above all people: there shall not be male or female barren among you, or among your cattle. 15 And the LORD will take away from thee all sickness, and will put none of the evil diseases of Egypt, which thou knowest, upon thee; but will lay them upon all *them* that hate thee. 16 And thou shalt consume all the people which the LORD thy God shall deliver thee; thine eye shall have no pity upon them: neither shalt thou serve their gods; for that *will be* a snare unto thee. 17 If thou shalt say in thine heart, These nations *are* more than I; how can I dispossess them? 18 Thou shalt not be afraid of them: *but* shalt well remember what the LORD thy God did unto Pharaoh, and unto all Egypt; 19 The great temptations which thine eyes saw, and the signs, and the wonders, and the mighty hand, and the stretched out arm, where*by* the LORD thy God brought thee out: so shall the LORD thy God do unto all the people of whom thou *art* afraid. 20 Moreover the LORD thy God will send the hornet among them, until they that are left, and hide themselves from thee, be destroyed. 21 Thou shalt not be affrighted at them: for the LORD thy God *is* among you, a mighty God and terrible. 22 And the LORD thy God will put out those nations before thee by little and little: thou mayest not consume them at once, lest the beasts of the field increase upon thee. 23 But the LORD thy God shall deliver them unto thee, and shall destroy them *with* a mighty destruction, until they be destroyed. 24 And he shall deliver their kings into thine hand, and thou shalt destroy their name from under heaven: there shall no man *be able to* stand before thee, until thou have destroyed them. 25 The graven images of their gods shall ye burn with fire: thou shalt not desire the silver or gold *that is* on them, nor take *it* unto thee, lest thou be snared therin: for it *is* an abomination to the LORD thy God. 26 Neither shalt thou bring an abomination into thine house, lest thou be a cursed thing like it: *but* thou shalt utterly detest it, and thou shalt utterly abhor it; for it *is* a cursed thing."

Rabbi Dr. J. Loeph wrote in an article entitled, "Jüdischer Volksbegriff", in the *Central-Verein Zeitung*, Volume 1, Number 2, (11 May 1922), p. 29,

"Jüdischer Volksbegriff.
Von Rabbiner Dr. J. Loeph.

Der Begriff des „Jüdischen Volkes" leidet in seiner Bedeutung unter derselben Unklarheit, die in der Regel mit dem Begriffe „Volk" überhaupt verbunden ist. Man muß hier scharf zwischen sprachlicher Herleitung und dem herausgebildeten, mit Synonymen arbeitenden Sprachgebrauch unterscheiden, obwohl nicht zu leugnen ist, daß häufig im sprachlichen Ursprung schon der scheinbar weit davon entfernte Sinn des späteren Sprachgebrauchs verdeckt enthalten ist.

Beim Herausschälen der ursprünglichen Bedeutung von „Jüdischem Volk" tut man am besten, auf die hebräischen Bezeichnungen für „Volk" zurückzugehen. Es scheiden zunächst aus als Sammelbegriffe engerer Art *Mischpacha*=Familie, *Beth-aboth*=Sippe, *Schebet*=Stamm. Für „Volk" hat die hebräische Sprache zwei Bezeichnungen, die häufig als Synonyma miteinander abwechseln, im Grunde aber ganz verschieden in ihrer Herleitung und rechten Anwendung sind: *Goj* und *Am*. *Goj* hängt mit der Wurzel *Gew*=Rücken, Rückgrat, aram. *Gew*.=das Innerste zusammen. Wies dieses ein von Natur fest zusammenhängendes homogenes Ganzes ist (Skelett), als wenig veränderlicher Halt für das angeschlossene, ständig Veränderliche, so stellt das Wort *Goj* zweifellos in seiner ursprünglichen Bedeutung den Begriff des von einem Ahnherrn ausgehenden, in fortlaufender Geschlechtsfolge sich ausbreitenden und abzweigenden Stammes dar, der zum Volke sich weitet. Das Kennzeichnende ist die Abstammung oder gemeinsamer ererbter Landbesitz, letzteres besonders in der Mehrzahl. Die Zusammengehörigkeit ist eine natürliche und braucht nicht bewußt zu sein. Es ist das griechische *Ethnos*—Volksstamm, Menschenklasse, wie die Septuaginta *Goj* stets übersetzt.

Am hängt grammatisch mit *Im*—„mit" zusammen und bedeutet einen bewußten, auf Kultur und Schicksalsgemeinschaft beruhenden Zusammenschluß stammlich oft ganz verschiedener Individuen und Körperschaften. Die Septuaginta übersetzt es regelmäßig mit *laos*—Volkshaufe, Masse, Menge von zusammengekommenen Menschen. Daher nennt Gott Israel selten *Goj*, wenn er nämlich den seinem Dienste geweihten Stamm (*Kadosch*) meint oder ihn als solchen mit anderen Völkerschaften vergleicht, meistens aber *Am*, wenn er sein persönliches Verhältnis zu der freiwillig ihm sich anschließenden, seinem Schutze anvertrauten, seiner Liebe oder Strafe im Schicksal zugewiesenen Gemeinschaft hervorheben will. Die jüdische Religions- und Schicksalsgemeinschaft „Israel" wird nie als *Goj*, sondern stets als *Am* bezeichnet, weshalb auch Gott sein Volk niemals *Goji* (die einzige widersprechende Stelle im Zephanja, II, 9 ist ohne Bedeutung, da es hier ganz deutlich nicht auf die Bedeutung, sondern lediglich auf die Herstellung des Parallelismus ankommt), sondern stets *Ammi*, „mein Volk", nennt, weil die Zugehörigkeit zu Gott weniger auf der Abstammung von Abraham beruht—wenn diese auch nicht ganz außer acht gelassen ist—, als auf dem Wandel in Gottes Wegen, der durch den Gehorsam gegen seine besonderen, dem Volke Israel gegebenen Gebote zum Ausdruck kommt.

Im gegenwärtigen Sprachgebrauch verstehen die verschiedenen jüdischen Richtungen unter „Jüdischem Volk" je nach ihrer Stellungnahme zum Rasse-, Glaubens- und jüdisch-politischen Standpunkt verschiedenes. Man muß also immer wissen, wer der Sprecher ist, um zu wissen, was mit „Jüdischem Volk" gemeint ist."

6.2 Political Zionism is a Form of Racism

Political Zionism has often been condemned as a form of racism by Jew and Gentile alike. The United Nations General Assembly passed a resolution number 3379 condemning Zionism as racism on 10 November 1975:

"3379 (XXX). Elimination of all forms of racial discrimination
The General Assembly,
Recalling its resolution 1904 (XVIII) of 20 November 1963, proclaiming the United Nations Declaration on the Elimination of All Forms of Racial Discrimination, and in particular its affirmation that 'any doctrine of racial differentiation or superiority is scientifically false, morally condemnable, socially unjust and dangerous' and its expression of alarm at 'the manifestations of racial discrimination still in evidence in some areas in the world, some of which are imposed by certain Governments by means of legislative, administrative or other measures',
Recalling also that, in its resolution 3151 G (XXVIII) of 14 December 1973, the General Assembly condemned, *inter alia*, the unholy alliance between South African racism and zionism,
Taking note of the Declaration of Mexico on the Equality of Women and Their Contribution to Development and Peace, 1975,[4] proclaimed by the World Conference of the International Women's Year, held at Mexico City from 19 June to 2 July 1975, which promulgated the principle that 'international co-operation and peace require the achievement of national liberation and independence, the elimination of colonialism and neo-colonialism, foreign occupation, zionism, *apartheid* and racial discrimination in all its forms, as well as the recognition of the dignity of

peoples and their right to self-determination',

Taking note also of resolution 77 (XII) adopted by the Assembly of Heads of State and Government of the Organization of African Unity at its twelfth ordinary session,[5] held in Kampala from 28 July to 1 August 1975, which considered 'that the racist régime in occupied Palestine and racist régimes in Zimbabwe and South Africa have a common imperialist origin, forming a whole and having the same racist structure and being organically linked in their policy aimed at repression of the dignity and integrity of the human being',

Taking note also of the Political Declaration and Strategy to Strengthen International Peace and Security and to Intensify Solidarity and Mutual Assistance among Non-Aligned Countries,[6] adopted at the Conference of Ministers for Foreign Affairs of Non-Aligned Countries held at Lima from 25 to 30 August 1975, which most severely condemned zionism as a threat to world peace and security and called upon all countries to oppose this racist and imperialist ideology,

Determines that Zionism is a form of racism and racial discrimination.

2400th plenary meeting
10 November 1975"[245]

This resolution was revoked in 1991, when the Zionist influence increased in the United Nations, in part due to the fall of the Soviet Union.

When confronted with the facts some racist Zionists and some of their advocates, including Einstein and many of Einstein's advocates, too often resort to smear tactics in lieu of reasoned arguments. The *Executive Council of the International Organization for the Elimination of All Forms of Racial Discrimination* stated,

"On 10 November 1975 the General Assembly of the United Nations

---

245. "3379 (XXX). Elimination of All Forms of Racial Discrimination", General Assembly—Thirtieth Session, Resolutions adopted on the reports of the Third Committee, 2400th Plenary Meeting, (10 November 1975), pp. 83-84. URL:

http://www.un.org/documents/ga/res/30/ares30.htm

Confer: Zionism & Racism: Proceedings of an International Symposium, International Organization for the Elimination of All Forms of Racial Discrimination, Tripoli, (1977), pp. 249-250. *Cf.* F. A. Sayegh, *Zionism: A Form of Racism And Racial Discrimination" Four Statements Made at the U.N. General Assembly*, Office of the Permanent Observer of the Palestine Liberation Organization to the United Nations, (1976), pp. 40-41. URL:

http://www.ameu.org/uploads/sayegh_march1_03.pdf

After the fall of the Soviet Union, which had long sponsored racial integration (*see:* "Circus" a motion picture released in 1936 directed by Grigori Alexandrov starring Lyubov Orlova), the U. N. withdrew this resolution under great pressure from Zionists.

adopted resolution 3379 (XXX) determining 'that Zionism is a form of racism and racial discrimination.' The response of Zionists and their supporters to this resolution was, not to attempt to demonstrate that the finding was in error, but to mount a campaign designed to discredit the UN and to impugn the motives of the 72 member states voting in support of it."[246]

Dr. Fayez A. Sayegh stated,

"[...]I am not chagrined by verbal abuse—by the insolent railing, the name-calling, to which the Delegation of the United States has resorted, both inside and outside the United Nations, ever since 3 October. 'Perverse,' 'obscene,' 'indecent,' 'lies'—these words have graced and punctuated the statements of the representatives of the United States. I am not chagrined and I am not disconcerted. Long, long ago, in my first elementary course in philosophy, I was told by my professors: Only he who has no argument resorts to name-calling.[47] Name-calling is no substitute for rational discourse; name-calling is an admission of intellectual bankruptcy."[247]

6.3 Most Jews Opposed Zionism

Following the Russian revolution and other Bolshevist takeovers, there was a strong backlash against Zionists and Bolshevists, who avowed segregationist and revolutionary stances that would render obliging persons disloyal to the nations in which they lived. Some successfully and unfairly portrayed all Jews, Bolshevists, Zionists, Anarchists and Social Democrats as if one body, though nothing could have been further from the truth. Those who wanted to stigmatize all Jews based upon the actions and beliefs of some Jews had an easier time of it, because the Zionists presumed to speak for all Jews. Of course, these radical speeches by radical Zionist "political Messiahs" only presumed to speak for all Jews, while in reality most Jews opposed this ultra-nationalistic ancient bigotry; as even the Zionist Bolshevist Adolf Hitler was forced to concede. Hitler, though pretending to doubt what he observed, wrote,

"[T]his was the *Zionists*. It looked, to be sure, as though only a part of the Jews approved this viewpoint, while the great majority condemned and

---

[246]. F. A. Sayegh, *Zionism: A Form of Racism And Racial Discrimination" Four Statements Made at the U.N. General Assembly*, Office of the Permanent Observer of the Palestine Liberation Organization to the United Nations, (1976), pp. 51-52.

<http://www.ameu.org/uploads/sayegh_march1_03.pdf>

[247]. Preface, *Zionism & Racism: Proceedings of an International Symposium*, International Organization for the Elimination of All Forms of Racial Discrimination, Tripoli, (1977), p. vii.

inwardly rejected such a formulation."[248]

In 1910, the eleventh edition of the *Encyclopædia Britannica* stated in its article on Theodor Herzl, the most successful advocate of political Zionism,

> "[Herzl] unexpectedly gained the accession of many Jews by race who were indifferent to the religious aspect of Judaism, but he quite failed to convince the leaders of Jewish thought, who from first to last remained (with such conspicuous exceptions as Nordau and Zangwill) deaf to his pleading."

and in its article on "Zionism",

> "Dr Herzl was joined by a number of distinguished Jewish literary men, among whom were Dr Max Nordau and Mr Israel Zangwill, and promises of support and sympathy reached him from all parts of the world. The *haute finance* and the higher rabbinate, however, stood aloof."

Political Zionism has always been a racist doctrine. Moses Hess' Zionist book *Rome and Jerusalem: A Study in Jewish Nationalism* of 1862 was blatantly racist and mirrored Herbert Spencer's "Social Darwinism".[249] In 1895, before the appearance of Herzl's *The Jewish State*, Zionist Yehiel Michael Pines stated,

> "The Jewish people is a race that is not by its nature capable of absorbing such an alien implantation."[250]

Zionists spoke in racist terms throughout the history of the movement in its various forms and incarnations.[251] Gerhard Holdheim stated in 1930,

> "The Zionist programme encompasses the conception of a homogenous, indivisible Jewry on a national basis. The criterion for Jewry is hence not a confession of religion, but the all-embracing sense of belonging to a racial community that is bound together by ties of blood and history and which is determined to keep its national individuality."[252]

---

248. A. Hitler, English translation by Ralph Manheim, *Mein Kampf*, Houghton Mifflin, Boston, New York, (1971), pp. 56-57.
249. M. Hess, *Rom und Jerusalem: die letzte Nationalitätsfrage*, Eduard Wengler, Leipzig, (1862); English: *Rome and Jerusalem: A Study in Jewish Nationalism*, Bloch, New York, (1918).
250. Y. M. Pines, quoted and translated in A. Hertzberg, *The Zionist Idea*, Harper Torchbooks, New York, (1959), p. 412.
251. *See, for example:* A. Hertzberg, *The Zionist Idea*, Harper Torchbooks, New York, (1959), pp. 120-121, 187, 188, 271, 356, 422, 481, 498-499, 560-561.
252. G. Holdheim, "Der Zionismus in Deutschland", *Süddeutsche Monatshefte*, Volume 12, (1930), p. 855; English translation in: K. Polkehn, "The Secret Contacts: Zionism and Nazi Germany, 1933-1941", *Journal of Palestine Studies*, Volume 5, Number 3/4, (Spring-Summer, 1976), pp. 54-82, at 57.

In 1930 the *Central-Verein*, an institution devoted to protecting Jews from anti-Semitism, publicly confronted the Zionists' allegiance to the Nazis and to Nazi ideology in the Party's official organ.[253] The Zionists called for the extermination of the *CV*, and the *CV* declared that it would fight back on behalf of the vast majority of German Jews, who wished to remain German. Most German Jews fought against the Zionists in the war the Zionist Jews, whom they likened to Hitler, had declared on patriotic German Jews. In their war against patriotic German Jews, whom the Zionist Jews considered abnormal, unreasonable and improper Jews,[254] the Zionists openly allied themselves with the the anti-Semites in the German Zionist Party's official organ the *Jüdische Rundshau*, on 13 June 1933, shortly after Hitler assumed power,

> "Zionism recognizes the existence of the Jewish question and wants to solve it in a generous and constructive manner. For this purpose, it wants to enlist the aid of all peoples; those who are friendly to the Jews as well as those who are hostile to them, since according to its conception, this is not a question of sentimentality, but one dealing with a real problem in whose solution all peoples are interested."[255]

The eleventh edition of the *Encyclopædia Britannica* in its article on "Zionism" of 1911 states,

> "Mendelssohnian culture, by promoting the study of Jewish history, gave a fresh impulse to the racial consciousness of the Jews. The older nationalism had been founded on traditions so remote as to be almost mythical; the new race consciousness was fed by a glorious martyr history, which ran side by side with the histories of the newly adopted nationalities of the Jews, and was not unworthy of the companionship. From this race consciousness came a fresh interest in the Holy Land. It was an ideal rather than a politico-nationalist interest—a desire to preserve and cherish the great monument of the departed national glories. It took the practical form of projects for improving the circumstances of the local Jews by means of schools, and for reviving something of the old social condition of Judea by the establishment of agricultural colonies. In this work Sir Moses Montefiore, the Rothschild family, and the Alliance Israélite Universelle were conspicuous. More or less passively, however, the older nationalism still lived on—especially in

---

253. *Central-Verein Zeitung*, Volume, 9, Number 28, (11 July 1930); and Volume 9, Number 37, (12 September 1930); and Volume 9, Number 38, (19 September 1930). K. Polkehn, "The Secret Contacts: Zionism and Nazi Germany, 1933-1941", *Journal of Palestine Studies*, Volume 5, Number 3/4, (Spring-Summer, 1976), pp. 54-82.
254. *Central-Verein Zeitung*, Volume, 9, Number 28, (11 July 1930), p. 2.
255. English translation in: K. Polkehn, "The Secret Contacts: Zionism and Nazi Germany, 1933-1941", *Journal of Palestine Studies*, Volume 5, Number 3/4, (Spring-Summer, 1976), pp. 54-82, at 59.

lands where Jews were persecuted—and it became strengthened by the revived race consciousness and the new interest in the Holy Land."

and in its article on "Anti-Semitism", the eleventh edition of the *Encyclopædia Britannica* wrote,

> "In the first place there is the so-called Zionist movement, which is a kind of Jewish nationalism and is vitiated by the same errors that distinguish its anti-Semitic analogue (see ZIONISM)."

Constantin Brunner stated in 1918 that political Zionists were a worse enemy to Jews than were political anti-Semites. Political Zionists and political anti-Semites were close allies who paradoxically found common ground in bigoted segregation and who demonstrated the universality of human weaknesses. Constantin Brunner stated,

> "Wer vertritt ihre Interessen, wer spricht denn überhaupt über die deutschen Juden außer den Judenhassern und — außer solchen, die in der Wahrheit ganz anderes vertreten als die wirklichen Interessen der deutschen Juden: die aber für die Vertreter der deutschen Juden genommen werden und damit deren Lage noch verschlimmern. Die lautesten Sprecher nämlich sind die aus andern, aus den östlichen Ländern eingewanderten Juden, die natürlich nicht sogleich ins deutsche Wesen hinein umwachsen: es bedarf (wovon später mehr) dreier Generationen, bis die Erziehung zur Nation vollendet ist, — zum Gentleman gar sind, wie die Engländer sagen, vier Generationen glücklicher Bedingungen nötig. Unmöglich können die neu eingewanderten Juden als zur deutschen Nation gehörig sich ansehen (so wenig wie Kants Großvater sich so ansehen konnte: Abstammung aus demselben Lande, Gemeinsamkeit der Geburt verbindet am leichtesten zur Nation, welches Wort von dem Worte natus, Geburt sich herleitet — das ist aber etwas ganz anderes als gemeinsamer Rassenursprung!), und sie dürfen sich nicht wundern, wenn sie von den Deutschen als Fremde angesehen werden. Auch den Deutschen jüdischer Abstammung sind sie fremd, ja ich sage nicht zu viel, wenn ich sage, sie sind manchem von diesen genau so fremd und unsympathisch, wie sie manchen Nichtjuden und wie manchen Nichtjuden die Juden überhaupt sind. Juden, die sich keinerlei Antisemitismus anders denn als Niederträchtigkeit vorstellen können, möchte ich raten, diese hier berührte Abneigung von Juden gegenüber Juden zu studieren: eine menschliche Schwäche, ein menschlicher Fehler, aber niederträchtig darf das nicht genannt werden, oder es sind alle die vielen Juden mit dieser Abneigung ebenso niederträchtig — als Nichtjuden geboren, wären sie Antisemiten. Die meisten jüdischen Deutschen hegen ein Vorurteil, manche ein sehr häßliches, gegen die neu eingewanderten Juden, und auch wo dies nicht der Fall ist, das bleibt doch immer: jene neu Eingewanderten haben nicht das Vaterland mit ihnen gemein und nicht das Sprachvaterland, und, selbst soweit sie Deutsch reden, nicht das

Aussprachvaterland (was so viel ausmacht schon zwischen Nord- und Süddeutschen — wo leider noch so manches ausmacht!). Diese neu eingewanderten Juden vertreten einseitig das Religiöse, oder sie versinken schnell in den unter uns grassierenden Ästhetismus und die entkräftende Nietzscherei (weil sie, ohne die Tradition unsrer Kultur, bei starker Anpassungsfähigkeit und Heißhunger, sich anzupassen, urteilslos der herrschenden Mode verfallen); und sie, die Unglücklichen, die kein Vaterland haben, weder dort wo ihre Wiege stand, noch unter uns, wo ihre Gräber stehen werden, sie sind die Träger der zionistischen Sehnsucht. Durch diese Juden fremder Länder fast ebenso sehr, wenn nicht noch mehr wie durch die Judenhasser, werden viele unter uns konfus gemacht und beeinträchtigt in ihrer deutschen Haltung.

Der Zionismus und der Judenhaß hängen aber aufs engste zusammen, wie Wirkung und Ursache. Der Zionismus ist die verkehrte Reaktivität der Juden, der Hereinfall der Juden auf den rassentheoretischen Judenhaß, — solcher Juden, die nicht einsehen können, daß es mit der Emanzipation langsam geht und unmöglich ohne Rückfälle vorangehen kann; welche Rückfälle also, bei der Natur der Menschheit und ihrer Geschichte, von psychologischer und historischer Berechtigung und Notwendigkeit sind. Historisch und psychologisch natürlich und unausbleiblich waren die politischen Rückschritte, die es bis zum Jahre 1869 gab, und ist auch — da seitdem, seit der damals ausgesprochenen verfassungsmäßigen völligen Emanzipation ein politischer Rückschritt nicht mehr möglich —, ist um so eher der gesellschaftliche Rückschritt, wie wir ihn jetzt erleben. Die staatlich anerkannte Freiheit und die gesellschaftlich anerkannte Freiheit sind zweierlei, trotzdem Staat und Gesellschaft im Grunde dasselbe sind und, was der Staat tut, die Gesellschaft tut. Aber jegliches Tun hat zweierlei Gesichter: bevor es getan und nachdem es getan ist; sowohl das rechte wie das verkehrte Tun hat diese zweierlei Gesichter. Die staatliche Emanzipation der Juden war das Tun der Gesellschaft vor der Verwirklichung: die eigentliche Emanzipation ist erst die der Wirklichkeit in der Gesellschaft; diese Emanzipation kann unmöglich so schnell in Gestaltung der Freiheit und alles Leben sich umsetzen, wie sie auf dem Papier der Verfassung vollständig geschrieben steht, aber sie hat doch bereits begonnen sich umzusetzen, das andre Gesicht der vollzogenen Emanzipation zeigt sich, und dagegen reagiert nun die Gesellschaft, als hätte sie gar nicht gewollt, was sie getan hat. Sie versteht sich selber nicht, sie hat wohl gewollt, sie will auch weiter (weil sie muß): sie kann nur noch nicht. Sie wird immer besser können, je mehr sie muß, und je mehr man ihr von dem abkämpft, was sie „geschenkt" hatte. Hier von Geschenk zu reden, das gehört zur Selbstglorifikation der Menschen — Geschenke haben oftmals gute Gründe anderswoher als aus Zucker und Freiheit, und gar Freiheit?! Freiheit wird niemals geschenkt und kann niemals geschenkt werden, sie will erkämpft sein in langem Kampfe, darin es nicht immer nur Siege geben kann; und wie selber das Siegen immer auch ein Stück Unterliegen und Verlieren mit sich bringt, so haben ebenfalls die Niederlagen ihr Wertvolles.

Was läßt sich Tröstlicheres und Wahreres sagen als das Sprichwort: „Ein Unglück ist besser als alle Ratschläge." Gut auch liest man bei Beaconsfield: „Ein Fehlschlag ist nichts, er kann verdient sein oder man kann ihm abhelfen: im ersten Falle bringt er Selbsterkenntnis, im zweiten ruft er eine neue Kombination hervor, die gewöhnlich siegreich ist." Aber die Menschen im allgemeinen, und also auch die Juden im allgemeinen, haben kürzere Gedanken und sind gar zu bald entmutigt; hinzu kommt noch der große Tiefstand der Emanzipationsidee in einigen Ländern, wo noch die Juden in mittelalterlichem Elend leben; dadurch wurden viele Juden unter uns vollends niedergeschlagen und verwirrt. So sind sie hereingefallen auf die Rassentheorie der Judenhasser weit schlimmer als andre Deutsche; kopfunten stürzten sie in den Abgrund [*Footnote:* Das ist kein erfundener Scherz, sondern man kann es bei Zollschan, „Das Rassenproblem" nachlesen, wie der Zionismus den Chamberlain zum Lehrmeister nimmt und dessen unsinnwüsteste Offenbarungen nachlallt.] Die übrigen Deutschen sind beinah ohne Rassenerinnerung, abgerechnet die Adligen, die aber gleichfalls allesamt immer noch tausendmal besser als mit ihren Vorfahren, mit Abraham, Isaak und Jakob Bescheid wissen — das sind Vorfahren, mit denen alle Deutschen Bescheid wissen, und mit Christus wissen alle Deutschen Bescheid: statt der Überlieferung von ihrer eigenen Rasse haben die Deutschen, haben überhaupt unsre Völker die Überlieferung von der jüdischen Rasse, wie unser Kulturzustand es mit sich bringt. Unter den übrigen Deutschen also, deren Rasse nicht so viel von sich selber spricht wie die Träger der Rassentheorie, konnte diese nichts andres hervorrufen als einen törichten, bald wieder verschwindenden Modeunfug: aber bei den Juden hat sie, wegen der Stärke der tatsächlich vorhandenen Rassenerinnerung, tatsächlich eine noch größere Steigerung des Rassenbewußtseins zur Folge gehabt; und einige Juden konnten auf die Konfundierung des Rassenbewußtseins mit der Nationalität derart konfus hereinfallen, daß sie aus ihrer wirklichen Nationalität herausfielen. Das heißt eine Tür aufmachen, um ein Fenster zu schließen. Der Zionismus führt nicht nach Zion, sondern ins Ghetto, wenn auch nicht korporaliter, so doch mentaliter; ins Ghetto ohne Mauern, in die Absonderung nach Leben und Lebensgefühl. Wie konnten Deutsche jüdischer Abstammung von einer jüdischen Nation zu reden beginnen und aus der bösesten Verleumdung den Traum ihres größten Unsinns machen! Wie konnten überhaupt Juden, die geschichtlichsten aller Menschen, mit der am höchsten hinaufreichenden geschichtlichen Erinnerung und mit dem lebendigsten geschichtlichen Wollen, wie konnten sie aus der Melodie geraten und so weit abirren zu derartigem geschichtslosen Pseudoideal! Die Juden eine Nation! Der österreicher Herzl hat sie gewiß verwechselt mit den nach nationaler Selbständigkeit ringenden österreichischen Völkern, und andre haben Zionsehnsucht der frommgläubigen Juden mit politischem Heimweh, mit politischem Zionismus verwechselt; die doch aber nichts miteinander gemein haben. Ernsthaft nehmen läßt sich nicht einmal die Schwärmerei osteuropäischer Juden, die auf alle Weise verhindert werden, das Land, in welchem sie leben, als ihr Vaterland zu betrachten, und deren Herz denn immer noch in Jerusalem und Zion ist — nicht einmal diese Schwärmerei kann man ernsthaft nehmen, und sie hat noch weniger Aussicht als die gleiche Schwärmerei der Kreuzfahrer hatte, oder als die gleiche Schwärmerei so mancher noch bestehender christlicher chiliastischer Sekten hat. Gar aber unsre frommgläubigen Juden, die auf die Tage des Messias harren, wo die Völker ihre Schwerter zu Pflugscharen und ihre Spieße zu Sicheln machen, der Löwe Stroh ißt wie ein Rind und Säuglinge ihre Lust haben werden am Loch der Otter, — ach, schließt nicht unser Wachen Träume in sich wie unser Schlafen? Jene frommgläubigen, jene traumgläubigen Juden mit ihrem Vertrauen auf die Verheißungen, mit ihrer Bibel, „dem aufgeschriebenen Vaterland der Kinder Gottes,'' sie harren wahrlich nicht auf ihr politisches Reich, sondern auf ein Wunder — das die Zionisten nimmer vollbringen werden, vielmehr heißt es von diesen Meschichim en masse und Verlockern zu einer falschen historischen Tat: „Deine Tröster verführen dich und zerstören den Weg, den du gehen sollst ''; sie sind „Diener der Zerschneidung '', und der Zionismus ist wahrlich eher Antimessias als Messias zu nennen. Die Juden eine Nation!? In den verschiedenen Häusern der Stadt die zerschnittenen Stücke Braten auf den Tellern will ich eher einen lebendigen Ochsen nennen als die Juden eine Nation! Aber wären sie tausendmal eine Nation — ließe sich darum diese Nation in Palästina wieder einsetzen? Ein Nagel haftet in der Wand, ist er aber einmal herausgerissen, dann nützt kein ihn wieder in das alte Loch Stecken; er hält da nicht mehr. — Wie es mit den Deutschen jüdischer Abstammung hinsichtlich der Nation steht, das wollen wir später betrachten, wo wir betrachten, wie es mit den übrigen Deutschen hinsichtlich der Nation steht. Das können wir erst, nachdem wir über den Staat und die politischen Parteien uns auseinandergesetzt haben.

    Mit den Worten gegen den Zionismus möchte ich nicht mißverstanden werden — doch muß ich das Gesagte gesagt sein lassen auf die Wahrscheinlichkeit hin, mißverstanden zu werden. Davor bleibe ich wohl nicht bewahrt, trotz der ausdrücklich hinzugefügten Erklärung, daß ich eine jüdische Siedelung von osteuropäischen Juden, eine Siedelung mit Selbstverwaltung unter Staatshoheit eines der bestehenden Staaten als ein mit allen Mitteln und mit allen Opfern zu erstrebendes Ziel ansehe — von osteuropäischen Juden, weil sie entrechtet, entehrt und entmenscht werden, aus keinem andern Grunde, und nicht der osteuropäischen Juden; denn man kann überzeugt sein, daß auch für Osteuropa die Judenemanzipation kommen wird wie für Westeuropa. Aber was hat eine derartige jüdische Siedelung mit der Pseudonationalidee der Zionisten zu schaffen? die ebenso närrisch und gefährlich ist, wie es unter diesen Zionisten bereits unleidliche Chauvinisten gibt, deren zionistische Betätigungen gegen die Nichtzionisten manchmal nicht besser sind als Antisemitismus. Die Zionisten haben sich das Dogma Rasse und Nation auf die allerärgste Weise angeeignet und sind, als Assimilanten dieses Antisemitendogmas mit ihrer verhängnisvollen Agitation dafür, Feinde nicht allein der Emanzipation der

Juden, sondern auch der Emanzipation der Menschheit oder der Kultur und damit auch der Grundidee des Judentums. (Ich meine hier nicht Manner wie Herzl, Nordau, Zangwill, die von ganz andrem Schlage sind und da niemals mitgingen — edle Manner, denen man bis in die letzten Ecken und Tiefen der Natur trauen kann, und die edel geirrt haben.) Der Zionismus ist die Traufe des Regens Antisemitismus, und die Zionisten sind den Juden gefährlicher als die Antisemiten. Indem die Zionisten den ungeheuersten aller Fehler begehen, die Juden zu isolieren und ihnen den lächerlichsten Nationalismus, den anationalen und antinationalen Traumnationalismus aufzureden, bringen sie tatsächlich die Juden zu dem, weswegen die Antisemiten sie nur verleumdeten; es gibt nun Juden, von denen wahr ist, was Antisemiten behaupten, und gilt nicht länger: Antisemiten sagens, es ist Lüge. Die Antisemiten bestritten nur den Juden ihre Nationalität, die Zionisten aber machen sie derselben unwürdig und unfähig und morden sie in ihnen. Die Zionisten bilden eine Gefahr und Schwierigkeit, deren Größe von den Deutschen jüdischer Abstammung nicht verkannt werden darf; aber unser Grundsatz laute: Es gibt keine Gefahren! Sie sind dazu da, überwunden zu werden, jede Gefahr ist zu überwinden — Feuer kann nicht verbrennen, aber ertrinken. Es gibt keine Gefahren und Schwierigkeiten, oder es gibt kein Leben! Hindurch durch Judenhaß hier, Zionismus dort; wir werden immer kräftig genug sein, zu überwinden und auch noch die um uns herum zu stärken und mit uns emporzuführen. Der Zionismus wird unter uns um so weniger Boden gewinnen und um so schneller den gewonnenen wieder verlieren, je weniger Einfluß und Macht wir den Judenhassern über uns zulassen."[256]

The dangerous rhetoric of racist Zionist Supreme Court Justice Louis D. Brandeis, and his Jewish racist compatriots, was rejected by most Jews, who were more sophisticated and enlightened in their beliefs. *The New York Times*, on 25 November 1917 (shortly after the Balfour Declaration was issued), in Section 9 on page 3, published an anti-Zionist appeal from Rabbi Dr. Samuel Schulman, which had first appeared in *The American Hebrew*, in which the Rabbi argued that Israel consisted of an international religious movement, not an individual nation. *The New York Times* published Professor Ralph Philip Boas' lecture "Youth and Judaism" on 23 April 1917 on page 9, in which Boas, who was Jewish, condemned Jewish racism,

"Racialism the Great Danger.
'The greatest danger, however, that stands in the way of the attempt of Judaism to reach its largest usefulness is racialism, that blind and unquestioning admission of one's superiority. The basic idea of racialism seems to be this, that, since one is born a Jew, it is his duty to develop his

---

256. C. Brunner, *Der Judenhass und die Juden*, Third Enlarged Edition, Oesterheld & Co., Berlin, (1919), pp. 192-197.

Jewishness to the fullest extent without reference to the fact that his particular race may not have absorbed all the good gifts of God. To put the case as brutally as possible, racialism is uncritical egotism.

'I am no assimilationist: I have no desire to see the Jews lose the qualities which they add to the commonwealth of nations, but I cannot but feel that there is something mechanical in the everlasting emphasis upon the things which make Jews different from other people. Why should a man always pride himself upon all the qualities, good and bad, which differentiate him from other people? There is something abhorrent in the throwing overboard of standards of value. Let us cultivate the good qualities of the Jewish race because they are good and refuse to cultivate the bad qualities just because they are Jewish.

'Four dangers, then, do their best to prevent the Jew from vitalizing in the hearts of young men an ideal of discipline and responsibility, and so making a valuable contribution toward the settlement of the central problem of democracy—religious romanticism, humanitarian materialism, formalism, and racialism. What then is left? What can the Jew offer which will make against the mere expression of impulse and make for the concentration of energy? What can the Jew offer? I can answer in one word—Judaism.

'Judaism, genuine and vital, Judaism freed from extravagance and excess. Judaism freed alike from formalism and false mysticism. Judaism, that religion the heart and soul of which is law, but law, magnetized by magnificent humanity."

The New York Times published Professor Ralph Philip Boas' statement directly condemning political Zionism as anti-American and dangerous, on 16 December 1917, Section 4, page 4:

"PROGRAM OF ZIONISM MENACES JEWISH UNITY"

BY RALPH P. BOAS

THE fall of Jerusalem is one of the most romantic events of the great war, for it gives reality to what has for a generation been a dream—Zionism. That Zion with its memories and its romance should once more pass into the keeping of its ancient people, that God should once again be praised in the ancient liturgy in a new temple, that the ancient culture which gave Europe its religion should once more flourish—here are possibilities the realization of which might well stir the most pedestrian mind.

It is just because the Zionist program is near fulfillment that honest criticism must not be stifled. This is no time for a comfortable and easy acquiescence to what is after all a matter involving the future, not of a few thousand colonists, but of the whole Jewish world. For Zionism is not merely a proposal to erect a new State in Palestine; it is a program of life for Jews everywhere. The Zionists maintain that Judaism is a way of life.

Judaism, they hold, presupposes a complete round of human activity. It presupposes certain theological beliefs and certain spiritual activities consequent upon these beliefs. It presupposes also a submission to the traditional discipline of Jewish law. It holds that such spiritual activity and such submission to discipline are impossible in countries which cannot make allowance for them, countries in which custom, prejudice, and convenience are all against separatism and individuality in everyday life.

Zionism maintains, therefore, that the only possible way for a man to be a complete Jew is to believe in Jewish theology, to order his spiritual life as that theology dictates, to obey faithfully the minute prescription of the traditional Jewish law, to speak a Jewish language, to cultivate the Jewish arts, to live in a Jewish land under a Jewish government. The Zionist maintains, moreover, that Judaism is now confronted with a very real issue, preservation or extinction.

With the last of the compact European Jewries in process of dissolution, Judaism has no longer any central home. The result is gradual but inevitable assimilation, which can have only one end, the extinction of Judaism as a religion and of the Jews as a group.

Assimilation is the crux of the Jewish problem as the Zionists see it. Zionism demands, therefore, that Jews regard themselves as a nationality forming with a dozen other nationalities a union under the Stars and Stripes. It would consider America not as a 'melting pot,' but as a magnified Balkan peninsula. It would, if consistently interpreted, regard the individualism shown by the Germans in the United States and by the French Canadians in the Dominion as entirely justified, since these groups refuse to allow their individuality to be fused with others into a single national group. Zionism is therefore more than a romantic adventure; it is a very practical and momentous issue.

That Zionism has its dangers is obvious. And these dangers are the more vital, since it is likely that men carried away by sympathy with and admiration for success may fail to recognize them.

The gravest danger lies in a concept of German pseudo-science, the 'Jewish race.' The world sees only dimly that the riot of national romanticism which is upon us is the child of Kultur. The idea of a primitive unspoiled German race which in all respects was like an individual and which, therefore, was under the biological necessity of living its life as the plant lives, tirelessly and remorselessly—this idea is the father of the present insistence of self-styled nationalities on independent existence. Some men, instead of finding out by what right they ask for the creation of a new state, assume that the world should recognize their yearning for peculiarity as inspired by a kind of zoological necessity. The mere blind impulse to be one's self, to remain on earth as an individual, fundamental though it be, is after all a characteristic which one shares with the cabbage.

The fact is that there is no pure Jewish blood. The whole record of Judaism is a record of constant intermarriage and assimilation. Every one knows that Jews differ among themselves as much as Frenchmen, and that the class-concept 'Jew' is the product of loose observation of particular groups. All talk of race necessity in connection with Zionism is misleading. The only possible justification for Zionism is that it will enable Jews to live better lives. Zionists are continually maintaining that only in Palestine can Jews live nobly; that Judaism as a religion can live only where Jews have political autonomy. There is a causal relationship assumed here which needs to be proved. Even Ahad Ha'Am, perhaps the greatest of the Zionists, sometimes despaired because many Zionists could see only the political side of their movement, and it therefore paid no attention to its truly valuable aspects, the Jewish culture, the Jewish religion, the Jewish ethics. What assurances have we that Jews, when tangled in the problems of political administration, will automatically become nobler and finer men. There is every assurance that they will not, for they must necessarily shift the burden of effort from religious and ethical achievement to political achievement.

Moreover, Zionism is continually emphasizing the breach between Jew and Christian which most of us are trying to bridge. As the child of anti-Semitism, it thrives on persecution. Its central argument is that Jews can never be at home in a 'foreign' land. It makes capital of every instance of petty intolerance and nourishes itself upon the ill-will which Jews are prone to fancy even when it is not present. The chip which many Jews bear more or less ostentatiously now that the yellow badge has been removed, some Zionists magnify into a veritable Pilgrim's burden which can drop from the bent back only upon the soil of Palestine. Zionists are continually heaping abuse upon the non-separatist, upon the man who has no desire to be different from other human beings and is very grateful that he does not have to be a marked man among men. Many of us do not believe that peculiarity is the most desirable thing in life. We honestly believe that the separation of church and state is one of the great blessings of life, and that among some Jews there is altogether too much inbreeding of ideas and sentiment. We honestly feel that Jews have still a few things to learn from others. We realize that we must continually make efforts to retain our Judaism, but if Judaism is so far gone that its only salvation lies in becoming a little State it had much better die and be done with the unequal struggle. As a mere survival it has only the value of a sentimental curiosity.

But one may grant all these things and still ask: If there are Jews who can be happy only in Palestine, why not let them go there and be happy? But such is not the real issue. Zionists want political independence. They want to speak as the Jewish people. In short, they want to arrogate a supremacy which non-Zionists can never dream of giving them without a struggle. Whether we will or no, the world insists upon looking at Jews as a unit. For what one Jew does, in the eyes of the world all Jews are responsible. We bear upon our backs the burden of many a Jew who is disgrace to the air he breathes. With such a spirit abroad, Zionists would, consciously or unconsciously, dragoon us into a citizenship and a nationality which we do not want. Every rash act of a Jewish politician would be the rash act of our brother. We will not be dragooned out of America into Palestine. It is all very well for Zionists to say that non-Zionists will not be affected by what goes on in a new Jerusalem, but they know that they are not facing the facts. Who of us Jews can escape being drafted into whatever is done by a 'Jewish people' under a 'Jewish flag'?

In its attempt to force unity upon all Jews, whether they want it or no, Zionism is on the brink of splitting Judaism irreconcilably. There are men who urge that now is the time for a new peace in Judaism, that with an approaching consummation Zionism ought to receive new confidence and encouragement. Such a wish is far from fulfillment. Not harmony but disruption is in sight. It is inconceivable that American Jews should allow their future to be determined by the group of men who will control the Zionist State. They would have but one resource, to cast off their bonds and convince the world that Jews, truly American Jews, could not take the responsibility for men who attempted to reconcile loyalty to America with a foreign nationality.

Who knows what the future may bring forth? Who knows what entangling alliances an independent Palestine might form? One must remember Trotzky, Dernburg, and Hillquit. They, too, are Jews. No country now can escape international association. Those dreamers who think that Zion could occupy a splendid isolation in international politics have no sense of history. They make the same mistake as the dreamers who think that the puny protests of a Government at Jerusalem could end Jewish persecution everywhere. Just so long as genuinely active anti-Semitism which would call forth protest is possible, just so long will little States have no power. A condition of international good-will which will make the voice of a little State heard in the council of nations, will make of anti-Semitism an impossibility.

The future is clear. The complete Zionist program means a complete disruption of Jewish unity. With Zion an independent state every American Jew must become a Zionist and take responsibility for the acts of Zionists, or find some other name than Jew. No one, of course, can object to colonies of Jews in Palestine, or anywhere else. But, every Jew who values his independence and the Americanism of which he has become a part will object as never before to the complete Zionist program. What was once a dream has now become almost a reality. And as it becomes real it becomes, just because of its romantic associations, insidiously dangerous."

In 1948, Zionist Mordecai Menahem Kaplan stated,

"Similarly American Jewry will for a long time have to give moral, political, and economic support to the Eretz Israel enterprise, which is the deciding factor in Israel's struggle for survival in the modern world. [***] Judaism cannot function in a vacuum. It has to be geared to a living community. In that community all who wish to be known as Jews should be registered, and expulsion from it should deprive one of the right to use the name Jew."[257]

In 1953, shortly after the State of Israel became a reality, Alfred M. Lilienthal, who is Jewish, reacted to the pressure placed on American Jews to support something foreign to them as if it were an unavoidable obligation,

> "During the events which altered the relationship between the Kremlin and Israel the reaction in this country was to treat the Israeli crisis as if it were the crisis of the Jewish people all over the world. But if the political problems of Israel continue to be the political responsibility of Jews in the United States, disaster must follow. Innumerable situations will involve Israel in policies and politics which nationals of no other country may dare underwrite. Next time, the enemy of Israel may not be the enemy of the United States. [***] This book has been written, against the concerned counsel of many who are close and dear to me, because I feel I owe a duty to my country above any duty I owe to my family and friends. [***] I have received innumerable admonitions 'not to say anything that might harm the Jewish people.' But, indeed, my efforts are intended to benefit American Jewry. Criticism expressed in these pages and directed against guilty leadership could involve all Jews only by the process of generalizing—the favorite weapon of anti-Semites. And yet, I do not underestimate for one moment the wrath that will descend upon me for having written this book. Every conceivable kind of pressure will be exerted, I am afraid, to prevent a fair consideration of the material set forth in its pages. [***] I have written this book because I, an American of Jewish faith, have not the slightest doubt that American Jewry, too, has a free choice—and must face the consequences of whatever it will choose. [***] In this one sense, the establishment of the State of Israel may yet prove to have been a providential blessing: now that those Jews who crave their separate nationhood can go to Israel, the last reason has been removed for the pernicious Jewish duality outside the Holy Land. Now each American Jew has been given a free choice to be either an American of Jewish faith or a nationalist Israeli in his own Middle East State. He can not be both. For him who cherishes the clannishness of particularism above everything else, there is only one honorable course—to emigrate to Israel. And the American Jew, who desires to harmonize his special religious beliefs with the universal pattern of American existence, will now have to cut all political ties with Zionism and the State of Israel. For American Judaism can survive only when it is so completely divorced from Israel as American Protestantism is divorced from England."[258]

---

[257]. M. M. Kaplan, *The Future of the American Jew*, Macmillan, New York, (1948); quoted in A. Hertzberg, *The Zionist Idea*, Harper Torchbooks, New York, (1959), pp. 536-544, at 537, 544.

[258]. A. M. Lilienthal, *What Price Israel*, Henry Regnery Company, Chicago, (1953), pp. vi-viii, 239. *See also:* "Israel's Flag Is Not Mine", *Reader's Digest*, (September, 1949),

In 1965, Moshe Menuhin published an exposé of Jewish racism and tribalism entitled *The Decadence of Judaism in Our Time*. Menuhin knew that he would probably be attacked for revealing many truths and sought to shield his son from any potential Zionist retaliation. He wrote, among other things, in the preface of his book,

> "It is not an easy or a pleasant job to perform; yet my very strong sense of duty and my anxiety compel me to undertake it. I am absolutely convinced of the truthfulness of my studies, observations and conclusions. I serve nobody's interests, and I am paid by no one. Yet, though I carefully and honestly stick to facts, I know that I am bound to antagonize the fanatical and professional idealists among the 'Jewish' nationalists. Therefore, please remember this: my son Yehudi Menuhin is in no way responsible for any opinion expressed here on Jewish life. In fact, he knows nothing about this spiritual adventure of mine. He has not read my manuscript. At this stage of our lives we are two wholly independent persons, fully emancipated from each other, intellectually and spiritually. Neither of us is answerable for the other. If the 'father has eaten sour grapes... the son shall not bear the sin of the father...'"[259]

Menuhin concluded his book,

> "Those, however, who cannot make the indispensable adjustment in the new post-World Wars nuclear age, and who feel that they must withdraw from the general community in America and live apart as 'Jewish' nationals—let them be honest enough with themselves and withdraw completely by going to live in Israel. Above all, they must leave us alone as integrated Americans.
>
> I have made my position witheringly clear. The time is immutably coming when we will have to face the awful question the 'Jewish' nationalists have imposed upon us: Are we American nationals, or Israeli nationals? We cannot and will not be both!
>
> I can hear some ask naïvely or bitterly: Is it nice to wash dirty linen in public? Well, shall we wait helplessly until a catastrophe overtakes us here, when a few of us might have the hollow satisfaction of saying: 'I told you so'? It will be much too late then. Must one contribute to the delinquency of presumptuous, fanatical and retrograde professional Jews who are running

---

pp. 49-53. "The State of Israel and the State of the Jew", *Vital Speeches of the Day*, Volume 16, Number 13, (15 April 1950).

<http://www.alfredlilienthal.com>

[259]. M. Menuhin, "Preface", *The Decadence of Judaism in Our Time*, Exposition Press, New York, (1965).

away with themselves? Must one be blind and join the complacent and silent Jews who help the destructive forces by sheer default?

The time has come to air and publicly expose this uncalled-for, self-engendered 'Jewish problem' that is being recklessly foisted upon us by 'Jewish' nationalists of the Old World. They are simply exploiting the goodness and kindness, as well as the sorrows and sympathies, of innocent, ignorant but warm-hearted Jews who feel that but for the grace of God, they too might have been turned into lampshades and soap in the crematoriums of Hitler's Germany. The 'Jewish' nationalists now want us, American and English and other Western Jews, to become 'refugees,' manpower in a greater 'Jewish homeland.'

My conscience had been bothering me ever since the Balfour Declaration came out in 1917 to undo the normal course of evolution of the Jews and of Judaism. I felt then that I could no longer belong to the 'gang' of which I was a dedicated member by indoctrination and brainwashing. I hope that this book will contribute to healthier and more independent thinking by innocent but misguided American and English Jews, as well as by Jews in other countries. I hope that it will also contribute to a better and more sympathetic understanding by the Gentile world of that great majority of innocent, loyal, grateful but confused Jews who now must win a new war of emancipation—an emancipation this time from their benighted fellow Jews, the 'Jewish' nationalists, who have perverted and degenerated the noble heritage of universal Judaism."[260]

Theodor Herzl gives evidence to the fact that from its very inception, most Jews vigorously opposed Zionism.[261] The only way for Herzl, a self-appointed Messiah, to be successful was for him to increase and generate anti-Semitism and in so doing force the Jews from their homes in Europe, which they did not wish to leave.

Professor Arnold J. Toynbee was quoted in an article entitled, "Toynbee Predicts Gains by Judaism: Historian Assails Zionism as Akin to Anti-Semitism", in *The New York Times* on 7 May 1961, on page 37,

"Zionism and anti-Semitism are expressions of an identical point of view[.] The assumption underlying both ideologies is that it is impossible for Jews

---

[260].M. Menuhin, *The Decadence of Judaism in Our Time*, Exposition Press, New York, (1965), pp. 488-489.

[261]. *Cf.* F. A. Sayegh, *Zionism: A Form of Racism And Racial Discrimination" Four Statements Made at the U.N. General Assembly*, Office of the Permanent Observer of the Palestine Liberation Organization to the United Nations, (1976).

<http://www.ameu.org/uploads/sayegh_march1_03.pdf>

Sayegh cites: T. Herzl, *Zionist Writings: Essays and Addresses*, Volume 1 Covering 1896-1898, Herzl Press, New York, (1973), pp. 62-70, 89-97, 119-124, 148, 232-239.

and non-Jews to grow together into a single community, and that therefore a physical separation is the only practical way out. The watchword of anti-Semitism is 'back to the medieval apartheid;' the watchword of Zionism is 'back to the medieval ghetto.' All far-flung ghettos in the world are to be gathered into one patch of soil in Palestine to create a single consolidated ghetto there."

The report on the First Zionist Congress in *The Jewish Chronicle* on 3 September 1897 on page 10 opened with the statement,

"The event which has been looked forward to with so much interest in a large section of the Jewish people and severely criticised in anticipation by another section has at length arrived—the Zionist Congress hasmet."

In fact the "section" opposed was immensely greater than the "section" that approved of political Zionism, and the majority of Jews hated the Zionists and considered them mad.[262] Both the Old Testament (*Leviticus* 26. *Deuteronomy* 4:24-27; 28:15-68; 30:1-3. II *Chronicles* 7:19-22. *Jeremiah* 29:1-7) and the Babylonian Talmud, *Tractate Kethuboth* (*also:* "Ketubot"), 111a, make it clear that the Jews must not hasten the coming of the Messiah and must wait for the Messiah to establish a Jewish state, before emigrating to Palestine in large numbers. Israel Shahak and Norton Mezvinsky wrote in their book *Jewish Fundamentalism in Israel*,

"The Haredi objection to Zionism is based upon the contradiction between classical Judaism, of which the Haredim are the continuators, and Zionism. Numerous Zionist historians have unfortunately obfuscated the issues here. Some detailed explanation is therefore necessary. In a famous talmudic passage in *Tractate Ketubot*, page 111, which is echoed in other parts of the Talmud, God is said to have imposed three oaths on the Jews. Two of these oaths that clearly contradict Zionist tenets are: 1) Jews should not rebel against non-Jews, and 2) as a group should not massively emigrate to Palestine before the coming of the Messiah. (The third oath, not discussed here, enjoins the Jews not to pray too strongly for the coming of the Messiah, so as not to bring him before his appointed time.) During the course of post-talmudic Jewish history, rabbis extensively discussed the three oaths. Of major concern in this discussion was the question of whether or not specific Jewish emigration to Palestine was part of the forbidden massive emigration. During the past 1,500 years, the great majority of traditional Judaism's most important rabbis interpreted the three oaths and the continued existence of the Jews in exile as religious obligations intended to

---

[262]. J. B. Agus, *The Meaning of Jewish History*, Volume 2, Abelard-Schuman, New York, (1963), pp. 412-413. *See also:* A. Hertzberg, *The Zionist Idea*, Harper Torchbooks, New York, (1959), pp. 509, 577, 581-582.

expiate the Jewish sins that caused God to exile them."[263]

Christians believe that the Jews had broken the Covenant and that a new Covenant had been made (*Matthew* 12:30; 21:43-45. *Romans* 9; 11:7-8. *Galatians* 3:16. *Hebrews* 8:6-10). Thomas A. Kolsky wrote in his book *Jews Against Zionism*,

> "The first vehement outburst by anti-Zionists occurred in 1897 at the time of the first Zionist congress. When news reached the German community in May 1897 that the Zionists were planning to hold their congress in Munich, German rabbis representing all shades of opinion, whom Herzl contemptuously dubbed *Protestrabbiner*, objected angrily and forced the Zionists to shift their gathering to Basle, Switzerland. In public protests in the Jewish *Allgemeine Zeitung des Judentums* on 11 June 1897 and in a number of German newspapers, including the *Berliner Tageblatt*, on 6 July 1897, the rabbis denounced Zionism as fanaticism, contrary to the teachings of the Jewish scriptures, and affirmed their undivided loyalty to Germany.[31]
> In 1917 the negotiations between the British cabinet and the Zionists over the Balfour Declaration stirred up a sharp reaction against the Zionist movement among British Jews. Prominent Jewish communal leaders, such as Lucien Wolf, Claude Montefiore, and Laurie Magnus, denounced Zionism as an ally of anti-Semitism, warning that it undermined the security of Jews throughout the world. In a letter to the London *Times* on 14 May 1917, the prestigious Conjoint Foreign Committee, the recognized representative body of British Jews in matters affecting Jews abroad, declared that the emancipated Jews of England considered themselves a religious community without any separate national aspirations. In fact, the foremost anti-Zionist within the British government during the deliberations over the Balfour Declaration was Sir Edwin Montagu, a Jewish member of the cabinet, who equated support for Zionism with anti-Semitism and characterized Zionism as 'a mischievous creed.' This anti-Zionist agitation and especially Montagu's influence undoubtedly contributed to diluting the final version of the Balfour Declaration: the change of the central phrase from 'Palestine as the National Home,' which the Zionists had suggested, to 'in Palestine as a National Home'; and the inclusion of a safeguard clause providing for the protection of the civil and religious rights of the 'non-Jewish communities in Palestine' as well as 'the rights and political status enjoyed by Jews in any other country.'[32]"[264]

The New York Times reported on 22 August 1897 on page 12,

---

[263]. I. Shahak and N. Mezvinsky, *Jewish Fundamentalism in Israel*, Pluto Press, London, (1999), p. 18.
[264]. T. A. Kolsky, *Jews Against Zionism: The American Council for Judaism, 1942-1948*, Temple University Press, Philadelphia, (1990), p. 17.

"Jews Against Zionism.

A violent split has taken place among the Jews all over the world. It is caused by the new course which Zionism recently adopted, and which is to find practical expression at the Zionist Congress to be held at Basle on the 29th, 30th, and 31st inst.

The meaning of Zionism hardly needs explanation. Up till recently Zionism, as is known, had only a religious and philanthropic tendency, and found many adherents also among believing Christians in England. But since the persecution of the Jews began in Russia and Roumania some ten years ago, and since anti-Semitism in Austria and Germany made the social position of the Jews more intolerable than it was before, the thought of establishing a Jewish state, if possible in old Judea, has gained ground, not only among the Jews of those countries mentioned, but also among the Jews in the rest of the world. Many of them thought that a purely philanthropic movement would always be but a palliative, and would never lead to a solution of the Jewish question. The many millions spent by Barons Hirsch and Rothschild on colonizing have produced only very slight results. And so arose the idea of political independence. The once philanthropic Jewish party of Zionists adopted a wider programme, so that now Zionism actually connotes the revival of the Jewish nationality by the establishment of a Jewish state. In short, Zionism has become a political and social movement. The idea had its origin among the Jews of Eastern Europe, where they are more or less persecuted or oppressed, as in Russia and Roumania, and even, too, in Austria; but curiously enough the movement has been eagerly fostered by many Jews in America and England, where Jewish citizens enjoy full and equal liberty with their Gentile fellow-citizens. The consolidation of nationalities is a characteristic feature of our century. Italy, Greece, Roumania, Servia, and Bulgaria owe their existence to the principle of nationality, a principle no less powerful and perhaps no less erroneous than was that of the Crusades, but one that urges on the people of this century with an irresistible force. National integration is carried so far in Europe that petty peoples, whose very names have hardly reached your shores, peoples so insignificant that they have neither a literary nor a spiritual past, which have possessed a grammar and a dictionary of their own languages only for the last ten to twenty years, are struggling for their national individuality with greater zeal and passion than those displayed in their pursuit of all other worldly goods, important as the latter may be.

Jewish Students Hold Aloof.

Considering that tendency, it is no wonder that the idea of political resurrection has taken possession of that race which produced monotheism, and which during 2,000 years of oppression has given numerous proofs of intellectual vigor and vitality. Only a few years ago no educated Jew in England, Germany, or Russia would have dreamed of calling himself anything but an Englishman, a German, or a Russian. To-day many are heard to say that they are only Jews. It is chiefly young people that has been seized with this unpatriotic tendency, which is so hurtful to them. At the universities of Berlin, Vienna, and other German towns the Jewish students have almost entirely given up the intercourse with the students of other creeds, which was carried on so pleasantly for decades. They have left all the common students' societies. In Vienna alone for some years there have been five academical societies for Jews only. With positively phantastic enthusiasm the young men cling to the dreamed-of ideal of a Jewish state. No doubt anti-Semitism in Austria and Germany has done a great deal to drive young Jews to this senseless and dangerous course. But it is not surprising that it has come to that. The Gentile university students are mostly anti-Semitic. But youth soon exaggerates and overflows. Formerly the brotherly understanding among the students of different races and religions was complete; creed and race were never thought of in their social intercourse. But now matters have utterly changed. Jewish students are met with blind racial hatred. No distinction is made; they are all socially banished. All noble qualities are denied them. They are declared to be nobodies. An insult from a Jew is no insult. At German universities dueling is still very usual. Jewish students are never challenged, or, it is said, a Jew is an unworthy individual, incapable of giving satisfaction.

Leaders of the Movement.

Political Zionism has been awakened and promoted chiefly by Dr. Herzl's book, 'Der Judenstaat,' (the Jewish State,) a clever but rather Utopian book, which was translated into all European languages immediately after its publication. Dr. Herzl and Dr. Max Nordau in Paris are the chief literary exponents of this new movement. Dr. Herzl has told me that the leaders are endeavoring first of all to organize a wholesale emigration of Jews from all countries. A 'Society of Jews' is to be formed; in London there are already considerable funds at its disposal. A plan has been formed for acquiring part of Palestine from Turkey and settling the immigrants there. Out of its funds the society would pay the Sultan a considerable annual tribute, on the strength of which he could raise a loan for the purpose of consolidating the disordered funds of his empire. In return he would protect the Jewish state, which would have complete self-government. The leaders hope for resolutions are to come which will lead to the execution of the scheme.

Meanwhile, however, a serious counter-movement has arisen, especially among the German Jews. Originally it was intended to hold the congress at Munich, but the German Jews protested against it. It has now become apparent that only a small number of the Jews in all countries favor these fantastic plans. As long as the projects were only on paper these objectors held their peace. But now that attempts are being made to carry them out, the great majority of thoughtful and serious Jews throughout the world have commenced a decided opposition to the unrealizable and damaging schemes. This majority emphatically denies the existence of a Jewish nationality, and condemns the new Zionist postulates. The majority holds that creed alone unites the Jews of divers countries. They are quite different from one another in language, manners, customs, thought, and culture. Such heterogeneous elements could never be welded together into a state; they simply would not understand one another. But, apart from that, they feel themselves modern citizens of those lands in which their ancestors lived for centuries.

A Split Over the Congress at Basle.

The new theories are only likely to compromise their patriotism toward the countries they live in without helping the Jews of Eastern Europe. For even were it possible to found a state artificially and in a barren country such as Palestine now is, and which would require many years of hard labor to restore its former fruitfulness, they characterize it as madness that honest men should subject themselves to the protection of the 'Great Assassin,' whose conscience is so little troubled by the blood of tens of thousands. Could any state that began by connecting itself with the blackest crimes, the most barbarous and villainous maladministration, prosper? The German Jews have accordingly been the first to declare against Zionism. The rabbis of Berlin, Frankfort, Munich, Dresden, and Hamburg have issued a manifesto to their co-religionists to the effect that the establishment of a Jewish state would be contrary to the Messianic prophecies, and that Judaism lays upon its adherents the obligation to support and foster with all devotion and with all their might the state they live in. Accordingly the rabbis call upon the Jews to oppose the Zionist ideas as contrary to Judaism, but especially to keep away from the Basle Congress. Similar declarations are likely soon to be made by the Jews of other countries. Consequently it is doubtful if the congress, which is to be attended by Zionist Jews from all parts of the world, will in face of that split be able to proceed to the realization of these Utopian schemes."

Note that the German Jews, the first Jews attacked by the Zionists' Nazism, were the Jews who were most strongly opposed to Zionism, and that it is very strange that Albert Einstein, a German Jew, should have aligned himself with Eastern Jewish Zionists. One suspects his motives were opportunism. One further suspects that the Zionists targeted German Jews to suffer from especially harsh attacks of anti-Semitism.

Apostate Jews like *Hofprediger* Adolf Stöcker promoted German anti-Semitism in the highest places, and Austrian anti-Semitism arose in Austria under Karl Lueger, who had very close ties to the Jewish community and who asserted that he had the self-declared right to determine who was, and who was not, a Jew—meaning that he could protect those Jews who protected and promoted him, while destroying assimilatory Jewry.

The self-segregation of Jewish students in the West came more at the instigation of the newly emigrated Eastern Jews from Galicia and Russian, than from anti-Semites. The Jewish press fanned the flames of the Protestant versus Catholic struggles of the *Kulturkampf*, and their hypocritical, outrageous and vicious attacks on Catholics and Christianity caused a surge in anti-Semitism even among Protestants, who could not help but be offended. Though the Jewish press felt free to express its intolerance towards Christians, they recoiled in horror should any criticism be directed at them and their campaign to discredit Catholicism and defame the priesthood and the Pope. Most Jews in Germany wanted only to assimilate and leave the intolerance of Jewish racism and tribalism, as well as the intolerance of anti-Semitism, in the past. The Zionists were determined to not let them succeed. The Zionists did not care about the survival of individuals. Rather they only cared about the survival of the "Jewish race". Most Jews knew this and they feared and hated the Zionists. Though Christians focus on the immortality of the soul, Judaism places more emphasis on the immortality of the Jewish people, should they be righteous. The Zionists transitioned this belief system into pure racism and massive imposed martyrdom on assimilatory Jewry. The Zionists revived the ancient system of holocaust, of human sacrifices in the form of burnt offerings.

Herzl's Zionism lacked broad appeal among Jews. Jew had to be forced into Palestine through violence. In 1897, after the first Zionist Conference, spiritual Zionist Ahad Ha'am wrote,

> "We must admit to ourselves that the 'ingathering of the exiles' is unattainable by natural means."[265]

In 1897, Ha'am acknowledged the belief of the "western" "political Zionists" that the "spiritual problem"—read "racial problem"—of remaining a Jew was,

> "a product of anti-Semitism, and [was] dependent on anti-Semitism for its existence[.]"[266]

Klaus Polkehn wrote in his article "The Secret Contacts: Zionism and Nazi Germany, 1933-1941", *Journal of Palestine Studies*, Volume 5, Number 3/4, (Spring-Summer, 1976), pages 54-82, at 55 and 57,

> "To the Zionist leaders, Hitler's assumption of power held out the possibility of a flow of immigrants to Palestine. Previously, the majority of German Jews, who identified themselves as Germans, had little sympathy with Zionist endeavours. [***] These German Jews were overwhelmingly non or anti-Zionist, and prior to 1937, the Zionist Union for Germany (Zionistische Vereinigung für Deutschland (henceforth ZVFD) experienced great difficulty in gaining a hearing. [***] The attitude of the Zionists towards the encroaching menace of fascist domination in Germany was determined by some common ideological assumption: the fascists as well as the Zionists believed in unscientific racial theories, and both met on the same ground in

---

265. A. Ha-Am, "The Jewish State and the Jewish Problem", in A. Hertzberg, *The Zionist Idea*, Harper Torchbooks, New York, (1959), pp. 262-269, at 264.
266. A. Ha-Am, "The Jewish State and the Jewish Problem", in A. Hertzberg, *The Zionist Idea*, Harper Torchbooks, New York, (1959), pp. 262-269, at 266.

their beliefs in such mystical generalizations as 'national character *(Volkstum)* and 'race', both were chauvinistic and inclined towards 'racial exclusiveness.'"

Given the opposition of the majority of Jews to Zionism, the conclusion the political Zionists of Einstein's ilk (who like Ha'am were mostly atheistic) drew was that since there was no supernatural means to force Jews to gather in Palestine, the unnatural means—which they sophistically called natural and good—the unnatural means of *anti-Semitism* was their only option to remove personal choice from individual Jews as to their own fate and force them to the desert. Many of these Zionist agitators survived the Holocaust in great comfort in America or in Switzerland, while their Jewish victims suffered and died. When eventually given the choice, many of these Zionist cheerleaders elected to live outside of Israel.

6.4 The Brotherhood of Anti-Semites and Zionists

The racism of political Zionists was often used to justify the racism of anti-Semites, and vice versa. The Nazis were quick to point out that the Zionists thought of Jews as a nation, not a religion.

After the Bolshevik Revolution and the First World War rocked the world, many criticized Jewish nationalism and Jewish racism. THE DEARBORN INDEPENDENT wrote on 16 October 1920:

"Jewish Testimony on 'Are Jews a Nation?'

> 'I will give you my definition of a nation, and you can add the adjective 'Jewish.' A Nation is, in my mind, an historical group of men of a recognizable cohesion held together by a common enemy. Then, if you add to that the word 'Jewish' you have what I understand to be the Jewish Nation.'—THEODOR HERZL.

> 'Let us all recognize that we Jews are a distinct nationality of which every Jew, whatever his country, his station, or shade of belief, is necessarily a Member.'—LOUIS D. BRANDEIS *Justice of the United States Supreme Court.*

THIS article is designed to put the reader in possession of information regarding the Jew's own thought of himself, as regards race, religion and citizenship. In the last article we saw the thought which Jewish representatives wish to plant in Gentile minds concerning this matter. The Senate committee which was to be convinced was made up of Gentiles. The witnesses who were to do the convincing were Jews.

Senator Simon Guggenheim said: 'There is no such thing as a Jewish race, because it is the Jewish religion.'

Simon Wolf said: 'The point we make is this * * * that Hebrew or

Jewish is simply a religion.'

Julian W. Mack said: 'Of what possible value is it to anybody to classify them as Jews simply because they adhere to the Jewish religion?'

The object of this testimony was to have the Jews classified under various national names, such as Polish, English, German, Russian, or whatever it might be.

Now, when the inquirer turns to the authoritative Jewish spokesmen who speak not to Gentiles but to Jews about this matter, he finds an entirely different kind of testimony. Some of this testimony will now be presented.

The reader will bear in mind that, as the series is not written for entertainment, but for instruction in the facts of a very vital Question, the present article will be of value only to those who desire to know for themselves what are the basic elements of the matter.

It should also be observed during the reading of the following testimony that sometimes the term 'race' is used, sometimes the term 'nation.' In every case, it is recognized that the Jew is a member of a separate people, quite aside from the consideration of his religion.

First, let us consider the testimony which forbids us to consider the term 'Jew' as merely the name of a member of a religious body only.

Louis D. Brandeis, Justice of the Supreme Court of the United States and world leader of the Zionist movement, says:

'Councils of Rabbis and others have undertaken at times to prescribe by definition that only those shall be deemed Jews who professedly adhere to the orthodox or reformed faith. But in the connection in which we are considering the term, it is not in the power of any single body of Jews—or indeed of all Jews collectively—to establish the effective definition. The meaning of the word 'Jewish' in the term 'Jewish Problem' must be accepted as co-extensive with the disabilities which it is our problem to remove * * * Those disabilities extend substantially to all of Jewish blood. The disabilities do not end with a renunciation of faith, however sincere * * * Despite the meditations of pundits or decrees of councils, our own instincts and acts, and those of others, have defined for us the term 'Jew.'' ('Zionism and the American Jews.')

The Rev. Mr. Morris Joseph, West London Synagogue of British Jews: 'Israel is assuredly a great nation * * * The very word 'Israel' proves it. No mere sect or religious community could appropriately bear such a name. Israel is recognized as a nation by those who see it; no one can possibly mistake it for a mere sect. To deny Jewish nationality you must deny the existence of the Jew.' ('Israel a Nation.')

Arthur D. Lewis, West London Zionist Association: 'When some Jews say they consider the Jews a religious sect, like the Roman Catholics or Protestants, they are usually not correctly analyzing and describing their own feelings and attitude. * * * If a Jew is baptized, or, what is not necessarily the same thing, sincerely converted to Christianity, few people think of him as no longer being a Jew. His blood, temperament and spiritual peculiarities are unaltered.' ('The Jews a Nation.')

Bertram B. Benas, barrister-at-law: 'The Jewish entity is essentially the entity of a People. 'Israelites,' 'Jews,' 'Hebrews'—all the terms used to denote the Jewish people bear a specifically historical meaning, and none of these terms has been convincingly superseded by one of purely sectarian nature. The external world has never completely subscribed to the view that the Jewish people constitute merely an ecclesiastical denomination. * * *' ('Zionism—The National Jewish Movement.')

Leon Simon, a brilliant and impressive Jewish scholar and writer, makes an important study of the question of 'Religion and Nationality' in his volume, 'Studies in Jewish Nationalism.' He makes out a case for the proposition that the Religion of the Jews is Nationalism, and that Nationalism is an integral part of their Religion.

'It is often said, indeed, that Judaism has no dogmas. That statement is not true as it stands.' He then states some of the dogmas, and continues—'And the Messianic Age means for the Jew not merely the establishment of peace on earth and good will to men, but the universal recognition of the Jew and his God. It is another assertion of the eternity of the nation. Dogmas such as these are not simply the articles of faith of a church, to which anybody may gain admittance by accepting them; they are the beliefs of a nation about its own past and its own future.' (p. 14.)

'For Judaism has no message of salvation for the individual soul, as Christianity has; all its ideas are bound up with the existence of the Jewish nation.' (p. 20.)

'The idea that Jews are a religious sect, precisely parallel to Catholics and Protestants, is nonsense.' (p. 34.)

Graetz, the great historian of the Jews, whose monumental work is one of the standard authorities, says that the history of the Jews, even since they lost the Jewish State, 'still possesses a national character; it is by no means merely a creed or church history. * * * Our history is far from being a mere chronicle of literary events or church history.'

Moses Hess, one of the historic figures through whom the whole Jewish Program has flowed down from its ancient sources to its modern agents, wrote a book entitled 'Rome and Jerusalem' in which he stated the whole matter with clearness and force.

'Jewish religion,' he says, 'is, above all, Jewish patriotism.' (p. 61.)

'Were the Jews only followers of a certain religious denomination, like the others, then it were really inconceivable that Europe, and especially Germany, where the Jews have participated in every cultural activity, 'should spare the followers of the Israelitish confession neither pains, nor tears, nor bitterness.' The solution of the problem, however, consists in the fact that the Jews are something more than mere 'followers of a religion,' namely, they are a race brotherhood, a nation * * *' (p. 71.)

Hess, like other authoritative Jewish spokesmen, denies that forsaking the faith constitutes a Jew a non-Jew. '* * * Judaism has never excluded anyone. The apostates severed themselves from the bond of Jewry. 'And not even them has Judaism forsaken,' added a learned rabbi in whose presence

I expressed the above-quoted opinion.'

'In reality, Judaism as a nationality has a natural basis which cannot be set aside by mere conversion to another faith, as is the case with other religions. A Jew belongs to his race and consequently also to Judaism, in spite of the fact that he or his ancestors have become apostates.' (pp. 97-98.)

Every Jew is, whether he wishes it or not, solidly united with the entire nation.' (p. 163.)

Simply to indicate that we have not been quoting outworn opinions, but the actual beliefs of the most active and influential part of Jewry, we close this section of the testimony with excerpts from a work published in 1920 by the Zionist Organization of America, from the pen of Jessie E. Sampter:

'The name of their national religion, Judaism, is derived from their national designation. An unreligious Jew is still a Jew, and he can with difficulty escape his allegiance only by repudiating the name of Jew.' ('Guide to Zionism,' p. 5.)

It will be seen that none of these writers—and their number might be multiplied among both ancients and moderns—can deny that the Jew is exclusively a member of a religion without at the same time asserting that he is, whether he will or not, the member of a nation. Some go so far as to insist that his allegiance is racial in addition to being national. The term 'race' is used by important Jewish scholars without reserve, while some, who hold the German-originated view that the Jews are an offshoot of the Semitic race and do not comprise that race, are satisfied with the term 'nation.' Biblically, in both the Old Testament and the New, the term 'nation' or 'people' is employed. But the consensus of Jewish opinion is this: the Jews are a separate people, marked off from other races by very distinctive characteristics, both physical and spiritual, and they have both a national history and a national aspiration.

It will be noticed how the testimony on the point of 'race' combines the thought of race and nationality, just as the previous section combined the thought of nationality with religion.

Supreme Justice Brandeis, previously quoted, appears to give a racial basis to the fact of nationality.

He says: 'It is no answer to this evidence of nationality to declare that the Jews are not an absolutely pure race. There has, of course, been some intermixture of foreign blood in the three thousand years which constitute our historic period. But, owing to persecution and prejudice, the intermarriages with non-Jews which have occurred have resulted merely in taking away many from the Jewish community. Intermarriage has brought few additions. Therefore the percentage of foreign blood in the Jews of today is very low. Probably no important European race is as pure. But common race is only one of the elements which determine nationality.'

Arthur D. Lewis, a Jewish writer, in his 'The Jews a Nation,' also bases nationality on the racial element:

'The Jews were originally a nation, and have retained more than most nations one of the elements of nationality—namely, the race element; this may be proved, of course, by the common sense test of their distinguishability. You can more easily see that a Jew is a Jew than that an Englishman is English.'

Moses Hess is also quite clear on this point. He writes of the impossibility of Jews denying 'their racial descent,' and says: 'Jewish noses cannot be reformed, and the black, wavy hair of the Jews will not turn through conversion into blond, nor can its curves be straightened out by constant combing. The Jewish race is one of the primary races of mankind that has retained its integrity, in spite of the continual change of its climatic environment, and the Jewish type has conserved its purity through the centuries.'

Jessie E. Sampter, in the 'Guide to Zionism,' recounting the history of the work done for Zionism in the United States, says: 'And this burden was nobly borne, due partly to the commanding leadership of such men as Justice Louis D. Brandeis, Judge Julian W. Mack, and Rabbi Stephen S. Wise, partly to the devoted and huge labors of the old-time faithful Zionists on the Committee, such as Jacob de Haas, Louis Lipsky, and Henrietta Szold, and partly to *the aroused race consciousness of the mass of American Jews.*'

Four times in the brief preface to the fifth edition of 'Coningsby,' Disraeli uses the term 'race' in referring to the Jews, and Disraeli was proud of being racially a Jew, though religiously he was a Christian.

In The Jewish Encyclopedia, 'the Jewish race' is spoken of. In the preface, which is signed by Dr. Cyrus Adler as chief editor, these words occur: 'An even more delicate problem that presented itself at the very outset was the attitude to be observed by the Encyclopedia in regard to those Jews who, while born within the Jewish community, have, for one reason or another, abandoned it. As the present work deals with the Jews as a race, it was found impossible to exclude those who were of that race, whatever their religious affiliations might have been.'

But as we are not interested in ethnology, the inquiry need not be contained further along this line. The point toward which all this trends is that the Jew is conscious of himself as being more than the member of a religious body. That is, Jewry nowhere subscribes in the persons of its greatest teachers and its most authoritative representatives, to the theory that a Jew is only 'a brother of the faith.' Often he is not of the faith at all, but he is still a Jew. The fact is insisted upon here, not to discredit him, but to expose the double minds of those political leaders who, instead of straightforwardly meeting the Jewish Question, endeavor to turn all inquiry aside by an impressive confusion of the Gentile mind.

It may be argued by the small body of so-called 'Reformed Jews' that the authorities quoted here are mostly Zionists. The reply is: there may be, and quite possible are, two Jewish programs in the world—one which it is intended the Gentiles should see, and one which is exclusively for the Jews. In determining which is the real Program, it is a safe course to adopt the one that is made to succeed. It is the Program sponsored by the so-called Zionists which is succeeding. It was made to succeed through the Allied Governments, through the Peace Conference, and now through the League of Nations. That, then, must be the true Jewish program, because it is hardly possible that the Gentile governments could have been led as they are being led, were they not convinced that they are obeying the behests of the real Princes of the Jews. It is all well enough to engage the plain Gentile people with one set of interesting things; the real thing is the one that has been put over. And that is the program whose sponsors also stand for the racial and national separateness of the Jews.

The idea that the Jews comprise a nation is the most common idea of all—among the Jews. Not only a nation with a past, but a nation with a future. More than that—not only a nation, but a Super-Nation.

We can go still further on the authority of Jewish statements: we can say that the future form of the Jewish Nation will be a kingdom.

And as to the present problems of the Jewish Nation, there is plenty of Jewish testimony to the fact that the influence of American life is harmful to Jewish life; that is, they are in antagonism, like two opposite ideas. This point, however, must await development in the succeeding article.

Israel Friedlaender traces the racial and national exclusiveness of the Jews from the earliest times, giving as illustrations two Biblical incidents— the Samaritans, 'who were half-Jews by race and who were eager to become full Jews by religion,' and their repulse by the Jews, 'who were eager to safeguard the racial integrity of the Jews'; also, the demand for genealogical records and for the dissolution of mixed marriages, as recorded in the Book of Ezra. Dr. Friedlaender says that in post-Biblical times 'this racial exclusiveness of the Jews became even more accentuated.' Entry into Judaism 'never was, as in other religious communities, purely a question of faith. Proselytes were seldom solicited, and even when ultimately admitted into the Jewish fold they were so on the express condition that they surrender their racial individuality.'

'For the purposes of the present inquiry,' says Dr. Friedlaender, 'it is enough for us to know that the Jews have always *felt* themselves as a separate race, sharply marked off from the rest of mankind. Anyone who denies the racial conception of Judaism on the part of the Jews in the past is either ignorant of the facts of Jewish history or *intentionally misrepresents them.*'

Elkan N. Adler says: 'No serious politician today doubts that our people have a *political future.*'

This future of political definiteness and power was in the mind of Moses Hess when he wrote in 1862—mark the date!—in the preface of his 'Rome and Jerusalem,' these words:

'No nation can be indifferent to the fact that in *the coming European struggle* for liberty, it *may have another people* as its friend or foe.'

Hess had just been complaining of the inequalities visited upon the Jews. He was saying that what the individual Jew could not get because he was a Jew, the Jewish Nation would be able to get because it would be a

Nation. Evidently he expected that nationhood might arrive before the 'coming European struggle,' and he was warning the Gentile nations to be careful, because in that coming struggle there might be another nation in the list, namely, the Jewish Nation, which could be either friend or foe to any nation it chose.

Dr. J. Abelson, of Portsea College, in discussing the status of 'small nations' as a result of the Great War, says: 'The Jew is one of these 'smaller nations,'' and claims for the Jew what is claimed for the Pole, the Rumanian, and the Serbian, and on the same ground—that of nationality.

Justice Brandeis voices the same thought. He says:

'While every other people is striving for development by asserting its nationality, and a great war is making clear the value of small nations * * * Let us make clear to the world that we too are a nationality clamoring for equal rights * * *'

Again says Justice Brandeis: 'Let us all recognize that we Jews are a distinct nationality, of which every Jew, whatever his country, his station, or shade of belief, is necessarily a member.'

And he concludes his article, from which these quotations are made, with these words:

'Organize, organize, organize, until every Jew must stand up and be counted—counted with us, or prove himself, wittingly or unwittingly, of the few who are against their own people.'

Sir Samuel Montagu, the British Jew who has been appointed governor of Palestine under the British mandate, habitually speaks of the Jewish Kingdom, usually employing the expression 'the restoration of the Jewish Kingdom.' It may be of significance that the native population already refer to Sir Samuel as 'The King of the Jews.'

Achad Ha-Am, who must be regarded as the one who has most conclusively stated the Jewish Idea as it has always existed, and whose influence is not as obscure as his lack of fame among the Gentiles might indicate, is strong for the separate identity of the Jews as a super-nation. Leon Simon succinctly states the great teacher's views when he says:

'While Hebraic thought is familiar with the conception of a Superman (distinguished, of course, from Nietzsche's conception by having a very different standard of excellence), yet its most familiar and characteristic application of that conception is not to the individual *but to the nation*—to *Israel as the Super-Nation* or 'chosen people.' In fact, the Jewish nation is presupposed in all characteristically Jewish thinking, just as it is presupposed in the teaching of the Prophets.'

'In those countries,' says Moses Hess, 'which form a dividing line between the Occident and the Orient, namely, Russia, Poland, Prussia and Austria, there live millions of our brethren who earnestly believe in the restoration of *the Jewish Kingdom* and pray for it fervently in their daily services.'

This article, therefore, at the risk of appearing tedious, has sought to summon from many sides and from many periods the testimony which should be taken whenever the subject of Jewish nationalism comes under discussion. Regardless of what may be said to Gentile authorities for the purpose of hindering or modifying their action, there can be no question as to what the Jew thinks of himself: He thinks of himself as belonging to a People, united to that People by ties of blood which no amount of creedal change can weaken, heir of that People's past, agent of that People's political future. He belongs to a race; he belongs to a nation; he seeks a kingdom to come on this earth, a kingdom which shall be over all kingdoms, with Jerusalem the ruling city of the world. That desire of the Jewish Nation may be fulfilled; it is the contention of these articles that it will not come by way of the Program of the Protocols nor by any of the other devious ways through which powerful Jews have chosen to work.

The charge of religious prejudice has always touched the people of civilized countries in a tender spot. Sensing this, the Jewish spokesmen chosen to deal with non-Jews have emphasized the point of religious prejudice. It is therefore a relief to tender and uninstructed minds to learn that Jewish spokesmen themselves have said that the troubles of the Jew have never arisen out of his religion, the Jew is not questioned on account of his religion, but on account of other things which his religion ought to modify. Gentiles know the truth that the Jew is not persecuted on account of his religion. All honest investigators know it. The attempt to shield the Jews under cover of their religion is, therefore, in face of the facts and of their own statements, an unworthy one.

If there were no other evidence, the very evidence which many Jewish writers cite, namely, the instant siding of the Jews one with another upon any and every occasion, would constitute evidence of racial and national solidarity. Whenever these articles have touched the International Jew Financier, hundreds of Jews in the lower walks of life have protested. Touch a Rothschild, and the revolutionary Jew from the ghetto utters his protest, and accepts the remark as a personal affront to himself. Touch a regular old-line Jewish politician who is using a government office exclusively for the benefit of his fellow Jews as against the best interests of the nation, and the Socialist and anti-government Jew comes out in his defense. Most of these Jews, it may be said, have lost a vital touch with the teachings and ceremonials of their religion, but they indicate what their real religion is by their national solidarity.

This in itself would be interesting, but it becomes important in view of another fact, with which the next article will deal, namely, the relation between this Jewish nationalism and the nationalism of the peoples among whom the Jews dwell."

The issue of Jewish racism was raised in THE DEARBORN INDEPENDENT on 14 May 1921, quoting Leo N. Levi from *Memorial Volume: Leo N. Levi. I. O. B. B.*, Hamburger Print. Co., Chicago, (1905); and Louis Dembitz Brandeis' *The Jewish Problem; How to Solve It*, Zionist Essays Pub. Committee, New York, (1915):

"B'nai B'rith Leader Discusses the Jews

TO THE pro-Jewish spokesmen who have filled the air with cries of 'lies' and 'slander,' to those self-appointed guardians of 'American ideals' who rule out with rare finality all those who would dare suggest that possibly there is a hidden side of the Jewish Question, it must come as something of a jolt to be reminded that in this series there is scarcely a line that is without high Jewish authority.

The Protocols themselves are written for centuries in Jewish authoritative teachings and records. All the plans that have been described from time to time in these articles are written in the fundamental laws of the Jews. And all that the ancients have taught, the modern Jews have reaffirmed.

The writer of these articles has had to take constant counsel of prudence in his selection of material, for the Jews have always counted confidently on the fact that if the whole truth were told in one comprehensive utterance, no one would believe it. Thus, bigots and minds bursting with the discoveries they have made, have never been feared by the Jews. They counted on the incapacity of the non-Jews to believe or receive certain knowledge. They know that facts are not accepted on proof, but only on understanding. Non-Jews cannot understand why human beings should lend themselves to certain courses. They are, however, beginning to understand, and the proof is therefore becoming more significant.

There are yet more important revelations to be made, always following closely the best Jewish sources, and when these revelations are made, it will be impossible for the Jewish leaders to keep silent or to deny. The time is coming for American Jewry to slough off the leadership which has led it and left it in the bog. Leadership knows that. Indeed, it is amazing to discover the number of indications that the attempts made to suppress THE DEARBORN INDEPENDENT have been made principally *to prevent the Jews reading it*. The leaders do not care how many non-Jews read these articles; but they do not desire their own people to read them. The Jewish leaders do not desire their people's eyes to be opened.

Why? Because, just now, only Jews can truly know whether the statements made in these articles are true or not. Non-Jews may know here and there, as their observations may confirm the printed statements. But informed Jews really *know*. And large numbers of the masses of the Jews really know. When they see the truth in all its relationships in these articles, the hitherto 'led' Jew may not be so tractable. Hence the effort to keep the non-Jewish point of view away from him.

In support of the statements that these articles have been based on Jewish authority, we quote today a series of declarations by one of the most able of the presidents of the B'nai B'rith, Leo N. Levi. Mr. Levi was American-born and died in 1904. He was a lawyer of distinction and attained the presidency of the international Jewish order, B'nai B'rith, in 1900. He took part in the international politics of his people and is credited with collaborating with Secretary of State John Hay on several important matters. The utterances here quoted were for the most made while he was president of B'nai B'rith, but all of them were published the year after his death under B'nai B'rith auspices. There is therefore no question of their Jewishness.

Non-Jewish defenders of the Jewish program have pretended to much indignation because of references that have been made to the Oriental character of certain Jewish manifestations. The references in these articles have been two in number, once regarding Oriental sensuality as it has been introduced to the American stage by Jewish theatrical pandlerers, and again in quoting Disraeli, the Jew who became premier of Britain, to the effect that the Jews—his people—were 'Mosaic Arabs.'

But it never seemed to have occurred to Leo N. Levi to deny the Oriental character of his race. Instead, he asserted it. On page 104 of the B'nai B'rith memorial, he excuses certain social crudities of the Jew on the ground 'that hailing originally from the Orient and having been compelled for twenty centuries to live in a society of his own, he has preserved in his tastes much that is characteristically Oriental.' Again on page 116, he excused the multiplicity of religious rites as being due to the fact that the Jew 'drew upon his Oriental imagination for a symbolism that appealed to his ideal emotions.' On page 312, he speaks of the Jews' 'Oriental devotion to their parents.' This easy recognition of the fact is commended to those bootlicking editors who, out of the vastness of their ignorance of the Jewish Question, have seen in the reference to Orientalism an 'insult' to the Jews and an unfailing indication of anti-Semitism.

The Jewish Question! Ah, that is another point which pro-Jewish spokesmen hasten to deny, but they will be somewhat disturbed by the candor with which true Jewish spokesmen admit the Question.

In a strong passage on page 101, Mr. Levi says:

'If I have dwelt so long upon this subject, it is because I recognize that if the Jew has been denied so much that is rightfully his, he often claims more than is his due. One of these claims, most persistently urged, is that there is no Jewish Question; that a Jew is a citizen like any other citizen and that as long as he abides by the law and does not subject himself to criminal prosecution or civil action, his doings are beyond legitimate inquiry by the public at large.

'This contention on his part would certainly be well based if he claimed nothing further than the right to live in peace, but when he demands social recognition the whole range of his conduct is a legitimate subject of inquiry against which no technical demurrers can be interposed .... nor must the Jew be over-sensitive about the inquiry.

'The inconsistencies and the unwisdom exhibited in the consideration of the Jewish Question are not to be found altogether on the side of those who are hostile to the Jews.'

'Since then the refugees from Russia, Galicia and Rumania have raised the Jewish Question to commanding importance. Since then it has dawned on the world that *we are witnessing another exodus which promises soon to change the habitat of the Jews to the Western Hemisphere.'* (Page 59)

'The Jewish Question cannot be solved by tolerance. There are thousands of well-meaning people who take to themselves great credit for exhibiting a spirit of tolerance toward the Jews.' (Page 98)

Mr. Levi also lays down rules for 'the study of the Jewish Question,' and he says that if they were followed the result 'would be startling at once to the Jews and the general public.' (Page 93) How far present Jewish leadership has departed from that frank and broad view taken by Mr. Levi, is everywhere evident.

Not that Mr. Levi was a critic of his people, but he was a lawyer who was accustomed to weighing facts, and he saw facts that weighed against his people. But he was pro-Jewish even in his most severe observations. He could make an attack on the rabbis, taunting them with the saying that 'many of you are 'rabbis for revenue only,'' but he could also insist on Jewish solidarity and exclusiveness.

In this connection it may be interesting to see how strongly Mr. Levi supports the contention of Jewish leaders (as outlined in THE DEARBORN INDEPENDENT of October 9 and 16, 1920) that the Jews are a *race* and not merely a *religion*, a nation and not merely a church, and that the term 'Jew' is biological rather than theological. This is specially commended to the attention of those dim-minded shouters of 'religious prejudice,' who come into action whenever the Jewish Question is mentioned. (Of 'religious prejudice' there are many examples to give in future articles.)

'Certain it is that thus far the race and the religion have been so fused, as it were, that none can say just where the one begins and the other leaves off.' (Page 116)

Attacking the contention of the 'liberals' or 'reformed Jews' to the effect that 'Jew' is the name of a member of religious denomination, and not of a member of a certain race, Mr. Levi says:

'Nothing to my mind is more pregnant with error than this postulate of unreason. (Page 185) It is not true that the Jews are only Jews because of their religion.' (Page 189)

'The Jews are not simply an indiscriminate lot of people who hold to a common belief.' (Page 190)

'A native Eskimo, an American Indian might conscientiously adopt every tenet of the Jewish church, might practice every form and ceremony imposed by the Jewish laws and the Jewish ritual, and as far as the religion is concerned, be a Jew, but yet, no one who will reflect for a moment would class them with the Jews as a people. If the truth were known, a very large percentage of so-called Christians would be found to be believers in the essentials of the Jewish religion, and yet, they are not Jews.

'It requires not only that men should believe in Judaism, but that they should be the descendants in a direct line of that people who enjoyed a temporal government and who owned a country up to the time of the destruction of the second commonwealth.

'That great event took away from the Jews their country and their temporal government; it scattered them over the face of the earth, *but it did not destroy the national and race idea* which was a part of their nature and of their religion.'

'Who shall say, then, that the Jews are no longer a race ? .... Blood is the basis and sub-stratum of the race idea, and no people on the face of the globe can lay claim with so much right to purity of blood, and unity of blood, as the Jews.'

'If I have reasoned to any purpose, the inquiry of rights in the premises is not to be limited to Jews as exponents of a particular creed, but *to the Jews as a race.*' (Pages 190-191)

'The religion alone does not constitute the people. As I have already maintained, a believer in the Jewish faith does not by reason of that fact become a Jew. On the other hand, however, *a Jew by birth remains a Jew, even though he abjures his religion.*' (Page 200)

This is the view of such men as Justice Brandeis, the Jew who sits on the Supreme Court of the United States. Justice Brandeis says, 'Let us all recognize that we Jews are a distinct nationality *of which every Jew, whatever his country, his station, his shade of belief, is necessarily a member.*'

Believing all this, Mr. Levi subscribes to the Jewish law and practice of exclusiveness.

Describing the state of the Jews, Mr. Levi says (page 92): 'The Jews have not materially increased or diminished in numbers for 2,000 years. They have made no proselytes to their religion ... . They have imbibed the arts, the literature and the civilization of successive generations, but have abstained very generally from intermixture of blood ... . They have infused their blood into that of other peoples but have taken little of other peoples into their own.'

As to intermarriage between the Jew and non-Jew, Mr. Levi calls it miscegenation. 'In remote countries, sparsely populated, the choice may lie between such marriages and a worse relation.' Those are his words on page 249. He does not advise the worse relation, but he has said quite enough to indicate the Jewish view of the case. He continues:

'It seems clear to me that Jews should avoid marriages with Gentiles and Gentiles with Jews, *upon the same principle that we avoid marrying the insane, the consumptive, the scrofulitic or the Negro.*' (Page 249)

This exclusiveness goes down through all human relations. The Jew has one counsel for non-Jews and another for himself in these matters. Of the non-Jew he demands as a right what he looks down upon as shady privilege. He uses the Ghetto as a club with which to bludgeon the non-Jew for his 'bigotry,' when as a fact he chooses the Ghetto for well-defined racial reasons. He condemns the non-Jew for the exclusion of the Jew from certain sections of society, when as a Jew his whole care is to keep himself unspotted from that very society to which he seeks entrance. The Jew insists on breaking down non-Jewish exclusiveness while keeping his own. The non-Jewish world is to be public and common, the Jewish world is to be kept sacrosanct. Read the teachings of this enlightened leader of Jewry as published by the B'nai B'rith.

He favors the public school for non-Jewish children, not for Jewish children; they are to be kept separate: they are the choice stock of the earth:

'Because the government tenders free education, it does not follow that it must he accepted if education be made compulsory, it does not follow that government schools must be attended ... . As a citizen I favor free schools, because the education they afford, imperfect as it is, is better than none, and society is benefitted thereby; but as an individual I prefer to pay to support free schools and send my children to more select places.' (Page 253) He speaks of the fact that 'all classes of children frequent the public schools' as an argument against Jewish children going there.

'In my judgment, Jewish children should be educated in Jewish schools.' (page 254) 'Not only is it a positive and direct advantage to educate our children as Jews, but it is absolutely necessary to our preservation. Experience has shown that our young people will be weaned from our people if allowed indiscriminately to associate with the Gentiles.' (Page 255)

Discussing the possibility of Jews losing their crudeness, Mr. Levi asks, 'How shall we best accomplish that end ?' Then he quotes the frequent answer: 'Since the exemplars of gentility most abound among the Gentiles, we should associate with them as much as possible, in order to wear our own rudeness away.' He meets the suggestion this way:

'If gentlemen were willing to meet all Jews on a parity because they are Jews, we should doubtless derive much benefit from such association. But, while it is true that no gentleman refuses association with another because that other is a Jew, he will not, as a rule, associate with a Jew unless he be a gentleman. As we are far from being all gentlemen, we cannot reasonably expect to be admitted as a class into good society. So, better keep by ourselves,' concludes Mr. Levi. (Page 260)

That is, Mr. Levi admits the willingness of society to meet Jews on equal terms, as with all others, but not on unequal terms. And this being so, Mr. Levi holds they had better meet as little as possible, they had better keep apart; in the formative years, certainly, Jewish young people should be kept rigidly apart from non-Jews. The exclusiveness of which the Jews complain is their own. The Ghetto is not a corner into which the non-Jews have herded the Semites; the Ghetto is a spot carved out of the community and consecrated to the Chosen People and is therefore the best section of the city in Jewish eyes, the rest being 'the Christian quarter,' the area of the heathen. Mr. Levi himself admits on page 220 that there is no prejudice against the Jew in this country.

Certain wild-eyed objectors to the series of studies on the Jewish Question have made the assertion that THE DEARBORN INDEPENDENT has declared cowardice to be a Jewish trait. That the statement is false as regards this paper does not change the fact that the subject has been generally discussed in and out of army circles. If it ever becomes necessary to discuss it in these studies, the facts will be set forth as far as they are obtainable. But the point just now is that Mr. Levi has had somewhat to say which may repay reading:

'Physical courage has always been an incident, not an element, of Jewish character. It has no independent existence in their make-up, and always depended on something else. With some exceptions this may be said of all Oriental people. The sense and fear of danger is highly developed in them, and there is no cultivation of that indifference to it which has distinguished the great nations of Western Europe.' (Page 205)

Were a non-Jew to call attention to this difference between the Jews and others, he would be met with the cry of 'anti-Semitism' and he would be twitted with the fact that all his relatives may not have served in the war. Loudest to twit him would be those who served in what our soldiers called 'the Jewish infantry,' the quartermasters corps in the late National Army.

It is to this aversion to danger, however, that Mr. Levi attributes the Jews' greatness among the nations. Other nations can fight, the Jews can *endure*, and that, he says, is greater. Note his words (the italics are his own)

'Other nations may boast conquests and triumphs born of aggression, but though the fruits of victory have been manifold, they have not been enduring; *and it may be truly said that the nation whose greatness grows out of valor* passes through the stages of discord and degeneracy to decay ... . In the virtue of endurance I believe the Jews have a safeguard against the decay that has marked the history of all other peoples.'

It appears, therefore, that the draft-dodger, if he can *endure* long enough, may yet come to own the country.

Jewish leaders have lately tried to minimize as 'wild words' the disclosures made by Disraeli with reference to the Jews' participation in European revolutions. What Disraeli said can be found in his 'Coningsby,' or in the quotations made therefrom in THE DEARBORN INDEPENDENT of December 18, 1920. With reference to the German Revolution of 1848, Disraeli wrote—before it had taken place:

'You never observe a great intellectual movement in Europe in which the Jews do not greatly participate ... . That mysterious Russian Diplomacy which so alarms Western Europe is organized and principally carried on by Jews. That mighty revolution which is at this moment preparing in Germany, and which will be, in fact, a second and greater Reformation, and of which so little is yet known in England, is entirely developing under the auspices of Jews.'

It is interesting, therefore, to hear Mr. Levi confirming from the American side those significant statements made by Disraeli.

'The revolution of 1848 in Germany, however, influenced a great many highly educated Jews to come to America.' (Page 181) 'It is unnecessary to review the events of 1848; suffice it to say, that not a few among the revolutionists were Jews, and that a considerable number of those who were proscribed by the government at home, fled to the United States for safety.'

(Page 182) These German Jews are now the arch-financiers of the United States. They found here complete liberty to exploit peoples and nations to the full extent of their powers. They still maintain their connections with Frankfort-on-the-Main, the world capital of International financial Jewry.

With these quotations from the speeches and writings of Leo N. Levi, a famous president of the B'nai B'rith, it would seem to be a fair question as to the reason for the denial and denunciation which have followed the making of these statements in the course of this series of studies. Leo N. Levi studied the Jewish Question because he knew a Jewish Question to exist. He knew that the Jewish Question was not a non-Jewish creation but appeared wherever Jews began to appear in numbers. They brought it with them. He knew the justice of many of the charges laid against the Jews. He knew the impossibility of disproving them, the futility of shrieking 'anti-Semitism' at them. He knew, moreover, that for the Jews to solve the Jewish Question by departing from the peculiar racial traditions of superiority, would be to cease to be Jews. Therefore, he threw his whole influence on the side of the Jews remaining separate, maintaining their tradition of The Chosen Race, looking upon themselves as the coming rulers of the nations, and there he left the Question just about where he found it.

But in the course of his studies he gave other investigators the benefit of his frank statements. He did not put lies into the mouths of his people. He was not endeavoring to maintain himself in position by prejudiced racial appeals. He looked certain facts in the face, made his report, and chose his side. Several times in the course of his argument, his very logic led him up to the point where, logically, he would have to cast aside his Jewish idea of separateness. But with great calmness he discarded the logic and clung to the Jewish tradition. For example:

'The better to facilitate such happiness in every country and in every age, various kinds of organizations have existed as they exist today. The Jews have theirs.

'For many reasons they are exclusive. In theory they should not be so. In our social organizations we should, in deference to the argument which I have already named, admit any congenial and worthy Gentile who honors us with his application. But what may be theoretically correct may be found practically wrong. It certainly is a wrong to exclude a worthy person because he does not happen to be a Jew; but on the other hand, where are you to draw the line?'

This is frankness to a fault. Of course, it is wrong, but the right is impractical! Logic goes by the boards in the face of something stronger. Mr. Levi is not to be blamed for having gone to his tribe. Every man's place is with his tribe. The criticism belongs to the lick-spittle Gentile Fronts who have no tribe and become hangers-on around the outskirts of Judah, racial mongrels who would be better off if they had one-thousandth of the racial sense which the Jew possesses.

This brief survey of the philosophy which Mr. Levi both lived and taught, amid which is shared by the leaders of American Jewry, is in strict agreement with Jewish principles all down the centuries. In his published addresses Mr. Levi does not touch upon all the implications of the separateness which he enjoins upon his nation. Why do they keep by themselves? What is it that keeps them distinct? Is it their religion? Very well; let us regard them as a sect of religious recluses and wish them well in their endeavors to keep themselves unspotted of the world. Is it their race? So their leaders teach. Race and nationality are strictly claimed. If this is so, there must be a political outlook. What is it? Palestine? Not that any one can notice. A great deal may be read about it in the newspapers, the newspapers in turn being supplied through the Associated Press with the Jewish Telegraph Agency's propaganda dispatches; but no one in Palestine notices the Land becoming more Jewish. Jewry's political outlook is world rule in the material sense. Jewry is an international nation. It is this, and nothing else, which gives significance to its financial, educational, propagandist, revolutionary and immigration programs."[267]

Even some leading "anti-Zionist Jews of America"[268] saw the Jews as a distinct race and a people united by a common religion. Henry Morgenthau wrote in 1921,

"The proudest boast of all these men, and my proudest boast, is: 'I am an American.' None of us would deny our race or faith. We are Jews by blood. We are Jews, though of various sects, by religion."[269]

Returning to the racist segregationist Zionists of America, blackmailer Supreme Court Justice Louis Dembitz Brandeis stated in 1915:

"The Jewish Problem; How to Solve It

THE SUFFERING of the Jews due to injustices continuing throughout nearly twenty centuries is the greatest tragedy in history. Never was the aggregate of such suffering larger than today. Never were the injustices more glaring. Yet the present is pre-eminently a time for hopefulness. The current of world thought is at last preparing the way for our attaining justice. The war is developing opportunities which make possible the solution of the Jewish problem. But to avail ourselves of these opportunities we must understand both them and ourselves. We must recognize and accept facts. We must consider our course with statesmanlike calm. We must pursue resolutely the course we shall decide upon; and be ever ready to make the sacrifices which

---

267. "B'nai B'rith Leader Discusses the Jews", *The Dearborn Independent*, (14 May 1920); reprinted in "Jewish Influences in American Life", *The International Jew: The World's Foremost Problem*, Volume 3, (1921), pp. 167-178.
268. H. Morgenthau, "Zionism a Surrender, Not a Solution", *The World's Work*, Volume 42, Number 3, (July, 1921), pp. i-viii, at vii.
269. H. Morgenthau, "Zionism a Surrender, Not a Solution", *The World's Work*, Volume 42, Number 3, (July, 1921), pp. i-viii, at vii.

a great cause demands. Thus only can liberty be won.

For us the Jewish Problem means this: How can we secure for Jews, wherever they may live, the same rights and opportunities enjoyed by non-Jews? How can we secure for the world the full contribution which Jews can make, if unhampered by artificial limitations?

The problem has two aspects: That of the individual Jew and that of Jews collectively. Obviously, no individual should be subjected anywhere, by reason of the fact that he is a Jew, to a denial of any common right or opportunity enjoyed by non-Jews. But Jews collectively should likewise enjoy the same right and opportunity to live and develop as do other groups of people. This right of development on the part of the group is essential to the full enjoyment of rights by the individual. For the individual is dependent for his development (and his happiness) in large part upon the development of the group of which he forms a part. We can scarcely conceive of an individual German or Frenchman living and developing without some relation to the contemporary German or French life and culture. And since death is not a solution of the problem of life, the solution of the Jewish Problem necessarily involves the continued existence of the Jews as Jews.

Councils of Rabbis and others have undertaken at times to prescribe by definition that only those shall be deemed Jews who professedly adhere to the orthodox or reformed faith. But in the connection in which we are considering the term, it is certainly not in the power of any single body of Jews, or indeed of all Jews collectively, to establish the effective definition. The meaning of the word Jewish in the term Jewish Problem must be accepted as co-extensive with the disabilities which it is our problem to remove. It is the non-Jews who create the disabilities and in so doing give definition to the term Jew. Those disabilities extend substantially to all of Jewish blood. The disabilities do not end with a renunciation of faith, however sincere. They do not end with the elimination, however complete, of external Jewish mannerisms. The disabilities do not end ordinarily until the Jewish blood has been so thoroughly diluted by repeated inter-marriages as to result in practically obliterating the Jew.

And we Jews, by our own acts, give a like definition to the term Jew. When men and women of Jewish blood suffer, because of that fact, and even if they suffer from quite different causes, our sympathy and our help goes out to them instinctively in whatever country they may live and without inquiring into the shades of their belief or unbelief. When those of Jewish blood exhibit moral or intellectual superiority, genius or special talent, we feel pride in, them, even, if they have abjured the faith like Spinoza, Marx, Disraeli or Heine. Despite the meditations of pundits or the decrees of council, our own instincts and acts, and those of others, have defined for us the term Jew.

Half a century ago the belief was still general that Jewish disabilities would disappear before growing liberalism. When religious toleration was proclaimed, the solution of the Jewish Problem seemed in sight. When the so-called rights of man became widely recognized, and the equal right of all citizens to life, liberty and the pursuit of happiness began to be enacted into positive law, the complete emancipation of the Jews seemed at hand. The concrete gains through liberalism were indeed large. Equality before the law was established throughout the western hemisphere. The Ghetto walls crumbled; the ball and chain of restraint were removed in central and western Europe. Compared with the cruel discrimination to which Jews are now subjected in Russia and Romania, their advanced condition in other parts of Europe seems almost ideal.

But the anti-Jewish prejudice was not exterminated even in those countries of Europe in which the triumph of civil liberty and democracy extended fully to Jews 'the rights of man.' The anti-Semitic movement arose in Germany a year after the granting of universal suffrage. It broke out violently in France, and culminated in the Dreyfus case, a century after the French Revolution had brought 'emancipation.' It expressed itself in England through the Aliens Act, within a few years after the last of Jewish disabilities had been there removed by law. And in the United States the Saratoga incident reminded us, long ago, that we too have a Jewish question.

The disease is universal and endemic. There is, of course, a wide difference between the Russian disabilities with their Pale of Settlement, their denial of opportunity for education and of choice of occupation, and their recurrent pogroms, and the German disabilities curbing university, bureaucratic and military careers. There is a wide difference also between these German disabilities and the mere social disabilities of other lands. But some of those now suffering from the severe disabilities imposed by Russia and Romania are descendants of men and women who in centuries before our modern liberalism enjoyed both legal and social equality in Spain and Southern France. The manifestations of the Jewish Problem vary in the different countries, and at different periods in the same country, according to the prevailing degrees of enlightenment and other pertinent conditions. Yet the differences, however wide, are merely in degree and not in kind. The Jewish Problem is single and universal. But it is not necessarily eternal. It may be solved.

Why is it that liberalism has failed to eliminate the anti-Jewish prejudice? It is because the liberal movement has not yet brought full liberty. Enlightened countries grant to the individual equality before the law; but they fail still to recognize the equality of whole peoples or nationalities. We seek to protect as individuals those constituting a minority; but we fail to realize that protection cannot be complete unless group equality also is recognized.

Deeply imbedded in every people is the desire for full development, the longing, as Mazzini phrased it, 'To elaborate and express their idea, to contribute their stone also to the pyramid of history.' Nationality like democracy has been one of the potent forces making for man's advance during the past hundred years. The assertion of nationality has infused whole peoples with hope, manhood and self-respect. It has ennobled and made purposeful millions of lives. It offered them a future, and in doing so revived and capitalized all that was valuable in their past. The assertion of nationality raised Ireland from the slough of despondency. It roused Southern Slavs to heroic deeds. It created gallant Belgium. It freed Greece. It gave us united Italy. It manifested itself even among the free peoples, like the Welsh, who had no grievance, but who gave expression to their nationality through the revival of the old Cymric tongue. Each of these peoples developed because, as Mazzini said, they were enabled to proclaim 'to the world that they also live, think, love, and labor for the benefit of all.'

In the past it has been generally assumed that the full development of one people necessarily involved its domination over others. Strong nationalities are apt to become convinced that by such domination only does civilization advance. Strong nationalities assume their own superiority, and come to believe that they possess the divine right to subject other people to their sway. Soon the belief in the existence of such a right becomes converted into a conviction that duty exists to enforce it. Ways of aggrandizement follow as a natural result of this belief.

This attitude of certain nationalities is the exact correlative of the position which was generally assumed by the strong in respect to other individuals before democracy became a common possession. The struggles of the eighteenth and nineteenth centuries both in peace and in war were devoted largely to overcoming that position as to individuals. In establishing the equal right of every person to development, it became clear that equal opportunity for all involves this necessary limitation: Each man may develop himself so far, but only so far, as his doing so will not interfere with the exercise of a like right by all others. Thus liberty came to mean the right to enjoy life, to acquire property, to pursue happiness in such manner and to such extent as the exercise of the right in each is consistent with the exercise of a like right by every other of our fellow-citizens. Liberty thus defined underlies twentieth century democracy. Liberty thus defined exists in a large part of the western world. And even where this equal right of each individual has not yet been accepted as a political right, its ethical claim is gaining recognition. Democracy rejected the proposal of the superman who should rise through sacrifice of the many. It insists that the full development of each individual is not only a right, but a duty to society; and that our best hope for civilization lies not in uniformity, but in wide differentiation... .

The difference between a nation and a nationality is clear; but it is not always observed. Likeness between members is the essence of nationality; but the members of a nation may be very different. A nation may be composed of many nationalities, as some of the most successful nations are. An instance of this is the British nation, with its division into English, Scotch, Welsh, and Irish at home; with the French in Canada; and throughout the Empire, scores of other nationalities. Other examples are furnished by the Swiss nation with its German, French and Italian sections; by the Belgian nation composed of Flemings and Walloons; and by the American nation which comprises nearly all the white nationalities. The unity of a nationality is a fact of nature; the unification into a nation is largely the work of man. The false doctrine that nation and nationality must be made co-extensive is the cause of some of our greatest tragedies. It is, in large part, the cause also of the present war. It has led, on the one hand, to cruel, futile attempts at enforced assimilation, like the Russianizing of Finland and Poland, and the Prussianizing of Posen, Schleswig-Holstein, and Alsace-Lorraine. It has led, on the other hand, to those Panistic movements which are a cloak for territorial ambitions. As a nation may develop though composed of many nationalities, so a nationality may develop though forming parts of several nations. The essential in either case is recognition of the equal rights of each nationality.

W. Allison Philips recently defined nationality as, 'An extensive aggregate of persons, conscious of a community of sentiments, experiences, or qualities which make them feel themselves a distinct people.' And he adds: 'If we examine the composition of the several nationalities we find these elements: race, language, religion, common habitat, common conditions, mode of life and manners, political association. The elements are, however, never all present at the same time, and none of them is essential... . A common habitat and common conditions are doubtless powerful influences at times in determining nationality; but what part do they play in that of the Jews or the Greeks, or the Irish in dispersion?'

See how this high authority assumes without question that the Jews are, despite their dispersion, a distinct nationality; and he groups us with the Greeks or the Irish, two other peoples of marked individuality. Can it be doubted that we Jews, aggregating 14,000,000 people, are 'an extensive aggregate of persons'; that we are 'conscious of a community of sentiments, experiences and qualities which make us feel ourselves a distinct people,' whether we admit it or not?

It is no answer to this evidence of nationality to declare that the Jews are not an absolutely pure race. There has, of course, been some intermixture of foreign blood in the 3000 years which constitute our historic period. But, owing to persecution and prejudice, the intermarriages with non-Jews which occurred have resulted merely in taking away many from the Jewish community. Intermarriage has brought few additions. Therefore the percentage of foreign blood in the Jews of today is very low. Probably no important European race is as pure.

But common race is only one of the elements which determine nationality. Conscious community of sentiments, common experiences, common qualities are equally, perhaps more, important. Religion, traditions and customs bound us together, though scattered throughout the world. The similarity of experience tended to produce similarity of qualities and community of sentiments. Common suffering so intensified the feeling of brotherhood as to overcome largely all the influences making for diversification. The segregation of the Jew was so general, so complete, and so long continued as to intensify our 'peculiarities' and make them almost ineradicable.

We recognize that with each child the aim of education should be to develop his own individuality, not to make him an imitator, not to assimilate him to others. Shall we fail to recognize this truth when applied to whole peoples? And what people in the world has shown greater individuality than the Jews? Has any a nobler past? Does any possess common ideas better worth expressing? Has any marked traits worthier of development? Of all the peoples in the world those of two tiny states stand preeminent as contributors to our present civilization, the Greeks and the Jews. The Jews gave to the world its three greatest religions, reverence for law, and the highest conceptions of morality. Never before has the value of our contribution been so generally recognized. Our teaching of brotherhood and righteousness has, under the name of democracy and social justice, become the twentieth century striving of America and of western Europe. Our conception of law is embodied in the American constitution which proclaims this to be a 'government of laws and not of men.' And for the triumph of our other great teaching, the doctrine of peace, this cruel war is paving the way.

While every other people is striving for development by asserting its nationality, and a great war is making clear the value of small nations, shall we voluntarily yield to anti-Semitism, and instead of solving our 'problem' end it by noble suicide? Surely this is no time for Jews to despair. Let us make clear to the world that we too are a nationality striving for equal rights to life and to self-expression. That this should be our course has been recently expressed by high non-Jewish authority. Thus Seton-Watson; speaking of the probable results of the war, said:

'There are good grounds for hoping that it [the war] will also give a new and healthy impetus to Jewish national policy, grant freer play to their splendid qualities, and enable them to shake off the false shame which has led men who ought to be proud of their Jewish race to assume so many alien disguises and to accuse of anti-Semitism those who refuse to be deceived by mere appearances. It is high time that the Jews should realize that few things do more to foster anti-Semitic feeling than this very tendency to sail under false colors and conceal their true identity. The Zionists and the orthodox Jewish Nationalists have long ago won the respect and admiration of the world. No race has ever defied assimilation so stubbornly and so successfully; and the modern tendency of individual Jews to repudiate what is one of their chief glories suggests an almost comic resolve to fight against the course of nature.'

Standing against this broad foundation of nationality, Zionism aims to give it full development. Let us bear clearly in mind what Zionism is, or rather what it is not.

It is not a movement to remove all the Jews of the world compulsorily to Palestine. In the first place there are 14,000,000 Jews, and Palestine would not accommodate more than one-third of that number. In the second place, it is not a movement to compel anyone to go to Palestine. It is essentially a movement to give to the Jew more, not less freedom; it aims to enable the Jews to exercise the same right now exercised by practically every other people in the world: To live at their option either in the land of their fathers or in some other country; a right which members of small nations as well as of large, which Irish, Greek, Bulgarian, Serbian, or Belgian, may now exercise as fully as Germans or English.

Zionism seeks to establish in Palestine, for such Jews as choose to go and remain there, and for their descendants, a legally secured home, where they may live together and lead a Jewish life, where they may expect ultimately to constitute a majority of the population, and may look forward to what we should call home rule. The Zionists seek to establish this home in Palestine because they are convinced that the undying longing of Jews for Palestine is a fact of deepest significance; that it is a manifestation in the struggle for existence by an ancient people which has established its right to live, a people whose three thousand years of civilization has produced a faith, culture and individuality which enable it to contribute largely in the future, as it has in the past, to the advance of civilization; and that it is not a right merely but a duty of the Jewish nationality to survive and develop. They believe that only in Palestine can Jewish life be fully protected from the forces of disintegration; that there alone can the Jewish spirit reach its full and natural development; and that by securing for those Jews who wish to settle there the opportunity to do so, not only those Jews, but all other Jews will be benefited, and that the long perplexing Jewish Problem will, at last, find solution.

They believe that to accomplish this, it is not necessary that the Jewish population of Palestine be large as compared with the whole number of Jews in the world; for throughout centuries when the Jewish influence was greatest, during the Persian, the Greek, and the Roman Empires, only a relatively small part of the Jews lived in Palestine; and only a small part of the Jews returned from Babylon when the Temple was rebuilt.

Since the destruction of the Temple, nearly two thousand years ago, the longing for Palestine has been ever present with the Jew. It was the hope of a return to the land of his fathers that buoyed up the Jew amidst persecution, and for the realization of which the devout ever prayed. Until a generation ago this was a hope merely, a wish piously prayed for, but not worked for. The Zionist movement is idealistic, but it is also essentially practical. It seeks to realize that hope; to make the dream of a Jewish life in a Jewish land come true as other great dreams of the world have been realized, by men working with devotion, intelligence, and self-sacrifice. It was thus that the dream of Italian independence and unity, after centuries of vain hope, came true through the efforts of Mazzini, Garibaldi and Cavour; that the dream of Greek, of Bulgarian and of Serbian independence became facts.

The rebirth of the Jewish nation is no longer a mere dream. It is in process of accomplishment in a most practical way, and the story is a wonderful one. A generation ago a few Jewish emigrants from Russia and from Romania, instead of proceeding westward to this hospitable country where they might easily have secured material prosperity, turned eastward for the purpose of settling in the land of their fathers.

To the worldly-wise these efforts at colonization appeared very foolish. Nature and man presented obstacles in Palestine which appeared almost insuperable; and the colonists were in fact ill-equipped for their task, save in their spirit of devotion and self-sacrifice. The land, harassed by centuries of misrule, was treeless and apparently sterile; and it was infested with malaria. The Government offered them no security, either as to life or property. The colonists themselves were not only unfamiliar with the character of the country, but were ignorant of the farmer's life which they proposed to lead; for the Jews of Russia and Romania had been generally denied the opportunity of owning or working land. Furthermore, these colonists were not inured to the physical hardships to which the life of a pioneer is necessarily subjected. To these hardships and to malaria many succumbed. Those who survived were long confronted with failure. But at last success came. Within a generation these Jewish Pilgrim Fathers, and those who followed them, have succeeded in establishing these two fundamental propositions:

First: That Palestine is fit for the modern Jew.

Second: That the modern Jew is fit for Palestine.

Over forty self-governing Jewish colonies attest to this remarkable achievement.

This land, treeless a generation. ago, supposed to be sterile and hopelessly arid, has been shown to have been treeless and sterile because of man's misrule. It has been shown to be capable of becoming again a land 'flowing with milk and honey.' Oranges and grapes, olives and almonds, wheat and other cereals are now growing there in profusion.

This material development has been attended by a spiritual and social development no less extraordinary; a development in education, in health and in social order; and in the character and habits of the population. Perhaps the most extraordinary achievement of Jewish nationalism is the revival of the Hebrew Language, which has again become a language of the common intercourse of men. The Hebrew tongue, called a dead language for nearly two thousand years, has, in the Jewish colonies and in Jerusalem, become again the living mother tongue. The effect of this common language in unifying the Jew is, of course, great; for the Jews of Palestine came literally from all the lands of the earth, each speaking, excepting those who used Yiddish, the language of the country from which he came, and remaining, in the main, almost a stranger to the others. But the effect of the renaissance of the Hebrew tongue is far greater than that of unifying the Jews. It is a potent factor in reviving the essentially Jewish spirit.

Our Jewish Pilgrim Fathers have laid the foundation. It remains for us to build the superstructure.

Let no American imagine that Zionism is inconsistent with Patriotism. Multiple loyalties are objectionable only if they are inconsistent. A man is a better citizen of the United States for being also a loyal citizen of his state, and of his city; for being loyal to his family, and to his profession or trade;

for being loyal to his college or his lodge. Every Irish American who contributed towards advancing home rule was a better man and a better American for the sacrifice he made. Every American Jew who aids in advancing the Jewish settlement in Palestine, though he feels that neither he nor his descendants will ever live there, will likewise be a better man and a better American for doing so.

Note what Seton-Watson says:

'America is full of nationalities which, while accepting with enthusiasm their new American citizenship, nevertheless look to some centre in the old world as the source and inspiration of their national culture and traditions. The most typical instance is the feeling of the American Jew for Palestine which may well become a focus for his *déclassé* kinsmen in other parts of the world.'

There is no inconsistency between loyalty to America and loyalty to Jewry. The Jewish spirit, the product of our religion and experiences, is essentially modern and essentially American. Not since the destruction of the Temple have the Jews in spirit and in ideals been so fully in harmony with the noblest aspirations of the country in which they lived.

America's fundamental law seeks to make real the brotherhood of man. That brotherhood became the Jewish fundamental law more than twenty-five hundred years ago. America's insistent demand in the twentieth century is for social justice. That also has been the Jews' striving for ages. Their affliction as well as their religion has prepared the Jews for effective democracy. Persecution broadened their sympathies. It trained them in patience and endurance, in self-control, and in sacrifice. It made them think as well as suffer. It deepened the passion for righteousness.

Indeed, loyalty to America demands rather that each American Jew become a Zionist. For only through the ennobling effect of its strivings can we develop the best that is in us and give to this country the full benefit of our great inheritance. The Jewish spirit, so long preserved, the character developed by so many centuries of sacrifice, should be preserved and developed further, so that in America as elsewhere the sons of the race may in future live lives and do deeds worthy of their ancestors.

But we have also an immediate and more pressing duty in the performance of which Zionism alone seems capable of affording effective aid. We must protect America and ourselves from demoralization, which has to some extent already set in among American Jews. The cause of this demoralization is clear. It results in large part from the fact that in our land of liberty all the restraints by which the Jews were protected in their Ghettos were removed and a new generation left without necessary moral and spiritual support. And is it not equally clear what the only possible remedy is? It is the laborious task of inculcating self-respect, a task which can be accomplished only by restoring the ties of the Jew to the noble past of his race, and by making him realize the possibilities of a no less glorious future. The sole bulwark against demoralization is to develop in each new generation of Jews in America the sense of *noblesse oblige*. That spirit can be developed in those who regard their people as destined to live and to live with a bright future. That spirit can best be developed by actively participating in some way in furthering the ideals of the Jewish renaissance; and this can be done effectively only through furthering the Zionist movement.

In the Jewish colonies of Palestine there are no Jewish criminals; because everyone, old and young alike, is led to feel the glory of his people and his obligation to carry forward its ideals. The new Palestinian Jewry produces instead of criminals, scientists like Aaron Aaronsohn, the discoverer of wild wheat; pedagogues like David Yellin, craftsmen like Boris Schatz, the founder of the Bezalel; intrepid *Shomrim,* the Jewish guards of peace, who watch in the night against marauders and doers of violent deeds.

And the Zionist movement has brought like inspiration to the Jews in the Diaspora, as Steed has shown in this striking passage from 'The Hapsburg Monarchy':

'To minds like these Zionism came with the force of an evangel. To be a Jew and to be proud of it; to glory in the power and pertinacity of the race, its traditions, its triumphs, its sufferings, its resistance to persecution; to look the world frankly in the face and to enjoy the luxury of moral and intellectual honesty; to feel pride in belonging to the people that gave Christendom its divinities, that taught half the world monotheism, whose ideas have permeated civilization as never the ideas of a race before it, whose genius fashioned the whole mechanism of modern commerce, and whose artists, actors, singers and writers have filled a larger place in the cultured universe than those of any other people. This, or something like this, was the train of thought fired in youthful Jewish minds by the Zionist spark. Its effect upon the Jewish students of Austrian universities was immediate and striking. Until then they had been despised and often ill-treated. They had wormed their way into appointments and into the free professions by dint of pliancy, mock humility, mental acuteness, and clandestine protection. If struck or spat upon by 'Aryan' students, they rarely ventured to return the blow or the insult. But Zionism gave them courage. They formed associations, and learned athletic drill and fencing. Insult was requited with insult, and presently the best fencers of the fighting German corps found that Zionist students could gash cheeks quite as effectually as any Teuton, and that the Jews were in a fair way to become the best swordsmen of the university. Today the purple cap of the Zionist is as respected as that of any academical association.

'This moral influence of Zionism is not confined to university students. It is quite as noticeable among the mass of the younger Jews outside, who also find in it a reason to raise their heads, and, taking their stand upon the past, to gaze straightforwardly into the future.'

Since the Jewish Problem is single and universal, the Jews of every country should strive for its solution. But the duty resting upon us of America is especially insistent. We number about 3,000,000, which is more than one fifth of all the Jews in the world, a number larger than comprised within any other country except the Russian Empire. We are representative of all the Jews in the world; for we are composed of immigrants, or descendants of immigrants coming from every other country, or district. We include persons from every section of society, and of every shade of religious belief. We are ourselves free from civil or political disabilities; and are relatively prosperous. Our fellow-Americans are infused with a high and generous spirit, which insures approval of our struggle to ennoble, liberate, and otherwise improve the condition of an important part of the human race; and their innate manliness makes them sympathize particularly with our efforts at self-help. America's detachment from the old world problem relieves us from suspicions and embarrassments frequently attending the activities of Jews of rival European countries. And a conflict between American interests or ambitions and Jewish aims is not conceivable. Our loyalty to America can never be questioned.

Let us therefore lead, earnestly, courageously and joyously, in the struggle for liberation. Let us all recognize that we Jews are a distinctive nationality of which every Jew, whatever his country, his station or shade of belief, is necessarily a member. Let us insist that the struggle for liberty shall not cease until equality of opportunity is accorded to nationalities as to individuals. Let us insist also that full equality of opportunity cannot be obtained by Jews until we, like members of other nationalities, shall have the option of living elsewhere or of returning to the land of our forefathers.

The fulfillment of these aspirations is clearly demanded in the interest of mankind, as well as in justice to the Jews. They cannot fail of attainment if we are united and true to ourselves. But we must be united not only in spirit but in action. To this end we must organize. Organize, in the first place, so that the world may have proof of the extent and the intensity of our desire for liberty. Organize, in the second place, so that our resources may become known and be made available. But in mobilizing our force it will not be for war. The whole world longs for the solution of the Jewish Problem. We have but to lead the way, and we may be sure of ample cooperation from non-Jews. In order to lead the way, we need not arms, but men; men with those qualities for which Jews should be peculiarly fitted by reason of their religion and life; men of courage, of high intelligence, of faith and public spirit, of indomitable will and ready self-sacrifice; men who will both think and do, who will devote high abilities to shaping our course, and to overcoming the many obstacles which must from time to time arise. And we need other, many, many other men, officers commissioned and non-commissioned and common soldiers in the cause of liberty, who will give of their efforts and resources, as occasion may demand, in unfailing and ever strengthening support of the measures which may be adopted. Organization, thorough and complete, can alone develop such leaders and the necessary support.

Organize, Organize, Organize, until every Jew in America must stand up and be counted, counted with us, or prove himself, wittingly or unwittingly, of the few who are against their own people."[270]

There are too many instances of hypocrisy and of sophistry in Brandeis' statement to address them all. I will mention the call for democracy while imposing a tyrannous Jewish minority on a majority Moslem and Christian native population in Palestine in hopes of someday outnumbering the native population with a deliberate and orchestrated demographic shift to a majority Jewish population. Brandeis' arbitrary statement that there could never be any conflict of interest between loyalty to a Palestinian Jewish state and loyalty to the United States of America proves his disloyalty to one, the other, or both; and statements such as his gave ammunition to the anti-Semites of Russia and Germany who sought to mischaracterize all Jews as if disloyal, even though most Jews were not Zionists.

Fellow Zionists like Jakob Klatzkin were more honest than Brandeis and openly declared their disloyalty to their present home states—Brandeis was a notorious liar and a mediocre sophist. Numerous Israeli agents, many, if not most of whom were American Jews, have been investigated and prosecuted by the Government of the United States of America for espionage against America. Israel has stolen American weapons and weapons technology and sold them to enemies of the United States. Zionist Jewish bankers have financed America's worst enemies including Great Britain, the Confederacy, Imperial Japan, Bolshevik Russia, Nazi Germany, etc. and have consistently agitated for grossly destructive wars. Zionist Jewish bankers are responsible for more American war casualties than any other group. They have deliberately cost Americans oceans of blood and mountains of treasure. Michael Collins Piper argues that Mossad agents were involved in the assassination of United States President John Fitzgerald Kennedy and that they wanted him dead because Kennedy opposed the Israeli nuclear weapons program, a program which is not in the best interests of the United States.[271] Zionist Jewish bankers have deliberately caused America's worst recessions and depressions. They have corrupted the American media and American politics.

Among the countless instances where Jews have been disloyal to their native countries, Egyptian Jews in collusion with the Israeli Government in "Operation Susannah" bombed American and British interests in Egypt and tried to make it appear that Egyptian Moslems had committed the terrorist atrocities that these Jews had committed at the behest of, and with the support of, the Israeli Government.[272] Israel has officially celebrated the disloyal Jews of the "Lavon

---

270. L. D. Brandeis, *The Jewish Problem; How to Solve It*, Zionist Essays Pub. Committee, New York, (1915).
271. M. C. Piper, *Final Judgment: The Missing Link in the JFK Assassination Conspiracy*, Wolfe Press, Washington, D.C., (1993).
272.A. Golan, *Operation Susannah*, Harper & Row, New York, (1978). **See also:** D. Raviv, *Every Spy a Prince: The Complete History of Israel's Intelligence Community*, Houghton Mifflin, Boston, (1990). **See also:** V. Ostrovsky and C. Hoy, *By Way of Deception: A Devastating Insider's Portrait of the Mossad*, Stoddart, Toronto, (1990). V.

Affair", who, without provocation, viciously attacked innocent Americans, Brits and Egyptians. In 1967, the State of Israel again committed an act of war against the United States of America and attempted to sink the *U. S. S. Liberty*, and to blame Egypt for the attack, so as to draw America into a broad war against its own best interests.[273]

Former Mossad agent Victor Ostrovsky wrote in his book *The Other Side of Deception: A Rogue Agent Exposes the Mossad's Secret Agenda* (note that a "Sayanim" is a disloyal and deceitful Jew, who is prepared to betray his or her neighbors at any time in order to advance a perceived Israeli interest),

> "The American Jewish community was divided into a three-stage action team. First were the individual *sayanim* (if the situation had been reversed and the United States had convinced Americans working in Israel to work secretly on behalf of the United States, they would be treated as spies by the Israeli government). Then there was the large pro-Israeli lobby. It would mobilize the Jewish community in a forceful effort in whatever direction the Mossad pointed them. And last was B'nai Brith. Members of that organization could be relied on to make friends among non-Jews and tarnish as anti-Semitic whomever they couldn't sway to the Israeli cause. With that sort of one-two-three tactic, there was no way we could strike out."[274]

In 2006, Professors John J. Mearsheimer and Stephen M. Walt wrote in their

---

Ostrovsky, *The Other Side of Deception: A Rogue Agent Exposes the Mossad's Secret Agenda*, Harper Paperbacks, New York, (1994). *See also:* I. Black and B. Morris, *Israel's Secret Wars: A History of Israel's Intelligence Services*, Grove Weidenfeld, New York, (1991). *See also:* S. Teveth, *Ben-Gurion's Spy: The Story of the Political Scandal That Shaped Modern Israel*, Columbia University Press, New York, (1996). *See also:* J. Beinin, *The Dispersion of Egyptian Jewry: Culture, Politics, and the Formation of a Modern Diaspora*, University of California Press, Berkeley, (1998).

[273]. W. Beecher, "Israel, in Error, Attacks U. S. Ship; 10 Navy Men Die, 100 Hurt in Raids North of Sinai Israelis, in Error, Attack U.S. Navy Ship 10 Navy Men Die and 100 Are Hurt Communications Vessel Is Raided From Air and Sea North of Sinai Peninsula", *The New York Times*, (9 June 1967), p. 1. N. Sheehan, "Sailors Describe Attack on Vessel; Israelis Struck So Suddenly U. S. Guns Were Unloaded", *The New York Times*, (11 June 1967), p. 27. "Israel Offers Compensation", *The New York Times*, (11 June 1967), p. 27. McClure M. Howland, "Families of Sailors" Letter to the Editor, *The New York Times*, (15 June 1967), p. 46. "Israel Accused at Hearing on U. S. Ship", *The New York Times*, (18 June 1967), p. 20. "U. S. Again Accused", *The New York Times*, (20 June 1967), p. 13. N. Sheehan, "Order Didn't Get to U. S. S. Liberty; Pentagon Reports Message Directing Ship Off Sinai to Move Arrived Late Ship 15.5 Miles Offshore", *The New York Times*, (29 June 1967), p. 1. *See also:* J. M. Ennes, *Assault on the Liberty: The True Story of the Israeli Attack on an American Intelligence Ship*, Random House, New York, (1979). *See also:* BBC Documentary, *Dead in the Water*, (21 August 2004, 7:00-8:10pm; rpt 1:50-3:00am)
<http://www.bbc.co.uk/bbcfour/documentaries/features/dead_in_the_water.shtml>

[274]. V. Ostrovsky, *The Other Side of Deception: A Rogue Agent Exposes the Mossad's Secret Agenda*, Harper Collins, New York, (1994), p. 32.

paper, "The Israel Lobby and U. S. Foreign Policy",

> "The U.S. national interest should be the primary object of American foreign policy. For the past several decades, however, and especially since the Six Day War in 1967, the centerpiece of U.S. Middle East policy has been its relationship with Israel. The combination of unwavering U.S. support for Israel and the related effort to spread democracy throughout the region has inflamed Arab and Islamic opinion and jeopardized U.S. security.
>
> This situation has no equal in American political history. Why has the United States been willing to set aside its own security in order to advance the interests of another state? One might assume that the bond between the two countries is based on shared strategic interests or compelling moral imperatives. As we show below, however, neither of those explanations can account for the remarkable level of material and diplomatic support that the United States provides to Israel.
>
> Instead, the overall thrust of U.S. policy in the region is due almost entirely to U.S. domestic politics, and especially to the activities of the 'Israel Lobby.' Other special interest groups have managed to skew U.S. foreign policy in directions they favored, but no lobby has managed to divert U.S. foreign policy as far from what the American national interest would otherwise suggest, while simultaneously convincing Americans that U.S. and Israeli interests are essentially identical."[275]

In America, many ethnicities have resolved their sense of heritage in different ways. Stephen Steinlight wrote, *inter alia*,

> "I'll confess it, at least: like thousands of other typical Jewish kids of my generation, I was reared as a Jewish nationalist, even a quasi-separatist. Every summer for two months for 10 formative years during my childhood and adolescence I attended Jewish summer camp. There, each morning, I saluted a foreign flag, dressed in a uniform reflecting its colors, sang a foreign national anthem, learned a foreign language, learned foreign folk songs and dances, and was taught that Israel was the true homeland. Emigration to Israel was considered the highest virtue, and, like many other Jewish teens of my generation, I spent two summers working in Israel on a collective farm while I contemplated that possibility. More tacitly and subconsciously, I was taught the superiority of my people to the gentiles who had oppressed us. We were taught to view non-Jews as untrustworthy outsiders, people from whom sudden gusts of hatred might be anticipated, people less sensitive, intelligent, and moral than ourselves. We were also taught that the lesson of our dark history is that we could rely on no one. I am of course simplifying a complex process of ethnic and religious identity

---

[275]. J. J. Mearsheimer and S. M. Walt, *The Israel Lobby and U. S. Foreign Policy*, Faculty Research Working Papers Series, Harvard University, John F. Kennedy School of Government, (March, 2006), p. 1.

formation; there was also a powerful counterbalancing universalistic moral component that inculcated a belief in social justice for all people and a special identification with the struggle for Negro civil rights."[276]

Brandeis intentionally confuses abstract ideals with peoples and circumstances. Peoples must interpret ideals and apply them to their interpretation of their circumstances. Different peoples with the same bill of ideals can come into conflict due to various circumstances and various interpretations of the same set of written ideals and different interpretations of the same set of circumstances. Different peoples can stress different ideals over others when choosing among the same set, etc. One group may change or abandon ideals as circumstances change due to demographic changes, wars, economic conditions, etc.; or by being misled, through mistakes or improvements, or because of perceived self-interest. When selecting that which constitutes the greater good, or the greater right, one side must prevail over the other.

Brandeis soon proved his own disloyalty to America by blackmailing the President of the United States and unnecessarily bringing America into a grossly destructive war, which was against America's best interests. His "you're either with us or against us" attitude toward assimilationist Jews, who were genuinely loyal to America, was typical of political Zionists, and is one reason why the Zionists were rejected by the vast majority of Jews, and the Zionists aided the anti-Semites in order to leave all persons of Jewish descent no option but to flee to a foreign state of the Zionists' making, or perish at the hands of the political Zionists and their political anti-Semitic allies in their home countries.

Brandeis, like many political Zionists, saw the "melting pot", to use Zangwill's term, of America as an alleged degeneration of the Jewish race and longed for the segregation of the Ghetto in a "World Ghetto"[277] for Jews in Palestine, to use Herzl's term. Many Zionists lamented the loss of the forced segregation of the Ghetto. As but one example of many, anti-Semitic Zionist Aaron David Gordon stated,

> "Our condition has changed strikingly in recent times, since the crumbling of the ghetto. The limited amount of independent life that still survived inside its walls has been destroyed while we, together with all mankind, have increased in knowledge, but at the expense of the spirit and of real life."[278]

---

[276]. S. Steinlight, "The Jewish Stake in America's Changing Demography: Reconsidering a Misguided Immigration Policy", *Backgrounder*, (October, 2001), p. 10.
[277]. T. Herzl, English translation by H. Zohn, R. Patai, Editor, *The Complete Diaries of Theodor Herzl*, Volume 1, Herzl Press, New York, (1960), p. 172.
[278]. A. D. Gordon, *Kitve A. D. Gordon*, In Five Volumes, Tel-Aviv, ha-Va'ad ha-merkazi shel mifleget ha-Po'el ha-tsa'ir, (1927-1930), parts translated to English in A. Hertzberg, *The Zionist Idea*, Harper Torchbooks, New York, (1959), pp. 371-386, at 384-385.

In 1924, racist Zionist Israel Zangwill wrote that Zionists wanted to segregate Jews in "Russia" in order to form an autonomous Jewish State. In his article, he points out that the only salvation to be had for Jews other than the "solution of the Jewish question" through "dissolution" was to absolutely segregate Jews in Ghettoes, be it in "Russia", New York, or Palestine. Given that the Nazis fulfilled this Zionist's objectives of segregating the Jews in large Ghettoes meant for deportation and took away their national rights, as did the Communists, one has a right to ask if the Zionists were behind, or influential members of, both the Hitler and Stalin régimes. Were the Nazis and Communists herding up the Jews of Europe for involuntary deportation to Palestine, Madagascar or a segregated state in Russia or China at the request of Zionists, or as part of a Zionist plan? Were Hitler and Stalin, in their messiah complexes, hacking apart European civilization, killing off the Gentiles and destroying religion and culture, in fulfillment of Communist objectives—those of genocidal Judaism? Had the political Zionists come to the conclusion that just as politics would play the role of Jewish Messiah in the modern world, it would fulfill that role to the letter of the Torah by making good on Jewish prophesies of the destruction of the Gentiles and the horrific punishment of "disobedient" Jews? Zangwill wrote in 1924,

> "It is true the situation may be modified if the Jewish republic now adumbrated in Russia, in the districts of Homel, Witebsk, and Minsk, really brings my own organization's ideal of an autonomous Jewish territory into being. [***] Of this species of nationalism, however, no pure example exists; it is only a tendency. New York's East Side comes nearest to it. But unless the East Side nationalists could be absolutely segregated from the general life in a close-barred Ghetto, they, and still more their children educated in the public schools, would be found responding to all the mass-emotions of the majority. [***] It was formally repudiated by Dr. Weizmann in a speech at Boston, but as even he cannot control the hot-heads or the muddle-heads of his movement, let me say here to any Diaspora nationalists that may happen to be in America that if they mean seriously that they are not merely sentimental sympathizers with the Palestine Jewry, as Irish-Americans are with Ireland, but that they are actual subjects of the Jewish National Home, they must naturally give up their American citizenship and all rights save those appertaining to resident aliens; a status which when proposed by a Belloc they are the first to cry out against. [***] The world's contempt for the Jew is not wholly undeserved. A people, a faith, in so parlous a situation, lives not under peace conditions but under war conditions, and the standard of duty exacted from every Jew is not a peace standard but a war standard. [***] That is why Zionism cannot afford to become the blind and obsequious agent of any Power."[279]

---

[279]. I. Zangwill, "Is Political Zionism Dead? Yes", *The Nation*, Volume 118, Number 3062, (12 March 1924), pp. 276-278.

Even in its infancy, the First World War was seen as an opportunity by the Zionists to grab land, and Brandeis called for soldiers to appropriate land—allegedly in the defense of liberty and democracy. There has been an oft repeated charge that both world wars only served the cause of the Zionists, and that a third is coming on their behalf. Perhaps one day history will record the world wars as a second wave of crusades to take Palestine for Zionists, instead of Christians, and to force Jews to emigrate there. Millions of completely innocent lives were lost due to political Zionist agitation.

Political Zionist leader Theodor Herzl, defamed Jews around the world as had John Chrysostom,[280] Raymund Martin,[281] Porchetus de Salvaticis,[282] Antonius Margaritha,[283] Martin Bucer,[284] Johann Eck,[285] Martin Luther,[286]

---

[280]. J. Chrysostom, translated by P. W. Harkins, "Discourses Against Judaizing Christians", *The Fathers of the Church*, Volume 68, Catholic University of America Press, Washington, D. C., (1979).

[281]. R. Martin, *Pugio fidei adversus Mauros et Judaeos*, Gregg Press, Farnborough, Hants, England, (1687/1967).

[282]. Porchetus de Salvaticis, *Victoria Porcheti aduersus impios Hebreos, in qua tum ex sacris literis, tum ex dictis Talmud, ac Caballistaru, et alioru omniu authoru, quos, HebrQei recipiut, monstratur veritas catholice fidei*, Jmpressit Guillerm[us] Desplains impensis Egidij Gourmõtij & Francisci Regnault, Parrhisijs, (1520).

[283]. A. Margaritha, *Der gantz Jüdisch Glaub mit sampt ainer gründtlichen vnd warhafften anzaygunge: aller Satzungen, Ceremonien, Gebetten, Haymliche vnd offentliche Gebreüch, deren sich dye Juden halten, durcdas ganzt Jar. Mit schönen und gegrundten Argumenten wyder jren Glauben*, H. Steyner, Augspurg, (1530).

[284]. M. Bucer, *Von den Jude ob, vn[d] wie die vnder den Christe zu halten sind, ein Rathschlag, durch die Gelerte am Ende dis Büchlins verzeichnet, zugericht. Item Ein weitere Erklerung vnd Beschirmung des selbigen Rahtschlags*, Wolfgang Köpfel, Strassburg, (1539); **and** M. Bucer and J. Kymeus, *Ratschlag jetz von den Corfürsten vnnd Fürsten, zü Franckfort gehalten, Ob Christlicher Obrigkeit gebüren müge, das sie die Jüden, vnter denn Christen zü wonen gedulden, vn[d] wo sie zü gedulden, welcher gestalt vnd mass. Durch die gelerten am ende dis Büchlins verzeichnet zü gericht*, Laurens vann der Müllen, Cöllen, (1539).

[285]. J. Eck, *Ains Juden büechlins verlegung: darin ain Christ, gantzer Christenhait zu schmach, will es geschehe den Juden vnrecht in bezichtigung der Christen Kinder Mordt*, Alexander Weissenhorn, Ingoldstat, (1541).

[286]. M. Luther, *Von den Juden und ihren Lügen*, Hans Lufft, Wittenberg, (1543); Reprinted, Ludendorffs, München, (1932); English translation by Martin H. Bertram, "On the Jews and Their Lies", *Luther's Works*, Volume 47, Fortress Press, Philadelphia, (1971), pp. 123-306.

Diderot,[287] Voltaire,[288] Ludolf Holst,[289] Richard Wagner,[290] Wilhelm Marr,[291] Eugen Karl Dühring,[292] and Edouard Adolphe Drumont.[293] Herzl declared the

---

287. D. Diderot, "Juif", *Encyclopédie, ou Dictionaire Raisonné des Sciences, des Arts et des Métiers*, Volume 9, A. Neufchastel, (1765).
288. F. M. Arouet de Voltaire, *Histoire de Charles XII, Roi de Suède*, (1731); **and** *Dictionnaire Philosophique*, Multiple Editions; multiple English translations, including: W. F. Flemming, *A Philosophical Dictionary*, Volume 6, Dingwall-Rock, New York, (1901), pp. 266-313; **and** *Essai sur les Moeurs et l'Esprit des Nations, et sur les Principaux faits de l'Histoire Depuis Charlemagne Jusqu'à Louis XIII*, Chapter 104, (1769); **and** *Philosophie Génerale: Métaphysique, Morale et Théologie*, Chez Sanson et Compagnie, Aux Deux-Ponts, (1792).
289. L. Holst, *Das Judentum in allen dessen Teilen. Aus einem staatswissenschaftlichen Standpunkt betrachtet*, Mainz, (1821).
290. R. Wagner under the *nom de plume* K. Freigedank, "Das Judenthum in der Musik", *Neue Zeitschrift für Musik*, (3 and 6 September 1850); Reprinted with revisions and an appendix, R. Wagner, *Das Judenthum in der Musik*, J. J. Weber, Leipzig, (1869); the original unrevised 1850 article was reprinted in *Gesammelte Schriften und Dichtungen*, Volume 5; English translation of the 1869 version by E. Evans, *Judaism in Music (Das Judenthum in der Musik) Being the Original Essay Together with the Later Supplement*, W. Reeves, London, (1910); Also, "Judaism in Music", *Richard Wagner's Prose Works*, Volume 3, Broude Brothers, New York, (1966), pp. 75-122.; which is a reprint of the original "Judaism in Music", *Richard Wagner's Prose Works*, Volume 3, Kegan Paul, Trench, and Trübner, London, (1894).
291. W. Marr, *Der Judenspiegel*, Im Selbstverlage des Verfassers, (1862); **and** *Der Sieg des Judenthums über das Germanenthum*, Rudolph Costenoble, Bern, (1879); **and** *Vom jüdischen Kriegsschauplatz: eine Streitschrift*, R. Costenoble, Bern, (1879); **and** *Wählet keinen Juden: Der Weg zum Siege des Germanenthums über das Judenthum*, O. Hentze, Berlin, (1879); **and** *Vom jüdischen Kriegsschauplatz eine Streitschrift*, R. Costenoble, Bern, (1879); **and** *Jeiteles teutonicus. Harfenklänge aus dem vermauschelten Deutschland*, Leipzig, (1879); **and** *Der Judenkrieg, seine Fehler und wie er Zu organisiren ist*, Richard Oschatz, Chemnitz, (1880); **and** *Goldene Ratten und rothe Mäuse*, E. Schmeitzner, Chemnitz, (1880); **and** *Wo steckt der Mauschel? oder, Jüdischer Liberalismus und wissenschaftlicher Pessimismus*, W. Raich, New York, (1880); **and** *Der Weg zum Siege des Germanenthums über das Judenthum*, O. Hentze, Berlin, (1880); and *Wählet keinen Juden: ein Mahnwort an die deutschen Wähler*, O. Hentze, Berlin, (1881).
292. E. K. Dühring, *Die Judenfrage als Racen-, Sitten- und Culturfrage: mit einer weltgeschichtlichen Antwort*, H. Reuther, Karlsruhe, (1881); English translation by A. Jacob, *Eugen Dühring on the Jews*, Nineteen Eighty Four Press, Brighton, England, (1997); **and** *Der Werth des Lebens: Eine philosophische Betrachtung*, Eduard Trewendt, Breslau, (1865); **and** *Kritische Geschichte der Philosophie von ihren Anfängen bis zur Gegenwart*, Heimann, Berlin, (1869); **and** *Kritische Geschichte der Nationalökonomie und des Socialismus*, T. Grieben, Berlin, (1871); **and** *Die Ueberschätzung Lessing's und Dessen Anwaltschaft für die Juden*, H. Reuther, Karlsruhe, (1881); **and** *Sache, Leben und Feinde: Als Hauptwerk und Schlüssel zu seinen sämmtlichen Schriften*, H. Reuther, Karlsruhe, Leipzig, (1882); **and** *Der Ersatz der Religion durch Vollkommeneres und die Ausscheidung alles Judenthums durch den modernen Völkergeist*, H. Reuther, Karlsruhe, (1883); **and** *Die Parteien in der Judenfrage*, München, (1907).
293. E. A. Drumont, *Richard Wagner, l'Homme et le musicien, à propos de Rienzi*, E. Dentu, Paris, (1869); **and** *Le dernier des Trémolin, : Société générale de librairie*

Jews a separate "race" incapable of assimilating into other groups of human beings. Theodor Herzl declared that Jews must leave European nations, and if they refused, they should be forced to do so by any and all means including: deliberate deception, provoked anti-Semitism, blackmail of Christians, and Zionist sponsored forced governmental expulsion. Herzl also libeled Jews by declaring that they should become disloyal to any nation other than the Jewish nation, because of alleged "racial" incompatibility.

Herzl advocated the forced expulsion of the Jews from Europe, which he sought to provoke, to the Jewish financiers he wanted to finance it; by asserting that they could profiteer from racism, panic and the slave labor of Eastern European Jews.

The opening line of Herzl's racist Zionist manifesto *The Jewish State* is a statement about economics and the rest reads like an archaic colonialistic business plan from the Roman Empire, promising exponential profits for all those who would invest in the scheme to segregate the Jews in a "World Ghetto".[294] Herzl was not timid in his declarations that these financiers could

---

catholique, Paris, (1879); **and** *La France juive; essai d'histoire contemporaine*, C. Marpon & E. Flammarion, Paris, (1886); German translation: *Das verjudete Frankreich: Versuch einer Tagesgeschichte*, A. Deubner, Berlin, (1889); **and** *La France juive devant l'opinion: La "France juive" et la critique, la conquête juive, le système juif et la question sociale, l'escrime sémitique, ce qu'on voit dans un tribunal*, Marpon & Flammarion, Paris, (1886); **and** *La fin d'un monde: étude psychologique et sociale*, A. Savine, Paris, (1889); **and** *La dernière bataille nouvelle étude psychologique et sociale*, E. Dentu, Paris, (1890); and *La daeniére batailla*, E. Dentu, Paris, (1890); **and** *Le testament d'un antisémite*, E. Dentu, Paris, (1891); **and** *Le testament d'un antisémite*, E. Dentu, Paris, (1891); **and** *Le secret de Fourmies: avec un plan de la place de l'église*, A. Savine, Paris, (1892); **and** *Les juifs contre la France une nouvelle Pologne*, Librairie Antisémite, Paris, (1899); **and** *Nos maîtres la tyrannie maçonnique*, Paris : Librairie antisémite, (1899); **and** *Pour la République! revue politique mensuelle*, Paris, (1899-1900); **and** *Socialismo Cattolico, con prefazione di Arturo Labriola*, Societa editrice partenopea, Napoli, (1911). ***See also:*** L. B. and E. A. Drumont, *Juif! quelques vers en réponse à La France juive*, A. Lanier, Paris, (1886). ***See also:*** H. Desportes and E. A. Drumont, *Le mystère du sang chez les juifs de tous les temps*, Albert Savine, Éditeur, Paris, (1889). ***See also:*** A. Rohling and E. A. Drumont, *Le juif selon le Talmud*, Albert Savine, Paris, (1889); German translation: *Prof. Dr. Aug. Rohling's Talmud-Jude*, T. Fritsch, Leipzig, (1891). ***See also:*** E. A. Drumont, A. de Rothschild and A. L. Burdeau, *Burdeau-Rothschild contre Drumont; Le proces de la libre parole, debats complets*, Paris, (1892). ***See also:*** J. Aron and E. A. Drumont, **Lettre ouverte à monsieur Édouard Drumont**, Paris, (1896). ***See also:*** A. Blanchard and E. A. Drumont, *Lettre de Benjamin Israël à Edouard Drumont la vérité sur la question juive en France*, C. Poinsignon, Neuilly-Plaisance, (1899). ***See also:*** M. L'Hermite and E. A. Drumont, *Drumont-démon, l'anti-pape: l'insurrection des Congrégations, le Belluaire; Moines et soldats, Alger-Milan*, L. Sery, Issoudun, (1899). ***See also:*** A. B. Monniot, *Les gouvernants contre la nation la trahison du ministère Waldeck*, Librairie antisémite, Paris, (1900). ***See also:*** P. Vergnet and E. A. Drumont, *Edouard Drumont, intime*, Libre parole, Paris, (1912). ***See also:*** The journals: *La libre parole illustrée / Libre parole / Almanach de la libre parole.*
294. T. Herzl, English translation by H. Zohn, R. Patai, Editor, *The Complete Diaries of*

control the trade between East and West by occupying Palestine and taking over the banking interests of the Sultan of Turkey, as well as forming new banks from smaller banks, potentially to put other larger banks out of business. Herzl promised the richest Jews of Western Europe palatial estates built with Eastern European Jews' blood and sweat. He shamelessly appealed to the anti-Semites' desire to see the Jews go, in order to encourage the governments of Europe to force the Jews of Europe out, so that these Jewish financiers could profit from the forced expulsion of the Jews.

In many respects, Herzl copied from Leon Pinsker's *Auto-Emancipation*. Pinsker claimed that Jews were incapable of assimilation, were an advanced race unlike some others, and were a parasite people with a "surplus" of untouchables, whom Herzl thought could be put to slave labor for the benefit of rich Western Jews. Pinsker wrote in 1882,

> "This is the kernel of the problem, as we see it: *the Jews comprise a distinctive element among the nations under which they dwell, and as such can neither assimilate nor be readily digested by any nation.* [***] The Jews are aliens who can have no representatives, because they have no country. Because they have none, because their home has no boundaries within which they can be entrenched, their misery too is boundless. The *general law* does not apply to the Jews as true aliens, but there are everywhere *laws for the Jews*, and if the general law is to apply to them, a special and explicit by-law is required to confirm it. Like the Negroes, like women, and unlike all free peoples, they must be *emancipated*. If, unlike the Negroes, they belong to an advanced race, and if, unlike women, they can produce not only women of distinction, but also distinguished men, even men of greatness, then it is very much the worse for them. [***] It is precisely the great misfortune of our race that we do not constitute a nation, but are merely Jews. [***] Such being the situation, we shall forever remain a burden to the rest of the population, parasites who can never secure their favor. The apparent fact that we can mix with nations only slightly offers a further obstacle to the establishment of amicable relations. Therefore, we must see to it that the *surplus*, the unassimilable residue, is removed and elsewhere provided for. [***] Our greatest and ablest forces—men of finance, of science, and of affairs, statesmen and publicists—must join hands with one accord in steering toward the common destination. They would aim chiefly and especially at creating a secure and inviolable home for the *surplus* of those Jews who live as proletarians in the different countries and are a burden to the native citizens. [***] The wealthy may also remain even where the Jews are not willingly tolerated. But, as we have said before, there is a certain point of saturation beyond which their numbers may not increase, if the Jews are not to be exposed to the dangers of persecution as in Russia, Roumania, Morocco and elsewhere. It is this surplus which, a

---

Theodor Herzl, Volume 1, Herzl Press, New York, (1960), p. 172.

burden to itself and to others, conjures up the evil fate of the entire people. It is now high time to create a refuge for this surplus. We must occupy ourselves with the foundation of such a lasting refuge, not with the meaningless collection of donations for emigrants or refugees who forsake, in their consternation, an *unhospitable home* to perish in the abyss of a strange and unknown land."[295]

Herzl was no friend to the Jews of Europe. Herzl advocated asking Christians to pay for the forced expulsion of the Jews of Europe, which Herzl strove to bring about. He assured the Christians that no such economic disasters would occur as happened when the Jews fled the Czar's pogroms and took with them their wealth. In statements certain to have provoked greed, Herzl promised that the expulsion of the Jews would be an economic boon and that Christians would profit by taking the jobs vacated by Jews and by managing the systems needed to expel them. The minutes of the Wannsee Conference of 1942[296] parallel many of the statements and proposals found in Herzl's book *The Jewish State* of 1896. Herzl appealed to all the most base and simplistic sensibilities later manifest in Zionist Fascist propgandists like Adolf Hitler. Like Hitler, Herzl wrote in absolutes of: *us* versus *them*, fatalistic inevitabilities, total self-assuredness, the common enemy, the alleged impossibility of different races living together in harmony, the mythologies of immutable conspiring forces in history demanding segregation, the allegedly feeble nature of democracy, etc.— all the Darwinistic and Hegelian cliches of the day meant to justify inhumane Colonialism and Imperialism. Herzl, Syrkin, and other anti-Semitic Zionists believed the racial mythology that,

"Competition from the Jew was all the harder to face, because natural selection had made him an especially fierce adversary in business."[297]

Herzl delighted in deceiving people and appealed to their greed in order to induce them into actions they would not otherwise take. Like Zionists in general, Herzl had little regard for informed personal choice. Herzl wrote,

"I am absolutely convinced that I am right—though I doubt whether I shall live to see myself proved to be so. [***] The Jewish State is essential to the world, it will therefore be created. [***] We are a people—one people. We have honestly endeavored everywhere to merge ourselves in the social life

---

295. L. Pinsker, translated by D. S. Blondheim, *Auto-Emancipation: An Admonition to His Brethren by a Russian Jew*, Federation of American Zionists, New York, (1916).

296. An English translation of the minutes appears in: "Wannsee Conference on the Final Solution of the Jewish Question", R. S. Levy, *Antisemitism in the Modern World: An Anthology of Texts*, D.C. Heath, Toronto, (1991), pp. 252-258; *see also:* pp. 250-252.

297. N. Syrkin, under the nom de plume "Ben Elieser", *Die Judenfrage und der socialistische Judenstaat*, Steiger, Bern, (1898); English translation in A. Hertzberg, *The Zionist Idea*, Harper Torchbooks, New York, (1959), pp. 333-350, at 339.

of surrounding communities, and to preserve only the faith of our fathers. It has not been permitted to us. In vain are we loyal patriots, our loyalty in some places running to extremes; in vain do we make the same sacrifices of life and property as our fellow-citizens; in vain do we strive to increase the fame of our native land in science and art, or her wealth by trade and commerce. In countries where we have lived for centuries we are still cried down as strangers, and often by those whose ancestors were not yet domiciled in the land where Jews had already made experience of suffering. The majority may decide which are the strangers; for this, as indeed every point which arises in the commerce of nations, is a question of might. I do not here surrender any portion of our prescriptive right, for I am making this statement merely in my own name as an individual. In the world of today, and for an indefinite period it will probably remain so, might precedes right. Therefore it is useless for us to be loyal patriots, as were the Huguenots who were forced to emigrate. If we could only be left in peace... . But I think we shall not be left in peace. [***] Every nation in whose midst Jews live is, either covertly or openly, Anti-Semitic. The common people have not, and indeed cannot have, any historic comprehension. They do not know that the sins of the Middle Ages are now being visited on the nations of Europe. We are what the Ghetto made us. We have attained pre-eminence in finance, because mediæval conditions drove us to it. The same process is now being repeated. Modern conditions force us again into finance, now the stock-exchange, by keeping us out of all other branches of industry. Being on the stock-exchange, we are therefore again considered contemptible. At the same time we continue to produce an abundance of mediocre intellects which find no outlet, and this endangers our social position as much as does our increasing wealth. Educated Jews without means are now fast becoming Socialists. Hence we are certain to suffer very severely in the struggle between classes, because we stand in the most exposed position in the camps of both Socialists and capitalists. [***] In the principal countries where Anti-Semitism prevails, it does so as a result of the emancipation of the Jews. [***] When we sink, we become a revolutionary proletariat, the subordinate officers of the revolutionary party; when we rise, there rises also our terrible power of the purse. [***] Thus, whether we like it or not, we are now, and shall henceforth remain, a historic group with unmistakable characteristics common to us all. We are one people—our enemies have made us one in our despite, as repeatedly happens in history. [***] The Governments of all countries scourged by Anti-Semitism will serve their own interests in assisting us to obtain the sovereignty we want. [***] This pamphlet will open a general discussion on the Jewish Question, avoiding, if possible, the creation of an opposition party. Such a result would ruin the cause from the outset, and dissentients must remember that allegiance or opposition are entirely voluntary. Who will not come with us, may remain. [***] Palestine is our ever-memorable historic home. The very name of Palestine would attract our people with a force of marvellous potency. Supposing His Majesty the Sultan were to give us Palestine, we could in return pledge ourselves to regulate the whole finances of Turkey. We should there form a portion of a rampart of Europe against Asia, an outpost of civilization as opposed to barbarism. The sanctuaries of Christendom would be safeguarded by assigning to them an extra-territorial status, such as is well known to the law of nations. We should form a guard of honor about these sanctuaries, answering for the fulfillment of this duty with our existence. This guard of honor would be the great symbol of the solution of the Jewish Question after eighteen centuries of Jewish suffering. [***] The Jewish Company is partly modelled on the lines of a great trading association. It might be called a Jewish Chartered Company, though it cannot exercise sovereign power, and has duties other than the establishment of colonial commerce. The Jewish Company will be founded as a joint-stock company subject to English jurisdiction, framed according to English laws, and under the protection of England. Its principal centre will be London. I cannot tell yet how large the Company's capital should be; I shall leave that calculation to our numerous financiers. But to avoid ambiguity, I shall put it at a thousand million marks (about £50,000,000); it may be either more or less than that sum. The form of subscription, which will be further elucidated, will determine what fraction of the whole amount must be paid in at once. The Jewish Company is an organization with a transitional character. It is strictly a business undertaking, and must be carefully distinguished from the Society of Jews. The Jewish Company will first of all see to the realisation of vested interests left by departing Jews. The method adopted will prevent the occurrences of crises, secure every man's property, and facilitate that inner migration of Christian citizens which has already been indicated. [***] At the same time the Company will buy estates, or, rather, exchange them. For a house it will offer a house in the new country, and for land, land in the new country; everything being, if possible, transferred to new soil in the same state as it was in the old. And this transfer will be a great and recognised source of profit to the Company. 'Over there' the houses offered in exchange will be newer, more beautiful, and more comfortably fitted, and the landed estates of greater value than those abandoned; but they will cost the Company comparatively little, because it will have bought the ground at a very cheap rate. [***] All the immense profits of this speculation in land will go to the Company, which is bound to receive this indefinite premium in return for having borne the risk of the undertaking. When the undertaking involves any risk, the profits must be freely accorded to those who have borne it. But under no other circumstances will profits be permitted. In the co-relation of risk and profit is comprehended financial justice. [***] I said before that the Company would not have to pay these unskilled labourers. [***] The principle is: the furnishing of every necessitous man with easy, unskilled work, such as chopping wood, or cutting faggots used for lighting stoves in Paris households. This is a kind of prison-work *before* the crime, done without loss of character. It is meant to prevent men from taking to crime out of want, by providing them with work and testing their willingness to do it.

Starvation must never be allowed to drive men to suicide; for such suicides are the deepest disgrace to a civilisation which allows rich men to throw tit-bits to their dogs. [***] But the Jewish Company will not lose one thousand millions; it will draw enormous profits from this expenditure. [***] Further and direct profit will accrue to Governments from the transport of passengers and goods, and where railways are State property the returns will be immediately recognisable. Where they are held by companies, the Jewish Company will make favourable terms for transport, in the same way as does every transmitter of goods on a large scale. [***] The capital required for establishing the Company was previously put at what seemed an absurdly high figure. The amount actually necessary will be fixed by financiers, and will in any case be a very considerable sum. There are three ways of raising this sum, all of which the Society will take under consideration. This Society, the great 'Gestor' of the Jews, will be formed by our best and most upright men, who must not derive any material advantage from their membership. Although the Society cannot at the outset possess any but moral authority, this authority will yet suffice to establish the credit of the Jewish Company in the nation's eyes. [***] The easiest, most rapid, and safest would be by 'la haute finance.' The required sum would then be raised in the shortest possible time by our great body of financiers, after they had discussed the advisability of the cause. The great advantage of this method would be that it would avoid the necessity of paying in the thousand millions (to keep to the original cipher) immediately in its entirety. A further advantage would be, that the unlimited credit of these powerful financiers would be of considerable value to the Company in its transactions. Many latent political forces lie in our financial power, that power which our enemies assert to be actually and now as effective as we know it might be if we exercised it. Poor Jews feel only the hatred which this financial power provokes; its use in alleviating their lot as a body, they have not yet felt. The credit of our great Jewish financiers would have to be placed at the service of the National Idea. But should these gentlemen, who are naturally satisfied with their lot, decline to do anything for their co-religionists who are unjustly held responsible for the large possessions of certain individuals—should these great financiers refuse to co-operate—then the realisation of this plan will afford an opportunity for drawing a clear line of distinction between them and the rest of Judaism. The great financiers, moreover, will certainly not be asked to raise an amount so enormous out of pure philanthropy; that would be expecting too much. The promoters and stock-holders of the Jewish Company are, on the contrary, intended to do a good piece of business, and they will be able to calculate beforehand what their chances of success are likely to be. For the Society of Jews will be in possession of all documents and references which may serve to define the prospects of the Jewish Company. The Society will also undertake the special duty of investigating with exactitude the extent of the new Jewish movement, so as to provide the Company promoters with thoroughly reliable information on the amount of support they may expect. The Society will also supply the Jewish Company with comprehensive modern Jewish statistics, thus doing the work of what is called in France a 'société d'études,' which undertakes all preliminary research previous to the financing of a great undertaking. Even so, the enterprise may not receive the valuable assistance of our money magnates. These might, perhaps, even try to oppose the Jewish movement by means of their secret servitors and agents. Such opposition we shall meet fairly and bravely. [***] The Company's capital might be raised without the assistance of a syndicate, by the direct imposition of a subscription on the public. Not only poor Jews, but also Christians who wanted to get rid of them, would subscribe their small quota to this fund. [***] The middle classes will involuntarily be drawn into the outgoing current, for their sons will be the Company's officials and employés over there.' [***] Great exertions will not be necessary to spur on the movement. Anti-Semites provide the requisite impetus. They need only do what they did before, and then they will create a love of emigration where it did not previously exist, and strengthen it where it existed before. Jews who now remain in Anti-Semitic countries do so chiefly because, even those among them who are most ignorant of history, know that numerous changes of residence in bygone centuries never brought them any permanent good. Any land which welcomed the Jews to-day, and offered them even fewer advantages than the future Jewish State would guarantee them, would immediately attract a great influx of our people. The poorest, who have nothing to lose, would drag themselves there. But I maintain, and every man may ask himself whether I am not right, that the pressure weighing on us arouses a desire to emigrate even among prosperous strata of society. Now our poorest strata alone would suffice to found a State; for these make the most vigorous conquerors, because a little despair is indispensable to the formation of a great undertaking. But when our desperadoes increase the value of the land by their presence and by the labour they expend on it, they make it at the same time increasingly attractive as a place of settlement to people who are better off. Higher and yet higher strata will feel tempted to go over. The expedition of the first and poorest settlers will be conducted by conjoint Company and Society, and will probably be additionally supported by existing emigration and Zionist societies. How may a number of people be concentrated on a particular spot without being given express orders to go there? There are certain Jews, benefactors on a large scale, who try to alleviate the sufferings of their co-religionists by Zionist experiments. To them this problem also presented itself, and they thought to solve it by giving the emigrants money or means of employment. Thus the philanthropists said: 'We pay these people to go there.' Such a procedure is utterly at fault, and all the money in the world will not achieve its purpose. On the other hand, the Company will say: 'We shall not pay them, we shall let them pay us. We shall merely offer them some inducements to go.' A fanciful illustration will make my meaning more explicit: One of those philanthropists (whom we will call 'The Baron') and myself both wish to get a crowd of people on to the plain of Longchamps near Paris, on a hot Sunday afternoon. The Baron, by promising them 10 francs each, will, for 200,000 francs, bring out 20,000 perspiring and miserable people, who will curse him for having given them so much annoyance. Whereas I will offer these 200,000 francs as a prize for the swiftest race-horse—and then I shall have to put up barriers to keep the people off Longchamps. They will pay to go in: 1 franc, 5 francs, 20 francs. The consequence will be that I shall get the half a million of people out there; the President of the Republic will drive *a la* Daumont; and the crowds will enjoy and amuse themselves. Most of them will think it an agreeable walk in the open air, spite of heat and dust; and I shall have made by my 200,000 francs about a million in entrance money and taxes on gaming. I shall get the same people out there whenever I like; but the Baron will not—not on any account. I will give a more serious illustration of the phenomenon of multitudes where they are earning a livelihood. Let any man attempt to cry through the streets of a town: 'Whoever is willing to stand all day long through a winter's terrible cold, through a summer's tormenting heat, in an iron hall exposed on all sides, there to address every passer-by, and to offer him fancy wares, or fish, or fruit, will receive 2 florins, or 4 francs, or something similar.' How many people would go to the hall? How many days would they hold out when hunger drove them there? And if they held out, what energy would they display in trying to persuade passers-by to buy fish, fruit, and fancy wares? We shall set about it in a different way. In places where trade is active, and these places we shall the more easily discover, since we ourselves forms channels for trade to various localities; in these places we shall build large halls, and call them markets. These halls might be worse built and more unwholesome than those above mentioned, and yet people would stream towards them. But we shall use our best efforts, and we shall build them better, and make them more beautiful than the first. And the people, to whom we had promised nothing, because we cannot promise anything without deceiving them, these brave, keen business men will gaily create most active commercial intercourse. They will harangue the buyers unweariedly; they will stand on their feet, and scarcely think of fatigue. They will hurry off day after day, so as to be first on the spot; they will make agreements, promises, anything to continue bread-winning undisturbed. And if they find on Sabbath night that all their hard work has produced only 1 florin, 50 kreutzer, or 3 francs, or something similar, they will yet look forward hopefully to the next day, which may, perhaps, bring them better luck. We have given them hope. Would any one ask whence the demand comes which creates the market? Is it really necessary to tell them again? I pointed out before that the labour-test increased our gain fifteenfold. One million produced fifteen millions; and one thousand millions, fifteen thousand millions. This may be the case on a small scale; is it so on a large one? Capital surely yields a return diminishing in inverse ratio to its own growth? Inactive capital yields this diminishing return, but active capital brings in a marvellously increasing return. Herein lies the social question. Am I stating a fact? I call on the richest Jews as witnesses of my veracity.

Why do these carry on so many different industries? Why do they send men to work underground and to raise coal amid terrible dangers for miserable pay? I cannot imagine this to be pleasant, even for the owners of the mines. For I do not believe that capitalists are heartless, and I do not take up the attitude of believing it. My desire is not to accentuate, but to smooth differences. Is it necessary to illustrate the phenomenon of multitudes, and their concentration on a particular spot, by references to pious pilgrimages? I do not want to hurt any one's religious sensibility by words which might be wrongly interpreted. I shall merely refer quite briefly to the Mohammedan pilgrimages to Mecca, the Catholic pilgrimages to Lourdes and to many other spots whence men return comforted by their faith, and to the holy Coat at Trier. Thus we shall also create a centre for the deep religious needs of our people. Our ministers will understand us first, and will be with us in this. [***] I imagine that Governments will, either voluntarily or under pressure from the Anti-Semites, pay certain attention to this scheme; and they may perhaps actually receive it here and there with a sympathy which they will also show to the Society of Jews. For the emigration which I suggest will not create any economic crises. Such crises as would follow everywhere in consequence of Jew-baiting would rather be prevented by the carrying out of my plan. A great period of prosperity would commence in countries which are now Anti-Semitic. For there will be, as I have repeatedly said, an intermigration of Christian citizens into the positions slowly and systematically evacuated by the Jews. If we are not merely suffered, but actually assisted to do this, the movement will have a generally beneficial effect. [***] Universal brotherhood is not even a beautiful dream. Antagonism is essential to man's greatest efforts. But the Jews, once settled in their own State, would probably have no more enemies, and since prosperity enfeebles and causes them to diminish, they would soon disappear altogether. I think the Jews will always have sufficient enemies, much as every nation has. But once fixed in their own land, it will no longer be possible for them to scatter all over the world. The diaspora cannot take place again, unless the civilization of the whole earth is destroyed; and such a consummation could be feared by none but foolish men. Our present civilization possesses weapons powerful enough for its self-defence. Innumerable objections will be based on low grounds, for there are more low men than noble in this world. I have tried to remove some of these narrow-minded notions; and whoever is willing to fall in behind our white flag with its seven golden stars must assist in this campaign of enlightenment. Perhaps we shall have to fight first of all against many an evil-disposed, narrow-hearted, short-sighted member of our own race. Again, people will say that I am furnishing the Anti-Semites with weapons. Why so? Because I admit the truth? Because I do not maintain that there are none but excellent men amongst us? Again, people will say that I am showing our enemies the way to injure us. This I absolutely dispute. My proposal could only be carried out with the free consent of a majority of Jews. Individuals or even powerful bodies of Jews might be attacked, but

Governments will take no action against the collective nation. The equal rights of Jews before the law cannot be withdrawn where they have once been conceded; for the first attempt at withdrawal would immediately drive all Jews rich and poor alike, into the ranks of the revolutionary party. The first official violation of Jewish liberties invariably brings about economic crisis. Therefore no weapons can be effectually used against us, because these cut the hands that wield them. Meantime hatred grows apace. The rich do not feel it much, but our poor do. Let us ask our poor, who have been more severely persecuted since the last renewal of Anti-Semitism than ever before. Our prosperous men may say that the pressure is not yet severe enough to justify emigration, and that every forcible expulsion shows how unwilling our people are to depart. True, because they do not know where to go; because they only pass from one trouble into another. But we are showing them the way to the Promised Land; and the splendid force of enthusiasm must fight against the terrible force of habit."[298]

Herzl proposed that,

"Supposing His Majesty the Sultan were to give us Palestine, we could in return pledge ourselves to regulate the whole finances of Turkey."[299]

Both sides of this bargain would appear to benefit the Zionists and take from Turkey. The Sultan of Turkey was in a financial crisis bought on by Jewish bankers, just as the Egyptian Khedive Ismail Pasha was in a financial crisis brought on by Jewish bankers when Disraeli purchased shares in the Suez Canal with the Bank of Rothschild.[300] But there were many reasons why the Zionists did not simply buy the land and end Turkey's humiliation, as Herzl had proposed at the Zionist Congress of 1897, and as the Rothschilds had proposed long before.[301]

---

298. T. Herzl, *A Jewish State: An Attempt at a Modern Solution of the Jewish Question*, The Maccabæan Publishing Co., New York, (1904), pp. xviii-xix, 5, 18, 21-25, 27, 29, 31-34, 37, 41-42, 50, 55-59, 67-73, 93-94, 98-100.

299. T. Herzl, *A Jewish State: An Attempt at a Modern Solution of the Jewish Question*, The Maccabæan Publishing Co., New York, (1904), p. 29.

300. *See:* "Letters to the Editor" with respect to the Memorandum to the Protestant monarchs regarding the "Restoration of the Jews", *The London Times*, (26 August 1840), p. 6. *See also:* L. Wolf, "The Story Of The Khedive's Shares", *The London Times*, (26 December 1905), p. 11; "The Story Of The Khedive's Shares. (Letters to the Editor)", *The London Times*, (28 December 1905), p. 4; "Story Of The Khedive's Shares.... (Letters to the Editor)", *The London Times*, (29 December 1905), p. 5; "The Story Of The Khedive's Shares. (Letters to the Editor)", *The London Times*, (18 January 1906), p. 4; "The Story Of The Khedive's Shares. (Letters to the Editor)", *The London Times*, (29 January 1906), p. 8.

301. "The Zionist Congress: Full Report of the Proceedings", *The Jewish Chronicle*, (3 September 1897), pp. 10-15, at 11. *See also:* E. Kedourie, "Young Turks, Freemasons and Jews", *Middle Eastern Studies*, Volume 7, Number 1, (January, 1971), pp. 89-104;

Herzl knew that the Jewish financiers who had caused the Turkish Empire's financial crisis were willing to cure it in exchange for the land of Palestine, and that the Sultan would agree to that deal. The Zionists had additional leverage on the Sultan due to the Turkish attacks on Armenian Christians. Though Jewish bankers were ultimately responsible for these attacks, they threatened to inflame the Christian world against the Turks. Herzl promised that he could improve the Sultan's public image, and prevent a Christian backlash against the Turkish Empire, through his contacts in the Jewish press. Herzl pledged that warm Jews in the media would bury the story of the Armenian attacks, and praise the Sultan and the Turkish Empire, if the Sultan would agree to sell Palestine to the Rothschilds. In 1902, an article published in *The American Monthly Review of Reviews* addressed some of the problems facing the Turkish Empire at the time Herzl tried to blackmail the Sultan,

"WHERE THE SULTAN FAILED.

Corrupt these pashas were. Many had come from low, and some were of ignoble, origin. Their birth was as varied as the races of the empire they administered but did not rule. The weakest Ottoman sultan does that. But they were undeniably able. They have disappeared. They have no successors. Palace has supplanted ministerial rule. Personal secretaries have taken the place of pashas. The grand vizierate has become an empty shade, unless Said Pasha change it. Nor is this likely. Able, shrewd, consummate diplomat, Abdul Hamid, for a decade and more equal to the task of inflicting on the European concert a fatal paralysis, until Austria acted alone in 1897, has proved unable to organize administration or to depute authority. The army he turned over to Goltz Pasha, and it is efficient, as the Greek war proved. The men are unpaid, but their cartridge-boxes are never empty. They are unshod, but their arms are serviceable. There are few or no ambulances, but the artillery is well horsed. The navy has disappeared. But in civil administration no man is secure. Imperial orders go above, below, and around. Some negro eunuch or palace underling may palsy the administration of a province or bring to disgrace by a secret order the ablest of valis, or provincial governors. Despotism in strong hands may prove both able and beneficent by organizing administration. But personal rule, smitten with a mania of fear of conspiracy, trusting no one, filling the empire with espionage, and selecting as instruments ignorant and ignoble personal attendants was certain to end in the collapse now clear.

For a season it prospered. In 1895, all held Abdul Hamid, doubtless, the subtlest schemer of his long line in generations, but in the broad sense successful. In twenty years, 1879-99, the population of the empire,

---

reprinted: E. Kedourie, *Arabic Political Memoirs and Other Studies*, Chapter 16, Frank Cass, London, (1974), pp. 243-263. ***See also:*** M. R. Buheiry, "Theodor Herzl and the Armenian Question", *Journal of Palestine Studies*, Volume 7, Number 1, (Autumn, 1977), pp. 75-97.

excluding tributary states, grew from 21,000,000 to 25,000,000— above the average of West Europe. The value of real estate advanced down to 1895 in all Turkish cities. In those with which I am most familiar in eastern Turkey, a fair 25 per cent. increase or more, in twenty years. There was no Turkish city, and I met residents from all, where building was not in progress in this period. All complained of taxes and oppression, and in all population, buildings, and realty values were growing. Imports, 1878 to 1898, rose from (estimated) $60,000,000 to $120,000,000, and exports from $35,000,000 to $68,750,000, an increase which stands for prosperity. The principal railroad in Asiatic Turkey, Smyrna-Aidin, 318 miles, increased its gross earnings from £140,538 to £354,406 from 1883 to 1893, and later lost its dividends.

But while figures of this character could be multiplied, the government itself was passing from one abyss of bankruptcy to another, if the imperial revenue only, averaging, 1892-95, $106,500,000—say $4 per capita—were collected in taxes, the burden would not be heavy. A semi-civilized country can easily raise a pound sterling a head, and a country like the United States averaged $16 in 1890, and did not feel the burden. But by universal consent, the Turkish revenue is extorted manifold by a system of farming the taxes and official peculation. The old government, by pashas, was ill. The new, by palace favorites, is worse. After wholesale repudiation in 1875,—the Porte compounded with its creditors in 1881,—Iradé, December 8-20, 1881, admitted the bailiff in the shape of a debt commission, and paid 1 per cent. on the unsecured debt. The nominal amount of the debt in 1875 was $1,200,000,000. It was scaled to $530,000,000 in 1881. In 1900, it was $682 000,000,—no great increase as national debts go. It is all held abroad,—77 per cent. in France, 10 per cent, in Germany, 9 per cent. in England, and 4 per cent. in Austria.[*Footnote:* London *Statist*, October3, 1896.] The aggregate national mortgage is not large—in all, in 1896, government, railroad, and other stocks, $792,370,000 at par, $397,125,000 quoted value, two thirds (67 per cent.) in France, 17 per cent. in Germany, 12 per cent. in England, and 4 percent, in Austria. A fair measure this of time pressure the diplomacy of these lands will on a pinch exert.

The debt commission collected $12,876,207 in 1900, against $9,998,230 in 1885—a fairly elastic revenue. An Oriental country whose salttax receipts grew in fifteen years from $3,071,502 to $3,729,721—twice as fast as population—plainly only needs decent administration for a prosperous budget. Instead, time treasury has wallowed for years in irretrievable deficits averaging $5,000,000 to $7,000,000, according to Sir Edward Vincent's last report. The treasury, a few weeks ago, borrowed a small sum for the most sacrosanct of all Moslem expenditures, the carpet and its escort, which the Sultan yearly, as caliph, sends to the Kaaba, at Mecca. it is as though the Pope had to raise a floating loan for the wine and wafer of the Easter eucharist. Every inquiry shows how easily the Turkish treasury might be solvent. Every week finds it unable to meet any expenditure.

ARMENIAN MASSACRE AND ITS CAUSE.

The Sultan's policy five years ago had, therefore, greatly reduced European interference in Turkish affairs, and greatly increased imperial authority, without securing either a stable budget or an efficient administration. Nothing is, perhaps, so dangerous in the affairs of state as unlimited power joined to none of the machinery which gives certainty to taxation or ordered action to authority. Such prosperity as had come was little felt by Moslems. There is that about the Moslem creed, code, and character which incapacitates for all practical affairs but war and rule. Turkish treasury accounts have always been kept by Greeks and Armenians. If a Turk owns land, some Christian keeps its rent-roll. If he has a business, Christian clerks manage it, If he owns mines or works the richer placer of official extortion, some Christian engineer or scribe manages and manipulates his accounts. Such prosperity as there was through the twenty years of Abdul Hamid's reign, which seemed prosperous, went to Christians. In all the cities where massacre came, it was the Christian and Armenian quarter that was thriving and rising in value. Armenian villages were waxing rich, buying hand and renting it. Armenian bankers were making loans. When massacre fell in one city, not a signature was left known to Constantinople bankers. Western manufactures, which were ruining native handicrafts, were all handled by Armenians. Economic strain and stress produced by this disproportionate prosperity of the small Christian fraction, gaining in wealth, education, and political aspiration, was a perilous irritant to add to the pride of a ruling and soldier caste and the fanaticism kindled by Moslem renaissance. The match of administrative order, or even administrative suggestion, had only to be touched to these explosive conditions to bring the Armenian massacres.

Into their history, it is no purpose of mine to enter. Beyond all refutation, the Sultan successfully prevented European interference or the punishment that was due. But great crimes of state bring their own inexorable penalty. For five years, since time last of the massacres, the Sultan visibly lost ground. Awful as is massacre, communities recover, if order is restored. Over the Armenian plateau this has never come. In all the empire a blight has fallen on trade. The fall in wool ruined southeastern Turkey and it is estimated there are 40,000,000 sheep between the Black Sea and the Persian Gulf. The silk collapse laid North Syria in ruins, and brought Beirut to beggary. The capital has never recovered from the mere business shock of massacre. The Greek war broke credits on the Levantine coast. From the Greek revolution to Bulgarian independence, 1828-78, the dismemberment of the Turkish empire had been accompanied by the appearance of communities capable of self-governnment. Even Algeria-Tunis and Egypt, which have passed under foreign contiol, had not done so until a separate, albeit despotic, autonomy had been gained. Driven back to its Moslem limits, nothing like this has appeared in the empire, in twenty years. Crete is separated, the hardships of its going being a measure of the relatively larger Moslem population. In Turkey proper, neither improvement in the central administration nor provinces capable of autonomy appear. Without either, the empire sinks in the slough of difficulties created by racial and physical problems. For a season these and all reforms were held at bay. Macedonian autonomy, Armenian protection, equitable taxation, improved administration—all these pledges of the Berlin treaty in 1878 remain unperformed through twenty years of Europe and an empire both without initiative, and both controlled by the inertia of events, the fear of a general war, and the address at intrigue of Abdul Hamid.

But the lack of sound government and an honest ruler nothing compensates—not even material prosperity, increasing trade, growing population, schools, museums, revived Islamism, and all the fruits of the reign marshaled by court journals when the quarter-century of the Sultan was celebrated. Instead, when collapse comes, as collapse has, and the powers, one by one, demonstrate the weakness of the empire, problems long postponed appear, as creditors haunt lesser lives in days of disaster."[302]

In the 1840's, the Rothschilds considered buying Palestine from the Turkish Empire. The real difficulties the Rothschilds faced did not come from the Turks, but rather from the Arabs, especially the Egyptians, and from the Christians, especially the Catholics. The Jews feared that the Arabs would swarm over them if the Jews took over Palestine, which had been managed by the Egyptians. The Jews expected that a Jewish migration *en masse* to Palestine, and especially to Jerusalem, and most especially if followed in short succession by the anointment of a Jewish King—no doubt a Rothschild, and the destruction of the Dome of the Rock and Al Aqsa Mosque in order to "rebuild" the Jewish Temple; as Jewish prophecy demanded, would provoke the Moslems to attack the Jews and wipe them out.

The reason the Rothschilds did not move more aggressively on Palestine, though they had the financial might to buy it, was that whenever they tested the world's reaction to their designs, they discovered that the Jews did not want to go, that the Arabs opposed them (as opposed to the Turks), and that the Catholics thought of them as the embodiment of the anti-Christ. The Rothschilds feared that the Christians would recognize the Biblical implications of Jewish financiers using their corruptly gotten gains to purchase Jerusalem, as the manifestation of the anti-Christ. The Jewish financiers feared that the Christians would join forces with Islam to crush the anti-Christ and the Jews, that is to say smite the Rothschilds and sack the Jews.

I *John* 2:18, 22 states:

"18 Little children, it is the last time: and as ye have heard that antichrist shall come, even now are there many antichrists; whereby we know that it is the last time. [***] 22 Who is a liar but he that denieth that Jesus is the Christ? He is antichrist, that denieth the Father and the Son."

---

[302]. "The Turkish Situation by One Born in Turkey", *The American Monthly Review of Reviews*, Volume 25, Number 2, (February, 1902), pp. 182-191, at 186-188.

I *John* 4:2-3 states:

"2 Hereby know ye the Spirit of God: Every spirit that confesseth that Jesus Christ is come in the flesh is of God: 3 And every spirit that confesseth not that Jesus Christ is come in the flesh is not of God: and this is *that spirit* of antichrist, whereof ye have heard that it should come; and *even* now already is it in the world."

II *John* 1:7 states:

"7 For many deceivers are entered into the world, who confess not that Jesus Christ is come in the flesh. This is a deceiver and an antichrist."

Hosea 8:14 states:

"For Israel hath forgotten his Maker, and buildeth temples; and Judah hath multiplied fenced cities: but I will send a fire upon his cities, and it shall devour the palaces thereof."

Matthew 5:9 states:

"Blessed *are* the peacemakers: for they shall be called the children of God."

Matthew 10:16-18 states:

"16¶ Behold, I send you forth as sheep in the midst of wolves: be ye therefore wise as serpents, and harmless as doves. 17 But beware of men: for they will deliver you up to the councils, and they will scourge you in their synagogues; 18 And ye shall be brought before governors and kings for my sake, for a testimony against them and the Gentiles."

Matthew 12:30 states:

"He that is not with me is against me; and he that gathereth not with me scattereth abroad."

Matthew 21:43-45 states:

"43 Therefore say I unto you, The kingdom of God shall be taken from you, and given to a nation bringing forth the fruits thereof. 44 And whosoever shall fall on this stone shall be broken: but on whomsoever it shall fall, it will grind him to powder. 45 And when the chief priests and Pharisees had heard his parables, they perceived that he spake of them."

Matthew 23:31-39 states:

"Wherefore ye be witnesses unto yourselves, that ye are the children of them which killed the prophets. Fill ye up then the measure of your fathers. Ye serpents, ye generation of vipers, how can ye escape the damnation of hell? Wherefore, behold, I send unto you prophets, and wise men, and scribes: and *some* of them ye shall kill and crucify; and *some* of them shall ye scourge in your synagogues, and persecute *them* from city to city: That upon you may come all the righteous blood shed upon the earth, from the blood of righteous Abel unto the blood of Zacharias son of Barachias, whom ye slew between the temple and the altar. Verily I say unto you, All these things shall come upon this generation. O Jerusalem, Jerusalem, *thou* that killest the prophets, and stonest them which are sent unto thee, how often would I have gathered thy children together, even as a hen gathereth her chickens under *her* wings, and ye would not! Behold, your house is left unto you desolate. For I say unto you, Ye shall not see me henceforth, till ye shall say, Blessed *is* he that cometh in the name of the LORD."

Matthew 27:25 states:

"Then answered all the people, and said, His blood *be* on us, and on our children."

Mark 8:15 states:

"And he charged them, saying, Take heed, beware of the leaven of the Pharisees, and *of* the leaven of Herod."

John 3:15-18 states:

"15 That whosoever believeth in him should not perish, but have eternal life. 16 For God so loved the world, that he gave his only begotten Son, that whosoever believeth in him should not perish, but have everlasting life. 17 For God sent not his Son into the world to condemn the world; but that the world through him might be saved. 18 He that believeth on him is not condemned: but he that believeth not is condemned already, because he hath not believed in the name of the only begotten Son of God."

John 5:15-18 states:

"15 The man departed, and told the Jews that it was Jesus, which had made him whole. 16¶ And therefore did the Jews persecute Jesus, and sought to slay him, because he had done these *things* on the sabbath day. 17 But Jesus answered them, My Father worketh hitherto, and I work. 18 Therefore the Jews sought the more to kill him, because he not only had broken the sabbath, but said also that God was his Father, making himself equal with God."

John 5:41-47 states:

"41 I receive not honour from men. 42 But I know you, that ye have not the love of God in you. 43 I am come in my Father's name, and ye receive me not: if another shall come in his own name, him ye will receive. 44 How can ye believe, which receive honour one of another, and seek not the honour that *cometh* from God only? 45 Do not think that I will accuse you to the Father: there is *one* that accuseth you, *even* Moses, in whom ye trust. 46 For had ye believed Moses, ye would have believed me: for he wrote of me. 47 But if ye believe not his writings, how shall ye believe my words?"

John 7:1 states:

"After these *things* Jesus walked in Galilee: for he would not walk in Jewry, because the Jews sought to kill him."

John 7:13 states:

"Howbeit no *man* spake openly of him for fear of the Jews."

John 8:37-40 states:

"37 I know that ye are Abraham's seed; but ye seek to kill me, because my word hath no place in you. 38 I speak *that* which I have seen with my Father: and ye do *that* which ye have seen with your father. 39 They answered and said unto him, Abraham is our father. Jesus saith unto them, If ye were Abraham's children, ye would do the works of Abraham. 40 But now ye seek to kill me, a man that hath told you the truth, which I have heard of God: this did not Abraham."

John 8:44-45 states:

"Ye are of *your* father the devil, and the lusts of your father ye will do. He was a murderer from the beginning, and abode not in the truth, because there is no truth in him. When he speaketh a lie, he speaketh of his own: for he is a liar, and the father of it. And because I tell *you* the truth, ye believe me not."

John 10:19-38 states:

"19 There was a division therefore again among the Jews for these sayings. 20 And many of them said, He hath a devil, and is mad; why hear ye him? 21 Others said, These are not the words of him that hath a devil. Can a devil open the eyes of the blind? 22¶ And it was at Jerusalem *the feast of* the dedication, and it was winter. 23 And Jesus walked in the temple in Solomon's porch. 24 Then came the Jews round about him, and said unto him, How long dost thou make us to doubt? If thou be the Christ, tell us plainly. 25 Jesus answered them, I told you, and ye believed not: the works that I do in my Father's name, they bear witness of me. 26 But ye believe not, because ye are not of my sheep, as I said unto you. 27 My sheep hear my voice, and I know them, and they follow me: 28 And I give unto them eternal life; and they shall never perish, neither shall any *man* pluck them out of my hand. 29 My Father, which gave *them* me, is greater than all; and no *man* is able to pluck *them* out of my Father's hand. 30 I and *my* Father are one. 31 Then the Jews took up stones again to stone him. 32 Jesus answered them, Many good works have I shewed you from my Father; for which of those works do ye stone me? 33 The Jews answered him, saying, For a good work we stone thee not; but for blasphemy; and because that thou, being a man, makest thyself God. 34 Jesus answered them, Is it not written in your law, I said, Ye are gods? 35 If he called them gods, unto whom the word of God came, and the scripture cannot be broken; 36 Say ye *of him*, whom the Father hath sanctified, and sent into the world, Thou blasphemest; because I said, I am the Son of God? 37 If I do not the works of my Father, believe me not. 38 But if I do, though ye believe not me, believe the works: that ye may know, and believe, that the Father *is* in me, and I in him."

John 19:38 states:

"And after this Joseph of Arimathaea, being a disciple of Jesus, but secretly for fear of the Jews, besought Pilate that he might take away the body of Jesus: and Pilate gave *him* leave. He came therefore, and took the body of Jesus."

John 20:19 states:

"Then the same day at evening, being the first *day* of the week, when the doors were shut where the disciples were assembled for fear of the Jews, came Jesus and stood in the midst, and saith unto them, Peace *be* unto you."

In contradiction to "Christian Zionists", who pretend to know the dates of "End Times" prophecy and who collaborate with Israel to artificially and deliberately bring them about, *Matthew* 24:34-36 states:

"34 Verily I say unto you, This generation shall not pass, till all these *things* be fulfilled. 35 Heaven and earth shall pass away, but my words shall not pass away. 36 But of that day and hour knoweth no *man*, no, not the angels of heaven, but my Father only."

Acts 1:6-7 states:

"6 When they therefore were come together, they asked of him, saying,

Lord, wilt thou at this time restore again the kingdom to Israel? 7 And he said unto them, It is not for you to know *the* times or *the* seasons, which the Father hath put in his own power."

The New Testament, which literally means "New Covenant", repeatedly reminds Christians that the Jews broke the Old Covenant. *Romans* 2:28-29 states:

"28 For he is not a Jew, which is one outwardly; neither *is that* circumcision, which is outward in the flesh: 29 But he *is* a Jew, which is one inwardly; and circumcision *is that* of the heart, in the spirit, *and* not *in* the letter; whose praise *is* not of men, but of God."

Galatians 3:16-29 states:

"16 Now to Abraham and his seed were the promises made. He saith not, And to seeds, as of many; but as of one, And to thy seed, which is Christ. 17 And this I say, that the covenant, that was confirmed before of God in Christ, the law, which was four hundred and thirty years after, cannot disannul, that it should make the promise of none effect. 18 For if the inheritance be of the law, it is no more of promise: but God gave it to Abraham by promise. 19 Wherefore then serveth the law? It was added because of transgressions, till the seed should come to whom the promise was made; and it was ordained by angels in the hand of a mediator. 20 Now a mediator is not a mediator of one, but God is one. 21 Is the law then against the promises of God? God forbid: for if there had been a law given which could have given life, verily righteousness should have been by the law. 22 But the scripture hath concluded all under sin, that the promise by faith of Jesus Christ might be given to them that believe. 23 But before faith came, we were kept under the law, shut up unto the faith which should afterwards be revealed. 24 Wherefore the law was our schoolmaster to bring us unto Christ, that we might be justified by faith. 25 But after that faith is come, we are no longer under a schoolmaster. 26 For ye are all the children of God by faith in Christ Jesus. 27 For as many of you as have been baptized into Christ have put on Christ. 28 There is neither Jew nor Greek, there is neither bond nor free, there is neither male nor female: for ye are all one in Christ Jesus. 29 And if ye be Christ's, then are ye Abraham's seed, and heirs according to the promise."

Galatians 4:9-11 states:
"9 But now, after that ye have known God, or rather are known of God, how turn ye again to the weak and beggarly elements, whereunto ye desire again to be in bondage? 10 Ye observe days, and months, and times, and years. 11 I am afraid of you, lest I have bestowed upon you labour in vain."

Philippians 3:2-3 states, in reference to Judaizers:

"2 Beware of dogs, beware of evil workers, beware of the concision. 3 For we are the circumcision, which worship God in the spirit, and rejoice in Christ Jesus, and have no confidence in the flesh."

The New Testament repeatedly warned against Judaizers of Christianity (*John* 8:37-45. *Acts* 15:1-12; 16:3. *Romans* 2:21-29; 6:3-11, 9:6-8; 14:14,20; 16:18. I *Corinthians* 2:2; 3; 7:18-19; 10:18. II *Corinthians* 3:18-4:6; 11:12-12:10. *Galatians* 1:7-8; 2:12, 19; 3:1, 16-29; 4; 5:2-3, 11, 24; 6:11-18. *Philippians* 3:2-3. *Colossians* 1:12-13; 2:8, 16, 20. I *Thessalonians* 2:14-16. I *John* 4:2-3. *Revelation* 2:9; 3:9), whom Paul called "dogs" and "evil workers" (*Philippians* 3:2-3). Judaizers have since made Christians their sword with which to destroy humanity.

I *Thessalonians* 2:14-16 states:

"For ye, brethren, became followers of the churches of God which in Judaea are in Christ Jesus: for ye also have suffered like things of your own countrymen, even as they *have* of the Jews: Who both killed the Lord Jesus, and their own prophets, and have persecuted us, and they please not God, and are contrary to all men: Forbidding us to speak to the Gentiles that they might be saved, to fill up their sins alway: for the wrath is come upon them to the uttermost."

Revelation 2:9 states:

"I know thy works, and tribulation, and poverty, (but thou art rich) and I *know* the blasphemy of them which say they are Jews, and are not, but *are* the synagogue of Satan."

Revelation 3:9 states:

"Behold, I will make them of the synagogue of Satan, which say they are Jews, and are not, but do lie; behold, I will make them to come and worship before thy feet, and to know that I have loved thee."

Zionist financiers realized that it would be a enormous risk to finance the endeavor, which would likely end up in a holy war they could not win. Though they prodded and probed over the course of many centuries, Jewish financiers made no move into the desert until the Holocaust of the Second World War primed the pump by making the Jews appear to be meek victims and no threat to the world in the form of the anti-Christ.

Over the centuries, Jews put out a tremendous amount of propaganda meant to undermine Christian beliefs and to make the Christians into the slavish guardians of the Jews, at the expense of the Christians' and the Moslems' own interests. The practice continues to this very day. Two Letters to the Editor of *The London Times* published on 26 August 1840 on page 6, evince the challenges the Rothschilds faced should they have bought Palestine outright, and these letters evince the Jewish propaganda meant to subvert Christianity and Islam, and to create an artificial enmity between the two religions, so that the Christians would slavishly guard the Jews against the Moslems when the Jews stole the Palestinians' homes and defiled their religion,

"TO THE EDITOR OF THE TIMES.

Sir,—Every right-minded person must feel gratified at the general expression of interest in the Jewish nation which has been elicited by the recent sufferings of their brethren at Damascus. It is to be hoped that the public feeling will not be allowed to evaporate in the mere expression of sympathy, but that some effectual measures may be adopted to prevent a recurrence of these atrocities, not merely in our own times, but in generations yet to come. We must not forget, when giving utterance to our indignation at the late transactions in the east, that but a few centuries have passed since our own country was the scene of similar enormities on a far larger scale. What reader of English history does not recall with shame and sorrow the wholesale tortures, executions, and massacres of the Jews who had sought shelter here, or who can estimate the amount of property seized and confiscated, or the number of hearts wrung by the endless repetition of cruelty and injustice? If in England they have till lately been thus treated, how can they look for more security elsewhere? Instead of wondering that they should become sordid and debased, the only cause for surprise is that any should rise to intelligence and respectability. Subject to the caprice and cruelty of any nation among whom they may dwell, fleeing from the persecutions of one only to meet with like treatment from another, having no city of refuge where they can be in safeguard, no single spot to call their own, they are in a more pitiable condition than the Indian of the forest, or the Arab of the desert.

'The wild bird hath her nest, the fox his cave,
'Mankind their country, Israel but the grave.'

Is this state of things always to continue? They think not. Though many hundreds of years of hope deferred might have been enough to quench the anticipations of the most sanguine, they still hope on, and turn with constant and earnest longing to the land of their forefathers. Their little children are taught to expect that they shall one day see Jerusalem. They purchase no landed property, and hold themselves in readiness at a few hours' notice to revisit what they and we tacitly agree to call 'their own land.' It is theirs by a right which no other nation can boast, for God gave it to them, and though dispossessed of it for so many ages, it is still but partially peopled, and held with a loose hand and a disputed title by a hostile power, as if in readiness for their return.

There are political reasons arising from the present aspect of affairs in Russia, Turkey, and Egypt, which would make it to the interest not only of England but of other European nations, either by purchase or by treaty, to procure the restoration of Judea to its rightful claimants. About a year since,

I heard it it said by a German Jew, that a proposal had some time before been made by our (then) Government to the late Baron Rothschild, that he should enter into a negotiation for this purpose, and that he declined, assigning as a reason, 'Judea is our own; we will not buy it, we wait till God shall restore it to us.' The desirableness as well as the possibility of such a step seems daily to become more evident, but England has lately proved that she needs no selfish motives to induce her to discharge a debt of national honour and justice, or to perform an act of pure benevolence. The one now suggested would not, judging from appearances, cost 20,000,000*l.* of money, or be unaccomplished after 50 years of exertion, or be so vast and so laborious an undertaking as the extinction of slavery throughout the world. It would be a noble thing for a Christian nation to restore these wanderers to their homes again. It would be a crowning point in the glory of England to bring about such an event. The special blessings promised in the Scriptures to those who befriend the Jews would rest upon her, and her sons and daughters would sit down with purer enjoyment to their domestic comforts when they thought that the persecuted outcasts of so many ages had, through their agency, been replaced in homes as happy and secure as theirs.

Hoping that some master mind may be led to take up this subject in all its bearings, and to form some tangible plan for its accomplishment, and that some Wilberforce may be raised up to plead for it by all the powerful and heart-stirring arguments of which it is capable,

I am, Sir, your obedient servant,

AN ENGLISH CHRISTIAN.

TO THE EDITOR OF THE TIMES.

Sir,—The extraordinary crisis of Oriental politics has stimulated an almost universal interest and investigation, and the fate of the Jews seems to be deeply involved with the settlement of the Syrian dilemma now agitating every Court of Christendom.

You have well and wisely recommended that a system of peaceful umpirage and arbitration should be adopted as the proper *role* of Britain, France, Austria, Prussia, and Russia, and you have exposed the extreme absurdity which these Powers would commit if in their zeal for accommodating the quarrels of the Ottomans they should stir up bloody wars among themselves.

The peace of Europe and the just balance of its powers being therefore assumed as the grand desideratum, as the consummation most devoutly to be wished, I peruse with particular interest a brief article in your journal of this day relative to the restriction of the Jews in Jerusalem, because I imagine that this event has become practicable through an unprecedented concatenation of circumstances, and that moreover it has become especially desirable, as the exact expedient to which it is the interest of all the belligerent parties to consent.

The actual feasability of the return of the Jews is no longer a paradox;

the time gives it proof. That theory of the restoration of the Jewish kingdom, which a few years ago was laughed at as the phantasy of insane enthusiasm, is now calculated on as a most practical achievement of diplomacy.

Let us view the question more nearly. It is granted that the Jews were the ancient proprietors of Syria; that Syria was the proper heart and centre of their kingdom. It is granted that they have a strong conviction that Providence will restore them to this Syrian supremacy. It is granted that they have entertained for ages a hearty desire to return thither, and are willing to make great sacrifices of a pecuniary kind to the different parties interested, provided they can be put in peaceful and secure possession.

It is likewise notorious, that since the Jews have been thrust out of Syria, that land has been a mere arena of strife to neighbouring Powers, all conscious that they had no legitimate right there, and all jealous of each other's intrusion.

Such having been the case, why, it may be asked, have not the Jews long ago endeavoured to regain possession of Syria by commercial arrangements? In reply it may be said, that though they have evidently wished to do so, and have made overtures of the kind, hitherto circumstances have mainly opposed their desires. For instance, they could not expect to purchase a secure possession of Syria from Turkey, while that empire, in the pride of insolent despotism, could have suddenly revoked its stipulations, and have seized on Jewish treasuries, none venturing to call it to account. Nor could the Jews have ventured to purchase Syria while the right to that country was vehemently disputed between Turkey and Egypt, without any powerful arbitrators to arrange the right at issue, and lend sanction and binding authority to diplomatic documents.

Now, however, these obstacles and hindrances are in a great measure removed; all the strongest Powers in Europe have come forward as arbitrators and umpires to arrange the settlement of Syria.

Under such potent arbitrators, pledged to the performance of any conditions finally agreed on, I have reason to believe that the Jews would readily enter into such financial arrangements as would secure them the absolute possession of Jerusalem and Syria.

If such an arrangement were formed, one great cause of dissension between France and England would be at once removed; for both the Porte and Egypt are decidedly in want of money, and will gladly sell their respective rights in the Syrian territory. They themselves begin to see the folly of enacting the part of the dog in the manger; they will drop the apple of discord if they can get fair compensation for their trouble.

I know no reason, under such powerful umpires, why the Hebrews should not restore an independent monarchy in Syria, as well as the Egyptians in Egypt, or the Grecians in Greece.

As a practical expedient of politics, I believe it will be easier to secure the peace of Europe and Asia by this effort to restore the Jews, than by any allotment of Syrian territories to the Turks or Egyptians, which will be sure to occasion fresh jealousies and discords.

In offering these remarks, I have viewed the question merely as a lawyer and a politician, and proposed the restoration of the Jews as a sort of *tertium quid*, calculated to win the votes of several of the parties at issue. But, Sir, there is a higher point of view from which many of the readers of *The Times* may wish to regard this topic of investigation. Whichever way the restoration of the Jews may finally be brought about, there is no doubt that it is a subject frequently illustrated by Biblical prophecies.

I will, therefore, if I may do so without the vain and presumptuous curiosity which some of the neologists have manifested, endeavour to detail the opinion of the church on this subject in the words of some of her most respectable writers.

It is generally supposed by Newton, Hales, Faber, and others, that the great prophetical period of 1,260 years is not very far from its termination. If they are right in this supposition, the period of the restoration of the Jews cannot be very remote.

These two contingencies are evidently connected by the prophet Daniel, who distinctly states that at the time of the end of this period there shall be great contests among the Eastern nations in Syria. And at that time (continues Daniel) shall Michael stand up, even the great Prince who standeth up for the children of the Jews, and there shall be a time of trouble such as never was since there was a nation, and at that time the Jews shall be delivered. (Daniel xii.)

Whatever this mysterious passage may imply, all the most learned expositors agree that it refers to the same crisis indicated by the author of the Apocalypse (Chapter xvi., verses 12, 16.) Most of these expositors seem to think that by the phrase 'drying up the great river Euphrates, that the way of the Kings of the East might be prepared,' we are to understand the diminution of the Turkish empire, that the Jews may regain their long lost kingdom of Syria.

I will not detain you by quoting a host of learned authorities in confirmation of this interpretation; but it may be important to hint, that the moral and intellectual position of the Jews in the present day, as well as their commercial connexions, has enabled them to assume a political sphere of activity at once lofty and extensive.

As to religion, they have of late years realized many of the predictions of Mendelssohn and D'Israeli. They have thrown off the absurd bigotry which once rendered them contemptible, and begin to give the New Testament and the writings of Christian divines that attention to which they are every way entitled among truth-searching and philosophic men. Though, perhaps, fewer positive conversions to Christianity have taken place than were expected by the clergy, still the Hebrew intellect has made within a few years past a wonderful approximation to that temper of impartial inquiry in which such books as *Grotius de Veritate* produce an indeliable impression.

I believe that the cause of the restoration of the Jews is one essentially generous and noble, and that all individuals and nations that assist this world-renounced people to recover the empire of their ancestors will be rewarded by Heaven's blessing. [It was and is commonplace for Zionists to appeal to the superstitions of Christians and others with the myth that Jews have supernatural connections which will bless those who help Jews and punish those who do not. The real forces at work are generally control over public opinion through media, planted rumor and gossip; sophisticated intelligence networks; and the might of higher education and investment capital, or lack thereof, which can raise a nation above others, or destroy it. Whoever controls news outlets and financial institutions is the first to learn of events and investments, and to profit from them, or prevent them.—CJB] Everything that is patriotic and philanthropic should urge Great Britain forward as the agent of prophetic revelations so full of auspicious consequence.

I dare not allow my mind to run into the enthusiasm on this subject which I find predominant among religious authors. I will, therefore, conclude with one quotation from *Hale's Analysis of Chronology*:—

'The situation of the new Jerusalem,' says this profound mathematician, 'as the centre of Christ's millennary kingdom in the Holy Land, considered in a geographical point of view, is well described by Mr. King in a note to his *Hymns to the Supreme Being*. How capable Syria is of a more universal intercourse than any other country with all parts of the world is most remarkable, and deserves to be well considered, when we read the numerous prophecies which speak of its future grandeur, when its people shall at length be gathered from all nations among whom they have wandered, and Sion shall be the joy of the whole earth.'

Your very obedient servant,
Aug. 17. F. B."

Many Christians were foolish and childish enough to be taken in by the Zionist propaganda promising them the joys of the apocalypse and their wonderful martyrdom for the sake of the Jews, but the Jews themselves wanted no part of it. The majority of Jews wanted nothing of the pseudo-Protestant movement, led by crypto-Jews, to banish them to the deserts of Palestine in the hopes that Jesus might return in the form of Rothschild. The Rothschilds were constantly testing to see if the Jews wanted to go to Palestine and consistently discovered that they did not. *The London Times* published the following set of queries on 17 August 1840 on page 3,

"SYRIA.—RESTORATION OF THE JEWS.
(From a Correspondent.)

The proposition to plant the Jewish people in the land of their fathers, under the protection of the five Powers, is no longer a mere matter of speculation, but of serious political consideration. In a Ministorial paper of the 31st of July an article appears bearing all the characteristics of a feeler on this deeply interesting subject. However, it has been reserved for a noble lord opposed to Her Majesty's Ministers to take up the subject in a practical and statesmanlike manner, and he is instituting inquiries, of which the following is a copy:—

QUERIES.

'1. What are the feelings of the Jews you meet with respect to their return to the Holy Land?

'2. Would the Jews of station and property be inclined to return to Palestine, carry with them their capital, and invest it in the cultivation of the land, if by the operation of law and justice life and property were rendered secure?

'3. How soon would they be inclined and ready to go back?

'4. Would they go back entirely at their own expense, requiring nothing further than the assurance of safety to person and estate?

5. Would they be content to live under the Government of the country as they should find it, their rights and privileges being secured to them under the protection of the European powers?

'Let the answers you procure be as distinct and decided and detailed as possible: in respect as to the inquiries as to property, it will of course be sufficient that you should obtain fair proof of the fact from general report.'

The noble Lord who is instituting these inquiries has given deep attention to the matter, and is well known as the writer of an able article in the *Quarterly* on the subject, in December, 1838.

In connexion with this, a deeply interesting discovery has been made on the south-west shores of the Caspian, enclosed in a chain of mountains, of the remnant of the Ten Tribes, living in the exercise of their religious customs in a primitive manner, distinct from the customs of modern Judaism. The facts which distinguish them as the remnant of that branch of the Jewish family are striking and incontrovertible, and are about to be given to the world. An intrepid missionary, the Rev. Mr. Samuel, of Bombay, has made the discovery, and resided amongst this people several months, under permission from the Russian Government, who directed him to institute inquiry concerning them."

The Christians were led by crypto-Jews and their agents, and the Jews controlled the press, but there was still the risk that a Christian movement might arise which was true to the Christian faith and unwilling to destroy the Gentiles for the sake of the Jews. Given that the Jews did not want to go, and given the risks of a holy war that could result in the extermination of the Jews and with them the Rothschilds, the Rothschilds decided to wait for more favorable circumstances before purchasing Palestine and chasing out the Palestinians.

Since the Jews themselves did not wish to go to Palestine, and the Zionists' potential financial backers feared that their investment would be lost due to a lack of Jewish interest and given the possibility that the Sultan would renege on the deal and take their money while the rest of the world stood idly by, the Rothschilds and their agents saved face by making it appear in the press that the Sultan wanted more than the Zionists were willing to give, and had recognized the value of Palestine to the British, the Germans, the Egyptians, the Russians and to Islam. Note that the Zionists' offer in Herzl's day was "to regulate the finances" of Turkey in exchange for Palestine, not to buy the territory. By managing the finances of the Sultan, as opposed to simply paying off his debts or transferring funds to him, the Zionists would have some means to retaliate against him, should the Sultan breach the contract—or if it simply suited their purposes—they were, after all, the cause of his financial difficulties in the first place.

Moses Hess quoted Ernest Laharanne's *La nouvelle question d'Orient: Empires d'Egypte et d'Arabie. Reconstitution de la nationalité juive*, E. Dentu, Paris, (1860), whose prose reveals why the Rothschilds were forced to propagandize the Christians and subvert Christianity, before moving into Palestine—which was also the primary cause of the Jews' ancient war on Catholicism,

> "I may, therefore, recommend this work, written, not by a Jew, but by a French patriot, to the attention of our modern Jews, who plume themselves on borrowed French humanitarianism. I will quote here, in translation, a few pages of this work, *The New Eastern Question*, by Ernest Laharanne.[*Footnote:* See note IX at end of book.]
>
> 'In the discussion of these new Eastern complications, we reserved a special place for Palestine, in order to bring to the attention of the world the important question, whether ancient Judæa can once more acquire its former place under the sun.
>
> 'This question is not raised here for the first time. The redemption of Palestine, either by the efforts of international Jewish bankers, or the nobler method, of a general subscription in which all the Jews should participate, has been discussed many times. Why is it that this patriotic project has not as yet been realized? It is certainly not the fault of pious Jews that the plan was frustrated, for their hearts beat fast and their eyes fill with tears at the thought of a return to Jerusalem.
>
> [*Footnote:* My friend, Armond L., who traveled for several years through the Danube Principalities, told me that the Jews were moved to tears when he announced to them the end of their suffering, with the words 'The time of the return approaches.' The more fortunate Occidental Jews do not know with what longing the Jewish masses of the East await the final redemption from the two thousand year exile. They know not that the patriotic Jew cannot suppress his cry of anguish at the length of the exile, even in the midst of his festive songs, as, for instance, the patriotic poem which is read on Chanukah, closes with the mournful call:
> 'For salvation is delayed for us and there is no end to the days of evil.'
> 'They asked me,' continued my friend, 'what are the indications that the end of the exile is approaching?' 'These,' I answered, 'that the Turkish and the papal powers are on the point of collapse.']
>
> 'If the project is still unrealized, the cause is easily cognizable. The

Jews dare not think of the possibility of possessing again the land of their fathers. Have we not opposed to their wish our Christian veto? Would we not continually molest the legal proprietor when he will have taken possession of his ancestral land, and in the name of piety make him feel that his ancestors forfeited the title to their land on the day of the Crucifixion?

'Our stupid Ultramontanism has destroyed the possibility of a regeneration of Judæa, by making the present of the Jewish people barren and unproductive. Had the city of Jerusalem been rebuilt by means of Jewish capital, we would have heard preachers prophesying, even in our progressive nineteenth century, that the end of the world is at hand and predictions of the coming of the Anti-Christ. Yes, we have lived to see such a state of affairs, now that Ultramontanism has made its last stand in oratorical eloquence. In the sacred beehive of religion, we still hear a continuous buzzing of those insects who would rather see a mighty sword in the hands of the barbarians, than greet the resurrection of nations and hail the revival of a free and great thought inscribed on their banner. This is undoubtedly the reason why Israel did not make any attempt to become master of his own flocks, why the Jews, after wandering for two thousand years, are not in a position to shake the dust from their weary feet. The continuous, inexorable demands that would be made upon a Jewish settlement, the vexatious insults that would be heaped upon them and which would finally degenerate into persecutions, in which fanatic Christians and pious Mohammedans would unite in brotherly accord—these are the reasons, more potent than the rule of the Turks, that have deterred the Jews from attempting to rebuild the Temple of Solomon, their ancient home, and their State."[303]

Hess, himself, wrote,

"It seems that extracts from the French pamphlet which I quoted to you, have awakened in you new thoughts. You think that the Christian nations will certainly not object to the restoration of the Jewish State, for they will thereby rid their respective countries of a foreign population which is a thorn in their side. Not only Frenchmen, but Germans and Englishmen, have expressed themselves more than once in favor of the return of the Jews to Palestine. You quote an Englishman who endeavored to prove, by Biblical evidence, the ultimate return of the Jews to Palestine and simultaneously also the conversion of the Jews to Christianity. Another Englishman attempts to prove that the present English dynasty is directly descended from the house of David and that the stone which plays such an important rôle in the coronation of English kings is the same on which Jacob's head rested when he dreamt of the famous ladder. A third magnanimously offers all the English ships for the purpose of conveying to Palestine, free of

---

[303]. M. Hess, "Eleventh Letter", *Rom und Jerusalem: die letzte Nationalitätsfrage*, Eduard Wengler, Leipzig, (1862); English: *Rome and Jerusalem: A Study in Jewish Nationalism*, Bloch, New York, (1918/1943), pp. 141-159, at 150-152.

charge, all the Jews who want to return there. These sentiments, however, seem to be, according to you, only a milder form of the desire, which in former ages expressed itself in frequent banishments of the Jews from Christian lands, for which mildness our people ought to be thankful. On the other hand, you see in such projects only a piece of folly which, in its final analysis, leads either to religious or secular insanity, and should not be taken into consideration. Such desires, moreover, if they come from pious Christians, would be opposed by all Jews. On the other hand, if pious Jews were the projectors, all Christians would object to the restoration; for as the latter would only consent to a return to Palestine on condition that the ancient sacrificial cult be reintroduced in the New Jerusalem, so would the former give its assistance to the plan, only on condition that we Jews would bring our national religion as a sacrifice to Christianity at the 'Holy Sepulchre.' And thus, you conclude, all the national aspirations of the Jews must inevitably founder on the rock of differences of opinion.

Now if rigid Christian dogma and inflexible Jewish orthodoxy could never be revived by the living current of history, they would certainly place an insurmountable obstacle to the realization of our patriotic aspirations. The thought of repossessing our ancient fatherland can, therefore, be taken under serious consideration, only when this rigidity of orthodox Jews and Christians alike, will have relaxed. And it is beginning to relax already, not only with the progressive elements, but even with pious Jews and Christians. Moreover, the Talmud, which is the corner-stone of modern Jewish orthodoxy, long ago counseled obedience to the dictates of life."[304]

The Christians believed that the Jews had only one way to save themselves from ultimate annihilation, and that was to convert to Christianity. Even those Gentiles willing to help the Jews take Palestine from the Turks and the Palestinians knew that the Jews would be attacked unless they pretended to convert to Christianity. Hence the countless books calling for the "restoration to Palestine" that were published by Christians, and by crypto-Jewish Zionists pretending to be Christians, concurrently called for the conversion of the Jews. They knew that this was the only safe way to establish a Jewish colony in Palestine without provoking the Christians into a holy war. This also had the benefit of allying the Christians with the Jews against the Moslems. In addition, Frankist Jews believed that the Messiah would only come when all Jews embraced heresy, and Christianity is a heresy to Jews. Ironically and paradoxically, a major part of the Jewish religious plan is the objective of making Jews heretical. Jewish leaders believe that the prophets commanded that Jews fall away from God and that they are duty bound to see to it that two thirds of Jews perish as a result (*Isaiah* 48:10. *Ezekiel* 5:12. *Zechariah* 13:8-9). They believe that the Messiah will only come when Jews have embraced heresy and

---

304. M. Hess, "Eleventh Letter", *Rom und Jerusalem: die letzte Nationalitätsfrage*, Eduard Wengler, Leipzig, (1862); English: *Rome and Jerusalem: A Study in Jewish Nationalism*, Bloch, New York, (1918/1943), pp. 160-162.

have made the world evil (*Sanhedrin* 97*a*).

Again, the problem the Jewish Zionists faced was that the Jews did not want to go to Palestine, let alone pretend to convert to Christianity and then go to Palestine.

Very early on, Cyprian stated in his Twelfth Treatise, "Three Books of Testimonies Against the Jews", First Book, Testimony 24, that the Jews had but one option, other than extermination, to atone for the death of Christ,

> "24. That by this alone the Jews can receive pardon of their sins, if they wash away the blood of Christ slain, in His baptism, and, passing over into His Church, obey His precepts.
>
> In Isaiah the Lord says: 'Now I will not release your sins. When ye stretch forth your hands, I will turn away my face from you; and if ye multiply prayers, I will not hear you: for your hands are full of blood. Wash you, make you clean; take away the wickedness from your souls from the sight of mine eyes; cease from your wickedness; learn to do good; seek judgement; keep him who suffers wrong; judge for the orphan, and justify the widow. And come, let us reason together, saith the Lord: and although your sins be as scarlet, I will whiten [*Footnote:* 'Exalbabo.'] them as snow; and although they were as crimson, I will whiten [*Footnote:* 'Inalbabo.'] them as wool. And if ye be willing and listen to me, ye shall eat of the good of the land; but if ye be unwilling, and will not hear me, the sword [Esau] shall consume you; for the mouth of the Lord hath spoken these things. [*footnote:* Isa. i. 15-20.]"[305]

The Zionists had to carefully nurture an antagonism over the course of centuries in Europe against the Pope and depict him as the anti-Christ, and against Catholicism as the evil ecumenical Church of the Apocalypse, and against Islam and the Turks as heathens; so that Christians would not see the Jews and Judaism as the prophesied evil ecumenical Church of the Apocalypse headed by the anti-Christ—the Jews' false Messiah; and so that the English Esau, or some other European force, would take Palestine from the Turks and give it to the Jews, who could then regulate the trade of the world. The best means to accomplish this monumental feat was to create anti-Catholic "reformations" and "second reformations" creating the Protestant and Puritan Churches, and for the Jews to pretend to convert to these Judaised Churches and form an alliance with Judaized "Christians" against Islam, while destroying Christianity along with Islam.

Herzl recalled the rôles of Esau and Jacob in his book *The Jewish State*, when he called on Europe, Esau, to guard Israel, Jacob,

---

[305]. Cyprian, Twelfth Treatise, "Three Books of Testimonies Against the Jews", First Book, Testimony 24, *The Anti-Nicene Fathers: Translations of the Writings of the Fathers down to A.D. 325*, Volume 5, Christian Literature Publishing Company, New York, (1886), p. 514-515.

"We should as a neutral State remain in contact with all Europe, which would have to guarantee our existence."[306]

Anti-Popism had a history in England dating back at least to Jon Wycliffe, who anticipated many aspects of the Protestant Reformation and Communism—the modern Utopian substitute for original Christian mythology. The wanton corruption of the Popes—especially the Spanish Borgia Popes (Pope Callixtus III, who waged war on the Turks, and his nephew Alexander VI, who waged war on Catholicism), made for fertile ground for the reformers who would convert Catholicism to Judaism and eventually atheism. This ground was tilled by Cabalist Jews and supposedly anti-Semitic Jews who claimed to have converted to Christianity, like: Konrad Mutian (a. k. a. Conradus Mutianus Rufus), Johann Reuchlin, Pico della Mirandola, Jakob Questenberg, Jakob ben Jehiel Loans, Obadja Sforno of Cesena, Johann Pfefferkorn, etc. Note that in the dualistic and dialectical terms of the Cabalah, anti-Semites and the defenders of Judaism serve the same purpose—the segregation of Jews.[307]

For centuries, the British and the Jews did what they could to diminish the power of Turkey and Egypt, fully achieving their Apocalyptic vision by the end of the First World War. As a supposedly Protestant English Zionist stated in a letter to the Editor of 17 August 1840, published in *The London Times*, on 26 August 1840, on page 6,

"Whatever this mysterious passage may imply, all the most learned expositors agree that it refers to the same crisis indicated by the author of the Apocalypse (Chapter xvi., verses 12, 16.) Most of these expositors seem to think that by the phrase 'drying up the great river Euphrates, that the way of the Kings of the East might be prepared,' we are to understand the diminution of the Turkish empire, that the Jews may regain their long lost kingdom of Syria."

Joseph Mede, of Cambridge, iterated this call for war against the Turks for the benefit of the Jews—under the guise of scripture—in the 1600's, and countless others echoed his call.[308] The Euphrates of Moslem might in the Middle East continues to evaporate under the influence of our present day Zionists, and with it the dignity of humankind is lost to the night.

The first act of the First Zionist Conference in 1897 was to pass a resolution

---

306. T. Herzl, *The Jewish State: An Attempt at a Modern Solution of the Jewish Question*, Zionist Organization of America, New York, (1943), p. 43.

307. M. A. Hoffman II, *Judaism's Strange Gods*, Independent History and Research, Coeur d'Alene, Idaho, (2000), pp. 108-109.

308. *Cf.* S. Snobelen, "'The Mystery of the Restitution of All Things': Isaac Newton on the Return of the Jews", in J. E. Force and R. H. Popkin, Editors, *The Millenarian Turn: Millenarian Contexts of Science, Politics, and Everyday Anglo-American Life in the Seventeenth and Eighteenth Centuries*, Chapter 7, Kluwer Academic Publishers, Dordrecht, Boston, (2001), pp. 95-118, at 112, note 11.

thanking the Sultan of Turkey, who, at the instigation of Jews and crypto-Jews, was committing atrocities against the Armenians. Crypto-Jews were the motive force behind the Sultan's atrocities against Armenian Christians. Jewish bankers, and crypto-Jewish bankers posing as Greek and Armenian Christians, managed the Sultan's accounts and led him into bankruptcy, while they, themselves, became immensely wealthy at the expense of the Turkish Empire. Jews prompted the Sultan to retaliate against innocent Armenian Christians, falsely blaming them for the theft, and diverting attention from the criminal Jews. The willingness of the political Zionists to fund and forgive (with their admitted corruption of the press) Jewish-Turkish atrocities began with their beginning and culminated in the genocide of the Armenians after the Sultan's Government was overthrown by the "Young Turks" in 1915—a group led by crypto-Jewish[309] positivist revolutionaries whose philosophies stemmed from Henri de Saint-Simon and Auguste Comte—philosophies which were popular among Jewish intellectuals, especially in Vienna. Thomas R. Bransten wrote in his compilation of David Ben-Gurion's *Memoirs*,

> "*No Messiah but nineteenth-century positivism as coupled to Biblical affirmation of the Jews' historical place in the land of Israel prompted their massive return.*"[310]

The Armenians are among the most ancient group of Christians—Christians whom some Jews have long sought to destroy. The Armenians were unwise enough to sponsor the Zionist venture in Palestine and publicly endorsed the Balfour Declaration in hopes that it would protect them from the Turks, not realizing that the Young Turks were massacring the Armenians in the millions at the instigation of their crypto-Jewish leadership. The Armenian leaders were corrupted by Zionist Jews and betrayed the Armenian People.

Herzl makes clear his evil intentions in his diaries. Herzl's deceit was earlier exposed in the eleventh edition of the *Encyclopædia Britannica* in 1911 in an article on "Zionism". The Zionists had cut a deal with the Sultan through Newlinsky to use their influence in the news media to control public opinion concerning the atrocities the Turks had committed against the Armenians at the instigation of the Jews,

> "The most encouraging feature in Dr Herzl's scheme was that the Sultan of Turkey appeared favourable to it. The motive of his sympathy has not hitherto been made known. The Armenian massacres had inflamed the whole of Europe against him, and for a time the Ottoman Empire was in

---

309. I. Zangwill, *The Problem of the Jewish Race*, Judaen Publishing Company, New York, (1914), pp. 9, 11; which was first published as an article, "The Jewish Race", *The Independent*, Volume 71, Number 3271, (10 August 1911), pp. 288-295, at 290-291. J. Prinz, *The Secret Jews*, Random House, New York, (1973), pp. 111-112.

310. D. Ben-Gurion, *Memoirs*, The World Publishing Company, New York, Cleveland, (1970), not paginated.

very serious peril. Dr Herzl's scheme provided him, as he imagined, with a means of securing powerful friends. Through a secret emissary, the Chevalier de Newlinsky, whom he sent to London in May 1896, he offered to present the Jews a charter in Palestine provided they used their influence in the press and otherwise to solve the Armenian question on lines which he laid down. The English Jews declined these proposals, and refused to treat in any way with the persecutor of the Armenians. When, in the following July, Dr Herzl himself came to London, the Maccabaean Society, though ignorant of the negotiations with the Sultan, declined to support the scheme. None the less, it secured a large amount of popular support throughout Europe, and in 1910 Zionism had a following of over 300,000 Jews, divided into a thousand electoral districts. The English membership is about 15,000. [***] Modern Zionism is vitiated by its erroneous premises. It is based on the idea that anti-Semitism is unconquerable, and thus the whole movement is artificial. Under the influence of religious toleration and the naturalization laws, nationalities are daily losing more of their racial character. The coming nationality will be essentially a matter of education and economics, and this will not exclude the Jews as such. With the passing away of anti-Semitism, Jewish nationalism will disappear. If the Jewish people disappear with it, it will only be because either their religious mission in the world has been accomplished or they have proved themselves unworthy of it."

Marwan R. Buheiry investigated these issues and made plain Herzl's treachery in an article, "Theodor Herzl and the Armenian Question", *Journal of Palestine Studies*, Volume 7, Number 1, (Autumn, 1977), pp. 75-97. As Buheiry proves, there were indeed very influential British press Jews who obliged Herzl's schemes, including Lucien Wolf.

Note the self-imposed pressure on early political Zionists to promote anti-Semitism, which was not considered by most Jews at the time to be nearly so unconquerable as Herzl had portrayed it, for without a dramatic increase of anti-Semitism brought about by the Zionists themselves, political Zionism, which was founded on the premise that Jews were incapable of assimilation, had no *raison d'être* and no hope of success. Political Zionism was premised on the success of anti-Semitism; which gave the Zionists the incentive to spread, not eliminate, anti-Semitism. The political Zionists became fanatical in this mission to generate anti-Semitism—unprecedentedly virulent anti-Semitism—because they convinced themselves that the survival of their divine race depended upon their ability to make the world hate and persecute Jews.

No one loved Herzl and his pamphlet more than the anti-Semites. Herzl wrote in his diary,

"the Pressburg anti-Semite Ivan von Simonyi [***] Loves me!"[311]

---

[311]. T. Herzl, English translation by H. Zohn, R. Patai, Editor, *The Complete Diaries of Theodor Herzl*, Volume 1, Herzl Press, New York, (1960), p. 317.

and,

> "In the beginning we shall be supported by anti-Semites through a *recrudescence** of persecution (for I am convinced that they do not expect success and will want to exploit their 'conquest.')"[312]

and,

> "The anti-Semites will become our most dependable friends, the anti-Semitic countries our allies."[313]

Herzl declared the virtue and justice, in his perverse mind, of anti-Semitism,

> "[W]e want to let respectable anti-Semites participate in our project [***] Present-day anti-Semitism can only in a very few places be taken for the old religious intolerance. For the most part it is a movement among civilized nations whereby they try to exorcize a ghost from out of their own past. [***] The anti-Semites will have carried the day. Let them have this satisfaction, for we too shall be happy. They will have turned out to be right because they *are* right. They could not have let themselves be subjugated by us in the army, in government, in all of commerce, as thanks for generously having let us out of the ghetto. Let us never forget this magnanimous deed of the civilized nations. [***] Thus, anti-Semitism, too, probably contains the divine Will to Good, because it forces us to close ranks, unites us through pressure, and through our unity will make us free."[314]

Herzl wrote in 1893,

> "What would you say, for example, if I did not deny there are good aspects of anti-Semitism? I say that anti-Semitism will educate the Jews. In fifty years, if we still have the same social order, it will have brought forth a fine and presentable generation of Jews, endowed with a delicate, *extremely sensitive* feeling for honor and the like."[315]

In 1897, Herzl told the First Zionist Congress,

---

312. T. Herzl, English translation by H. Zohn, R. Patai, Editor, *The Complete Diaries of Theodor Herzl*, Volume 1, Herzl Press, New York, (1960), p. 56.
313. T. Herzl, English translation by H. Zohn, R. Patai, Editor, *The Complete Diaries of Theodor Herzl*, Volume 1, Herzl Press, New York, (1960), p. 84.
314. T. Herzl, English translation by H. Zohn, R. Patai, Editor, *The Complete Diaries of Theodor Herzl*, Volume 1, Herzl Press, New York, (1960), pp. 143, 171, 182, 231.
315. T. Herzl, as quoted by A. Elon, *Herzl*, Holt, Rinehart and Winston, New York, (1975), pp. 114-115.

> "The feeling of communion, of which we have been so bitterly accused, had commenced to weaken when anti-Semitism attacked us. Anti-Semitism has restored it. We have, so to speak, gone home. Zionism is the return home of Judaism even before the return to the land of the Jews."[316]

Max Nordau wrote in 1905,

> "Anti-Semitism has also taught many educated Jews the way back to their people."[317]

Benjamin Disraeli, who was to become the Prime Minister of England, wrote in 1844, referring to Jews as the "superior race" and the "pure persecuted race",

> "And every generation they must become more powerful and more dangerous to the society which is hostile to them. Do you think that the quiet humdrum persecution of a decorous representative of an English university can crush those who have successively baffled the Pharaohs, Nebuchadnezzar, Rome, and the Feudal ages? The fact is, you cannot destroy a pure race of the Caucasian organization. It is a physiological fact; a simple law of nature, which has baffled Egyptian and Assyrian Kings, Roman Emperors, and Christian Inquisitors. No penal laws, no physical tortures, can effect that a superior race should be absorbed in an inferior, or be destroyed by it. The mixed persecuting races disappear; the pure persecuted race remains."[318]

The Zionists sponsored anti-Semitism: 1) By raising the issue wherever and whenever they could promoting the idea of the "common enemy" to Jews to lead them into panic and segregation. 2) By smearing famous figures of all ethnic groups including those Jewish financiers who would not fund them. 3) By smearing and intimidating assimilationists. 4) By promoting racial segregation as if it had a scientific basis. 5) By censoring ideas contrary to their own and otherwise manipulating the press and politicians as best they could. 6) By promoting the massive emigration of Eastern European Jews to the West believing it would provoke and agitate anti-Semitism—as is reflected in Einstein's actions and speeches in the late Teens through the Twenties of the Twentieth Century. The political Zionists even used *agents provocateur* to spread anti-Semitism and the political Zionists founded anti-Semitic societies, societies which produced the Nazis.

---

316. "The Zionist Congress: Full Report of the Proceedings", *The Jewish Chronicle*, (3 September 1897), pp. 10-15, at 11.
317. M. Nordau and G. Gottheil, *Zionism and Anti-Semitism*, Fox, Duffield & Company, (1905), p. 19.
318. B. Disraeli, *Coningsby; or, The New Generation*, The Century Co., New York, (1904), pp. 231.

Herzl was a corrupt journalist, and he established the precedent of the political Zionists' frequent attempts to corrupt the mass media, which has continued through to Robert Maxwell[319] and beyond. Benjamin Harrison Freedman's writings and speeches document the political Zionists' tactics of smear and distraction, which are manifest in abundance in their shameless and dishonest promotion of Einstein, who is for them not merely a national hero, but a saint. The Jewish industrialist Benjamin Harrison Freedman warned Americans in the immediate post-World War II period, as quoted by Douglas Reed in 1951,

> "Mr. Freedman, some time before Mr. Truman's 'proudest moment,' wrote: 'The threat of Political Zionism to the welfare and security of America is little realized.... There may soon take place in Palestine an explosion which will set off another world war.... The influence of the Zionist organization reaches into the inner policy-making groups of nearly every government in the world—particularly into the Christian West. This influence causes these groups to adopt pro-Zionist policies which are often in conflict with the real interests of the peoples they govern. This condition exists in the United States... New York, Pennsylvania, Illinois, Massachusetts and California control 151 electoral votes out of a total of 531. In these states are concentrated the overwhelming majority of Americans of Jewish faith. In these states Jews hold the balance of power. Zionists claim that they can 'deliver' this vote. Although a great majority of American Jews are not Zionists ... the Zionist minority has found means to silence them, and to convince nearly everybody that anti-Zionism means being anti-Jewish. In the light of this, and in the light of past elections, the present administration, with its eye on the next elections (which President Truman's supporters won) 'has been strongly pro-Zionist. The pro-Zionist, politically motivated declarations of the President have been accepted throughout the world as official statements of American foreign policy. Yet it has always been a cardinal principle of American policy that all civilized peoples have a right to enjoy their own freedom.... Soviet Communism will succeed in its attempt to conquer the world in direct proportion to the support which America gives to Zionism.... It will take courage for Americans of whatever origin to think these facts through and take public positions upon them. They will be smeared. They will be slandered. Already, Zionists have been able to bring about economic ruin of many Christians and Jews who have dared challenge the right to claim Palestine for a Jewish national State.'"[320]

In 1955, James Rorty called Benjamin Freedman a "Jewish anti-Semite",

---

319. G. Thomas and M. Dillon, *Robert Maxwell, Israel's Superspy*, Carroll and Graf, New York, (2002).
320. B. Freedman as quoted in D. Reed, *Somewhere South of Suez*, Devin-Adair Company, U. S. A., (1951), pp. 331-332.

"One of McGinley's angels is the Jewish anti-Semite Benjamin Freedman, who told the Armed Services Committee on December 12, 1950 that he had given $15,000 to *Common Sense*."[321]

Arnold Forster wrote in his book *Square One* of 1988,

"And I said that we knew the purpose of the trip was to seek documentation for the case against Mrs. Rosenberg from one Benjamin Freedman, an affluent, self-hating apostate Jew who had spent untold thousands of dollars purchasing, reprinting and disseminating widely the anti-Jewish materials produced by the nation's worst professional anti-Semites. I told, too, how the two men, the one from Fulton Lewis' office, the other on the senator's staff, had carried a letter of introduction to Freedman from Gerald L. K. Smith, one of the most notorious bigots in the United States."[322]

Retired Congressman Paul Findley has written extensively on political Zionism's undue influence in shaping American public opinion and of its interference in American politics.[323] Douglas Reed published many scathing indictments of political Zionism and of political Zionism's negative impact on the world.[324]

Here are but a few of examples of the corruption of the press and of Herzl's intended manipulation of the press to smear those who disagreed with him, to cover-up and forgive atrocities, and to corruptly control public opinion—but a small sample taken from the *many* to be found in Herzl's diaries:

"But if he does, I shall smash him, incite popular fanaticism against him, and demolish him in print [***] I shall probably make enemies of the big Jews. Well, this is going to be apparent from the attacks or the silence of the servile part of the press. [***] I am writing de Haas a few compliments for Mr. Prag, and am authorizing him at the same time to publish the Turkish ambassador's denial in the press—only the substance, not the wording. [***] I must endeavor to gain influence over a newspaper. I can have such influence only as an owner of shares. [***] Let the gentlemen found or buy one large daily paper in London and one in Paris. There are papers that yield a good profit and on which the Fund would not lose anything. The politics

---

321. J. Rorty, "Storm Over the Investigating Committees", *Commentary*, Volume 19, Number 2, (February, 1955), pp. 128-136, at 131.
322. A. Forster, *Square One*, D. I. Fine, New York, (1988), p. 121.
323. P. Findley, *They Dare to Speak Out: People and Institutions Confront Israel's Lobby*, Lawrence Hill, Westport, Connecticut, (1985); **and** *Deliberate Deceptions: Facing the Facts about the U.S.-Israeli Relationship*, Lawrence Hill Books, Chicago, (1993); **and** *Silent No More: Confronting America's False Images of Islam*, D : Amana Publications, Beltsville, Maryland, (2001).
324. D. Reed, *Somewhere South of Suez*, Devin-Adair Company, U. S. A., (1951); **and** *The Controversy of Zion*, Bloomfield, Sudbury, (1978).

of the Jews should be conducted through these papers, for or against Turkey, depending on circumstances, etc. On the outside, the papers need not be recognizable as Jewish sheets. [***] Here I wish to insert *incidemment* something that will show how easily we can transplant many of our customs. The newspapers which are now being hawked as Jewish sheets—and rightly so, I believe—will have editions over there, like the Paris edition of the *New York Herald*. The news will be exchanged between both sides by cable. After all, we shall remain in contact with our old homelands. Gradually the demand for newspapers will increase, the colonial editions will grow, the Jewish editors will move overseas, leaving the Gentile ones by themselves. Little by little and imperceptibly, the Jewish papers will turn into Gentile papers, until the overseas editions are as independent as the European ones. It is an amusing thought in this serious plan that many a government will be willing to help us for that reason alone. [etc. etc. etc.]"[325]

Amos Elon wrote in his book *Herzl*,

"Herzl had a long talk with Badeni (in French) but did not commit himself. He was as reluctant to leave the *Presse* as Bacher and Benedikt were to lose him. A few days after his talk with Badeni he approached Benedikt, the *Presse*'s senior publisher, with his plan. Beneclikt, one of the most influential men in Austria, married to a Gentile woman, was the embodiment of Jewish pan-Germanism.* A man of great erudition, with an exceptionally keen mind, his editorials resounded from one end of the empire to another. Unlike Bacher, he realized how serious Herzl was, and he feared that the newspaper might lose one of its stars. Benedikt and Herzl took a long walk in the country. For three hours they argued back and forth, talking and walking themselves to exhaustion.

Herzl said he would like best to be able to launch his plan 'in and with' the newspaper, but he would launch it even if they refused.

'You are confronting us with a monstrous choice,' Benedikt said. 'The entire newspaper would take on a different complexion. We have always

---

325. T. Herzl, English translation by H. Zohn, R. Patai, Editor, *The Complete Diaries of Theodor Herzl*, Volumes 1 and 2, Herzl Press, New York, (1960), pp. 37, 97, 170, 455, 457, 480. Racist political Zionist Theodor Herzl wrote on 12 June 1895,

"Jewish papers! I will induce the publishers of the biggest Jewish papers (*Neue Freie Presse, Berliner Tageblatt, Frankfurter Zeitung*, etc.) to publish editions over there, as the *New York Herald* does in Paris."—T. Herzl, English translation by H. Zohn, R. Patai, Editor, *The Complete Diaries of Theodor Herzl*, Volume 1, Herzl Press, New York, (1960), p. 84.

THE DEARBORN INDEPENDENT, praised the *New York Herald*."When Editors Were Independent of the Jews", THE DEARBORN INDEPENDENT, (5 February 1921). ***See also:*** *The Collected Papers of Albert Einstein*, Volume 7, Document 35, Princeton University Press, (2002), pp. 296-297, note 8.

been considered a Jewish newspaper. But we have never admitted it. Now we are supposed to suddenly remove our cover.'

'But you won't need that cover anymore,' said Herzl. 'The very moment my idea is publicized, the entire Jewish question is solved honestly. After all, we can stay in all those countries that accept our good citizenship and loyalty to the fatherland. Only where they do not want us shall we move away.'

Herzl asked Benedikt to place a Sunday edition of the newspaper at his disposal. On the front page would be printed *The Solution to the Jewish Question, by Dr. Theodor Herzl*. Herzl would invite all of Jewry to contribute questions and comments; the ensuing debate would be more interesting than anything ever published by a newspaper."[326]

Samuel Landman wrote in 1936,

"In the early years of the War great efforts were made by the Zionist Leaders, Dr. Weizmann and Mr. Sokolow, chiefly through the late Mr. C. P. Scott of the *Manchester Guardian*, and Sir Herbert Samuel, to induce the Cabinet to espouse the cause of Zionism."[327]

Herbert Samuel was a highly religious Jew and an ardent Zionist who must have known the significance of the Messianic prophecies. His family were Jewish bankers and bullion merchants. P. W. Wilson wrote in 1922,

"For many years, I have known Sir Herbert Samuel and watched his career. He and his family belong to the stricter and more orthodox section of the Jewish community in Britain. In business, they are bankers and bullion merchants, an enterprise which depends for its success upon a meticulous accuracy of method and reliability of character. It is this high standard of personal responsibility that Herbert Samuel has applied to all his conduct as a British Minister in England and as the executive in Palestine."

Many of the leaders and ambassadors that Americans and the British have sent to predominantly Moslem lands have been Jewish—notably, but by no means limited to, the appointment of the racist political Zionist and Orthodox Jew Herbert Louis Samuel as the High Commissioner of Palestine in 1920.[328]

There is yet another odd aspect to Herzl's book *The Jewish State*. Why did someone as intelligent as Herzl say such foolish racist things, and why did he so heavily stress financial incentives? Herzl had earlier spoken far more rationally.

---

[326]. A. Elon, *Herzl*, Holt, Rinehart and Winston, New York, (1975), pp. 167-168.
[327]. S. Landman, *Great Britain, the Jews and Palestine*, New Zionist Press, London, (1936), p. 3
[328]. A. T. Clay, "Political Zionism", *The Atlantic Monthly*, Volume 127, Number 2, (February, 1921), pp. 268-279, at 277-279. B. L. Brasol, *The World at the Cross Roads*, Small, Mayhard & Co., Boston, (1921), pp. 371-379.

Such irrational reversals as Herzl's usually derive from insanity, a desire for revenge or from greed. Herzl focused on money in his pamphlet *Judenstaat* and in his book *Altneustadt*, and may have been a mouthpiece for a few of the financiers who stood to profit from the "scheme"—though many are known to have opposed him. The Anglicans had been trying to finance Zionism at least since the 1830's. Herzl received the early support of the Jewish financier Baron Hirsch and desperately sought the support of the Rothschild family, where he apparently was initially not so well received. However, the Balfour Declaration was addressed to Lord Rothschild and the Rothschilds had been trying to take Palestine for a very long time. Perhaps they sensed that Herzl was after their money. Perhaps the relationship between Herzl and the Rothschilds was indirect, or perhaps it was better than we have yet learned.

Another possibility is that Herzl was suddenly struck with a Messiah complex, and he does speak in his diaries of how famous he will become and does reveal that he was obsessed with his cause—but all this can also be attributed to greed. Messiahs don't usually proselytize to the checkbook, nor deny their alleged divinity. However, Herzl did once dream of the Messiah, and the Anglican Zionist William Henry Hechler tried to lead Herzl to be believe that he was the Messiah—but Herzl resisted any personal association with Messianic prophecy. Again, the Anglicans had been trying to finance Zionism for quite some time and sought assurances for the Jewish financiers that should the Jews buy Palestine from the Sultan, the Sultan would be unable to renege on the deal. Herzl was very careful to promote his venture as a secular enterprise so as to alleviate any Christian concerns that he was the anti-Christ. Given the pressure on Herzl to conceal any religious motivations he may have had, it is difficult to discern if he or his backers were not in fact motivated to fulfill Jewish Messianic prophecy, or if he was simply a greedy opportunist who took advantage of the religious aspirations of those wealthier than he.

The usual explanation for Herzl's change of attitude is that the success of some anti-Semitic politicians, like Karl Lueger in Austria, and the crisis of the Dreyfus Affair, prompted the change, but if Herzl genuinely believed that this converted all Jews into one people and rendered impossible the coexistence of this people with others, he was alone in his delusion. Herzl often mentions Eugen Karl Dühring's racist book *Die Judenfrage als Racen-, Sitten- und Culturfrage: mit einer weltgeschichtlichen Antwort*, H. Reuther, Karlsruhe, (1881); which profoundly affected him, and perhaps inspired his racism, or at least gave him a source to copy.[329] Nathan Birnbaum, the Zionist, accused Herzl of profiteering from Zionism, which appears to be the most plausible explanation for Herzl's sudden interest in raising money and casting the Jews out of Europe.[330]

---

[329]. M. Samuel, "Diaries of Theodor Herzl", in: M. W. Weisgal, *Theodor Herzl: A Memorial*, The New Palestine, New York, (1929), pp. 125-180, at 129. T. Herzl, English translation by H. Zohn, R. Patai, Editor, *The Complete Diaries of Theodor Herzl*, Volume 1, Herzl Press, New York, (1960), pp. 4, 111.

[330]. T. Herzl, English translation by H. Zohn, R. Patai, Editor, *The Complete Diaries of Theodor Herzl*, Volume 1, Herzl Press, New York, (1960), p. 307.

Herzl, who is seen by some as a prophet—Herzl, who congratulated anti-Semites on their supposed wisdom—Herzl, the fool, believed that he could provoke governments to expel Jews with complete impunity—Herzl stated on 14 June 1895,

> "They cannot throw us into the sea, at least not all of us, nor burn us alive. After all, there are societies for the prevention of cruelty to animals everywhere. What, then? They would finally have to find us some piece of land on the globe—a world ghetto, if you please."[331]

and in his book *The Jewish State* of 1896, Herzl's fatal hubris was again unleashed,

> "Again, people will say that I am furnishing the Anti-Semites with weapons. Why so? Because I admit the truth? Because I do not maintain that there are none but excellent men amongst us? Again, people will say that I am showing our enemies the way to injure us. This I absolutely dispute. My proposal could only be carried out with the free consent of a majority of Jews. Individuals or even powerful bodies of Jews might be attacked, but Governments will take no action against the collective nation. The equal rights of Jews before the law cannot be withdrawn where they have once been conceded; for the first attempt at withdrawal would immediately drive all Jews rich and poor alike, into the ranks of the revolutionary party. The first official violation of Jewish liberties invariably brings about economic crisis. Therefore no weapons can be effectually used against us, because these cut the hands that wield them."[332]

Leon Pinsker had stated in 1882,

> "We waged the most glorious of all guerrilla struggles with the peoples of the earth, who with one accord wished to destroy us. But the war we have waged—and God knows how long we shall continue to wage it—has not been for a fatherland, but for the wretched maintenance of millions of 'Jew peddlers.' [***] When an individual finds himself despised and rejected by society, no one wonders if he commits suicide. But where is the deadly weapon to give the *coup de grace* to the scattered limbs of the Jewish nation, and then who would lend his hand to it! The destruction is neither possible nor desirable. Consequently, we are bound by duty to devote all our remaining moral force to re-establishing ourselves as a living nation, so that we may ultimately assume a more fitting and dignified role among the

---

331. T. Herzl, English translation by H. Zohn, R. Patai, Editor, *The Complete Diaries of Theodor Herzl*, Volume 1, Herzl Press, New York, (1960), p. 172.
332. T. Herzl, *A Jewish State: An Attempt at a Modern Solution of the Jewish Question*, The Maccabæan Publishing Co., New York, (1904), p. 99.

family of the nations."³³³

The Zionists later used Einstein, then a celebrity, as an attraction to lure in crowds, and with them, cash, just as Herzl had planned. In return, Einstein, Herzl's proposed prize horse, was able to bask in the limelight he so loved. Einstein, as a political personality, was especially vulnerable to Herzl's racist belief system. Einstein generally hated Gentile Germans and was an impressionable and simplistic absolutist, who sought his opinions in the writings of others, and who formed generalized, stereotypical opinions expressed in absolutes. Einstein spoke of the "common destiny" of Jews in all of the countries of the world, of "our race", of Jews "sticking together", of ties of "blood", of the "Gentile world", of the "whole Jewish people", of the "salvation for the race", etc.³³⁴ While asserting his Zionist racism, Einstein would sometimes soften his statements, and mask his Jewish racism and supremacism, by asserting that he would prefer a world in which all human beings were brothers in the spirit of internationalism, but such a world did not exist because of anti-Semitism and he had to face facts and so practiced racism in order to protect himself from racism. Some anti-Semites had already justified segregation in the same terms as Einstein. Some anti-Semites claimed that they would prefer a Utopian world with universal brotherhood in the true Christian spirit, but that Zionist racism made such a world impossible and they just had to face facts and protect themselves from Jewish racists.³³⁵

Weizmann, Blumenfeld and Ginsberg ordered Einstein around, and he dutifully followed them until tensions and divisions arose among the Zionists. It is a myth that all Zionists were Communists or, alternatively, that all were right-wing extremists, though many did tend towards extremes as was natural for a fledgling movement caught in the tumult of turbulent times. There was a great deal of infighting among the political Zionists. The most common theme among Zionists was racism. Ber Borochov, a Marxist Zionist, cited Marx and Engel's materialistic racism in an effort to justify Zionism.³³⁶ Racist Zionist Moses Hess, who was condemned to death in the German Revolution of 1848 and who had worked with Marx and Engels, opposed the dogmatic approach of communistic materialistic determinism, and preferred nationalistic Socialism—like the Nazis later would.

The Zionists were able to corrupt the press and to promote anti-Semitism, so that the anti-Semites would force European governments to force the Jews to

---

333. L. Pinsker, translated by D. S. Blondheim, *Auto-Emancipation: An Admonition to His Brethren by a Russian Jew*, Federation of American Zionists, New York, (1916).
334. A. Einstein, *The World As I See It*, Citadel, New York, (1993), pp. 92-97, 103, 106, 108.
335. P. W. Massing, *Rehearsal for Destruction: A Study of Political Anti-Semitism in Imperial Germany*, Howard Fertig, New York, (1967), pp. 303-305.
336. B. Borochov, *Nationalism and Class Struggle*, New York, (1935), pp. 135-136, 183-205; quoted in A. Hertzberg, *The Zionist Idea*, Harper Torchbooks, New York, (1959), pp. 355-360, at 356.

leave Europe and assist in the expulsion of Jews to Palestine. Herzl, even before the Russian revolution, but after the French Revolution and the revolutions of 1848, played on the fear European governments had of the Jewish mission to rule the world by deposing monarchies through revolution, and in so doing the political Zionists reinforced anti-Semitism. Herzl unwisely believed that he could threaten the governments of the world,

> "The governments will give us their friendly assistance because we relieve them of the danger of a revolution which would start with the Jews—and stop who knows where!"[337]

Herzl wrote in his book *The Jewish State*,

> "When we sink, we become a revolutionary proletariat, the subordinate officers of the revolutionary party; when we rise, there rises also our terrible power of the purse. [***] Again, people will say that I am furnishing the Anti-Semites with weapons. Why so? Because I admit the truth? Because I do not maintain that there are none but excellent men amongst us? Again, people will say that I am showing our enemies the way to injure us. This I absolutely dispute. My proposal could only be carried out with the free consent of a majority of Jews. Individuals or even powerful bodies of Jews might be attacked, but Governments will take no action against the collective nation. The equal rights of Jews before the law cannot be withdrawn where they have once been conceded; for the first attempt at withdrawal would immediately drive all Jews rich and poor alike, into the ranks of the revolutionary party. The first official violation of Jewish liberties invariably brings about economic crisis. Therefore no weapons can be effectually used against us, because these cut the hands that wield them."[338]

However, it is clear from Herzl's book *The Jewish State* of 1896, that Herzl knew that the Jews of various nations were loyal to their homelands and would never leave Europe and America in large enough numbers of their own volition. Herzl took it upon himself, as self-appointed pseudo-Messiah, to generate political conditions whereby the Jews would have no choice but to leave. This political Zionist policy of provoking anti-Semitism fit in well with Einstein's desire to avoid criticism by dangerously stigmatizing scientific disagreement as if anti-Semitism, *per se*.[339] In this way Einstein accomplished two ends with one tactic. He generated and increased anti-Semitic sentiments in academia and he

---

337. T. Herzl, English translation by H. Zohn, R. Patai, Editor, *The Complete Diaries of Theodor Herzl*, Volume 1, Herzl Press, New York, (1960), p. 183.
338. T. Herzl, *A Jewish State: An Attempt at a Modern Solution of the Jewish Question*, The Maccabæan Publishing Co., New York, (1904), pp. 23, 99.
339. "Prof. Einstein Here, Explains Relativity", *The New York Times*, (3 April 1921), pp. 1, 13, at 1.

publicly smeared anyone who disagreed with him or threatened to expose him.

6.5 Albert Einstein Becomes a Cheerleader for Racist Zionism

Albert Einstein actively campaigned for Herzl's racism and traveled to America in April of 1921 in order to promote it. Einstein brought a "secretary", Salomon Ginzberg, the son of the famous Zionist leader Ha-Am. Ginzberg apparently had little respect for Einstein. He ridiculed Einstein for one of Einstein's "speeches"—a pre-Goebbels-like plea for ethnic unity behind a lone *Führer*,[340]

"You have one leader — Weizmann. Follow him and no other!"[341]

Ginzberg and Einstein's second wife failed to persuade Albert to return to his rehearsed lines, when Einstein was interviewed by *The New York Times Book Review* quoted herein. Note that Einstein's "secretary" repeated lines from Einstein's Zionist arrival speech—much to Einstein's annoyance. This speech was covered in *The New York Times* in a story which began on the front page and spilled over onto page 13, on 3 April 1921, reprinted herein. The interview in the *New York Times Book Review* was arranged for Einstein to promote his book, and to raise money for Zionists, not for Einstein to babble and boast.

But why, in contrast to his pro-American attitude in that interview, was Einstein so bitter after he had left America? The Zionists quibbled among themselves in America and the trip turned out to be a disappointment for them. The American Zionists wanted to proceed slowly and to maintain the bonds Jews had to the many nations of the world. Few wanted to venture from their comfortable mansions in America to tame the deserts of Palestine. European Zionists were more militant and isolationist, and resented the fact that masses of Jews could not be persuaded to voluntarily emigrate to Palestine.

6.5.1 While Zionists and Sycophants Hailed Einstein, Most

---

340. *See, for example,* J. Goebbels, "Der Führer", *Aufsätze auf der Kampfzeit*, Zentralverlag der NSDAP, Munich, (1935), pp. 214-216; **and** "Goldene Worte für einen Diktator und für solche, die es werden wollen", *Der Angriff*, (1 September 1932); reprinted in: *Wetterleuchten: Aufsätze aus der Kampfzeit*, Zentralverlag der NSDAP., Franz Eher Nachf., München, (1939), pp. 325-327. On the Zionists' quest to find a "great man" to be their "dictator", *see:* N. Goldman, "Zionismus und nationale Bewegung", *Der Jude*, Volume 5, Number 4, (1920-1921), pp. 237-242, at 240-242; which was part of a series including: "Zionismus und nationale Bewegung", *Der Jude*, Volume 5, Number 1, (1920-1921), pp. 45-47; and "Zionismus und nationale Bewegung", *Der Jude*, Volume 5, Number 7, (1920-1921), pp. 423-425.

341. *Cf.* Schlomo Ginossar, a. k. a. Simon Ginsburg, a. k. a. Salomon Ginzberg, "Early Days", *The Hebrew University of Jerusalem, 1925-1950*, Universitah ha-'uvrit bi-Yerushalayim, Jerusalem, (1950), pp. 71-74, at 73. *See also:* J. Stachel, *Einstein from 'B' to 'Z'*, Birkhäuser, Boston, (2002), p. 79, note 41.

Scientists Rejected Him and "His" Theories

In addition to Zionist strife and infighting, which caused Einstein problems during his trip to America, Einstein's scientific work was not so well-received, nor so perfect, as his present day advocates would have us believe. As a result, Albert Einstein had quite a rough time in America, where he was again and again challenged for his plagiarism and for his irrationality.[342] The same was true in Germany. The same was true in England. Louis Essen wrote,

> "But there have always been its critics: Rutherford treated it as a joke: Soddy called it a swindle: Bertrand Russell suggested that it was all contained in the Lorentz transformation equations and many scientists commented on its contradictions. These adverse opinions, together with the fact that the small effects predicted by the theory were becoming of significance to the definition of the unit of atomic time, prompted me to study Einstein's paper. I found that it was written in imprecise language, that one assumption was in two contradictory forms and that it contained two serious errors."[343]

John T. Blankart stated in 1921,

> "The 'Kinertia' articles offer food for thought when considered in connection with the colossal claims made by Einstein's supporters concerning his almost super-human originality. In fact, one begins to doubt the justice of these claims and to wonder if the charges made by a fast growing group of German scientists who, like E. Gehrcke, P. Lenard, and Paul Weyland, hold that Einstein is both a plagiarist and a sophist, are not, after all, true. We have done little justice in the above to the rare dialectic skill with which Dr. Einstein has applied his intellectual anæsthesia to the minds of his readers. All intellectual obstructions have been removed, and the reader is prepared to venture forth boldly into the mysterious realm of 'curved' space *whose geometrical properties depend upon the matter present*. This most curious inference of Einstein is the master stroke in his

---

[342]. *See, for example:* A. Lynch, *The Case Against Einstein*, P. Allan, London, (1932). H. Dingler, *Die Grundlagen der Physik; synthetische Prinzipien der mathematischen Naturphilosophie*, Second Edition, Walter de Gruyter & Co., Berlin, (1923); **and** *Physik und Hypothese Versuch einer induktiven Wissenschaftslehre nebst einer kritischen Analyse der Fundamente der Relativitätstheorie*, Walter de Gruyter & Co., Berlin, Leipzig, (1921); **and** "Kritische Bemerkungen zu den Grundlagen der Relativitätstheorie", *Physikalische Zeitschrift*, Volume 21, (1920), pp. 668-669. H. Nordenson, *Relativity, Time and Reality: A Critical Investigation of the Einstein Theory of Relativity from a Logical Point of View*, Allen and Unwin, London, (1969).
[343]. L. Essen, "Relativity — Joke or Swindle?", *Electronics and Wireless World*, (February, 1988), pp. 126-127. URL:
<http://www.cfpf.org.uk/articles/scientists/essen.html>

skillful massing of inconsistent sophistries."[344]

Einstein once asked,

"Do I have something of a charlatan or a hypnotist about me that draws people like a circus clown?"[345]

Paul Weyland[346] and Ernst Gehrcke[347] proved that Einstein's rise to fame was a "mass suggestion" fed by the insecurities of some of the authorities, and by the press, who would frequently misrepresent the facts, and misrepresented the views of many leading authorities, who were in reality mostly opposed to relativity theory. Weyland pointed out that Einstein obviously could not defend himself or "his" theories, because Einstein relied upon the *ad hominem* attack of calling his opponents "anti-Semites", instead of refuting their arguments in a rational manner.

Ernst Gehrcke and Stjepan Mohorovičić pointed out that Einstein rose to prominence, not because "his" theories were sound, but rather because his hangers-on, his connections in the press, and his racist smears intimidated the scientific community and deliberately inhibited the debate, with their frenzied personal attacks and their proven threats of violence, smears and career infringement against any who would question Einstein. Bruno Thüring, in 1941, stated that the acceptance of the theory of relativity resulted from a "mass psychosis" brought about by Jewish led propaganda, intimidation and the career infringement of anyone who opposed the dogmatism of Einstein.[348] Ernst Mach

---

344. J. T. Blankart, "Relativity or Interdependence", *Catholic World*, Volume 112, (February, 1921), pp. 588- 610, at 606.
345. K. Sugimoto, translated by B. Harshav, *Albert Einstein, A Photographic Biography*, Schocken Books, New York, (1989), p. 74.
346. P. Weyland, "Einsteins Relativitätstheorie—eine wissenschaftliche Massensuggestion", *Tägliche Rundschau*, (August 6, 1920).
347. E. Gehrcke, "Die gegen die Relativitätstheorie erhobenen Einwände", *Die Naturwissenschaften*, Volume 1, Number 3, (1 January 1913), pp. 62-66; republished *Kritik der Relativitätstheorie*, Hermann Meusser, Berlin, (1924), pp. 20-28, "Massensuggestion" appears at page 28; Gerhcke also delivered a lecture, which Einstein attended, on 24 August 1920 in the Berlin Philharmonic, *Die Relativitätstheorie eine wissenschaftliche Massensuggestion*, Arbeitsgemeinschaft Deutscher Naturforscher zur Erhaltung reiner Wissenschaft, Berlin, (1920); republished in *Kritik der Relativitätstheorie*, pp. 54-68. See also: *Die Massensuggestion der Relativitätstheorie*, Hermann Meusser, Berlin, (1924). The editors of *The Collected Papers of Albert Einstein*, Volume 7, Princeton University Press, (2002), p. 102; cite the earliest appearance by Gehrcke of this charge as: E. Gehrcke, Ed., P. Drude, *Lehrbuch der Optik*, Third Edition, S. Hirzel, Leipzig, (1912), p. 470. Reference is also had to this fact in H. Goenner, "The Reaction to Relativity Theory. I: The Anti-Einstein Campaign in Germany in 1920", *Science in Context*, Volume 6, Number 1, (1993), pp. 107-133.
348. B. Thüring, "Albert Einsteins Umsturzversuch der Physik und seine inneren Möglichkeiten und Ursachen", *Forschungen zur Judenfrage*, Volume 4, (1940), pp. 134-162. Republished as: *Albert Einsteins Umsturzversuch der Physik und seine inneren*

considered Einstein a charlatan, and Mach, too, categorized the theory of relativity as a "mass suggestion"—even before the terrible hype of the 1919 eclipse observations.

We know Mach's opinion from a letter which Čeněk Dvořák wrote to Mach on 19 August 1915,

> "The best contemporary physicists would agree with you about the exaggerated speculation, mass suggestion, and modish tendencies in modern physics."[349]

Arvid Reuterdahl was quoted in the *Minneapolis Sunday Tribune* on 20 November 1921, after Einstein's humiliating departure from America,

"Einstein Foes Prove Theory False Claim

Twin Cities Mathematical Association Hears Talk on Relativity.

Former Exponents Are Now Sorry, Says St. Thomas Engineering Dean.

Einstein's theory of relativity, which created a stir in the scientific world when first promulgated, is rapidly being rejected by the leading scholars of Europe and America. Prof. Arvid Reuterdahl told members of the Twin Cities Mathematical association last night at the Minnesota Union, University of Minnesota.

Professor Reuterdahl, who has been a vigorous opponent of Einsteinism since its inception, is dean of the department of engineering and architecture at St. Thomas college.

'Seething in Revolt.'

'It is literally true that Europe is seething in revolt against the yoke of Einsteinism,' Professor Reuterdahl declared. 'The eminent thinkers of Europe emphatically object to the steam roller methods used by the Einsteinian propagandists.

'The affair of Einstein was overdone and as a result the entire world is united, not only against a palpable fallacy, but also against the questionable methods by which this fallacy was flaunted before an unsuspecting public

---

Möglichkeiten und Ursachen, Dr. Georg Lüttke Verlag, Berlin, (1941).
[349]. Č. Dvořák quoted in J. T. Blackmore, *Ernst Mach: His Work, Life, and Influence*, University of California Press, Berkeley, Los Angeles, London, (1972), p. 279.

as a super-truth.'

A score of eminent scientists of both Europe and America were named by Professor Reuterdahl as actively opposed to the Einstein theory.

'Even in England where Einsteinism has been firmly entrenched since the findings of the English polar expedition were made known, the rebellion is gaining strength,' he said. 'In the front rank of the English expedition we find Prof. W. D. Ross of Oxford, and the celebrated mathematicians Gaynor and Whitehead.'

Professor Reuterdahl asserted that the leading astronomers of the United States are now either directly denying the truth of Einstein's theory or openly doubting the correctness of its contentions.

Majority Opposed.

'It is no longer an intellectual misdemeanor to doubt the validity of his speculations,' he said, "Undoubtedly the great majority of American scientists are today solidly opposed to the theories of Einstein. Many of those scientists who succumbed to the mass psychology of his trumpet blasts now sincerely wish that they had remained discreetly neutral.

Doctor T. J. J. See, professor of mathematics, United States navy, and director of the Mare island observatory, California, was said by Professor Reuterdahl to be one of the leading opponents of the theory in America.

'It is truly a sad ending to a perfect Einsteinian day,' he said, 'A camouflaged formula successfully used to gather renown is finally shown by an American scientist to be contrary to that great law which serves as the basic foundation of the entire structure of science.'"

The *Minneapolis Evening Tribune* of 5 May 1921 wrote,

"Scientists Rally to Support Reuterdahl in Fight on Einstein

Mysterious 'Kinertia' Attacks Theory and Thanks Minnesota Man.

'Fantastic Jazz of Mathematical Symbols,' Says Dr. S. P. Skidmore.

American scientists are rallying to the support of Professor Arvid Reuterdahl of St. Thomas college in his fight against Doctor Albert Einstein, including the mysterious 'Kinertia,' to whom Professor Reuterdahl gives credit for originating the theory of relativity.

Professor Reuterdahl has received a statement signed by 'Kinertia,' through an intermediary in New York, in which the scientist again attacks Einsteinism and thanks the St. Thomas dean for his efforts to prove the theory false.

All Write to Reuterdahl.

Doctor Sydney P. Skidmore of Philadelphia, Dr. W. E. Glanville, noted astronomer of Baltimore, and Dr. Robert P. Browne, author of 'Mystery of Space,' are others who have communicated with Professor Reuterdahl.

Doctor Skidmore says:

'Einsteinism is a fantastic jazz of mathematical symbols, devoid of quanta, in a dance hall floored by a parquetry of ifs, supposings and assumptions, and has no application to anything in the realm of objective truth.'

Doctor Glanville likens the Einstein theory to a newly discovered drug which is brought forth and acclaimed as a universal scientific panacea. He also compares Einsteinism to a great deflated scientific bubble. Doctor Brown assures Professor Reuterdahl that he will be allied in the fight 'against the mathematical usurpations of Einstein and relativity.'

Doubts Efficiency of Test.

'In the critical portion of the article just sent me by 'Kinertia' he points out some of the outstanding errors in Einstein's theory,' said Reuterdahl today. 'He expresses serious doubt that the solar spectrum test proposed by Einstein to prove his theory will be confirmative in its result. 'Kinertia' states:

"'In dynamics, acceleration and weight are not forces or physical causes. This is the dangerous ground Einstein assumes in his apparent anxiety to relegate forces to the waste basket because they disappear in the parallelogram law; he proposes to substitute uniform antecedents in place of natural causation.'

"'Kinertia,' moreover, demands that Einstein be consistent in his application of the motion of acceleration. In order to be consistent, 'Kinertia' holds, Einstein must develop a law which provides that bodies at the earth's surface be pushed from its center with the same acceleration with which falling bodies are apparently drawn toward it.

What Differentials Show.

"'Kinertia' further states:

'Einstein's differentials only show that either case would suffice if the acceleration was the same.'

He concludes his article with this pertinent statement:

'Science wants more than agnosticism; it wants to know the absolute truth, before accepting any such theory; even if d'Alembert's static ghost is dressed in Hamiltonian functions.'"

Hubert Goenner contended that,

"Also, a majority of theoretical physicists in Germany moved away from a theory with little potential for experiments and testable consequences."[350]

This view is supported by the record, for example the *St. Paul Dispatch* wrote on 3 April 1921, that Einstein had run away from Germany to America to hide from his critics,

"EINSTEIN ON RUN, SAYS LETTER TO REUTERDAHL
 Albert Einstein, denounced by the opponents of his alleged 'discoveries,' is on the run, according to a letter received from Dr. Hermann Fricke, physicist and astronomer of Berlin, dated August 19, by Prof. Arvid Reuterdahl, dean of the department of engineering and architecture at St. Thomas college, and author of 'Einstein and the New Science,' an attack on the Einstein theory, recently published.
 Einstein's popularity has waned, the Berlin scientist writes, and he says also that a large edition of Prof. Reuterdahl's book is to be published in the German capital.
 Dr. J. G. A. Goedhart, astronomer at Amsterdam, writes that Einstein has left Germany and has taken a professorship at the university in Leiden Holland. Circulation of Prof. Reuterdahl's book in Holland, and also in Sweden, is to he undertaken by foreign scientists opposed to the Einstein theory."

Nobel Prize laureate Johannes Stark wrote in 1922,

"Vorwort
 Die deutsche Physik macht gegenwärtig eine Krisis durch. Es kämpfen in ihr zwei Richtungen miteinander. Einsteins und durch den Dogmatismus der Quantentheorie hat eine theoretische Richtung einen beherrschenden Einfluß gewonnen, welcher die physikalische Wissenschaft grundsätzlich zu schädigen begonnen hat, indem sie deren Quellen mehr in der gedanklichen Konstruktion als in der Erfahrung sucht und diese zur Dienerin der Formel machen will. Ihr gegenüber findet sich die experimentelle Richtung in der Verteidigungsstellung; sie sieht die Quelle der Physik in der Beobachtung und Messung und in der Theorie ein heuristisches und systematisches Hilfsmittel für die Gewinnung und Darstellung der physikalischen Erkenntnis. Es kann kein Zweifel darüber bestehen, welche Richtung schließlich die Oberhand gewinnen wird. Die vorliegende Schrift hat die Aufgabe, durch die rückhaltlose Kritik von experimenteller Seite her die Entwicklung der gegenwärtigen Krisis in der deutschen Physik zu beschleunigen.
 [***]

---

[350]. H. Goenner, "The Reaction to Relativity Theory in Germany, III: 'A Hundred Authors against Einstein'", J. Earman, M. Janssen, J. D. Norton, Editors, *The Attraction of Gravitation: New Studies in the History of General Relativity*, Birkhäuser, Boston, Basel, Berlin, (1993), p. 249.

Die vorstehenden Ausführungen über das Verhältnis der physikalischen Theorie zur Erfahrung enthalten nichts Neues und in späterer Zeit mag einem Leser ihre Wiederholung als überflüssig erscheinen. In der gegenwärtigen Zeit ist es aber gegenüber dem anspruchsvollen Auftreten moderner Theorien notwendig, an sie zu erinnern. Für die Überschätzung der Theorie und die Unterschätzung der Beobachtung ist ein Ausspruch Einsteins, des Schöpfers der allgemeinen Relativitätstheorie, kennzeichnend. Anfangs dieses Jahres hielt Herr Einstein in Berlin vor einem auserwählten Kreis von Wissenschaftern, Wirtschaftern und Politikern einen Vortrag über die neuere Entwicklung der physikalischen Forschung. Gegen den Schluß desselben äußerte er sich zusammenfassend über die Quantentheorie des Atoms folgendermaßen: man dürfe erwarten, daß die Theorie bald imstande sein werde, die Eigenschaften der chemischen Atome und ihre Reaktionen vorauszuberechnen, so daß sich die mühevollen zeitraubenden experimentellen Arbeiten der Chemiker erübrigen würden. Als ich diese lobpreisende Überschätzung der Theorie mitanhörte, mußte ich aus Höflichkeit gegen den Gastgeber an mich halten, um nicht in Lachen auszubrechen. Aber danach war ich über die Leichtfertigkeit empört, mit welcher Herr Einstein, der von dem breiten Publikum herausgestellt wird, eine Auffassung verbreitet, welche auf die Dauer großen Schaden stiften muß. Herr Einstein sollte sich einmal eingehender mit der Erfahrung der anorganischen und organischen Chemie befassen, dann würde ihm klar werden, wie ungeheuer übertrieben seine theoretischen Erwartungen im Gebiete der Chemie sind und wie wenig gerade diese Wissenschaft die immer erneute Erfahrung entbehren kann. Es würde auch lehrreich für ihn sein, zu sehen, wie erstaunlich weit sich diese Wissenschaft fast allein auf Grund der Erfahrung ohne die mathematische Hilfe der Theorie entwickelt hat.

II. Die Stellung der allgemeinen Relativitätstheorie Einsteins in der Physik und die Propaganda für sie.

Wenn die Bedeutung einer Theorie proportional der Zahl der Abhandlungen, Bücher und Vorträge über sie wäre, so müßte die allgemeine Relativitätstheorie Einsteins als die weitaus bedeutendste Theorie aller Zeiten gewertet werden. Denn über keine Theorie in der Physik ist bisher von berufener und unberufener Seite soviel geschrieben und geredet worden wie über sie; es ist für sie seit Jahren in aller Welt sowohl in wissenschaftlichen Zeitschriften wie in Flugschriften und in der Tagespresse eine Propaganda getrieben worden, wie sie bisher unbekannt in der physikalischen Wissenschaft war. Diese Propaganda und der Einfluß des Einsteinschen Kreises ist in erster Linie für das Überwuchern der Theorie in der gegenwärtigen Physik, für die Unterschätzung der experimentellen Forschung und für die Vernachlässigung der angewandten Physik in Unterricht und Forschung verantwortlich zu machen. Mit Recht haben bereits Lenard [*Footnote:* P. Lenard, Über Relativitätstheorie, Äther, Gravitation, S. Hirzel, Leipzig 1921.] und Gehrcke Einspruch gegen die Fiktionen der Einsteinschen Relativitätstheorie erhoben und auch W. Wien [*Footnote:* W. Wien, Die Relativitätstheorie, Joh. Ambr. Barth, Leipzig 1921.] hat zu physikalischer Besinnung in dem Für und Wider um sie gemahnt. Aber Lenards und Gehrckes Kritik wurde von der Seite Einsteins als persönliche Beleidigung aufgefaßt und in unsachlicher Weise beantwortet. Und trotzdem die Auseinandersetzungen über die Einsteinsche Theorie auf der Nauheimer Naturforscherversammlung in persönlicher Hinsicht höchst unerquicklich und in sachlicher Hinsicht unfruchtbar waren, und obwohl seitdem kein unbestrittener Fortschritt in der experimentellen Prüfung der Theorie erfolgt ist, soll auf der diesjährigen Naturforscherversammlung in Leipzig die Einsteinsche Theorie wieder einem Kreise vorgeführt werden, der nur zu einem kleinen Teile aus Physikern besteht.

Bei dieser Lage der Dinge ist eine kritische Auseinandersetzung mit der allgemeinen Relativitätstheorie hinsichtlich ihrer physikalischen Bedeutung und der Propaganda für sie dringend geboten.

Von einer physikalischen Theorie ist zu verlangen, daß sie an ihre Spitze eine grundlegende Aussage über eine Beziehung zwischen physikalischen Größen stellt. So liegt der mechanischen Wärmetheorie der Gedanke von der wechselseitigen Umwandelbarkeit von Wärme und Arbeit zugrunde, der Maxwellschen Theorie der Gedanke der raumzeitlichen Verknüpfung von elektrischer und magnetischer Feldstärke. Welche grundlegende Aussage über eine zahlreiche Erscheinungen umfassende Beziehung zwischen physikalischen Größen stellt nun Einstein an die Spitze seiner allgemeinen Relativitätstheorie? Er selbst versteht unter „allgemeinem Relativitätsprinzip" die Behauptung: „Alle Bezugskörper func usw. sind für die Naturbeschreibung (Formulierung der allgemeinen Naturgesetze) gleichwertig, welches auch deren Bewegungszustand sein mag." An einer anderen Stelle derselben Schrift bezeichnet Einstein als exakte Formulierung seines allgemeinen Relativitätsprinzips folgende Aussage: „Alle Gaussschen Koordinatensysteme sind für die Formulierung der allgemeinen Naturgesetze prinzipiell gleichwertig."

Wie selbst der Nichtphysiker erkennt, macht das so formulierte allgemeine Relativitätsprinzip keine Aussage über eine Beziehung zwischen physikalischen Größen, sondern über die formal-mathematische Darstellung von physikalischen Gesetzen. Entsprechend seinem formal-mathematischen Grundgedanken ist es darum überhaupt nicht unter die physikalischen Theorien in dem oben umschriebenen Sinne zu rechnen, sondern gehört in das Grenzgebiet zwischen Physik, Mathematik und Erkenntnistheorie. In dem formal-mathematischen Grundgedanken der allgemeinen Relativitätstheorie ist es denn auch gelegen, daß Nichtphysiker, vor allem Erkenntnistheoretiker und Mathematiker, sie mit Eifer aufgegriffen und in zahlreichen Abhandlungen und dicken Schriften auf ihre

Weise ausgearbeitet haben. Wenn ich dieser Art Relativitätsliteratur, welche vorzügliche philosophische oder mathematische Leistungen darstellen mögen, jeglichen Wert für die physikalische Wissenschaft abspreche, so werde ich zwar von den Einsteinianern als armseliger Banause abgetan werden, dies kann mich aber nicht hindern, meinerseits als Physiker mein Urteil über die physikalische Bedeutung des allgemeinen Relativitätsprinzips zu bekennen und sogar folgende Blasphemie auszusprechen: Wäre Einstein mit seiner Theorie doch von Anfang unter die Mathematiker und Philosophen gegangen! Die deutsche Physik wäre dann vielleicht von dem lähmenden Gift des Gedankens verschont geblieben, man könne aus geistreichen Fiktionen („Gedankenexperimenten") mit Hilfe mathematischer Operationen physikalische Erkenntnisse oder, wie es in der Regel heißt, das „Weltbild" gewinnen.

Der Vorwurf der physikalischen Inhaltslosigkeit trifft die allgemeine Relativitätstheorie Einsteins ins Herz und diejenigen ihrer Verteidiger, welchen mein Urteil nicht von vornherein gleichgültig ist, werden sich beeilen mir entgegenzuhalten, daß die Relativitätstheorie doch zu bestimmten Folgerungen von sachlich-physikalischem Inhalt gelange, so zu einer Aussage über den Einfluß des Gravitationsfeldes auf die Lichtpflanzung und auf die optischen Eigenfrequenz chemischer Atome. Ist bis jetzt der Ausgangspunkt der Relativitätstheorie vom physikalischen Standpunkt aus beurteilt worden, so kommen wir mit der Antwort auf den vorstehenden Einwand zur physikalischen Beurteilung der methodischen Seite der Theorie. Ich gebe vorweg zu, daß die allgemeine Relativitätstheorie zu sachlich-physikalischen Folgerungen gelangt. Indes haben diese ihre Wurzel nicht allein in ihrem formal-mathematischen Grundgedanken, sondern auch in den sachlich-physikalischen Zutaten bei seiner mathematischen Verarbeitung, so vor allem in der Verknüpfung des Gravitationsfeldes mit der beschleunigten Bewegung und in der Verwertung der Tatsache von Proportionalität der schweren und der trägen Masse.

Die Art der Verarbeitung des Grundgedankens der Relativitätstheorie entspricht ebensowenig den an eine physikalische Theorie zu stellenden Forderungen wie ihr Grundgedanke selbst. Wie oben dargelegt wurde, ist eine physikalische Theorie in erster Linie für den experimentellen Physiker bestimmt; sie soll da, wo sie nicht seine Messungen zusammenfassend beschreibt, sondern Vorhersagen macht, auch für denjenigen Experimentalphysiker verständlich sein, welcher nicht die Kenntnisse des Fachmathematikers besitzt. Wie steht es in dieser Hinsicht mit Einsteins allgemeiner Relativitätstheorie? Zwar Einstein glaubte seine Theorie selbst dem Nichtphysiker verständlich machen zu können; seiner „gemeinverständlicher" Schrift [*Footnote:* A. Einstein, Über die spezielle und die allgemeine Relativitätstheorie, 51.-55. Tausend. F. Vieweg & Sohn, Braunschweig.] über sie, die in mehr als $50\,''00$ Stück verbreitet ist, schickt er nämlich folgende Sätze voraus: „Das vorliegende Büchlein soll solchen eine möglichst exakte Einsicht in die Relativitätstheorie vermitteln, die sich vom allgemein wissenschaftlichen, philosophischen Standpunkt für die Theorie interessieren, ohne den mathematischen Apparat der theoretischen Physik zu beherrschen. Die Lektüre setzt etwa Maturitätsbildung und — trotz der Kürze des Büchleins — ziemlich viel Geduld und Willenskraft beim Leser voraus."

Einstein war also der Meinung, daß für das Verständnis seiner Relativitätstheorie die Kenntnis der höheren Mathematik nicht nötig sei. In Wirklichkeit ist wohl noch keine Theorie in der physikalischen Literatur mitgeteilt worden, welche so schwer verständlich gewesen wäre wie die Einsteinsche Relativitätstheorie. Hierfür zeugt schon die Tatsache, daß man es für nötig hielt, sie durch zahlreiche Bücher selbst dem physikalischen und mathematischen Fachmann verständlich zu machen. Und auf der Seite ihrer Verteidiger hat man sich heute gegenüber der Kritik von Experimentalphysikern hinter die bequeme Ausrede zurückgezogen, sie besäßen nicht die höhere mathematische Bildung, welche zum Verständnis der allgemeinen Relativitätstheorie notwendig sei. Diejenigen Physiker, welche an ihr Kritik üben, verfügen nach ihnen nicht über dies nötige mathematische Begabung, um sie zu verstehen; sie werden gegenüber den Relativitätstheoretikern in eine tiefere Klasse verwiesen. Diese Behandlung ist selbst einem Physiker von den experimentellen Leistungen und mathematischen Kenntnissen Lenards von Seite Einsteins und seiner Anhänger widerfahren. Indes sprechen diese Theoretiker, welche so überlegen nicht bloß die höhere, sondern die höchste mathematische Bildung für das Verständnis der allgemeinen Relativitätstheorie fordern, dieser selbst das Urteil. Sie vergessen in Selbsteingenommenheit, daß die Theorie in der Physik nicht Selbstzweck, nicht allein für den Theoretiker und den Mathematiker da ist, sondern daß sie eine Hilfe für den Experimentalphysiker sein, seine Arbeit anregen oder formal abschließen soll. Für eine Theorie, welche dieser Forderung nicht genügt, sollte in physikalischen Zeitschriften kein Platz sein.

Die Übertreibung ins Abstrakte und Formale, die Beschränkung auf das intellektuelle Spiel mit den mathematischen Definitionen und Formeln kommt in der Einsteinschen Relativitätstheorie vor allem in der absichtlichen Ignorierung des Äthers zum Ausdruck. Gewiß kann man physikalische Beziehungen zwischen materiellen Körpern in mathematischen Formeln unter Absehen vom Äther zwischen ihnen darstellen. Wird aber damit der Begriff des Äthers überflüssig, wird damit die Tatsache der Existenz des Äthers aus der Welt geschafft? In einer der Ansprachen auf der Nauheimer Naturforscherversammlung wurde es von einem Nichtphysiker als eine naturwissenschaftliche Großtat Einsteins gefeiert, daß er den Äther abgeschafft habe. Soll man lachen über diese Wertschätzung einer vermeintlichen Großleistung Einsteins, oder soll man empört sein über die von seinen Fiktionen angerichtete Verwüstung. Nein, die gefeierte Abschaffung des Äthers durch Einstein ist nicht eine Großtat, sondern der Versuch zu einem verheerenden Rückschritt in der physikalischen Wissenschaft. Die Einführung des Äthersbegriffes in die Optik und in die Elektrodynamik, das anschauliche Denken mit ihm hat sich in der Physik als außerordentlich fruchtbar erwiesen; der Äther ist durch die physikalische Forschung eines Jahrhunderts aus einer Hypothese zu einer Tatsache geworden. Eine Physik ohne den Äther ist keine Physik. Einstein ist wohl selbst ob seiner Großtat der Abschaffung des Äthers bange geworden; denn in neuerer Zeit scheint er in einem Vortrag den Äther wieder einführen zu wollen, freilich ist es nicht der alte abgeschaffte Äther, sondern eine Art Einsteinscher Relativitätsäther.

Man mag nun zugeben, daß die allgemeine Relativitätstheorie weder in ihrem Grundgedanken noch in ihrer Entwicklung den Anforderungen genügt, welche von physikalischer Seite an eine physikalische Theorie zu stellen sind. Es könnte aber doch sein, daß ihr das große Verdienst zuzusprechen wäre, die Entdeckung neuer Erscheinungen veranlaßt zu haben und daß ihre Folgerungen experimentell bestätigt worden sind. Es ist darum zu prüfen, ob dies für die allgemeine Relativitätstheorie zutrifft. Drei Erscheinungen sind es, welche in dieser Hinsicht in Betracht kommen.

Da ist zunächst die Anomalie in der Perihelbewegung des Merkurs; sie war bereits vor Aufstellung der Relativitätstheorie aus der astronomischen Beobachtung bekannt. Ihr Betrag schien früher genau mit der Rechnung nach der Relativitätstheorie übereinzustimmen; dies ist indes nach einer kürzlich erschienenen Nachprüfung durch Großmann zum mindesten fraglich geworden. Aber selbst wenn die Übereinstimmung vorhanden wäre, könnte durch sie die Richtigkeit der Relativitätstheorie noch nicht als erwiesen gelten. Denn es gibt noch eine andere Möglichkeit (Annahme interplanetarer Massen) zur Deutung jener Anomalie.

Denn soll die Relativitätstheorie durch den Nachweis der Ablenkung des Fixsternlichtes beim Vorbeigang an der Sonne bestätigt worden sein. Es muß zugestanden werden, daß der Gedanke eines Einflusses des Gravitationsfeldes auf die Lichtbewegung ursprünglich und wertvoll ist. Es erfordert allerdings die geschichtliche Gerechtigkeit, die Priorität dieses Gedankens Soldner zuzuerkennen, der ihn bereits vor hundert Jahren, wenn auch auf Grund einer anderen Annahme über das Wesen des Lichtes zur Grundlage einer theoretischen Abhandlung in den Annalen der Physik und Chemie gemacht hat. Wie steht es aber mit der experimentellen Bestätigung dieser zweiten Folgerung der Relativitätstheorie? Bisher liegen nur Beobachtungen bei einer einzigen Sonnenfinsternis vor. Wer die für derartige Messungen notwendige Meßtechnik zu beurteilen und den Wert von Meßdaten, welche nahe der Grenze der Meßgenauigkeit liegen, abzuwägen versteht, der wird erklären, daß durch jene Beobachtungen lediglich wahrscheinlich gemacht ist, daß Lichtstrahlen, wenn sie nahe bei der Sonne verlaufen, aus ihrer anfänglichen Richtung etwas abgelenkt werden. Von einer quantitativen Bestätigung der allgemeinen Relativitätstheorie durch jene Beobachtungen kann jedoch nicht die Rede sein. Die Ablenkung von Lichtstrahlen in der Nähe der Sonne kann einen anderen Grund haben, als in der Relativitätstheorie angenommen wird.

Die dritte Folgerung der Relativitätstheorie behauptet, daß durch die Wirkung eines Gravitationsfeldes, z. B. durch dasjenige an der Sonne, die optischen Eigenfrequenzen der chemischen Atome etwas verkleinert, also die ihnen entsprechenden Spektrallinien etwas nach Rot verschoben werden. Die bis jetzt in dieser Hinsicht vorliegenden Messungen widersprechen sich in ihrem Ergebnis hinsichtlich der Relativitätstheorie. Amerikanische und deutsche Beobachter, welche mit einer guten Technik arbeiteten, erklären, daß eine Rotverschiebung von Sonnenlinien in dem von der Theorie geforderten Betrag nicht vorhanden ist. Wieder andere deutsche Beobachter und ein französischer behaupten, sie hätten die von Einstein gefolgerte Rotverschiebung der Sonnenlinien gefunden. Es steht also Behauptung wider Behauptung und es kann nur durch neue, mit besonderer Umsicht durchgeführte Messungen die Entscheidung gebracht werden. Diese neuen Messungen sollten ohne jegliche Voreingenommenheit für und wider die Theorie unternommen werden. Bei dem Lesen des Berichtes über sie sollte man nicht den Eindruck haben, daß sie in der Absicht durchgeführt und zurechtgemacht wurden, um die Theorie zu bestätigen. Und der spektralanalytische Fachmann wird mit Zurückhaltung und theoretischer Skepsis an die Deutung einer geringen Verschiebung von Linien im Sonnenspektrum gegenüber ihrer Lage im Spektrum irdischer Lichtquellen herangehen. Weiß er doch, daß es eine Reihe von Wirkungen gibt, welche geringe Verschiebungen von Spektrallinien hervorbringen, und da uns die Bedingungen an der Oberfläche der Sonne nicht genügend bekannt sind, so wird er an die Beweisführung zugunsten einer besonderen Wirkung hohe Anforderungen stellen.

In keinem der drei Fälle, welche in der Regel als Beweise für die Richtigkeit der allgemeinen Relativitätstheorie angeführt werden, liegen also die Verhältnisse so, daß ein vorsichtiger Physiker anerkennen könnte, daß die Richtigkeit der Einsteinschen Relativitätstheorie erwiesen sei; er kann höchstens zugeben, daß es nicht ausgeschlossen ist, daß weitere verfeinerte Messungen eine Übereinstimmung zwischen den Folgerungen der Theorie und den Beobachtungen ergeben. Und es kann der Relativitätstheorie darum noch nicht das Verdienst zugesprochen werden, neue Entdeckungen veranlaßt zu haben.

Bedenkt man, daß die „Bestätigung" der Einsteinschen Theorie noch aussteht, nimmt man dazu, daß ihr Grundgedanke formal-mathematisch ist und das Verständnis ihrer Entwicklung hohe mathematische Kenntnisse erfordert, so versteht man nicht, wie mit einer solchen Theorie eine so unerhörte Propaganda getrieben werden konnte, wie es bisher mit keiner anderen Theorie der Fall gewesen ist. Weit über den Kreis der wenigen physikalischen und mathematischen Fachleute hinaus, welche sie zu beurteilen vermögen, wurde sie dem urteilslosen Publikum in angeblich gemeinverständlichen Schriften, in der Tagespresse, in öffentlichen Vorträgen und im Salon als höchste und tiefste naturwissenschaftliche Weisheit angepriesen. Und zuletzt scheute man nicht einmal vor dem Unfug zurück, Illustrationen zur Relativitätstheorie im Film dem Kinopublikum vorführen zu lassen. Diese Propaganda fand in der Zeit der politischen und sozialen Revolution einen fruchtbaren Boden, redete sie doch von dem

Umsturz unserer bisherigen Anschauungen von Raum und Zeit und von einer die Welt umspannenden Theorie. Sie lag auch insofern dem Geiste der letzten Jahre, als ihre jüngsten Jünger mit großen Worten ihre Weisheit vortragen konnten, ohne auf die Wirklichkeit Rücksicht nehmen zu brauchen.

Einstein ist der Vorwurf nicht zu ersparen, daß er sich dem Hinauszerren seiner Theorie auf den Jahrmarkt nicht entgegengesetzt hat, die Propaganda seiner Freunde und Anhänger gewähren ließ, ja Schriften von Dilettanten zum Ruhme seiner Theorie ermunterte. Er mag es entrüstet zurückweisen, mit seinen Vortragsreisen ins Ausland selbst Propaganda für seine Theorie getrieben zu haben. Gut. Aber hinsichtlich seiner Auslandsreisen halte ich es für notwendig, daß ihm bei dieser Gelegenheit folgender Hinweis gegeben wird.

In einem Artikel im Berliner Tageblatt hat sich Einstein zu internationaler Gesinnung bekannt. Gleichwohl ist es nicht zu verstehen, daß er ohne Rücksicht auf die furchtbare Bedrückung des deutschen Volkes durch die Franzosen einer französischen Einladung zu einem Vortrag in Paris in diesem Frühjahre Folge geleistet, ja im Anschluß daran sogar darauf gehalten hat, auf einer Automobilfahrt sich die „verwüsteten" Gegenden (les régions dévastées) zeigen zu lassen. Einstein lebt doch in Deutschland, und ist Mitglied amtlicher deutscher Ausschüsse, vor allem Direktor eines Kaiser-Wilhelm-Instituts; da hätte er mit Rücksicht darauf soviel Takt haben müssen, die Reise nach Paris zu einer Zeit zu unterlassen, wo der französische Druck besonders stark war. Und wenn er dies nicht von selbst einsah, so hätten es ihm seine Freunde, die ihm sonst so rasch beispringen, bedeuten sollen. Daß über die Franzosenreise Einsteins große deutsche Tageszeitungen telegraphisch berichteten, daß sie nicht von selbst daran Kritik übten, ja nicht einmal einen Einspruch dagegen aus physikalischen Kreisen aufnahmen, ist ein trauriges Zeichen von dem deutschen Verfall.

Doch zurück zur Propaganda für die Relativitätstheorie! Während sie sich selbst keine Schranken setzte, nahmen Einstein und seine Anhänger sogar eine Kritik aus Fachkreisen sehr übel auf. So warf er Lenard, einem unserer tiefsten und gewissenhaftesten Denker, im Berliner Tageblatt (27. Aug. 1920) Oberflächlichkeit vor und Gehrckes [*Footnote:* Gehrcke ist der Kampf gegen die Relativitätstheorie übel bekommen; trotz seiner zahlreichen hervorragenden experimentellen Arbeiten wird er von Fakultäten nicht für ein physikalisches Ordinat vorgeschlagen.] Kritik unterstellte er unsachliche Motive.

Auch der Fernerstehende erkennt beim Lesen der vorstehenden Ausführungen, daß durch die allgemeine Relativitätstheorie ein Zwiespalt zwischen den Physikern aufgerissen worden ist. Experimentell gerichtete Physiker lehnen sich gegen den nach ihrer Meinung unphysikalischen Geist der Relativitätstheorie und gegen die maßlose Propaganda für sie auf; deren Anhänger werfen ihnen dafür Beschränktheit, Mangel an mathematischer Bildung oder gar unsachliche Motive vor. Ferner fühlt selbst der Fernerstehende, daß die experimentelle Begründung einer so umstrittenen

Theorie noch nicht gesichert sein kann und daß es unangebracht ist, eine Theorie, über welche selbst die physikalischen und mathematischen Fachleute noch im Streit liegen, vor den weiten Kreis der Laien bis herab zum Kinopublikum zu bringen.

Bei dieser Lage der Dinge muß es auf physikalischer Seite als ein bedauerlicher Mißgriff bezeichnet werden, daß für die Hundertjahr-Feier der Gesellschaft deutscher Naturforscher und Ärzte in diesem Jahre in Leipzig als Thema für die erste allgemeine Sitzung die Relativitätstheorie in Aussicht genommen wurde. Daß dies nach den Auseinandersetzungen in Nauheim geschehen konnte, ist, wie ich bereits bemerkte, ein Zeichen für das Überwuchern der Theorie. Man lasse uns Physiker endlich eine Zeitlang in Ruhe mit der bis zum Überdruß abgehandelten Einsteinschen Relativitätstheorie! Man warte endlich einige Jahre mit der Propaganda für sie, bis ihre Folgerung durch zuverlässige Beobachtungen geprüft sind!"[351]

The New York Times stated in 1923,

"It was reported in January from Berlin that fifty German physicists, mathematicians and philosophers were 'seriously grieved' to see public opinion misled by the suggestion that the Theory of Relativity is the solution of the problems of the universe, and by the concealment of the fact that many savants, 'including the most distinguished,' do not accept this theory as a proved hypothesis, but look upon it as fiction."[352]

This was quoted in a press release Thomas Jefferson Jackson See issued to the *Associated Press* on 18 April 1923. It appears to paraphrase a flier distributed at the meeting of the *Gesellschaft Deutscher Naturforscher und Ärtze* in Leipzig in 1922.[353] T. J. J. See concluded his press release with the rhetorical question,

"Under the circumstances is it any wonder that some of us who owe a duty of Truth to the Public, should be obliged to vigorously contest the unauthorized and indefensible conclusion that the observed refraction of starlight near the Sun is a confirmation of the discredited Doctrine of Relativity?"

See later published similar statements in *The San Francisco Journal* on 20 May 1923 in an article entitled, "Einstein a Second Dr. Cook?"

Ones sees that it wasn't just the Germans who were disgusted with Einstein, his theories, his self-promotion and his plagiarism. As Einstein himself

---

351. J. Stark, *Die gegenwärtige Krisis in der Deutschen Physik*, Johann Ambrosius Barth, Leipzig, (1922), Forward and pp. 6-16.
352. "Einstein's Triumph", *The New York Times*, (13 April 1923), p. 16.
353. *See:* C. Schönbeck, "Albert Einstein und Philipp Lenard", *Schriften der mathematisch-naturwissenschaftliche Klasse der Heidelberger Akademie der Wissenschaften*, Volume 8, (2000), pp. 1-42, at 37.

professed, it was only in America that his theories were generally accepted and where he was loved, a fact he found comical. Einstein made a scathing, ethnocentric, misogynist and hateful denouncement of America and American scientists.[354] However, in America, See, Reuterdahl and Poor wrote several articles exposing Einstein as a fraud. Each complained of censorship of their efforts to expose Einstein.

French savants had little love for Einstein. *The New York Times* reported on 4 April 1922 on page 21:

> "*Einstein Breaks Engagement In Paris, Fearing Hostility*
> PARIS, April 3.—Professor Albert Einstein of the University of Berlin, who recently delivered his first lecture here under the auspices of the College of France and had a notable reception, canceled an engagement to attend the session of the Academy of Sciences today in order to avoid a hostile manifestation.
> Some of the members of the academy had decided as a protest against his presence to rise and leave the hall as soon as he entered."

The New York Times reported on 5 April 1922 on page 21:

> "PLEDGED TO SNUB EINSTEIN.
> 30 French Scientists Would Have Left if He Had Attended Meeting.
> PARIS, April 4.—The failure of Professor Albert Einstein to pay his formal visit to the French Academy of Sciences yesterday was due to the fact that he had received a friendly warning that the occasion would be made embarrassing by a certain element of that distinguished body. This statement is in L'Oeuvre. Scoring French scientists for their unbelievable narrowness, L'Oeuvre declares that thirty members had pledged themselves, if Professor Einstein made his appearance, to leave the hall in a body."

The New York Times reported on 16 November 1922 on the front page that the Russians had condemned Einstein's theory:

> "*Einstein Theory 'Bourgeois' And Dangerous, Say Russians*
> PARIS, Nov. 15.—A message from Moscow to the Echo de Paris says that Professor Albert Einstein has been solemnly excommunicated by the Russian Communists.
> At a special council meeting held in order to examine the question the Russian Communist Party condemned the Einstein theory as being 'reactionary of nature, furnishing support for counter-revolutionary ideas'; also as being 'the product of the bourgeois class in decomposition.'
> Professor Timirazeff presented a long report to the council in which he discussed whether Einstein's theories could be reconciled with the theory of

---

[354].C. Brown, *The New York Times*, (8 July 1921), p. 9.

materialism. He decided that they could not, and because, in his opinion, they led to 'pure idealism,' the council pronounced condemnation."

Irving Levy published the following comment in *The New York Times*, on 2 March 1936, page 16,

"The relativity theory advanced by Professor Einstein is held in such uncomprehending awe by the vast majority of people that it is not generally known there exists a far from unanimous acceptance of it in the scientific world."

So we see that, contrary to the popular history told today, Einstein was internationally known as a sophist and a plagiarist when he came to America in 1921. Einstein tried to head off any criticism he might face in America by stigmatizing any criticism of him, or of the theory of relativity as if "anti-Semitism" *per se* before he even stepped off the boat onto America's shores.[355] He was a coward who hid behind the power of Jewish tribalism.

6.5.2 Hypocritical and Cowardly Einstein Plays the "Race Card" and Cripples Scientific Progress

Like his cowardly Zionist comrades, hypocritical Einstein "played the race card." In an effort to change the subject from his plagiarism and fallacious theories, which subject was beginning to destroy his fame, Einstein smeared anyone and everyone who would dare question him or the theory of relativity as if an anti-Semite *per se* in *The New York Times* on 3 April 1921 on pages 1 and 13, and bear in mind that *The New York Times*, itself, reported that relativity theory was "much-debated",

"PROF. EINSTEIN HERE, EXPLAINS RELATIVITY

'Poet in Science' Says It Is a
Theory of Space and Time,
But It Baffles Reporters.

SEEKS AID FOR PALESTINE

Thousands Wait Four Hours to
Welcome Theorist and His
Party to America.

---

[355]. "Prof. Einstein Here, Explains Relativity", *The New York Times*, (3 April 1921), pp. 1, 13, at 1.

A man in a faded gray raincoat and a flopping black felt hat that nearly concealed the gray hair that straggled over his ears stood on the boat deck of the steamship Rotterdam yesterday, timidly facing a battery of cameramen. In one hand he clutched a shiny briar pipe and with the other clung to a precious violin. He looked like an artist—a musician. He was.

But underneath his shaggy locks was a scientific mind whose deductions have staggered the ablest intellects of Europe. One of his traveling companions described him as an 'intuitive physicist' whose speculative imagination is so vast that it senses great natural laws long before the reasoning faculty grasps and defines them.

The man was Dr. Albert Einstein, propounder of the much-debated theory of relativity that has given the world a new conception of space, and time and the size of the universe.

Dr. Einstein comes to this country as one of a group of prominent Jews who are advocating the Zionist movement and hope to get financial aid and encouragement for the rebuilding of Palestine and the founding of a Jewish university. He is of medium height, with strongly built shoulders, but an air of fragility and self-effacement. Under a high, broad forehead are large and luminous eyes, almost childlike in their simplicity and unworldliness.

Thousands Welcome Him.

With him as fellow-travelers were Professor Chaim Weizmann, President of the Zionist World Organization, discoverer of trinitrotoluol, and head of the British Admiralty laboratories during the war; Michael Ussichkin, a member of the Zionist delegation to the Paris Peace Conference and now Resident Chairman of the Zionist Commission in Palestine, and Dr. Benzion Mossinson, President of the Hebrew Teachers Organization in Palestine.

The party was welcomed at the Battery by thousands of fellow-Jews who had waited there for hours.

The crowds were packed deeply along the Battery wall, waving Jewish flags of white with two blue bars, wearing buttons with Zionist inscriptions, and cheering themselves hoarse as the police boat John F. Hylan drew near. Dozens of automobiles were parked near the landing, and when the welcoming committee and the visitors had entered them they started uptown to the Hotel Commodore, preceded by a police escort. They turned into Second Avenue, where the sidewalks were lined nearly all the way uptown with thousands who waved hands and handkerchiefs and shouted welcome to the visitors.

Professor Einstein was reluctant to talk about relativity, but when he did speak he said most of the opposition to his theories was the result of strong anti-Semitic feeling. He was amused at attempts by reporters to get some idea of his theory by questioning him, and he did his best to make his answers as simple as possible. He spoke through an interpreter.

A Theory of Space and Time.

The interview took place in the Captain's cabin, where Professor Einstein was almost surrounded by seekers after knowledge. He was asked to define his theory.

'It is a theory of space and time, so far as physics are concerned,' he said.

'How long did it take you to conceive your theory?" he was asked.

'I have not finished yet,' he said with a laugh. 'But I have worked on it for about sixteen years. The theory consists of two grades or steps. On one I have been working for about six years and on the other about eight or nine years.

'I first became interested in it through the question of the distribution and expansion of light in space; that is, for the first grade or step. The fact that an iron ball and a wooden ball fall to the ground at the same speed was perhaps the reason which prompted me to take the second step.'

He was asked about those who oppose his theory, and said:

'No man of culture or knowledge has any animosity toward my theories. Even the physicists opposed to the theory are animated by political motives.'

When asked what he meant, he said he referred to anti-Semitic feeling. He would not elaborate on this subject, but said the attacks in Berlin were entirely anti-Semitic.

Dr. Einstein said the theory was a step in the further development of the Newtonian theory. He hoped to lecture at Princeton on relativity before he left the country, he said, as he felt grateful to the Faculty of Princeton, which was the first college to become interested in his work.

Poses for Moving Picture Men.

As the questioners gave up their attempts to seek further elucidation of the Einstein principles, the professor laughed and said:

'Well, I hope I have passed my examination.'

Professor Einstein's interview came soon after he had escaped the moving picture men. As they ground away at their machines, ordering him about, he seemed at first bewildered, then amused. He posed with other members of his party and with Mrs. Einstein for nearly half an hour, and then almost ran away, shaking his head in exasperation and refusing to do any more.

'Like a prima donna,' he exclaimed.

'He does not like to be, what you call it, a showcase,' said Mrs. Einstein. 'He does not like society, for he feels that he is on exhibition. He would rather work and play his violin and walk in the woods.'

'Do you understand his theory?' Mrs. Einstein was asked.

'Oh, no,' she said, laughing, 'although he has explained it to me so many times. I understand it in a general way, but in its details it is too much for a woman to grasp. But it is not necessary for my happiness.'

Dr. Einstein was an inspirational worker, she said. When he was engaged on some problem, 'there was no day and no night,' but in his periods of relaxation he went for weeks without doing anything in particular but dream and play on his violin. Whenever he became weary in the midst of his work he went to the piano or picked up his violin and rested his mind with music.

'He improvises,' she explained. 'He is really an excellent musician.'

Mozart and Brahms His Favorites.

On the ship, when a concert was held Dr. Einstein played selections from Mozart, of whose work he is particularly fond, on the violin. Brahms is another of his favorites.

'I never met Professor Einstein before this voyage,' said Professor Weizmann, who is a great admirer of his fellow-scientist. 'He has a singularly sweet and lovable nature, and is exceedingly simple in his habits of life. I have talked with him many times about his work, and he is glad to speak of it when he can find some one who is interested and at least partly capable of understanding it. I do not entirely, for when I get beyond the atom I am lost.

'When he was called 'a poet in science' the definition was a good one. He seems more an intuitive physicist, however. He is not an experimental physicist, and although he is able to detect fallacies in the conceptions of physical science, he must turn his general outlines of theory over to some one else to work out. That would be readily understandable to a man of science. He first became interested in mathematics when he was 14 years old, and his work is his life. He spends most of his time reading and thinking when he is not playing his violin.'

Professor Weizmann also is accompanied by his wife. He and the other Zionist visitors, during their visit of several weeks, will endeavor to interest American Jews in the Zionist movement and obtain money and moral support for both the national Zionist idea and for the university.

Dr. Weizmann Explains Mission.

'It is a great satisfaction to me as President of the Zionist Organization to find myself for the first time in the Union States,' said Dr. Weizmann. 'The cause of the Jewish national home in Palestine has from the first appealed to the generous instincts of the American people and owes much to the sympathetic support it has consistently received from leaders of public opinion in the United States.

'Our primary object is to confer with the American Zionists who have, under the distinguished leadership of Justice Brandeis, Judge Mack and other representative American Jews, rendered invaluable services to the Zionist movement during the past few critical years. In the task of reconstruction in Palestine, for which the time has now arrived, it is confidently expected that the American Zionists will play an equally conspicuous and honorable part. In this connection we hope to enlist the active interest of American Jews in the Keren Hayesod, or Foundation Fund, the central fund for the building up of the Jewish National Home, to which Jews throughout the world are being called upon to contribute to the utmost limit of their resources.

'Professor Einstein has done us the honor of accompanying us to America in the interest of the Hebrew University of Jerusalem. Zionists have long cherished the hope of creating in Jerusalem a centre of learning in which the Hebrew genius shall find full self-expression and which shall play its part as interpreter between the Eastern and Western worlds. Professor Einstein attaches the utmost importance to the early inauguration of the Jerusalem university and is prepared when the time arrives personally to associated himself within its activities—a course in which there is reason to hope he will be followed by other Jewish scholars and scientists of world-wide reputation.'

Einstein to Work for University.

Professor Einstein will devote most of his time while here to advocating support of the university by American Jews.

'The establishment of such a university has been for a long time one of the most cherished plans of the Zionist organization,' he said. 'But for the outbreak of the war it would have materialized in 1914, when a site was actually purchased on the Mount of Olives. In 1918 the foundation stone was laid by Dr. Weizmann. Since then the university site has been extended and a building purchased in which it will be possible for a beginning to be made. There is also a library of 30,000 volumes which is rapidly growing.

'Plans have been worked out both for the complete university of the future and for a comparatively modest beginning. The time has now come to insure the immediate realization of the latter. Such is the importance attached by the Zionist Organization to the spiritual values in the Zionist national home that even at this moment, when the organization is faced with tremendous tasks of immigration and colonization, and is concentrating all efforts upon the Palestine Foundation Fund, an exception is made in favor of the university to which a special branch of the fund is devoted.

'I know of no public event which has given me such delight as the proposal to establish a Hebrew university in Jerusalem. The traditional respect for knowledge which Jews have maintained intact through many centuries of severe hardship made it particularly painful for us to see so many talented sons of the Jewish people cut off from higher education and study, and knocking vainly at the doors of universities of Eastern and Central Europe.

Home For Spiritual Life.

'Others who have gained access to the regions of free research only did so by undergoing a painful, even dishonoring, process of assimilation which crippled and robbed them again and again of their cultural leaders. The time has now come when our spiritual life will find a home of its own. Distinguished Jewish scholars in all branches of learning are waiting to go to Jerusalem, where they will lay the foundation of a flourishing spiritual life and will promote the intellectual and economic development of Palestine.

'Notwithstanding the crude political realism of our times and the materialistic atmosphere in which it has enveloped us, there are visible none the less glimmerings of a nobler conception of human aspirations, such as were expressed in the part played by the American people in world politics. And so we come from sick and suffering Europe with feelings of hope, being convinced that our spiritual aims will command the full sympathy of the American nation and will receive enthusiastic approval and powerful support from our Jewish brethren in the United States.'

The Zionists were met down the bay by a delegation from the Mayor's committee of welcome, Captain Abraham Tulin, who served as American liason officer with General Mangin's army in the war; Dr. Schmarya Levin, who was member of the first Russian Duma and of the Cadet Party in Russia, and Magistrate Bernard Rosenblatt. They were delayed by the quarantine examination and were not able to board the Rotterdam until nearly 1 o'clock. On the way up the bay they had lunch with Professor Einstein, Professor Weizmann and others in the party, and remained with them on the ship until sundown. As it was the Sabbath their religion prevented them from leaving until that time.

Crowd Waits Four Hours at Pier.

At the pier were several hundred welcomers, although the ship was more than four hours late in reaching her pier. They gave the Zionists a rousing welcome before they went aboard the police boat John F. Hylan, which landed them at the Battery. The boat flew the Jewish flag in honor of the party. On board were L. Lipsky, Secretary of the Zionist Organization of America; L. Robison of the National Executive Committee; B. G. Richards, Secretary of the American Jewish Congress; M. Rothenberg, Chairman of the American Jewish Congress; J. Fishman, managing editor of The Jewish Morning Journal; W. Edlin, editor of The Day; Rabbi M. Berlin; David Pinski, editor of Die Zeit; John F. Sinnott, Secretary to Mayor Hylan; Henry H. Klein, Commissioner of Accounts; Judge Gustave Hartman, the Rev. H. Masliansky, Judge Jacob S. Strahl and many others.

An official meeting of welcome will be held at the City Hall on Tuesday at which Mayor Hylan, Frank L. Polk, George W. Wickersham, Magistrate Rosenblatt, Professor Einstein and Professor Weizmann will speak.

Among those on the Committee of Welcome are Nathan Straus, Arthur Brisbane, Chancellor E. E. Brown, Judge Benjamin Cardoza, Abram I. Elkus, James A. Foley, F. H. LaGuardia, Justice Samuel Greenbaum, William D. Guthrie, Mrs. William R. Hearst, Adolph Lewisohn, Alfred E. Smith, Leon Kaimaky, Judge Otto A. Rosalsky, Benjamin Schlessinger, Oscar S. Straus, Senator Nathan Straus Jr., Marcus Loew, Dr. Bernard Flexner, Colonel Robert Grier Monroe, Herman Bernstein, Samuel Koenig and George Gordon Battle.

A meeting also will be held at the Metropolitan Opera House on April 10. Professor Einstein will not touch on relativity at these meetings, but it is expected that before he leaves the city he will speak before some scientific gathering, at which he will discuss his discovery."

Einstein prevented an uninhibited debate over the merits of the theory of relativity. His shrill cries of "anti-Semitism" had a chilling effect, which froze Twentieth Century Physics in a mythology of metaphysical "Space-Time" and physical gravitation via mathematical abstraction and imaginary dimensions.

The Chicago Tribune reported on 3 April 1921 on page 5 (and note that Einstein was careful to not offend the lovers of Newton as was done in 1919),

"EINSTEIN IN N. Y.; EVEN WIFE CAN'T GRASP THEORIES

Hopes to Lecture at Princeton, He Says.

New York, April 2.—[Special]—A man in a faded gray raincoat, topped off by a flopping black felt hat, which nearly concealed straggling gray hair that fell over his ears, stood on the boat deck of the steamship Rotterdam today, timidly facing a battery of camera men. In one hand he clutched a shiny briar pipe and the other clung to a violin.

Dr. Albert Einstein, discoverer of the famous theory of relativity, which has given the world a new conception of space and time, looks like a musician, and he is.

Dr. Einstein comes to this country as one of a group of prominent Jews, advocating the Zionist movement. They hope to get financial aid and encouragement for the rebuilding of Palestine and the founding of a Jewish university.

Amused by Questions.

The scientist was reluctant to talk about relativity. He was greatly amused at the attempts of reporters to search out by their questions some idea of what his theory is, and did his best to make his answers as simple as possible. He does not speak English and answered through an interpreter.

'It is a theory of space and time, so far as physics are concerned,' he said.

'How long did it take you to conceive your theory?' he was asked.

'I have not finished yet,' he said with a laugh. 'But I have worked on it for about sixteen years. The theory consists of two grades or steps. On one I have been working for about six years and on the other about eight or nine years.'

Iron and Wooden Balls.

'I first became interested in it through the question of the distribution and expansion of light in space. That is, for the first grade or step. The fact that an iron ball and a wooden ball fall to the ground at the same speed was perhaps the reason which prompted me to take the second step.'

He was asked about those who opposed his theory, and said:

'No man of culture or knowledge has any animosity toward my theories. Even the physicists opposed to the theory are animated by political motives.'

Asked what he meant, he said he referred to anti-semitic feeling. He would not elaborate on this subject, but said that the attacks in Berlin were entirely anti-semitic.

Develops Newton's Theory.

Dr. Einstein said that the theory is a step in the further development of the Newtonian theory. He hopes to lecture at Princeton on relativity before he leaves the country, as he feels grateful to the faculty of Princeton, which was the first college to become interested in his work.

'Do you understand his theory,' Mrs. Einstein was asked.

'O, no,' she said, 'although he has explained it to me so many times. I understand it in a general way, but it is too subtle for a woman to grasp. Still it is not necessary for my happiness.'"

Einstein called anti-Semitic, among other things, the thesis of Gehrcke and Weyland that: Einstein's promotion mirrored Hans Christian Andersen's fairy tale *The Emperor's New Clothes*; that, the overblown public reaction to the theory of relativity was a "mass suggestion" and a "mass psychosis"; and Gehrcke and Weyland's criticism that theory of relativity had not been proven correct and was instead contradicted by St. John's experiments; and Gehrcke and Weyland's accusation that Einstein's theory, while promoted as a radically new development, was not a new idea, but was derived from Lorentz and others. Einstein, himself, had complained to Heinrich Zangger on 24 December 1919,

"[T]his business reminds one of the tale of 'The Emperor's New Clothes,' but it is harmless tomfoolery."[356]

Einstein endorsed and plagiarized Gehrcke and Weyland's other views, which he had called anti-Semitic in 1920, on 3 April 1921, and would again plagiarize Gehrcke and Weyland's ideas when Einstein returned to Europe and was again interviewed by the press. *The Chicago Tribune* reported on 4 April 1921 on page 6,

"EINSTEIN, TOO, IS PUZZLED; IT'S AT PUBLIC INTEREST

Can't See Why Theories
Are Widely Discussed.

New York, April 3.—[Special]—Prof. Albert Einstein, the German scientist, who is visiting this country, today discussed his famous 'relativity' theory with reporters.

Before going into details with the reporters, Prof. Einstein exploded the accepted story that he had said only twelve men in the world were capable of understanding it. He thinks most scientists understand his theories and

---

[356]. A. Einstein to H. Zangger of 24 December 1919, English translation by A. Hentschel, *The Collected Papers of Albert Einstein*, Volume 9, Document 233, Princeton University Press, (2004), pp. 197-198, at 197.

added that his students in Berlin understand them perfectly.

No theory can be susceptible of absolute proof, he added, and mentioned that an American scientist, St. John, is now conducting experiments which seem to give results at variance with the Einstein theory.

'The two theories, that of St. John and my own, have not yet been brought into harmony,' Prof. Einstein said. 'The subject dealt with is that of the wave lengths in the spectrum. It is impossible at the present stage of the experiments to say what the result will be.'

Calls for Psychologist.

Prof. Einstein was rather puzzled to account for the public interest in his conception of time and space, and said the public attitude seemed to call for a psychologist who could determine why persons who are not generally interested in scientific work should be interested in him.

'It seems psycho-pathological,' he said, with a laugh.

When it was suggested that perhaps people were interested because he seemed to give a new conception of the universe, which, next to the idea of God, has been the subject of the most fascinating speculations of the mind, he agreed that such might be the case.

'The theory has a certain bearing in a philosophical sense on the conception of the universe,' he said, 'but not from the scientific point of view. Its great value lies in the logical simplicity with which it explains apparently conflicting facts in the operation of natural law. It provides a more simple method. Hitherto science has been burdened by many general assumptions of a complicated nature.'

Not a Radical Departure.

Two of the great facts explained by the theory are the relativity of motion and the equivalence of mass of inertia and mass of weight, said Prof. Einstein.

'There has been a false opinion widely spread among the general public,' he said, 'that the theory of relativity is to be taken as differing radically from the previous developments in physics from the time of Galileo and Newton—that it is violently opposed to their deductions. The contrary is true. Without the discoveries of every one of the great men of physics, those who laid down preceding laws, relativity would have been impossible to conceive and there would have been no basis for it. Psychologically, it is impossible to come to such a theory at once without the work which must be done before. The four men who laid the foundations of physics on which I have been able to construct my theory are Galileo, Newton, Maxwell, and Lorenz.'

Man in Street Needn't Worry.

Whatever the value of relativity, it will not necessarily change the conceptions of the man in the street, said Prof. Einstein.

'The practical man does not need to worry about it,' he said. 'From the philosophical aspect, however, it has importance, as it alters the conceptions of time and space which are necessary to philosophical speculations and conceptions. Up to this time the conceptions of time and space have been such that if everything in the universe were taken away, if there was nothing left, there would still be left to man time and space. But under this theory even time and space would cease to exist, because they are unalterably bound up with the conceptions of matter.'

The reporters did not argue the point."

The New York Times responded to Einstein's "PSYCHOPATHIC RELATIVITY" on 5 April 1921 on page 18, and quoted Einstein on 8 July 1921 on page 9,

"'You ask whether it makes a ludicrous impression on me to observe the excitement of the crowd for my teaching and my theory, of which it, after all, understands nothing? I find it funny and at the same time interesting to observe this game.

'I believe quite positively that it is the mysteriousness of what they cannot conceive which places them under a magic spell. One tells them of something big which will influence all future life, of a theory which only a small group, highly learned, can comprehend. Big names are mentioned of men who have made discoveries, of which the crowd grasps nothing. But it impresses them, takes on color and the magic power of mystery, and thus one becomes enthusiastic and excited."

Einstein wrote to Max Born on 15 September 1950, in the context of politics,

"And the idiotic public can be talked into anything."[357]

Among those who actively opposed relativity theory, as it was expressed by Einstein—who, according to Einstein's assertions, must have been uncultured, ignorant anti-Semites—we find Hendrik Antoon Lorentz, Max Abraham, Alfred North Whitehead, Ernst Mach, Albert Abraham Michelson, Friedrich Adler, Henri Bergson, Oskar Kraus, Melchior Palágyi, [etc. etc. etc.]. Clearly, Einstein lied about a very serious matter, and, what is worse, Einstein was himself a racist instigator and a political agitator; and, therefore, a hypocrite and a deliberate inciter of "racial" discord.

6.5.3 What is Good for the Goose is not Good for the Goyim

The political Zionists emphasized their mistaken belief that Jews are a distinct race incapable of assimilation, and that Jews constitute a foreign nation within Germany. Einstein's anti-assimilationist rhetoric would later find its match in Philipp Lenard's segregationist belief in "Aryan Physics". Nobel Prize laureate Philipp Lenard was reacting to the Jews' bigoted assertions of their distinct racial

---

[357]. A. Einstein, *The Born-Einstein Letters*, Walker and Company, New York, (1971), p. 188.

characteristics and the Zionists' open declarations of their disloyalty to Germany.[358] Many Jews viewed Physics in expressly racist terms long before Lenard joined their ranks.[359]

Following the racial mythologies of Gobinaeu and Renan, Philipp Lenard joined the Jewish movement to segregate science and wrote of "Aryan Physics". Like many Jews before him, Lenard artificially distinguished between "German Physics" and "Jewish Physics" in 1936. Johannes Stark and Wilhelm Müller adopted this nomenclature in 1941 at the behest of the Zionist Nazis.[360]

Racist Jews provided the segregationist dogma. For example, there was the segregated "Jüdisch-Russisch Wissenschaftliches Verein" (Russian-Jewish Scientific Society) which participated in the foundation of the modern Zionist movement with its leaders Shmarya Levin, Leo Motzkin, Nachman Syrkin, Victor Jacobson, Arthur Hantke, Heinrich Löwe, Zelig Soskin, Willi Bambus, and many others.[361] In the late 1800's men like Theodor Mommsen and Anatole Leroy-Beaulieu were criticizing segregated Jewish associations, which they rejected as bigoted and segregated institutions.[362]

Just as some Christians felt uncomfortable around Eastern Jews, some Eastern Jews felt uncomfortable around Christians and found them dirty and disgusting. These Jews refused to eat at the same table with Christians, who did not oblige their Kosher laws.

In the early 1800's there was an influential movement to promote "Jewish

---

358. J. Klatzkin, *Krisis und Entscheidung im Judentum; der Probleme des modernen Judentums*, Jüdischer Verlag, Berlin, (1921). Heinrich Class under the pseudonym Daniel Frymann, *Wenn ich der Kaiser wär': politische Wahrheiten und Notwendigkeiten*, Dieterich, Leipzig, (1912); English translation, R. S. Levy, "If I were the Kaiser / Daniel Freymann", *Antisemitism in the Modern World: An Anthology of Texts*, Chapter 14, D.C. Heath, Toronto, (1991).
359. T. Lessing, *Der jüdische Selbsthaß*, Zionistischer Bücherbund, Berlin, (1930).
360. P. Lenard, *Deutsche Physik in vier Bänden*, J. F. Lehmann, München, (1936-1937); **and** *Philipp Lenard, der deutsche Naturforscher: sein Kampf um nordische Forschung: Reichssiegerarbeit*, J. F. Lehmanns Verlag, München, (1937). J. Stark and W. Müller, *Jüdische und deutsche Physik, Vorträge zur Eröffnung des Kolloquiums für theoretische Physik an der Universität München*, Helingsche Verlagsanstalt, (1941). **See also:** P. Lenard and J. Stark, "Hitlergeist und Wissenschaft", *Großdeutsche Zeitung. Tageszeitung für nationale und soziale Politik und Wirtschaft*, Volume 1, Number 81, (8 May 1924), pp. 1-2; reprinted *Nationalsozialistische Monatshefte*, Volume 7, Number 71, (February, 1936), pp. 110-111; annotated English translation, K. Hentschel, "Philipp Lenard & Johannes Stark: The Hitler Spirit and Science [May 8, 1924]" *Physics and National Socialism: An Anthology of Primary Sources*, Basel, Boston, Birkhäuser, (1996), pp. 7-10; which book also contains numerous other Nazi-era texts in English translation.
361. *Cf.* C. Weizmann, *Trial and Error: The Autobiography of Chaim Weizmann*, Harper & Brothers, New York, (1949), p. 35.
362. J. B. Agus, *The Meaning of Jewish History*, Volume 2, Abelard-Schuman, New York, (1963), pp. 407-408. Agus cites: A. Leroy-Beaulieu, *Israel among the Nations: A Study of the Jews and Antisemitism*, G.P. Putnam's sons, New York, W. Heinemann, London, (1895), pp. 45, 131, 134; and T. Mommsen, *Auch ein Wort über unser Judenthum*, Weidmannsche Buchhandlung, Berlin, (1880), pp. 5, 7, 15-16.

science". At the time, some Jews were forced to feign Christian conversion if they wished to become university professors. In 1822, Gans, Zunz and Moser created the *Verein für Kultur und Wissenschaft der Juden*, a segregated Jewish institution which offered Jews an alternative to an insincere and degrading baptism. They published a journal on "Jewish science", the *Zeitschrift für die Wissenschaft des Judenthums* published from 1822-1823.[363] There was also the *Jeschurun. Zeitschrift für die Wissenschaft des Judenthums* published from 1856-1870 by Joseph Kobak in German and Hebrew; and the *Monatsschrift für Geschichte und Wissenschaft des Judenthums* published from 1851-1939 by Rudolf Kuntze of the *Gesellschaft zur Förderung der Wissenschaft des Judentums*, Dresden.

Albert Einstein traveled to America in order to raise money[364] for an ethnically segregated "Jewish university"[365] or "Hebrew University" in Jerusalem. Many Zionists asserted that Jews had to be segregated in order to manifest their superior Jewish racial characteristics, which had lain dormant in the Diaspora. In accord with Jewish Messianic prophecy, they asserted that the Jewish race would again shine and lead the world of thought if only they could be "restored" to Palestine and segregated and at long last be permitted to be "Jews" and be relieved of the burden of being pseudo-Gentiles. Even after the Holocaust, Einstein was still calling for the segregation of Jewish students from Gentile students, which he argued was the only solution to the problem of anti-Semitism. Peter A. Bucky quoted Albert Einstein,

> "I think that Jewish students should have their own student societies. [***] One way that it won't be solved is for Jewish people to take on Christian fashions and manners. [***] In this way, it is entirely possible to be a civilized person, a good citizen, and at the same time be a faithful Jew who loves his race and honors his fathers."[366]

Shortly after World War One, Zionist Shmuel Hugo Bergmann wrote to Einstein,

> "[...]whether you, Professor, whom the world rightly calls the greatest Jewish scientist, but above all whom we love and value also as a person— whether you would be willing to participate in this conference and help us

---

363. M. Hess, English translation by M. Waxman, *Rome and Jerusalem: A Study in Jewish Nationalism*, Bloch, New York, (1918/1943), pp. 17-18, 24.
364. A. Einstein, *The World As I See It*, Citadel, New York, (1993), p. 98.
365. Letter from A. Einstein to H. Bergman of 5 November 1919, English translation by A. Hentschel, *The Collected Papers of Albert Einstein*, Volume 9, Document 155, Princeton University Press, (2004), pp. 132-133, at 132. *See also:* H. N. Bialik, "Bialik on the Hebrew University", in A. Hertzberg, *The Zionist Idea*, Harper Torchbooks, New York, (1959), pp. 281-288, at 284-285.
366. P. A. Bucky, Einstein, and A. G. Weakland, *The Private Albert Einstein*, Andrews and McMeel, Kansas City, (1992), p. 88.

with its preparation. I do not need to say how happy the Jewish people would be if *you* could be appointed to its university, but that is a question for the future."[367]

Albert Einstein stated,

"Antisemitism must be seen as a real thing, based on true hereditary qualities, even if for us Jews it is often unpleasant. I could well imagine that I myself would choose a Jew as my companion, given the choice. On the other hand I would consider it reasonable for the Jews themselves to collect the money to support Jewish research workers outside the universities and to provide them with teaching opportunities."[368]

and,

"The psychological root of anti-Semitism lies in the fact that the Jews are a group of people unto themselves. Their Jewishness is visible in their physical appearance, and one notices their Jewish heritage in their intellectual works, and one can sense that there are among them deep connections in their disposition and numerous possibilities of communicating that are based on the same way of thinking and of feeling. The Jewish child is already aware of these differences as soon as it starts school. Jewish children feel the resentment that grows out of an instinctive suspicion of their strangeness that naturally is often met with a closing of the ranks. [***] [Jews] are the target of instinctive resentment because they are of a different tribe than the majority of the population."[369]

Maja Winteler-Einstein wrote in her biography of her brother Albert Einstein,

"His later advocacy of Zionism and his activities on its behalf came from this impulse: less in accordance with and on the basis of Jewish teachings than from an inner sense of obligation toward those of his race for whom an independent working place for scholarly activity in the sciences should be created, where they would not be discriminated against as Jews."[370]

---

[367]. Letter from A. Einstein to H. Bergman of 5 November 1919, English translation by A. Hentschel, *The Collected Papers of Albert Einstein*, Volume 9, Document 155, Princeton University Press, (2004), pp. 132-133, at 132.

[368]. M. Born, *The Born-Einstein Letters*, Walker and Company, New York, (1971), p. 16.

[369]. A. Einstein, English translation by A. Engel, *The Collected Papers of Albert Einstein*, Volume 7, Document 35, Princeton University Press, (2002), pp. 156-157.

[370]. M. Winteler-Einstein, English translation by A. Beck, "Albert Einstein—A Biographical Sketch", *The Collected Papers of Albert Einstein*, Volume 1, Princeton University Press, (1987), pp. *xv-xxii*, at *xx*.

Albert Einstein wrote to Paul Ehrenfest on 8 November 1919,

"This university will contribute toward making less Jewish talent, particularly in Poland and Russia, have to go wretchedly to waste."[371]

6.5.3.1 Supremacist and Segregationist Jewish "Neo-Messianism"

After emancipation, Jews had initially faced the dilemma that if they sought to become a professor they had to convert, at least on paper, to Christianity. In 1822, Gans, Zunz and Moser created a segregated Jewish institution in order to offer an alternative to the often insincere baptisms of Jews. They called their society the *Verein für Kultur und Wissenschaft der Juden*, which published the *Zeitschrift für die Wissenschaft des Judenthums*. The association attracted Heinrich Heine, but soon disbanded. Heine, Gans and countless others took the baptismal plunge and the integration of Jews into Christian society began—some would later say in effort to undermine Gentile society.

Several articles appeared in *La Revue de Paris* in 1928 under the title "Les Origines Secrètes du Bolchevisme: Henri Heine et Karl Marx", in which "Salluste", a pseudonym, argued that Communism was a "neo-Messianic" scheme created by Heinrich Heine and Karl Marx, by means of which they meant to fulfill Judaic Messianic prophecies.[372] Salluste further argued that the Jewish societies which grew out of Moses Mendelssohn's reformation movement, had organized to subvert Gentile cultures and religions and replace them with Judaized culture and world revolution—fulfilling Judaic Messianic myth under the pretext of a secular movement for progress. This movement also manifested itself in the arrogance of the "Jewish Mission" of "reformed" Judaism.

On 22 September 1922, a Jewish Bolshevist Zionist apologist who published under the pseudonym "Mentor",[373] on pages 13 and 14 of *The Jewish Chronicle*, confirmed that the "Jewish Mission" was to subvert other nations, cultures and religions in the name of "peace" and to force Gentiles to comply to

---

371. Letter from A. Einstein to P. Ehrenfest of 8 November 1919, English translation by A. Hentschel, *The Collected Papers of Albert Einstein*, Volume 9, Document 160, Princeton University Press, (2004), pp. 135-136, at 136.
372. "Salluste", "Henri Heine et Karl Marx. Les Origines Secrètes du Bolchevisme", *La Revue de Paris*, Volume 35, Number 11, (1 June 1928), pp. 567-589; **and** "Henri Heine et Karl Marx II. Les Origines Secrètes du Bolchevisme", *La Revue de Paris*, Volume 35, Number 12, (15 June 1928), pp. 900-923; **and** "Henri Heine et Karl Marx III. Les Origines Secrètes du Bolchevisme", *La Revue de Paris*, Volume 35, Number 13, (1 July 1928), pp. 153-175; **and** "Henri Heine et Karl Marx IV. Les Origines Secrètes du Bolchevisme", *La Revue de Paris*, Volume 35, Number 14, (15 July 1928), pp. 426-445. *See also, Rabbi Liber's Response:* "Judaïsm et Socialisme", *La Revue de Paris*, Volume 35, Number 15, (1 August 1928), pp. 607-628; *To which "Salluste" Replied:* "Autour d'une Polémique: Marxism et Judaïsm", *La Revue de Paris*, Volume 35, Number 16, (15 August 1928), pp. 795-834.
373. *Cf.* Mentor, "Peace, War—and Bolshevism", *The Jewish Chronicle*, (4 April 1919), p. 7; **and** "From My Note Book", *The Jewish Chronicle*, (11 April 1919), p. 9.

the will of the Jews,

"What are the Jews Doing?"

By MENTOR.

WHEN I wrote in this column last week, I had no idea that the premonitions to which I alluded, of another great catastrophe of like sort to the war that began in 1914, would so soon be justified. Within a few hours of my words appearing in print a document was issued by the British Government, threatening the beginning of a war of which, once started, no man could foretell the end. Hardly was the last issue of the *Jewish Chronicle* published than we seemed whirled back in a sudden instant to the time eight years ago that preluded the terrible world-struggle that lasted through nearly five years. There were rumours of war; there were ominous movements of politicians from the four corners of the kingdom, which newspapers interpreted as meaning all sorts of things. The evil birds of Militarism were foregathering. Like vultures they flew to gather their prey. Stories were bruited abroad, craftily designed to work upon the sentiments and the emotions of the people. Reasons and excuses, arguments and assurances, were cleverly designed, so that when the dogs of war were unleashed, proof of the inevitability and the justification for starting wholesale murder, for man going out to kill his fellow man, might be prudently provided beforehand. As I write, the situation—as it is termed—seems, if anything, a good deal less dangerous than it did at the beginning of the week. That is because those who were for war, those who were willing if not anxious to resort to arms in order to fight about a dispute instead of adjusting it by negotiation, have not received the encouraging response from the country which they had evidently hoped would come to them. Once bit twice shy! All the conventional paraphernalia of diplomats and politicians were again employed by the men of war as they were used eight years ago. Then their assurances were accepted, and men believed they could by war accomplish a great deal. Now, some of the public at least are wiser, and recollect the fraud, the chicanery, the double-dealing, the falsity, and the two-facedness which were so largely responsible for the determination of this country to enter into war eight years ago. They know that the same people are up to the same dodges, that the like people are bent on the like wiles, and the country this time has put a large discount upon all the mongering for War. The experience of the Great War has thus not been wholly lost, and there seems a healthy disposition, in more than one quarter, to regard the Minister who leads this country into war as *ipso facto* unfitted to hold the trust he has dishonoured by muddlement. There is proved to be now a looking upon war as the crowning disaster of any nation, not as its glory, as a visitation and not as a proud happening.

Jewish Doctrine and Christian.

If war is averted, if those responsible for the Government of the country finding war 'no go,' because the people will have none of it, have to seek other means for adjusting international differences, then the incident which looked so grave at the beginning of the week will have been of advantage.

For it will have shown at least one Government that the way of war is not the easiest at hand for them for settling any disputes that may arise. So far, so good; and if that spirit of antagonism to and hatred and—if you will— fear of war be maintained, so that men, beginning by disliking it, will go on to loathe and detest it, then we shall have made a long stride to the abolition of war and the arbitrament of the sword, and towards that condition which is the Jewish ideal; when man shall no longer lift up sword against man, nor learn war any more. [*Isaiah* 2:4] I call that the Jewish ideal, but we Jews have not a monopoly of it. Peace is a Christian ideal, too. Indeed, Christianity goes much farther, and is a doctrine of non-resistence to evil. Judaism does not teach that; it is far more practical and far more human. But if Christianity were really practised and the Christian spirit were truly in the souls of those who profess Christianity, war would be impossible. But a Jew is here writing for Jews, and it is because peace is a Jewish ideal that I revert to this question here and now—now, because we are on the threshold of the most sacred days in the Jewish calender, when the Jew, if ever, is brought into close contact with the Almighty, when, if ever, he feels strong upon him the duty which is his as a Jew.

The Jewish Mission.

And I ask: What are the Jews doing in the war against war, the war which the King himself the other day said is the only war worth while; the war for Civilisation, for salving Humanity, for making the life of men and women in the world tolerable and bearable; the war against one of the most fertile roots of poverty with its fruits of hunger, and vice, and disease—what are the Jews doing in the war for which the King of Kings long ago conscripted certainly every Jew? I suppose the answer will reach me that Jews ought not, as such and of themselves, to be expected to take any definite part in such a campaign. I shall be told that war is really a political matter, and that Jews have no politics of their own, they share in the politics of the nations of which they are citizens. But this argument, carried to its logical conclusion, would place the Jew in such a position that the whole of the claim which he has made concerning his place in the world, and in respect to the Judaism he professes, would have to be seriously overhauled. How can a Jew be true to Jewish teachings, to the teachings of the Prophets, to Rabbinical teaching, to all that Judaism connotes for the Jew, unless Peace on earth and Goodwill among men be believed in by him and hoped for by him? How can he pray, as he constantly prays, from year end to year end, and day by day, for peace, and yet not mean it and not wish it? And if he means it and wishes it, then how can he place even his duty to the State (if it is conceivable that his duty to the State can involve war as a principle) before his duty to his God? The Christian does it. He worships a Divinity that he hails as the emblem of peace. He invokes the one whom he regards as Messiah, the harbinger of peace. He subscribes to the doctrine of Peace enunciated by the great Founder of his faith, and yet he contrives instruments of violence, engines of slaughter, and all the hellish devices for maintaining War on earth and illwill towards men. But that is a matter for

Christians. That they do thus is no reason, and assuredly no justification for Jews doing likewise. Following the multitude to do evil is not Jewish work. And so I ask again, just as we are slipping into yet another New Year: What are the Jews doing so that war shall cease from the earth, so that peace may reign and goodwill prevail among the children of men?

Our Separateness.

What are the Jews doing? It is a pertinent and not an impertinent question; because it asks, though not in those words, how is the Jew justifying his existence? We elect to remain a separate people. In every country and in every land we segregate ourselves from our fellow-citizens, and throughout the ages we have obstinately (as our enemies term it), faithfully as we believe, kept ourselves apart as a separate people. For what? Some Jews will tell you that we have refused to assimilate in the sense of losing ourselves in the multitudes surrounding us, because we have all along been conscious of being a separate national entity. So we have maintained our separateness in the hope that some day our national being would be restored. This, put very broadly, is the attitude of Zionists and Jewish Nationalists. But all Jews are not one or the other. The majority are neither, or at least care not at all for either striving. Their idea of Jewish separateness is altogether another. They say that we Jews have kept apart in order to carry on, amid the nations of the world, a Jewish Mission. That mission, so it is claimed, comprises our weaning other peoples away from error of thought and sin of action to a true conception of God. It means that we have to urge the breaking up of all idols and securing allegiance alone to the Almighty Governor of the universe. Very well, let us accept, for the purpose of argument, the contention of these fellow Jews that their separateness is maintained alone for the Mission potentialities of our people. Then I would ask: What are they doing in the way of propagating that Mission? Some of them argue that although it is true they are not actively engaged in spreading the message of Israel, or in preaching its truths to those of other faiths, they are doing service to the mission passively in the living of their lives. Their example, they say, is even better than precept. Surely this is a paltering with the question; it is an excuse, a subterfuge, and it makes the whole idea of the Mission of Israel not alone the sham that it is with those who thus argue, but a ridiculous parody of every idea of the purpose and the object which any mission worthy of the name must have.

The Jew's Contribution.

This paltry excuse for neglect of the call of the Mission of Israel does not rob us of the right to ask: What is the Jew doing in pursuance of what he believes to be his mission to Mankind? The answer must be: precious little. We are standing at the dawn of a New Year. We are about to reach another milestone in our history. Is the Jew to go on year by year in the same meaningless, chaotic existence, just living, just existing without a worthy purpose as Jew; for mere material selfish objects, as a people without an ideal, without an aspiration? Broadly speaking, there are only two possible ideals for Jews, the National ideal and the Mission of Israel ideal. They are not antagonistic or even mutually exclusive. For the Jewish Nationalist also believes—believes very strongly—in the Mission of Israel, but believes, too, that it is impossible of accomplishment without national existence in a Jewish land. But taking the Jewish position as it is, either aspiration, if the Jew be true to it, will justify his separateness among the nations of the world. But if he nourish neither of those ideals, as is the way with thousands and thousands of Jews, then the *raison d'être* of his existence is *nil*, the part he plays in the world is a mirage. He is a mere parasite, and he justifies nothing so much as the indictment that is made by some enemies of our people. They denounce us because we remain separate as a people, and yet take no count of any service which we should do as Jews for the common benefit of Mankind. Well, if there be any reality in the Mission of Israel ideal, then I ask again: What are the Jews doing? What part are they taking in the war against war, in leading men from violence and slaughter and murder in the wholesale, back or rather forward to ways of peace, to ways of goodwill and happiness among men. We are doing precious little, even as individual Jews. **As a Jewish people, we are doing nothing.**

Here surely, as I have more than once suggested, is a great and glorious opportunity for the Jewish People. They do not want to be a separate nation. They wish to be separate among the nations of the world. Very well, then let them justify that aspiration. All the trouble Jews encounter is traceable to nothing so surely as to the fact that they are despised. And they are despised, not as individuals—as individuals even anti-Semites respect Jews—but because, however commendable individual Jews may be, whatever service individual Jews may have done for the world and for civilisation—and Dr. Joseph Jacobs left a posthumous work showing how great had been the service of individual Jews in that respect—as a people Jews contribute nothing to the service of mankind. We do not cultivate a Jewish culture; we are not known for any great or enduring office which we perform. But suppose we carried on our mission, our God-given mission as the bringers and the promoters of peace, as the bearer of that great ideal, is it not palpable that there would be something we should be doing by which we should win the respect of mankind? Because sooner or later, after misunderstanding had passed away and misrepresentation and vituperation had evaporated, the world would come to acknowledge itself our debtors for the good we should have effected. It seems to me that in the times in which we live—with the constant menace and danger of war, with the ineffable wickedness which allows great talent and scientific attainment to be misused and misapplied, as they are being misused and misapplied in devising means for carnage, for bloodshed, for violence, for all the indescribable horror comprised in war—and particularly at this hour when we are entering into the most solemn moments of conclave—the Jew with his God—it is not inapt to ask: What are the Jews doing in the war that alone matters, the war against war? I ask it here and now, because the hearts of my fellow-Jews, attuned at this season to higher thoughts and loftier aspirations, may bethink themselves that there is a great evil in the world, the greatest evil that mankind and civilisation have to contend against. And mayhap there will arise in their souls a determination, each one as he can and where he can, to do what he can—thus making it a Jewish mission—so as to roll away the menace of war from the path that humanity is treading."

If the "Jewish Mission" were truly to convince the Peoples of the world that monotheism is the most rational choice among extant religions, then Jews would be applying themselves to this task, but they are not. Instead, it appears that where Jews involve themselves in religious questions, they are most often ridiculing other religions. Far from inviting other Peoples to join Judaism, Jewish leaders instead attempt through their disproportionate control of media and education to destroy all religious beliefs in other Peoples, including the monotheism of Christianity and Islam—save the false beliefs they have instilled in Dispensationalist Christian Zionists who serve as their slavish and gleefully suicidal "Esau" to their "Jacob". The true nature of the "Jewish Mission" is made obvious by the actions of Jewish leaders and is spelled out in Jewish religious literature. It is to destroy other cultures, religions, nations and "races". It is not a mission of peace and tolerance, rather it is a mission of segregation, "race" hatred, Jewish supremacy, war and death. As the Jewish book of *Exodus* 34:11-17 states, the "Jewish Mission" is to:

"11 Observe thou that which I command thee *this* day: behold, I drive out before thee the Amorite, and the Canaanite, and the Hittite, and the Perizzite, and the Hivite, and the Jebusite. 12 Take heed to thyself, lest thou make a covenant with the inhabitants of the land whither thou goest, lest it be for a snare in the midst of thee: 13 But ye shall destroy their altars, break their images, and cut down their groves: 14 For thou shalt worship no other god: for the LORD, whose name *is* Jealous, *is* a jealous God: 15 Lest thou make a covenant with the inhabitants of the land, and they go a whoring after their gods, and do sacrifice unto their gods, and *one* call thee, and thou eat of his sacrifice; 16 And thou take of their daughters unto thy sons, and their daughters go a whoring after their gods, and make thy sons go a whoring after their gods. 17 Thou shalt make thee no molten gods. [King James Version]"

The Jewish book of *Obadiah* states,

"1 The vision of Obadiah. Thus saith the Lord GOD concerning Edom: We have heard a message from the LORD, and an ambassador is sent among the nations: 'Arise ye, and let us rise up against her in battle.' 2 Behold, I make thee small among the nations; thou art greatly despised. 3 The pride of thy heart hath beguiled thee, O thou that dwellest in the clefts of the rock, thy habitation on high; that sayest in thy heart: 'Who shall bring me down to the ground?' 4 Though thou make thy nest as high as the eagle, and though thou set it among the stars, I will bring thee down from thence, saith the LORD. 5 If thieves came to thee, if robbers by night—how art thou cut off!—would they not steal till they had enough? If grape-gatherers came to thee, would they not leave some gleaning grapes? 6 How is Esau searched out! How are his hidden places sought out! 7 All the men of thy confederacy have conducted thee to the border; the men that were at peace with thee have beguiled thee, and prevailed against thee; they that eat thy bread lay a snare under thee, in whom there is no discernment. 8 Shall I not in that day, saith the LORD, destroy the wise men out of Edom, and discernment out of the mount of Esau? 9 And thy mighty men, O Teman, shall be dismayed, to the end that every one may be cut off from the mount of Esau by slaughter. 10 For the violence done to thy brother Jacob shame shall cover thee, and thou shalt be cut off for ever. 11 In the day that thou didst stand aloof, in the day that strangers carried away his substance, and foreigners entered into his gates, and cast lots upon Jerusalem, even thou wast as one of them. 12 But thou shouldest not have gazed on the day of thy brother in the day of his disaster, neither shouldest thou have rejoiced over the children of Judah in the day of their destruction; neither shouldest thou have spoken proudly in the day of distress. 13 Thou shouldest not have entered into the gate of My people in the day of their calamity; yea, thou shouldest not have gazed on their affliction in the day of their calamity, nor have laid hands on their substance in the day of their calamity. 14 Neither shouldest thou have stood in the crossway, to cut off those of his that escape; neither shouldest thou have delivered up those of his that did remain in the day of distress. 15 For the day of the LORD is near upon all the nations; as thou hast done, it shall be done unto thee; thy dealing shall return upon thine own head. 16 For as ye have drunk upon My holy mountain, so shall all the nations drink continually, yea, they shall drink, and swallow down, and shall be as though they had not been. 17 But in mount Zion there shall be those that escape, and it shall be holy; and the house of Jacob shall possess their possessions. 18 And the house of Jacob shall be a fire, and the house of Joseph a flame, and the house of Esau for stubble, and they shall kindle in them, and devour them; and there shall not be any remaining of the house of Esau; for the LORD hath spoken. 19 And they of the South shall possess the mount of Esau, and they of the Lowland the Philistines; and they shall possess the field of Ephraim, and the field of Samaria; and Benjamin shall possess Gilead. 20 And the captivity of this host of the children of Israel, that are among the Canaanites, even unto Zarephath, and the captivity of Jerusalem, that is in Sepharad, shall possess the cities of the South. 21 And saviours shall come up on mount Zion to judge the mount of Esau; and the kingdom shall be the LORD'S. [version of the Jewish Publication Society]"

"Salluste" alleged that Heinrich Heine pulled out of these movements which grew from Moses Mendelssohn's Jewish reformation, not because Heine sincerely wished to disassociate from the Jewish destruction of Western Civilization, but because these groups had begun to draw attention to themselves and Heine wanted his views to be kept secret, and shied away from the political pressure placed on these subversive organizations. Salluste later republished his articles in book form, *Les Origines Secrètes du Bolchevisme: Henri Heine et Karl Marx*, Jules Tallandier, Paris, (1930); and his ideas were championed by Denis Fahey in his book *The Mystical Body of Christ in the Modern World*, Browne and Nolan Limited, London, (1935); and later by Robert H. Williams, *The Ultimate World Order—As Pictured in "The Jewish Utopia"*, CPA Book Publisher, Boring, Oregon, (1957?).

In 1845, *The North American Review* wrote of the Frankist-style forces in the Jewish community Salluste later described,

> "We might confidently look for reformers under such a system as Rabbinism; and, even without the name of reformation, for wide departures from the Talmud, either towards the 'old paths,' or to infidelity. The man who in modern times exerted the most commanding influence on Judaism was Moses Mendelssohn. He was born at Dessau, in 1729, was carefully educated in the Bible and Talmud, but was thrown upon Hebrew charity in Berlin, at the age of thirteen. Following the bent of his own genius, and stimulated by various associations, he left the dreary paths of tradition, to pursue the intricate but flowery ways of Gentile philosophy. He even improved the German language, in which he wrote with great taste. The influence of his works and his example was soon manifest. An enthusiasm for German literature and science was awakened among the Jewish people, when they beheld their kinsman ranking with the first scholars of the age. 'Parents wished to see their children like Mendelssohn. Rashi and Kimchi, the Shulchan, Aruch, and Josaphoth, were laid on the shelf. Schiller and Wieland, Wolff and Kant, were the favorite books of the holy nation.' Mendelsshon was very strict in Talmudical observances, and did not in his works directly oppose them; yet he certainly intended to undermine Rabbinism, and covertly labored to obliterate superstitions and prejudices, and to render his religion consistent with free intercourse between Jew and Gentile, and with the palpable benefits of modern progress in letters and refinement in manners. After all, he was probably at best but a deist; and he certainly lacked that directness, candor, and earnestness of purpose, which true-hearted reformers have usually manifested. Christians must deny to Judaism that vitality which is essential to its maintenance upon the true basis even of a pure pre-Messianic creed. As a system, though not indeed strictly in each individual, it must ever oscillate between Rabbinism, or the like, and rationalism,—finding no stable, middle, spiritual ground.
>
> Mendelssohn died in 1786; but others arose to carry out his innovations. A Jewish literary and philosophical society was formed at Königsberg, in 1783, which supported the first Jewish periodical ever published,—a journal devoted to the cause of reform. The 'new light' rapidly spread; and now Mendelssohnism, in different varieties, inclined more or less to the Talmud, or to infidelity, is the religion of a great majority of the Jews in all Europe west of Poland, into which country itself, especially Austrian Poland, the revolution has in some degree extended. The 'Jews of the New Temple,' or ' Rational' or 'Reformed Jews,' as they are called, where their numbers have

not secured peaceable ascendency, have generally seceded from the Talmudists; who, on their own part, where the so-called reformation has made good progress, adhere to the Talmud scarcely even in name.

The creed of the new sect has never appeared in an authoritative shape, but may be gathered from their writings and practices. The believers in it agree, that the Jews are no longer a chosen people, in the sense hitherto commonly received. They reject the Talmud, professing to receive the Hebrew Scriptures as the true basis of religious belief, and as a divine revelation; though after explaining away their inspiration, and the miracles recorded in them, on rationalistic principles. Regarding the Mosaic institutions as never abrogated, they consider, however, that most of their requirements are applicable only to a state of national establishment in Palestine; and therefore hold, that, until the unknown period of the Messiah's advent, and Israel's restoration, such laws only are to be observed as are necessary to preserve the essence of religion, or useful to form pious ecclesiastical communities, and which do not interfere with Gentile governments, with any of the existing relations of life, or with intellectual culture. The synagogue service has been remodelled; and the modern languages have been generally substituted for the Hebrew. A weekly lecture has taken the place of the semi-annual sermons of the Rabbinists. Contrary to the precept of the Talmud, instrumental music is introduced into public worship. 'The question of organ or no organ,' says a late journal devoted to the Jews, 'divides Judaism on both sides of the Atlantic.'

Before long, the latitudinarian views of the leaders in this movement clearly discovered themselves; and there was a temporary reaction in favor of Rabbinism, to which the more devout among their converts receded. Yet the new system has signally prevailed and flourished. It is in France, perhaps, that the Jews have thrown off most completely the trammels of Judaism,—indeed, of all religion. They now style themselves *French Israelites*, or *Israelitish Frenchmen*, according to the doctrine of Napoleon's Sanhedrim; and seem anxious to amalgamate themselves more and more with the nation at large. Most of their leaders are infidels, undisguisedly aiming to obliterate all the common notions about a Messiah, as utterly superstitious; referring the prophecies of his advent—which they still nominally treat as prophecies—to the political emancipation of the Jews in the various lands of their sojourn. 'The Regeneration,' a journal published at Paris by some of their most learned and influential men, has represented the French Revolution as the coming of the Messiah, bringing, first, judgment, then, liberty and peace. The grand rabbi of Metz, a few years ago, in addressing the Jews of his district, spoke thus:—

'God has permitted different religions, according to the different necessities of men, in the same way as he has created different plants, different animals, and men of different characters, genius, constitutions, physiognomies, and colors. Consequently, all religions are salutary for those who are born in these religions; consequently, we must respect all religions. All men, without distinction of religion, will be partakers of eternal beatitude, provided they have practised virtue in this life.'

On the 12th of June last, a voluntary Jewish synod met at Brunswick, composed of twenty-five eminent rabbins, from various parts of the continent. It was the first of a proposed succession of annual synods, to deliberate on Jewish affairs. They sat eight days, passed various resolutions proposing important changes, and declared their concurrence in all the decisions of Napoleon's Sanhedrim. The Jews of England, though visibly influenced by residence in so enlightened a kingdom, were all nominally Rabbinists, until, within the last four or five years, a reforming party seceded in London whence their principles and denomination—' British Jews—have since gradually spread. Even among those who remained, great difference of opinion prevails as to Talmudical observances. Both there and in this country, the Portuguese Jews seem most active in the work of revolution. The tide of Jewish emigration to the United States is rapidly swelling; and as it comes from many lands, it exhibits a variety of hue. But the voluntary emigrant is ever and characteristically a lover of change; and here the Talmud has little sway, and that rapidly declining. Mr. Leeser represents the Bible alone as the basis of the Jewish faith and in the whole article already referred to, does not so much as mention the Talmud. He edits, at Philadelphia, 'The Occident and American Jewish Advocate,' the first Jewish periodical established in this country. Soon after its establishment, 'The Israelite,' a weekly German paper, devoted to the same cause, and also published in Philadelphia, was announced; whether this still survives, we know not. Mr. Leeser expects a literal Messiah, —not God, or a son of God, but a mere man, eminently endowed, like Moses, to accomplish all that is foretold of him. He protests against some of the decisions of the late Brunswick synod, particularly the one reaffirming the *dictum* of the French Sanhedrim, that Jews might intermarry with Gentiles. He has long had in his congregation a Sabbath school, or a school for religious instruction, held, not on the seventh day, but on the Christian Sabbath, which Christian observance makes necessarily a day of convenient leisure for the purpose.

Among the stricter Jews, all over the world, the expectation of Messiah's advent is becoming more and more anxious. They not unfrequently talk, though without serious purpose, of embracing Christianity, should he not appear within a certain time. Migration to the Holy Land is visibly increasing. Multitudes from all parts of the world would hasten thither, could they become possessors of the dear soil, and enjoy reasonable protection. Mr. Noah proposes, that Christian societies and governments interested in the welfare of the Jews should exert their influence to procure these advantages for them in their native land of promise. The suggestion deserves notice.

Of modern efforts for the conversion of Israel to Christianity we can speak but briefly. The chief extraordinary obstacles which have hitherto opposed such efforts have been, a bigotry which treated the bare thought of investigating Christianity as a heinous sin, and which was ever prepared to stifle free inquiry by persecution; the character of Talmudical education, which disqualified the pupil for independent judgment; and accumulated prejudices against a religion too often exemplified only by profligate persecutors. But all these obstacles are gradually sinking away; nor does growing infidelity appear so formidable as the superstition and fanaticism which have given place to it. Moreover, the spirit of inquiry, and the dissensions kindled by the progress of the revolution which Mendelssohn commenced, are favorable to Christian effort. We shall speak only of what Protestants have done."[374]

Salluste quoted a rather famous letter which had for decades been attributed to the "Neo-Messianist" Baruch Lévy (a pseudonym?), which was allegedly written to Karl Marx, and which mirrors many of racist Zionist Moses Hess' statements, and which further anticipates Michael Higger's philo-Semitic Messianic book *The Jewish Utopia*. The Lévy letter stated,

"The Jewish people as a whole will itself be the Messiah. It will reign over the world by intermixing the other races of mankind and by eliminating borders and monarchies, which are a defense against particularism, and by the establishment of a world-wide Republic, which will universally grant the Jews the right of citizenship. In this new organization of humanity, the children of Israel who are now spread over the entire surface of the globe, all of the same race and the same traditions—without, however, forming a distinct nationality—will exclusively become the leaders, without ever meeting opposition, especially if they manage to set some segment of the working masses on a stable course. The governments of the Nations which form the Universal Republic will all pass into the hands of the Jews without any effort, as a reward for the victory of the proletariat. The ruling Judaic race will then be able to eliminate personal property, and will control all of the public's wealth. Thus the promise of Talmud will have been fulfilled, which states that in the messianic era the Jews will hold the wealth of all the people of the world under their lock and key."

"Le peuple juif pris collectivement sera lui-même son Messie. Son règne sur l'Univers s'obtiendra par l'unification des autres races humaines, la suppression des frontières et des monarchies, qui sont le rempart du particularisme, et l'établissement d'une République Universelle qui reconnaîtra partout les droits de citoyens aux Juifs. Dans cette organisation nouvelle de l'Humanité, les fils d'Israël répandus dès maintenant sur toute la surface du globe, tous de même race et de même formation traditionnelle sans former cependant une nationalité distincte, deviendront sans opposition l'élément partout dirigeant, surtout s'ils parviennent à imposer aux masses ouvrières la direction stable de quelques-uns d'entre eux. Les

---

[374]. "The Modern Jews", *The North American Review*, Volume 60, Number 127, (April, 1845), pp. 329-368, at 361-365.

gouvernements des Nations formant la République Universelle passeront tous, sans effort, dans des mains israélites, à la faveur de la victoire du prolétariat. La propriété individuelle pourra alors être supprimée par les gouvernants de race judaïque qui administreront partout la fortune publique. Ainsi se réalisera la promesse du Talmud que, lorsque les Temps du Messie seront venus, les Juifs tiendront sous leurs clefs les biens de tous les peuples du monde."[375]

Rabbi Liber doubted the authenticity of this letter and published a polemic against "Salluste", stating, *inter alia*,

"Salluste quotes but one letter, which is enough to impress the novice, from the 'neo-messianist' Baruch Lévy to Karl Marx. Who is this Baruch Lévy? From where is this text taken? It is a mystery. Until proven otherwise, I hold this letter to be a forgery. Let me assure the reader. There exists, in the antisemitic literature, a whole series of false letters of the same tone, manufactured in more or less clandestine dispensaries, to say nothing of the 'Protocols of the Learned Elders of Zion', that forgery by officers of the Czar's police, whose origin was definitively unmasked[1]."

"Salluste cite seulement une lettre, assez impressionnante pour les novice, du «néo-messianiste» Baruch Lévy à Karl Marx. Qui est ce Baruch Lévy? D'où est tiré ce texte? Mystère. Jusqu'à preuve du contraire, je tiens cette lettre pour *un faux*. Que le lecteur ne se récrie pas. Il existe, dans la littérature antisémitique, toute une série de fausses lettres du même ton, fabriquées dans des officines plus ou moins clandestines, sans parler des «Protocles des Sages de Sion», cette forgerie de policiers tsaristes dont l'origine a été définitivement démasquée[1]."[376]

Salluste responded by quoting from numerous Jewish sources which justified his conclusions, though apparently without directly touching upon the provenance of the "Baruch Lévy" letter—which reappeared in Salluste's book of 1930.[377] Morris Kominsky argued that the letter was a hoax.[378] Kominsky rather unconvincingly relies upon Herbert Aptheker's conclusion that the letter

---

[375]. "Salluste", "Henri Heine et Karl Marx. Les Origines Secrètes du Bolchevisme", *La Revue de Paris*, Volume 35, Number 11, (1 June 1928), pp. 567-589, at 574. An alternative English translation appears in D. Fahey, *The Mystical Body of Christ in the Modern World*, Browne and Nolan Limited, London, (1935), p. 83.

[376]. Rabbi Liber, "Judaïsm et Socialisme", *La Revue de Paris*, Volume 35, Number 15, (1 August 1928), pp. 607-628, at 623-624.

[377]. "Salluste", "Autour d'une Polémique: Marxism et Judaïsm", *La Revue de Paris*, Volume 35, Number 16, (15 August 1928), pp. 795-834. "Salluste", *Les Origines Secrètes du Bolchevisme: Henri Heine et Karl Marx*, Jules Tallandier, Paris, (1930), at pp. 33-34.

[378]. M. Kominsky, *The Hoaxers: Plain Liars, Fancy Liars, and Damned Liars*, Branden Press, Boston, (1970), pp. 189-191.

is a hoax on its face and does not appear to Aptheker to resemble anything else attributed to Marx, his correspondents or Marxism, in Aptheker's experience.[379] In any event, it is interesting that such accusations should be attributed to Marx's correspondent. It is even more interesting that Michael Higger's book of 1932, *The Jewish Utopia*,[380] unabashedly advocates nearly the exact same plan as the letter from "Lévy" to Marx. These, however, have a common source in the Hebrew Bible, and in the virulently anti-Christian and anti-Gentile Talmud and Cabalistic literature.

When Salluste republished the Lévy letter in 1930, Salluste added the following notation, which did address the provenance of the letter and which states that though letter might be of dubious origin it had been in circulation for almost half a century without raising a protest, that Marx's correspondence was purged of unflattering materials before it was published, and that the letter agreed with common sentiments among current authors and fit the current situation perfectly,

> "(1) Ce texte d'une lucidité prodigieuse, et dont chaque phrase paraît s'appliquer à la situation politique et sociale du monde à l'époque où nous écrivons (1928), est connu depuis près d'un demi-siècle. Il a été cité pour la première fois au Congrès Antisémite de Berlin, en 1888, puis reproduit à plusieurs reprises en France, et pour la dernière fois à notre connaissance, en 1919. Son insertion dans notre étude n'en a pas moins provoqué une véritable fureur chez nos contradicteurs, et l'on verra plus loin que le rabbin Liber nous accuse carrément de faux à ce sujet...
>
> Nous nous permettons d'observer: 1° que le fait que cette lettre ne figure pas dans la Correspondance de Karl Marx ne prouve rien contre l'authenticité de la pièce, les gendres du prophète judéo-communiste, Paul Lafargue et Charles Longuet, n'ayant livré à l'impression les lettres de leur beau-père et de ses correspondants qu'après les avoir soigneusement expurgés; 2° que la lettre ci-dessus a été citée à plusieurs reprises, depuis quarante ans, sans soulever la moindre protestation de la part d'autorités juives tout aussi qualifiées que M. le rabbin Liber; 3° que les idées contenues dans cette lettre sont absolument conformes à celles exprimées, sous une forme très voisine, par d'autres écrivains juifs contemporains, tels que MM. Edmond Fleg, Barbusse, André Spire, etc., etc.; 4° qu'en admettant même que le document soit d'origine incertaine, tout ce qui se passe dans le monde quarante ans après sa production, spécialement au point de vue de la judaïsation des partis révolutionnaires, montre que son auteur était admirablement renseigné."[381]

---

[379]. M. Kominsky, *The Hoaxers: Plain Liars, Fancy Liars, and Damned Liars*, Branden Press, Boston, (1970), pp. 190-191.
[380]. M. Higger, *The Jewish Utopia*, Lord Baltimore Press, Baltimore, (1932).
[381]. "Salluste", *Les Origines Secrètes du Bolchevisme: Henri Heine et Karl Marx*, Jules Tallandier, Paris, (1930), pp. 34-35.

This "Neo-Messianism" of Communism, which manifested itself in the French Revolution as a political Messiah, is truly the Paleo-Messianism of Deutero-Isaiah (*Isaiah*, Chapters 40-66, or 40-55 if one accepts the theory of Trito-Isaiah). Today "Neo-Messianism" bears the title of "political Zionism". Moses Hess based his racist political Zionism and his Communism on the ancient Jewish Messianic prophecies, not those of a personal messiah, but of the Jewish People as Messiah, the Jewish People as the master race. Before Moses Hess, Mordecai Manuel Noah called upon the Jewish People to act as the Messiah and "restore" themselves to Palestine.[382]

But Communism was only one side of the Jewish Messianic coin. Jewish Capitalists sought to control all the wealth of the world by accumulating it through corrupt means, and by hoarding gold—even by melting down the coins of the nations. Like the Communists, whom they funded, the Jewish Capitalists sought to ruin the nations with wars and with debt, and by destroying their cultures, religions and educational institutions. They also sought to Judaize them.

Joseph Klausner wrote of the concept of the Jewish People as Messiah and master race—the usurper of the nations, and of the wealth of the world—in his book *The Messianic Idea in Israel: From Its Beginnings to the Completion of the Mishnah*. In his analysis, we can find the dogmatic Judaic basis for the persecution of the Jews who chose not to be political Zionists by the Zionist Nazis; and the persecution and oppression of the Gentiles by the Jewish People. Klausner wrote, *inter alia*,

> "And kings shall be thy foster-fathers,
> And queens thy nursing mothers;
> They shall bow down to thee
>   with their face to the earth,
> And lick the dust of thy feet (49:23).
>
> For all the enemies of Judah will be cut off, and all who lift themselves up against her will not succeed (49:17-19, 25-26; 54:17)— just as the prophet had said in his prophecies of the first period. So great will be the *political success*. And *material prosperity* will not be less. 'O thou afflicted one, storm-tossed, uncomforted!'—the prophet turns toward the beloved

---

[382]. M. M. Noah, *Call to America to Build Zion*, Arno Press, New York, (1814/1977); **and** *Discourse Delivered at the Consecration of the Synagogue of [K. K. She'erit Yisra'el] in the City of New-York on Friday, the 10th of Nisan, 5578, Corresponding with the 17th of April, 1818*, Printed by C.S. Van Winkle, New-York, (1818); **and** *Discourse on the Evidences of the American Indians Being the Descendants of the Lost Tribes of Israel: Delivered Before the Mercantile Library Association*, Clinton Hall, J. Van Norden, New York, (1837); **and** *Discourse on the Restoration of the Jews: Delivered at the Tabernacle, Oct. 28 and Dec. 2., 1844*, Harper, New York, (1845); **and** *The Jews, Judea, and Christianity: A Discourse on the Restoration of the Jews*, Hugh Hughes, London, (1849).

homeland in great compassion—

> Behold, I will set thy bases with beryl,[18]
> And lay thy foundations with sapphires.
> And I will make thy pinnacles of rubies,
> And thy gates of carbuncle stones,
> And all thy border of jewels (54:15—12).

At the same time spiritual blessings will multiply:

> And all thy children shall be taught of the LORD;
> And great shall be the peace of thy children (54:13).

For Zion will be established in righteousness (54:14), Jerusalem will be 'the Holy City,' and the uncircumcised and unclean will no more enter it (52:1)—just as Ezekiel had said. In spite of all the universalism of the prophet, which we shall soon see in all its glory, his nationalism is not diminished, just as in spite of all his spirituality his political and worldly hopes are not impaired. The Gentiles will exalt Israel as the Chosen People, as their kings bow down to the earth before him and lick the dust of his feet. The Gentiles, therefore, will not be equal to Israel in glory and honor,[19] although all of them will become sons of God because all of them will be called by the name of the LORD. Israel will remain the center, while the Gentiles will be only points on the circumference.

On what basis should Israel have an advantage over the rest of the nations? The answer could have been only this: because Israel will teach the knowledge of the LORD and ethical insight to all peoples. But this answer was the result of a long evolution of ideas and the cause of a new chain of profound ideas closely bound together by their own nature.

Not *all* the people of Israel have acknowledged the LORD; among this Chosen People are evil ones and sinners, who do not wish to know the LORD and to walk in His ways. Only the prophets and their disciples are the servants of the LORD, and only they have spread His teaching in Israel—and for this they have been persecuted by their own people, slain like Uriah the son of Shemaiah, or cast into cisterns and into prison like Jeremiah. Thus the one attempting to spread the knowledge of the LORD and the love of the good, that is, to *benefit* the people, is forced to endure many *evils* for the LORD's sake and to take comfort in the hope that finally the sinful people will acknowledge and understand that the servant of the LORD was in the right.

The people Israel is the only nation within which is the knowledge of the LORD and the recognition of the good; therefore it must disseminate these two things among the other peoples, as the prophets disseminate them within it. This ethical demand was already made by the pre-Exilic prophets from Amos to Zephaniah and Jeremiah. And if the Exilic and post-Exilic prophets saw that Israel was suffering greatly, that its land was laid waste and its Temple ruined, that it had gone into exile among the Gentiles and become in its political weakness an object of mockery and derision among them, verily—unless the prophet and his disciples were willing to conclude that the God of Israel had no power or ability to save His people and that the whole idea of the choice of the people Israel is only vanity and emptiness— there was left to them only the conclusion that just as the prophet suffers without having committed a fault, suffers from the transgressors among his own people whom he is seeking to benefit, that is to say, *takes upon himself the iniquity of others*, so suffers also the people Israel from other peoples more sinful than Israel, because Israel seeks to benefit them. In other words, *the people Israel takes upon itself the iniquity of all the rest of the peoples, the iniquity of the whole world.* What the prophet is to Israel, Israel becomes to all the world: the servant of the LORD, holding up the standard of the highest righteousness in the world and suffering for his pursuit of good.

This is the profound conception that lies hidden in 42:1-7; 49:1-9; 50:4-9; 52:13-15 plus 53:1-12. The ancient Jewish interpreters were divided as to whether these passages refer to the prophet alone or to the whole people Israel. The early Christians, from Paul the Apostle (Acts 8:32-35) onward, saw in them a reference to the sufferings and death of Jesus of Nazareth. (As a matter of fact, some of his career did resemble what is described in Chapter 53; and the rest of his career is *intentionally* portrayed in the Gospels in such a manner that the events appear to have happened in fulfillment of the words in this chapter.) Some modern Christian scholars wish to see in these passages a description of the fate of Zerubbabel or Jehoiachin (Sellin), or of some other great man of Israel who lived in the middle of the sixth century B.C.E. (Duhm).[20] After what I said above by way of explanation, it should now be clear that the prophet could not separate his own fate, as one persecuted for his pursuit of good, from the fate of his disciples and of all the servants of the LORD, whom he considered to be the real nucleus of the people Israel, the Israel in whom the LORD 'will be glorified' (49:3). Thus everything said in these chapters can and must be related in one process both to the prophet and to the whole Jewish nation: the servants of the LORD are this nation's chosen remnant, to which alone belongs the future.[21]

Nevertheless, there is a kernel of Messianism—not Christian, but completely Jewish Messianism—in these chapters.

I have already said in a number of places that the Jewish Messiah is composite in his nature: in him are some of the politico-worldly virtues of the king and some of the ethico-spiritual virtues of the prophet. In the period of the Second Isaiah there was no place for an individual political Jewish Messiah, as was said above; and apart from the reference to 'the sure mercies of David' we do not find this subject mentioned at all by the prophet of consolation to Zion. But precisely because the ethico-spiritual virtues of the Messiah were exalted and became the shining symbols of Messianism, the bearer of Messianism came to be either the individual 'servant of the LORD,' the prophet, or the collective 'servant of the LORD,' the best of the people Israel. Thus the *whole* people Israel *in the form of the elect of the nation* gradually became *the Messiah of the world, the redeemer of mankind*. This Messiah must suffer just as the prophet suffers. Here also punishment precedes redemption; but this punishment is unique: it comes as a penalty *for the sin of others*. And it redeems the world; for if Israel had not been willing to suffer and to spread the knowledge of God and of pure morality in the earth, the world would have remained sunk in sin against religion and morality. And for this punishment, bringing good to all peoples except Israel, this people receives a worthy reward in 'the end of days' [future age], in that it becomes 'a light to the Gentiles,' [*Isaiah* 42:6; 49:6] in that it is placed in the center of mankind.

This, *in its broadest aspects*, is the content of those chapters which treat of the servant of the LORD. In it are included the spiritual, the universalistic, Messianic expectations of the people Israel, expectations which serve to supplement the nationalistic, the worldly, and the political expectations of which I have already spoken above. Therefore it is impossible to pass over them in silence; they must be presented as completely as possible here, since the greatness of their value for the development of Messianism in the future is incalculable.

The servant of the LORD, 'Israel in whom He will be glorified,' suffers, and it seems that he has labored in vain and spent his strength for nothing; but actually his accomplishment is great and his reward is with the LORD, who says to him:

> It is too light a thing that thou shouldest be My servant
> To raise up the tribes of Jacob,
> And to restore the survivors of Israel (the nationalistic expectation);
> *I will also give thee for a light of the nations,*
> *That My salvation may be unto the end of the earth* (49:1-6).

And not only for 'a light of the nations' but also for 'a covenant of the people' (49:8)—as Deutero-Isaiah had said in his earlier prophecies (42:6). 'The Redeemer and the Holy One of Israel' promises His servant, whom He has chosen and who was despised and abhorred and 'a slave of rulers' (from this it seems clear that even here the meaning does not apply to the prophet alone) ,[22] that 'kings shall see and arise (before him), princes, and they shall prostrate themselves (49:7)—something which the prophet had already promised to the whole people Israel (49:23)."[383]

6.5.3.2 It is Alright for Jews to Claim that "Einstein's Theories" are "Jewish", but Goyim Dare Not Say It

---

[383]. J. Klausner, *The Messianic Idea in Israel, from its Beginning to the Completion of the Mishnah*, Translated from the third Hebrew edition by W. F. Stinespring, Macmillan, New York, (1955), pp. 160-164.

As the Twentieth Century arrived, the situation of the Jews had greatly improved in Germany. Relations between Jews and Christians were quite amicable and Jews frequently married Christians. The political Zionists saw the rapidly increasing process of assimilation as a threat to their racial heritage. The political Zionists had few qualms about forwarding their goals of racial segregation by corrupt means. They learned from the Dreyfus affair that Jews could be unified by the charge of anti-Semitism. It immediately became their favored means to unite and organize their members, to raise funds, and to segregate. It was also their favorite means to censor their critics, which was nothing new. The Jews attempted to censor the Egyptians, Romans and Greeks with false claims of "anti-Semitism" more than two thousand years ago.

There are often political forces involved in the appointment of professorships and the rejection of literature antagonistic to the agenda of any given publication. Ethnically biased institutions inhibit the progress of science; whether they are forced into segregation, as was often the case in Russia, or elect to be segregated, as was also often the case in Russia. Graduates streaming out of ethnically oriented schools sometimes obtain positions of power throughout the world and carry their bigotry with them. Jews were the victims of ethnic bias throughout the Nineteenth Century. It taught them to organize and to act as a unified force and in particular instances, the tables were turned. Few other groups were as successful at creating and maintaining societies, hospitals, associations and charities as the Jews of the early Twentieth Century—no one had the power of the press or as much money at their disposal as the Jews. Some Germans became resentful and felt that they were being pushed out of their own institutions. They tended to blame the *Ostjuden* who had immigrated from the East to cities like Berlin.[384]

The political Zionists thrived on the tension that existed between *Ostjuden* and the traditional Gentile Germans following the German loss in the First World War. German Jews found themselves caught in the middle of this struggle for the national identity of Germany. Following the Second World War, after the Zionists had had their revenge on assimilatory German Jewry, they continued to ridicule German Jews in Israel, as reported in *Time Magazine* in 1948,

> "In other lands the German Jews tend to look upon themselves as the aristocrats of Jewry (although they give precedence to the Sephardic families from Spain and Portugal). In Palestine the recent German aliyah is looked down upon and made the butt of the same kind of joke that German Jews in the U.S. used to hurl at their Russian brethren.
>
> Israel calls the German Jew a *yecki* (roughly: squarehead), laughs at his naiveté. Many of the *yecki* are physicians (of that great, devoted band of German-Jewish doctors) and they have a hard time adjusting to the land. Many try chicken farming, going about it in that highly scientific Teuton

---

384. B. Haase, "Die Ostjudenfrage", *Central-Verein Zeitung*, Volume 1, Number 1, (4 May 1922), p. 6. "Ostjudenfrage", *Central-Verein Zeitung*, Volume 1, Number 1, (4 May 1922), p. 5.

way which makes the Polish and Russian Israelis guffaw. They say that when one *yecki* found a sick chicken he sent all the way to India for a serum, inoculated every one of his flock. They tell of a *yecki* with an old dry cow who asked a Polish Jew to sell it for him. The Pole found a Russian Jew to whom he said: 'This is a fine young cow; she gives six liters of milk every day.' The *yecki*, standing by, said: 'Well, well, that I didn't know; I'd like to buy her back.' To new arrivals the Eastern Jews say: 'Did you come here from conviction—or from Germany?'"[385]

There are allegations that Ashkenazi Jews later practiced genocide against the Sephardic Jews in Israel, by irradiating them with x-ray machines under the pretext that they were treating them for ringworm. This allegedly occurred under David Ben-Gurion's leadership.

There is terrible enmity between the Sephardim and Ashkenazim in Israel.[386] The article "Israel" in the *Great Soviet Encyclopedia: A Translation from the Third Edition*, Volume 10, Macmillan, New York, (1976), pp. 477-484, at 478, states,

> "Jews make up more than 85 percent of the population (1970); Arabs (14.6 percent) and a small number of Armenians make up the rest. Arabs are subjected to harsh racial discrimination. More than half of the Jewish population is made up of immigrants from Europe, Asia, Africa, and America. The various ethnic groups of the Jewish population of Israel are unequal in terms of social position. The sabras (Jews born in Israel) enjoy the special confidence of the chauvinist ruling circles: next in position are the Ashkenazim (immigrants from Europe). Jewish immigrants from the countries of Asia and Africa are subjected to discrimination. The official language is Hebrew; however, some Jews do not know it, and Yiddish, Ladino (close to Spanish), Arabic, English, and other languages are used in everyday life. Jewish believers practice Judaism. The Arabs are Sunni Muslims, although some are Druze and Christians. The Armenians are Christians."

Ethnic pride (and insecurity resulting from both fairminded and unfair attacks) often resulted in pro-Jewish ethnic mythologies, which anti-Semites used as examples to criticize Jews in general, much to the delight of the political Zionists. Bruno Thüring wrote, citing Salomon Wininger's *Grosse jüdische National-Biographie mit mehr als 8000 Lebensbeschreibungen namhafter jüdischer Männer und Frauen aller Zeiten und Länder, ein Nachschlagewerk für das jüdische Volk und dessen Freunde*, in seven volumes, Druck "Orient", Cernauti, (1925-1936); and Theodor Lessing's *Der jüdische Selbsthass*, Zionistischer Bücher-Bund, Berlin, (1930):

---

[385]. *Time Magazine*, Volume 52, Number 7, (16 August 1948), p. 27.
[386]. J. Jiladi, *Discord in Zion: Conflict Between Ashkenazi & Sephardi Jews in Israel*, Scorpion Pub., London, (1990).

"So können wir also verstehen, wenn der betreffende Referent in der großen jüdischen Nationalbiographie (Wininger) in die Worte ausbricht: „Ptolemäus und Kopernikus waren als Forscher Waisenknaben gegen Einstein, der Raum und Zeit ins Wanken bringt. Kopernikus stürzte die absolute Ruhe der Erde, Einstein aber stürzte den Absolutismus überhaupt. Nichts ist ‚wirklich', für jeden Beobachter ist das Weltbild ein anderes, aber jeder hat recht."

Daß aber das Judentum sich auch bewußt war, in diesen Dingen das eigene Selbst zum Ausdruck gebracht zu sehen, zeigt eine Stelle aus dem Buche: „Der jüdische Selbsthaß" von dem Juden Theodor Lessing:

„Die durch das Wachstum der nichteuklidischen Geometrien möglich gewordenen neuen Wissensgebiete, die Anzahlen-, die Mengen-, die reine Mannigfaltigkeitslehre, das Auflösen der mit dem Unendlichen auf jenen Wissensgebieten verknüpften Paradoxien und die Relativierung auch der letzten Konkretheit und Anschaulichkeit zugunsten des absoluten Kalküls, das war das Werk eigentlich jüdischer Intelligenzen wie Georg Cantor, Alfred Fränkel, Alfred Pringsheim, Arthur Schoenfließ, Felix Hausdorff, Ludwig Kronecker, Alfred[1]) Sommerfeld, bis schließlich durch Michelson, Minkowski und Einstein die Weltwende, die Überwindung des Aristoteles, Newton und Kant erzwungen wurde. Es ist, als ob diese Kohorte sich verschworen hätte, das letzte arme Restchen sinnfälliger Gestaltlichkeit zu verflüchtigen.""[387]

Zionist Theodor Lessing also stated that,

"Vor nahezu einem Menschenalter, etwa um die gleiche Zeit, da das Werk Weiningers erschien, veröffentlichte ich eine Abhandlung zur Psychologie der Mathematik, welche zu zeigen suchte, daß die damals mächtig einsetzende Geometrisierung der Physik und Arithmetisierung der Geometrie und der schon damals sich ankündigende Aufstieg der „Relativitätslehre" eng zusammenhänge mit der Seele jüdischer Menschen."[388]

In 1850 and 1869, the German composer Richard Wagner published a scathing indictment of Jews as, in his view, tending to be inherently incompetent in the arts—as, in Wagner's view, too often mere poseurs and cultural parasites lacking natural talent.[389] Wagner's essays have doubtless unnecessarily led to

---

[387]. B. Thüring, "Albert Einsteins Umsturzversuch der Physik und seine inneren Möglichkeiten und Ursachen", *Forschungen zur Judenfrage*, Volume 4, (1940), p. 154. Republished as: *Albert Einsteins Umsturzversuch der Physik und seine inneren Möglichkeiten und Ursachen*, Dr. Georg Lüttke Verlag, Berlin, (1941).
[388]. T. Lessing, *Der jüdische Selbsthass*, Matthes & Seitz, München, (1984), pp. 84-84.
[389]. R. Wagner under the *nom de plume* K. Freigedank, "Das Judenthum in der Musik", *Neue Zeitschrift für Musik*, (3 and 6 September 1850); Reprinted with revisions and an

lingering insecurity in the German Jewish community and its decedents. Zionist Leon Pinsker disdainfully referred to the accusation in 1882, "to reproach us with a lack of men of genius!"[390] Burton J. Hendrick wrote in his 1923 defense of Jewish Americans from the accusation that they dominated finance,

> "Wagner, in his essay on 'The Jews and Music' denies them creative power in this art. They have lesser lights—a Mendelssohn, a Meyerbeer, an Offenbach; they have no Beethoven, no Mozart, or—he might have added—no Wagner. In poetry they have a Heine, but no Milton, no Byron, no Keats, no Wordsworth. In the drama they possess several figures of minor importance, but where is the Jewish Shakespeare or Molière or Schiller? In statesmanship they have a Disraeli, but no Cromwell or Pitt or Washington or Lincoln. What Jewish orator is there to put in the same class with Burke or Fox or Sheridan or Webster? What Jewish jurist ranks with Blackstone, Lord Mansfield, or John Marshall? In philosophy indeed the Jews do possess one man of the very first rank, Spinoza, and that exception to the generalization made above must be noted; but in science is there any Jewish name to put beside Copernicus or La Place or Galileo or Newton or Darwin—unless, indeed, the recent work of Einstein may ultimately include him in these exalted ranks? Even in that branch in which the Jews have been especially active and in which they have demonstrated great ability, medicine and surgery, their names by no means occupy the first place. Run over the list of the great medical discoveries of the last three centuries from that of the circulation of the blood to that of bacteriology; the most impressive fact is that the vast majority of the preeminent brains are Gentiles. Even in the Nineteenth and Twentieth Centuries, when Jewish scholarship in this country and in Europe has had free scope, the great accomplishments have been made by non-Jews. Probably the greatest medical achievements of modern times were the discovery of vaccination, of anæsthetics, and of bacteriology; the first was English, the second American, the third French. Indeed it would probably be possible to mention half a dozen American achievements—such as anæsthetics, ovariotomy, Marion Sims' work in gynecology, Dr. Beaumont's discovery of the laws

---

appendix, R. Wagner, *Das Judenthum in der Musik*, J. J. Weber, Leipzig, (1869); the original unrevised 1850 article was reprinted in *Gesammelte Schriften und Dichtungen*, Volume 5; English translation of the 1869 version by E. Evans, *Judaism in Music (Das Judenthum in der Musik) Being the Original Essay Together with the Later Supplement*, W. Reeves, London, (1910); Also, "Judaism in Music", *Richard Wagner's Prose Works*, Volume 3, Broude Brothers, New York, (1966), pp. 75-122.; which is a reprint of the original "Judaism in Music", *Richard Wagner's Prose Works*, Volume 3, Kegan Paul, Trench, and Trübner, London, (1894).

390. L. Pinsker, "Auto-Emancipation", quoted in A. Hertzberg, *The Zionist Idea*, Harper Torchbooks, New York, (1959), pp. 181-198, at 190. Hertzberg cites: English translation by D. Blondheim in B. Netanyahu, *Road to Freedom*, New York, (1944), pp. 74-95, 105-106.

of digestion, Dr. Holmes's discovery of the contagiousness of child bed fever, Dr. Walter Reed's work in yellow fever—to which Jewish medical science can present few parallels. In this department, as in the arts, the Jewish minds lack the great faculty of creation: Jewish medical scientists, such as Metchnikoff, Ehrlich, and Wasserman, have important achievements to their credit, but their work consists in elaborating principles discovered by other men; the work of the three mentioned, for example, is all based upon the original investigations of Pasteur. Nor is it any sufficient answer to point to the comparatively small number of Jews, for one of the most certain teachings of history is that the genius of a people, and the proportion of great men it produces has no relation to its numbers. The genius of the English people had its finest flowering in the days of Elizabeth, when the population of the little island was less than two million. The genius of the Greeks reached its most eloquent expression in the days of Pericles when the population was only a few hundred thousand. The small numbers of the Jews as compared with Gentiles is therefore no reason why they should not have produced a great array of geniuses of the first class if, as we have been taught to believe, we are dealing with a race of supermen."[391]

It is quite probable that such Wagnerian venom played no small rôle in the psychological need of some Jews to deceitfully promote Einstein as if greater than Copernicus, Kepler, Galileo, Huyghens and Newton; and to promote the deceit that Einstein's work was unprecedented and of an exclusively "Jewish" character. The desire to discredit the Wagnerian view also provided racist and tribalistic Jews with an incentive to Judaize Gentile culture, and to take over university departments so as to promote their own interests and encourage Jews to achieve and fulfill their sensibilities, while discouraging the advancement of Gentiles, thereby inhibiting the progress of the Jews' perceived competition.

Immediately after Einstein's humiliating retreat from America and at a critical moment in the Zionist movement, *The London Times* wrote on 14 June 1921 on page 8, referring to the occasion of Einstein's lecture at King's College:

"LORD HALDANE, who presided, said they were there to give a British welcome to a man of genius. (Cheers.) The highest knowledge was a possession of which the world at large was proud, and genius knew no frontier. That morning he had been touched to observe that his distinguished guest had left his house to gaze on the tomb of Newton in Westminster Abbey. What Newton was to the 18th century Einstein was to the 20th century. In the lecture they were about to hear they would find a new point in the theory of relativity which had never been so definitely stated. Einstein had given a new conception of the universe, a conception, he thought, more revolutionary than that of Galileo, Copernicus, or Newton. He had taught

---

391. B. J. Hendrick, "The Jews in America: II Do the Jews Dominate American Finance?", *The World's Work*, Volume 44, Number 3, (January, 1923), pp. 266-286, at 269-271.

them to think of the universe of externality as relative in its reality to knowledge. Reality was relative, not merely our knowledge of it. He had given a view which brought us back to the deeper meaning of knowledge itself.

The new doctrine, added Lord Haldane, had come from a man distinguished by his desire, if possible, to efface himself and yet impelled by the unmistakable power of genius which would allow the individual of whom it had taken possession to rest for one moment. Professor Einstein had two great qualities for his task. He had a command of the tremendous instrument of mathematics as complete, at least, as that of any man alive. He had something more, a creative imagination akin to that of the poet. He fashioned creations apparently out of nothing in the way that genius alone could do. He was, too, a musician who played with a feeling and insight not always found in even the very best professionals. He was a master of the violin as well as of mathematics. The 20th century had produced one of the greatest thinkers that the last 500 years had seen and they were proud to be there to welcome him. (Cheers.)"

Einstein, himself, admitted that he was no mathematician. He was an absolutist and his ideas were not original and others expressed these ideas far more cogently than he was ever able to express them.

Even before Wagner, long before Wagner, Jews suffered under false accusations that they were incapable of creative thought. Josephus wrote in his ancient polemic in defense of the Jews, *Against Apion*,

> "Hence hath arisen that accusation which some make against us, that we have not produced men that have been the inventors of new operations, or of new ways of speaking; for others think it a fine thing to persevere in nothing that has been delivered down from their forefathers, and these testify it to be an instance of the sharpest wisdom when these men venture to transgress those traditions; whereas we, on the contrary, suppose it to be our only wisdom and virtue to admit no actions nor supposals that are contrary to our original laws; which procedure of ours is a just and sure sign that our law is admirably constituted; for such laws as are not thus well made are convicted upon trial to want amendment."[392]

Jews did suffer from the rigid dogmatism of Judaism, which inhibited their progress in the ancient world and during the Enlightenment (see, for example, *Babylonian Talmud*, Tractate Menahoth, folio 99*b*, which discourages the study of "Greek science"). The uncreative indoctrination of Jewish scholars in the beliefs of the Talmud and in the learning of the Hebrew language also tended to

---

[392]. Josephus, "Flavius Josephus Against Apion", *The Works of Flavius Josephus: Comprising the Antiquities of the Jews; a History of the Jewish Wars; and Life of Flavius Josephus, Written by Himself*, Book 2, S. S. Scranton Co., Hartford, Connecticutt, (1916), pp. 917-918.

destroy their ability to think independently and creatively.

Adolf Hitler attacked Jews as if uncreative and parasitic in many of his speeches. Following Rathenau's murder in 1922, Hitler spent a month in jail. When he was released, he stated,

> "That is the lurking danger, and the Jew can meet it in one way only—by destroying the hostile national intelligentsia. That is the inevitable ultimate goal of the Jew in his revolution. And this aim he must pursue; he knows well enough his economics brings no blessing: his is no master-people: he is an exploiter: the Jews are a people of robbers. He has never founded any civilization, though he has destroyed civilizations by the hundred. He possesses nothing of his own creation to which he can point. Everything that he has is stolen. Foreign peoples, foreign workmen build him his temples, it is foreigners who create and work for him: it is foreigners who shed their blood for him. He knows no 'people's army': he has only hired mercenaries who are ready to go to death on his behalf. He has no art of his own: bit by bit he has stolen it all from the other peoples or has watched them at work and then made his copy. He does not even know how merely to preserve the precious things which others have created: as he turns the treasures over in his hand they are transformed into dirt and dung. He knows that he cannot maintain any State for long. [***] All that the Jew cannot do. And because he cannot do it, therefore all his revolutions must be 'international'. They must spread as a pestilence spreads. He can build no State and say 'See here! Here stands the State, a model for all. Now copy us!' He must take care that the plague does not die, that it is not limited to one place, or else in a short time this plague-hearth would burn itself out. So he is forced to bring every mortal thing to an international expansion. For how long? Until the whole world sinks in ruins and brings him down with it in the midst of the ruins."[393]

At the Nuremberg *Parteitag* in 1937, Hitler stated,

> "The people which has thus through Jewish agitators been driven into madness, reinforced by non-social elements liberated from the prisons, now destroys its own national intelligentsia on the scaffold and the Jew without scruple and without conscience is supreme. The Jew is himself completely uncreative: he may in many countries hold 90 per cent. of the positions in the intellectual world, but he never discovered, formed, or conceived the elements of knowledge, culture, or art, and the same is true in trade. Therefore of necessity, if he wishes to hold power for any length of time in a country, he must proceed to a bloody annihilation of the former intellectual upper class; otherwise he would soon be conquered once more by this superior intelligence."[394]

---

[393]. A. Hitler, *The Speeches of Adolf Hitler April 1922-August 1939*, Volume 1, Howard Fertig, New York, (1969), pp. 30-31.
[394]. A. Hitler, *The Speeches of Adolf Hitler April 1922-August 1939*, Volume 1, Howard

The Jewish Bolsheviks made it a priority to mass murder the intellectual elite of the Gentiles in the nations they conquered, while elevating educated and intelligent Jews into positions of power and comparative wealth. Hitler, as a good Bolshevist, did much to destroy the intellectual class of Germany, and to ruin its educational institutions.

The charge that Jews are incapable of producing great minds in the arts and science has resulted in an unnecessary insecurity among Jews. This may be why some have a pro-Einstein ethnic bias and so violently oppose the exposure of the truth which results in the loss of one of "their" supposed greats. This is not only a mistake on their part, it is unnecessary, as there have been many great minds of Jewish descent in history, and even were there not, the insecurity which results in zealous hero worship is artificial and destructive and ultimately results in arrogance and cultural stagnation, as was recognized even before the time of Josephus.

Albert Einstein realized that the cult of personality surrounding him was destructive to science and to progress. He was very much aware of the fact that people believed in what he said out of blind faith—not because it was true or because it was logical, but merely because the miraculous "Einstein" had said it and the press had applauded it. It worried him that people had surrendered their individuality, their ability to make their own judgements, to his authority; but he worried privately and enjoyed the limelight.[395]

The shameless hype of Einstein as if equal to, or greater than, Copernicus, Galileo, Kepler, Huyghens and Newton,[396] was begun by Alexander Moszkowski, who was familiar with Eugen Karl Dühring's work, and favored by Einstein,[397] who was also familiar with Dühring's work. Dühring's book *Robert Mayer, der Galilei des neunzehnten Jahrhunderts. Eine Einführung in seine Leistungen und Schicksale*, E. Schmeitzner, Chemnitz, (1880); provided Moszkowski with the inspiration to call Einstein the Galileo of the Twentieth Century. Moszkowski's shameless hype was likely a direct response to Dühring's accusation of 1881,

"If one surveys the history of the Jewish tribe as a whole, one finds

---

Fertig, New York, (1969), p. 699.

[395]. P. A. Bucky, Einstein, and A. G. Weakland, *The Private Albert Einstein*, Andrews and McMeel, Kansas City, (1992), pp. 32, 110, 116-117.

[396]. It should be noted that Arnold Sommerfeld stated at the end of 1915, that Einstein was the greatest spirit since Gauss and Newton. *See:* Letter from A. Sommerfeld to K. Schwarzschild of 28 December 1915, Niedersächsische Staats- und Universitätsbibliothek Göttingen, Cod. Ms. K. Schwarzschild 743. Sommerfeld was one of the first to profit by promoting Einstein while demeaning the deceased Poincaré in Sommerfeld's annotations for *Das Relativitätsprinzip*, B. G. Teubner, Berlin, Leipzig, (1913)—a book published shortly after Poincaré died, which failed to include any of Poincaré's works, but obviously helped to promote Sommerfeld.

[397]. P. A. Bucky, Einstein, and A. G. Weakland, *The Private Albert Einstein*, Andrews and McMeel, Kansas City, (1992), p. 31.

immediately how it has not managed a fibre of real science in its national existence. [***] Where, however, is — to recall only the development of science since Copernicus, Kepler, Galilei, Huyghens, etc. — the Jew, to whom, in these significant centuries too, even a single natural scientific discovery is due?"[398]

Houston Stewart Chamberlain later repeated the insult. Ironically, Jewish litterateurs countered the charge that Jews were uncreative, by plagiarizing their critics.

Paul Ehrenfest, who opposed the dishonest promotion of Einstein as the "Jewish Newton", wrote to him on 9 December 1919,

"I hear, for ex., that your accomplishments are being used to make propaganda, with the 'Jewish Newton, who is simultaneously an ardent Zionist' (I personally haven't *read* this yet, but only *heard* it mentioned). [***] But I cannot go along with the propagandistic fuss with its *inevitable* untruths, precisely *because* Judaism is at stake and *because* I feel myself so thoroughly a Jew."[399]

Communist Zionist Nachman Syrkin thought that Jews had an innate "national character". He wrote in 1898,

"The peculiar literature, thought, and sentiment of the Jewish masses, which stamp them unmistakably with a well-defined national character, are clearly reflected in Jewish socialism."[400]

The pro-Jewish promoter A. A. Roback wrote in his book *Jewish Influence in Modern Thought*, that racial characteristics happily gave Jews the edge in creating the theory of relativity; which, according to Roback, was a Jewish creation, one might even say, according to Roback, a racially predetermined Jewish physics resulting from uniquely Jewish biological forces. Roback even thought it a shame that Lorentz was not Jewish, made much of the fact that most everyone considered Lorentz to be Jewish, stated that Lorentz looked Jewish, and then demeaned Lorentz' contribution to the theory. Roback wrote in 1929,

---

398. E. K. Dühring, *Die Judenfrage als Racen-, Sitten- und Culturfrage: mit einer weltgeschichtlichen Antwort*, H. Reuther, Karlsruhe, (1881); English translation by A. Jacob, *Eugen Dühring on the Jews*, Nineteen Eighty Four Press, Brighton, England, (1997), pp. 96-97.

399. Letter from P. Ehrenfest to A. Einstein of 9 December 1919, English translation by A. Hentschel, *The Collected Papers of Albert Einstein*, Volume 9, Document 203, Princeton University Press, (2004), pp. 173-175, at 174.

400. N. Syrkin, under the nom de plume "Ben Elieser", *Die Judenfrage und der socialistische Judenstaat*, Steiger, Bern, (1898); English translation in A. Hertzberg, *The Zionist Idea*, Harper Torchbooks, New York, (1959), pp. 333-350, at 344.

> "It is common knowledge that the man whose name is most intimately associated with the theory of relativity is a Jew of unmistakable Semitic origin and avowedly nationalistic tendencies. Albert Einstein has already taken his place with Galileo, Kepler, Copernicus and Newton in the forefront of scientific achievement. But it is not generally known that the doctrine of relativity has been reared, so to speak, on a Jewish foundation. [***] If Michelson, Minkowski, Levi-Civita, and other Jews all had a hand with Einstein in the establishment of the great principle, only as a result of chance or coincidence, then the line between a coincidence and a miracle almost vanishes. In self-defense for broaching this delicate subject, I may call attention to the fact that the issue between the House of Israel and the principle of relativity has already been picturesquely and good-humoredly brought up by a non-Jew. [***] It is my belief that a theory, principle or even law, *must be in us before we can discover it in nature.* [***] In the development of the relativity theory, it is perhaps significant that the Jewish stamp is found at almost every turn. Were Einstein, alone of all Jewry, responsible for the vast physical transformation, the connection between relativity and the Jews could be regarded as wholly fortuitous, but where the names of Michelson, Levi-Civita, Minkowski, Born, and Silberstein are all associated, in a more or less intimate way, with Einstein's achievement, one begins to feel that the 'Elders of Zion' have unwittingly conspired to explain the world's most baffling phenomena, and apparently have met with success."[401]

Roback and Lenard were kindred spirits. Some have asserted that Lenard was a crypto-Jew.[402]

Roback was inspired by L. Roth, who also went too far in 1927 in his essay *Jewish Thought in the Modern World*,

> "In the same way, what is perhaps the most remarkable of modern intellectual movements, the development in mathematical physics, is largely the result of the labours of the Jews Michelson, Minkowski, Einstein, and Weyl, while its philosophical interpretation (as a part of a vast body of other fruitful work in the general history and evaluation of the sciences) is being furthered by the insight of Cassirer, Brunschvicg, and Meyerson. Yet truth is its own witness and its own judge, and it is absurd to discuss it in terms of its discoverers. Like many other pioneers these men are of Israel, but their work is for the whole world."[403]

---

401. A. A. Roback, *Jewish Influence in Modern Thought*, Sci-Art Publishers, Harvard Square, Cambridge, Massachusetts, (1929), pp. 237-238, 245-246, 250-251.

402. D. Bronder, *Bevor Hitler kam: Eine historische Studie*, Hans Pfeiffer Verlag, Hannover, (1964), p. 204 (p. 211 in the 1974 edition). H. Kardel, *Adolf Hitler, Begründer Israels*, Verlag Marva, Genf, (1974); English translation *Adolf Hitler: Founder of Israel*, Modjeskis' Society Dedicated to Preservation of Cultures, San Diego, (1997), pp. 4, 73.

403. L. Roth, "Jewish Thought in the Modern World", *The Legacy of Israel*, Clarendon

These statements were made at a time when Jews were characterized by anti-Semitic Jewish Zionists as "parasites" feeding off of the nations. Many Jews began to doubt their ability to live independently of a "host".[404] One can certainly understand the need to correct that injustice and self-doubt, but it would more than have sufficed to have simply told the truth without distorting and exaggerating the facts in a way that did gross injustice not only to history and to the public which was lied to, but also to the many philosophers, mathematicians and scientists whose legacies were stolen and whose good reputations were destroyed for the sake of promoting mediocre Jewish minds.

Racist Jews tried to justify themselves by claiming that if race is the standard, then the Jews are a superior race. For example, Ignatz Zollschan stated at least as early as 1914, referring to his book, *Das Rassenproblem unter besonderer Berücksichtigung der theoretischen Grundlagen der jüdischen Rassenfrage*, W. Braumüller, Wien, (1910),

"JEWISH QUESTIONS

I.
The Cultural Value of the Jewish Race

The cultural value of the Jewish race has long been established by students of history and philosophy. A race whose genius has created all prevailing religions among all civilized nations, a race whose spiritual heroes have given to the world the principles of freedom and justice, a race whose sons have for thousands of years made vast contributions to the advance of civilization—such a race unquestionably represents a useful member in the family of nations. And yet a minute, scientific investigation of this problem, from the point of view of anthropology and biology, is urgently needed.

For, at the present time, some writers are busily engaged in disseminating the view that the Jews are no race at all; that modern Jews are not descendants of the ancient Hebrews, and are accordingly no Jews, but merely adherents of the Mosaic creed. Should this opinion prove to be correct, we would naturally have no right to appeal to the achievements of the Jewish intellect in ancient times. If this view is right, then all the facts

---

Press, Oxford, (1927), pp. 433-463 at 463.

404. L. Pinsker, "Auto-Emancipation", in A. Hertzberg, *The Zionist Idea*, Harper Torchbooks, New York, (1959), pp. 181-198, at 193. H. N. Bialik, "Bialik on the Hebrew University", in A. Hertzberg, *The Zionist Idea*, Harper Torchbooks, New York, (1959), pp. 281-288, at 284. A. D. Gordon, *Kitve A. D. Gordon*, In Five Volumes, Tel-Aviv, ha-Va'ad ha-merkazi shel mifleget ha-Po'el ha-tsa'ir, (1927-1930), parts translated to English in A. Hertzberg, *The Zionist Idea*, Harper Torchbooks, New York, (1959), pp. 371-386, at 376. S. H. Landau, *Kithe*, Warsaw, (1935), pp. 36-43; translated to English in A. Hertzberg, *The Zionist Idea*, Harper Torchbooks, New York, (1959), pp. 434-439, at 437-438.

enumerated above must be eliminated, when we consider the cultural value of the Jewish race. This opinion, however, can easily be refuted by anthropological arguments. But far more serious and more dangerous are the theories of a different kind, which pretend to be the result of strictly scientific research.

These theories do not deny that the Jews of to-day are the descendants of the Jews of ancient times, but assert that both modern and ancient Jews represent an inferior racial element, and that they are injurious to the State and Society in whose midst they dwell. The anti-Semitic theories, of which H. Stewart Chamberlain is now the foremost exponent, are as follows:

The Jewish race has developed its characteristics on lines diametrically opposed to those of the rest of mankind. The inoculation of the characteristics of the Jewish race in other nations would be a great menace to the latter. Above all, however, the Jews deserve to be contemned and despised for their spiritual inferiority. The Semites have never created anything great and comprehensive. They never founded a great organized State. Loyalty, respect for the great, and nobility of character in general, are entirely unknown among the Jews. In all these thousands of years they have not rendered any exceptionally great service in the domain of philosophy, science and art. There are a number of talented Jews, but they have no surpassing genius. The Semitic race, accordingly, is far below the Aryan race. Even the religious genius, which has been, ascribed to the Jews, does not exist, according to Chamberlain. It is just the Jews, he maintains, who are the least gifted in matters of religion. Even the Negro is above them in this respect.

Now anyone familiar with modern tendencies and with the latest literature, will recognize the reality of these disgraceful attacks, and will understand that should such theories be allowed to remain unanswered, they would become a great political danger. It is very desirable, therefore, that we should employ the same weapons as our opponents: that is to say, the weapons of anthropology, sociology and natural science, to investigate the social value of the Jews.

It is unfortunately impossible, you realize, to solve this problem in a single lecture. In the short time allotted to me, I can only give a rough outline of a sketch, to show the manner in which our opponents argue in order to attain such results, and to point out the method we are to choose in our refutation.

It has hitherto been the commonly accepted theory, that in remote antiquity all the nations, from the East Indians to the Britons, from the Greeks to the Norwegians, formed one common race—the Aryan. The great historians of human culture, and especially Renan, propounded the theory, that all great things that were achieved by German industry, British energy, Roman power, Greek art and Indian philosophy, were due to this common Aryan spirit. With these they compared the cultural achievements of the Semites, and arrived at the conclusion that the Semites have indeed achieved much in the field of religion, but have been surpassed by far by the Aryans, in all other domains. To this Aryan theory, which was important enough in itself, there has, in the course of the last decade, been added another one, which is of infinitely greater significance. What is the purport of this new theory, and what relation does it bear to our subject?

The well-known migration of natives, which entirely devastated the south of Europe at the end of classical antiquity was, according to this theory, not an isolated event, but the last link of a chain of such migrations from the Germanic North. These migrations were the consequence of the overcrowded population of these countries, the soil of which became diminished on account of the encroachment of the sea and through glaciation. The severity of the glacial period made the struggle for existence very strenuous, and only the fittest survived. This struggle made it necessary to exert all bodily and mental power. And thus arose in these cold regions a blond, well-built nation, endowed to the highest degree with vitality and mental activity.

When the population became overcrowded, part of this race crossed the Alps, and inhabited in prehistoric times all countries in Southern Europe, the northern coast of Africa, and the western and southern parts of Asia. Some of these stocks even came to China and Japan, and even further. We indeed find to-day in all these countries, men of high stature, blue eyes, blond hair, and long heads. These men are considered the descendants of those men of the prehistoric migrations.

Many problems now appear to be solved. In the first place, we understand why the Aryan speech is so widely spread. For these wanderers brought their language along with them. Hence all the languages, of all the kindred nations from India to the Atlantic Ocean, are related. But this is not the only problem that is solved. It was discovered that the blood-relationship reaches much further. A reason was finally found for the phenomenon that there are so many blond and dolichocephalic, that is, longheaded people, in the South. The explanation was simple. Anthropologically, they belonged to the nations that hailed from the North. This newly won experience is even applied to the Jews. For instance, Esau was red; King David was blond; Jesus, too, as it is sometimes claimed, was blond—hence those men, as well as modern blond Jews, were not pure Semites, but descendants of the Amorites; that is to say, of a race that hailed from the North and which, according to Chamberlain, had a great share in the composition of the Jewish race.

It is claimed, that scientific inquiry has succeeded in demonstrating that great achievements, which history ascribes to the Jews, are due to these non-Jewish elements. Furthermore, that scientific inquiry appears to establish the fact that many of these great achievements were not at all produced by the Jews, but were borrowed by them from the neighboring nations. Thus the most important elements of Jewish culture are supposed to be derived from Babylon and Egypt; and the bulwarks of their religion are supposed to be borrowed from the Sumero-Accadians. But, according to Chamberlain and the politico-anthropological school, these Sumero-Accadians were dolichocephalic—longheaded—and hence of Aryan; of Northern origin.

All these Aryan Germanic natives, according to this theory, had in common, certain characteristics of soul and mind, as well as of creative genius. And in consequence of those creative characteristics, all the enumerated nations had already, in remotest antiquity, attained their high classical culture. To-day, however, all these Oriental countries are almost entirely excluded from cultural creations. The historian of human culture has often occupied himself with this question. But the solution of this problem is only apparently difficult. For in our own times also, only the Germanic nations are politically, economically, spiritually and artistically, the standard-bearers of idealism and progress. These anthropologists find that all the great and important achievements have proceeded from men of Germanic extraction. An explanation was thus found for nearly all striking phenomena.

For through these migrations in remote antiquity, not only Germanic blood, but also Germanic power and energy, and Germanic intellectual productivity were imported to the South. Along with their blood and language these Northern hordes, also brought, according to this theory, to the South, their high and gigantic cultural ability; while the primitive inhabitants of the latter countries had lived in an intellectual lethargy. Thanks to these invasions, all the oriental nations of antiquity were enabled to attain the loftiest summit of civilization. But as the northern blood of that uncultured primitive population was slowly and gradually waning, these primitive nations fell back to their present-day inactivity and sluggishness. Their cultural value was reduced, in proportion to the dilution of the quality of their blood. The decline of Greece and Rome is thus easily explained by the anthropologists, through the waning of the fair-complexioned race elements. For the cultural value of a nation stands in direct relations to its racial value. And this racial value depends on the quantity of northern blood which still flows in its veins. Hence the racial value of the Jews is very insignificant, according to the teaching of Gobineau, the politico-anthropological school and Chamberlain.

According to Chamberlain, the Jews are, apart from this, a bastard nation, which arose through the mingling of racially different nations: Semitic Arabs, Aryan Amorites and Syrian Hittites. It is this bastard character which is responsible for the unusual inferiority of the Jewish race.

I am extremely sorry that I am not in a position to discuss here in detail the anthropology of the Semites. For, although theories explained here appear far-fetched at first sight, they are, nevertheless, important. It would by far lack due emphasis, were I merely to explain to you that these theories are incorrect. It is necessary to enter deeply into this question, in order to see how furidamentally wrong these theories are, and that in many cases just the opposite is true. But one must enter into linguistic and pre-historical, as well as into sociological and anthropological investigation, and into a study of the laws of heredity, if one wishes even to begin to criticize this system. By investigating the history of human cultute we find, to take only a single example, that no Aryans ever existed at all, and that identity of language does not permit us to draw any conclusions about identity of race. For, according to this language theory, all negroes in South America would be pure Spaniards and all negroes of North America would be pure Anglo-Saxons! Languages are altered and transformed through political and social influences, so that two neighboring and kindred nations may by chance speak different languages. Thus the Jews of to-day collectively speak all the languages of the world except their own. And thus, also, the Persian or the Armenian, who is supposedly Aryan, is, according to all anthropological characteristics without any doubt, more akin to the Semitic Syrian than to the Iberian or Norwegian. For this reason alone, it is impossible to speak of the contrast between the Semites and the Aryans.

But more significant than these linguistic considerations are the anthropological investigations themselves, of too technical a nature to be discussed here in detail, concerning which I must refer you to my book on this subject. The researches about this matter force upon us the conclusion that the Germanic race theory is from beginning to end untenable and without foundation.

All this is, however, only a part of that which an impartial investigation into the material reveals, but even this is sufficient to prove the whole proud edifice of these theorists to be only a house of cards, which can offer no resistance to a keen critic. But anthropological inquiry yields still more important results. For the division of the races of man, according to their historical development—and this is the only division possible to-day—arrives at conclusions diametrically opposed to those maintained by these theorists.

When we enter into the study of anthropology, we find an entirely different grouping of nations. On account of the glaciation of the Alps, the entire white Caucasian race was, for many thousand years after the glacial period, divided into two unequal groups of nations differing, therefore, from each other, in their development and physiognomy; the land in the cold regions north of the Alps was inhabited by the fair-complexioned group — the Xanthochroic or light-haired—and the land south of the Alps was populated by a darker-haired group—the Melanochroic. To the Xanthochroic belong the Slavonic-Keltic-Germanic nations; while to the Melanochroic, south of the Alps, belong the nations of Southern Europe, North Africa, and the white nations of Asia. To the southern group belong, accordingly, the Jews and other Semites, as well as the East Indians, Persians, Sumero-Accadians, Egyptians, Greeks, Romans, etc.

According to the dogma of the race theorists, innate ability is determined by birth, and nations of the same race must necessarily be equally gifted. The Jews, according to this division, are of the same race as the nations enumerated above, and hence their innate ability must in no respect differ from that of the Indians, Sumero-Accadians and Greeks. The racial value of the Jews must, therefore, be the same as that of those nations of which the race theory treated; namely, of all Aryan nations except those of the Germanic group. For just as the Germanic nations distinguished themselves among the Xanthochroic group, so did the Jews excel among the Melanochroic types. That group to which the Germans belong, entered the stage of civilization only as late as the 13th century, and it is only in the very late periods that it assumed a leading role in the advance of European culture. The nations of the other group had a high state of civilization in remotest ages, and some of them, for instance, the Egyptians and Babylonians, stood thousands of years ago, at the highest stage of classical development. As Greeks and Romans they created the classical culture, and as Moors, Byzantines and Italians, they were the authors of post-classical civilization.

Is it, however, at all true, that innate ability depends upon race, and that every race has its specific racial peculiarities which invariably adhere to it forever, under all conditions and circumstances? Is there an innate racial soul which never changes? Are the psychical bases of various races really fundamentally different? It is true that there are different racial characteristics and abilities. But do these fundamental racial peculiarities remain the same throughout all ages or are they subject to the laws of change?

This is a problem with which Science has interested itself for more than a century. Formerly it was merely a subject for philosophic speculation, but it has now entered into the field of experimental investigation. In the field of heredity two views are now current, that of Lamarck, who insists upon the adaptability and changeability of characteristics in the entire organic world, and that of Weissmann, who maintains that the specific character always remains the same. However interesting it may be to pursue this theme in detail, I must confine myself to a brief resume of the results obtained from a historico-philosophical analysis and further study of the laws of heredity. The theory that acquired characteristics are not transmissible and that the specific character is absolutely constant, can now be regarded as exploded. As it is impossible to give details on this point in a single lecture, I must again refer you to my book for a fuller discussion. What applies to the entire organic world applies with greater force to man. It is therefore not true that we are justified in assuming specific racial psychical powers for each race. It is indeed true that the Greeks distinguished themselves by their artistic sense and the Romans by their energy, and that the peculiarities of the Italians differed from those of the Scandinavians. But the reason for these differences are to be found in their historical and social environment. The inductive method of historical investigation shows that the internal character of these nations changed, when the external conditions altered fundamentally. Thus the so-called innate family virtues of the Jews may be lost, when they come in disturbing environments. It is equally untrue that the essential psychical differences of the various races can be demonstrated by natural science, in the sense that all pre-eminent Frenchmen must distinguish themselves by their *esprit*, and Germans can only excel as poets and thinkers, and that the specific ability of the ancient Greeks lay only in art, that of the ancient Indians only in philosophy, that of the Romans only in conquest and control, and that of the Jews only in Commerce.

The psychology of a people changes at the various stages of culture through which it passes. Most people pass through the same stages of 'Volkspsychologie,' at one stage or another of their existence, and this 'Volkspsychologie' is the product of the particular stage. There is a peculiar psychology of hunters and husbandmen, of scholars and merchants; a distinct psychology of the inhabitants of the country and of the inhabitants of the city. This is the same among all races. There would accordingly be more justification to speak of a psychology of stages than of a psychology of races. The quality of the capability of a nation does not depend upon its race, but upon environment, the stage of development through which it at the moment happens to be passing, and upon the influences of tradition.

And yet when we consider the capacity and psychical intellectual ability of a nation, we cannot say that it is immaterial from which race it descended. The descendants of one race may indeed be more gifted than those of another. The explanation is to be found in the past experience of that stock. In the entire organic world, we find that every being developed and perfected those organs which were mostly employed. The limb which is most exercised, grows best. When it was necessary, therefore, for a certain species to develop its brain to the highest perfection—when a certain race, by its own free-will or by force of circumstances, devoted itself to work which required it to perfect the brain, it necessarily follows that the descendants of such a race have the advantage over the descendants of another race. The quality of their ability, as was remarked above, depends upon environment, the stage of development and the influences of tradition; but the quantity of their capacity, the magnitude and intensity of their ability does not depend upon environment, but upon race, or rather upon the cultural activity of their ancestors. This is, therefore, a factor of heredity.

Now with what people and with what race was the cultural activity of their ancestors greater than with the Jews? For with the Jews study was a religious duty, and those among them who did not possess a high degree of intellectual activity were not fit for the struggle for existence. In consequence of the intensive cultural activity of their ancestors, the Jews must possess the maximum sum of innate ability.

This result is obtained from the theory of heredity. Anthropology, as we have shown, points to the contrast between the Xanthochroic and Melanochroic. But this contrast also led us to a conclusion different from that taught in the schools. All those nations which achieved the great things, and created the intellectual monuments, belong to the same groups of races to which the Jews belong. This would be the inference from the mode of distributing the intellectual ability, if we are to maintain with the race theorists, that nations derived from the same race are equally gifted. I merely wish to hint at this conclusion.

But the racial pride of the Semites does not require them to employ any speculative demonstration and logical deductions, which may perhaps be considered as sophistry. The simple, but forceful historical facts in themselves render all other demonstrations unnecessary. The principal reproach cast upon the Jews by their foes, that the Semitic race lacks creative genius, stands self-condemned in the light of the result of modern research, which considers Mesopotamia, the cradle of all the Semites, as the place where civilization originated. And furthermore, no period of history is more neglected by these theorists than the golden age of Semitic culture in Spain. They pass over in silence the influence that that period had on the development of modern Europe. There is an unbroken chain of evidence to prove that the origin of Humanism and of the Renaissance of which Europe is so proud, can be traced to the Semites, Jews and Arabs, in Spain The Jew indeed among the nations, who draw upon his resources and in whose midst he lives, is only one of the heirs of his own past achievements.

There is, however, another important question which waits an answer. We have seen that the Jews and the other Semitic nations were the torch-bearers of civilization. In ancient times the Babylonians, Phoenicians, and Carthaginians took an active part in advancing human culture, while in mediaeval times the Arabs achieved wonders, and were the leading and creative genius of all that is great. How is it that now, as it seems, the Jews are merely receptive and reproductive, but do not produce anything really new? An explanation of this phenomenon is to be found in the social structure of presentday Jewry.

In Mesopotamia, Palestine, and finally in Spain, these nations lived in accordance with their own culture. They did not confine themselves to one branch of industry, but, like all other nations of the earth, cultivated all sorts of trades. But the unnatural historical development of the Jews, and the quite unnatural distribution of professions of to-day must inevitably produce unnatural results. The social structure of present-day Jewry is unsound. The keen struggle for existence stifles much that is really great and profound, so that for the most part only those that are commercially fit are able to rise. In consequence of the present-day development, which is contrary to the law of natural selection, Judaism of to-day cannot fully bring out its dormant powers, and its cultural energies cannot be brought into complete action.

The development of great talents finds a favorable field among such nations, ashaving grown to fruition with their soil—owing to their calm and stable pursuits, have the necessary leisure to think and contemplate for its own sake. But in a commercial community where the struggle for existence is still more intensified by political and economic conditions, such talents are crippled or lie fallow and rusty. It is due to this influence, which is contrary to the law of natural selection, that the Jews are extremely ambitious. Prof. Werner Sombart erroneously takes this as the principal characteristic of the Jewish race. In addition to those disadvantages, we must take account of the destruction of the old religious and Ghetto environment, in which the people were at least complete after their fashion. Ours is a period of hollow and empty transition. The inner distraction and disruption of our people in this transition, have caused this characteristic to be considered as the principal feature of the Jewish race. It is very unfortunate that, owing to exceedingly superficial reasoning, the noble personalities are left out of account. The mediocre and obtrusive Jews are in evidence, and they form the criterion for the entire Jewry. The gross, misleading picture which arose through the social structure of Jewry in the diaspora depicted the Jew as the type opposed to all that is lofty in humanity.

The peculiar environment brought it about, that the actual conditions could not have been different from what they are to-day. Under the conditions existing at present, the Jews cannot attain that richly productive activity which in remote antiquity their ancestors developed in Mesopotamia, and later on in the Pyrhenean peninsula. And yet even to-day, under the most discouraging circumstances, the Jews have created not only the modern system of capital, or not only a large number of prominent workers in purely intellectual domains, but they are also the creators of the new currently dominant tendencies of knowledge. One at once thinks—to mention only a few—of Hertz and Ehrlich, of Marx and Stahl, of Spinoza and Bergson, and of Georg Kantor in mathematics. One sees that your profound thinkers have very often created also in heterogeneous cultures, a transvaluation of all intellectual, ethical and religious values, a radical change and renewal of the whole spiritual life. One wonders what their cultural value would be under healthy and normal circumstances. We fear to draw any definite conclusion on this point, lest it should sound exaggerated and speculative, to say the least.

Through the conscious efforts of numerous generations of thinkers and statesmen and through the influence of religion, a nation of pure blood, not tainted by diseases of excess or immorality, of a highly developed sense of family purity, and of deeply rooted, virtuous habits, would develop an exceptional intellectual activity. Furthermore, the prohibition against mixed marriage provided that these highest ethnical treasures should not be lost, through the admixture of less carefully bred races. This prohibition brought it about that heredity, which is the first factor in the formation of a race, should exercise its power in a most beneficial way, and thus the racial qualities are not only transmitted from generation to generation, but are gradually heightened.

Thus from the striving after eternal existence (which was likewise a commandment of the Deity), there resulted that natural selection which has no parallel in the history of the human race. In the struggle for existence imposed upon this nation, which was shaken by fire and sword, by the hardest economic and moral oppression, and by constant enticements to fall away, only those individuals who were morally and physically strong could survive and propagate.

Thus the Jews form an ancient, chaste race of a maximum cultural value. If a race that is so highly gifted were to have the opportunity of again developing its original power, nothing could equal it as far as cultural value is concerned.

We thus admit that, despite the extraordinary share that the modern Jews contribute to the advance of civilization, their achievements are only an insignificant part of that which they could have produced under normal conditions. The philosopher Eduard von Hartmann, who can by no means be regarded as a friend of the Jews, has admirably expressed himself on this point when he says:

The conflicting position of Judaism makes it impossible for the Jews to produce anything new in the field of a Jewish national culture, which does not exist, or in the field of the national culture of other nations. But the versatility of Judaism and the originality of its comprehension are sufficiently large to enable it to adapt itself to alien national cultures of various kinds, and by good fortune sometimes to reach as far as that borderline, which divides talent from genius.' This proves, at least, there is nothing against the assumption, that should a Jewish national culture exist, the old productivity of Judaism would manifest itself once more.

I have made no reference in this lecture to the enormous influence of the religions to which Judaism gave birth. There is hardly any parallel for such activity in the cultural world. Nor have I spoken of the Jewish spirit, that is to say, Judaism in a broader sense, that lies hidden in these religions and in the most important intellectual movements of modern times, as, for instance, in Philosophy and Socialism. I have purposely confined myself to the services rendered by the Sernites in other domains, to the material culture, and to the investigation of our problem from the point of view of pure Natural Science.

I am satisfied if I have been able to show you, that even if the Jewish people should prove itself unequal to the task of carrying out its wonderful mission, namely, to realize its dormant potentialities, no stigma of belonging to an inferior race can be attached to it in the name of Science."[405]

In agreement with Philipp Lenard's later view that "Jewishness" could be seen in intellectual works published by Jews, Einstein stated sometime "after 3 April 1920",

"The psychological root of anti-Semitism lies in the fact that the Jews are a group of people unto themselves. Their Jewishness is visible in their physical appearance, and one notices their Jewish heritage in their intellectual works, and one can sense that there are among them deep connections in their disposition and numerous possibilities of communicating that are based on the same way of thinking and of feeling. The Jewish child is already aware of these differences as soon as it starts school. Jewish children feel the resentment that grows out of an instinctive suspicion of their strangeness that naturally is often met with a closing of

---

405. I. Zollschan, "The Cultural Value of the Jewish Race", *Jewish Questions: Three Lectures*, New York, Bloch Pub. Co., (1914), pp. 3-19.

the ranks. [***] [Jews] are the target of instinctive resentment because they are of a different tribe than the majority of the population."[406]

Viktor G. Ehrenberg, Hedwig Born's father, wrote to Einstein on 23 November 1919,

"So it uplifts the heart and strengthens one's faith in the future of mankind when one sees the researchers of all nations prostrating themselves before a man of Jewish blood, who thinks and writes in the German language, in full recognition of his greatness."[407]

The Zionist United States Supreme Court Justice Louis Dembitz Brandeis wrote in a letter dated 1 March 1921,

"You have doubtless heard that the Great Einstein is coming to America soon with Dr. Weizmann, our Zionist Chief. Palestine may need something more now than a new conception of the Universe or of several additional dimensions; but it is well to remind the Gentile world, when the wave of anti-Semitism is rising, that in the world of thought the conspicuous contributions are being made by Jews."[408]

Brandeis' racist views were, in part, a reaction to the views of the ancients who asserted that the Judeans produced nothing new, and men like Bauer, Marx, Wagner, Dühring and Houston Stewart Chamberlain, who asserted that Jews, with their dogmatic and obedient monotheism, detested anything new and repressed science and art. Ada Sterling wrote in *The Jew and Civilization* published in 1924,

"NOTWITHSTANDING the honor which the world of scientists yields to their Jewish confrères, the Messrs. Michaelson, Bergson, Einstein, and a host of lesser men, who, nevertheless, have made and are making continually, great discoveries toward improving conditions of life for humanity, there have been published, and recently, a vast amount of deliberate mispraisement of the Jew in science as in other departments of life, and ingenious arguments, the purpose of which is to minimize his present-day worth, and to deny his race a position among the pioneers in the field of physics. It is not surprising if the uninformed, overwhelmed by the dogmatic positiveness of such a rabid foe to the Jews as Mr. Chamberlain—

---

[406]. A. Einstein, English translation by A. Engel, *The Collected Papers of Albert Einstein*, Volume 7, Document 35, Princeton University Press, (2002), pp. 156-157.
[407]. Letter from V. G. Ehrenberg to A. Einstein of 23 November 1919, English translation by A. Hentschel, *The Collected Papers of Albert Einstein*, Volume 9, Document 173, Princeton University Press, (2004), p. 145.
[408]. L. D. Brandeis, M. I. Urofsky and D. W. Levy, Editors, *Letters of Louis D. Brandeis* Volume 4, State University of New York Press, Albany, New York, (1975), pp. 536-537.

who angrily deplores that 'Walhalla and Olympus became depopulated because the Jewish priests wished it so'—should take on similar prejudice and beliefs; or that they should accept his violent assertions when he declares that it was the Jews' scorn of science which long retarded the spread of knowledge along scientific lines. Nor is it to be wondered at if the uneducated, seeing in a news-sheet a belittling allusion to the uselessness of star-measuring should find themselves repeating such idle estimates of the scientific seekers, especially in connection with the measuring of Betelgueuse.

Mr. Chamberlain's statement is an interesting admission in more ways than one. It ascribes to a people strictly 'inferior', and he so names them over and over again, powers which only a distinctly superior people could possess. This contradiction is a common characteristic, as has been pointed out in another connection, of the resolute anti-Semite; but few so often display it as does the writer just referred to. He pronounces the Jews 'mentally sterile', and presently shows them to be the most mentally active people in the world, dangerously creative in fact; he undertakes to prove them the most money-worshipping race in the world—by means of a characteristic with which he invests the matriarch Rebekah—and denies them wit enough to invent numerals.

To prove their stupidity he says that in sharp business transactions 'one Armenian is a match for three Jews.' He resorts, as well, to quoting Apion's time-worn accusation—ascribing it to Wellhausen—that 'the Jews never invented anything,' and he attaches a deal of indexed learning to prove that the race has never even been near to 'grasping the eel of science'; to prove, as well, that all the Jewish people knew—they with a known history of three thousand years, and a traditional one of many thousands more—they borrowed, he says, from their young neighbors, the Greeks, who came into existence 800 B. C."[409]

The dangers of a racist Jewish reaction to any criticism of Jews are many, and the racism of Brandeis and his ilk only serve to inhibit the progress of science and the uninhibited criticism of scientific theories. Brandeis and Sterling are wrong to make a "racial" defense and to assume that all criticisms are completely false, merely because they are false in their "racial" aspects. Ironically, Brandeis and Sterling reinforce the racism they ought to have attempted to discredit. Sterling continued, arrogantly parroting the lies many racist Jews told to promote Einstein,

"As each new ascent in knowledge is made possible by the plane attained by our predecessors, so it has been said that the Morley-Michelson experiments are the starting-point whence arises the Einstein theories on Equivalents and Relativity, which latest discovery of the Jewish mind,

---

[409]. A. Sterling, *The Jew and Civilization*, Aetco, New York, (1924), pp. 202-203.

though yet to be proven, have been greeted by the scientists of the age as 'probably the most profound and far- reaching application of mathematics to the phenomena of the material universe that the world has ever known'; one which 'takes us behind our present ideas about space, time and matter to the primitive reality out of which we have built up those ideas'. Professor Thomson says Mr. Wells had a pretty clear idea of it all before Einstein's theory appeared; but, he adds, Einstein takes us a big step farther. He asked a question which nobody had asked before him: 'Is the space and time interval which separates two events the same for everybody'?

[***]

The Einstein Theory, while still, in part, under experiment, nevertheless has already solved problems that had worried great mathematicians for generations. To test it, England sent out an important expedition, for the purpose of photographing the stars whose light passed near the sun, when it was in eclipse. The 'Theories' stood the test; more, strikingly verified them. 'Einstein's Theory', say the editors of 'The Outline of Science', shows, further, 'that there is something in the nature of an ultimate entity in the universe' though even yet we know nothing intelligible about it; but, these authorities believe it will presently be made clear through the Einstein discoveries that the whole universe has been .... created by the mind itself.

To what insignificant proportions do fanatical critics shrink before the blaze of scientific accomplishment which haloes the modern Jew, and this, not alone because of his exploration of the spaces of the sky, not alone for setting back of the horizon to take in undreamed of worlds, but because, too, great men of the race, regardless alike of fame, and of profit, work on in the secret quiet of 'Science's holy cell', seeking tirelessly and often finding panaceas for the relief of humanity's ills!

But, great as are the findings of the race in the broader fields of physics, to the individual they are of less instant value than are the mysteries of life which chemists, physicians and other scientists of the race may be credited with. At these, too, we will now glance."[410]

---

[410]. A. Sterling, *The Jew and Civilization*, Aetco, New York, (1924), pp. 221-222.

7 NAZISM IS ZIONISM

Zionists have always employed the bogey of "anti-Semitism" to force Jews into segregation. After thousands of years of planning and work, the Jewish bankers had finally accumulated enough wealth to buy Palestine and destroy the Gentile world in fulfillment of Jewish Messianic prophecy. They only lacked one resource needed to become King of the Jews, the Holy Messiah. That one last necessary ingredient for fulfillment of the prophecies of the End Times was the Jewish People, the majority of whom rejected Zionism. The Jewish bankers had an ancient solution for that problem. They manufactured an anti-Semitic dictator who segregated the Jews and filled them with the fear of God. Palestine was for the fearful remnant. Those who would not obey were to have their necks broken and be thrown into the well.

> "The way I see it, the fact of the Jews' racial peculiarity will necessarily influence their social relations with non-Jews. The conclusions which—in my opinion—the Jews should draw is to become more aware of their peculiarity in their social way of life and to recognize their own cultural contributions. First of all, they would have to show a certain noble reservedness and not be so eager to mix socially—of which others want little or nothing. On the other hand, anti-Semitism in Germany also has consequences that, from a Jewish point of view, should be welcomed. I believe German Jewry owes its continued existence to anti-Semitism."—ALBERT EINSTEIN[411]

> "Hitler will be forgotten in a few years, but he will have a beautiful monument in Palestine. You know the coming of the Nazis was rather a welcome thing. So many of our German Jews were hovering between two coasts; so many of them were riding the treacherous current between the Scylla of assimilation and the Charybdis of a nodding acquaintance with Jewish things. Thousands who seemed to be completely lost to Judaism were brought back to the fold by Hitler, and for that I am personally very grateful to him.'"—EMIL LUDWIG[412]

> "[H]ad I been a Jew, I would have been a fanatical Zionist. I could not imagine being anything else. In fact, I would have been the most ardent Zionist imaginable."—ADOLF EICHMANN[413]

7.1 Introduction

The Old Testament's solution to the Jewish question was two-fold. If the Jews obeyed God and remained segregated, God would give them the land from the

---

411. A. Einstein, A. Engel translator, "How I became a Zionist", *The Collected Papers of Albert Einstein*, Volume 7, Document 57, Princeton University Press, (2002), pp. 234-235, at 235.
412. M. Steinglass, "Emil Ludwig before the Judge", *American Jewish Times*, (April, 1936), p. 35; *as quoted in:* L. Brenner, *Zionism in the Age of the Dictators*, Chapter 6, Croom Helm, London, L. Hill, Westport, Connecticut, (1983), p. 59.
413. A. Eichmann, "Eichmann Tells His Own Damning Story", *Life Magazine*, Volume 49, Number 22, (28 November 1960), pp. 19-25, 101-112; at 22.

Nile to the Euphrates. Note that the Jews were not the original inhabitants of the land and that they promised it to themselves. If the Jews did not obey God and assimilated into the Gentile world, they would be laid to waste in the lands in which they dwelt, and the righteous remnant—the most racist Jews—would steal the Promised Land from its original inhabitants. Note that racist Jews created this religious mythology and only racist Jews feel obliged to fulfill it.

The Old Testament taught racist Jews to subvert Gentile society. They suppressed the advancement of the Gentile world as best they could by taxing the people with wars and perpetual strife, which left the nations in debt. They fomented highly destructive revolutions, often by scapegoating fellow Jews. They did whatever they could to prevent Gentiles from accumulating wealth and dominating politics, the universities, the professions and the press. Always a minority, racist Jews have no regard for democratic principles. They are religious fanatics, who pretend that they are the master race elected by God to rule the world (*Isaiah* 65; 66).

7.2 *Blut und Boden*—A Jewish Ideal

Judaism is a racist and genocidal world view; in that it creates the mythology of a master race, the "chosen people", tied to a specific "Holy Land", who are after world domination following their deliberate destruction of other nations. This warmongering tribe believes in a God that exterminates their enemies in order to enthrone them as rulers of the Earth. Many who have spoken out against Judaism are repelled by what they consider to be a slavish loyalty to a mythology which denigrates the nobility and the dignity of the individual human being—especially the non-believer and all who are not "racially" allied with the supposedly master race of Judeans. They argue that it is irrational to utilize an imaginary "God" as the fundamental source of logical deduction for all that is said to be moral. They sought something more synthetic and rational. When searching for a religion to replace Judaic and Christian mythologies, Schopenhauer and Wagner adopted the mythologies of metempsychosis reincarnation of the "Aryans" as the supposedly true product of the "racial instincts" of the "Aryan race"—as opposed to the slavish and destructive Judaic and Christian mythologies, which many anti-Semites believed stemmed from the inferior and corrosive "racial instincts" of the Jews.

Many modern Zionists embraced these racist systems of thought and spoke in Nazi-like terms of "the end justifies the means", or "hevlei Mashiah", in order to justify their deliberate destruction of the Earth for the sake of Israel. Zionist Judah Leon Magnes criticized their "Joshua methods" and arrogance. Magnes captured their prevalent beliefs,

> "There is the *Wille zur Macht*, the state, the army, the frontiers. We have been in exile; now we are to be masters in our own Home. We are to have a Fatherland, and we are to encourage the feelings of pride, honor, glory that are part of the paraphernalia of the ordinary nationalistic patriotism. In the face of such danger one thinks of the dignity and originality of that passage

in the liturgy which praises the Lord of all things that our portion is not like theirs and our lot not like that of all the multitude."[414]

The farce of Jewish Zionist nationalistic supremacy reached a very low point when Martin Buber argued that the Zionist beliefs in their master race were superior to the Nazis' assertions that they constituted a master race of the "chosen"; because God chose the Jews, and the Nazis just chose themselves; and, therefore, the Jews were entitled to their racism, while the Nazis were not,

> "It is asserted that every great people regards itself as the chosen people; in other words, awareness of peculiarity is interpreted as a function of nationalism in general. Did not the National Socialists believe that Destiny had elected the German people to rule the entire world? [***] Our doctrine is distinguished from their theories in that our election is completely a demand. [***] Israel was chosen to become a true people, and that means God's people."[415]

The origins of the Judeans lie in the Canaanites, and others. There never were "Israelites". The mythology of a man named Moses, who led "his people" out of Egypt, lacks evidentiary archeological support. It is likely that the Judeans' Torah originates with the Egyptian monotheism of Pharaoh Akhenaton IV.

Judaism and Christianity were the products of fancy, not fact. The philosophy of Philo "Judæus" is in many respects strikingly similar to the philosophy of Jesus, and one need only mix in some of Æsop's legend and fables with Heraclitean and Platonic philosophy to arrive at the teachings of Christ as the recasting of Mosaic (Egyptian) and Greek monotheism in Greek dialectic terms as expressed by Essenian Jews—which is especially clear in the oldest known texts in their original languages. The life story of Jesus mirrors many of the much older myths of Mithras,[416] whom the Roman soldiers worshiped as their Sun god. The apocalyptic myths of Christianity were first stated by the Jewish apocalyptic writers not long before Christianity emerged from the wreckage of Judaism. Some argue that the entire gospels may have been fabrications written by Alexandrian and Essenian Jews in their efforts to

---

414. J. L. Magnes, *Like All Nations?*, Jerusalem, (1930), quoted in A. Hertzberg, *The Zionist Idea*, Harper Torchbooks, New York, (1959), pp. 443-449, at 447.

415. M. Buber, *Israel and the World*, Schocken Books, New York, (1948), quoted in A. Hertzberg, *The Zionist Idea*, Harper Torchbooks, New York, (1959), pp. 457-463, at 460-461.

416. Friedrich Wilhelm Ghillany, under the pseudonym Richard von der Alm, *Die Urtheile heidnischer und jüdischer Schriftsteller der vier ersten christlichen Jahrhunderte über Jesus und die ersten Christen: eine Zuschrift an die gebildeten Deutschen zur weiteren Orientirung in der Frage über die Gottheit Jesu*, Otto Wigand, Leipzig, (1864); **and** *Theologische briefe an die gebildeten der deutschen Nation*, In Three Volumes, Otto Wigand, Leipzig, (1862-1863).

incorporate Greek philosophy, sayings and superstitions into Judaism in a form of Christianity which would appeal to the Romans. Some of the ancient Greeks argued that Judaism is itself a plagiarism of the Greek philosophers.

When the Romans rejected this highly Helenized Judaism of Christianity and campaigned against the Jews, some of the Jews themselves may have then changed the stories that they had written, in order to make it appear that they, too, had opposed the Heretic "Jesus", which "name" may be interpreted simply as "Jew" or "Judean". Agrippa and Alexander Lysimachus are then said to have persecuted the early Christians.[417] Those who clung to Christianity took the Helenization of Judaism even further in their apologies to the Romans by trying to convince them that this form of Judaism was Greek in nature. The Romans greatly admired and thoroughly copied Greek culture and religion and it was a sound stratagem to attempt to convince the Romans that Christianity was Greek in nature, and, after all, it was. The early Christians also took on the Judaic penchant for religious fanaticism, proselytizing and the Judaic love of martyrdom. Jews of the Diaspora in general tended to segregate and considered themselves an independent nation, regardless of where they happened to live at the time. This troubled many who worried where their loyalties lay. Many have argued that the Jews were dispersed throughout the civilized world long before the Romans tossed them out of Palestine.

According to some, the Jewish segregationist habit of forming a "nation within a nation" predated the Roman imposed Diaspora. Though the story of the Egyptian captivity is fabricated, it evinces a segregationist spirit in the earliest Jewish works. Schiller wrote in his *Die Sendung Moses*,

"Die Hebräer kamen, wie bekannt ist, als eine einzige Nomadenfamilie, die nicht über 70 Seelen begriff, nach Ägypten und wurden erst in Ägypten zum Volk. Während eines Zeitraums von ungefähr vierhundert Jahren, die sie in diesem Land zubrachten, vermehrten sie sich beinahe bis zu zwei Millionen, unter welchen sechshunderttausend streitbare Männer gezählt wurden, als sie aus diesem Königreich zogen. Während dieses langen Aufenthalts lebten sie abgesondert von den Ägyptern, abgesondert sowohl durch den eigenen Wohnplatz, den sie einnahmen, als auch durch ihren nomadischen Stand, der sie allen Eingebornen des Landes zum Abscheu machte und von allem Anteil an den bürgerlichen Rechten der Ägypter ausschloss. Sie regierten sich nach nomadischer Art fort, der Hausvater die Familie, der Stammfürst die Stämme, und machten auf diese Art einen Staat im Staat aus, der endlich durch seine ungeheure Vermehrung die Besorgnis der Könige erweckte."

Exodus 1:8-14 and 3:2 taught the Jews that oppression strengthened their "race" and ultimately increased their numbers, and note the ancient declaration made by the Jews themselves (the story is a fabrication) that the Jews were a dangerously disloyal nation within a nation, note also the image of enduring a

---

417. *Acts*, Chapter 4, Josephus' *Antiquities of the Jews*, Book 20, Chapter 5.

holocaust,

> "8 Now there arose up a new king over Egypt, which knew not Joseph. 9 And he said unto his people, Behold, the people of the children of Israel *are* more and mightier than we: 10 Come on, let us deal wisely with them; lest they multiply, and it come to pass, that, when there falleth out any war, they join also unto our enemies, and fight against us, and *so* get them up out of the land. 11 Therefore they did set over them taskmasters to afflict them with their burdens. And they built for Pharaoh treasure cities, Pithom and Raamses. 12 But the more they afflicted them, the more they multiplied and grew. And they were grieved because of the children of Israel. 13 And the Egyptians made the children of Israel to serve with rigour: 14 And they made their lives bitter with hard bondage, in morter, and in brick, and in all manner of service in the field: all their service, wherein they made them serve, *was* with rigour. [***] 3:2 And the angel of the LORD appeared unto him in a flame of fire out of the midst of a bush: and he looked, and, behold, the bush burned with fire, and the bush was not consumed."

While it is true that the myth of the Hebrews in Egypt is the story of a nation within a nation, the relevant point is that the Jews were spread out across the ancient world long before the Romans sacked Jerusalem. They tended to live in highly segregated communities, which hypocritically insisted upon religious freedom for Jews, while proscribing the exercise of all other religions in their communities. Jewish intolerance and hypocrisy was typified in the age old refrain that the "Jews are a nation within a nation." Therefore, they are a disloyal and potentially treacherous force in all other nations. Zionists played on this common conception to create animosity towards Jews, which would leave them with no choice but to found an independent nation on conquered soil.

Zionist Napoleon Bonaparte stated,

> "The Jews must be considered as a nation, and not as a sect. They are a nation within a nation."[418]

Thomas Jefferson declared that Jews constitute a nation within a nation. Daniel J. Boorstein wrote in his book *The Americans: The Colonial Experience*,

> "The Society of Friends had become a kind of international conspiracy for Peace and for primitive Christian perfection. Some years after the Revolution, Thomas Jefferson called them 'a religious sect ... acting with one mind, and that directed by the mother society in England. Dispersed, as the Jews, they still form, as those do, one nation, foreign to the land they live in. They are Protestant Jesuits, implicitly devoted to the will of their superior, and forgetting all duties to their country in the execution of the

---

[418]. S. Schwarzfuchs, *Napoleon, the Jews, and the Sanhedrin*, Routledge & Kegan Paul, London, Boston, (1979), p. 49.

policy of their order.'"[419]

Adolf Stoecker stated in 1879,

> "The Jews are and remain a people within a people, a state within a state, a separate tribe within a foreign race."[420]

Richard Gottheil, while not taking the harsh position that a Jew must be loyal only to Palestine, stated in 1898,

> "We believe that the Jews are something more than a purely religious body; that they are not only a race, but also a nation; though a nation without as yet two important requisites — a common home and a common language."[421]

Gottheil repeated this in a pamphlet and also stated in 1898,

> "Zionism has sought and has found for us a basis which is a broader one than the religious one (and on which all religious distinctions vanish), that of race and of nationality. And even though we do not know it, and even though we refuse to recognize it, there are forces which are unconsciously making for the same end, working out in spite of us the will of Almighty God. Never before has such intelligent interest been taken by the Jews in their own past history. Germany has become honeycombed with societies for the study of Jewish history. Vienna, Hamburg, and Frankfurt have associations for the preservation of Jewish art. The Société des Etudes Juives, the American Jewish Historical Society, the Anglo-Jewish Historical Society, the Maccabeans in London, the Judaeans in New York, the Council of Jewish Women, the Chautauqua Assembly meetings—all of these and many others are working in the same direction. They are welding the people of Israel together once more. They are not religious societies, mark you. They rest upon the solid basis of common racial and national affinity. [***] Nay! it would seem to me that just those who are so afraid that our action will be misinterpreted should be among the greatest helpers in the Zionist cause. For those who feel no racial and national communion with the life from which they have sprung should greet with joy the turning of Jewish immigration to some place other than the land in which they dwell. They must feel, e.g., that a continual influx of Jews who are not Americans is a continual menace to the more or less complete absorption for which they are

---

[419]. D. J. Boorstein, *The Americans: The Colonial Experience*, Vintage Books, New York, (1958), pp. 64-65.
[420]. P. W. Massing, *Rehearsal for Destruction: A Study of Political Anti-Semitism in Imperial Germany*, Howard Fertig, New York, (1967), pp. 285-285.
[421]. R. Gottheil, "The Jews as a Race and as a Nation", *The World's Best Orations*, Volume 6, F. P. Kaiser, St. Louis, (1899), pp. 2294-2298, at 2296, at 2297.

striving. But I must not detain you much longer. Will you permit me to sum up for you the position which we Zionists take in the following statements: We believe that the Jews are something more than a purely religious body; that they are not only a race, but also a nation; though a nation without as yet two important requisites—a common home and a common language."[422]

Zionist racist Max Nordau declared at least as early as 1905,

"The first Zionist congress solemnly proclaimed in the face of the attentive world that the Jews are a nation, and that they do not desire to be absorbed by other nations."[423]

Zionist Rabbi Abraham Isaac Kook proclaimed,

"After our race was weaned [***] This people was fashioned by God to speak of His Glory; it was granted the heritage of the blessing of Abraham so that it might disseminate the knowledge of God, and it was commanded to live its life apart from the nations of the world. [***] It is a grave error to be insensitive to the distinctive unity of the Jewish spirit, to imagine that the Divine stuff which uniquely characterizes Israel is comparable to the spiritual content of all the other national civilizations. [***] It is a fundamental error to [***] discard the concept that we are a chosen people. We are not only different from all the nations, set apart by a historical experience that is unique and unparalleled, but we are also of a much higher and greater spiritual order."[424]

Zionist leader Nachum Goldman stated,

"Diaspora Jewry (all Jews outside Palestine) has to overcome the conscious or subconscious fear of so-called double loyalty. It has to be convinced that it is fully justified in tying up its destiny with Israel's. It has to have the courage to reject the idea that Jewish communities owe loyalty only to the states where they live."[425]

and,

---

422. R. Gottheil, *The Aims of Zionism*, Publication No. 1 of the American Federation of Zionists, New York, (1899); quoted in A. Hertzberg, *The Zionist Idea*, Harper Torchbooks, New York, (1959), pp. 496-500, at 498-500.
423. M. Nordau and G. Gottheil, *Zionism and Anti-Semitism*, Fox, Duffield & Company, New York, (1905), p. 27.
424. A. I. Kook, *Orat*, Second Edition, Jerusalem, (1950); English translation in A. Hertzberg, *The Zionist Idea*, Harper Torchbooks, New York, (1959), pp. 419-431, at 422, 425, 427.
425. English translation from K. A. Strom, Editor, *The Best of Attack! and National Vanguard Tabloid*, National Alliance, Arlington, Virginia, (1984), p. 60.

"Judaism can have nothing in common with Germanism, if we go by the standards of race, history and culture, and the Germans have the right to prevent the Jews from intruding into the affairs of their folk. [***] The same demand I raise for the Jewish folk, as against the German. The tragedy of the situation consists in the fact that it is not yet possible to establish the rule whereby the Jews should be assisted to move toward their state in Palestine. The Jews are divided into two categories, those who admit that they belong to a race distinguished by a history thousands of years old and those who don't. The latter are open to the charge of dishonesty. [***] *It is true that the participation of Jews in subversive movements and in the overthrow of the German government in November, 1918, was extraordinarily strong.* This is to be regretted since as a consequence of these activities, the Jewish people lost forces which could have been useful in its own folkist affairs."[426]

Before political Zionism took root, the cry that Jews form a nation within a nation was long considered an anti-Semitic outburst. Haman was quoted in *Esther* 3:8,

"And Haman said unto king Ahasuerus, There is a certain people scattered abroad and dispersed among the people in all the provinces of thy kingdom; and their laws *are* diverse from all people; neither keep they the king's laws: therefore it *is* not for the king's profit to suffer them."

Zionist Edmond Flegg tells us that he was initially surprised that some Jews had adopted the rhetoric of anti-Semites,

"It was then that, for the first time, I heard of Zionism. You cannot imagine what a light that was, my child! Remember that, at the period of which I am writing, this word Zionism had never yet been spoken in my presence. The anti-Semites accused the Jews of forming a nation within the nations; but the Jews, or at any rate those whom I came across, denied it. And now here were the Jews declaring: 'We are a people like other peoples; we have a country just as the others have. Give us back our country.'"[427]

The Zionist ideologist Jakob Klatzkin stated, among other things, in his book of 1921 *Krisis und Entscheidung im Judentum; der Probleme des modernen Judentums*, Second Enlarged Edition, Jüdischer Verlag, Berlin, pages 61-63, and 118:

---

[426]. N. Goldman, quoted by C. Bloch, *Mi Natan Li-meshisah Ya'akov ve-Yisrael Le-vozezim? Meharsim ve-maharivim mi-Hertsl ve-Nordoy derekh Goldman ve-Klatskin el ha-shehitah ve-hurban ha-Yahadut, mishak dam 'im goral Yisra'el*, Bronx, (1957). English translation from J. B. Agus, *The Meaning of Jewish History*, Volume 2, Abelard-Schuman, New York, (1963), p. 427-428.
[427]. E. Flegg, "Why I am a Jew", English translation by V. Gollancz in A. Hertzberg, *The Zionist Idea*, Harper Torchbooks, New York, (1959), pp. 480-485, at 481.

" [I applaud] the contribution of our enemies in the continuance of Jewry in eastern Europe. [***] We ought to be thankful to our oppressors that they closed the gates of assimilation to us and took care that our people were concentrated and not dispersed, segregatedly united and not diffusedly mixed [***] One ought to investigate in the West and note the great share which antisemitism had in the continuance of Jewry and in all the emotions and movements of our national rebirth . [***] Truly our enemies have done much for the strengthening of Judaism in the diaspora. [***] Experience teaches that the liberals have understood better than the antisemites how to destroy us as a nation. [***] We are, in a word, naturally foreigners; we are an alien nation in your midst and we want to remain one."[428]

"Man vergegenwärtige sich, wie groß der Anteil unserer Feinde am Fortbestand des Judentums im Osten ist. [...] Wir müßten beinahe unseren Bedrängern dankbar sein, wenn sie die Tore der Assimilation vor uns schlossen und dafür Sorge trugen, daß unsere Volksmassen konzentriert und nicht zerstreut, abgesondert geeint und nicht zerklüftet vermischt werden[....] Man untersuche es im Westen, welchen hohen Anteil der Antisemitismus am Fortbestand des Judentums und an all den Regungen und Bewegungen unserer nationalen Wiedergeburt hat. [...] Wahrlich, unsere Feinde haben viel zur Stärkung des Judentums in der Diaspora beigetragen. [...] Und die Erfahrung lehrt, daß die Liberalen es besser als die Antisemiten verstanden haben, uns als Volk zu vernichten. [...] Wir sind schlechthin Wesensfremde, sind — wir müssen es immer wiederholen — ein Fremdvolk in eurer Mitte und wollen es auch bleiben."

Klatzkin gave credence to the accusations of Friedrich Wilhelm Ghillany in the 1840's (Ghillany,[429] among other things, accused Jews of ritual murder near

---

428. K. J. Herrmann, "Historical Perspectives on Political Zionism and Antisemitism", *Zionism & Racism: Proceedings of an International Symposium*, International Organization for the Elimination of All Forms of Racial Discrimination, Tripoli, (1977), pp. 197-210, at 204-205. A lengthy quotation from Klatzkin, in English translation, appears in: M. Menuhin, *The Decadence of Judaism in Our Time*, Exposition Press, New York, (1965), pp. 482-483.

429. F. W. Ghillany, *Die Menschenopfer der alten Hebräer. Eine geschichtliche Untersuchung...* , Nürnberg, (1842); **and** *Die Judenfrage: Eine Beigabe zu Bruno Bauers Abhandlung über dieser Gegenstand*, Nürnberg, (1843); **and** *Das Judenthum und die Kritik; odor, Es bleibt dei den Menschenopfern der Hebräer und bei der Nothwendigkeit einer zeitgemaessen Reform des Judenthums*, J.A. Stein, Nurnberg, (1844). Friedrich Wilhelm Ghillany, under the pseudonym Richard von der Alm, *Die Urtheile heidnischer und jüdischer Schriftsteller der vier ersten christlichen Jahrhunderte über Jesus und die ersten Christen: eine Zuschrift an die gebildeten Deutschen zur weiteren Orientirung in der Frage über die Gottheit Jesu*, Otto Wigand, Leipzig, (1864); **and** *Theologische briefe an die gebildeten der deutschen Nation*, In Three Volumes, Otto Wigand, Leipzig, (1862-1863).

the time of the Damascus Affair, and published important critical texts on the history and divinity of Jesus),

> "[The Jews] have been an alien, foreign element within Germany for more than a thousand years. We must either help them towards the land of their fathers, or fuse completely with them.... But it would be best for Europe if they were to emigrate... to Palestine... or to America."[430]

Prominent Zionist and author of the *Encyclopaedia Judaica; das Judentum in Geschichte und Gegenwart*, Jakob Klatzkin wrote in Hebrew in a work published in 1925,

> "The national viewpoint taught us to understand the true nature of antisemitism, and this understanding widens the horizons of our national outlook. [***] In the age of enlightenment antisemitism was included among the phenomena that are likely to disappear along with other forms of prejudice and iniquity. The antisemites, so the rule stated, were the laggard elements in the march of progress. Hence, our fate is dependent on the advance of human culture, and its victory is our victory. [***] In the period of Zionism, we learned that antisemitism was a psychic-social phenomenon that derives from our existence as a nation within a nation. Hence, it cannot change, until we attain our national end. But if Zionism had fully understood its own implications, it would have arrived, not merely as a psycho-sociological explanation of this phenomenon, but also as a justification of it. It is right to protest against its crude expressions, but we are unjust to it and distort its nature so long as we do not recognize that essentially it is a defense of the integrity of a nation, in whose throat we are stuck, neither to be swallowed nor to be expelled. [***] And when we are unjust to this phenomenon, we are unfair to our own people. If we do not admit the rightfulness of antisemitism, we deny the rightfulness of our own nationalism. If our people is deserving and willing to live its own national life, then it is an alien body thrust into the nations among whom it lives, an alien body that insists on its own distinctive identity, reducing the domain of their life. It is right, therefore, that they should fight against us for their national integrity. [***] Know this, that it is a good sign for us that the nations of the world combat us. It is proof that our national image is not yet utterly blurred, our alienism is still felt. If the war against us should cease or be weakened, it would indicate that our image has become indistinct and our alienism softened. We shall not obtain equality of rights anywhere save at the price of an explicit or implied declaration that we are no longer a

---

[430]. P. L. Rose, *Revolutionary Antisemitism in Germany from Kant to Wagner*, Princeton University Press, (1990), p. 260. Rose cites: F. W. Ghillany, "Dedication", *Die Judenfrage: Eine Beigabe zu Bruno Bauers Abhandlung über dieser Gegenstand*, Nürnberg, (1843); **and** R. W. Stock, *Die Judenfrage durch fünf Jahrhunderte*, Verlag Der Stürmer, Nürnberg, (1939), pp. 406ff.

national body, but part of the body of the host-nation; or that we are willing to assimilate and become part of it. [***] Instead of establishing societies for defense against the antisemites, who want to reduce our rights, we should establish societies for defense against our friends who desire to defend our rights. [***] When Moses came to redeem the children of Israel, their leaders said to him, 'You have made our odor evil in the eyes of Pharaoh and in the eyes of his servants, giving them a sword with which to kill us.' Nevertheless, Moses persisted in worsening the situation of the people, and he saved them."[431]

7.3 Zionism is Built on Lies and Hatred

Journalist Lucien Wolf decried the Herzlian belief in the unassimilability of Jews and Herzl's cry for a "World Ghetto". Wolf made it clear that the Zionists must aim for Palestine, and use any other relocation of Eastern Jews merely as an opportunity to train these Jews for emigration to Palestine. Wolf also prayed for the Jews to succeed in their plans to destroy Eastern Europe, and to create thereby conditions for the Jews to overtake Palestine, and history shows that the Jews did just that with the Jewish takeovers and deliberate destruction of the Russian and Turkish Empires, and their manufacture of the First World War. Note that Wolf was warned that it was hopeless to attempt to argue with the fanatical Zionists. Wolf wrote in *The London Times*, on 8 September 1903, on page 5,

"THE ZIONIST PERIL.

TO THE EDITOR OF THE TIMES.

Sir,—Mr. Zangwill is mistaken when he endeavours to account for my hostility to the East African scheme of the Zionists by the theory that I have 'not the faintest understanding of the new forces developing in Jewry.' It is because I recognize but too clearly the bearings of the new Zionism, because I feel that its teaching is fraught with the gravest peril to the Jewish people, that I have ventured to lift up my voice against this scheme.

What is that teaching? Mr. Zangwill throughout his eloquent speech on Saturday carefully abstained from formulating it, but the Zionists in congress in Basel have not been so reticent. Here it is in a few words as laid down by Dr. Max Nordau, amid the cheers of the congress, on the 24th ult. :—

'I claim it as a great service rendered by Zionism that it has put an end to the humbug about being happy and contented... . We are not contented. We regard a fundamental change in our situation as a vital necessity. After

---

[431]. J. Klatzkin, *Tehumim: Ma'amarim*, Devir, Berlin, (1925). English translation in J. B. Agus, *The Meaning of Jewish History*, Volume 2, Abelard-Schuman, New York, (1963), pp. 425-426.

the humiliating attempts we have made at assimilation with other peoples we have taken counsel with ourselves and we desire to live in our own way.'

Were this mere rhetoric I should attach no importance to it. In the heat of campaigning one says much that one does not dream of acting upon when the time of responsibility arrives. But the Zionists have been as good as their crazy word. They have prevailed upon his Majesty's Government to promise favourable consideration to a scheme of colonization by which a Jewish settlement is to be established in a British colony on a self-governing basis, 'with,' to quote Mr. Zangwill, 'Jewish Home Rule, Jewish national customs, and a Jewish Governor.'

I can conceive of no more serious set-back to Jewish history than this scheme. With the original idea of Zionism I have always sympathized. That the Jews, who are refused their legitimate place in the national life of countries like Russia and Rumania, and who find the alternative of emigration to more liberal countries closing against them, should dream of reviving an independent Hebrew State on the consecrated soil of Palestine is not unreasonable. If I have ventured to throw cold water upon it my reason has been that I have doubted its practicability, and, in view of the terrible consequences of the Sabbetai Zevi fiasco in the 17th century, I have refused to be an accomplice in a disillusionment which would probably result in widespread demoralization. But I have always urged that, if through any political convulsion in Eastern Europe, it should become possible to secure the Holy Land for a Jewish State under the protection of the Powers, then, no matter what the difficulties might be, the whole Jewish people should strain their utmost endeavour to establish the unassimilated Jewish population of Europe in such a State and to make it a social and political success.

This, however, is not the scheme that is now before us. The starting point of the Zionists to-day is not the political situation of the unemancipated Jewish communities of Russia and Rumania, but a comprehensive capitulation to the calumnies of the Anti-Semites. It is not enough for them that we must turn tail in our fight for citizenship in the lands in which we live, but we are now asked admit that the arguments and pledges with which we won our rights in the West and with which we have been claiming them in the East are so many 'conventional lies,' that we are in truth an unassimilable element among Christian peoples, and that the Jewish problem can only be solved by the restoration of the Ghetto on a new model.

This is the inner meaning of the East African scheme. It is the Hirsch idea *plus* a theory of the invincible unassimilability of the Jew. I quite recognize that the Zionists could not act otherwise after all their contemptuous and predatory talk of the Jewish Colonization Association's policy of 'mitigation.' It would have been too much of a stultification to have copied the Argentine example with a mere change of *locale*. Hence this baseless assertion of unassimilability, this disingenuous invention of national customs which do not exist, and this demand for Home Rule in a country where equal rights have long ago solved the Jewish question. In future what can we Jews reply when another Goldwin Smith talks of our 'tribalism,' and asks, 'Can Jews be patriots?'? What can we answer when the theory of *perpetuus inimicus* of the medieval jurists is cast in our teeth by the Pobiedonostzeffs, the Drumonts, and the Stoeckers of modern Anti-Semitism, when we are accused of undue separatism and clannishness, and when public writers like Mr. Hamilton Fyfe in this month's *Nineteenth Century* declare that the anti-Semitic agitation is due to the feeling that 'Jews put their original nationality before their adopted nationality'? What, indeed, can we reply when in a British colony Jewish hands will have reared a monument to our civic incapacity?

Nor will the evil effects be limited to these controversial embarrassments. If Anti-Semitism has hitherto been a miserable failure in Western Europe it will take heart of grace now that the truth of its central doctrine is avowed by the Jews themselves. Already in Germany this is the case; for during the last fortnight German newspapers have been publishing long articles about *Jüdische Abschliessungsbestrebungen*, complaining that a tendency to separatism has set in among the German Israelites. When we find a journal like the *Kölnische Zeitung* warning the Jews in large type that 'Der Staat hat Euch zu Bürgern gemacht um Euch zu Deutschen zu erziehen,' prudent Jews will do well to halt and reflect. And what about Russia and Rumania? Will not the struggle for emancipation in those countries become more hopeless than ever? How can people claim to form part of the Russian and Rumanian nationalities when even under tolerant shadow of the British flag they ask for a separate political existence? If there were a chance of balancing these evils by making the East African colony a real training ground for the political independence of unassimilated Jews, to be followed by a wholesale emigration to some legalized home within a reasonable period, I should be less perturbed; but this is not the case. The only compensation that is offered us is the settling of some 800 Jews—not Russians—in the vicinity of Uganda at a cost of £240,000. Can reckless levity go further than this?

I have been warned that it is a thankless task to expostulate with the Zionists because they are idealists. I am getting a little tired of the cant of idealism. The fact is that people confuse idealism with chimera. The true ideal, like true art, is only the real better understood. Were it otherwise our lunatic asylums would be empty. There is nothing of this sort of idealism in the present scheme of the Zionists. It is based on a false conception of Judaism, on a travesty of Jewish history, and on a low estimate of Jewish character. It is a libel on the past and a peril for the future. For these reasons I oppose it, and I trust most earnestly that the good sense and public spirit of the Jewish people will be found on my side. Yours obediently,

London. Sept. 7. LUCIEN WOLF."

Bernard Lazare tells us of the plagiarism and forgeries of some Alexandrian Jews in their efforts to lay claim to the contributions of Greek minds and of the hatred directed by some toward the Jews and from some Jews toward the Gentiles,

> "Why were the Jews hated in all those countries, in all those cities? Because they never entered any city as citizens, but always as a privileged class. Though having left Palestine, they wanted above all to remain Jews, and their native country was still Jerusalem, i.e., the only city where God might be worshiped and sacrifices offered in His Temple. They formed everywhere republics, as it were, united with Judea and Jerusalem, and from every place they remitted monies to the high priest in payment of a special tax for the maintenance of the Temple the didrachm.
>
> Moreover, they separated themselves from other inhabitants by their rites and their customs; they considered the soil of foreign nations impure and sought to constitute themselves in every city into a sort of a sacred territory. They lived apart, in special quarters, secluded among themselves, isolated, governing themselves by virtue of privileges which were jealously guarded by them, and excited the envy of their neighbours. They intermarried amongst themselves and entertained no strangers, for fear of pollution. The mystery with which they surrounded themselves excited curiosity as well as aversion. Their rites appeared strange and gave occasion for ridicule; being unknown, they were misrepresented and slandered.
>
> At Alexandria they were quite numerous. According to Philo, Alexandria was divided into five wards. Two were inhabited by the Jews. The privileges accorded to them by Caesar were engraved on a column and guarded by them as a precious treasure. They had their own Senate with exclusive jurisdiction in Jewish affairs, and they were judged by an ethnarch. They were ship-owners, traders, farmers, most of them wealthy; the sumptuousness of their monuments and synagogues bore witness to it. The Ptolemies made them farmers of the revenues; this was one of the causes of popular hatred against them. Besides, they had a monopoly of navigation on the Nile, of the grain trade and of provisioning Alexandria, and they extended their trade to all the provinces along the Mediterranean coast. They accumulated great fortunes; this gave rise to the invidia auri Judaici. The growing resentment against these foreign cornerers, constituting a nation within a nation, led to popular disturbances; the Jews were frequently assaulted, and Germanicu, among others, had great trouble protecting them.
>
> The Egyptians took revenge upon them by deriding their religious customs, their abhorrence of pork. They once paraded in the city a fool, Carabas by name, adorned with a papyrus diadem, decked in a royal gown, and they saluted him as king of the Jews. Under Philadelphus, one of the first Ptolemies, Manetho, the high-priest of the Temple at Heliopolis, lent his authority to the popular hatred; he considered the Jews descendants of the Hyksos usurpers, and said that that leprous tribe had been expelled for sacrilege and impiousness. Those fables were repeated by Chaeremon and Lysimachus. It was not only popular animosity, however, that persecuted

the Jews; they had also against them the Stoics and the Sophists. The Jews, by their proselytism, interfered with the Stoics; there was a rivalry for influence between them, and, notwithstanding their common belief in divine unity, there was opposition between them. The Stoics charged the Jews with irreligiousness, judging by the sayings of Posidonius and Apollonius Molo; they had a very scant knowledge of the Jewish religion. The Jews, they said, refuse to worship the gods; they do not consent to bow even before the divinity of the emperor. They have in their sanctuary the head of an ass and render homage to it; they are cannibals; every year they fatten a man and sacrifice him in a grove, after which they divide among themselves his flesh and swear on it to hate strangers. 'The Jews, says Apollonius Molo, are enemies of all mankind; they have invented nothing useful, and they are brutal.' To this Posidonius adds: 'They are the worst of all men.'

Not less than the Stoics did the Sophists detest the Jews. But the causes of their hatred were not religious, but, I should say, rather literary. From Ptolemy Philadelphus, until the middle of the third century, the Alexandrian Jews, with the intent of sustaining and strengthening their propaganda, gave themselves to forging all texts which were capable of lending support to their cause. The verses of Aeschylus, of Sophocles, of Euripides, the pretended oracles of Orpheus, preserved in Aristobulus and the Stromata of Clement of Alexandria were thus made to glorify the one God and the Sabbath. Historians were falsified or credited with the authorship of books they had never written. It is thus that a History of the Jews was published under the name of Hecataeus of Abdera. The most important of these inventions was the Sibylline oracles, a fabrication of the Alexandrian Jews, which prophesied the future advent of the reign of the one God. They found imitators, however, for since the Sibyl had begun to speak, in the second century before Christ, the first Christians also made her speak. The Jews would appropriate to themselves even the Greek literature and philosophy. In a commentary on the Pentateuch, which has been preserved for us by Eusebius,[17] Aristobulus attempted to show that Plato and Aristotle had found their metaphysical and ethical ideas in an old Greek translation of the Pentateuch. The Greeks were greatly incensed at such treatment of their literature and philosophy, and out of revenge they circulated the slanderous stories of Manetho, adapting them to those of the Bible, to the great fury of the Jews; thus the confusion of languages was identified with the myth of Zeus robbing the animals of their common language. The Sophists, wounded by the conduct of the Jews, would speak against them in their teaching. One among them, Apion, wrote a Treatise against the Jews. This Apion was a peculiar individual, a liar and babbler, to a degree uncommon even among rhetors, and full of vanity, which earned him from Tiberius the nickname 'Cymbalum mundi.' His stories were famous; he claimed to have called out, by means of magic herbs, the shade of Homer, says Pliny.

Apion repeated in his Treatise against the Jews the stories of Manetho, which had been previously restated by Chaeremon and Lysimachus, and supplemented them by quoting from Posidonius and Apollonius Molo.

According to him, Moses was 'nothing but a seducer and wizard,' and his laws contained 'nothing but what is bad and dangerous.'[18]

As to the Sabbath, the name was derived, he said, from a disease, a sort of an ulcer, with which the Jews were afflicted, and which the Egyptians called sabbatosim, i.e., disease of the groins.

Philo and Josephus undertook the defence of the Jews and fought the Sophists and Apion. In Contra Apionem, Josephus is very severe on his adversary. 'Apion,' says he, 'is as stupid as an ass and as imprudent as a dog, which is one of the gods of his nation.' Philo, on the other hand, prefers to attack the Sophists in general, and if he mentions Apion at all, in his Legatio ad Caium, it is merely because Apion was sent to Rome to prefer charges against the Jews before Caligula."[432]

The Old Testament makes many references to the diseases of the Egyptians afflicting the Jews, which probably included leprosy, syphilis and other sexually transmitted diseases. This tends to indicate that the Judeans learned monotheism from an expelled group of Egyptians, or perhaps from the Hyksos.

The Russian Jewess Helena Petrovna Blavatsky wrote of the ancient Jews' enmity towards the rest of humanity and spoke of Jewish phallic worship, which was also practiced by the Turkic Khazars, who adopted Judaism and became one of the major bloodlines of today's Ashkenazi Jews—the Orthodox and Ultra-Orthodox of whom worship the Talmud,[433]

---

432. B. Lazare, *Antisemitism: Its History and Causes*, (1894), pp. 20-23. *L'Antisémitisme, son Histoire et ses Causes*, L. Chailley, Paris, (1894).
433. B. J. Hendrick, "The Jews in America: I How They Came to This Country", *The World's Work*, Volume 44, Number 2, (December, 1922), pp. 144-161; **and** "The Jews in America: II Do the Jews Dominate American Finance?", *The World's Work*, Volume 44, Number 3, (January, 1923), pp. 266-286; **and** "The Jews in America: III The Menace of the Polish Jew", *The World's Work*, Volume 44, Number 4, (February, 1923), pp. 366-377; **and** "Radicalism among the Polish Jews", *The World's Work*, Volume 44, Number 6, (April, 1923), pp. 591-601. *See also:* D. M. Dunlop, *The History of the Jewish Khazars*, Princeton University Press, (1954). *See also:* A. Koestler, *The Thirteenth Tribe: The Khazar Empire and Its Heritage*, Random House, New York, (1976). P. B. Golden, *Khazar Studies: An Historico-Philological Inquiry into the Origins of the Khazars*, In Two Volumes, Akadémiai Kiadó, Budapest, (1980). *See also:* N. Golb and O. Pritsak, *Khazarian Hebrew Documents of the Tenth Century*, Cornell University Press, Ithaca, (1982). *See also:* ha-Levi Judah (12th Century) and M. Lazar, *Book of the Kuzari: A Book of Proof and Argument in Defense of a Despised Faith : a 15th Century Ladino Translation (Ms. 17812, B.N. Madrid)*, Labyrinthos, Culver City, California, (1990). ha-Levi Judah (12th Century) and N. D. Korobkin, *The Kuzari: In Defense of the Despised Faith*, J. Aronson, Northvale, New Jersey, (1998). *See also:* P. Wexler, The Ashkenazi Jews: A Slavo-Turkic People in Search of a Jewish Identity, Slavica Publishers, Columbus, Ohio, (1993). *See also:* K. A. Brook, *The Jews of Khazaria*, Jason Aronson, Northvale, New Jersey, (1999). *See also:* M. F. Hammer, *et al.*, Jewish and Middle Eastern Non-Jewish Populations Share a Common Pool of Y-Chromosome Biallelic Haplotypes", *Proceedings of the National Academy of Sciences*, Volume 97, (2000), pp. 6769-6774. *See also:* M. P. Stumpf and D. B. Goldstein, "Genealogical and Evolutionary

"But phallic worship has developed only with the gradual loss of the keys to the inner meaning of religious symbols; and there was a day when the Israelites had beliefs as pure as the Âryans have. But now Judaism, built *solely* on phallic worship, has become one of the latest creeds in Asia, and theologically a religion of hate and malice toward everyone and everything outside themselves. Philo Judæus shows what was the genuine Hebrew faith. The Sacred Writings, he says, prescribe what we ought to do, *commanding us to hate the heathen and their laws and institutions*. They did hate Baal or Bacchus worship publicly, but left its worst features to be followed secretly. It is with the Talmudic Jews that the grand symbols of nature were the most profaned."[434]

Jews took great offense at the claim often made in antiquity that they were decedents of the Egyptians and that Moses was an Egyptian priest. The Egyptian Pharaoh Akhenaton IV, is considered by many today to have been the father of monotheism and the Biblical stories of Moses' background appear rather implausible, are apparently close copies of other ancient myths, and tend to indicate that Moses was an Egyptian by birth and descent, perhaps even the Egyptian Pharaoh Akhenaton IV.

Among many such "blasphemies", Apion claimed that the Jews practiced human sacrifices, and would fatten up a Greek for the slaughter each year in the Temple, sacrifice the Greek in the woods, feast on the Greek's viscera and swear a curse of hatred upon the Greeks. The Jews, especially Jewish royalty, had practiced Baal worship, and the worship of Baal and of Moloch entailed human sacrifices. Christianity is a Jewish tale of human sacrifice not unlike Moloch worship—though Jesus was sacrificed on the cross, not the funeral pyre. The Old Testament contains many verses which make reference to human sacrifices, for example *Genesis* 22:1-18; *Exodus* 8:26;13:2. *Joshua* 13:14. *Judges* 11:29-40. I *Kings* 13:1-2. II *Kings* 16:3-4; 17:17; 21:6; 23:20-25. II *Chronicles* 28:1-4; *Jeremiah* 7:3; 19:5; 32:35. *Ezekiel* 16:20-21; 20:26, 31; 23:37. and *Leviticus* 27:28-29:

Inference with the Human Y Chromosome", *Science*, Volume 291, (2001), pp. 1738-1742. **See also:** M. F. Hammer, *et al.*, "Hierarchical Patterns of Global Human Y-Chromosome Diversity", *Molecular Biology and Evolution*, Volume 18, (2001), pp. 1189-1203. **See also:** D. M. Behar, *et al.*, "Multiple Origins of Ashkenazi Levites: Y Chromosome Evidence for Both Near Eastern and European Ancestries", *The American Journal of Human Genetics*, Volume 73, Number 4, (October, 2003), pp. 768-779. **See also:** D. M. Behar, *et al.*, "The Matrilineal Ancestry of Ashkenazi Jewry: Portrait of a Recent Founder Event", *The American Journal of Human Genetics*, Volume 78, Number 3, (March, 2006), pp. 487-497.

434. H. P. Blavatsky, "Who the Jews Really Are", *The Secret Doctrine: The Synthesis of Science, Religion, and Philosophy*, Volume 2, Part 2, Section 3, The Theosophical Publishing Society, (1893), pp. 493-494.

"28 Notwithstanding, no devoted thing, that a man shall devote unto the LORD of all that he hath, *both* of man and beast, and of the field of his possession, shall be sold or redeemed: every devoted thing *is* most holy unto the LORD. 29 None devoted, that shall be devoted of men, shall be redeemed; *but* shall surely be put to death."

And *Leviticus* 20:1-7 admonishes Jews not to allow sacrifices of their own children to Moloch, indicating that Jewish child sacrifices were occurring, and perhaps prescribing that they should be made to the "Lord", Baal, instead of to Moloch:

"1 And the LORD spake unto Moses, saying, 2 Again, thou shalt say to the children of Israel, Whosoever *he be* of the children of Israel, or of the strangers that sojourn in Israel, that giveth *any* of his seed unto Molech; he shall surely be put to death: the people of the land shall stone him with stones. 3 And I will set my face against that man, and will cut him off from among his people; because he hath given of his seed unto Molech, to defile my sanctuary, and to profane my holy name. 4 And if the people of the land do any ways hide their eyes from the man, when he giveth of his seed unto Molech, and kill him not: 5 Then I will set my face against that man, and against his family, and will cut him off, and all that go a whoring after him, to commit whoredom with Molech, from among their people. 6 And the soul that turneth after such as have familiar spirits, and after wizards, to go a whoring after them, I will even set my face against that soul, and will cut him off from among his people. 7 Sanctify yourselves therefore, and be ye holy: for I *am* the LORD your God."

Beyond this, Judaism, as a religious doctrine which revolves around a fabricated history, depends upon repeated instances, and prophecies, of the sacrifice of masses of human beings in genocides directed by God—ultimately the mass murder of all Gentiles and apostate Jews as a human sacrifice to God for the sake of Zionism and the remnant of "righteous" Jews.

Josephus recounts some of Apion's tales, among them,

"Now, although I cannot but think that I have already demonstrated, and that abundantly more than was necessary, that our fathers were not originally Egyptians, nor were thence expelled, either on account of bodily diseases, or any other calamities of that sort; yet will I briefly take notice of what Apion adds upon that subject; for in his third book, which relates to the affairs of Egypt, he speaks thus:—'I have heard of the ancient men of Egypt, that Moses was of Heliopolis, and that he thought himself obliged to follow the customs of his forefathers, and offered his prayers in the open air, toward the city walls; but that he reduced them all to be directed toward sunrising, which was agreeable to the situation of Heliopolis: that he also set up pillars instead of gnomons, under which was represented a cavity like that of a boat, and the shadow that fell from their tops fell down upon that cavity, that it might go round about the like course as the sun itself goes round in the other.' [***] He then assigns a certain wonderful and plausible occasion for the name of Sabbath, for he says, that 'when the Jews had travelled a six days' journey, they had swellings on their groins; and that on this account it was that they rested on the seventh day, as having got safely to that country which is now called Judea; that then they preserved the language of the Egyptians, and called that day the Sabbath, for that malady of swellings on their groin was named Sabbatosis by the Egyptians.' [***] And as for this grammatical translation of the word Sabbath, it either contains an instance of his great impudence or gross ignorance; for the words Sabbo and Sabbath are widely different from one another; for the word Sabbath in the Jewish language denotes rest from all sorts of work; but the word Sabbo, as he affirms, denotes among the Egyptians the malady of a swelling in the groin. [***] He adds another Grecian fable, in order to reproach us. In reply to which, it would be enough to say, that they who presume to speak about divine worship, ought not to be ignorant of this plain truth, that it is a degree of less impurity to pass through temples, than to forge wicked calumnies of its priests. Now, such men as he are more zealous to justify a sacrilegious king than to write what is just and what is true about us, and about our temple; for when they are desirous of gratifying Antiochus, and of concealing that perfidiousness and sacrilege which he was guilty of, with regard to our nation, when he wanted money, they endeavour to disgrace us, and tell lies even relating to futurities. Apion becomes other men's prophet upon this occasion, and says, that 'Antiochus found in our temple a bed, and a man lying upon it, with a small table before him, full of dainties, from the [fishes of the] sea, and the fowls of the dry land; that this man was amazed at these dainties thus set before him; that he immediately adored the king, upon his coming in, as hoping that he would afford him all possible assistance; that he fell down upon his knees, and stretched out to him his right hand, and begged to be released; and that when the king bade him sit down, and tell him who he was, and why he dwelt there, and what was the meaning of those various sorts of food that were set before him, the man made a lamentable complaint, and with sighs, and tears in his eyes, gave him this account of the distress he was in; and said that he was a Greek, and that as he went over this province, in order to get his living, he was seized upon by foreigners, on a sudden, and brought to this temple, and shut up therein, and was seen by nobody, but was fattened by these curious provisions thus set before him: and that truly at the first such unexpected advantages seemed to him matter of great joy; that after a while they brought a suspicion upon him, and at length astonishment, what their meaning should be; that at last he inquired of the servants that came to him, and was by them informed that it was in order to the fulfilling a law of the Jews, which they must not tell him, that he was thus fed; and that they did the same at a set time every year: that they used to catch a Greek foreigner, and fatten him thus up every year, and then lead him to a certain wood, and kill him, and sacrifice with their accustomed solemnities, and taste of his entrails, and take an oath upon this sacrificing a Greek, that they would ever be at enmity with the Greeks; and that then they threw the remaining parts of the miserable wretch into a certain pit.' Apion adds further, that 'the man said there were but a few days to come ere he was to be slain, and implored of Antiochus that, out of the reverence he bore to the Grecian gods, he would disappoint the snares the Jews laid for his blood, and would deliver him from the miseries with which he was encompassed.'"[435]

Many such charges of "ritual murder" were made against Jews throughout history. They have become known as "blood libels". One of the more famous of these is the story of Saint Simon of Trent. In Rev. S. Baring-Gould, *The Lives of the Saints*, New and Revised Edition, John Grant, Edinburgh, (1914), pp. 447-449, it states:

"S. SIMON, BOY M.
(A.D. 1475.)

[Roman Martyrology. Authority :—The Acts of Canonization by Benedict XIV., and the Acts published in the Italian immediately after the event took place.]

THROUGH the Middle Ages, in Europe the Jews were harshly treated, suffering from sudden risings of the people, or from the exactions of princes and nobles. Tales of murder of Christian children were trumped up against them. This was, perhaps, the case in Trent, where on Tuesday in Holy Week, 1475, the Jews met to prepare for the approaching Passover, in the house of one of their number named Samuel, and it was agreed between three of them—Samuel, Tobias, and Angelus—that a child should be crucified, as an act of revenge against their tyrants, and of hatred against Christianity. The difficulty, however, was how to get one. Samuel sounded his servant Lazarus, and attempted to bribe him into procuring one, but the suggestion so scared the fellow, that he packed up all his traps and ran away. On the Thursday, Tobias undertook to get the boy, and going out in the evening, whilst the people were in church during the singing of Tenebræ, he prowled about till he found a child sitting on the threshold of his father's door in the Fossati Street, aged twenty-nine months, and named Simon. The Jew began to coax the little fellow to follow him, and the boy did so, and he conducted him to the house of Samuel, where he was put to bed, and given raisins and apples to amuse him.

In the mean time the parents, Andrew and Mary, missing their child, began to seek him everywhere, but not finding him, and night falling darkly upon them, they returned, troubled and alarmed to their home.

---

435. Josephus, "Flavius Josephus Against Apion", *The Works of Flavius Josephus: Comprising the Antiquities of the Jews; a History of the Jewish Wars; and Life of Flavius Josephus, Written by Himself*, Book 2, S. S. Scranton Co., Hartford, Connecticutt, (1916), pp. 905-906, 911.

During the night, when all was still, a Jew named Moses took the child from its bed, and carried it into the vestibule of the synagogue, which formed a part of the house of Samuel, and sitting down on a bench began to strip the infant; a handkerchief being twisted round its throat to prevent it from crying. Then stretching out his limbs in the shape of a cross they began the butchery of the child, cutting the body in several places, and gathering his blood in a basin. The child being half dead, they raised him on his feet, and whilst two of them held him by the arms, the rest pierced his body on all sides with their awls.

When the child was dead, they hid the body in a cellar behind the barrels of wine.

All Friday the parents sought their son, but found him not, and the Jews, alarmed at the proceedings of the magistrates, who had taken the matter up, and were making investigations in all quarters, consulted what had better be done. They could not carry the body away, as every gate was watched, and the perplexity was great. At length they determined to dress the body again and throw it into the stream which ran under Samuel's window, but which was there blocked by an iron cage in which the refuse was caught. Tobias was to go to the bishop and chief magistrates and tell them that there was a child's body entangled in the grate, and he hoped that by thus drawing attention to it all suspicion of having been implicated in the murder would be diverted from him and his co-religionists.

This was done, and when John de Salis, the bishop, and James de Sporo, the governor, heard the report of the Jew, they at once went, and the body was removed before their eyes, and conveyed to the cathedral, followed by a crowd. As, according to a popular mediæval superstition, blood is supposed to flow from the wound when the murderer approaches, the officers of justice examined the body as the crowds passed it; and they noticed that blood exuded as Tobias approached. On the strength of this the house of Samuel and the synagogue were examined, and blood and other traces of the butchery were found in the cellar, and in the place where the deed had been done, and the bowl of blood was discovered in a cupboard. The most eminent physicians were called to investigate the condition of the corpse, and they unanimously decided that the child could not have been drowned, as the body was not swollen, and as there were marks on the throat of strangulation. The wounds they decided were made by sharp instruments like awls and knives, and could not be attributed to the gnawing of water-rats. The popular voice now accusing the Jews, the magistrates seized on the Jews and threw them into prison, and on the accusation of a renegade Jew named John, who had been converted to Christianity seven years before, and who declared that the Jews had often sought to catch and kill a child, and had actually done this elsewhere, more than five of the Jews were sentenced to be broken on the wheel, and then burnt.

The blood found in the basin is preserved in the cathedral of Trent, and the body of the child is also enshrined there in a magnificent mausoleum. Such is the story. A boy was drowned, and his body gnawed by rats. This was worked up into a charge against the Jews, to excuse a massacre and plunder of the unfortunate Hebrews."

Josephus' denials of the charge that the Jews fabricated the "history" of the Israelites include attempted refutations of many ancient scholars, but Josephus has proven less than reliable and there is no ancient evidence yet found to support his views on the origins of the Israelites and Jews. Among the ancients who saw the Jews as Egyptians was Strabo, who wrote,

> "[I]t is inhabited in general, as is each place in particular, by mixed stocks of people from Aegyptian and Arabian and Phoenician tribes; for such are those who occupy Galilee and Hiericus [*Footnote:* Jericho] and Philadelphia and Samaria, which last Herod surnamed Sebastê. [*Footnote: i. e.* in Latin, 'Augusta,' in honour of Augustus Caesar.] But though the inhabitants are mixed up thus, the most prevalent of the accredited reports in regard to the temple at Jerusalem represents the ancestors of the present Judaeans, as they are called, as Aegyptians. [***] Moses, namely, was one of the Aegyptian priests, and held a part of Lower Aegypt, as it is called, but he went away from there to Judaea, since he was displeased with the state of affairs there, and was accompanied by many people who worshipped the Divine Being. [***] Now Moses, saying things of this kind, persuaded not a few thoughtful men and led them away to this place where the settlement of Jerusalem now is[.]"[436]

If these many voices from antiquity (Herodotus, Ptolemy of Medes, Apion, Chaeremo, Diodorus, Lysimachus, Tacitus and Strabo) speak the truth, yet another tenet of racist Zionist mythology is proven false.

Not only ancient voices are raised against Zionist myth. Philip R. Davies also discredits Zionist folklore in his book *In Search of "Ancient Israel"*, JSOT Press, Sheffield, England, (1992). In addition, A. Arnaiz-Villena, N. Elaiwa, C. Silvera, A. Rostom, J. Moscoso, E. Gómez-Casado, L. Allende, P. Varela, and J. Martínez-Laso, published an important paper, "The Origin of Palestinians and Their Genetic Relatedness with other Mediterranean Populations", *Human Immunology*, Volume 62, Number 9, (2001), pp. 889-900; which discredited Zionist legend, but which has been removed from the publication's website (*see also:* A. Nebel, *et al.*, "High-Resolution Y Chromosome Haplotypes of Israeli and Palestinian Arabs Reveal Geographic Substructure and Substantial Overlap with Haplotypes of Jews", *Human Genetics*, Volume 107, Number 6, (December, 2000), pp. 630-641.). In an article by Robin McKie, in *The Observer* (Guardian Unlimited), "Journal Axes Gene Research on Jews and Palestinians", (25 November 2001), McKie claims,

---

436. Strabo, translated by H. L. Jones, *Geography*, Volume 7, Book 16, Chapter 2, Sections 34-36, Harvard University Press, Cambridge, Massachusetts, (2000), pp. 281, 283.

"Academics who have already received copies of Human Immunology have been urged to rip out the offending pages and throw them away. Such a drastic act of self-censorship is unprecedented in research publishing and has created widespread disquiet, generating fears that it may involve the suppression of scientific work that questions Biblical dogma."[437]

The exodus is a painful topic for some political Zionists, because, in addition to the loss of another pillar in the temple of their racist beliefs, the story of the exodus has long been a source of inspiration in times of crisis for the Jews in general and has come to symbolize the many long struggles many Jews have endured in their efforts to maintain their "race", their religion and their "nation". The reassuring story gives a promise of hope which many Jews believe carried them through to the promised land. It is perhaps this romantic and sentimental love of the promised land that enabled the political Zionists, who were motivated in no small part by greed, to gain some support in their attempts to risk all and join forces with the anti-Semites in a racist call for segregation in the first half of the Twentieth Century, which ultimately had horrific consequences. George Henry Borrow wrote in *The Zincali* in 1841,

"If there be one event in the eventful history of the Hebrews which awakens in their minds deeper feelings of gratitude than another, it is the exodus, and that wonderful manifestation of olden mercy still serves them as an assurance that the Lord will yet one day redeem and gather together his scattered and oppressed people. 'Art thou not the God who brought us out of the land of bondage?' they exclaim in the days of their heaviest trouble and affliction. He who redeemed Israel from the hand of Pharaoh is yet capable of restoring the kingdom and sceptre to Israel."[438]

Like the ancients, Sigmund Freud again questioned the origins of Judaism in 1938, arguing that Moses was an Egyptian, in his book *Moses and Monotheism*,[439] and he was attacked for it by some fellow Jews, who were concerned that if this question were further explored, the answers might profit the anti-Semites. The Nazis had already begun to inflict their cowardly violence against defenseless Jews, and Freud himself wisely left his beloved Vienna. Many have stated that the Old Testament contains nothing new and was plagiarized from many sources including the Monotheism of the Egyptian Pharaoh Amenhotep IV, a. k. a. Akhenaton. Many believe that David's *Psalm 104* is plagiarized from Akhenaton's *Great Hymn to the Aten*.

The theory that the "Israelites" were Egyptians, who were expelled from Egypt due to their diseases and depravity, fits some of the plague stories and

---

[437]. http://www.observer.co.uk/international/story/0%2C6903%2C605798%2C00.html
[438]. G. H. Borrow, *The Zincali, or, An Account of the Gypsies of Spain*, Volume 1, John Murray, London, (1841), pp. 157-158.
[439]. S. Freud, *Der Mann Moses und die monotheistische Religion*, (1938); English translation, *Moses and Monotheism*, Knopf, New York, (1939).

their related counterparts in Egyptian texts as contagions among the Hyksos, which stories otherwise rely upon divine intervention. The Hyksos conquerors were known as lepers, and may be the source of many of the legends of "Israelites". The dietary and other laws of the Jews, and their practice of circumcision, burial of the dead, etc. are carry-overs from an Egyptian heritage—Egyptian priests emphasized cleanliness and completely shaved their bodies. Egyptians were circumcised and did not eat swine flesh, which was considered to be a source of leprosy. Disease may have been brought in by the Hyksos, who then infected other Egyptians—though the existence of the Hyksos is established, the existence of the ancient Israelites is not, and they were clearly fictions created by the Judeans, who may have received Egyptian lore from expelled Hyksos travellors, or from expelled Egyptian priests.

Publius Cornelius Tacitus wrote in his *Histories*, Book V, Chapters 2-8, of 70AD—the year Jerusalem is thought by some to have been destroyed, that the Jews were lepers among the Egyptians, who were expelled. Tacitus tells us that Moses rejected Egyptian gods due to the affliction his people suffered (there are recurring accusations of worship of the golden calf), and out of spite Moses sacrificed the animals the Egyptians worshiped and turned to monotheism as a means of uniting his group behind a belief system foreign to all others. Tacitus claimed that Jews do not eat swine flesh, because this animal carried the leprosy which afflicted them. Tacitus wrote, among other things,

> "Most writers, however, agree in stating that once a disease, which horribly disfigured the body, broke out over Egypt; that king Bocchoris, seeking a remedy, consulted the oracle of Hammon, and was bidden to cleanse his realm, and to convey into some foreign land this race detested by the gods. The people, who had been collected after diligent search, finding themselves left in a desert, sat for the most part in a stupor of grief, till one of the exiles, Moyses by name, warned them not to look for any relief from God or man, forsaken as they were of both, but to trust to themselves, taking for their heaven-sent leader that man who should first help them to be quit of their present misery."[440]

Tacitus concludes,

> "5. This worship, however introduced, is upheld by its antiquity; all their other customs, which are at once perverse and disgusting, owe their strength to their very badness. The most degraded out of other races, scorning their national beliefs, brought to them their contributions and presents. This

---

[440]. Tacitus, translated by A. J. Church and W. J. Brodribb, *The Complete Works of Tacitus*, The Modern Library, New York, (1942), p. 658. An alternative, and far more accessible translation appears in: Tacitus, *History of the Jews*, Book V, Chapters 2-8 in: "Dissertation III", *The Works of Flavius Josephus: Comprising the Antiquities of the Jews; a History of the Jewish Wars; and Life of Flavius Josephus, Written by Himself*, S. S. Scranton Co., Hartford, Connecticut, (1916), pp. 951-956.

augmented the wealth of the Jews, as also did the fact, that among themselves they are inflexibly honest and ever ready to shew compassion, though they regard the rest of mankind with all the hatred of enemies. They sit apart at meals, they sleep apart, and though, as a nation, they are singularly prone to lust, they abstain from intercourse with foreign women; among themselves nothing is unlawful. Circumcision was adopted by them as a mark of difference from other men. Those who come over to their religion adopt the practice, and have this lesson first instilled into them, to despise all gods, to disown their country, and set at nought parents, children, and brethren. Still they provide for the increase of their numbers. It is a crime among them to kill any newly-born infant. They hold that the souls of all who perish in battle or by the hands of the executioner are immortal. Hence a passion for propagating their race and a contempt for death. They are wont to bury rather than to burn their dead, following in this the Egyptian custom; they bestow the same care on the dead, and they hold the same belief about the lower world. Quite different is their faith about things divine. The Egyptians worship many animals and images of monstrous form; the Jews have purely mental conceptions of Deity, as one in essence. They call those profane who make representations of God in human shape out of perishable materials. They believe that Being to be supreme and eternal, neither capable of representation, nor of decay. They therefore do not allow any images to stand in their cities, much less in their temples. This flattery is not paid to their kings, nor this honour to our Emperors. From the fact, however, that their priests used to chant to the music of flutes and cymbals, and to wear garlands of ivy, and that a golden vine was found in the temple, some have thought that they worshiped father Liber, the conqueror of the East, though their institutions do not by any means harmonize with the theory; for Liber established a festive and cheerful worship, while the Jewish religion is tasteless and mean."[441]

Tacitus was familiar with the Old Testament which is filled with stories of mass murder and of genocides allegedly sanctioned—insisted upon—perpetrated—by God. King David, a great hero of the work, was a treacherous murderer. It is truly a brutal and bloody religious mythology and largely a fabricated and horrific[442] history of racism, misogyny, martyrdom, world

---

[441]. Tacitus, translated by A. J. Church and W. J. Brodribb, *The Complete Works of Tacitus*, The Modern Library, New York, (1942), pp. 659-660. An alternative, and far more accessible translation appears in: Tacitus, *History of the Jews*, Book V, Chapters 2-8 in: "Dissertation III", *The Works of Flavius Josephus: Comprising the Antiquities of the Jews; a History of the Jewish Wars; and Life of Flavius Josephus, Written by Himself*, S. S. Scranton Co., Hartford, Connecticutt, (1916), pp. 951-956.

[442]. F. Delitzsch, *Die grosse Täuschung: kritische Betrachtungen zu den alttestamentlichen Berichten über Israels Eindringen in Kanaan, die Gottesoffenbarung vom Sinai und die Wirksamkeit der Propheten*, In Two Volumes, Deutsche Verlags-Anstalt, Berlin, Stuttgart, (1921).

domination, slavery, rape, genocide, infanticide, cannibalism and human sacrifice.[443] The disgustingly low level of the Talmud is vividly displayed in the Rabbis' pedophilia in *Kethuboth* 11*a* and 11*b*, where the learned discuss their conclusion that a girl under the age of three who has sexual intercourse with a grown man can later lawfully claim to be a virgin for the purposes of marriage, because her hymen will have since grown back. According to these religious leaders, a grown woman who fornicates with a prepubescent boy can likewise claim to be a virgin. *Kethuboth* 11*b*,

> "A small boy who has intercourse with a grown-up woman makes her [as though she were] injured by a piece of wood. [*Footnote:* Although the intercourse of a small boy is not regarded as a sexual act, nevertheless the woman is injured by it as by a piece of wood.] [***] It means[5] this: When a grown-up man has intercourse with a little girl it is nothing, for when the girl is less than this,[*Footnote:* Lit., 'here', that is, less than three years old.] it is as if one puts the finger into the eye;[*Footnote:* I.e., tears come to the eye again and again, so does virginity come back to the little girl under three years. Cf. Nid. 45*a*.] but when a small boy has intercourse with a grown-up woman he makes her as 'a girl who is injured by a piece of wood,'"[444]

The *Sanhedrin* 69*a* states that a girl of three years and one day can enter marriage if an adult male rapes her, and that she can be passed to her husband's brother, should her husband die, by his having raped her,

> "A maiden aged three years and a day may be acquired in marriage by coition, and if her deceased husband's brother cohabited with her, she becomes his. The penalty of adultery may be incurred through her; [if a *niddah*,] she defiles him who has connection with her, so that he in turn defiles that upon which he lies, as a garment which has lain upon [a person afflicted with gonorrhoea]. If she married a priest, she may eat of *terumah;* if any unfit person cohabits with her, he disqualifies her from the priesthood. If any of the forbidden degrees had intercourse with her, they are executed on her account, but she is exempt.[3] [69*b*] But why so: may she not prove to be barren, her husband not having married her on such a condition?[5] Hence it must be that we take into account only the majority, and the majority of

---

443. *Genesis* 3:13-16; 9:25-27; 19:24-29; 22:1-18. *Exodus* 8:26; 12:12-13, 29-30; 13:2; 15:3. *Leviticus* 25:44-46; 26:29; 27:28-29. *Numbers* 11:33-34; 15:35-36; 25:1-9; 31:1-54. *Deuteronomy* 7:1-26; 28:1- 68; 29:20-24; 32:19-43. *Joshua* 6:17-21; 13:14, 21-22. *Judges* 6:1-40; 8:18; 11:29-40. *I Samuel* 6:19. I *Kings* 13:1-2. II *Kings* 2:24; 6:26 -30; 16:3-4; 17:17; 21:6; 23:20-25. *I Chronicles* 13:10; 21:14-15. II *Chronicles* 28:1-4. *Psalms* 2:8-9; 78:31-51; 83:9-10; 137:8-9. *Isaiah* 10:6; 13:1-22; 19:2-4; 34:1-17; 61:5-9. *Hosea* 13:16. *Joel* 3:8. *Zechariah* 14:2-3. *Malachi* 2:2-3. *Jeremiah* 7:3; 19:5; 32:35. *Ezekiel* 16:20-21; 20:26, 31; 23:37.

444. I. Epstein, Editor, Kethuboth 11*b*, *The Babylonian Talmud*, Volume 17 (Kethuboth I), The Soncino Press, London, (1936), pp. 57-58.

women are not constitutionally barren! No. The penalty incurred on her account is a sacrifice, [but not death]. But it is explicitly stated, 'They are executed on her account?'—That refers to incest by her father. But the statement is, If *any* of the forbidden degrees had intercourse with her?⁶—Hence this [Mishnah] refers to a husband who explicitly accepted her under all conditions."⁴⁴⁵

Shabbath 133 commands a mohel who performs a circumcision to suck on the wounded penis with his mouth and draw blood into his mouth—a process referred to as *metzitzah b'peh*. This tradition continues to this day and may result in the transmission of the herpes virus and other diseases from the mohel to the baby.⁴⁴⁶ *Shabbath* 133*a* states,

"*MISHNAH*. WE PERFORM ALL THE REQUIREMENTS OF CIRCUMCISION ON THE SABBATH. WE CIRCUMCISE,¹ UNCOVER [THE CORONA],² SUCK [THE WOUND],³ AND PLACE A COMPRESS AND CUMMIN UPON IT.⁴ IF ONE DID NOT CRUSH [THE CUMMIN] ON THE EVE OF THE SABBATH, HE MUST CHEW [IT] WITH HIS TEETH AND APPLY [IT TO THE WOUND]; IF HE DID NOT BEAT UP WINE AND OIL ON THE EVE OF THE SABBATH,⁵ EACH MUST BE APPLIED SEPARATELY. WE MAY NOT MAKE A HALUK⁶ FOR IT IN THE FIRST PLACE, BUT MUST WRAP A RAG ABOUT IT. IF THIS WAS NOT PREPARED FROM THE EVE OF THE SABBATH, ONE WINDS IT ABOUT HIS FINGER⁷ AND BRINGS IT, AND EVEN THROUGH ANOTHER COURTYARD."⁴⁴⁷

---

445. I. Epstein, Editor, *The Babylonian Talmud*, Sanhedrin II, Volume 28, The Soncino Press, London, (1935), pp. 469-470.
446. J. Siegel-Itzkovich, "Ritual Circumcisers Can Pass Infection to Babies", *The Jerusalem Post*, (29 August 2004), p. 7. *See also:* M. Haberman, "Fear Rabbi Gave Tots Herpes Probe Death of Baby after Circumcision", *Daily News* (New York), (2 February 2005), p. 7. *See also:* "Rabbi's Circumcision Rites Linked to Herpes Death", The Houston Chronicle, (3 February 2005), p. A9. *See also:* "Baby 'Herpes Death' Inquiry", *The Daily Telegraph* (Sydney, Australia), 3 February 2005), p. 26. *See also:* "Circumcision Rite Poses 'Inherent Risks,' City Sez", *Daily News* (New York), (3 February 2005), p. 42. *See also:* J. Siegel-Itzkovich, "The Kindest Cut of All", *The Jerusalem Post*, (18 September 2005), p. 7. *See also:* (Rev) Michael Plaskow Mbe, Richard Rinberg, Glenn Woiceshyn, Colin L. Leci, Henry Kaye, Joe Frankl, Yoel Tamari, Shifra Tarem, Raymond Cannon, Shakil Khan, Ellie Morris, Leon Harris, Naomi Cohen, "Letters", *The Jerusalem Post*, (19 September 2005), p. 14. *See also:* J. Rutenberg, "Metro Briefing New York: Manhattan: Infection Connected To Religious Rite", *The New York Times*, (14 December 2005), p. B9. *See also:* D. Saltonstall, "2 More Tots Infected with Herpes", *Daily News* (New York), (14 December 2005), p. 20. *See also:* D. Saltonstall, "New Cases of Tot Herpes Infections", *Daily News* (New York), (14 December 2005), p. 20. *See also:* J. Rutenberg and A. Newman, "City Officials and Rabbis Clash Over Rite", *The New York Times*, (5 January 2006), p. B5; **and** "Mayor Balances Hasidic Ritual Against Fears for Babies' Health", *The New York Times*, (6 January 2006), pp. A1, B4. *See also:* J. Purnick, "Taking a Stand On a Rite With Hazards", *The New York Times*, (9 January 2006), p. B1.
447. I. Epstein, Editor, "Shabbath", *The Babylonian Talmud*, Volume 8, Part 2, The Soncino Press, (1938), pp. 666-669, at 668-669.

Shabbath 133*b* states,

> "WE SUCK OUT, etc. R. Papa said: If a surgeon does not suck [the WOUND], it is dangerous and he is dismissed. It is obvious? Since we desecrate the Sabbath for it, it is dangerous?[6]—You might say that this blood is stored up, therefore he informs us that it is the result of a wound, and it is like a bandage and cummin: just as when one does not apply a bandage and cummin there is danger, so here too if one does not do it there is danger.""[448]

The Jewish Encyclopedia in its article "Circumcision" states,

> "Mez,iz,ah: By this is meant the sucking of the blood from the wound. The mohel takes some wine in his mouth and applies his lips to the part involved in the operation, and exerts suction, after which he expels the mixture of wine and blood into a receptacle (see Fig. 4, below) provided for the purpose. This procedure is repeated several times, and completes the operation, except as to the control of the bleeding and the dressing of the wound."[449]

The Talmud contains numerous obvious lies and fables, as Johannes Buxtorf pointed out.[450]

This ancient mythology and fabricated history, Judaism, of course, does not reflect upon the majority of modern ethnic Jews, most of whom have not read it, let alone believed in it; and, speaking anecdotally, the vast majority of those who genuinely believed in these works and prophesies whom your author has encountered, were fundamentalist Dispensationalist Christians, not ethnic Jews. Moderns enjoy the inheritance of a great many loving and wise Jewish and Christian philosophers, who have tried to construct a dignified and beautiful religion from the brutal past, as has been the case with many human groups and their various ancient religions. However, there are even today large and influential Jewish and Christian political movements which still adhere to the ancient bigotry and genocidal designs of Judaism, and they pose a genuine and substantial threat to humanity, for when the means are attained for achieving their ends, it is likely that they will win over many more adherents and carry out their ancient religious mandates with the religious fanaticism and disregard for

---

448. I. Epstein, Editor, "Shabbath", *The Babylonian Talmud*, Volume 8, Part 2, The Soncino Press, (1938), pp. 666-669, at 668-669.
449. "Circumcision", *The Jewish Encyclopedia*, Volume 4 CHAZARS-DREYFUS CASE, Funk and Wagnells Company, New York, London, (1903), pp. 92-102, at 99.
450. J. Buxtorf, "Touching the Jews Messias who is yet for to come", Chapter 36, *Synagoga Judaica: Das ist Jüden Schul ; Darinnen der gantz Jüdische Glaub und Glaubensubung... grundlich erkläret*, Basel, (1603); as translated in the 1657 English edition, *The Jewish Synagogue: Or An Historical Narration of the State of the Jewes, At this Day Dispersed over the Face of the Whole Earth*, Printed by T. Roycroft for H. R. and Thomas Young at the Three Pidgeons in Pauls Church-Yard, London, (1657).

human life and individual freedom they have displayed throughout their treacherous history.

The history of anti-Semitism to the time of political Zionism, as it was understood at the time political Zionism was formulated, is documented in brief in the eleventh edition of *Encyclopædia Britannica* (1910) in its article "Anti-Semitism" and in greater detail in Bernard Lazare's *Antisemitism: Its History and Causes* of 1894. Both contain detailed references to the literature of the period and are more apologetic to the views of anti-Semites than we are today in the post-Holocaust world. Lazare's work is especially noteworthy for its rejection of the mythology of the distinction of separate human races, which has since been bolstered by genetic research, and which foreshadowed Franz Boas' cultural anthropology.

In very early Christian times, Marcion declared that the God of the Old Testament, the Creator God of the Jews, was a hateful and genocidal maniac, as revealed in the Old Testament—was in essence the Devil, and that Christ was of a different, loving supreme God; which view was, of course, greatly offensive to Jews. Tertullian[451] and others slandered Marcion, and his views, though initially popular, eventually fell out of favor—thought they were in part revived by Friedrich Delitzsch in 1920, who openly criticized the God of the Old Testament and the Jews who created Him.[452] St. John Chrysostom accused the Jews with the alleged selfaccusation iterated in Matthew 27:25, "[Jesus'] blood be upon us [Jews] and upon our children."[453] Other such passages appear in the New Testament and were perhaps added to generate and/or justify animosity against Jews. Matthew 23:31-39 states:

> "Wherefore ye be witnesses unto yourselves, that ye are the children of them which killed the prophets. Fill ye up then the measure of your fathers. *Ye serpents, ye* generation of vipers, how can ye escape the damnation of hell? Wherefore, behold, I send unto you prophets, and wise men, and scribes: and *some* of them ye shall kill and crucify; and *some* of them shall ye scourge in your synagogues, and persecute *them* from city to city: That upon you may come all the righteous blood shed upon the earth, from the blood of righteous Abel unto the blood of Zacharias son of Barachias, whom ye slew between the temple and the altar. Verily I say unto you, All these things

---

451. Tertullian, *Adversus Marcionem*, Clarendon Press, Oxford, (1972); English translation: *The Five Books of Quintus Sept. Flor. Tertullianus Against Marcion*, T. & T. Clark, Edinburgh, (1868). "Marcionites", *The Catholic Encyclopedia*, Volume 9, Robert Appleton Company, New York, (1910), pp. 645-649.

452. F. Delitzsch, *Die grosse Täuschung: kritische Betrachtungen zu den alttestamentlichen Berichten über Israels Eindringen in Kanaan, die Gottesoffenbarung vom Sinai und die Wirksamkeit der Propheten*, In Two Volumes, Deutsche Verlags-Anstalt, Berlin, Stuttgart, (1921).

453. J. Chrysostom, translated by P. W. Harkins, "Discourses Against Judaizing Christians", *The Fathers of the Church*, Volume 68, Catholic University of America Press, Washington, D. C., (1979), p. 18.

shall come upon this generation. O Jerusalem, Jerusalem, *thou* that killest the prophets, and stonest them which are sent unto thee, how often would I have gathered thy children together, even as a hen gathereth her chickens under *her* wings, and ye would not! Behold, your house is left unto you desolate. For I say unto you, Ye shall not see me henceforth, till ye shall say, Blessed *is* he that cometh in the name of the LORD."

Luke 11 gives a somewhat different account. I *Thessalonians* 2:14-16 states,

> "For ye, brethren, became followers of the churches of God which in Judaea are in Christ Jesus: for ye also have suffered like things of your own countrymen, even as they *have* of the Jews: Who both killed the Lord Jesus, and their own prophets, and have persecuted us; and they please not God, and are contrary to all men: Forbidding us to speak to the Gentiles that they might be saved, to fill up their sins alway: for the wrath is come upon them to the uttermost."

John 8:44-45 states,

> "Ye are of *your* father the devil, and the lusts of your father ye will do. He was a murderer from the beginning, and abode not in the truth, because there is no truth in him. When he speaketh a lie, he speaketh of his own: for he is a liar, and the father of it. And because I tell *you* the truth, ye believe me not."

Revelation 2:9:

> "I know thy works, and tribulation, and poverty, (but thou art rich) and I *know* the blasphemy of them which say they are Jews, and are not, but *are* the synagogue of Satan."

Revelation 3:9:

> "Behold, I will make them of the synagogue of Satan, which say they are Jews, and are not, but do lie; behold, I will make them to come and worship before thy feet, and to know that I have loved thee."

St. Chrysostom placed the blame for Christ's crucifixion on the Jews as a religious group and employed scripture to defame them, calling them evil tempters and proselytizers, and instructed all Christians to refrain from contact with Jews; and most importantly, in his view, to refrain from practicing Jewish rituals and to avoid the synagogues, which he claimed were places of demon worship, brothels and theaters—much like the Canaanite temples. The cries for segregation sounded throughout the Dark Ages, and Jews were forced into "Ghettos" and were denied many rights. This lasted well into the 1800's and in the 1900's became the means for segregation and mass expulsion at the instigation of the political Zionists, who sought after a "world ghetto"[454] in order to preserve the alleged racial purity of Jews.

Christians were forbidden to charge usurious interest rates when loaning money, because, *inter alia*, Aristotle had declared that the practice of Usury was unethical and a form of stealing.[455] This view was adopted by Christians. Hence, Christians did not often loan money to those who were most desperate for it, because there was no reward to compensate the risk. Jews were not so inhibited against Gentiles, though usury against fellow Jews was discouraged (*Deuteronomy* 15; 23:20), and Jews had always been skilled financiers long before Christianity emerged, though the Christians are most often blamed today for the usury of Jews given that they sometimes limited the ability of Jews to own land or work in agriculture or certain industries.[456] Exercising self-discipline and freed from the burdens of funding armies and instead profiteering from them, these financiers accumulated great wealth, compounding their fortunes with exorbitant interest rates that led Gentile societies into ruin. When governments wished to conduct wars, or when they ran into financial difficulties, or when they needed money to build palaces, or buy ships, or bribe other nations, or to pay ransoms, they often turned to these Jewish financiers for funds.[457] No one loans such large sums of money without considering the risks involved and applying a rate of interest commensurate to the risk that will make the transaction profitable. As such, some Jews, who were otherwise segregated, gained access to the ruling classes of Europe and naturally exercised a tremendous measure of influence over the destiny of the European powers—which involvement by those who considered themselves foreigners was very much resented.

During their forced (and in many instances voluntary) segregation in the Ghettos, the Jews evolved a strong Bourgeoisie class, which thrived on city life, and, following emancipation, became a competitive threat to the underdeveloped Bourgeoisie class of the Gentiles, who had not emerged from the feudal system as advanced and prepared for city living as had the Jews—which was especially true in Russia. From the most ancient times of their history, the Jews stressed the value of educating of their young. Josephus noted long ago in his *Against Apion*,

"Our principal care of all is this, to educate our children well[.]"[458]

---

[454]. T. Herzl, English translation by H. Zohn, R. Patai, Editor, *The Complete Diaries of Theodor Herzl*, Volume 1, Herzl Press, New York, (1960), p. 172.

[455]. Aristotle, *Nichomachean Ethics* V, 5, 1133ab; **and** *Politics*, I, 10, 1258ab. **See also:** Plato, *Laws*, V, 742. **See also:** Luke 6:35.

[456]. B. J. Hendrick, "The Jews in America: II Do the Jews Dominate American Finance?", *The World's War*, Volume 44, Number 3, (January, 1923), pp. 266-286, at 275.

[457]. D. S. Jordan, *Unseen Empire; a Study of the Plight of Nations that Do Not Pay Their Debts*, American Unitarian Association, Boston, (1912).

[458]. Josephus, "Flavius Josephus Against Apion", *The Works of Flavius Josephus: Comprising the Antiquities of the Jews; a History of the Jewish Wars; and Life of Flavius Josephus, Written by Himself*, Book 1, S. S. Scranton Co., Hartford, Connecticutt, (1916), p. 886.

In the 1870's in the newly formed German nation, Jewish financiers were heavily involved in the many scandals which led to financial hardships for the general population, and this was generalized into a general hatred for Jews by Wilhelm Marr and Otto Glogau[459]—though it was Edward Lasker, who was Jewish, who on 14 January 1873 called the attention of the Prussian Diet to the crisis and indicted Bethel Henry Strousberg, who was also Jewish, on 7 February 1873.[460] These scandals were soon followed by the "Black Friday" of 9 May 1873 in Vienna, where the police closed the stock market, which had collapsed under the weight of widespread Jewish corruption. An article published under the heading, "Bernhardt in Russia", *The New York Times*, (24 January 1882), page 6, wrote, *inter alia*,

"[...]The anti-Semitic press sneered disdainfully at 'Sarah the Jewess,' the Slavophiles made her name the pretext for fresh controversy with the *Zapadniki*—the 'Occidentals'—as are termed those who are modest enough to admit that Russia has still a good deal to learn in the matter of civilization from Western Europe.

This polemic was at its height when she arrived at St. Petersburg; she was tired with her journey, out of spirits, suffering from a bad cold, and the reception at the depot was not calculated to bring up her morale, all of which suited the plans of her enemies, who prophesied a stupendous failure and an assault on the box-office by the public to get its money back before the end of the first act. The aristocracy took no part in this discussion, as the aristocracy affects supreme contempt for everything relating to the question of Jews versus Gentiles, simply ignoring the chosen people altogether, except when it wants its paper discounted. But, though caring naught for the stranger's religious creed, it was, all the same, not benevolently inclined to her. These and all the other details indeed I got from another private letter written by one of the Bernhardt troupe, where all is not peace and good-will,

---

459. W. Marr, *Der Judenspiegel*, Im Selbstverlage des Verfassers, (1862); **and** *Der Sieg des Judenthums über das Germanenthum*, Rudolph Costenoble, Bern, (1879); **and** *Vom jüdischen Kriegsschauplatz: eine Streitschrift*, R. Costenoble, Bern, (1879); **and** *Wählet keinen Juden: Der Weg zum Siege des Germanenthums über das Judenthum*, O. Hentze, Berlin, (1879); **and** *Vom jüdischen Kriegsschauplatz eine Streitschrift*, R. Costenoble, Bern, (1879); **and** *Jeiteles teutonicus. Harfenklänge aus dem vermauschelten Deutschland*, Leipzig, (1879); and *Der Judenkrieg, seine Fehler und wie er Zu organisiren ist*, Richard Oschatz, Chemnitz, (1880); **and** *Goldene Ratten und rothe Mäuse*, E. Schmeitzner, Chemnitz, (1880); **and** *Wo steckt der Mauschel? oder, Jüdischer Liberalismus und wissenschaftlicher Pessimismus*, W. Raich, New York, (1880); **and** *Der Weg zum Siege des Germanenthums über das Judenthum*, O. Hentze, Berlin, (1880); and *Wählet keinen Juden: ein Mahnwort an die deutschen Wähler*, O. Hentze, Berlin, (1881). *See also:* O. Glagau [Glogau???], *Die Bösen und Grunderqeschwindel in Berlin*, Leipzig, (1876); *Les besoins de l'Empire et le nouveau Kulturkampf*, Osnabruck, (1879).
460. Refer to the Eleventh Edition of the *Encyclopædia Britannica* (1910) in its article "Anti-Semitism". H. Blum, *Das deutsche Reich zur Zeit Bismarcks. Politische geschichte von 1871 bis 1890*, Bibliographisches Institut, Leipzig, Wien, (1893), pp. 153-181.

although Marie Colombier is not of it. Sarah is accused of being too fond of M. Angelo, of snubbing certain great personages on account of that good-looking actor, and such derogations are not readily pardoned in Muscovy, as Mme. Patti learned to her cost. And so the fatidical moment came; the house was literally crammed; nowhere was even standing room obtainable; the fronts of every box resembled a horticultural exhibition; but the silence was glacial. The curtain rose, not a Russian hand was clapped in welcome, the Nationals in the pit smiled with ill-concealed satisfaction, not venturing on a hiss, but sternly shutting up the few French spectators who might have been disposed to manifest favorably. The first act passed, so did the second, and always with the same hostile reserve; Sarah was seriously disheartened; she said nothing and kept the door of her dressing-room closed and locked during the *entr'actes*, but rumor affirms that the grinding of teeth inside was distinctly audible from without, and that Mr. Jarrett grew pale as he thought of the scene which would follow that performance, and of the other scene which he might expect when summoned by the Director of Police to explain why there was no second performance, as the showman felt quite sure that she who had left the Français because her interpretation of the 'Aventurière' was pronounced to be defective would never risk a second affront from a Russian audience. However, Sarah is a brave woman; whatever she thought she kept severely to herself; never did she play better, and at last, suddenly, in the third act, in the pathetic scene between Marguerite and Armand Duval's father, the ice melted as by enchantment; the artist had won her cause, and the whole audience broke into an enthusiastic tempest of applause, which was all the more passionate that it had been so long contained. No such triumph of talent over stupid prejudice is on record in the city of Peter; the stage became literally a parterre of flowers; even the Nationals cheered, and, forgetting their hatred of her origin, joined with the 'Occidentals' in sincere admiration of this 'queen of dramatic art.' Since then all has gone well with Mlle. Bernhardt, but it was so nearly not going well that she wants to come home, and, as I have already hinted as a strong probability, will not insist on M. Perrin's making *amende honorable* in his shirt with a lighted taper in his hand and barefooted, which were, says Mrs. Grundy, the original conditions demanded by her as the *sine qua non* of her return, the derelict manager and the *societarians* on the other hand, and as a *quid pro quo* for this gracious condescension, agreeing not to talk any more about those 100,000f. damages awarded by the court for breach of contract.

From a French actress on the Russian stage to a Russian play on the French stage, the transition is easy and natural, but before saying anything about 'Serge Panine,' I will just mention a few words concerning that tremendous anti-Semitic feeling which, first showing itself in semi-barbarous Southern Muscovy, has gradually spread westward, until not only Germany, but even skeptical, atheistical France begins to exhibit symptoms of the agitation. Far be it from me to justify or even palliate the atrocious acts of violence perpetrated at Odessa, at Kiev, at Elizabethgrad, and at a hundred other points upon the Israelites. Persecution is never justifiable, but truth and justice compel me to say that the Israelites have brought it upon themselves by over-zeal in acting up to the one command of the lawgiver which none of them have ever transgressed; they were bidden to spoil the Egyptians, and they have done so, confounding all who are not of their own communion in the same category as the original subjects of King Pharaoh, until the Gentiles have arisen in their wrath and smitten their oppressors. All through the East of Europe, except in Turkey, where, thanks to the superior cleverness of the Armenians, the Jews starve, these Pariahs prove that Shylock was not the mere creation of a poet's fancy. A vast deal of maudlin sympathy was expressed about their sufferings in Roumania by their coreligionists in England, whose lying publications were circulated on all sides and awakened even an echo in America. I wish some of their sympathizers would go to Roumania, to Galicia, to Bessarabia, and judge from what they see how those Provinces have flourished since they were overrun by this scum of mankind, more destructive than the locusts and the grasshoppers of Egypt; the travelers might be sincere Philo-Semites when they started on their journey, but not one on his return, if he spoke the truth, would profess other than sentiments of ultra-anti-Semitism. The hatred of Poles for Russians is proverbial, and yet the late riots at Warsaw have proved that this animosity is about to be forgotten in their still stronger detestation of the Jews, whom the population accuses of its ruin by their shameless speculations on corn and breadstuffs, and their disgraceful manœuvres on the Stock Exchanges of Frankfort, Berlin, and Vienna. I believe, and I think have already so stated, that Sarah Bernhardt is not a child of Israel, although she certainly is of that type, but the mere suspicion of her connection with 'the foul accursed race who crucified our Lord'—I quote literally from a Russian author—is quite enough to create hosts of enemies; if she had tried it on about Easter, or rather about the time of the Passover, when the Slav peasant is convinced that the Jews steal and murder Christian babes 'in their unholy rites,' Sarah would have been pelted with something even more disagreeable than cucumbers, and about one-half of the nation would have thought, if it did not say so in so many words: 'Served her right!' [...]"[461]

Such corruption also took place in the American stock market, where pools run by Jewish financiers would run up stock prices to bilk the comparatively poor, who had bought shares on margin. It was common knowledge that Jewish financiers kept specific reporters in their pockets. These corrupt cabals would bride newspaper correspondents to write favorable reviews of certain companies in order to lure in poorer investors. The rich would then sell the stock for a profit, and then sell it short, for an additional profit. These profits were stolen from the middle class and the comparatively poor. This eventually resulted in the Great

---

461. A Letter to the Editor by "Truth" in response to the above article appeared under the heading "Jews in Russia and the East" in *The New York Times*, (28 January 1882), p. 2.

Depression.

There were many accusations made in the 1800's of undue and disproportionate Jewish influence in the newspapers by Friedrich Wilhelm Ghillany;[462] and perhaps most notably by Richard Wagner, who charged that Jewish influences had destroyed his career after he had published an essay which was critical of undue Jewish influence; and by Eugen Karl Dühring, who, like Wagner, charged that Jews controlled the press and manipulated public opinion in a grossly destructive way, converting high culture into something cheap, base and banal; as well as by Adolf Stoecker,[463] who called for segregation, as the racist Zionist Albert Einstein later would. Dühring wrote in 1881 that the promotion of Spinoza, Heine, Lessing, Lassal, etc. by some Jews was overblown and degenerative and resulted from dishonest self-advertisement. Heinrich von Treitschke and Wilhelm Marr also alleged in the Nineteenth Century that Jews controlled the press. Treitschke, who most famously stated, "Die Juden sind unser Unglück!" wrote in 1879,

> "The little man can no longer be talked out of the fact that the Jews write the newspapers. Therefore, he won't believe them any longer. Our newspaper system owes a great deal to Jewish talents. From the first the trenchancy and acuity of the Jewish spirit found a fruitful field. But here, too, the effect was ambiguous. Börne was the first to introduce a characteristically shameless tone into our journalism."[464]

The accusation of undue Jewish influence in journalism again spiked when the *Protocols of the Learned Elders of Zion*, which advocated Jewish control over all important media outlets, became infamous in 1920.[465] Many pointed to the pride some Jews took of their disproportionate rôle in the media of New York and Berlin.

Dühring's attacks probably resulted in an insecurity that led many to dishonestly hype Einstein beyond all reasonable limits in order to establish in the public mind that Jews were not only capable of creative thought, but were the most significant thinkers. Dühring wrote, *inter alia*,

---

462. P. L. Rose, *Revolutionary Antisemitism in Germany from Kant to Wagner*, Princeton University Press, (1990), p. 260. Rose cites: F. W. Ghillany, *Die Judenfrage: Eine Beigabe zu Bruno Bauers Abhandlung über dieser Gegenstand*, Nürnberg, (1843), p. xxi; **and** R. W. Stock, *Die Judenfrage durch fünf Jahrhunderte*, Verlag Der Stürmer, Nürnberg, (1939), pp. 412ff.

463. P. W. Massing, *Rehearsal for Destruction: A Study of Political Anti-Semitism in Imperial Germany*, Howard Fertig, New York, (1967), pp. 278-287.

464. H. v. Treitschke, *Preussische Jahrbücher*, Volume 44, (1879), pp. 572-574; English translation in R. S. Levy, *Antisemitism in the Modern World: An Anthology of Texts*, D. C. Heath and Company, Toronto, (1991), pp. 69-73, at 72.

465. S. Nilus, *The Protocols and World Revolution Including a Translation and Analysis of the "Protocols of the Meetings of the Zionist Men of Wisdom"*, Small, Maynard & Co., Boston, (1920), pp. 120-122; **and** the V. E. Mardsen translation as republished as: *Protocols of the Learned Elders of Zion*, The Thunderbolt, Savannah, Georgia, p.73.

"For a century, the emergence of the Jews, as well as its precisely not modest propaganda for everything which belongs to their race, has had the greatest share in the fact that Spinoza has come more to the foreground. [***] Heine has formed something out of Romanticism and has moreover plagiarized great models like the British poet Byron down to his own level. [***] Those small people, like Mr. Marx, who conducted even from London, but under the name of Socialism, a so-called worker's association, in truth however a Jewish alliance, showed wherever they erred in science, noteworthy talent really only in their literary shamelessness. In this way Mr. Marx had drawn along his Jews so discreetly to a formless and eccentric fragmentary book which he produced from himself without any talent, after unmentionably long toils, that these people were soon speaking of a Marxist century. The humour however became complete only when rather similarly people spoke of a Jewish century; for this entire so-called science in which such propagandistic Jews made a business aims, in its way, also not at the supposed happiness of the nations but at the merging of all nations into a Jewish kingdom. [***] The Jews, who do not create anything even in science, but even there only conduct business with the products and the work of another may occasionally put on the market individual talents and especially acquisitive talents—the creative power and genius, however, remain always foreign to them. [***] One needs only to consider the advertisements with which the Jews seek at present, at any cost, to raise their Lessing up to a god after they have for a century raised his fame ten times more than what he is worth with all the arts of false praise. The business which the Jewish press and Jewish literature have always systematically made out of bringing a powerful overvaluation of Lessing into the public has recently been carried out indeed to the point of disgust. The Jewish newspaper writers have raised the author of that flat Jewish piece which is entitled *Nathan der Weise* over the greatest authors and poets and declared him to be, for example, the greatest German, to say something against whom would be a *lèse Majésté*. [***] The preceding hurried treatment is however based here only on the fact that the overvaluation of Lessing by the Jews forms the example lying closest at hand, and the most popular, of the effects of the most unashamed Jewish advertisement, and that Lessing himself, together with Börne and Heine, represents a group of literary renown which must be briefly characterized as a Jewish group and be separated from the really creative and truly original greats like Voltaire, Rousseau, Bürger, Byron, to a certain extent also Goethe, Schiller, and Shelley. If the Jews did not have the daily press in their hands, it would not have been possible to falsify the truth with so many tears before the eyes of the peoples, to displace the natural judgement and force everywhere an interested Jewish opinion in its place. [***] Their inherited lack of imagination is the cause of their aversion to clear illustration, and correspondingly also a reason of the religious statutes founded by them. [***] In this coarse and base material direction also lies a chief reason of the incapacity of the Jews to prove themselves creative in science and art. [***] Some talent, which however remains far removed from creative genius and mostly indeed only apes, is all that is found exceptionally among individual Jews. Almost always, however, this talent is, above all, one of appropriation and of trading with the intellectual accomplishments of others. [***] That is now the same Lasalle on whom the Jews later pride themselves in their lack of better racial comrades, and whom they glorify, in spite of everything which he has effected against the bourgeoisie and therewith also against themselves, with the most unashamed advertisements."[466]

Wagner may have had an axe to grind, due to his inability to live within his means, and as an historical witness, his claims need be scrutinized with an especial care. Dühring also believed that "Jewish elements" had interfered with his career. In 1882, Franz Mehring quoted a Jewish author who criticized Jews for, among other things, "the malicious gloating when veritable conspiracies deprived of their livelihoods people who were suspected of anti-Jewish feelings[.]"[467]

Some Germans, such as Wilhelm Marr, Eugen Karl Dühring and Friedrich Nietzsche, saw Christianity as an evolution of Jewish dogma, and wished to rid Germany of both "foreign" influences of a "slavish" philosophy that thrived not because of any intrinsic worth, but instead from the manipulation of human frailty and an appeal to the human will to everlasting life which renders us gullible. Dühring saw the "Jewish question" as a racial question—a viewpoint also taken by the political Zionists, most especially Theodor Herzl—who learned it from Dühring.[468] Before Dühring and Herzl was the racist Zionist Socialist writer Moses Hess, who advocated a biological and "racial" answer to the "Jewish question".

As early as Justin Martyr, Christians argued that God had given them the "prophetical gifts" promised to the Jews. Though the Jews saw their life eternal as the eternal life of the Jewish People, Christians personalized this prophetical gift to give each individual eternal life, thereby taking the nationalistic racism out of Judaism and greatly increasing the allure of the religion to Gentiles. The Jews' Talmud and *Zohar* considered Jesus an insane imposter who had brought

---

466. E. K. Dühring, *Die Judenfrage als Racen-, Sitten- und Culturfrage: mit einer weltgeschichtlichen Antwort*, H. Reuther, Karlsruhe, (1881); English translation by A. Jacob, *Eugen Dühring on the Jews*, Nineteen Eighty Four Press, Brighton, England, (1997), pp. 101, 104, 110-111, 113-114, 115, 120, 121, 124, 127. *See also:* E. K. Dühring, *Die Ueberschätzung Lessing's und Dessen Anwaltschaft für die Juden*, H. Reuther, Karlsruhe, (1881).
467. English translation from: P. W. Massing, *Rehearsal for Destruction: A Study of Political Anti-Semitism in Imperial Germany*, Howard Fertig, New York, (1967). p. 315.
468. M. Samuel, "Diaries of Theodor Herzl", in: M. W. Weisgal, *Theodor Herzl: A Memorial*, The New Palestine, New York, (1929), pp. 125-180, at 129. T. Herzl, English translation by H. Zohn, R. Patai, Editor, *The Complete Diaries of Theodor Herzl*, Volume 1, Herzl Press, New York, (1960), pp. 4, 111.

them great suffering, an alleged imposter who had been called the "whore's son" in antiquity, and they continued to ridicule Jesus calling him the "hanged rogue". Segregation among neighbors led to increasing suspicions, and many allegations accrued, some of which defamed the Jews, some of which defamed the Christians. The Jews were accused of murdering Christian babies in order to use their blood in Passover rituals, of desecrating the Host, etc. Conversely, some Jews accused Gentiles of murdering their babies.[469]

Martin Luther; angered by the Jews' refusal to convert to Christianity, and their Talmudic writings which call the Virgin Mary a whore (Mary Magdalene) and Jesus the "whore's son" and "the hanged man";[470] decried the Jews' racism and genocidal plans in the strongest of terms, and claimed that they were no longer the "chosen people" because a loving God could not have inflicted the misery upon them that they had suffered in the intervening 1,500 years since the death of Christ, and that God was punishing them for having rejected His Son. In the 1500's, Martin Luther wrote, among other things,

> "Further, they presume to instruct God and prescribe the manner in which he is to redeem them. For the Jews, these very learned saints, look upon God as a poor cobbler equipped with only a left last for making shoes. This is to say that he is to kill and exterminate all of us Goyim through their Messiah, so that they can lay their hands on the land, the goods, and the government of the whole world. And now a storm breaks over us with curses, defamation, and derision that cannot be expressed with words. They wish that sword and war, distress and every misfortune may overtake us accursed Goyim. They vent their curses on us openly every Saturday in their synagogues and daily in their homes. They teach, urge, and train their children from infancy to remain the bitter, virulent, and wrathful enemies of the Christians."[471]

Jewish Bolshevism and Nazism very nearly accomplished all these Messianic goals for the Jews in the Twentieth Century.

George Henry Borrow recorded in his book of 1841, *The Zincali*, that,

> "There are certainly some points of resemblance between the children of Roma [Gypsies, as in E*gyp*t, though in reality of Indian origin] and those of Israel. Both have had an exodus, both are exiles and dispersed amongst the gentiles, by whom they are hated and despised, and whom they hate and despise, under the names of Busnees and Goyim; both, though speaking the

---

469. Talmud, *Abodah Zarah* 26a.
470. G. Dalman, *Jesus Christ in the Talmud, Midrash, Zohar, and the Liturgy of the Synagogue*, Deighton Bell, Cambridge, (1893).
471. M. Luther, *Von den Juden und ihren Lügen*, Hans Lufft, Wittenberg, (1543); Reprinted, Ludendorffs, München, (1932); English translation by Martin H. Bertram, "On the Jews and Their Lies", *Luther's Works*, Volume 47, Fortress Press, Philadelphia, (1971), pp. 123-306, at 264.

language of the Gentiles, possess a peculiar tongue, which the latter do not understand, and both possess a peculiar cast of countenance, by which they may, without difficulty, be distinguished from all other nations; but with these points the similarity terminates. The Israelites have a peculiar religion, to which they are fanatically attached, the Romas have none, as they invariably adopt, though only in appearance, that of the people with whom they chance to sojourn; the Israelites possess the most authentic history of any people in the world, and are acquainted with and delight to recapitulate all that has befallen their race, from ages the most remote; the Romas have no history, they do not even know the name of their original country, and the only tradition which they possess, that of their Egyptian origin, is a false one, whether invented by themselves or others; the Israelites are of all people the most wealthy, the Romas the most poor; poor as a Gypsy being proverbial amongst some nations, though both are equally greedy of gain; and finally, though both are noted for peculiar craft and cunning, no people are more ignorant than the Romas, whilst the Jews have always been a learned people, being in possession of the oldest literature in the world, and certainly the most important and interesting."[472]

Max Nordau stated, *inter alia*, in his address to the First Zionist Congress in 1897 published in *The Jewish Chronicle* on 3 September 1897 on pages 7-9, at 8 and 9,

"In the Ghetto, the Jew had his own world; it was to him the sure refuge which had for him the spiritual and moral value of a parental home. Here were associates by whom one wished to be valued, and also could be valued; here was the public opinion to be acknowledged by which was the aim of the Jew's ambition. To be held in low esteem by that public opinion was the punishment for unworthiness. Here all specific Jewish qualities were esteemed, and through their special development that admiration was to be obtained which is the sharpest spur to the human mind. What mattered it that outside the Ghetto was despised that which within it was praised? The opinion of the outside world had no influence, because it was the opinion of ignorant enemies. One tried to please one's co-religionists, and their applause was the worthy contentment of his life. So did the Ghetto Jews live, in a moral respect, a real full life. [***] Before the emancipation the Jew was a stranger among the peoples, but he did not for a moment think of making a stand against his fate. He felt himself as belonging to a race of his own, which had nothing in common with the other people of the country. The emancipated Jew is insecure in his relations with his fellow-beings, timid with strangers, suspicious even toward the secret feeling of his friends. His best powers are exhausted in the suppression, or at least in the difficult

---

[472]. G. H. Borrow, *The Zincali, or, An Account of the Gypsies of Spain*, Volume 1, John Murray, London, (1841), pp. 159-160.

concealment of his own real character. For he fears that this character might be recognized as Jewish, and he has never the satisfaction of showing himself as he is in all his thoughts and sentiments. He becomes an inner cripple, and externally unreal, and thereby always ridiculous and hateful to all higher feeling men, as is everything that is unreal. All the better Jews in Western Europe groan under this, or seek for alleviation. They no longer possess the belief which gives the patience necessary to bear sufferings, because it sees in them the will of a punishing but yet loving God."

In 1898, Zionist Communist Nachman Syrkin wrote,

"This sense of their higher religious estate, rooted in the general cast of the Jewish spirit, was the source of their morale in their war with the world. [***] How did the Jews react to the world? The religious-psychological difference had already sown the seed of estrangement and hatred between Christian and Jew, and the many troubles the Jews had suffered added to their bitterness. Huddling together with his brethren in the ghetto, the Jew gritted his teeth, cursed the enemy, and dreamed of revenge, the vengeance of heaven and earth."[473]

The Talmud, *Shabbath* 89a, states that Jewish hatred of all other peoples proceeded from Mount Sinai. The Old Testament book of *Deuteronomy*, Chapter 7, instructs the Jews to commit genocide against the other peoples of the Earth and *Numbers*, Chapter 24, prophesies a Messiah from the seed of Jacob who will exterminate all the other nations, all of the descendants of Esau, all of the rest of humanity.

Albert Einstein was not above hypocrisy, hatemongering and smear campaigns to achieve his personal political ends. Einstein had a reputation as a rabid anti-assimilationist, which is to say that Einstein was a rabid racist segregationist. On 15 March 1921, Kurt Blumenfeld wrote to Chaim Weizmann,

"Einstein [***] is interested in our cause most strongly because of his revulsion from assimilatory Jewry."[474]

Einstein, who had himself married outside his religion, ethnicity and "race" over his mother's racist objections, avowed (in nationalistic and racist terms Hitler would later use),

"To deny the Jew's nationality in the Diaspora is, indeed, deplorable. If one adopts the point of view of confining Jewish ethnical nationalism to Palestine, then one, to all intents and purposes, denies the existence of a

---

[473]. N. Syrkin, under the nom de plume "Ben Elieser", *Die Judenfrage und der socialistische Judenstaat*, Steiger, Bern, (1898); English translation in A. Hertzberg, *The Zionist Idea*, Harper Torchbooks, New York, (1959), pp. 333-350, at 334-335.
[474]. J. Stachel, *Einstein from 'B' to 'Z'*, Birkhäuser, Boston, (2002), p. 79, note 41.

Jewish people. In that case one should have the courage to carry through, in the quickest and most complete manner, entire assimilation.

We live in a time of intense and perhaps exaggerated nationalism. But my Zionism does not exclude in me cosmopolitan views. I believe in the actuality of Jewish nationality, and I believe that every Jew has duties towards his coreligionists. [***] [T]he principal point is that Zionism must tend to strengthen the dignity and self-respect of the Jews in the Diaspora. I have always been annoyed by the undignified assimilationist cravings and strivings which I have observed in so many of my friends."[475]

Racist and alarmist Hitler sounded very much like racist and alarmist Einstein,

"Only today, when the same deplorable misery is forced on many millions of Germans from the Reich, who under foreign rule dream of their common fatherland and strive, amid their longing, at least to preserve their holy right to their mother tongue, do wider circles understand what it means to be forced to fight for one's nationality. [***] The elemental cry of the German-Austrian people for union with the German mother country, that arose in the days when the Habsburg state was collapsing, was the result of a longing that slumbered in the heart of the entire people — a longing to return to the never-forgotten ancestral home. [***] Gradually I began to hate them. All this had one good side: that in proportion as the real leaders or at least the disseminators of Social Democracy came within my vision, my love for my people inevitably grew. [etc. etc. etc.]"[476]

The racist legacy of political Zionism lingers. Israeli Supreme Court Justice Haim Cohn was quoted in *The London Times* on 25 July 1963 on page 8:

"It is one of the bitterest ironies of fate that the same biological or racist approach which was propagated by the Nazis and characterized the infamous Nuremberg laws should, because of an allegedly sacrosanct Jewish tradition, become the basis for the official determination or rejection of Jewishness in the state of Israel."[477]

When some Jews attempted to help the Falasha, the "black Jews" of Ethiopia, to emigrate to Israel in the early 1980's, they initially received little help from the Israeli Government.[478]

---

475. A. Einstein, "Jewish Nationalism and Anti-Semitism", *The Jewish Chronicle*, (17 June 1921), p. 16.
476. A. Hitler, English translation by Ralph Manheim, *Mein Kampf*, Houghton Mifflin, Boston, New York, (1971), pp. 11, 13, 63.
477. See also: J. Badi, *Fundamental Laws of the State of Israel*, Twayne Publishers, New York, (1961), p. 156.
478. Video Documentary, S. Jacobovici, *Falasha: Exile of the Black Jews*, New Yorker Video, Matara Film Productions, New York, (1983).

7.4 The Hypocritical Vilification of Caligula— Ancient Jewish Historians are not Credible

In the era of Philo the Jew, the Roman Emperor Gaius Caesar Augustus Germanicus (a. k. a. *Caligula*) had, pursuant to Roman custom, declared himself a god and demanded that the Jews worship him and instructed the Jews of Alexandria to erect statues to him and to swear by his name. This was an intolerable request and constituted sacrilege for the Jews. However, it was the normal practice of a Roman Emperor and the Jews themselves had often desecrated the temples of other religions and committed genocide against many other peoples, in order to spare the honor of their "jealous God",

> "Thou shalt not bow down thyself to them, nor serve them: for I the LORD thy God *am* a jealous God, visiting the iniquity of the fathers upon the children unto the third and fourth *generation* of them that hate me; [***] For thou shalt worship no other god: for the LORD, whose name is *Jealous*, *is* a jealous God: [***] For the LORD thy God *is* a consuming fire, *even* a jealous God. [***] (For the LORD thy God *is* a jealous God among you) lest the anger of the LORD thy God be kindled against thee, and destroy thee from off the face of the earth."[479]

and they believed, "The LORD *is* a man of war: the LORD *is* his name."[480] Among the many such acts we find in *Deuteronomy* 7:4-6,

> "4 For they will turn away thy son from following me, that they may serve other gods: so will the anger of the LORD be kindled against you, and destroy thee suddenly. 5 But thus shall ye deal with them; ye shall destroy their altars, and break down their images, and cut down their groves, and burn their graven images with fire. 6 ¶ For thou *art* a holy people unto the LORD thy God: the LORD thy God hath chosen thee to be a special people unto himself, above all people that *are* upon the face of the earth."

Many of the ancient Greeks and Egyptians of Alexandria hated the Jews who lived there;[481] because, in addition to their hypocritical religious intolerance, the Jews of Alexandria had exercised monopoly control over many markets, were tax collectors, and employed other corrupt means to accumulate vast fortunes and draw off the gold of other peoples.[482] Manetho and Apion

---

[479]. *Exodus* 20:5, 34:14, *Deuteronomy* 4:24, 5:9, 6:15.
[480]. *Exodus* 15:3.
[481].N. Bentwich, *Philo-Judæus of Alexandria*, The Jewish Publication Society of America, Philadelphia, (1910).
[482]. B. Lazare, *Antisemitism: Its History and Causes*, (1894); *L'Antisémitisme, son Histoire et ses Causes*, L. Chailley, Paris, (1894).

exposed Judaism as a vulgar and hateful religion, which was filled with lies, plagiarisms and historical deceptions. The *Septuagint* (which corrupted Judaism to somewhat soften and render less obvious the Jews' quest for world domination, their racism and their religious and historical fabrications) was in part a work meant to placate the Greeks and the Egyptians. In the early Christian Era, Marcion, Clement, Origen, and many others, again exposed the hateful nature of Judaism. The Jews were largely successful in attacking these early critics, through deceitful "Christian" reactionaries. They grew less successful at suppressing anti-Judaism as time went on and they faced Cyprian, Chrysostom, and many other Christians, who would not tolerate Jewish proselytizing, Jewish intolerance and Jewish anti-Christianism. The cruder attacks tended to be the more successful.

As is evident in Philo the Jew's writings, the Jews of Alexandria virulently defamed the Egyptians and Greeks, and treated all other religions with utter and expressed contempt and otherwise degraded their neighbors in the harshest terms—terms that would later be applied to Christians and then to Moslems in the Talmud and Cabalistic literature. Among many examples, Philo stated,

> "The greater portion of these men ere Egyptians, wicked, worthless men, who had imprinted the venom and evil disposition of their native asps and crocodiles on their own souls, and gave a faithful representation of them there."[483]

The Bible and Talmud also treat Egyptians as if sub-human animals.[484] Rabbi Meir Kahane quoted *Midrash Tehillim* 22:1, on 23 May 1986,

> "Each Jew took his dog and put his foot on the throat of a dead Egyptian and said to his dog: Eat of the hand that enslaved me; eat of the heart that showed me no pity."[485]

Philo the Jew's racist hatred and hypocrisy are even more apparent in his essay *Flaccus* than in his *On the Embassy to Gaius*—which will be addressed here in detail. He asserts in *Flaccus* that Jews have a right to their religion and the rights and privileges of all other countries, but that Judea is a holy place and cannot be violated by any other religion, while demanding that the Egyptians, whom he loathes, give up their lands to Jews and allow Jews to dominate their chief cities and political life through the Roman leaders they have bought, while

---

483. Philo the Jew, "On the Embassy to Gaius", *The Works of Philo*, Hendrick Publishing, U. S. A., (2000), pp. 757-790, at 772.
484. *Ezekiel* 23:20. *Berakoth* 25b, 58a. *Shabbath* 150a. *Yebamoth* 98a. *Niddah* 45a. *Arakin* 19b.
485. M. Kahane, *On Jews and Judaism: Selected Articles 1961-1990*, Volume 1, Institute for the Publication of the Writings of Rabbi Meir Kahane, Jerusalem, (1993), p. 130. An English translation of the entire text of *Midrash Tehillim* 22:1 appears in: W. G. Braude, *The Midrash on Psalms*, Volume 1, Yale University Press, New Haven, (1959), p. 297.

the Egyptians, whom he describes as subhuman, struggle in poverty and slavery in their own lands. In those superstitious times, Philo repeatedly threatens people with the power of his God, and concludes, "that the nation of the Jews is not left destitute of the providential assistance of God."[486] However, it is clear in his self-indulgent stories that the governing force "behind the scenes"[487] was the corrupt influence of Philo's money, not God, on the Romans—a fact well-known to Caligula, who allegedly became archenemy of the Jews.

According to the stories of the ancient Jews, Caligula, who, like all Roman Emperors of his era, believed himself to be a god, demanded that a statue of him be placed in the Jewish Temple in Jerusalem, due to an act of hypocrisy by some of the Jews of Jamnia, who demanded that other peoples obey their Jewish laws in the Jewish holy land, while the Jews refused to obey the laws of Rome.

The Jews hypocritically forbade Gentiles to practice Gentile religions in Judea, or in any of the sections of foreign cities with large Jewish populations, while demanding that Jews be given religious freedom throughout the world. Jewish hypocrisy and intolerance offended the Romans' sense of justice. Tacitus wrote,

> "Quite different is their faith about things divine. The Egyptians worship many animals and images of monstrous form; the Jews have purely mental conceptions of Deity, as one in essence. They call those profane who make representations of God in human shape out of perishable materials. They believe that Being to be supreme and eternal, neither capable of representation, nor of decay. They therefore do not allow any images to stand in their cities, much less in their temples. This flattery is not paid to their kings, nor this honour to our Emperors. From the fact, however, that their priests used to chant to the music of flutes and cymbals, and to wear garlands of ivy, and that a golden vine was found in the temple, some have thought that they worshiped father Liber, the conqueror of the East, though their institutions do not by any means harmonize with the theory; for Liber established a festive and cheerful worship, while the Jewish religion is tasteless and mean."[488]

According to the stories of the ancient Jews, the Jews of Jamnia violated a religious monument to Caligula by destroying it on the grounds that it violated

---

486. Philo the Jew, "Flaccus", *The Works of Philo*, Hendrick Publishing, U. S. A., (2000), pp. 725-741, at 741.
487. Philo the Jew, "Flaccus", *The Works of Philo*, Hendrick Publishing, U. S. A., (2000), pp. 725-741, at 737.
488. Tacitus, translated by A. J. Church and W. J. Brodribb, *The Complete Works of Tacitus*, The Modern Library, New York, (1942), pp. 659-660. An alternative, and far more accessible translation appears in: Tacitus, *History of the Jews*, Book V, Chapters 2-8 in: "Dissertation III", *The Works of Flavius Josephus: Comprising the Antiquities of the Jews; a History of the Jewish Wars; and Life of Flavius Josephus, Written by Himself*, S. S. Scranton Co., Hartford, Connecticutt, (1916), pp. 951-956.

Jewish religious laws, then were shocked to learn that Caligula would retaliate by violating the Jewish Temple with a statue of himself. Instead of realizing their hypocrisy, Jews like Philo and Josephus instead heaped defamation upon defamation on people who sought social justice and the equitable distribution of wealth—the people of Alexandria and other cities where the Jews were a privileged, segregated and intolerant class. Philo records,

"'You know the principal and primary cause of all; for that indeed is universally known to all men. [Caligula] desires to be considered a god; and he conceives that the Jews alone are likely to be disobedient; and that therefore he cannot possibly inflict a greater evil or injury upon them than by defacing and insulting the holy dignity of their temple; for report prevails that it is the most beautiful of all the temples in the world, inasmuch as it is continually receiving fresh accessions of ornament and has been for an infinite period of time, a never-ending and boundless expense being lavished on it. And as he is a very contentious and quarrelsome man, he thinks of appropriating this edifice wholly to himself. (199) And he is excited now on this subject to a much greater degree than before by a letter which Capito has sent to him.

'Capito is the collector of the imperial revenues in Judaea, and on some account or other he is very hostile to the nations of the country; for having come thither a poor man, and having amassed enormous riches of every imaginable description by plunder and extortion, he has now become afraid lest some accusation may be brought against him, and on this account he has contrived a design by which he may repel any such impeachment, namely, by calumniating those whom he has injured; (200) and a circumstance which we will now mention, has given him some pretext for carrying out his design.

'There is a city called Jamnia; one of the most populous cities in all Judaea, which is inhabited by a promiscuous multitude, the greatest number of whom are Jews; but there are also some persons of other tribes from the neighbouring nations who have settled there to their own destruction, who are in a manner sojourners among the original native citizens, and who cause them a great deal of trouble, and who do them a great deal of injury, as they are continually violating some of the ancestral national customs of the Jews. (201) These men hearing from travellers who visit the city how exceedingly eager and earnest Gaius is about his own deification, and how disposed he is to look unfavourably upon the whole race of Judaea, thinking that they have now an admirable opportunity for attacking them themselves, have erected an extemporaneous altar of the most contemptible materials, having made clay into bricks for the sole purpose of plotting against their fellow citizens; for they knew well that they would never endure to see their customs transgressed; as was indeed the case.

(202) 'For when the Jews saw what they had done, and were very indignant at the holiness and sanctity and beauty of the sacred place being thus obscured and defaced, they collected together and destroyed the altar;

so the sojourners immediately went to Capito who was in reality the contriver of the whole affair; and he, thinking that he had made a most lucky hit, which he had been seeking for a long time, writes to Gaius dilating on the matter and exaggerating it enormously; (203) and he, when he had read the letter, ordered a colossal statue gilt all over, much more costly and much more magnificent than the rich altar which had been erected in Jamnia, by way of insult to be set up in the temple of the metropolis, having for his most excellent and sagacious counsellors Helicon, that man of noble birth, a chattering slave, a perfect scum of the earth, and a fellow of the name of Apelles, a tragic actor, who when in the first bloom of youth, as they say, made a market of his beauty, and when he was past the freshness of youth went on the stage; (204) and in fact all those who go on the stage selling themselves to the spectators, and to the theatres, are not lovers of temperance and modesty, but rather of the most extreme shamelessness and indecency.

'On this account Apelles was taken into the rank of a fellow counsellor of the emperor, that Gaius might have an adviser with whom he might indulge in mocking jests, and with whom he might sing, passing over all considerations of the general welfare of the state, as if everything in every quarter of the globe was enjoying profound peace and tranquillity under the laws.

(205) 'Therefore Helicon, this scorpion-like slave, discharged all his Egyptian venom against the Jews; and Apelles his Ascalonite poison, for he was a native of Ascalon; and between the people of Ascalon and the inhabitants of the holy land, the Jews, there is an irreconcileable and neverending hostility although they are bordering nations.'

(206) When we heard this we were wounded in our souls at every word he said and at every name he mentioned; but those admirable advisers of admirable actions a little while afterwards met with the fit reward of their impiety, the one being bound by Gaius with iron chains for other causes, and being put to the torture and to the rack after periods of relief, as is the case with people affected with intermittent diseases; and Helicon was put to death by Claudius Germanicus Caesar, for other wicked actions, that, like a madman as he was, he had committed; but there occurrences took place at a later date."[489]

The book of *Esther*, Chapter 3, tells another story in which retribution was sought against Jews for their failure to abide by the laws of the nations in which they lived—though the Jews hypocritically forbade the practice of any other religion in their presence:

"1 After these things did king Ahasuerus promote Haman the son of

---

[489]. Philo the Jew, "On the Embassy to Gaius", *The Works of Philo*, Hendrick Publishing, U. S. A., (2000), pp. 757-790, at 775-776.

Hammedatha the Agagite, and advanced him, and set his seat above all the princes that *were* with him. 2 And all the king's servants, that *were* in the king's gate, bowed, and reverenced Haman: for the king had so commanded concerning him. But Mordecai bowed not, nor did *him* reverence. 3 Then the king's servants, which *were* in the king's gate, said unto Mordecai, Why transgressest thou the king's commandment? 4 Now it came to pass, when they spake daily unto him, and he hearkened not unto them, that they told Haman, to see whether Mordecai's matters would stand: for he had told them that he *was* a Jew. 5 And when Haman saw that Mordecai bowed not, nor did him reverence, then was Haman full of wrath. 6 And he thought scorn to lay hands on Mordecai alone; for they had shewed him the people of Mordecai: wherefore Haman sought to destroy all the Jews that *were* throughout the whole kingdom of Ahasuerus, *even* the people of Mordecai. 7 In the first month, that *is*, the month Nisan, in the twelfth year of king Ahasuerus, they cast Pur, that *is*, the lot, before Haman from day to day, and from month to month, *to* the twelfth *month*, that *is*, the month Adar. 8 And Haman said unto king Ahasuerus, There is a certain people scattered abroad and dispersed among the people in all the provinces of thy kingdom; and their laws *are* diverse from all people; neither keep they the king's laws: therefore it *is* not for the king's profit to suffer them. 9 If it please the king, let it be written that they may be destroyed: and I will pay ten thousand talents of silver to the hands of those that have the charge of the business, to bring *it* into the king's treasuries. 10 And the king took his ring from his hand, and gave it unto Haman the son of Hammedatha the Agagite, the Jews' enemy. 11 And the king said unto Haman, The silver *is* given to thee, the people also, to do with them as it seemeth good to thee. 12 Then were the king's scribes called on the thirteenth day of the first month, and there was written according to all that Haman had commanded unto the king's lieutenants, and to the governors that *were* over every province, and to the rulers of every people of every province according to the writing thereof, and *to* every people after their language; in the name of king Ahasuerus was it written, and sealed with the king's ring. 13 And the letters were sent by posts into all the king's provinces, to destroy, to kill, and to cause to perish, all Jews, both young and old, little children and women, in one day, *even* upon the thirteenth *day* of the twelfth month, which *is* the month Adar, and *to take* the spoil of them for a prey. 14 The copy of the writing for a commandment to be given in every province was published unto all people, that they should be ready against that day. 15 The posts went out, being hastened by the king's commandment, and the decree was given in Shushan the palace. And the king and Haman sat down to drink; but the city Shushan was perplexed."

To this day, there are Zionists who want to force all Gentiles, and especially Palestinians, out of Jerusalem (*Isaiah* 52:1. *Joel* 3:17), and even out of Israel and

"Greater Israel" from the Nile to the Euphrates.[490] They follow the ancient law of the *halakha* that only Jews and Judaism be permitted in the "Holy Land", while hypocritically insisting that Jews be permitted religious freedom, be permitted citizenship and be enfranchised in all other lands. Yehoshafat Harkabi noted that Rabbi Meir Kahane wrote in modern times,

> "The Arabs of Israel are a desecration of God's name. Their non-acceptance of Jewish sovereignty over the Land of Israel is a rejection of the sovereignty of the God of Israel and of his kingdom. Removing them from the land is therefore more than a political matter. It is a religious matter, a religious obligation to wipe out the desecration of God's name. Instead of worrying about the reactions of the Gentiles if we act, we should tremble at the thought of God's wrath if we do not act. Tragedy will befall us if we do not remove the Arabs from the land, since redemption can come at once in its full glory if we do, as God commands us.... Let us remove the Arabs from Israel and hasten the Redemption (*Thorns in Your Eyes*, pp. 244-245)."[491]

Yehoshafat Harkabi noted that Maimonides wrote long ago,

> "An affirmative precept is enjoined for the destruction of idolatry and its worshippers, and everything made for its sake.... In the Land of Israel, it is a duty actively to chase out idolatry until we have exterminated it from the whole of our country. Outside of the holy land, however, we are not so commanded; but only that whenever we acquire any territory by conquest, we should destroy all the idolatry found there (*Hilkhot Avodah Zara*, ch. 7:1)."[492]

and,

> "It is forbidden to show them mercy, as it was said, 'nor show mercy unto them' (Deut. 7:2). Hence, if one sees one of them who worships idols perishing or drowning, one is not to save him.... Hence you learn that it is forbidden to heal idolators even for a fee. But if one is afraid of them or apprehends that refusal might cause ill will, medical treatment may be given for a fee but not gratuitously.... The foregoing rules apply to the time when the people of Israel live exiled among the nations, or when the Gentiles' power is predominant. But when Israel is predominant over the nations of the world, we are forbidden to permit a gentile who is an idolator to dwell among us. He must not enter our land, even as a temporary resident; or even

---

490. Y. Harkabi, *Israel's Fateful Hour*, Harper & Row, New York, (1988), pp. 147-149, 156-158.
491. English translation in: Y. Harkabi, *Israel's Fateful Hour*, Harper & Row, New York, (1988), p. 148. Harkabi cites: M. Kahane, *Thorns in Your Eyes*, New York, Jerusalem, (1980?), pp. 244-245.
492. Y. Harkabi, *Israel's Fateful Hour*, Harper & Row, New York, (1988), p. 156.

as a traveler, journeying with merchandise from place to place, until he has undertaken to keep the seven precepts which the Noachides were commanded to observe (*Hilkhot Avodah Zara*, ch. 10:8)."[493]

Philo the Jew also records that attempts were planned to bribe Helicon, but that he could not be bribed, and to control Caligula through Macro, who was close friends with Flaccus of Alexandria—at that time friendly to the Jews of Alexandria; and Philo records conversations that must either have been the product of his imagination, or the result of the corrupt use of agents within the Roman government. It appears that the schemes and conspiracies alleged in Philo's writings did not occur, but that he and others were the ones attempting to exert their influence in Rome and sought to maintain a privileged status at the expense of others. Philo relentlessly smears Caligula and Gentile peoples. The ancient slanders against Caligula are actually quite similar to the life history of Mausolus, whose tomb was one of the seven wonders of the ancient world. It was an Old Testament habit for Jews to defame all other peoples and to plagiarize their beliefs and their historical stories.

Philo threatens that a Jewish God will punish all who contradict the wishes of the Jews and sees no hypocrisy in his demand that all people on Earth obey Jewish law, lest there be civil war instigated by Jews; while the Jews refused to obey the laws of Rome. Philo wrote, quoting Caligula,

> "'If any people in the bordering countries, with the exception of the metropolis itself, wishing to erect altars or temples, nay, images of statues, in honour of me and of my family are hindered from doing so, I charge you at once to punish those who attempt to hinder them, or else to bring them before the tribunal.' (335) Now this was nothing else but a beginning of seditions and civil wars, and an indirect way of annulling the gift which he appeared to be granting. For some men, more out of a desire of mortifying the Jews than from any feelings of loyalty towards Gaius, were inclined to fill the whole country with erections of one kind or another. But they who beheld the violation of their national customs practised before their eyes were resolved above all things not to endure such an injury unresistingly."[494]

Great enmity existed between the Egyptians, Greeks and Jews of Alexandria; in part because the Jews maintained themselves as a privileged and segregated class. As often happens in places of great wealth discrepancy and wealth condensation, Egyptians and Greeks looted Jewish estates in their quest for social justice. According to Philo, Flaccus reduced the two Jewish sectors of the city into one smaller parcel and quarantined the Jews in it. According to Josephus, three ambassadors of Alexandria called upon the Emperor. One of the

---

[493]. Y. Harkabi, *Israel's Fateful Hour*, Harper & Row, New York, (1988), p. 157.
[494]. Philo the Jew, "On the Embassy to Gaius", *The Works of Philo*, Hendrick Publishing, U. S. A., (2000), pp. 757-790, at 787.

ambassadors, Apion, criticized the Jews.[495] Another ambassador was Philo the Jew, who was the brother of the wealthiest man in the world, Alexander Lysimachus. Philo's own accounts, however, reveal that Josephus misrepresented the history. Though it appears that Philo was himself a liar, and Josephus simply corrupted Philo's fantasies. There is, however, redeeming value in analyzing their accounts, which exhibit Jewish double standards and duplicity in the ancient world.

There were, in Philo's accounts, two meetings of the embassies of Jewish Alexandria with Caligula (Gaius or Caius). Philo the Jew, who attended both meetings, makes no mention of Apion. Whereas Philo himself stated that the Emperor treated him cordially at the first meeting,

> "For it appeared good to present to Gaius a memorial, containing a summary of what we had suffered, and of the way in which we considered that we deserved to be treated; (179) and this memorial was nearly an abridgment of a longer petition which we had sent to him a short time before, by the hand of king Agrippa; for he, by chance, was staying for a short time in the city, while on his way into Syria to take possession of the kingdom which had been given to him, (180) but we, without being aware of it, were deceiving ourselves, for before also we had done the same, when we originally began to set sail, thinking that as we were going before a judge we should meet with justice; but he was in reality an irreconcilable enemy to us, attracting us, as far as appearance went, with favourable looks and cheerful address; (181) for, receiving us favourably at first, in the plains on the banks of the Tiber (for he happened to be walking about in his mother's garden), he conversed with us formally, and waved his right hand to us in a protecting manner, giving us significant tokens of his good will, and having sent to us the secretary, whose duty it was to attend to the embassies that arrived, Obulus by name, he said, 'I myself will listen to what you have to say at the first favourable opportunity.'
>
> So that all those who stood around congratulated us as if we had already carried our point, and so did all those of our own people, who are influenced by superficial appearances. (182) But I myself, who was accounted to be possessed of superior prudence, both on account of my age and my education, and general information, was less sanguine in respect of the matters at which the others were so greatly delighted. 'For why,' said I, after pondering the matter deeply in my own heart, 'why, when there have been such numbers of ambassadors, who have come, one may almost say, from every corner of the globe, did he say on that occasion that he would hear what we had to say, and no one else? What could have been his meaning?

---

495. Josephus, "Embassy of the Jews to Caius", *The Works of Flavius Josephus: Comprising the Antiquities of the Jews; a History of the Jewish Wars; and Life of Flavius Josephus, Written by Himself*, "Antiquities of the Jews", Book 18, Chapter 8, S. S. Scranton Co., Hartford, Connecticutt, (1916), pp. 563-567. Also: *Contra Apionem*, Book 2, Chapter 8, Section 95.

for he was not ignorant that we were Jews, who would have been quite content at not being treated worse than the others; (183) but to expect to be looked upon as worthy to receive especial privileges and precedence, by a master who was of a different nation and a young man and an absolute monarch, would have seemed like insanity. But it would seem that he was showing civility to the whole district of the Alexandrians, to which he was thus giving a privilege, when promising to give his decision speedily; unless, indeed, disregarding the character of a fair and impartial hearer, he was intending to be a fellow suitor with our adversaries and an enemy of ours, instead of behaving like a judge.'"[496]

Josephus misrepresented the story in his *Antiquities of the Jews*, Book 18, Chapter 8, and stated that the Emperor, angered by what Apion (a person Josephus wished to defame) had allegedly told him, refused to hear Philo the Jew, and only then ordered that a statue of Caligula be erected in the Temple. Josephus wrote,

"CHAPTER VIII.

Embassy of the Jews to Caius—Caius sends Petronius into Syria to make war against the Jews.

THERE was now a tumult arisen at Alexandria, between the Jewish inhabitants and the Greeks; and three ambassadors were chosen out of each party that were at variance, who came to Caius. Now one of these ambassadors from the people of Alexandria was Apion, (29) who uttered many blasphemies against the Jews; and, among other things that he said, he charged them with neglecting the honors that belonged to Caesar; for that while all who were subject to the Roman empire built altars and temples to Caius, and in other regards universally received him as they received the gods, these Jews alone thought it a dishonorable thing for them to erect statues in honor of him, as well as to swear by his name. Many of these severe things were said by Apion, by which he hoped to provoke Caius to anger at the Jews, as he was likely to be. But Philo, the principal of the Jewish embassage, a man eminent on all accounts, brother to Alexander the alabarch, (30) and one not unskillful in philosophy, was ready to betake himself to make his defense against those accusations; but Caius prohibited him, and bid him begone; he was also in such a rage, that it openly appeared he was about to do them some very great mischief. So Philo being thus affronted, went out, and said to those Jews who were about him, that they should be of good courage, since Caius's words indeed showed anger at them, but in reality had already set God against himself.

Hereupon Caius, taking it very heinously that he should be thus

---

496. Philo the Jew, "On the Embassy to Gaius", *The Works of Philo*, Hendrick Publishing, U. S. A., (2000), pp. 757-790, at 773-774.

despised by the Jews alone, sent Petronius to be president of Syria, and successor in the government to Vitellius, and gave him order to make an invasion into Judea, with a great body of troops; and if they would admit of his statue willingly, to erect it in the temple of God; but if they were obstinate, to conquer them by war, and then to do it. Accordingly, Petronius took the government of Syria, and made haste to obey Caesar's epistle. He got together as great a number of auxiliaries as he possibly could, and took with him two legions of the Roman army, and came to Ptolemais, and there wintered, as intending to set about the war in the spring. He also wrote word to Caius what he had resolved to do, who commended him for his alacrity, and ordered him to go on, and to make war with them, in case they would not obey his commands. But there came many ten thousands of the Jews to Petronius, to Ptolemais, to offer their petitions to him, that he would not compel them to transgress and violate the law of their forefathers; 'but if,' said they, 'thou art entirely resolved to bring this statue, and erect it, do thou first kill us, and then do what thou hast resolved on; for while we are alive we cannot permit such things as are forbidden us to be done by the authority of our legislator, and by our forefathers' determination that such prohibitions are instances of virtue.' But Petronius was angry at them, and said, 'If indeed I were myself emperor, and were at liberty to follow my own inclination, and then had designed to act thus, these your words would be justly spoken to me; but now Caesar hath sent to me, I am under the necessity of being subservient to his decrees, because a disobedience to them will bring upon me inevitable destruction.' Then the Jews replied, 'Since, therefore, thou art so disposed, O Petronius! that thou wilt not disobey Caius's epistles, neither will we transgress the commands of our law; and as we depend upon the excellency of our laws, and, by the labors of our ancestors, have continued hitherto without suffering them to be transgressed, we dare not by any means suffer ourselves to be so timorous as to transgress those laws out of the fear of death, which God hath determined are for our advantage; and if we fall into misfortunes, we will bear them, in order to preserve our laws, as knowing that those who expose themselves to dangers have good hope of escaping them, because God will stand on our side, when, out of regard to him, we undergo afflictions, and sustain the uncertain turns of fortune. But if we should submit to thee, we should be greatly reproached for our cowardice, as thereby showing ourselves ready to transgress our law; and we should incur the great anger of God also, who, even thyself being judge, is superior to Caius.'

When Petronius saw by their words that their determination was hard to be removed, and that, without a war, he should not be able to be subservient to Caius in the dedication of his statue, and that there must be a great deal of bloodshed, he took his friends, and the servants that were about him, and hasted to Tiberias, as wanting to know in what posture the affairs of the Jews were; and many ten thousands of the Jews met Petronius again, when he was come to Tiberias. These thought they must run a mighty hazard if they should have a war with the Romans, but judged that the transgression of the law was of much greater consequence, and made supplication to him, that he would by no means reduce them to such distresses, nor defile their city with the dedication of the statue. Then Petronius said to them, 'Will you then make war with Caesar, without considering his great preparations for war, and your own weakness?' They replied, 'We will not by any means make war with him, but still we will die before we see our laws transgressed.' So they threw themselves down upon their faces, and stretched out their throats, and said they were ready to be slain; and this they did for forty days together, and in the mean time left off the tilling of their ground, and that while the season of the year required them to sow it. (31) Thus they continued firm in their resolution, and proposed to themselves to die willingly, rather than to see the dedication of the statue.

When matters were in this state, Aristobulus, king Agrippa's brother, and Heleias the Great, and the other principal men of that family with them, went in unto Petronius, and besought him, that since he saw the resolution of the multitude, he would not make any alteration, and thereby drive them to despair; but would write to Caius, that the Jews had an insuperable aversion to the reception of the statue, and how they continued with him, and left of the tillage off their ground: that they were not willing to go to war with him, because they were not able to do it, but were ready to die with pleasure, rather than suffer their laws to be transgressed: and how, upon the land's continuing unsown, robberies would grow up, on the inability they would be under of paying their tributes; and that Caius might be thereby moved to pity, and not order any barbarous action to be done to them, nor think of destroying the nation: that if he continues inflexible in his former opinion to bring a war upon them, he may then set about it himself. And thus did Aristobulus, and the rest with him, supplicate Petronius. So Petronius, (32) partly on account of the pressing instances which Aristobulus and the rest with him made, and because of the great consequence of what they desired, and the earnestness wherewith they made their supplication, — partly on account of the firmness of the opposition made by the Jews, which he saw, while he thought it a terrible thing for him to be such a slave to the madness of Caius, as to slay so many ten thousand men, only because of their religious disposition towards God, and after that to pass his life in expectation of punishment; Petronius, I say, thought it much better to send to Caius, and to let him know how intolerable it was to him to bear the anger he might have against him for not serving him sooner, in obedience to his epistle, for that perhaps he might persuade him; and that if this mad resolution continued, he might then begin the war against them; nay, that in case he should turn his hatred against himself, it was fit for virtuous persons even to die for the sake of such vast multitudes of men. Accordingly, he determined to hearken to the petitioners in this matter.

He then called the Jews together to Tiberias, who came many ten thousands in number; he also placed that army he now had with him opposite to them; but did not discover his own meaning, but the commands of the emperor, and told them that his wrath would, without delay, be executed on such as had the courage to disobey what he had commanded, and this immediately; and that it was fit for him, who had obtained so great a dignity by his grant, not to contradict him in any thing: — 'yet,' said he, 'I do not think it just to have such a regard to my own safety and honor, as to refuse to sacrifice them for your preservation, who are so many in number, and endeavor to preserve the regard that is due to your law; which as it hath come down to you from your forefathers, so do you esteem it worthy of your utmost contention to preserve it: nor, with the supreme assistance and power of God, will I be so hardy as to suffer your temple to fall into contempt by the means of the imperial authority. I will, therefore, send to Caius, and let him know what your resolutions are, and will assist your suit as far as I am able, that you may not be exposed to suffer on account of the honest designs you have proposed to yourselves; and may God be your assistant, for his authority is beyond all the contrivance and power of men; and may he procure you the preservation of your ancient laws, and may not he be deprived, though without your consent, of his accustomed honors. But if Caius be irritated, and turn the violence of his rage upon me, I will rather undergo all that danger and that affliction that may come either on my body or my soul, than see so many of you to perish, while you are acting in so excellent a manner. Do you, therefore, every one of you, go your way about your own occupations, and fall to the cultivation of your ground; I will myself send to Rome, and will not refuse to serve you in all things, both by myself and by my friends.'

When Petronius had said this, and had dismissed rite assembly of the Jews, he desired the principal of them to take care of their husbandry, and to speak kindly to the people, and encourage them to have good hope of their affairs. Thus did he readily bring the multitude to be cheerful again. And now did God show his presence to Petronius, and signify to him that he would afford him his assistance in his whole design; for he had no sooner finished the speech that he made to the Jews, but God sent down great showers of rain, contrary to human expectation; (33) for that day was a clear day, and gave no sign, by the appearance of the sky, of any rain; nay, the whole year had been subject to a great drought, and made men despair of any water from above, even when at any time they saw the heavens overcast with clouds; insomuch that when such a great quantity of rain came, and that in an unusual manner, and without any other expectation of it, the Jews hoped that Petronius would by no means fail in his petition for them. But as to Petronius, he was mightily surprised when he perceived that God evidently took care of the Jews, and gave very plain signs of his appearance, and this to such a degree, that those that were in earnest much inclined to the contrary had no power left to contradict it. This was also among those other particulars which he wrote to Caius, which all tended to dissuade him, and by all means to entreat him not to make so many ten thousands of these men go distracted; whom, if he should slay, (for without war they would by no means suffer the laws of their worship to be set aside,) he would lose the revenue they paid him, and would be publicly cursed by them for all future ages. Moreover, that God, who was their Governor, had shown his power most evidently on their account, and that such a power of his as left no room for doubt about it. And this was the business that Petronius was now engaged in.

But King Agrippa, who now lived at Rome, was more and more in the favor of Caius; and when he had once made him a supper, and was careful to exceed all others, both in expenses and in such preparations as might contribute most to his pleasure; nay, it was so far from the ability of others, that Caius himself could never equal, much less exceed it (such care had he taken beforehand to exceed all men, and particularly to make all agreeable to Caesar); hereupon Caius admired his understanding and magnificence, that he should force himself to do all to please him, even beyond such expenses as he could bear, and was desirous not to be behind Agrippa in that generosity which he exerted in order to please him. So Caius, when he had drank wine plentifully, and was merrier than ordinary, said thus during the feast, when Agrippa had drunk to him: 'I knew before now how great a respect thou hast had for me, and how great kindness thou hast shown me, though with those hazards to thyself, which thou underwentest under Tiberius on that account; nor hast thou omitted any thing to show thy good-will towards us, even beyond thy ability; whence it would be a base thing for me to be conquered by thy affection. I am therefore desirous to make thee amends for every thing in which I have been formerly deficient; for all that I have bestowed on thee, that may be called my gifts, is but little. Everything that may contribute to thy happiness shall be at thy service, and that cheerfully, and so far as my ability will reach.' (34) And this was what Caius said to Agrippa, thinking be would ask for some large country, or the revenues of certain cities. But although he had prepared beforehand what he would ask, yet had he not discovered his intentions, but made this answer to Caius immediately: That it was not out of any expectation of gain that he formerly paid his respects to him, contrary to the commands of Tiberius, nor did he now do any thing relating to him out of regard to his own advantage, and in order to receive any thing from him; that the gifts he had already bestowed upon him were great, and beyond the hopes of even a craving man; for although they may be beneath thy power, [who art the donor,] yet are they greater than my inclination and dignity, who am the receiver. And as Caius was astonished at Agrippa's inclinations, and still the more pressed him to make his request for somewhat which he might gratify him with, Agrippa replied, 'Since thou, O my lord! declarest such is thy readiness to grant, that I am worthy of thy gifts, I will ask nothing relating to my own felicity; for what thou hast already bestowed on me has made me excel therein; but I desire somewhat which may make thee glorious for piety, and render the Divinity assistant to thy designs, and may be for an honor to me among those that inquire about it, as showing that I never once fail of obtaining what I desire of thee; for my petition is this, that thou wilt no longer think of the dedication of that statue which thou hast ordered to be set up in the Jewish temple by Petronius.'

And thus did Agrippa venture to cast the die upon this occasion, so great was the affair in his opinion, and in reality, though he knew how dangerous a thing it was so to speak; for had not Caius approved of it, it had tended to no less than the loss of his life. So Caius, who was mightily taken with Agrippa's obliging behavior, and on other accounts thinking it a dishonorable thing to be guilty of falsehood before so many witnesses, in points wherein he had with such alacrity forced Agrippa to become a petitioner, and that it would look as if he had already repented of what he had said, and because he greatly admired Agrippa's virtue, in not desiring him at all to augment his own dominions, either with larger revenues, or other authority, but took care of the public tranquillity, of the laws, and of the Divinity itself, he granted him what he had requested. He also wrote thus to Petronius, commending him for his assembling his army, and then consulting him about these affairs. 'If therefore,' said he, 'thou hast already erected my statue, let it stand; but if thou hast not yet dedicated it, do not trouble thyself further about it, but dismiss thy army, go back, and take care of those affairs which I sent thee about at first, for I have now no occasion for the erection of that statue. This I have granted as a favor to Agrippa, a man whom I honor so very greatly, that I am not able to contradict what he would have, or what he desired me to do for him.' And this was what Caius wrote to Petronius, which was before he received his letter, informing him that the Jews were very ready to revolt about the statue, and that they seemed resolved to threaten war against the Romans, and nothing else. When therefore Caius was much displeased that any attempt should be made against his government as he was a slave to base and vicious actions on all occasions, and had no regard to what was virtuous and honorable, and against whomsoever he resolved to show his anger, and that for any cause whatsoever, he suffered not himself to be restrained by any admonition, but thought the indulging his anger to be a real pleasure, he wrote thus to Petronius: 'Seeing thou esteemest the presents made thee by the Jews to be of greater value than my commands, and art grown insolent enough to be subservient to their pleasure, I charge thee to become thy own judge, and to consider what thou art to do, now thou art under my displeasure; for I will make thee an example to the present and to all future ages, that they. may not dare to contradict the commands of their emperor.'

This was the epistle which Caius wrote to Petronius; but Petronius did not receive it while Caius was alive, that ship which carried it sailing so slow, that other letters came to Petronius before this, by which he understood that Caius was dead; for God would not forget the dangers Petronius had undertaken on account of the Jews, and of his own honor. But when he had taken Caius away, out of his indignation of what he had so insolently attempted in assuming to himself divine worship, both Rome and all that dominion conspired with Petronius, especially those that were of the senatorian order, to give Caius his due reward, because he had been unmercifully severe to them; for he died not long after he had written to Petronius that epistle which threatened him with death. But as for the occasion of his death, and the nature of the plot against him, I shall relate them in the progress of this narration. Now that epistle which informed Petronius of Caius's death came first, and a little afterward came that which commanded him to kill himself with his own hands. Whereupon he rejoiced at this coincidence as to the death of Caius, and admired God's providence, who, without the least delay, and immediately, gave him a reward for the regard he had to the temple, and the assistance he afforded the Jews for avoiding the dangers they were in. And by this means Petronius escaped that danger of death, which he could not foresee."[497]

Note the Greek-like fairytale nature of Josephus' religious story, with its superstitious threats by God, its miracles and omens, its morals, its cunning heroes and villains, and its fatalistic resolution through divine wisdom making all right in the world, demonstrating the power of the Jewish God and the blessings received by those who obeyed His will, and the fall of those who disobeyed it through fatal hubris.

According to Josephus' spurious account, after being refused an audience with the Emperor, Philo the Jew then slandered the Emperor and threatened him, declaring that God would exact vengeance upon him. In his alleged anger, the Emperor ordered that Publius Petronius be made president of Syria and sent him to Judea with Roman armies to erect a statue of the Emperor in the Temple, demanding that the Jews worship Caligula as god.

However, Philo the Jew, quoting another, informs us that Caligula had already issued his order that his image appear in the Temple in the form of statue before meeting with the ambassadors, the first time,

> "And he with difficulty, sobbing aloud, and in a broken voice, spoke as follows: 'Our temple is destroyed! Gaius has ordered a colossal statue of himself to be erected in the holy of holies, having his own name inscribed upon it with the title of Jupiter!'"[498]

Josephus' account contradicts Philo the Jew's own account, that it was the Jews' destruction of a shrine to Caligula in Jamnia that provoked Caligula to retaliate by demanding that they place a statue of him in the Temple. In addition, whereas Philo accuses Helicon of putting thoughts into Caligula's head, Josephus changes the story to blame Apion—the same Josephus who made such a show of declaring his honesty and virtue as a historian, while defaming the Greeks, in his work *Against Apion*—in which Josephus recklessly defames Apion and the Egyptians with still more shrill lies. Of course, the fact that

---

497. Josephus, "Embassy of the Jews to Caius", *The Works of Flavius Josephus: Comprising the Antiquities of the Jews; a History of the Jewish Wars; and Life of Flavius Josephus, Written by Himself*, "Antiquities of the Jews", Book 18, Chapter 8, S. S. Scranton Co., Hartford, Connecticutt, (1916), pp. 563-567.
498. Philo the Jew, "On the Embassy to Gaius", *The Works of Philo*, Hendrick Publishing, U. S. A., (2000), pp. 757-790, at 774.

Josephus' account is dishonest does not render Philo's account accurate, though it is certainly more plausible.

The entire story appears to be a canard meant to artificially fulfil prophecies and may simply be a repetition of the story of Antiochus Epiphanes, who was also said to be mad, and who also desecrated the Temple with a statue of a foreign god, Zeus, and who had pigs sacrificed at the alter of the Temple which he converted to the worship of Zeus (I *Maccabees* 1:20-24, 54-64). Antiochus IV Epiphanes was said to be the first of a line of what Christians call the "anti-Christ"; and Jews fear that another anti-Messiah will rise in Syria (*Daniel* 11), which may well explain why the Neo-Conservative Jews in America are pushing America towards war with Syria, another obvious reason being their desire for a greater Israel from the Nile to the Euphrates, and perhaps most ominously *Isaiah* 17:1 states, "The burden of Damascus. Behold, Damascus *is* taken away from *being* a city, and it shall be a ruinous heap."

Some doubt whether the Temple actually was at the site of the Dome of the Rock and the Al Aqsa Mosque. The Roman Emperor Hadrian built a temple of Jupiter on the site claimed for the Jewish Temple, and the remains which are extant today of an ancient temple may be the remains of Hadrian's temple to Jupiter. The supposed first Temple of Solomon probably never existed. The Samaritans placed Solomon's Temple at Mount Gerizim (*John* 4:20).[499] Some hold that the Temple was at the Gihon Spring, which made a better location for a temple, given that it had a spring available to wash away the blood from sacrifices.[500]

According to Josephus, the Jews declared that they would die before they would see the Temple defiled. In his story, the Jews lay with their throats bare for forty nights aware that war with the Romans was futile and declared that they should be killed before their laws were defiled and their Temple desecrated—an account that is false on its face. According to Josephus, Publius Petronius, speaking as if a monotheistic Jew himself (Philo implies that Petronius, Augustus and Julia Augusta were Jewish converts), declared that he could not commit such an injustice against a people behaving so nobly and called upon the "power of God" to help him persuade the Emperor to change his mind. This speech is a figment of Josephus' imagination. According to Josephus, Petronius sent out a letter informing Caligula that he would rather die himself than follow such an unfair order. This, too, is a figment of Josephus' imagination. The letter instead attempted to persuade Caligula that it would be impractical to place the statue in the Temple before the crops were harvested and gave other justifications for delay.

In Philo's account, it was the Jewish King Agrippa who courted death, not the Roman Petronius. Philo reproduced Agrippa's supposed words,

---

[499]. "The Modern Jews", *The North American Review*, Volume 60, Number 127, (April, 1845), pp. 329-368, at 353.
[500]. E. L. Martin, *The Temples that Jerusalem Forgot*, ASK Publications, Portland, Oregon, (2000).

> "O master, that your Agrippa may not be driven wholly to forsake life; for I shall appear (if you do not do so) to have been released from bondage, not for the purpose of being saved, but for that of being made to perish in a more conspicuous manner. [***] command me at once to be put out of the way. For what advantage would it be to me to live, who place my whole hopes of safety and happiness in your friendship and favour?"[501]

Not only did Petronius not offer to commit suicide, Philo tells us that he planned to run away to Alexandria,

> "[Petronius] himself was intending, as is said, to sail to Alexandria in Egypt[.]"[502]

During this period of civil disobedience, the Jews could not attend to their occupations and so could not pay tribute to the Romans, and, should the Romans murder them, could never increase the wealth of Rome. While a sit-down strike may have occurred, it did not occur in the fashion of Josephus' fancy. The Jewish King Herod Agrippa was closely associated with, and literally indebted to, Alexander Lysimachus, Philo the Jew's wealthy brother, and to Caligula who had made him King.

According to Josephus, Herod Agrippa tricked Caligula, his friend and supporter. King Herod Agrippa, who oppressed the first followers of Christ (*Acts* 12) and ruled over the lands of Judea—intervened on behalf of the Jews by getting Caligula drunk and flattering him, compelling Caligula to offer Agrippa, the man he had made King, such graces in return, lest he appear less noble than Agrippa. According to Josephus' accounts, when Caligula asked Agrippa what it is that he might desire, Agrippa asked Caligula to free the Jews from the requirement of erecting a statue to the worship of Caligula in the Temple. Caligula allegedly agreed.

In the meanwhile, according to Josephus' fictional account, Publius Petronius' letter reached Caligula, who ordered that Publius Petronius must follow through on his pledge to kill himself before committing such an injustice against so many good people. Amazingly, Caligula was assassinated by Cassius Chaerea of the Prætorian Guard before his letter reached Publius Petronius, and so slow was the ship dispatched to deliver his order to Publius Petronius, that news of Caligula's assassination reached Publius Petronius before Caligula's letter, and both Publius Petronius and the sanctity of the Temple were spared by divine providence. This story of guile is a fabrication meant to save Agrippa's honor, the honor of the Temple and of the Jewish people; while concomitantly smearing the Romans and threatening all with the power of the Jewish God.

We must be on guard against the ethnic bias shown by "historians" like

---

501. Philo the Jew, "On the Embassy to Gaius", *The Works of Philo*, Hendrick Publishing, U. S. A., (2000), pp. 757-790, at 786.
502. Philo the Jew, "On the Embassy to Gaius", *The Works of Philo*, Hendrick Publishing, U. S. A., (2000), pp. 757-790, at 780.

Josephus, who fantasize and distort in order to embellish their image and the image of their people, while smearing others. (It would be inappropriate and cumbersome to name all of the discrepancies between Josephus' and Philo's story in this place.) Max Born and Philipp Frank lied about events in the early 1920's in much the same way Josephus lied about the history of Caligula—in order to protect the image of their humiliated saint and to smear those who opposed Einstein with deliberate lies. Jewish tribalism has many ill effects and the dishonesty of no small number of Jewish historians is one of them.

Contrary to the accounts of Josephus, Philo the Jew stated that Agrippa coincidently called upon Caligula and had no knowledge of the unfolding events. Caligula communicated to Agrippa that something was terribly wrong by staring through him, and then informed Agrippa of his plan to place a statue of himself in the Temple, at which point Agrippa went into shock and passed out. When he had somewhat recovered—with the help of drugs—Agrippa wrote a long and passionate letter to Caligula, pleading with him to not defile the Temple. Caligula appeared to relent somewhat and wrote to Petronius. However, Philo makes clear that this was a delaying tactic by Caligula, one made after Petronius' deceitful delaying tactics. Caligula constructed an even greater statue in Rome than that which he had ordered Petronius, governor of Syria, to construct in Sidon. In this way, Caligula was able to install the statue in the Temple before the Jews could in any way organize to obstruct him. He had used their tactics against them and outwitted them. Philo wrote,

> "What advantage, then, was gained? some one will say; for even when they were quiet, Gaius was not quiet; but he had already repented of the favour which he had showed to Agrippa, and had re-kindled the desires which he had entertained a little while before; for he commanded another statue to be made, of colossal size, of brass gilt over, in Rome, no longer moving the one which had been made in Sidon, in order that the people might not be excited by its being moved, but that while they remained in a state of tranquillity and felt released from their suspicions, it might in a period of peace be suddenly brought to the country in a ship, and be suddenly erected without the multitude being aware of what was going on.
>
> XLIII. (338) And he was intending to do this while on his voyage along the coast during the period which he had allotted for his sojourn in Egypt. For an indescribable desire occupied his mind to see Alexandria, to which he was eager to go with all imaginable haste, and when he had arrived there he intended to remain a considerable time, urging that the deification about which he was so anxious, might easily be originated and carried to a great height in that city above all others, and then that it would be a model to all other cities of the adoration to which he was entitled, inasmuch as it was the greatest of all the cities of the east, and built in the finest situation in the world. For all inferior men and nations are eager to imitate great men and

great states."[503]

Though Josephus would have us believe that the sanctity of the Temple was preserved through the wit and cunning of Agrippa, the humanity of Petronius who had converted to Judaism, and the will of God; Philo the Jew informs us that all the synagogues in the world were violated with statues of Caligula, and the Temple, the holy of holies, was indeed made a temple to Caligula, and that Philo the Jew and all the ambassadors of the Jews recanted and worshiped Caligula as the god Jupiter,

> "So great therefore was his inequality of temper towards every one, and most especially towards the nation of the Jews to which he was most bitterly hostile, and accordingly beginning in Alexandria he took from them all their synagogues there, and in the other cities, and filled them all with images and statues of his own form; for not caring about any other erection of any kind, he set up his own statue every where by main force; and the great temple in the holy city, which was left untouched to the last, having been thought worthy of all possible respect and preservation, he altered and transformed into a temple of his own, that he might call it the temple of the new Jupiter, the illustrious Gaius. [***] We have now related in a concise and summary manner the cause of the hatred of Gaius to the whole nation of the Jews; we must now proceed to make our palinode to Gaius."[504]

The fact that Philo the Jew's account differs from Josephus' account does not mean that either man told the truth. Josephus, however, has been discredited and his story is false on its face, with its physical impossibilities. Philo was a participant in the events. Artifacts from the period confirm Philo's story to the extent that Jewish religious artifacts and temple carvings from the period bear images of Roman gods.

It would be very interesting to discover that Jews, or Jewish Christians, not the Romans, were the ones to destroy the Jewish Temple, which had been desecrated by the statue, the idol, of a foreign god. Such an event would be in keeping with the story of Christ and would explain the dispersal of the Jews as their self-fulfillment of Judaic prophetic myth. The Jews were bound by God to destroy any and all idolatrous temples in Jerusalem, and if Philo's story is true, the Jews were duty bound to ruin the formerly Jewish Temple and all others like it. It would be like them to blame this on the Romans and use the hatred of the Romans as a means to preserve what was left of the unity of their disintegrating nation.

Caligula was not the only Roman Emperor who has been smeared by historians in order to preserve Jewish honor. Nero has been blamed for the

---

503. Philo the Jew, "On the Embassy to Gaius", *The Works of Philo*, Hendrick Publishing, U. S. A., (2000), pp. 757-790, at 787.
504. Philo the Jew, "On the Embassy to Gaius", *The Works of Philo*, Hendrick Publishing, U. S. A., (2000), pp. 757-790, at 788.

burning of Rome, which was more likely carried out by Jews, who resented being forced to worship Roman gods and the Emperor of Rome, and who resented the Roman occupation. Though it is claimed that Nero blamed the Christians for the fire in order to deflect suspicions that he had set it, others have claimed that Nero's Jewish wife Poppæa urged him to blame the Christians[505] in order to protect the Jews from retaliation against them, and is so doing not only burned Rome, but also put to death many Gentiles who had the audacity to pretend to be Jews. Pope St. Clement (88-97AD) recognized that the Jews were responsible for Nero's persecution of the Christians. The murders of the Christians often took the form of human sacrifices. Nero is said to have killed Poppæa by kicking her in the stomach while pregnant, which is perhaps symbolic of an attempted abortion. Their first child died as an infant and was declared a deity, almost as if the child were a sacrifice.

The intercession of Poppæa into Roman-Jewish relations[506] is interesting on another point. She allegedly persuaded Nero to allow the Jews to shield the Temple from the Romans' view, so that religious sacrifices could be made in complete privacy. Jews were often accused of human sacrifice and the Old Testament repeatedly mentions Jews sacrificing their children by "passing them through the fire". One bone of contention among more modern Jews who debated Zionism and the reconstruction of the Temple was whether or not the sacrifices of animals should be resumed. Among racist Jews, Gentiles have long been considered animals. Judaism has long called on Gentiles to sacrifice themselves in the name of God in wars fought on Israel's behalf. Esau was destined to soldier for Jacob.

Idolatry was nothing new to the Jews. Solomon was an idolater who supposedly constructed the Temple with magic, employing both demons and

---

505. Josephus, "Antiquities of the Jews", Book XX, Chapter 8, *The Works of Flavius Josephus: Comprising the Antiquities of the Jews; a History of the Jewish Wars; and Life of Flavius Josephus, Written by Himself*, S. S. Scranton Co., Hartford, Connecticutt, (1916), pp. 609-613, at 612-613. *See also:* Tacitus, *Annal*, Book XV, in: "Dissertation III", *The Works of Flavius Josephus: Comprising the Antiquities of the Jews; a History of the Jewish Wars; and Life of Flavius Josephus, Written by Himself*, S. S. Scranton Co., Hartford, Connecticutt, (1916), p. 960. *See also:* E. Gibbon, "The Conduct of the Roman Government towards the Christians, from the Reign of Nero to that of Constantine", *The History of the Decline and Fall of the Roman Empire*, Chapter 16, Volume 3, Fred De Fau and Company, New York, (1776).

506. Josephus, "Antiquities of the Jews", Book XX, Chapter 8, *The Works of Flavius Josephus: Comprising the Antiquities of the Jews; a History of the Jewish Wars; and Life of Flavius Josephus, Written by Himself*, S. S. Scranton Co., Hartford, Connecticutt, (1916), pp. 609-613, at 612-613. *See also:* Tacitus, *Annal*, Book XV, in: "Dissertation III", *The Works of Flavius Josephus: Comprising the Antiquities of the Jews; a History of the Jewish Wars; and Life of Flavius Josephus, Written by Himself*, S. S. Scranton Co., Hartford, Connecticutt, (1916), p. 960. *See also:* E. Gibbon, "The Conduct of the Roman Government towards the Christians, from the Reign of Nero to that of Constantine", *The History of the Decline and Fall of the Roman Empire*, Chapter 16, Volume 3, Fred De Fau and Company, New York, (1776).

angels to build it. This is an instance of Jewish Dualism, the belief that ultimate power can only be attained by the use of both good and evil forces, both of which stem from God. Posidonius and Apollonius Molo charged the Jews with worshiping the head of an ass. When Moses came down from Mount Sinai with the Commandments, he found his brother Aaron and their people worshiping a Golden Calf they had made from golden earrings molten in a pot (*Exodus* 32). Some believe that Jewish sects to this day worship the Golden Calf, which is symbolic of wealth accumulation, and in some minds, of the "Beast" Baal. The Hebrews called God "Baal" and Jacob called God "El", who was Baal's father. The Frankists openly advocated the deliberate practice of evil, accused fellow Jews of using Christian blood in their rituals, and taught their followers to, "acquire wealth even in deceitful and crooked ways."[507]

Jews ofttimes worshiped the earthly and devilish "Covenant of Baal" (*Exodus* 32. *Leviticus* 26:30. *Numbers* 22:41. *Judges* 6:25, 31; 8:33; 9:4; 11:31, 39. I *Kings* 14:22-24; 16:31-33; 18:18-19, 26; 19:10, 14, 18; 22:53. II *Kings* 3:2-3; 8:18, 27; 10:18-28; 11:18; 16:3-4; 17:10, 16-18, 23; 18:4-5; 21:6; 22:5; 23:5, 12, 32, 37; 24:9, 19. I *Chronicles* 12:5 ("Bealiah"); II *Chronicles* 23:17; 24:7; 28:1-4. *Jeremiah* 7:3, 9, 31; 11:12-13; 17:2; 19:5,13; 32:29, 35. *Ezekiel* 14:11. *Hosea* 2:16)—a. k. a. Baal-Berith (*Judges* 8:33, 9:4), also called El-Berith (*Judges* 9:46), Baal-Zebub (II *Kings*, 1:2, 3, 6, 16. *Shabbath* 83*b*. *Sanhedrin* 63b), Baal-Peor (*Numbers* 25:1-9, 18; 31:16. *Deuteronomy* 3:29. *Joshua* 22:17. *Hosea* 9:10. *Psalm* 106:28 [eating the sacrifices of the dead]), Baal-Habab, Baal-Moloch (II *Chronicles* 28:1-4)—the God of Flies, the Golden Calf, the religion of Devil worship and human sacrifices (*Genesis* 22:1-18. *Exodus* 8:26; 13:2. *Leviticus* 27:28-29. *Joshua* 13:14. *Judges* 11:31, 39. I *Kings* 13:1-2. II *Kings* 16:3-4; 17:17; 21:6; 23:20-25. II *Chronicles* 28:1-4. *Jeremiah* 7:3; 19:5; 32:35. *Ezekiel* 16:20-21; 20:26, 31; 23:37.)

Ancient Jews kept secrets hidden behind the screen of the Temple in the Holy of Holies. The Temple had a secret area where only the High Priest was allowed to enter and it contained the Ark of the Covenant. The Ark was covered in gold, and embellished with golden rings, which facilitated movement and which might have been symbolic of the Golden Calf and of the earrings used to make it. Might the Ark have contained the Golden Calf? Aaron, who had introduced the worship of the Golden Calf, was the first High Priest. The cover of the Ark had two Cherubim on it, which were forbidden by the very Commandments it was said to house:

> "3 Thou shalt have no other gods before me. 4 Thou shalt not make unto thee any graven image, or any likeness of *any thing* that *is* in heaven above, or that *is* in the earth beneath, or that *is* in the water under the earth: 5 Thou shalt not bow down thyself to them, nor serve them: for I the LORD thy God *am* a jealous God, visiting the iniquity of the fathers upon the children unto

---

[507]. H. Graetz, *Popular History of the Jews*, Volume 5, Fifth Edition, Hebrew publishing Company, New York, (1937), p. 247.

the third and fourth *generation* of them that hate me;"—*Exodus* 20:3-5

Solomon's Temple contained Solomon's Molten Sea—a giant pot perhaps symbolizing the smelting pot used to melt the gold Aaron used to cast the Golden Calf. The Molten Sea sat upon twelve oxen, three facing each of the four points of the compass, or the winds (I *Kings* 7:23-26. II *Chronicles* 4:2-5, 15) of Baal. God commanded that golden offerings be made to the Ark (I *Samuel* 6:8). Solomon's Temple was filled with carved Cherubim covered in Gold. The oracle and its *giant* Cherubim were covered in gold (I *Kings* 6). The forbidden images of angels might well have been erected in reverence of the fallen angels said to have bred with human females and to have introduced evil to the world (*Genesis* 6:1-5. I *Enoch*) and may have reflected the tradition of Dualism present in the story of Adam and Eve, Cain and Abel, and Jacob and Esau.

The New Testament ascribes Satanic aspects to some Jews. *John* 8:44-45 states,

> "Ye are of *your* father the devil, and the lusts of your father ye will do. He was a murderer from the beginning, and abode not in the truth, because there is no truth in him. When he speaketh a lie, he speaketh of his own: for he is a liar, and the father of it. And because I tell *you* the truth, ye believe me not."

Revelation 2:9 states:

> "I know thy works, and tribulation, and poverty, (but thou art rich) and I *know* the blasphemy of them which say they are Jews, and are not, but *are* the synagogue of Satan."

Revelation 3:9 states:

> "Behold, I will make them of the synagogue of Satan, which say they are Jews, and are not, but do lie; behold, I will make them to come and worship before thy feet, and to know that I have loved thee."

John 7:1 tells that,

> "After these *things* Jesus walked in Galilee: for he would not walk in Jewry, because the Jews sought to kill him."

John 7:13 states:

> "Howbeit no *man* spake openly of him for fear of the Jews."

John 19:38 states:

> "And after this Joseph of Arimathaea, being a disciple of Jesus, but secretly

for fear of the Jews, besought Pilate that he might take away the body of Jesus: and Pilate gave *him* leave. He came therefore, and took the body of Jesus."

John 20:19 states:

"Then the same day at evening, being the first *day* of the week, when the doors were shut where the disciples were assembled for fear of the Jews, came Jesus and stood in the midst, and saith unto them, Peace *be* unto you."

I *John* 4:2-3 states:

"2 Hereby know ye the Spirit of God: Every spirit that confesseth that Jesus Christ is come in the flesh is of God: 3 And every spirit that confesseth not that Jesus Christ is come in the flesh is not of God: and this is *that spirit* of antichrist, whereof ye have heard that it should come; and *even* now already is it in the world."

Caligula did unto the Jews as the Jews did unto others. Ancient Jews were religious zealots blind to their own arrogance and hypocrisy. Philo the Jew's family, in collusion with Agrippa and Claudius, murdered Caligula. Philo must have been deeply gratified to have served up this revenge upon Caligula—if any of these events actually took place.

Philo the Jew and Josephus went to great lengths to defame Caligula, and their characterizations have prevailed through history. There is, though, another side to the story. Jews hypocritically demanded religious freedom and religious tolerance from others, but forbid others from practicing their religions, from worshiping their gods and idols, from building their temples of perishable materials, in any districts predominantly inhabited by Jews. Alexandrian Jews were resented for their control of markets, tax collecting and corruption, as well as their segregationist religion and nationalism, and the Roman Emperor was not so disliked as they were in the nations—in fact, Caligula was loved and they were often hated. Philo the Jew decries the civil wars he alleges Caligula caused, but only hints at the social injustice and rampant corruption in those cities which prompted revolts against Jews, and Philo admits that the worship of Caligula was only a pretext for attempts to obtain social justice. Philo betrays the religious arrogance and bigotry of the Alexandrian Jews. Quoting Caligula, Philo wrote,

> "'If any people in the bordering countries, with the exception of the metropolis itself, wishing to erect altars or temples, nay, images of statues, in honour of me and of my family are hindered from doing so, I charge you at once to punish those who attempt to hinder them, or else to bring them before the tribunal.' (335) Now this was nothing else but a beginning of seditions and civil wars, and an indirect way of annulling the gift which he appeared to be granting. For some men, more out of a desire of mortifying the Jews than from any feelings of loyalty towards Gaius, were inclined to

fill the whole country with erections of one kind or another. But they who beheld the violation of their national customs practised before their eyes were resolved above all things not to endure such an injury unresistingly. [***] What is this that you say? Do you, who are a man, seek to take to yourself the air and the heaven, not being content with the vast multitude of continents, and islands, and nations, and countries of which you enjoy the sovereignty? And do you not think any one of the gods who are worshipped in that city or by our people worthy of any country or city or even of any small precinct which may have been consecrated to them in old time, and dedicated to them with oracles and sacred hymns, and are you intending to deprive them of that, that in all the vast circumference of the world there may be no visible trace or memorial to be found of any honour or pious worship paid to the true real living God? (348) Truly you are suggesting fine hopes to the race of mankind; are you ignorant that you are opening the fountains of evils of every kind, making innovations, and committing acts of audacious impiety such as it is wicked to do and even to think of? [***] For if he were to give us up to our enemies, what other city could enjoy tranquillity? What city would there be in which the citizens would not attack the Jews living in it? What synagogue would be left uninjured? What state would not overturn every principle of justice in respect of those of their countrymen who arrayed themselves in opposition to the national laws and customs of the Jews? They will be overthrown, they will be shipwrecked, they will be sent to the bottom, with all the particular laws of the nation, and those too which are common to all and in accordance with the principles of justice recognized in every city."[508]

And wealthy Philo betrays another of his motives, one less noble than the preservation of the honor of his religion and that of his nation, which he eventually betrayed,

"For [Caligula's] designs were prepared against all those in authority and all those possessed of riches, and especially against those in Rome and those in the rest of Italy, by whom such quantities of gold and silver had been treasured up that even if all the riches of all the rest of the habitable world had been collected together from its most distant borders, it would have been found to be very inferior in amount. On this account he began, he, this hater of the citizens, this devourer of the people, this pestilence, this destructive evil, began to banish all the seeds of peace from his country, as if he were expelling evil from holy ground [***] Is it fitting now to compare with these oracles of Apollo the ill-omened warning of Gaius, by means of which poverty, and dishonor, and banishment, and death were given premature notice of to all those who were in power and authority in any part of the

---

[508]. Philo the Jew, "On the Embassy to Gaius", *The Works of Philo*, Hendrick Publishing, U. S. A., (2000), pp. 757-790, at 787, 788, 790.

world?"509

After Caligula was assassinated, Claudius took the throne—with the assistance, one might even say, at the insistence of King Agrippa—and exacted vengeance upon the enemies of the Jews. Claudius was intimate friends with Philo the Jew's brother, Alexander Lysimachus, long before the assassination took place. Alexander Lysimachus was also steward to Claudius' mother Antonia. Caligula had imprisoned Alexander Lysimachus. Claudius set Alexander Lysimachus free. Claudius executed the assassins of Caligula, made Agrippa, the Great Herod's grandson, not only King of Judea, but also of Samaria, and Claudius issued an edict in two forms, as repayment to the Jews who had given him the throne,

"Now about this time there was a sedition between the Jews and the Greeks, at the city of Alexandria; for when Caius was dead, the nation of the Jews, which had been very much mortified under the reign of Caius, and reduced to very great distress by the people of Alexandria, recovered itself, and immediately took up their arms to fight for themselves. So Claudius sent an order to the president of Egypt to quiet that tumult; he also sent an edict, at the requests of King Agrippa and King Herod, both to Alexandria and to Syria, whose contents were as follows: 'Tiberius Claudius Cæsar Augustus Germanicus, high priest, and tribune of the people, ordains thus: Since I am assured that the Jews of Alexandria, called Alexandrians, have been joint inhabitants in the earliest times with the Alexandrians, and have obtained from their kings equal privileges with them, as is evident by the public records that are in their possession, and the edicts themselves; and that after Alexandria had been subjected to our empire by Augustus, their rights and privileges have been preserved by those presidents who have at divers times been sent thither; and that no dispute had been raised about those rights and privileges, even when Aquila was governor of Alexandria; and that when the Jewish ethnarch was dead, Augustus did not prohibit the making such ethnarchs, as willing that all men should be so subject [to the Romans] as to continue in the observation of their own customs, and not be forced to transgress the ancient rules of their own country religion; but that, in the time of Caius, the Alexandrians became insolent towards the Jews that were among them, which Caius, out of his great madness and want of understanding, reduced the nation of the Jews very low, because they would not transgress the religious worship of their country, and call him a god: I will therefore that the nation of the Jews be not deprived of their rights and privileges, on account of the madness of Caius; but that those rights and privileges which they formerly enjoyed be preserved to them, and that they may continue in their own customs. And I charge both parties to take very

---

509. Philo the Jew, "On the Embassy to Gaius", *The Works of Philo*, Hendrick Publishing, U. S. A., (2000), pp. 757-790, at 766-767.

great care that no troubles may arise after the promulgation of this edict.'

And such were the contents of this edict on behalf of the Jews that was sent to Alexandria. But the edict that was sent into the other parts of the habitable earth was this which follows: 'Tiberius Claudius Caesar Augustus Germanicus, high priest, tribune of the people, chosen consul the second time, ordains thus: Upon the petition of King Agrippa and King Herod, who are persons very dear to me, that I would grant the same rights and privileges should be preserved to the Jews which are in all the Roman empire, which I have granted to those of Alexandria, I very willingly comply therewith; and this grant I make not only for the sake of the petitioners, but as judging those Jews for whom I have been petitioned worthy of such a favor, on account of their fidelity and friendship to the Romans. I think it also very just that no Grecian city should be deprived of such rights and privileges, since they were preserved to them under the great Augustus. It will therefore be fit to permit the Jews, who are in all the world under us, to keep their ancient customs without being hindered so to do. And I do charge them also to use this my kindness to them with moderation, and not to show a contempt of the superstitious observances of other nations, but to keep their own laws only. And I will that this decree of mine be engraven on tables by the magistrates of the cities, and colonies, and municipal places, both those within Italy and those without it, both kings and governors, by the means of the ambassadors, and to have them exposed to the public for full 30 days, in such a place whence it may plainly be read from the ground.'"[510]

This edict continued a long tradition of governmental edicts which granted Jews special privileges and which ensured religious tolerance towards Jews, while granting the Jews the privilege of being intolerant. Ancient Jews did not accord the peoples they had vanquished religious freedom and were intolerant even of their neighbors' religious beliefs, which they forbade and sought to completely destroy. The book of *Ezra*, Chapter 6, provides an example both of the special privileges allegedly accorded to ancient Jews, and of the hatred of ancient Jews against the "heathens", as well as the use of the Jewish God as a superstitious granter of gifts to those who sponsor the Jews—the gods of various peoples were often used as a threat to curse or to bless enemies or friends in the ancient world, a mythology which continues today (Zionists curried, and curry, favor for their cause among Christians by promising them uninhibited access to the holy sites and the fulfilment of religious prophecy). Note also the use of the wealth of other nations for the construction of the Temple, and the hypocrisy lying in the fact that a monument to Caligula was destroyed by a people who demanded religious tolerance for themselves. It must be borne in mind that the

---

[510]. Josephus, "Claudius restores to Agrippa his grandfather's kingdoms—augments his dominions; and publishes an edict in behalf of the Jews", *The Works of Flavius Josephus: Comprising the Antiquities of the Jews; a History of the Jewish Wars; and Life of Flavius Josephus, Written by Himself*, "Antiquities of the Jews", Book 19, Chapter 5, S. S. Scranton Co., Hartford, Connecticutt, (1916), pp. 593-594.

accounts of *Ezra* may be fabrications as may the alleged "roll":

> "Then Darius the king made a decree, and search was made in the house of the rolls, where the treasures were laid up in Babylon. 2 And there was found at Achmetha, in the palace that *is* in the province of the Medes, a roll, and therein *was* a record thus written: 3 In the first year of Cyrus the king, *the same* Cyrus the king made a decree *concerning* the house of God at Jerusalem, Let the house be builded, the place where they offered sacrifices, and let the foundations thereof be strongly laid; the height thereof threescore cubits, *and* the breadth thereof threescore cubits; 4 *With* three rows of great stones, and a row of new timber: and let the expenses be given out of the king's house: 5 And also let the golden and silver vessels of the house of God, which Nebuchadnezzar took forth out of the temple which *is* at Jerusalem, and brought unto Babylon, be restored, and brought again unto the temple which *is* at Jerusalem, *every one* to his place, and place *them* in the house of God. 6 Now *therefore*, Tatnai, governor beyond the river, Shethar-boznai, and your companions the Apharsachites, which *are* beyond the river, be ye far from thence: 7 Let the work of this house of God alone; let the governor of the Jews and the elders of the Jews build this house of God in his place. 8 Moreover I make a decree what ye shall do to the elders of these Jews for the building of this house of God: that of the king's goods, *even* of the tribute beyond the river, forthwith expenses be given unto these men, that they be not hindered. 9 And that which they have need of, both young bullocks, and rams, and lambs, for the burnt offerings of the God of heaven, wheat, salt, wine, and oil, according to the appointment of the priests which *are* at Jerusalem, let it be given them day by day without fail: 10 That they may offer sacrifices of sweet savors unto the God of heaven, and pray for the life of the king, and of his sons. 11 Also I have made a decree, that whosoever shall alter this word, let timber be pulled down from his house, and being set up, let him be hanged thereon; and let his house be made a dunghill for this. 12 And the God that hath caused his name to dwell there destroy all kings and people, that shall put to their hand to alter *and* to destroy this house of God which *is* at Jerusalem. I Darius have made a decree; let it be done with speed. 13 ¶ Then Tatnai, governor on this side the river, Shethar-boznai, and their companions, according to that which Darius the king had sent, so they did speedily. 14 And the elders of the Jews builded, and they prospered through the prophesying of Haggai the prophet and Zechariah the son of Iddo. And they builded, and finished *it*, according to the commandment of the God of Israel, and according to the commandment of Cyrus, and Darius, and Artaxerxes king of Persia. 15 And this house was finished on the third day of the month Adar, which was in the sixth year of the reign of Darius the king. 16 ¶ And the children of Israel, the priests, and the Levites, and the rest of the children of the captivity, kept the dedication of this house of God with joy, 17 And offered at the dedication of this house of God a hundred bullocks, two hundred rams, four hundred lambs; and for a sin offering for all Israel, twelve he goats,

according to the number of the tribes of Israel. 18 And they set the priests in their divisions, and the Levites in their courses, for the service of God, which is at Jerusalem; as it is written in the book of Moses. 19 ¶ And the children of the captivity kept the passover upon the fourteenth *day* of the first month. 20 For the priests and the Levites were purified together, all of them *were* pure, and killed the passover for all the children of the captivity, and for their brethren the priests, and for themselves. 21 And the children of Israel, which were come again out of captivity, and all such as had separated themselves unto them from the filthiness of the heathen of the land, to seek the LORD God of Israel, did eat, 22 And kept the feast of unleavened bread seven days with joy: for the LORD had made them joyful, and turned the heart of the king of Assyria unto them, to strengthen their hands in the work of the house of God, the God of Israel."

Alexander Lysimachus was the wealthiest man in the world. He decorated the Temple with gold. Alexander Lysimachus' son, Philo the Jew's nephew, Marcus Julius Alexander married Agrippa's daughter Bernice. Ironically, Alexander Lysimachus' other son, Tiberius Julius Alexander, allegedly abandoned Judaism, became procurator of Judea and prefect of Egypt and took part in the attacks on Jerusalem. The facts tend to indicate that the family of Philo the Jew, the wealthiest family in the world, assassinated Caligula in the defense of the Temple, Jewish law and the Jews of Alexandria—truly speaking, also in defense of their wealth and privilege, and to free Alexander Lysimachus. Then the Jews did what they were bound to do by their religion.

Josephus' story of Publius Petronius is implausible, and it is far more likely that Caligula never intended to withdraw his order that a statue of him be placed in the Temple, and that Claudius and Philo the Jew's family conspired to murder him. Claudius attained the throne, and Philo the Jew received the edict favoring the Jews, the freedom of his brother, and corrupt influence in the Roman government, which enabled him to maintain a privileged status for the Jews of the ancient world—though this was short-lived.

Philo the Jew, also known as Philo of Alexandria, is most famous for Helenizing the Jewish faith with mystical writings on the Pentateuch, a Helenization carried out in earnest in Alexandria, with, among other things the translation of the Torah into Greek in the Septuagint with Heraclitean and Platonic language and overtones. Many ancients claimed that the Old Testament itself was a plagiarized fabrication by the Judeans, who had no known authentic ancient history of their own and instead cobbled one together *circa* 500-450 BC, copying the beliefs of the Egyptians, Greeks and others.

In the 1870's, Julius Wellhausen[511] set out to prove the contention that the

---

[511]. J. Wellhausen, "Israel", *Encyclopaedia Britannica*, Ninth Edition, (1881); **and** *Sketch of the History of Israel and Judah*, Third Edition, Adam and Charles Black, London, (1891); **and** *Prolegomena to the History of Ancient Israel: With a Reprint of the Article Israel from the Encyclopaedia Britannica*, Meridian Books, New York, (1957); **and** *Die Composition des Hexateuchs und der historischen Bücher des Alten Testaments*,

Old Testament was of comparatively recent origin and that the Pentateuch had multiple authors. He established that the Old Testament signifies the creation of the new religion of Judaism and not the history of Israel. His work was popular and well-received.

7.5 All the Best Zionists are Anti-Semites

The worst enemy of the common Jew has always been the Zionist.

In 1932, Einstein stated, referring to the "deplorably high development of nationalism everywhere"—his own rabid Zionism excepted,

> "The introduction of compulsory service is therefore, to my mind, the prime cause of the moral collapse of the white race, which seriously threatens not merely the survival of our civilization but our very existence. This curse, along with great social blessings, started with the French Revolution, and before long dragged all the other nations in its train."[512]

Einstein complained to Lorentz on 12 January 1920 that even well-educated persons fell victim to "the illiberal nationalistic standpoint."[513] Einstein called "Nationalism" an "ugly name".[514] Einstein's Zionist hypocrisy did not go unnoticed. He was asked why he stood firmly against Gentile nationalism, while making Zionist nationalism his primary purpose in life. According to Thüring, the *Jüdische Presse* reported on 29 May 1929,

> "Man fragte [Einstein], warum er als Verfechter aller internationalen Interessen, als Gegner aller nationalistischen Bestrebungen die jüdische nationale Sache zu seiner eigenen mache. Er erklärte seinen Standpunkt durch ein Gleichnis: Wer einen rechten Arm hat und davon spricht und immer davon spricht, ist ein Narr. Wem aber rechte Arm fehlt, der darf alles tun, um sich das fehlende Glied zu ersetzen. Daher sei er in einer Welt, in der jedes Volk die Bedingungen des nationalen Lebens hat, ein Feind des Nationalismus, als Jude aber ein Anhänger der jüdisch-nationalen Idee, weil den Juden die notwendige und natürliche Voraussetzung ihres nationalen Lebens fehlt."

This clearly elucidates Einstein's nationalistic perspective, which mirrored the Nazis' nationalistic perspective. The Nazis simply pursued the same false reasoning as Einstein and asserted that their right arm was infected with

---

Fourth Unaltered Edition, W. de Gruyter, Berlin, (1963).
512. A. Einstein, translated by A. Harris, "The Disarmament Conference of 1932. I." *The World As I See It*, Citadel, New York, (1993), pp. 59-60.
513. Letter from A. Einstein to H. A. Lorentz of 12 January 1920, English translation by A. Hentschel, *The Collected Papers of Albert Einstein*, Volume 9, Document 256, Princeton University Press, (2004), pp. 214-215, at 214.
514. A. Einstein, *The World As I See It*, Citadel, New York, (1993), p. 109.

Einstein's self-described foreign and disloyal nationalists. Einstein agreed with the Nazis and saw them as the salvation of the Jews.

Therein lies the potential danger of Einstein's segregationism. Segregationist nationalism is bound to lead to genocidal nationalism. Einstein's tacit premise that citizenry and nationhood be based on ancient territory, ethnicity, race and religion—on *Blut und Boden*, instead of the sovereignty of a group of living persons in a territory, whether homogenous or heterogeneous in its ethnicities and religions, was racist bigotry—commonly held bigotry, but bigotry nonetheless. Einstein's Zionist nationalism, which was no different from Nazi nationalism, would disconnect Jews around the world from the nations in which they were citizens. His racist nationalism definitely did not conform with his internationalist views, which were premised upon a community of nations, which implies a human family. In addition, Einstein voluntarily amputated his right arm, though he pretended that his self-inflicted wound was a congenital defect. Einstein was born a German, not a Palestinian. But Einstein's hypocrisy, his system of double standards, his desire that the Gentiles be consumed in wars and that the Jews reestablish a State and rule the world, were nothing new. They were Judaism.

Einstein was an advocate of world government and a segregated "Jewish State". While this seemed a contradiction to many, including many Jews, especially many secular Jews, Einstein was merely expressing his loyalty to Jewish Messianic myth. Given Einstein's racist Zionism, it is clear that Einstein wished for a day when Jews would rule a world devoid of Gentile government and that they would be segregated from, and reign over, the "Goyim", to use Einstein's term. "Internationalism" was a code word for a world devoid of Gentile government—a Jewish Messianic prophecy. "Zionism" was a code word for Jewish supremacy reigning over the world from Jerusalem in the Jewish Nation. Einstein's "Internationalism" and Einstein's "Zionism" need no reconciliation, they are one in the same objective—Judaism. Rather those who are confused by Einstein's apparent contradictions need only read the Hebrew Bible, where the Jewish prophets tell the Jews to reconstruct the Jewish State and at the same time destroy all the Gentile governments of the world.

After World War II had ended, Einstein's friend Peter A. Bucky also questioned the apparent contradiction in Einstein's political philosophy. Bucky asked Einstein how he reconciled his Zionism with his anti-Nationalism. As a good racist Zionist Jew was wont to do, Einstein exploited modern anti-Semitism to legitimize racist Jewish Nationalism which is at least 2,500 years old,

> "I think that [nationalism] is justified in this special case because the world has forced the Jews to entrench themselves with the continued existence of anti-Semitism."[515]

---

515. P. A. Bucky, Einstein, and A. G. Weakland, *The Private Albert Einstein*, Andrews and McMeel, Kansas City, (1992), pp. 88-89.

Einstein felt that Jews owed anti-Semites a great debt of appreciation for forcing Jews to "entrench themselves". He must also have known that the Zionists created the Nazis to force reluctant assimilating Jews to Palestine. Einstein dreaded a world without anti-Semitism, without segregation and without segregated racist Jews like himself. The incentive for Jews to create anti-Semitism is clear. There is abundant evidence that leading Jews have again and again down through history created and sponsored anti-Semitism. In the racist Zionist's view, racist segregationist Judaism and the Jewish tribe cannot continue to exist without manufacturing anti-Semitism to keep them alive.

Given that the vast majority of German Jews during Einstein's lifetime vehemently opposed his bigotry, it is especially odd that Einstein was so unenlightened and so racist. His own children were assimilated Jews, and he hated them for it.[516] Whereas most German Jews considered the racism of Zionist Eastern Jews primitive and uncivilized, Einstein considered assimilation uncivilized and inhuman, because Einstein believed that European Gentiles were sub-human and incapable of civilization. His Zionist sponsors created wars for, among other things, the purpose of discrediting Gentile government. Einstein owed his fame to Zionists, who used him to publicize their cause. Einstein was more loyal to the Zionists' racism, than he was to his own children. Racism buttered Einstein's bread, his children wanted eat it, though he wouldn't let them—they were sub-human. Fellow Jewish racists kept Einstein in the spotlight and shielded him from criticism.

Einstein, himself, echoed and endorsed the views of the anti-Semites in an interview in which he again revealed himself to be a racist and a segregationist. Zionists intentionally provoked and sought to inspire anti-Semitism, and anti-Semites welcomed the openly racist positions of the Zionists.[517] Einstein went along with the crowd of prominent political Zionists who openly stated that anti-Semitism is welcomed, encouraged and useful to the Zionists. They based their myth on Spinoza's declaration that emancipation leads to assimilation and that the Jews only exist in modern times because glorious anti-Semitism kept them segregated.[518]

Prominent Zionist and author of the *Encyclopaedia Judaica; das Judentum in Geschichte und Gegenwart*, Jakob Klatzkin stated in 1925,

> "The national viewpoint taught us to understand the true nature of antisemitism, and this understanding widens the horizons of our national outlook. [***] In the age of enlightenment antisemitism was included

---

516. P. A. Bucky, Einstein, and A. G. Weakland, *The Private Albert Einstein*, Andrews and McMeel, Kansas City, (1992), pp. 107-108.
517. *Cf.* K. J. Herrmann, "Historical Perspectives on Political Zionism and Antisemitism", *Zionism & Racism: Proceedings of an International Symposium*, International Organization for the Elimination of All Forms of Racial Discrimination, Tripoli, (1977), pp. 197-208.
518. J. Wellhausen, *Sketch of the History of Israel and Judah*, Third Edition, Adam and Charles Black, London, (1891), pp. 201-203.

among the phenomena that are likely to disappear along with other forms of prejudice and iniquity. The antisemites, so the rule stated, were the laggard elements in the march of progress. Hence, our fate is dependent on the advance of human culture, and its victory is our victory. [***] In the period of Zionism, we learned that antisemitism was a psychic-social phenomenon that derives from our existence as a nation within a nation. Hence, it cannot change, until we attain our national end. But if Zionism had fully understood its own implications, it would have arrived, not merely as a psycho-sociological explanation of this phenomenon, but also as a justification of it. It is right to protest against its crude expressions, but we are unjust to it and distort its nature so long as we do not recognize that essentially it is a defense of the integrity of a nation, in whose throat we are stuck, neither to be swallowed nor to be expelled. [***] And when we are unjust to this phenomenon, we are unfair to our own people. If we do not admit the rightfulness of antisemitism, we deny the rightfulness of our own nationalism. If our people is deserving and willing to live its own national life, then it is an alien body thrust into the nations among whom it lives, an alien body that insists on its own distinctive identity, reducing the domain of their life. It is right, therefore, that they should fight against us for their national integrity. [***] Know this, that it is a good sign for us that the nations of the world combat us. It is proof that our national image is not yet utterly blurred, our alienism is still felt. If the war against us should cease or be weakened, it would indicate that our image has become indistinct and our alienism softened. We shall not obtain equality of rights anywhere save at the price of an explicit or implied declaration that we are no longer a national body, but part of the body of the host-nation; or that we are willing to assimilate and become part of it. [***] Instead of establishing societies for defense against the antisemites, who want to reduce our rights, we should establish societies for defense against our friends who desire to defend our rights. [***] When Moses came to redeem the children of Israel, their leaders said to him, 'You have made our odor evil in the eyes of Pharaoh and in the eyes of his servants, giving them a sword with which to kill us.' Nevertheless, Moses persisted in worsening the situation of the people, and he saved them."[519]

Klaus J. Herrmann has collected a great deal of evidence related to Zionist racism in his presentation, "Historical Perspectives on Political Zionism and Antisemitism", *Zionism & Racism: Proceedings of an International Symposium*, International Organization for the Elimination of All Forms of Racial Discrimination, Tripoli, (1977), pp. 197-210. At page 197, Herrmann states, [quoting Constantin Brunner, *Der Judenhass und die Juden*, Berlin, (1918), p. 112; and Ernst Ludwig Pinner, "Meine Abkehr vom Zionismus", *Los vom*

---

519. J. Klatzkin, *Tehumim: Ma'amarim*, Devir, Berlin, (1925). English translation in J. B. Agus, *The Meaning of Jewish History*, Volume 2, Abelard-Schuman, New York, (1963), pp. 425-426.

Zionismus, J. Kauffmann, Frankfurt, (1928), pp. 32-33; and referencing Houston Stewart Chamberlain, *Die Grundlagen des neunzehnten Jahrhunderts*, F. A. Bruckmann, München, (1899), English translation by John Lees, *Foundations of the Nineteenth Century*, John Lane, New York, (1910)—*see:* F. Kahn, "H. St. Chamberlain (Eine Charakteristik)", *Jüdische Rundschau*, Volume 25, Nummer 63/64, (10 September 1920), pp. 499-500, for a contemporary view of the impact on Jews of Chamberlain's much-read book. His book was popular among Zionists and the English translation of it received a long and favorable review in the *Times Literary Supplement* of 15 December 1910, pp. 500-501.]:

> "Jews,' wrote Brunner, 'have been taken in by the racial theories of the Jew-haters;' and he accused the Zionists of having taken as their teacher the notorious racist and forger of scholarly documentation Houston Stewart Chamberlain, whose 'confused nonsense revelations' had been 'restammered' in a Zionist book on the subject of race. 'How could Germans of Jewish background begin to talk of a Jewish nation, and to fashion of the worst calumny the dream of their greatest nonsense!'[1]
>
> One of Brunner's disciples, Ernst Ludwig Pinner, who had been a Zionist earlier, bitterly accused the Zionists of having
>
>> taken up Europe's newest nonsense, namely racial theory as the justification of national emotion. Racial arrogance and racial hate poison national emotion, as did previously religious arrogance and religious hatred. Today it is race which is exalted as the banner in whose name everything is justified.
>
> Pinner also designated the Zionists as 'Jews infected by the sickness of racial insanity ... because, similar to the Jew-haters, they drew political consequences out of race-consciousness.'[2] Pinner did absolve Zionists of 'preaching arrogance and hatred;'[3] whether or not he would have done so in later years remains open to conjecture."

At pages 204-205, Klaus J. Herrmann quotes the Zionist ideologist Jakob Klatzkin who stated, among other things, in his book of 1921 *Krisis und Entscheidung im Judentum; der Probleme des modernen Judentums*, Second Enlarged Edition, Jüdischer Verlag, Berlin, pages 61-63, and 118:

> "[I applaud] the contribution of our enemies in the continuance of Jewry in eastern Europe. [***] We ought to be thankful to our oppressors that they closed the gates of assimilation to us and took care that our people were concentrated and not dispersed, segregatedly united and not diffusedly mixed [***] One ought to investigate in the West and note the great share which antisemitism had in the continuance of Jewry and in all the emotions and movements of our national rebirth . [***] Truly our enemies have done much for the strengthening of Judaism in the diaspora . [***] Experience teaches that the liberals have understood better than the antisemites how to

destroy us as a nation. [***] We are, in a word, naturally foreigners; we are an alien nation in your midst and we want to remain one."[520]

"Man vergegenwärtige sich, wie groß der Anteil unserer Feinde am Fortbestand des Judentums im Osten ist. [...] Wir müßten beinahe unseren Bedrängern dankbar sein, wenn sie die Tore der Assimilation vor uns schlossen und dafür Sorge trugen, daß unsere Volksmassen konzentriert und nicht zerstreut, abgesondert geeint und nicht zerklüftet vermischt werden[... .] Man untersuche es im Westen, welchen hohen Anteil der Antisemitismus am Fortbestand des Judentums und an all den Regungen und Bewegungen unserer nationalen Wiedergeburt hat. [...] Wahrlich, unsere Feinde haben viel zur Stärkung des Judentums in der Diaspora beigetragen. [...] Und die Erfahrung lehrt, daß die Liberalen es besser als die Antisemiten verstanden haben, uns als Volk zu vernichten. [...] Wir sind schlechthin Wesensfremde, sind — wir müssen es immer wiederholen — ein Fremdvolk in eurer Mitte und wollen es auch bleiben."

Some Jews, and some critics of the Jews, have for thousands of years asserted that Jews always form a separate state within the nations they inhabit. This they attribute to the Jewish religion, with its one God to rule over all—Jews being the chosen people who will one day receive the Messiah who will assist them in ruling the world after all other nations are destroyed, which fatalistic belief system inspires the nationalism many Jews have expressed in the Diaspora.

When Zionists like Herzl, Klatzkin and Rabbi Meir Bar-Ilan, who stated in 1922,

"We have no 'church' that is not also concerned with matters of state, just as we have no state which is not also concerned with 'church' matters—in Jewish life these are not two separate spheres."[521]

confirmed that these ancient religious, nationalistic and political aspirations were current in modern Europe, where Jews had been emancipated, it caused many to view Jews not only with suspicion, but with contempt, most especially so because radical revolutionary organizations were often led by, and populated with, Jews in disproportionate numbers to Gentiles. Many leading figures

---

520. K. J. Herrmann, "Historical Perspectives on Political Zionism and Antisemitism", *Zionism & Racism: Proceedings of an International Symposium*, International Organization for the Elimination of All Forms of Racial Discrimination, Tripoli, (1977), pp. 197-210, at 204-205. A lengthy quotation from Klatzkin, in English translation, appears in: M. Menuhin, *The Decadence of Judaism in Our Time*, Exposition Press, New York, (1965), pp. 482-483.
521. M. Bar-Ilan, *Kitve Rabi Me'ir Bar-Ilan*, Volume 1, Mosad ha-Rav Kuk, Jerusalem, (1950), pp. 5-16; English translation in A. Hertzberg, *The Zionist Idea*, Harper Torchbooks, New York, (1959), pp. 548-555, at 548.

warned the public that the Bolshevik Jews were seeking world domination. They wanted to end the immigration of Eastern European Jews to Germany, and to expel Eastern European Jews from Germany. The Bolshevik Jews had already conducted successful, though short-lived, revolutions in German territory. Many of the Jews emigrating to Germany from the East were the descendants of the Frankists, who had pledged themselves to destroy the Gentile nations by means of deception and revolution. Frankist Jews were often crypto-Jews who hid their Jewish ethnicity in order to deceive Christians who might not otherwise trust them, to place the blame for their actions on other peoples so as to cause an unjust hatred towards those innocent peoples, and to prevent a backlash against Jews for the vile actions Jews were taking against other peoples. The Talmud teaches the Jews that they can sin against others with immunity if they hide the fact that they are Jewish such that Jews will not be attacked in retaliation. *Moed Katan* 17a states,

> "R' IL'AI SAYS: [***] IF A PERSON SEES THAT HIS evil INCLINATION IS OVERWHELMING HIM, [***] HE SHOULD GO TO A PLACE WHERE THEY DO NOT RECOGNIZE HIM [***] AND CLOTHE HIMSELF IN BLACK AND WRAP HIMSELF IN BLACK, [***] AND HE SHOULD DO WHAT HIS HEART DESIRES, [***] AND HE SHOULD NOT DESECRATE THE NAME OF HEAVEN OPENLY."[522]

An alternative translation:

> "For R. Il'ai says, If one sees that his [evil] *yezer*[5] is gaining sway over him, let him go away where he is not known; let him put on sordid[6] clothes, don a sordid wrap and do the sordid deed that his heart desires rather than profane the name of Heaven openly.[7]"[523]

Albert Einstein stated in the *Berliner Tageblatt* on 30 December 1919,

> "It is quite likely that there are Bolshevist agents in Germany, but they undoubtedly hold foreign passports, have at their disposal ample funds and cannot be seized by any administrative measures. The big profiteers among the Eastern European Jews have certainly, long ago, taken precautions to elude arrest by officials. The only [Jews] affected would be *those poor and unfortunate ones*, who in recent months made their way to Germany under

---

522. *Moed Katan* 17a, Rabbi Y. S. Schorr, *et al.*, Editors, "Tractate Moed Katan", *Talmud Bavli: the Schottenstein edition: the Gemara: the classic Vilna edition, with an annotated, interpretive elucidation, as an aid to Talmud study / elucidated by a team of Torah scholars under the general editorship of Hersh Goldwurm and Nosson Scherman.*, Volume 21, Mesorah Publications, Ltd., Brooklyn, New York, (1999), p. 17a^2.
523. Mo'ed Katan, Rabbi I. Epstein, Editor, "Seder Mo'ed", *The Babylonian Talmud*, Volume 14, The Soncino Press, (1938), p. 107.

inhumane privations, in order to look for work here."[524]

Albert Einstein was himself a racist; and, therefore, a hypocrite when criticizing the racism of others. John Stachel wrote,

> "While he lived in Germany, however, Einstein seems to have accepted the then-prevalent racist mode of thought, often invoking such concepts as 'race' and 'instinct,' and the idea that the Jews form a race."[525]

On 8 July 1901, Einstein wrote to Winteler,

> "There is no exaggeration in what you said about the German professors. I have got to know another sad specimen of this kind — one of the foremost physicists of Germany."[526]

Einstein wrote to Besso sometime after 1 January 1914,

> "A free, unprejudiced look is not at all characteristic of the (adult) Germans (blinders!)."[527]

After the war Einstein and some of his friends alluded to much earlier conversations with Einstein where he had correctly predicted the eventual outcome of the war. In his diaries, Romain Rolland recorded his conversations with Einstein in Switzerland at their meeting of 16 September 1915,

> "What I hear from [Einstein] is not exactly encouraging, for it shows the impossibility of arriving at a lasting peace with Germany without first totally crushing it. Einstein says the situation looks to him far less favorable than a few months back. The victories over Russia have reawakened German arrogance and appetite. The word 'greedy' seems to Einstein best to characterize Germany. [***] Einstein does not expect any renewal of Germany out of itself; it lacks the energy for it, and the boldness for initiative. He hopes for a victory of the Allies, which would smash the power of Prussia and the dynasty... . Einstein and Zangger dream of a divided Germany—on the one side Southern Germany and Austria, on the other side Prussia. [***] We speak of the deliberate blindness and the lack of

---

[524]. A. Einstein, translated by A. Engel, *The Collected Papers of Albert Einstein*, Volume 7, Document 29, Princeton University Press, (2002), pp. 110-111.

[525]. J. Stachel, "Einstein's Jewish Identity", *Einstein from 'B' to 'Z'*, Birkhäuser, Boston, Basel, Berlin, (2002), pp. 57-83, at 68.

[526]. A. Einstein to J. Winteler, English translation by A. Beck, *The Collected Papers of Albert Einstein*, Volume 1, Document 115, Princeton University Press, (1987), pp. 176-177, at 177.

[527]. A. Einstein, English translation by A. Beck, *The Collected Papers of Albert Einstein*, Volume 5, Document 499, Princeton University Press, (1995), pp. 373-374, at 374.

psychology in the Germans."[528]

Jews often sought to Balkanize nations so as to weaken the power of any faction within a nation and to created perpetual agitation between the nations which could be exploited for profit and other Jewish gains. For example, the Rothschilds created the American Civil War and profited from the debts it generated. They hoped to divide America into two nations and to pit these against one another. They were successful. Jews had long been pitting North German Protestants against South German and Austrian Catholics. Jews were the motive force behind the *Kulturkampf*. After creating these divides and promoting perpetual agitations amongst neighbors, Jewry could then fund one side against the other to destroy it whenever Jewry decided to wreck a given nation.

Einstein's dreams during the First World War remind one of the "Carthaginian Peace" of the Henry Morgenthau, Jr. plan for the destruction of Germany following the Second World War. Morgenthau worked with Lord Cherwell (Frederick Alexander Lindemann), Churchill's friend and advisor, who planned to bomb German civilian populations into submission. Lindemann studied under Einstein's friend, Walther Nernst, who worked with Fritz Haber, a Jewish developer of poisonous gas. James Bacque argues that the Allies, under the direction of General Eisenhower, starved hundreds of thousands, if not millions of German prisoners of war to death. Dwight David Eisenhower was called "the terrible Swedish-Jew" in his yearbook for West Point, *The 1915 Howitzer*, West Point, New York, (1915), p. 80. He was also called "Ike", as in... Eisenhower? The Soviets also abused countless German POW's after the Second World War.[529]

Einstein often spoke in genocidal and racist terms against Germany, and for the Jews and England, and he betrayed Germany before, during and after the war. Einstein wrote to Paul Ehrenfest on 22 March 1919,

> "[The Allied Powers] whose victory during the war I had felt would be by far the lesser evil are now proving to be *only slightly* the lesser evil. [***] I get most joy from the emergence of the Jewish state in Palestine. It does seem to me that our kinfolk really are more sympathetic (at least less brutal) than these horrid Europeans. Perhaps things can only improve if only the Chinese are left, who refer to all Europeans with the collective noun

---

528. R. Romain, *La Conscience de l'Europe*, Volume 1, pp. 696ff. English translation from A. Fölsing, *Albert Einstein: A Biography*, Viking, New York, (1997), pp. 365-367. *See also:* Letter from A. Einstein to R. Romain of 15 September 1915, *The Collected Papers of Albert Einstein*, Volume 8, Document 118, Princeton University Press, (1998); **and** Letter from A. Einstein to R. Romain of 22 August 1917, *The Collected Papers of Albert Einstein*, Volume 8, Document 374, Princeton University Press, (1998).
529. J. Bacque, *Other Losses: An Investigation into the Mass Deaths of German Prisoners at the Hands of the French and Americans after World War II*, Stoddart, Toronto, (1989).

'bandits.'"530

While responsible people were trying to preserve some sanity in the turbulent period following World War I, Zionists like Albert Einstein sought to validate and encourage the racism of anti-Semites. The Dreyfus Affair taught them that anti-Semitism had a powerful effect to unite Jews around the world. The Zionists were afraid that the "Jewish race" was disappearing through assimilation. They wanted to use anti-Semitism to force the segregation of Jews from Gentiles and to unite Jews, and thereby preserve the "Jewish race". They hoped that if they put a Hitler into power—as Zionists had done in the past, they could use him to herd up the Jews and force the Jews into Palestine against their will. This would also help the Zionists to inspire distrust and contempt for Gentile government, while giving the Zionists the moral high-ground in international affairs, despite the fact that the Zionists were secretly behind the atrocities. Theodor Herzl wrote in his book *The Jewish State*,

> "Oppression and persecution cannot exterminate us. No nation on earth has survived such struggles and sufferings as we have gone through. Jew-baiting has merely stripped off our weaklings; the strong among us were invariably true to their race when persecution broke out against them. This attitude was most clearly apparent in the period immediately following the emancipation of the Jews. Later on, those who rose to a higher degree of intelligence and to a better worldly position lost their communal feeling to a very great extent. Wherever our political well-being has lasted for any length of time, we have assimilated with our surroundings. I think this is not discreditable. Hence, the statesman who would wish to see a Jewish strain in his nation would have to provide for the duration of our political well-being; and even Bismarck could not do that. [***] The Governments of all countries scourged by Anti-Semitism will serve their own interests in assisting us to obtain the sovereignty we want. [***] Great exertions will not be necessary to spur on the movement. Anti-Semites provide the requisite impetus. They need only do what they did before, and then they will create a love of emigration where it did not previously exist, and strengthen it where it existed before. [***] I imagine that Governments will, either voluntarily or under pressure from the Anti-Semites, pay certain attention to this scheme; and they may perhaps actually receive it here and there with a sympathy which they will also show to the Society of Jews."531

Albert Einstein wrote to Max Born on 9 November 1919, and encouraged anti-Semitism and advocated segregationism (one must wonder what rôle

---

530. Letter from A. Einstein to Paul Ehrenfest of 22 March 1919, English translation by A. Hentschel, *The Collected Papers of Albert Einstein*, Volume 9, Document 10, Princeton Univsersity Press, (2004), pp. 9-10, at 10.
531. T. Herzl, *A Jewish State: An Attempt at a Modern Solution of the Jewish Question*, The Maccabæan Publishing Co., New York, (1904), pp. 5-6, 25, 68, 93.

Albert's increasing racism played in his divorce from Mileva Marić—a Gentile Serb),

> "Antisemitism must be seen as a real thing, based on true hereditary qualities, even if for us Jews it is often unpleasant. I could well imagine that I myself would choose a Jew as my companion, given the choice. On the other hand I would consider it reasonable for the Jews themselves to collect the money to support Jewish research workers outside the universities and to provide them with teaching opportunities."[532]

In 1933, the Zionists publicly declared their allegiance to the Nazis. They wrote in the *Jüdische Rundshau* on 13 June 1933,

> "Zionism recognizes the existence of the Jewish question and wants to solve it in a generous and constructive manner. For this purpose, it wants to enlist the aid of all peoples; those who are friendly to the Jews as well as those who are hostile to them, since according to its conception, this is not a question of sentimentality, but one dealing with a real problem in whose solution all peoples are interested."[533]

On 21 June 1933, the Zionists issued a declaration of their position with respect to the Nazi régime, in which they expressed a belief in the legitimacy of the Nazis' racist belief system and condemned the anti-Fascist forces.[534]

Michele Besso wrote that it might have been Albert Einstein's racism and bigotry which caused him to separate from his first wife Mileva Marić in 1914. Besso wrote to Einstein on 17 January 1928,

> "[...]perhaps it is due in part to me, with my defense of Judaism and the Jewish family, that your family life took the turn that it did, and that I had to bring Mileva from Berlin to Zurich[.]"[535]

The hypocrisy of racist Zionists often manifested itself in this way. Many

---

[532]. M. Born, *The Born-Einstein Letters*, Walker and Company, New York, (1971), p. 16.

[533]. English translation in: K. Polkehn, "The Secret Contacts: Zionism and Nazi Germany, 1933-1941", *Journal of Palestine Studies*, Volume 5, Number 3/4, (Spring-Summer, 1976), pp. 54-82, at 59.

[534]. L. S. Dawidowicz, "The Zionist Federation of Germany Addresses the New German State", *A Holocaust Reader*, Behrman House, Inc., West Orange, New Jersey, (1976), pp. 150-155. *See also:* H. Tramer, Editor, S. Moses, *In zwei Welten: Siegfried Moses zum fünfundsiebzigsten Geburtstag*, Verlag Bitaon, Tel-Aviv, (1962), pp. 118.ff; cited in K. Polkehn, "The Secret Contacts: Zionism and Nazi Germany, 1933-1941", *Journal of Palestine Studies*, Volume 5, Number 3/4, (Spring-Summer, 1976), pp. 54-82, at 59.

[535]. English translation quoted from J. Stachel, "Einstein's Jewish Identity", *Einstein from 'B' to 'Z'*, Birkhäuser, Boston, Basel, Berlin, (2002), pp. 57-83, at 78. Stachel cites M. Besso, A. Einstein, *Correspondance, 1903-1955*, Hermann, Paris, (1972), p. 238.

had "intermarried". Racist Zionist Moses Hess was married to a Christian Gentile prostitute named Sybille Pritsch.

Einstein may have been affected by his mother's early racist opposition to his relationship with Marić. Another factor in the Einsteins' divorce was, of course, Albert's incestuous relationship with his cousin Else Einstein, and his desire to bed her daughters, as well as his general promiscuity. Albert Einstein opposed his sister Maja's marriage to Gentile Paul Winteler on racist grounds, and Albert thought they should divorce. Albert Einstein wrote to Michele Besso on 12 December 1919, "No mixed marriages are any good (Anna says: oh!)"[536] Besso, himself, was married to a Gentile, Anna Besso-Winteler. Denis Brian wrote,

> "When asked what he thought of Jews marrying non-Jews, which, of course, had been the case with him and Mileva, [Albert Einstein] replied with a laugh, 'It's dangerous, but then all marriages are dangerous.'"[537]

On 3 April 1920, Einstein wrote, criticizing assimilationist Jews,

> "And this is precisely what he does *not* want to reveal in his confession. He talks about religious faith instead of tribal affiliation, of 'Mosaic' instead of 'Jewish' because the latter term, which is much more familiar to him, would emphasize affiliation to his tribe."[538]

Albert Einstein often referred to Jews as "tribesmen" and Jewry as the "tribe". Fellow German Jew Fritz Haber was outraged at Albert Einstein's racist treachery and disloyalty. Einstein confirmed that he was disloyal and a racist, and was obligated,

> "[...] to step in for my persecuted and morally depressed fellow tribesmen, as far as this lies within my power[.]"[539]

Einstein bore no such loyalty to Germans, who had feed him and made him famous. In fact, Einstein wanted to exterminate the Germans.

After declaring that Jewish children segregate due to natural forces and that

---

[536]. Letter from A. Einstein to M. Besso of 12 December 1919, English translation by A. Hentschel, *The Collected Papers of Albert Einstein*, Volume 9, Document 207, Princeton University Press, (2004), pp. 178-179, at 179.

[537]. D. Brian, *The Unexpected Einstein: The Real Man Behind the Icon*, Wiley, Hoboken, New Jersey, (2005), p. 42.

[538]. A. Einstein, English translation by A. Engel, *The Collected Papers of Albert Einstein*, Volume 7, Document 34, Princeton University Press, (2002), pp. 153-155, at 153.

[539]. A. Einstein quoted in: H. Gutfreund, "Albert Einstein and the Hebrew University", J. Renn, Editor, *Albert Einstein Chief Engineer of the Universe: One Hundred Authors for Einstein*, Wiley-VCH, Berlin, (2005), pp. 314-318, at 316.

they are "different from other children",[540] not due to religion or tradition, but due to genetic features and "heritage", Einstein continued his 3 April 1920 statement,

> "With adults it is quite similar as with children. Due to race and temperament as well as traditions (which are only to a small extent of religious origin) they form a community more or less separate from non-Jews. [***] It is this basic community of race and tradition that I have in mind when I speak of 'Jewish nationality.' In my opinion, aversion to Jews is simply based upon the fact that Jews and non-Jews are different. [***] Where feelings are sufficiently vivid there is no shortage of reasons; and the feeling of aversion toward people of a foreign race with whom one has, more or less, to share daily life will emerge by necessity."[541]

Einstein made similar comments in a document dated sometime "after 3 April 1920". Einstein was in agreement with Philipp Lenard that a "Jewish heritage" could be seen in intellectual works published by Jews. Einstein stated,

> "The psychological root of anti-Semitism lies in the fact that the Jews are a group of people unto themselves. Their Jewishness is visible in their physical appearance, and one notices their Jewish heritage in their intellectual works, and one can sense that there are among them deep connections in their disposition and numerous possibilities of communicating that are based on the same way of thinking and of feeling. The Jewish child is already aware of these differences as soon as it starts school. Jewish children feel the resentment that grows out of an instinctive suspicion of their strangeness that naturally is often met with a closing of the ranks. [***] [Jews] are the target of instinctive resentment because they are of a different tribe than the majority of the population."[542]

In a draft letter of 3 April 1920, Einstein wrote that children are conscious of "racial characteristics" and that this alleged "racial" gulf between children results in conflicts, which instill a sense of foreignness in the persecuted child. Einstein wrote,

> "Unter den Kindern war besonders in der Volksschule der Antisemitismus lebendig. Er gründete [s]ich auf die den Kindern merkwürdig bewussten

---

540. A. Einstein, English translation by A. Engel, *The Collected Papers of Albert Einstein*, Volume 7, Document 34, Princeton University Press, (2002), pp. 153-155, at 153.
541. A. Einstein, English translation by A. Engel, *The Collected Papers of Albert Einstein*, Volume 7, Document 34, Princeton University Press, (2002), pp. 153-155, at 153-154.
542. A. Einstein, English translation by A. Engel, *The Collected Papers of Albert Einstein*, Volume 7, Document 35, Princeton University Press, (2002), pp. 156-157.

Rassenmerkmale und auf Eindrücke im Religionsunterricht. Thätliche Angriffe und Beschimpfungen auf dem Schulwege waren häufig, aber meist nicht gar zu bösartig. Sie genügten immerhin, um ein lebhaftes Gefühl des Fremdseins schon im Kinde zu befestigen."[543]

Einstein's racism was perhaps a defense mechanism to depersonalize the attacks he faced as a child and to counter the hurt with a sense of communal love, and communal hatred. Like Adolf Stoecker before him,[544] Albert Einstein advocated the segregation of Jewish students. Peter A. Bucky quoted Albert Einstein,

"I think that Jewish students should have their own student societies. [***] One way that it won't be solved is for Jewish people to take on Christian fashions and manners. [***] In this way, it is entirely possible to be a civilized person, a good citizen, and at the same time be a faithful Jew who loves his race and honors his fathers."[545]

Einstein stated,

"We must be conscious of our alien race and draw the logical conclusions from it. [***] We must have our own students' societies and adopt an attitude of courteous but consistent reserve to the Gentiles. [***] It is possible to be [***] a faithful Jew who loves his race and honours his fathers."[546]

On 5 April 1920, Einstein repeated what he had heard from his political Zionist friends, who believed that anti-Semitism was necessary to the preservation of the "Jewish race",

"Anti-Semitism will be a psychological phenomenon as long as Jews come in contact with non-Jews—what harm can there be in that? Perhaps it is due to anti-Semitism that we survive as a race: at least that is what I believe."[547]

and,

---

543. Letter from A. Einstein to P. Nathan of 3 April 1920, *The Collected Papers of Albert Einstein*, Volume 9, Document 366, Princeton University Press, (2004), p. 492. Also: *The Collected Papers of Albert Einstein*, Volume 1, Princeton University Press, (1987), p. lx, note 44.
544. P. W. Massing, *Rehearsal for Destruction: A Study of Political Anti-Semitism in Imperial Germany*, Howard Fertig, New York, (1967), pp. 278-294.
545. P. A. Bucky, Einstein, and A. G. Weakland, *The Private Albert Einstein*, Andrews and McMeel, Kansas City, (1992), p. 88.
546. A. Einstein, *The World As I See It*, Citadel, New York, (1993), pp. 107-108.
547. A. Einstein, English translation by A. Engel, *The Collected Papers of Albert Einstein*, Volume 7, Document 37, Princeton University Press, (2002), p. 159.

"I am neither a German citizen, nor is there in me anything that can be described as 'Jewish faith.' But I am happy to belong to the Jewish people, even though I don't regard them as the Chosen People. Why don't we just let the Goy keep his anti-Semitism, while we preserve our love for the likes of us?"[548]

This letter was published in the *Israelitisches Wochenblatt für die Schweiz*, on 24 September 1920, on page 10. It became famous and was widely discussed in newspapers and was used as a political issue. Einstein's racism had already become a weapon for critics of the Jews to wield against German Jews loyal to the Fatherland. Einstein ridiculed the *Central-Verein deutscher Staatsbürger jüdischen Glaubens*, an organization that combated anti-Semitism and vigorously defended and celebrated Jews, because Einstein sought to promote anti-Semitism and because Einstein believed that being "Jewish" was a racial, not a religious, condition. Einstein knew quite well that the letter had been published. The *C. V.* contacted him about it and published a statement regarding it in their periodical *Im deutschen Reich* in March of 1921,

"So wurde auch in einzelnen Versammlungen der bekannte Brief des Naturforschers Professor Einstein, den dieser an den Central-Verein gerichtet hat, und in welchem er die Bestrebungen des Central-Vereins ablehnt, weil sie zu national-deutsch und zu wenig jüdisch orientiert seien, zum Gegenstand der Erörterungen gemacht. Dieser Brief hat in der öffentlichen Erörterung der jüdischen und judengegnerischen Presse in den letzten Monaten und auch bei den Wahlen eine gewisse Rolle gespielt und Anlaß zu den verschiedenartigsten Betrachtungen je nach der Parteistellung der Versammlungsredner und der verschiedenen Zeitungen gegeben. So hat sich z. B. die jüdisch-nationale „Wiener Morgenzeitung" veranlaßt gesehen, den Central-Verein in wenig vornehmer Weise anzugreifen und ihn wegen seines nationaldeutschen Standpunktes zu verdächtigen. Diese Angriffe würden durch die Auffassung von Professor Einstein nicht gedeckt worden sein, wenn die „Wiener Morgenzeitung" gewußt hätte, daß Professor Einstein ohne nähere Kenntnis der Bestrebungen und der Arbeit des Central-Vereins seinen Brief geschrieben und keineswegs an eine Veröffentlichung, die nur durch eine Indiskretion erfolgt ist, gedacht hat. Erst nach der Veröffentlichung hat er von der Art und Weise der Tätigkeit des Central-Vereins Kenntnis erhalten und hat, wie mit gutem Grund

---

548. A. Einstein quoted in A. Fölsing, English translation by E. Osers, *Albert Einstein, a Biography*, Viking, New York, (1997), p. 494; which cites speech to the *Central-Verein Deutscher Staatsbürger Jüdischen Glaubens*, in Berlin on 5 April 1920, in D. Reichenstein, *Albert Einstein. Sein Lebensbild und seine Weltanschauung*, Berlin, (1932). This letter from Einstein to the Central Association of German Citizens of the Jewish Faith of 5 April 1920 is reproduced in *The Collected Papers of Albert Einstein*, Volume 9, Document 368, Princeton University Press, (2004).

versichert werden kann, infolge dieser Kenntnis eine wesentlich andere Auffassung vom Werte der Arbeit unseres Central-Vereins gewonnen. Auch dieser Vorfall sollte Anlaß geben, Urteile in der Oeffentlichkeit erst dann zu fällen, wenn die Sachlage einigermaßen geklärt ist."[549]

On 24 May 1931, the *Sunday Express* of London published an interview it claimed it had had with Einstein while he was visiting Oxford. The interview contained inflammatory statements similar to those published in the *Israelitisches Wochenblatt für die Schweiz* on 24 September 1920. These statements were repeated in several German language newspapers across Europe together with scathing editorial indictments of Einstein. Einstein claimed that no interview had taken place and the quotations were taken from a letter he had written eleven years prior. Einstein stated in a letter to Michael Traub of 22 August 1931 that this letter had never been published,[550] though it had been published and Einstein knew quite well that it had been published.

Einstein accused the *Central-Verein deutscher Staatsbürger jüdischen Glaubens e. V.* of instigating the "forgery". The *C.V.* denied that it was behind the publication in the *Sunday Express* and invited Einstein to respond in their official organ the *Central-Verein Zeitung*. Einstein took the opportunity and stated, "Es wurden mir schon wiederholt Auszüge aus einem Artikel der „Sunday Expreß" zugesandt, aus denen ich ersehe, daß es sich um eine glatte Fälschung handelt. Ich habe in Oxford überhaupt kein einziges Zeitungsinterview gegeben. Der Inhalt ist eine böswillige Entstellung eines vor elf Jahren geschriebenen, nicht für die Oeffentlichkeit bestimmten Briefes."[551] He affirmed in 1931 that he had made the statements and did not repudiate them.

In 1932, Einstein stated, referring to the "deplorably high development of nationalism everywhere"—his own rabid Zionism excepted,

> "The introduction of compulsory service is therefore, to my mind, the prime cause of the moral collapse of the white race, which seriously threatens not merely the survival of our civilization but our very existence. This curse, along with great social blessings, started with the French Revolution, and before long dragged all the other nations in its train."[552]

Einstein had a reputation as a rabid anti-assimilationist, which is to say that Einstein was a rabid racist segregationist. On 15 March 1921, Kurt Blumenfeld wrote to Chaim Weizmann,

---

549. "Zeitschau", *Im deutschen Reich*, Volume 27, Number 3, (March, 1921), pp. 90-97, at 92.
550. D. K. Buchwald, *et al.*, Editors, *The Collected Papers of Albert Einstein*, Volume 7, Document 37, Princeton University Press, (2002), p. 304, note 8.
551. "Professor Einstein erklärt das „Sunday Expreß"-Interview für gefälscht", *Central-Verein Zeitung*, Volume 10, Number 37, (11 September 1931), p. 443.
552. A. Einstein, translated by A. Harris, "The Disarmament Conference of 1932. I." *The World As I See It*, Citadel, New York, (1993), pp. 59-60.

"Einstein [***] is interested in our cause most strongly because of his revulsion from assimilatory Jewry."[553]

Einstein stated in 1921,

"To deny the Jew's nationality in the Diaspora is, indeed, deplorable. If one adopts the point of view of confining Jewish ethnical nationalism to Palestine, then one, to all intents and purposes, denies the existence of a Jewish people. In that case one should have the courage to carry through, in the quickest and most complete manner, entire assimilation. We live in a time of intense and perhaps exaggerated nationalism. But my Zionism does not exclude in me cosmopolitan views. I believe in the actuality of Jewish nationality, and I believe that every Jew has duties towards his coreligionists. [***] [T]he principal point is that Zionism must tend to strengthen the dignity and self-respect of the Jews in the Diaspora. I have always been annoyed by the undignified assimilationist cravings and strivings which I have observed in so many of my friends."[554]

In 1921, Einstein declared, referring to Eastern European Jews,

"These men and women retain a healthy national feeling; it has not yet been destroyed by the process of atomisation and dispersion."[555]

Einstein wrote in the *Jüdische Rundschau*, on 21 June 1921, on pages 351-352,

"This phenomenon [*i. e.* Anti-Semitism] in Germany is due to several causes. Partly it originates in the fact that the Jews there exercise an influence over the intellectual life of the German people altogether out of proportion to their number. While, in my opinion, the economic position of the German Jews is very much overrated, the influence of Jews on the Press, in literature, and in science in Germany is very marked, as must be apparent to even the most superficial observer. This accounts for the fact that there are many anti-Semites there who are not really anti-Semitic in the sense of being Jew-haters, and who are honest in their arguments. They regard Jews as of a nationality different from the German, and therefore are alarmed at the increasing Jewish influence on their national entity. [***] But in Germany the judgement of my theory depended on the party politics of the

---

[553]. J. Stachel, *Einstein from 'B' to 'Z'*, Birkhäuser, Boston, (2002), p. 79, note 41.
[554]. A. Einstein, "Jewish Nationalism and Anti-Semitism", *The Jewish Chronicle*, (17 June 1921), p. 16.
[555]. J. Stachel, "Einstein's Jewish Identity", *Einstein from 'B' to 'Z'*, Birkhäuser, Boston, (2002), p. 65. Stachel cites, *About Zionism: Speeches and Letters*, Macmillan, New York, (1931), pp. 48-49. For Zionist Ha-Am's use of the image of atomisation and dispersion, *see:* A. Hertzberg, *The Zionist Idea*, Harper Torchbooks, New York, (1959), p. 276.

Press[.]556

Einstein also stated,

"The way I see it, the fact of the Jews' racial peculiarity will necessarily influence their social relations with non-Jews. The conclusions which—in my opinion—the Jews should draw is to become more aware of their peculiarity in their social way of life and to recognize their own cultural contributions. First of all, they would have to show a certain noble reservedness and not be so eager to mix socially—of which others want little or nothing. On the other hand, anti-Semitism in Germany also has consequences that, from a Jewish point of view, should be welcomed. I believe German Jewry owes its continued existence to anti-Semitism."557

Nazi Zionist Joseph Goebbels, sounding very much like political Zionist Albert Einstein, was quoted in *The New York Times*, on 29 September 1933, on page 10,

"It must be remembered the Jews of Germany were exercising at that time a decisive influence on the whole intellectual life; that they were absolute and unlimited masters of the press, literature, the theatre and the motion pictures, and in large cities such as Berlin, 75 percent of the members of the medical and legal professions were Jews; that they made public opinion, exercised a decisive influence on the Stock Exchange and were the rulers of Parliament and its parties."

On 1 July 1921, Einstein was quoted in the *Jüdische Rundshau* on page 371,

"Let us take brief look at the *development of German Jews* over the last hundred years. With few exceptions, one hundred years ago our forefathers still lived in the Ghetto. They were poor and separated from the Gentiles by a wall of religious tradition, secular lifestyles and statutory confinement and were confined in their spiritual development to their own literature, only relatively weakly influenced by the forceful progress which intellectual life in Europe had undergone in the Renaissance. However, these little noticed, modestly living people had one thing over us: *Every one of them belonged with all his heart to a community*, into which he was incorporated, in which he felt a worthwhile member, in which nothing was asked of him which conflicted with his normal processes of thought. Our forefathers of that era were pretty pathetic both bodily and spiritually, but—in social relations—

---

556. A. Einstein, "Jewish Nationalism and Anti-Semitism", *The Jewish Chronicle*, (17 June 1921), p. 16.
557. A. Einstein, A. Engel translator, "How I became a Zionist", *The Collected Papers of Albert Einstein*, Volume 7, Document 57, Princeton University Press, (2002), pp. 234-235, at 235.

in an enviable state of mental equilibrium. Then came emancipation. It offered undreamt of opportunities for advancement. The isolated individual quickly found their way into the upper financial and social circles of society. They eagerly absorbed the great achievements of art and science which the Occidentals[558] had created. They contributed to the development with passionate affection, and themselves made contributions of lasting value. They thereby took on the lifestyle of the Gentile world, turning away from their religious and social traditions in growing masses—took on Gentile customs, manners and mentality. It appeared as if they were being completely dissolved into the numerically superior, politically and culturally better organized host peoples, such that no trace of them would be left after a few generations. The complete eradication of the Jewish nationality in Middle and Western Europe appeared to be inevitable. However, it didn't turn out that way. It appears that racially distinct nations have instincts which work against interbreeding. The adaptation of the Jews to the European peoples among whom they have lived in language, customs and indeed even partially in religious practices *was unable to eliminate all feelings of foreignes* which exist between Jews and their European host peoples. In short, this spontaneous feeling of foreignness is ultimately due to a loss of energy.[559] For this reason, *not even well-meant arguments can eradicate it*. Nationalities do not want to be mixed together, rather they want to go their own separate ways. A state of peace can only be achieved by mutual tolerance and respect."

Einstein stated that Jews should not participate in the German Government,

"I regretted the fact that [Rathenau] became a Minister. In view of the attitude which large numbers of the educated classes in Germany assume towards the Jews, I have always thought that their natural conduct in public should be one of proud reserve."[560]

Einstein merely parroted the Zionist Party line. Werner E. Mosse wrote,

"While the leaders of the CV saw it as their special duty to represent the interests of the German Jews in the active political struggle, Zionism stood

---

558. At the time Einstein made his statement, Jews and Gentiles often referred to Jews as "Orientals".
559. Einstein repeatedly spoke of the Germans as "greedy" to acquire territory and of the "loss of energy" when different "races" attempted to live together. He have been speaking literally. Georg Friedrich Nicolai wrote of the struggle of life to aquire the energy of the sun and he applied this struggle to humanity. G. Nicolai, *Die Biologie des Krieges, Betrachtungen eines deutschen Naturforschers*, O. Füssli, Zürich, (1917); English translation: *The Biology of War*, Century Co., New York, (1918), pp. 36-39, 44-53.
560. R. W. Clarck, *Einstein, the Life and Times*, World Publishing Company, USA, (1971), p. 292. Clarck refers to: *Neue Rundschau*, Volume 33, Part 2, pp. 815-816.

for... systematic Jewish non-participation in German public life. It rejected as a matter of principle any participation in the struggle led by the CV."[561]

In 1925, Einstein wrote in the official Zionist organ *Jüdische Rundschau*,

"By study of their past, by a better understanding of the spirit [Geist] that accords with their race, they must learn to know anew the mission that they are capable of fulfilling. [***] What one must be thankful to Zionism for is the fact that it is the only movement that has given many Jews a justified pride, that it has once again given a despairing race the necessary faith, if I may so express myself, given new flesh to an exhausted people."[562]

On 12 October 1929, Albert Einstein wrote to the *Manchester Guardian*,

"In the re-establishment of the Jewish nation in the ancient home of the race, where Jewish spiritual values could again be developed in a Jewish atmosphere, the most enlightened representatives of Jewish individuality see the essential preliminary to the regeneration of the race and the setting free of its spiritual creativeness."[563]

Einstein's public racism eventually waned, but he continued to publicly express his segregationist philosophy in the same terms as anti-Semites, as well as his belief that Jews "thrived on" and owed their "continued existence" to anti-Semitism.

Einstein stated in December of 1930 to an American audience,

"There is something indefinable which holds the Jews together. Race does not make much for solidarity. Here in America you have many races, and yet you have the solidarity. Race is not the cause of the Jews' solidarity, nor is their religion. It is something else—which is indefinable."[564]

Einstein's confusing public statement perhaps resulted from his desire to promote multi-culturalism in America, which had the benefit of freeing up

---

561. W. E. Mosse, "Die Niedergang der deutschen Republik und die Juden", *The Crucial Year 1932*, p. 38; English translation in: K. Polkehn, "The Secret Contacts: Zionism and Nazi Germany, 1933-1941", *Journal of Palestine Studies*, Volume 5, Number 3/4, (Spring-Summer, 1976), pp. 54-82, at 56-57.
562. English translation by John Stachel in J. Stachel, "Einstein's Jewish Identity", *Einstein from 'B' to 'Z'*, Birkhäuser, Boston, (2002), p. 67. Stachel cites, "Botschaft", *Jüdische Rundschau*, Volume 30, (1925), p. 129; French translation, *La Revue Juive*, Volume 1, (1925), pp. 14-16.
563. J. Stachel, "Einstein's Jewish Identity", *Einstein from 'B' to 'Z'*, Birkhäuser, Boston, (2002), p. 65. Stachel cites, *About Zionism: Speeches and Letters*, Macmillan, New York, (1931), pp. 78-79.
564. A. Einstein quoted in "Einstein on Arrival Braves Limelight for Only 15 Minutes", *The New York Times*, (12 December 1930), pp. 1, 16, at 16.

Jewish immigration to the United States.⁵⁶⁵ Einstein was also likely parroting, or trying to parrot, a fellow anti-assimilationist political Zionist whose pamphlet was well known in America, Solomon Schechter and his *Zionism: A Statement*, Federation of American Zionists, New York, (1906), in which Schechter states, among other things, "Zionism is an ideal, and as such is indefinable."

Einstein stated in 1938,

"JUST WHAT IS A JEW?"

The formation of groups has an invigorating effect in all spheres of human striving, perhaps mostly due to the struggle between the convictions and aims represented by the different groups. The Jews, too, form such a group with a definite character of its own, and anti-Semitism is nothing but the antagonistic attitude produced in the non-Jews by the Jewish group. This is a normal social reaction. But for the political abuse resulting from it, it might never have been designated by a special name.

What are the characteristics of the Jewish group? What, in the first place, is a Jew? There are no quick answers to this question. The most obvious answer would be the following: A Jew is a person professing the Jewish faith. The superficial character of this answer is easily recognized by means of a simple parallel. Let us ask the question: What is a snail? An answer similar in kind to the one given above might be: A snail is an animal inhabiting a snail shell. This answer is not altogether incorrect; nor, to be sure, is it exhaustive; for the snail shell happens to be but one of the material products of the snail. Similarly, the Jewish faith is but one of the characteristic products of the Jewish community. It is, furthermore, known that a snail can shed its shell without thereby ceasing to be a snail. The Jew who abandons his faith (in the formal sense of the word) is in a similar position. He remains a Jew.

[***]

WHERE OPPRESSION IS A STIMULUS
[***]

Perhaps even more than on its own tradition, the Jewish group has thrived on oppression and on the antagonism it has forever met in the world. Here undoubtedly lies one of the main reasons for its continued existence through so many thousands of years."⁵⁶⁶

Albert Einstein was parroting racist political Zionist leader Theodor Herzl, who wrote in his book *The Jewish State*,

---

565. E. A. Ross, *The Old World in the New: The Significance of past and Present Immigration to the American People*, Century Company, New York, (1914), p. 144.

566. A. Einstein, "Why do They Hate the Jews?", Collier's, Volume 102, (26 November 1938); reprinted in *Ideas and Opinions*, Crown, New York, (1954), pp. 191-198, at 194, 196. Einstein expressed himself in a similar way to Peter A. Bucky, P. A. Bucky, Einstein, and A. G. Weakland, *The Private Albert Einstein*, Andrews and McMeel, Kansas City, (1992), p. 87.

"Oppression and persecution cannot exterminate us. No nation on earth has survived such struggles and sufferings as we have gone through. Jew-baiting has merely stripped off our weaklings; the strong among us were invariably true to their race when persecution broke out against them. This attitude was most clearly apparent in the period immediately following the emancipation of the Jews. Later on, those who rose to a higher degree of intelligence and to a better worldly position lost their communal feeling to a very great extent. Wherever our political well-being has lasted for any length of time, we have assimilated with our surroundings. I think this is not discreditable. Hence, the statesman who would wish to see a Jewish strain in his nation would have to provide for the duration of our political well-being; and even Bismarck could not do that. [***] The Governments of all countries scourged by Anti-Semitism will serve their own interests in assisting us to obtain the sovereignty we want. [***] Great exertions will not be necessary to spur on the movement. Anti-Semites provide the requisite impetus. They need only do what they did before, and then they will create a love of emigration where it did not previously exist, and strengthen it where it existed before. [***] I imagine that Governments will, either voluntarily or under pressure from the Anti-Semites, pay certain attention to this scheme; and they may perhaps actually receive it here and there with a sympathy which they will also show to the Society of Jews."[567]

Einstein's statements and those of other like-minded racist Zionists threw fuel on the fire and were reflective of the spirit and tone enunciated in *Protocols of the Learned Elders of Zion*, Number 9, which states (no matter who wrote it),

"Nowadays, if any States raise a protest against us, it is only *pro forma* at our discretion, and by our direction, for their anti-Semitism is indispensable to us, for the management of our lesser brethren."[568]

Many Zionist leaders espoused racist nationalism, which made them the darlings of the Nazis, the Nazis they had put into power. Joachim Prinz wrote, among other things, a racist polemic against assimilation in his book published in Germany in the Hitler-era, *Wir Juden* of 1934,

"Die Theorie der Assimilation ist zusammengebrochen. Kein Schlupfwinkel birgt uns mehr. Wir wünschen an die Stelle der Assimilation das Neue gesetzt: *das Bekenntnis zur jüdischen Nation und zur jüdischen Rasse*. Ein Staat, der aufgebaut ist auf dem Prinzip der Reinheit von Nation und Rasse, kann nur vor dem Juden Achtung und Respekt haben, der sich

---

[567]. T. Herzl, *A Jewish State: An Attempt at a Modern Solution of the Jewish Question*, The Maccabæan Publishing Co., New York, (1904), pp. 5-6, 25, 68, 93.
[568]. L. Fry, *Waters Flowing Eastward: The War Against the Kingship of Christ*, TBR Books, Washington, D. C., (2000), p. 137.

zur eigenen Art bekennt. Nirgendwo kann er in diesem Bekenntnis mangelnde Loyalität dem Staate gegenüber erblicken. Er kann keine anderen Juden wollen, als die Juden des klaren Bekenntnisses zum eigenen Volk. Er kann keine liebedienerischen, kriecherischen Juden wollen. Er muß von uns das Bekenntnis zur eigenen Art fordern. Denn nur jemand, der *eigene* Art und *eigenes* Blut achtet, wird den Respekt vor dem *nationalen Wollen anderer Nationen* haben können.

In dem Bekenntnis des Juden zu seiner eigenen Nation, in seiner Gewißheit, in sich sein eigenes Blut zu tragen, seine eigene Vergangenheit und seine eigene Art — wird er erst beginnen, die Distanz vor den Erlebnissen der anderen Nationen zu wahren, die notwendig ist, um ein ehrliches Miteinander und eine anständige Nachbarschaft zu halten. In dem Augenblick, in dem dieses Bekenntnis zur jüdischen Nationalität die Mehrheit der Judenheit ergreift, beginnt *die erste ehrliche Aussprache zwischen Juden und Nichtjuden.*"[569]

Prinz wrote of the supposed suicide of the emancipated Jews through assimilation in *liberal* states, and he despised liberalism. His goal was to preserve the alleged purity of the Jewish race in a Jewish nation, *i. e.* the expulsion of the Jews to a new territory which allowed the Zionists to enforce racial segregation. Prinz wrote,

"The brochure of the baptized Jew Karl *Marx* on the Jewish question is an anti-Jewish pamphlet and an autobiographical entry in the chapter of Jewish self-hatred."

"Die Broschüre des getauften Juden Karl *Marx* über die Judenfrage ist ein antijüdisches Pamphlet und ein autobiographischer Beitrag zum Kapitel des jüdischen Selbsthasses."[570]

Prinz was not alone in his condemnation of Karl Marx's anti-Semitism.[571] Hitler and Prinz had much in common. Lenni Brenner documents Prinz' and the Zionists' *kinship* with the Nazis' nationalistic and racial views in his book

---

[569]. J. Prinz, *Wir Juden*, Erich Reiss, Berlin, (1934), pp. 154-155.
[570]. J. Prinz, *Wir Juden*, Erich Reiss, Berlin, (1934), p. 44.
[571]. E. Bernstein, "Jews and German Social Democracy", *Die Tukunft* (New York), Volume 26, (March, 1921), pp. 145ff.; English translation in: P. W. Massing, *Rehearsal for Destruction: A Study of Political Anti-Semitism in Imperial Germany*, Howard Fertig, New York, (1967), pp. 322-330. On Marx's alleged "self-hatred", *see:* H. Hirsch, "The Ugly Marx: Analysis of an 'Outspoken Anti-Semite'", *Philosophical Forum*, Volume 8, (1978), pp.150-162. *See also:* P. L. Rose, *Revolutionary Antisemitism in Germany from Kant to Wagner*, Princeton University Press, (1990), pp. 296-305. *See also:* R. Grooms, "The Racism of Marx and Engels", *The Barnes Review*, Volume 2, Number 10, (October, 1996), pp. 3-8. Communists have always been opportunistic Jew baiters.

Zionism in the Age of the Dictators.[572]

Dietrich Bronder and Hennecke Kardel[573] state that the top leadership of the Nazi Party and the orchestrators of the "final solution" were of Jewish descent, including Adolf Hitler,[574] Adolf Eichmann, Reinhard Heydrich, Rudolf Hess (member of the *Thule-Gesellschaft*, an organization Zionists created to promote anti-Semitism in order to force Jews to accept Zionism), Dietrich Eckart (member of the *Thule-Gesellschaft*), Alfred Rosenberg (member of the *Thule-Gesellschaft*), Julius Streicher (member of the *Thule-Gesellschaft*), Joseph Goebbels, and Hans Frank (member of the *Thule-Gesellschaft*). Dietrich Bronder wrote in 1964,

> "Aus den eigenen Untersuchungen des Verfassers über die führenden Nationalsozialisten sei hier nur mitgeteilt, daß sich unter 4000 Männern der Reichsführung 120 Ausländer von Geburt befanden, viele mit einem oder zwei Elternteilen ausländischer Herkunft und ein Prozent sogar jüdischer Abkunft — also im Sinne der NS-Rassengesetze „untragbar''.
> a)　So rechnen zu den Auslandsgeborenen:
> Reichsminister und Führerstellvertreter Rudolf Heß (Ägypten); Reichsminister Darré (Argentinien); Gauleiter und Staatssekretär E. W. Bohle und der Reichskommissar Herzog von Sachsen-Coburg (England); Generaloberst Löhr (Jugoslawien); General der Waffen-SS Phleps (Rumänien); Reichsärzteführer und Staatssekretär Dr. Conti und der Berliner Oberbürgermeister Lippert (Schweiz); NSKK-Obergruppenführer G. Wagner (Frankreich); sowie aus Rußland: Reichsminister und Reichsleiter Alfred Rosenberg und die NS-Reichshauptamtsleiter Brockhausen, Dr. von Renteln und Schickedanz, Reichsminister Backe, Präsident Dr. Neubert, Staatsrat Dr. Freiherr von Freytag-Loringhoven und Bischof J. Beermann.

---

572. See especially Chapter 5: L. Brenner, *Zionism in the Age of the Dictators*, Croom Helm, London, L. Hill, Westport, Connecticut, (1983).

<http://www.aaargh-international.org/engl/zad/zad.html>

573. D. Bronder, *Bevor Hitler kam: Eine historische Studie*, Hans Pfeiffer Verlag, Hannover, (1964), p. 204 (p. 211 in the 1974 edition). H. Kardel, *Adolf Hitler, Begründer Israels*, Verlag Marva, Genf, (1974); English translation *Adolf Hitler: Founder of Israel*, Modjeskis' Society Dedicated to Preservation of Cultures, San Diego, (1997).

574. H. Koehler, *Inside the Gestapo: Hitler's Shadow Over the World*, Pallas Pub. Co. Ltd., London, (1940). **See aslo:** H. Frank, *Im Angesicht des Galgens; Deutung Hitlers und seiner Zeit auf Grund eigener Erlebnisse und Erkenntnisse. Geschrieben im Nürnberger Justizgefängnis*, F. A. Beck, München-Gräfelfing, (1953), pp. 330-331. **See aslo:** D. Bronder, *Bevor Hitler kam: Eine historische Studie*, Hans Pfeiffer Verlag, Hannover, (1964), p. 204 (p. 211 in the 1974 edition). **See aslo:** H. Kardel, *Adolf Hitler, Begründer Israels*, Verlag Marva, Genf, (1974); English translation *Adolf Hitler: Founder of Israel*, Modjeskis' Society Dedicated to Preservation of Cultures, San Diego, (1997).

b) Darüber hinaus stammten von einem oder beiden ausländischen Elternteilen (u. v. a.):
Der Reichsjugendführer Baldur von Schirach, Generaloberst Rendulic sowie der Generaldirektor Gustav Krupp von Bohlen-Halbach.
c) Selbst jüdischer Abkunft bzw. mit jüdischen Familien verwandt waren: der Führer und Reichskanzler Adolf Hitler; seine Stellvertreter, die Reichsminister Rudolf Heß und Reichsmarschall Hermann Göring; die Reichsleiter der NSDAP Gregor Strasser, Dr. Josef Goebbels, Alfred Rosenberg, Hans Frank und Heinrich Himmler; die Reichsminister von Ribbentrop (der mit dem berühmten Zionisten Chaim Weizmann, dem 1952 verstorbenen ersten Staatsoberhaupt von Israel, einst Brüderschaft getrunken hatte) und von Keudell; die Gauleiter Globocznik (der Judenvernichter), Jordan und Wilhelm Kube; die hohen SS-Führer und z. T. in der Judenvernichtung tätigen Reinhard Heydrich, Erich von dem Bach-Zelewski und von Keudell II; die Bankiers und alten Förderer Hitlers vor 1933 Ritter von Stauß (Vizepräsident des NS-Reichstages) und von Stein; der Generalfeldmarschall und Staatssekretär Milch, der Unterstaatssekretär Gauß; die Physiker und Alt-Pg.'s Philipp von Lenard und Abraham Esau; die Uralt-Pg.'s Hanffstaengel (NS-Auslandspressechef) und Prof. Haushofer (s. S. 190)."[575]

Inferences can be drawn that these crypto-Jewish Nazi leaders were either motivated by self-hatred, or they were front men under the control of Herzlian political Zionists. Both may have been true of the genocidal Nazi Party leaders. Bryan Mark Rigg estimates the total number of Jewish soldiers and sailors in the Nazi military perhaps ranged upwards to 150,000.[576]

Many Zionists hated themselves and Jews in general and defamed Jews in their literature, especially the relatively impoverished and uneducated Jews of the East, whom the Zionists tried to bribe into migrating to Palestine, though they only largely succeeded in capturing ne'er-do-wells. Herzl considered himself to be a sleazy ultra-Jew in the most pejorative sense of which he could conceive to use the term "Jew". Herzl justified himself by generalizing his character flaws as if they were a racial "Jewish" trait. He hated the masses of poor Jews from the East and the rich Jews of the West, who wanted to assimilate.

In 1845, *The North American Review* wrote of the snobbish class hatred

---

575. D. Bronder, *Bevor Hitler kam: Eine historische Studie*, Hans Pfeiffer Verlag, Hannover, (1964), pp. 203-204 (pp. 210-211 in the 1974 edition).

576. "Who Were Hitler's Jewish Soldiers", *The Jewish Chronicle*, (6 December 1996), p. 1. *See also:* W. Hoge, "Rare Look Uncovers Wartime Anguish of Many Part-Jewish Germans", *The New York Times*, (6 April 1997), p. 16. *See also:* B. M. Rigg, *Hitler's Jewish Soldiers: The Untold Story of Nazi Racial Laws and Men of Jewish Descent in the German Military*, University Press of Kansas, Lawrence, Kansas, (2002); **and** *Rescued from the Reich: How One of Hitler's Soldiers Saved the Lubavitcher Rebbe*, Yale University Press, New Haven, (2004).

common among Jews, the inter-Jewish racism which has long plagued Jews, and the various dogmatic Jewish sects hatefully at odds with one another (note the misogyny and dogmatic indoctrination of Jews, which continues to this day,[577] and which manifests itself in, among other things, virulent Jewish censorship of others),

> "As the Jews were anciently divided into several religious sects,—the Pharisees, the Sadducees, the Essenes,—so we find them distinguished at the present day. Their chief modern denominations, some of which represent the more ancient, are the Caraites, the Zabathaites, the Chasidim, the Rabbinists, or Talmudists, and the Reformed Jews. The Samaritans
>
> [*Footnote:* Mixed descendants of a remnant of the ten tribes left in their own land, and of the Assyrians colonized among them. 2 Kings, xvii. 24, &c. In Christ's time they had a temple on Mount Gerizim, which they held more sacred than Mount Zion and its temple. They receive only the Pentateuch, and perhaps the Books of Joshua and Judges, which are found among them; but confidently wait for the Messiah, and observe the Mosaic laws more strictly than even the Jews. Wolff found fifty families of them at the foot of Gerizim, and they have also been met with in other parts of Palestine and in Egypt.]
>
> are not to be classed among them, though akin to them in many respects. The main point of difference between most of these sects, though not the only one, respects the Talmud. The *Talmud*—a word that means *doctrine*—is a voluminous work of two parts,—the *Mishna*, that is the *second law*, and the *Gemara*, or *completion*. The former, consisting of a divine interpretation of the written law, say the Talmudists, was given to Moses at the same time with that delivered on Mount Sinai, together with rules for its exegesis, all to be orally handed down; and by him it was made known to the whole people, and specially committed to his successors. These traditions were collected in the Mishna, a work ascribed to Judah Hannasi,—the Holy, as he is usually called,—about the middle of the second century. Many glosses upon this text soon accumulated, which the Rabbi Jochanan, about the year 230, threw together in the form of a perpetual commentary upon it, entitled the Gemara; and this, with the Mishna, is called the Jerusalem Talmud; though sometimes the Mishna, and sometimes the Gemara alone, is, by synecdoche, called the Talmud. About a century later, Ashi and Abhina, distinguished Babylonian rabbins, compiled a much larger collection of opinions, which, with the Mishna, is styled the Babylonian Talmud, a work held in much higher esteem than the other, and generally understood when the Talmud, without further specification, is mentioned. It has commonly

---

[577]. E. Kaye, *The Hole in the Sheet: A Modern Woman Looks at Orthodox and Hasidic Judaism*, L. Stuart Inc., Secaucus, New Jersey, (1987).

been published in twelve large folios. The other is comprised in a single folio. The Talmud has been justly described as 'containing things frivolous and superstitious, impieties and blasphemies, absurdities and fables.' As an example of all these in one,—God is represented as having contracted impurity by the burial of Moses, and as washing in fire to cleanse himself. These traditions, many of them the same by which, in Christ's time, the Jews 'made the commandment of God of none effect,' since then, in accumulated instances, have been used to destroy the force of the Old Testament Scriptures; which, indeed, Rabbinists consider of very little importance.

[***]

Rabbinism is the Catholic faith, from which all these sects are, in modern phrase, dissenters. It is the lineal descendant of Pharisaism, and distinguished by its blind adherence to the Talmud. The estimation in which strict Rabbinists hold this book is unbounded. 'He that has learned the Scripture, and not the Mishna,' says the Gemara, 'is a blockhead.' Isaac, a distinguished rabbi, says, 'Do not imagine that the written law is the foundation of our religion, which is really founded on the oral law.' The Rabbinical doctrine is, ' The Bible is like water, the Mishna like wine, and the Gemara like spiced wine.' [*Soferim* 13*b*] Some even say, that 'to study the Bible is but a waste of time.' [*Baba Mezia* 33*a*] For strict Rabbinism, a melancholy compound of superstition and fanaticism, we must look to Poland, Russia, Hungary, and Palestine, of which we speak, in describing the system. In those countries, the Rabbinists, or Talmudists, discountenance as profane all other study than that of the Bible and Talmud, but are very careful to educate their sons in their religious lore. The Talmud forbids teaching females more than their appropriate domestic arts. 'Whoever instructs his daughter in the Bible is as if he instructed her in abominations.' But it is a disgrace, if boys are not taught to read the Hebrew Bible. The rich provide teachers for their own children, and either permit the poorer to share this provision, or aid them in obtaining masters. So honorable is the office of teacher made, that a bare support is enough generally to secure a competent one. The ordinary method of instruction is very simple. The child, when four years old, is taught the Hebrew letters, and then to pronounce words, the meaning of which he afterwards learns from his tutor; and thus proceeds, without grammar or dictionary, until he can translate the Pentateuch with tolerable ease. Then he begins at Genesis to study exegetically, surrendering his mind, however, entirely to the guidance of some Jewish commentator; and, from first to last, never forming an independent judgment, but implicitly following tradition, and of course never detecting its gross perversions of the Bible. Some stop short of this commentary, with which others conclude their education; while others still, whose parents can afford it, especially if they display quickness in study and fondness for it, pass on to the Talmud,—first the Mishna, then the Gemara, each with its rabbinical commentaries. As an evidence of the ardor sometimes manifested in these studies, and of complete devotion to them, we are told, that a traveller, some years ago, met three young educated rabbins, who 'were born and lived to manhood in the middle of Poland, and yet knew not one word of its language.' A Jewish youth, distinguished for proficiency in Talmudical learning, is anxiously sought in marriage for the daughters of wealthy parents; who look not only at the certain honor of such an alliance, but also at the chance, thus increased, of the Messiah's coming in their line. On the other hand, the Talmud designates by the name of 'people of the land,' equivalent to *peasantry*, those educated in the Bible alone, or not at all; and represents them as an inferior class, fit only for servile labor, with whom others may not intermarry; applying Deut. xxvii. 21,—'Cursed be he that lieth with any manner of beast.' Indeed, the Talmud authorizes every species of oppression towards such, giving them the hope of heaven only if they submit. The Jewish 'peasant' is a servant of servants, ground down by those who have learned, by being oppressed, the art of oppression. In Russia and Poland, where the Jews collect the government taxes among themselves, the rabbins make the peasantry pay nearly the whole. This class, too, where the Jews regulate the conscription, must furnish all the soldiers required.

Some other characteristics of the strict Rabbinists may be briefly noticed. They are the lowest of the Jews in point of morals, and this is sufficiently accounted for by the gross immorality of many Talmudical precepts. On the great yearly *Day of Atonement*, complete absolution from all past sins is pronounced, and from all religious vows, bonds, and oaths taken since the last preceding, and until the next, atonement. This latter absolution, contained in a prayer denominated *col nidre*, being supposed by Christians to extend to all oaths and obligations, civil as well as religious, which the Jews deny, has caused them much trouble in some parts of Europe. The Talmud teaches, moreover, that no respect is due to a Gentile's, or an unlearned Jew's, rights of property; and it accumulates other abominable doctrines, too numerous, and some of them too vile, to mention. Indeed, the modern Rabbinical Jews are generally, in practice, superior to the precepts of the Talmud. They believe in a purgatory, and pray for the souls of the dead; they hold that all Hebrews will rise in the Holy Land, those dying elsewhere rolling painfully under ground until they reach that soil; and that 'all Israel hath part in eternal life.' The dead buried in the Holy Land are expected to be the first to rise in the Messiah's day; [*Kethuboth* 111*a-b*. *Yerushalmi Kilayim* 9, 3. *Ezekiel* 37. *Genesis* 50:25] and so strong has been the desire of burial there, that in the seventeenth century large quantities of Jewish bones were yearly sent thither to be interred. Ship-loads of this melancholy freight might often be seen at Joppa. They believe that a council properly constituted is infallible, and practically, by their implicit confidence in the Talmud, they make the ancient rabbins their 'fathers.' They place a high estimate on the merits of good works, especially those of a ceremonial kind. Thus, though the reading of the Bible is considered hardly a good act, and even as a positive waste of time, the act of taking out the Pentateuch from its depository in the synagogue, and the duty of standing on the left side of the reader, and of closing and removing the roll after service, are considered highly meritorious, and the privilege of performing them is often sold to the highest bidder. A pilgrimage to the Holy Land, much more to pass one's life there, is a superlative merit. They place great confidence in the supererogatory merits of their ancient saints, especially of Abraham, Isaac, and Jacob, for the males, and of Sarah, Rebekah, and Leah, for females. They have daily morning and evening prayer in the synagogue; and all the prayers for public and private devotion are prescribed, and in Hebrew; for the Talmud affirms, that the angels who receive them understand no other language. Women, servants, and children under twelve years of age, are not required to observe the hours of prayer. The Jews of the Holy Land are, perhaps, singular in praying to saints, and honoring and even worshipping relics. They never approach the supposed stones of the temple, some of which are much worn by kissing, without removing their shoes. Every spot where a saint is supposed to be buried is a place of prayer and pilgrimage. The Talmudists do not allow women to attend the synagogue, until they are married; and then, in Poland, Russia, and the East, they occupy a separate apartment.

Public worship among the Talmudical Jews is, for the most part, where the civil power has not interfered, very irreverent and disorderly. A missionary at Beyroot saw comfits thrown among the people in the synagogue, when particular portions of the service were read, *to show the sweetness of the law!* and the audience—some of the adults, and all the boys—tumbling over one another in the scramble for them on the floor. The Talmud declares, that, in observing the feast of Purim, 'Every man must get so drunk, that he cannot distinguish between the phrases, *Blessed be Mordecai*, and *Cursed be Haman*.' While the Talmud imposes many burdensome ceremonies in addition to the Mosaic institutions, it also furnishes multiplied expedients for lightening the latter; and a fertile ingenuity, newly exercised for each emergency, or perpetuated in legendary rules, has extended the dispensation to every desirable point. Stephens, in his travels in the Holy Land, lodged with a Jew, who would not suffer a lamp, lighted the day before, to be extinguished on the Sabbath; but 'described an admirable contrivance he had invented for reconciling appetite with duty;—an oven, heated the night before to such a degree, that the process of cooking was continued during the night, and the dishes were ready when wanted on the Sabbath.' Yet even the Talmudical Jews are generally superior in morals to their Christian neighbours, especially in the point of female purity. No wonder they hate the New Testament, reading it only through the profligate and intolerant conduct of their persecutors.

Hospitality and alms-giving to their brethren are sacred duties among all the Jews. A large majority of those in Palestine are paupers, and, for their support, contributions, averaging fourteen thousand dollars a year, are made in different parts of Europe, deposited at Amsterdam, and thence transmitted to Beyroot. Jerusalem, Hebron, Tiberias, and Saphet are holy cities in Jewish esteem, and in all the Italian synagogues money-boxes are kept, marked, 'For Jerusalem,' 'For Saphet,' &c. The largest collections are in

Amsterdam. Leghorn sends about four thousand dollars. But the poor unlearned Jews of Palestine are greatly oppressed by the rabbins, and generally defrauded, wholly or in part, of their share in these charities. When the Hebrew quarter at Smyrna was destroyed by fire, in 1841, Mr. Rothschild, of Vienna, gave 20,000 francs for the sufferers. He and his brothers have lately offered 100,000 francs for founding a Jewish hospital at Jerusalem. Sir Moses Montefiori, during his late visit to Palestine, contributed munificently to the wants of his poor brethren there."[578]

Lenni Brenner cites numerous examples of defamations against Jews by the Jewish Zionists Maurice Samuel, Ben Frommer, Micah Yosef Berdichevsky, Yosef Chaim Brenner[579] and Aaron David Gordon.[580] One could add Theodor Herzl's, Berl Katzenelson's[581] and Vladimir Jabotinsky's[582] names to the list. Mussolini called Jabotinsky a "Jewish Fascist" and David Ben-Gurion found Adolf Hitler's writings reminiscent of Jabotinsky's.[583] Lenni Brenner wrote, quoting Chaim Greenberg,

"In March 1942 Chaim Greenberg, then the editor of New York's Labour Zionist organ, *Jewish Frontier*, painfully admitted that, indeed, there had been:

> a time when it used to be fashionable for Zionist speakers (including the writer) to declare from the platform that 'To be a good Zionist one must be somewhat of an anti-Semite'... To this day Labor Zionist circles are under the influence of the idea that the Return to Zion involved a process of purification from our economic uncleanliness. Whosoever doesn't engage in so-called 'productive' manual labor is believed to be a sinner against Israel and against mankind."[584]

---

578. "The Modern Jews", *The North American Review*, Volume 60, Number 127, (April, 1845), pp. 329-368, at 353-354, 357-361.
579. J. H. Brenner, "Self-Criticism", in A. Hertzberg, *The Zionist Idea*, Harper Torchbooks, New York, (1959), pp. 307-312. Brenner cites Mendele Moher Sefarim's works (Mendele's real name was Shalom Jacob Abramowitz).
580. A. D. Gordon, *Kitve A. D. Gordon*, In Five Volumes, Tel-Aviv, ha-Va'ad ha-merkazi shel mifleget ha-Po'el ha-tsa'ir, (1927-1930), parts translated to English in A. Hertzberg, *The Zionist Idea*, Harper Torchbooks, New York, (1959), pp. 371-386, *see especially* pp. 372, 376, 377, 379.
581. B. Katzenelson, *Ba-Mivhan*, Tel-Aviv, (1935); parts translated to English in A. Hertzberg, *The Zionist Idea*, Harper Torchbooks, New York, (1959), pp. 390-395, *see especially* pp. 390-391.
582. V. Jabotinsky, *Evidence Submitted to the Palestine Royal Commission*, London, (1937), pp. 10-29; in A. Hertzberg, *The Zionist Idea*, Harper Torchbooks, New York, (1959), pp. 559-570, 1t 560-561.
583. M. Bar-Zohar, *Ben-Gurion: A Biography*, Delacorte Press, New York, (1978), p. 67.
584. L. Brenner, *Zionism in the Age of the Dictators*, Chapter 2, Croom Helm, London, L. Hill, Westport, Connecticut, (1983), p. 24. *Brenner cites*: C. Greenberg, "The Myth of

Martin Luther accused the Jews of not sharing the societal burden of manual labor. The ancients also made such accusations against the Jews. Zionists like Theodor Herzl emphasized that Jews must engage in manual labor in their proposed segregated society, so that their gene pool would not be corrupted by foreign laborers, and Herzl stressed his assertions that the poor Jews of Galicia and Eastern Europe were well accustomed to manual labor. Echoing the charges of anti-Semites, Herzl and other Zionists publicly accused the Jews of being "parasites"—to use their term—and the Zionists wanted Jews to take up farming and manual labor allegedly so as to cease to be "parasites". They had other ulterior motives. Racists, and there was no one more racist than the Zionists, had long argued that conquered peoples exacted a vengeance of the vanquished by outbreeding, and by overwhelming the cultures of nations which used them as slave labor. The Zionists wanted to avoid any such occurrence by using exclusively Jewish labor in the "Jewish Homeland". They also wanted to strengthen their gene pool, which they believed had been weakened by the ghetto system and urbanization. In addition, in the early 1920's, some, like Lord Sydenham, complained that the Zionists were exporting Jews of poor character from the East to Palestine, people who were not fit for, nor skilled in, the farmwork that was needed in Palestine. Beyond this, Jewish laws forbids non-Jews to live in Jerusalem, even in Greater Israel, and the racist Jews needed a Jewish slave labor force of Eastern European Jews to build them a new nation without violating Jewish law.

Indeed, one of the first objectives of the Jewish Bolshevists was to train Eastern European Jews to farm and perform the trades. A. Borisow wrote in *The Jewish Chronicle* on 22 September 1922 on page 16,

"'Nep and the Jews.

A New Element in Soviet Russia.

BY A. BORISOW.

A new persecutor has arisen to plague our long-suffering Russian Jewry in the form of the New Economic Policy, familiarly known in Russia as the 'Nep.'

Most people will look up in surprise when they hear me describe the 'Nep,' the far-famed and much-heralded New Economic Policy of the Soviet Government, as a persecutor. For does not 'Nep' mean the renunciation of the Communist illusions, liberation from the bureaucratic Soviet institutions, the reintroduction of trade into the country, the circulation of money, the right of possession of land and factories? All that

---

Jewish Parasitism", *Jewish Frontiers*, (March, 1942), p. 20. Brenner also refers to Yehezkel Kaufman, "Hurban Hanefesh: A Discussion of Zionism and Anti-Semitism", *Issues*, (Winter, 1967), p.106.

is surely a blessing to the Jewish population, mainly an urban and commercial element, and yet I stigmatise it as a persecutor!

Still I repeat that the 'Nep' in Russia is a persecutor of the Jews. During the whole of the last two years the Jews have not suffered economically so much as they have during the few months since the introduction of the 'Nep.' It is not for nothing that the Jews translate the initials of the 'Nep' as the 'Nestchastnaja' ('luckless') Economic Policy.

What is it that the 'Nep' has brought us?

To begin with, it has reduced the number of officials. Many of the Soviet institutions have been closed down. In most of the others, 50 to 60 per cent. of the staff has been dismissed. Viewed on its merits, this is most welcome. It will mean a decrease in the heavy taxation which went to keep all these officials. But for the Jewish population it is a terrible blow. It is no secret that the Soviet institutions, especially in the cities, were staffed almost entirely by Jews. About three-quarters of the total number of officials were Jews. Tens of thousands of Jewish intellectuals and semi-intellectuals, lawyers, journalists and doctors, managed to earn a crust of bread in the service of the Soviet institutions. They formed the majority of the lettered population. Now they are dismissed, driven out into the streets, condemned to unemployment and to starvation. That is the first blessing which the 'Nep' has brought to the Jews.

Trade in Russia has again become free. People are allowed to exchange commodities, to buy and sell. As that was the usual occupation of the majority of the Jews in pre-war Russia, it should be an excellent thing for the Jews. There is no need now to fear that the 'Cheka' will come down on the traders and have them shot for speculation.

But what is the result?

The reintroduction of trade has meant the annihilation of everything that has been done to foster productive work among the Jews. During the four years of Communism in Russia, the foundations of the old economic order were undermined.

With fire and sword the Communists wiped out every trace of trading in the country. They put a stop to what they called 'speculation.' The 'Cheka' drove our *Luftmenschen* by the fear of death into productive work. No one imagined there would ever be a return to the old conditions. Lest they died of hunger, they were compelled to adapt themselves to the new conditions. They learned some kind of handicraft, or they took to agriculture. Productive co-operatives sprang up in the towns. The younger generation, especially, took to establishing agricultural co-operatives. Thousands of young men and women joined the *Hechaluz*, joined together in a rigid discipline in order to take up agriculture as their life work. The Jewish population, under compulsion, became if not proletarianised, in the sense of becoming a factory population, at least labourised—engaged in direct labour. They did productive work instead of engaging in barter.

And now the 'Nep' has come, and stamped out all these hopeful signs, put a stop to all this new endeavour which has meant so much adaptation and hard work. It has killed the co-operatives, and the *Hechaluz* groups. People have left their handicrafts, their agricultural work, and they have again started their small trading—not only those who were traders in the pre-Revolution period, but also people who had never before in their lives had anything to do with commerce or barter. Men who were intellectuals, lawyers, writers, Government officials, have hailed the 'Nep' as the liberator. People who had grown tired of hungering, who had sold their last garment in order to get a dry crust of bread, who could no longer stand being herded together, ten of them in an unheated cellar, have become drunk with the lust of making money. Hundreds of millions of roubles, they heard, could be made by engaging in trade. So they went into trade. They are 'Nepping.'

It would be ridiculous to blame anyone for that. All we can do is to deplore it. But we must regret that the forced and unwilling, yet nevertheless healthy work of transition of a large part of the Jewish population to productive work has been brought to nothing.

If the 'Nep' at least provided the people with the means of livelihood, if those who have thrown away their handicraft and their agriculture, in order to engage in trade improved their economic position, there would be some sort of justification even for the loss.

Business in Russia to-day is conducted by the million. The slightest transaction involves tens of millions. Where are the people to obtain these huge initial sums with which to start their businesses? Nobody had any money. Most of those who have started in business have sold their effects down to their very last plate or spoon in order to get some sort of a starting capital. They buy up goods for several tens of millions and they sell them again at some hundreds per cent. profit. Splendid, it seems at first sight. But in the interval which the transaction takes to complete, the rate of exchange has generally fallen to such an extent that the total sum realised buys less than the original sum had purchased. Sixty million roubles, for example, to-day buy about as much as 20 to 25 millions bought a short while ago. Nominally, the 'Nep' man has become richer. Actually, he has become poorer.

There is an anecdote in circulation among these 'Nep' people which will serve as an apt illustration.

Somebody bought in the Urals five waggon-loads of nails, brought them to Moscow and sold them at a tremendous profit. He went back again to the Urals, but this time he was unable to get more than three waggon-loads of nails for his money. He came back to Moscow, sold them again at an immense profit, and went back to the Urals. This time he managed to buy no more than one waggon-load of nails. And so it went on and on, until at last he went to the Urals with a simply colossal sum, but all he could get for it was just one nail. So he took that nail and hitched a rope to it and hanged himself.

It is not difficult to earn money, but to become rich or even to make a decent livelihood is impossible, especially with the State shearing the 'Nep' people unmercifully. They are taxed to an enormous extent. And it is not to be wondered at that hundreds and thousands of Jews who at first petitioned for permits to become 'Neppists' are now returning their permits to the Government asking to be released from the honour of being among the builders of the New Economic Policy.

But it is not easy to give it up. When the shopkeepers in Homel, staggering under their heavy taxation, declared a sort of strike, refusing to open their shops and engage in business, they were denounced as counter-revolutionaries, and one of the leaders of the 'Yewsekzie,' the notorious Merejin, published two inflammatory articles in the 'Emess' denouncing the 'first attack by the Jewish bourgeoisie against the Soviet Government,' and demanding that they should be punished as traitors.

Naturally, not all the 'Nep' people go through the same kind of thing. As always, there are exceptions, and there are individuals who have made fortunes, especially in Moscow, which is to-day the greatest, perhaps the only trading centre in Russia. It is to Moscow that the Jews are flocking from every part of the country. Till recently things were not so bad in Minsk, where people managed to do well on contraband trade with Poland. But now there is a Customs office at the railway station in Minsk; all goods are thoroughly examined, and permission to bring goods back into Russia is given only to those who agree to smuggle illegal Communist literature into Poland.

Things are somewhat better for those families who have children over the age of twelve, able to travel round the villages, buy up goods and bring them home to their parents to sell. To have several grown-up children to-day in Russia means to be a rich man. Each child is a bread-winner. So from their earliest days children are being brought up to trade. Speculation is again becoming the forte of the Jews. All education is neglected, in order to train the children to become good business people. Of ideals it is better to say nothing at all.

But the most fortunate under the "Nep' are those families who have been down with typhus. That is an exceptional bit of luck. These people have no fear of again contracting the disease, so they travel about along the railway lines, and bring goods to their homes. There are very few others who will venture to set foot in a train, for the compartments are generally the homes of lice and contagion. They are consequently becoming monopolists. People who want to have things done for them in distant parts usually have to employ these typhus people, who get a good proportion of the profits.

A few individuals become rich, speculating in diamonds and in the exchange rate. The overwhelming majority, however, scuttle about the place like poisoned rats, buying and selling, working sixteen hours in the day, thinking of nothing in the world except their little businesses, and at the end of it all they have gained hardly anything.

Economically, 'Nep' has brought nothing but demoralisation into the life of Russian Jewry.

The moral degeneration is appalling. The mentality of the few new rich is disgusting. Everything is to their view concentrated within their little business transactions. The hunger for profits is stronger than anything else in the world, more potent than social or intellectual interests, for which there is no room left in their minds. A 'Nep' man who has really done well will never give anything away unless he is given a place of honour on some committee or other. The rule of the day in Jewish life in Russia is that 'he who has the money gets the honey.' The few new rich 'Nep' people are the rulers of Jewish life. The old social order has been broken up. The former communal workers have emigrated or have become the poorest of the poor. New people have taken their place.

The story of the 'Nep' is not finished yet. We will not venture to prophesy what it may bring to Russian Jewry in the future. But there is no doubt about what it is to-day. It is a persecution. It is not a New Economic Policy, but, as they say a 'luckless' economic policy."

In an age of Social Darwinism, the Zionists promoted the idea that only young and strong Jews should emigrate to Palestine and that they alone should avoid death at the hands of the Nazis. The infamous stories of the selection process of the *SS*, whereby healthy Jews of childbearing age were selected to survive, while others were selected to die, was, if true, most likely a Zionist directive meant to undue the supposed genetic damage of the ghettoes. The Nazis were also infamous for forcing Jews to perform strenuous manual labor, literally working the old and the weak to death.[585] This practice fulfilled several Zionist objectives—killing off the old and the weak—training Jews to do the dirty work that would be needed to be done in Palestine—fulfilling Jacob's Biblical rôle as an agrarian—and ensuring that the Holy Land would become predominantly Jewish, almost exclusively Jewish, which is also a Biblical goal and one the racist Israelis are still attempting to achieve today. The Nazis devoted a great deal attention to identification of "Jewish racial traits".

While the Zionist Nazis favored Zionist Jews and helped to usher them out of Nazi occupied lands, the Zionist Nazis targeted assimilatory Jewry and Orthodox Jewry, who were largely opposed to Zionism. The Zionists hoped to persuade both assimilated Jews and Orthodox Jews to violate their sensibilities and the Talmud and emigrate to Palestine *en masse* after the Second World War. The Zionists viciously punished these assimilated and Orthodox Jews who had opposed the Zionists after the First World War. The following article appeared in *The Jewish Chronicle* on 11 April 1919 on page 10,

"Jewish Factions in the Polish Parliament.

Co

PENHAGEN [F. O. C.]

---

585. *See: The Holocaust Chronicle*, Publications International, Ltd., Lincolnwood, Illinois, (2003), p. 177.

When the leaders of the Jewish factions in Poland rose in Parliament to explain the Jewish policy and demands in the new State, there was only one note of agreement struck by them, namely, the loyalty of the Jews to Poland and their goodwill towards the State. Otherwise, a sharp conflict of opinions manifested itself between the Jewish Nationalists (the Zionists and the People's Party) and the Orthodox Group (to which the Assimilation Party also leans on Jewish National questions). The former demanded national minority rights for the Jews, whereas the latter claimed equal rights only. Rabbi Perlmutter, on behalf of the Orthodox Party (speaking in the House without uncovering), declared that he desired to see a great Poland sweeping to the sea. M. Prilutzky, on behalf of the Jewish Nationalists, claimed National rights for the Jews, including special schools and the right to employ Yiddish in Courts of Justice and in State documents. He had a very hostile reception, members shouting at him: 'Let America grant such demands first, and we shall follow.'

Rabbi Halpern replied to M. Prilutzky that the Orthodox Party, which, as he believed, formed the preponderating group of Jews in Poland, only demanded equal rights. He stated that the Jews loved Poland, and that they believed the declarations of the Polish Party leaders that the Jews would get equal rights. He expressed the fear that the Nationalists would impair the relations between the Poles and the Jews."

Many Jews were aware of the fact that the Zionists were sponsoring anti-Semitism and that Zionists agreed with the precepts of anti-Semitism—were themselves anti-Semites. Some Zionists loudly protested against this truth. On 3 September 1897 on page 12, an article in *The Jewish Chronicle* paraphrased Dr. Birnbaum's statement at the First Zionist Congress,

"Dr. BIRNBAUM mentioned that it had often been contended that Zionism was but a reaction against anti-Semitism. It had not been denied that the growth of Zionism coincided with that of anti-Semitism, and, therefore, the conclusion was arrived at that the former only existed at the mercy of the latter. This was a complete mistake. It should be remembered that every movement had its causes and impetus. Through the former it obtained its pioneers, and through the latter its troops. Zionism could proudly say of itself that all who stood at its head had either long left the anti-Semitic impetus behind them, or that from the beginning their belief originated in the anomaly of the existence of a Jewish people. The want of a land of their own caused this anomaly to be the greater. There was a sentimental feeling in favour of Palestine, but sentiment would not suffice because the land whither they would go did not need special attraction; any country in which their distress would cease would be attractive; what they required was a land which would be able to keep them once they were there, till the grand process of converting them from a mercantile people into a people devoted to all callings, especially agriculture, had been completed, and they would no longer hanker after the flesh pots of Egypt [*Exodus* 16:3]. Palestine was the only country able to accomplish this. The second reason in favour of Palestine was the benefits that would be conferred not on Jews alone, but on mankind in general. A Jewish people in Palestine would not alone be the medium between the social-ethical and political-æsthetical elements of Europeism, but also the long-sought medium between the East and the West. No people is so apt for this as the Jews with their inherited Oriental qualities and their acquired European character. No country is so fitted to be the territorial medium as Palestine, with its proximity to Europe and to the Suez Canal, and as being the inevitable station on the railway to India. Fears had been expressed for the future of the Holy Sepulchre if Jews became the masters of Palestine, but by making the Christian holy places extra-territorial the difficulty would be overcome and all fears would be dissipated."

Much of what Birnbaum stated echoed the sophistry contained in Theodor Herzl's book *The Jewish State*. If the Zionists were genuinely interested in the best interests of humanity, they would have propped up the Turkish Empire, instead of trying to tear it down. It was the Turkish Empire which had the potential to fulfill the rôles the Zionist European Jews artificially claimed as their own. The Zionists knew that a large Jewish presence in Palestine would have the exact opposite effect of what they claimed. Instead of bringing peace to the region, it would inflame the Moslems and Catholics against the Jews and against one another. The Zionists tossed out the bait that the Suez Canal was of vital interest to European trade, and then falsely asserted that a Jewish presence would secure that interest, when in fact the Jews knew quite well that a Jewish presence would jeopardize European interests by instigating religious conflict. There was nothing that prevented the British and the French from maintaining productive relations with Moslems, other than Jewish Messianic designs.

The Jews did not want to secure the Suez for the sake of the Europeans, rather the Jews wanted the Northern European and British Protestants to secure Palestine for the Jews and protect them from the Catholics and the Muhammadans who would be inflamed by a Jewish colonization of their Holy Lands and shrines. Jewish intolerance of other religions remains a threat today, when Jewish Israelis attack Christians in Bethlehem, violate international law in Jerusalem, and seek to destroy the militaries and societies in Moslem countries, so that the Moslems will have no means with which to fight back when the Jews and Dispensationalist Christians destroy the Dome of the Rock and Al Aqsa Mosque and build a Jewish Temple in their place. The Israelis are also using the military of the United States to take over the territory of Greater Israel, which they know will eventually pass into their hands.

Beginning in the late 1800's, Jewish Zionists heavily promoted anti-Semitism and anti-Semites. Crypto-Jews founded and led anti-Jewish societies, which were financed with Jewish bankers' money. The most prominent Nazis were crypto-Jewish Frankists—Zionist propagandists in anti-Semite's clothing, *agents provocateur*, including Alfred Rosenberg, who took his political ideology from Houston Stewart Chamberlain, Liebenfels and List. Theodor Herzl took his racist political ideology from Eugen Karl Dühring, making Dühring an influence on both Herzlian Zionist racist mythology and Nazi racist mythology.[586] Before Dühring was the Jewish racist Zionist Moses Hess, who created National Socialist racism. Dietrich Eckart proposed that a demagogue should lead the Germans to drive out the Jews—long a Zionist objective.

In 1909, Zionist Max Nordau presented a character profile for the successful revolutionary that fit Hitler, Lenin and Stalin.[587] Nordau, though born and raised in Austro-Hungary, called himself a German, parroted the *Übermensch* philosophy, ridiculed Judaism and Christianity, copied the Germanic Hegelian dialectic, then called the modern world and those philosophers he was copying "degenerate"—a favorite word of Lombroso, who was Jewish,[588] and the Nazis.[589] Disraeli, Nordau and Zollschan promoted the alleged superiority of the "blonde Nordic race" in order to promote the segregation of their own "Jewish race". They asserted that the German and the Jew were superior races to the Slav and the Negro.

---

[586]. M. Samuel, "Diaries of Theodor Herzl", in: M. W. Weisgal, *Theodor Herzl: A Memorial*, The New Palestine, New York, (1929), pp. 125-180, at 129. T. Herzl, English translation by H. Zohn, R. Patai, Editor, *The Complete Diaries of Theodor Herzl*, Volume 1, Herzl Press, New York, (1960), pp. 4, 111.

[587]. M. Nordau, *Der Sinn der Geschichte*, C. Duncker, Berlin, (1909); English translation: *The Interpretation of History*, Willey Book Co., New York, (1910), pp. 309-315.

[588]. I. Zangwill, *The Problem of the Jewish Race*, Judaen Publishing Company, New York, (1914), p. 13; which was first published as an article, "The Jewish Race", *The Independent*, Volume 71, Number 3271, (10 August 1911), pp. 288-295, at 292.

[589]. J. B. Agus, *The Meaning of Jewish History*, Volume 2, Abelard-Schuman, New York, (1963), pp. 410-411, 420.

Cesare Lombroso,[590] who was Jewish,[591] advocated the extermination of

---

590.C. Lombroso, Italian: *L'Uomo di Genio*, Bocca, Torino, (1888); English: *The Man of Genius*, C. Scribner's Sons, New York,(1891); French: *L'Homme de Génie*, Alcan, Paris, (1889); **and** *Le Crime, Causes et Remèdes*, F. Alcan, Paris, (1907); English: *Crime: Its Causes and Remedies*, W. Heinemann, London, (1911); **and** *L'Uomo di Genio in Rapporto alla Psichiatria, alla Storia ed all'Estetica*, Fratelli Bocca, Torino, (1888); **and** *Applications de l'Anthropologie Criminelle*, Félix Alcan, Paris, (1892); **and** *Les Anarchistes*, E. Flammarion, Paris, (1896); **and** *Le Più Recenti Scoperte ed Applicazioni della Psichiatria ed Antropologia Criminale*, Bocca, Torino, (1893); **and** *Palimpsesti del Carcere; Raccolta Unicamente Destinata Agli Uomini di Scienza*, Bocca, Torino, (1888); **and** *L'Anthropologie Criminelle et ses Récents Progrès*, F. Alcan, Paris, (1890); **and** French: *L'Antisemitismo e le Scienze Moderne*, L. Roux e C., Torino, (1894); German: *Der antisemitismus und die Juden im lichte der modernen Wissenschaft*, G.H. Wigand, Leipzig, (1894); **and** French: *L'Antisémitisme*, V. Giard & E. Brière, Paris, (1899); Spanish: *El Antisemitismo*, Viuda de Rodríguez Serra, Madrid, (1900's); **and** *L'Uomo Delinquente, in Rapporto all'Antropologia, alla Giurisprudenza ed alle Discipline Carcerarie: Delinquente-Nato e Pazzo Morale*, Fratelli Bocca, Torino, (1884); **and** *Kerker-Palimpseste; Wandinschriften und Selbstbekenntnisse gefangener Verbrecher*, Verlagsanstalt und Druckerei a.g., Hamburg, (1899); **and** *Criminal Man According to the Classification of Cesare Lombroso*, Patterson Smith, Montclair, New Jersey, (1911); **and** *L'Homme Criminel; Criminel-né, Fou Moral, Épileptique; Étude Anthropologique et Médico-Légale*, Alcan, Paris, (1887) ; **and** *Nuovi Studii sul Genio*, R. Sandrom, Milano, (1902); **and** *Les Applications de l'Anthropologie Criminelle*, Alcan, Paris, (1892); **and** *Crime: Its Causes and Remedies*, Little, Brown, Boston, (1911); **and** *L'Homme Criminel; Criminel né, fou Moral, Épileptique, Criminel fou, Criminel d'Occasion*, F. Alcan, Paris, (1895); **and** *Palimpsestes des Prisons Recueillis*, A. Storck, Lyon, (1894); **and** *Der Verbrecher, in anthropologischer, ärztlicher und juristischer Beziehung*, Richter, Hamburg, (1887-1890); **and** Italian: *Genio e Degenerazione. Nuovi Studi e Nuove Battaglie*, Sandron, Palermo, (1897); German: *Entartung und Genie*, G.H. Wigand, Leipzig, (1894); **and** *Genie und Irrsinn in ihren Beziehungen zum Gesetz, zur Kritik und zur Geschichte*, P. Reclam, Leipzig, (1887); **and** *Neue Fortschritte in den Verbrecherstudien*, Leipzig, (1894); **and** *Neue Verbrecherstudien...* , a.S., Carl Marhold, Halle, (1907); **and** *Die Anarchisten; eine kriminalpsychologische und sociologische Studie*, J.F. Richter, Hamburg, (1895); **and** ; **and** *Die Ursachen und Bekämpfung des Verbrechens*, H. Bermühler, Berlin, (1902); **and** *Genio e Follia; in Rapporto alla Medicina Legale, alla Critica ed alla Storia*, Bocca, Roma, (1882); **and** *Aplicaciones Judiciales y Médicas de la Antropología Criminal*, La España Moderna, Madrid, (1892) ; **and** *Les Palimpsestes des Prisons*, A. Storck, Lyon, G. Masson, Paris, (1894); **and** *Der Selbstmord der Verbrecher insbesondere im Zellengefaengnis*, Berlin, (1901); **and** *L'Homme Criminel: Etude Anthropologique et Psychiatrique*, Félix Alcan Ed., Paris, (1895); **and** *Criminal Man According to the Classification of Cesare Lombroso*, G.P. Putnam's Sons, New York, (1911); **and** *Genie und Entartung: Autorisierte Ubersetzung aus dem Stalienischen*, P. Reclam, Leipzig, (1910); **and** *The Heredity of Acquired Characteristics* ; **and** *L'Uomo Bianco e l'Uomo di Colore Letture su l'Origine e la Varietà delle Razze Umane*, Fratelli Bocca, Torino, Firenze, (1892); **and** *Studien über Genie und Entartung*, P. Reclam, Leipzig, (1910); **and** *Problemes du Jour*, Libr. Universelle, Paris, (1906); **and** *Lombroso und die Criminal-Anthropologie von heute*, Hubertusburg, (1897) ; **and** *L'Amore nel Suicidio e nel Delitto*, Ermanno Loescher, Torino, (1881); **and** *Criminal Anthropology: Its Origin and Application*, Forum Pub. Co.,

alleged criminal phenotypes. His theories later became the model for the Nazis' gassing of political opponents, criminals, the insane and the infirm. Jewish Zionist Max Nordau was in many senses the archetype Nazi. Many Nazis and anti-Semites were of Jewish origin. Both the Zionists and the Nazis loathed the Slavic "race" and brought about its downfall in modern times. They also attempted to wipe out the blonde "race" of Nordics they pretended to admire.

The Old Testament is filled with stories of Jews massacring other Jews and of human sacrifice. The Old Testament and the Talmud instruct pious Jews to kill Jews who abandon Judaism, especially those who sincerely convert to other religions, as well as heathen priests.[592] While addressing the justifications given by religious Zionist terrorists for Yigal Amir's murder of Yitzhak Rabin, Jessica Stern wrote in her book *Terror in the Name of God: Why Religious Militants Kill*,

> "According to the halakah, the rulings of Din Mosser and Din Rodef apply to those Jews who have committed the most despicable crime imaginable—the betrayal of their fellow Jews. The punishment of the Mosser—a person

---

New York, (1895); **and** *The Physiognomy of the Anarchists*, Philadelphia, (1891/1993); **and** *Virchow und die Kriminalanthropologie*, (1896); **and** *The Physiology & Psychology of Crime*, American Institute for Psychological Research, Albuquerque, (1895/1980); **and** *Pazzi ed Anomali; Sággi*, Lapi, Città di Castello, (1886); **and** *Studj Clinici ed Esperimentali sulla Natura, Causa e Terapia della Pellagra*, G. Bernardoni, Milano, (1870); **and** *Études de sociologie: Les Anarchistes*, E. Flammarion, Paris, (1897). *See also:* C. Lombroso, E. Ferri, R. Garofalo, *Et al., Polemica in Difesa della Scuola Criminale Positiva*, N. Zanichelli, Bologna, (1886). *See also:* C. Lombroso, G. Regnier, and A. Bournet, *L'Homme Criminel; Étude Anthropologique et Médico-Légale*, F. Alcan, Paris,(1887).. *See also:* C. Lombroso, R. Laschi, *Il Delitto Politico e le Rivoluzioni in Rapporto al Diritto, all'Antropologia Criminale ed alla Scienza di Governo. Con 10 Tavole e 21 Figure nel Testo*, Fratelli Bocca, Torino, (1890). *See also:* C. Lombroso, R. Laschi, Rodolfo, and A. Bouchard, *Le Crime Politique et les Révolutions par Rapport au Droit, à l'Anthropologie Criminelle et à la Science du Gouvernement*, F. Alcan, Paris, (1892). *See also:* C. Lombroso, R. Laschi, Rudolfo, H. Kurella, *Der politische Verbrecher und die Revolutionen in anthropologischer, juristischer und staatswissenschaftlicher Beziehung*, Verlagsanstalt und Druckerei, Hamburg, (1891-1892). *See also:* C. Lombroso, G. Ferrero, Italian: *La Donna Delinquente: La Prostituta e la Donna Normale*, L. Roux, Torino, (1893); German: *Das Weib als Verbrecherin und Prostituirte: Anthropologische Studien, gegründet auf eine Darstellung der Biologie und Psychologie des normalen Weibes*,Verlagsanstalt und Druckerei, Hamburg, (1894); French: *La Femme Criminelle et la Prostituée*, F. Alcan, Paris, (1896); English: *The Female Offender*, T.F. Unwin, London, (1895). *See also:* P. Näcke, C. Lombroso, *Ein Willkommengruss von Herrn Lombroso*, Leipzig, (1894). *See also:* L. Fratiny, C. Lombroso, *Une Interview. Criminalité Génialité: C. Lombroso Jugé par Mignozzi-Bianchi*, Firenze, (1909).

591. I. Zangwill, *The Problem of the Jewish Race*, Judaen Publishing Company, New York, (1914), p. 13; which was first published as an article, "The Jewish Race", *The Independent*, Volume 71, Number 3271, (10 August 1911), pp. 288-295, at 292.

592. *Abodah Zarah* 26b.

who hands over sacred Jewish property to the gentile—as well as that of the Rodef—a person who murders or facilitates the murder of Jews—shall be death. Since the execution of the Mosser or the Rodef is aimed at saving the lives of other Jews, there is no need for a trial."[593]

The political Zionists considered non-Zionist Jews to be traitors and they believed assimilation would lead to the death of the mythical "Jewish race". Moses Hess wrote,

"The most touching point about these Hebrew prayers is, that they are really an expression of the collective Jewish spirit; they do not plead for the individual, but for the entire Jewish race. The pious Jew is above all a Jewish patriot. The 'new' Jew, who denies the existence of the Jewish nationality, is not only a deserter in the religious sense, but is also a traitor to his people, his race and even to his family. If it were true that Jewish emancipation in exile is incompatible with Jewish nationality, then it were the duty of the Jews to sacrifice the former for the sake of the latter. This point, however, may need a more elaborate explanation, but that the Jew must be above all a Jewish patriot, needs no proof to those who have received a Jewish education. Jewish patriotism is not a cloudy Germanic abstraction, which dissolves itself in discussions about being and appearance, realism and idealism, but a true, natural feeling, the tangibility and simplicity of which require no demonstration, nor can it be disposed of by a demonstration to the contrary."[594]

Anti-Semitism was very useful to both the Communists and the Zionists. Politically active anti-Semitic demagogues like Lueger, Ahlwardt, Treitschke and Stoecker had numerous Jewish connections, as did Adolf Hitler—some even had Jewish blood, as did Hitler. Anselm von Rothschild stated that Stoecker was an apostate Jew. The Rothschilds wanted desperately to buy Palestine and establish a Jewish state there, with a Rothschild sitting as king of the world, but the Rothschilds lacked broad Jewish support. The political Zionists later concluded that they could only obtain Jewish support in a climate of advanced anti-Semitism. *The Chicago Tribune*, on 12 December 1881 on page 6, reprinted a letter from Rothschild to Stoecker:

"BARON ROTHSCHILD.

The Letter Written by Him in Defense of the Jews.
Baron Anselm von Rothschild, of Vienna, wrote the annexed letter to

---

[593]. J. Stern, *Terror in the Name of God: Why Religious Militants Kill*, Ecco, New York, (2003), p. 91.
[594]. M. Hess, *Rom und Jerusalem: die letzte Nationalitätsfrage*, Eduard Wengler, Leipzig, (1862); English translation, *Rome and Jerusalem: A Study in Jewish Nationalism*, Bloch, New York, (1918/1943), pp. 62-63.

Hof-Prediger Stoecker, of Berlin, the instigator of the anti-Semitic agitation in Germany:

VIENNA, November, 1881.—*To the Court Preacher Stoecker*—SIR: If I am correctly informed, your physician once advised you to take plenty of exercise, and since then you have been almost constantly employed in anti-Semitic movements. This matter really concerns me very little, for, thank God, Austria has not yet advanced so far on the path of intelligence and refinement as to possess a 'Judenhetze,' such as the cultivated city of Berlin can boast of. But still I should like to call the attention of your reverence to certain grave errors which have crept into your speech recently delivered in the German Parliament.

You said in that address, 'Behind me stand the millions.' You are mistaken: the millions stand behind me, and if you doubt this you are respectfully invited to visit my counting-house, where ample proofs shall be given you. You contend that 'the Jewish usurers have ruined all classes of people.' Now, pray tell me, my dear Court-Preacher, who goes to the Jewish usurer? Is it not those whose credit is exhausted? And if their fellow-men will not trust them any longer, are they not already ruined before they seek their last resort—the Jewish usurer? This is only another of many cases where the Jew is made the scapegoat for the offenses of his neighbors. [A Gentile, or Gentile government, could be easily forced to seek loans from a Jewish usurer through the agitation of the Jewish press for war, or an infinite number of other corrupt means—as evinced by the Jewish financiers' destruction of Russia.]

You say further that the Jews, out of all proportion to their numbers, assisted by talent and capital, exercise a mighty influence in the community. I am really surprised that this should surprise you. As if talent had not, from time immemorial, held the sceptre. Would you rather that this world should be ruled by fools than by wise men? And as far as the disproportion of our numbers is concerned we Jews cannot help but feel highly flattered if we possess more talent than our Gentile countrymen. And as for our power as capitalists this is the result of our business genius and our economy. Why do not the Christians imitate us? Do we hinder them from earning money or from saving it? [The answer to this question is obviously yes. Jewish power, wealth and influence in the press result from Jewish racism and Jewish tribalism, and the Cabalist and Talmudic doctrines which encourage Jews to take advantage of Christian integrity in order to exploit Christians. Rich Jews promoted honesty and decentralized power among the Gentiles, while promoting dishonesty and tribalism among their own. This gave the Jews an advantage, which could only be overcome by the Gentiles' sinking to the debased level of the foe, or expelling it. Rothschild's racist arrogance is ample proof of the fact, and if the Gentiles had truly leveled the playing field by sinking to Rothschild's level, they would have quickly crushed the Jews.]

'The Jews should be more modest,' you say. It is true that modesty is a most desirable virtue, suited alike to Jew and Gentile, but as Goethe has it, 'Only scoundrels are modest.' Now, among the Jews there are so few scoundrels, and then really it is much easier for a Court Preacher to be modest than for a Jew. If a Court Preacher displays that commendable virtue, his flock will bow before him and exclaim, 'So mighty and yet so modest.' Let a Jew be modest and he is kicked and spurned, and the mob say, 'Serves him right.' [The reality is that the rich Jews concentrated their wealth and shared it with neither Gentile nor poor Jew. This concentration of wealth gave the Jews enormous power and the resentment this corrupt and undemocratic warmongering power caused was directed at poor Jews, often through the machinations of rich Jews, who sought to keep their poor brethren segregated.]

You aver that the Jew in Lessing's 'Nathan' is no Jew at all, but a Christian. With the same right I might say the Court Preacher Stoecker is no Christian, but an apostate Jew who has banded himself with some barbarous relics of the Middle Ages to prosecute a miserable anti-Semitic agitation. But I will not say this, as I would not desire to so grossly insult my co-religionists. [It is highly interesting that Rothschild called Stoecker an apostate Jew.]

Your friend Bachem is of the opinion that the people are backing him. I will admit that there is a people in his wake, but as a German philosopher once said: 'There are enough wretches in the world to back any bad cause.' Your friend also indulges in the crushing accusation that the Jews are grain speculators. Now, do you know who was the first speculator in grain? None other than Jew, Joseph, in Egypt, although at that time there was no Court Preacher to discover any crime in his action, and the people were grateful to him. [The Egyptian people were not grateful to Joseph, who brought them into slavery (*Genesis* 47).]

In one of your discourses you once exclaimed: 'Look at Herr von Bleichroeder. He has more money than all the evangelical preachers put together.' Now, I am sure that Herr von Bleichroeder has never said: 'Look at Court Preacher Stoecker. He earns more money by a single sermon than a hundred Jewish firms do in a whole year.'

In conclusion, if you will not admit that the Jews have any good qualities, you will at least not envy them for the little money they may possess. If in spite of their wealth they cannot prevent, in the year of 1881, the formation of an agitation against them, what would become of them if they had no money? It is true the Jews put some value on wealth, and I must say that I would rather be a rich Jew than a poor Christian. But then, are there not poor Christians who would rather be rich Jews? Even you, most reverent sir, might, perhaps, be willing to change positions with me (and I flatter myself that you would not make so bad a bargain). For myself, I can only say that if I were not Rothschild, I should still be very far from wishing myself the Court Preacher Stoecker. Very respectfully, A. VON ROTHSCHILD."

The essentially meaningless term "anti-Semitism", a term for anti-Jewish which is employed as a weapon by hypocritical racist Jews with which to smear and threaten, was coined by an "anti-Semitic" crypto-Jew named Wilhelm Marr after Jewish corruption imploded the stock markets of Europe. In its article entitled "ANTI-SEMITISM", *The Universal Jewish Encyclopedia*, Volume 1, The Universal Jewish Encyclopedia, Inc., New York, (1939), pp. 341-409, at 341; states:

> "The word was probably first used by Wilhelm Marr, said to have been a converted Jew, in Der Sieg des Judentums ueber das Germanentum, a pamphlet which he published in 1879, the same year in which he founded the Anti-Semitic League; two years later he began publication of Zwanglose antisemitische Hefte."

Under the heading "Foreign Articles", the following statement appeared in *Niles' Weekly Register*, Volume 17, Number 427, (13 November 1819), p. 169,

> "Mr. Rothschild, the great London banker, indignant at the persecution of his Jewish brethren in Germany, has refused to take bills upon any of the cities in which they are persecuted; and great embarrassments to trade have been experienced in consequence of his determination. ☞It is intimated that the persecution of the Jews is in part owing to the fact, that Mr. Rothschild and his brethren were among the chief of those who furnished the 'legitimates,' with money to forge chains for the people of Europe."

There would not have been agitations against the Jews, if the Jews in the press had not attacked Christianity and if the Jewish financiers had not attacked Europe with perpetual war[595] throughout the Nineteenth Century. That "little money" in the hands of the Rothschilds alone amounted to some, or one might say "sum" $3,400,000,000.00, acquired through deceitful and inhuman means. It is interesting, though not at all unusual, that Stoecker was a Jew and was behind the anti-Jewish agitation. The same could be said of Goebbels, Streicher, Heydrich, Frank, etc., and, no doubt, Rothschild.

The outspoken Mayor of Vienna Karl Lueger proclaimed that he decided who was, and who was not, a Jew, meaning that he could protect those Jews who helped him—those Jews who put him in power in order to spread anti-Semitism. He had Jewish backers and was an agent for their agenda. Anti-Semite Hermann Ahlwardt advocated the segregation of Jews in the Reichstag in 1895. The segregation of Jews was a Zionist objective. Ahlwardt spoke in anti-assimilationist terms Theodor Herzl would soon use in his book *The Jewish State*. Adolf Stoecker also raised his voice to advocate segregation in the schools, as did racist Zionist Albert Einstein. The dogma of segregation had both Zionist and anti-Semitic origins—for example racist Zionist Moses Hess' *Rom und*

---

595. G. E. Griffin, "The Rothschild Formula", *The Creature from Jekyll Island: A Second Look at the Federal Reserve*, Chapter 11, Fourth Edition, American Media, Westlake Village, California, (2002), pp. 217-234.

Jerusalem: die letzte Nationalitätsfrage of 1862 and anti-Semite Eugen Karl Dühring's *Die Judenfrage als Racen-, Sitten- und Culturfrage: mit einer weltgeschichtlichen Antwort* of 1881. Hermann Ahlwardt stated to the Reichstag in 1895,

> "A Jew who was born in Germany does not thereby become a German; he is still a Jew. Therefore it is imperative that we realize that Jewish racial characteristics differ so greatly from ours that a common life of Jews and Germans under the same law is quite impossible because the Germans will perish."[596]

Jewish Zionist Bernard Lazare wrote in 1894,

> "Everything is tending to bring about such a consummation. Such is the irony of things that antisemitism which everywhere is the creed of the conservative class, of those who accuse the Jews of having worked hand in hand with the Jacobins of 1789 and the Liberals and Revolutionists of the nineteenth century, this very antisemitism is acting, in fact, as an ally of the Revolution. Drumont in France, Pattai in Hungary, Stoecker and von Boeckel in Germany are co-operating with the very demagogues and revolutionists whom they believe they are attacking. This antisemitic movement, in its origin reactionary, has become transformed and is acting now for the advantage of the revolutionary cause. Antisemitism stirs up the middle class, the small tradesmen, and sometimes the peasant, against the Jewish capitalist, but in doing so it gently leads them toward Socialism, prepares them for anarchy, infuses in them a hatred for all capitalists, and, more than that, for capital in the abstract. And thus, unconsciously, antisemitism is working its own ruin, for it carries in itself the germ of destruction.
>
> Such, then, is the probable fate of modern antisemitism. I have tried to show how it may be traced back to the ancient hatred against the Jews; how it persisted after the emancipation of the Jews, how it has grown and what are its manifestations. In every way I am led to believe that it must ultimately perish, and that it will perish for the various reasons which I have indicated, because the Jew is undergoing a process of change; because religious, political, social, and economic conditions are likewise changing; but above all, because antisemitism is one of the last, though most long lived, manifestations of that old spirit of reaction and narrow conservatism, which is vainly attempting to arrest the onward movement of the Revolution."[597]

---

596. English translation from: P. W. Massing, *Rehearsal for Destruction: A Study of Political Anti-Semitism in Imperial Germany*, Howard Fertig, New York, (1967). p. 304.
597. B. Lazare, *Antisemitism: Its History and Causes*, (1894), pp. 182-183. *L'Antisémitisme, son Histoire et ses Causes*, L. Chailley, Paris, (1894).

Dietrich Eckart[598] promoted Adolf Hitler as a viable anti-Semitic demagogue, though many thought that Hitler appeared to be a Jewish actor or comedian spoofing an anti-Semitic demagogue, and they laughed at him. Eckart said,

> "The best would be a worker who knows how to talk.... . He doesn't need much brains, politics is the stupidest business in the world, and every marketwoman in Munich knows more than the people in Weimar. I'd rather have a vain monkey who can give the Reds a juicy answer, and doesn't run away when people begin swinging table legs, than a dozen learned professors. He must be a bachelor, then we'll get the women."[599]

Erik Jan Hanussen, a crypto-Jew whose real name was Hermann Steinschneider, coached Hitler on effective public speaking.[600] In 1933, the sick joke, the clown, the Bolshevik Zionist Adolf Hitler rose to power over Germany lifted on the golden purse strings of the Jewish bankers.

In 1934, Jacob R. Marcus incorrectly predicted that Nazis would not carry out the Holocaust, because so many of its prominent leaders were, by Nazism's own standards, "sub-human",

> "The present National Socialist government is too shrewd, in spite of its racial commitments, to lend itself to such extravaganzas. It wants no Brahmanic caste-system in which even the shadow of a low caste Hindu is a pollution. It knows that any attempt toward racial eugenics along purely Nordic lines would disrupt present day Germany with its half-dozen racial strains. Nordicization, if it were literally true to itself, would mean the exclusion from the German state of the following non-Nordic types: the late Paul von Hindenburg; Streicher, the rabid anti-Semitic Nuremberg journalist; Ley, the head of the National Socialist Labor Front; Goebbels, Minister of National Enlightenment and Propaganda; and finally, Hitler, himself. Here is a racial analysis of Hitler made in 1929 by the racial-hygienist, Professor von Gruber, then President of the Bavarian Academy of Sciences and a member of the racially-minded Pan-American Association:
>> 'For the first time I saw Hitler at close range. Face and head of poor race, mongrel, low slanting forehead, ugly nose, broad cheek bones, small eyes, dark hair. A short brushlike mustache, no broader than the nose, gives the face a defiant touch. The facial expression is not that of

---

598. A. Hitler and D. Eckart, *Der Bolschewismus von Moses bis Lenin*, Hoheneichen-Verlag, München, (1924); English translation by W. L. Pierce, *Bolshevism from Moses to Lenin: A Dialogue between Adolf Hitler and Me*, World Union of National Socialists, Arlington, Virginia, (1966).
599. D. Eckart, quoted in A. Hitler, translated by R. Manheim with an introduction by A. Foxman, *Mein Kampf*, Houghton Mifflin, Boston, New York, (1999), p. 687.
600. M. Gordon, *Erik Jan Hanussen: Hitler's Jewish Clairvoyant*, Feral House, (2001).

a man who has complete control of himself but of one who is aroused to frenzy. Repeated twitching of the facial muscles. When through, expression of contented self-reliance.' (*Essener Volkswacht*, Nov. 9, 1929.)"[601]

The exposure of the active involvement of Zionists with the Nazi hierarchy—even as instigators of the entire Nazi movement—is shocking, but one is reminded of the willingness of some Jewish religious fanatics to commit suicide and to submit to genocide in order to preserve the integrity of the holy land and of the "race" of the "chosen". Racists like the Jewish Zionist Meir Kahane thrived on conflict. Kahane asked Jews to rejoice at the United Nations Resolution which acknowledged that Zionism is a form of racism. He hoped that it would lead to strife between Gentiles and Jews, because he believed that this would ultimately lead to the destruction of the Gentile world, as Jewish prophecy foretold. Kahane hoped that the entire world would turn against Israel, and falsely tied all Jews to Israel, in the hopes that Jews would be humiliated and then the Gentiles would be destroyed by God, in the form of Zionist subversion. Kahane succinctly wrote, *inter alia*,

"The banding together by the nations of the world against Israel is the guarantee that their time of destruction is near and the final redemption of the Jew at hand."[602]

Jessica Stern, in her book *Terror in the Name of God: Why Religious Militants Kill*, writes of Jews who are willing,

"to risk a world war in pursuit of religious redemption for the Jewish people.30"[603]

Baruch Kimmerling wrote,

"At the center of this culture of death is the remembrance of martyrs—Jews who, in Zionist ideology, had to die so that the state might be born. [***] *A triumphal creed shadowed by death, Zionism transformed the catastrophes of Jewish history into nationalist fables of redemption.*"[604]

---

[601]. J. R. Marcus, *The Rise and Destiny of the German Jew*, The Union of American Hebrew Congregations, Cincinnati, (1934), p. 62.
[602]. M. Kahane, *On Jews and Judaism: Selected Articles 1961-1990*, Volume 1, Institute for the Publication of the Writings of Rabbi Meir Kahane, Jerusalem, (1993), p. 104.
[603]. J. Stern, *Terror in the Name of God: Why Religious Militants Kill*, Ecco, New York, (2003), p. 105. Stern cites E. Sprinzak, "From Messianic Pioneering to Vigilante Terrorism", in D. C. Rapoport, Editor, *Inside Terrorist Organizations*, Frank Cass, London, (1988), pp. 194-216.
[604]. B. Kimmerling, "Israel's Culture of Martyrdom", *The Nation*, (10-17 January 2005), pp. 29-30, 33-34, 36, 38, 40, at 29; which is a review of I. Zertal, *Israel's Holocaust and*

Though Kahane has been rejected by the vast majority of Jews, and by the majority of Israelis, his message is in keeping with Judaism. Kahanism has a romantic allure to some Jews of a promise of community and common enemy. That battles with their better natures and Kahanism threatens to become a broad movement if not checked and exposed again and again as the hateful mythology that it is. This lust for persecution and martyrdom in order to bring death upon the enemies of the Jews, real or imagined enemies, is an ancient tradition for Jews. It is clearly advocated in the writings of Philo the Jew and Josephus, as well as those of Theodor Herzl. Philo the Jew vilified the Egyptians and Caligula with lies—as did Josephus with even more outrageous lies. Josephus fabricated the myth of Masada. These legendary lies are ingrained in the psyches of those who see these lies as their history and who have a "Masada Complex" of imagined persecution and martyrdom.[605] Many modern Jews have created an unhealthy culture of death and persecution around the Holocaust—some say the Holocaust has become a new religion.[606]

In the book of *Numbers*, Chapter 25, Jews were commanded by God to commit genocide against Jews who had assimilated. According to the *Gospel of John* 11:47-53, Caiaphas chose to execute Jesus in order to preserve the nation of the Jews and to gather back its supposedly chosen people:

> "47 Then gathered the chief priests and the Pharisees a council, and said, What do we? for this man doeth many miracles. 48 If we let him thus alone, all men will believe on him: and the Romans shall come and take away both our place and nation. 49 And one of them, named Caiaphas, being the high priest that same year, said unto them, Ye know nothing at all, 50 Nor consider that it is expedient for us, that one man should die for the people,

---

the Politics of Nationhood, Cambridge University Press, (2005); and Y. Grodzinsky, *In the Shadow of the Holocaust: The Struggle Between Jews and Zionists in the Aftermath of World War II*, Common Courage Press, Monroe, Maine, (2004).

605. Philo the Jew, "Flaccus" and "On the Embassy to Gaius", *The Works of Philo*, Hendrick Publishing, U. S. A., (2000), pp. 725-741 and 757-790. Josephus, "Against Apion", *The Works of Josephus*.

606. R. Garaudy, *Les Mythes Fondateurs de la Politique Israélienne*, Samiszdat, Paris, (1996); English translations: *The Founding Myths of Israeli Politics*, and *The Mythical Foundations of Israeli Policy*, Studies Forum International, London, (1997) and *The Founding Myths of Modern Israel*, Institute for Historical Review, Newport Beach, California, (2000). *See also:* N. Finkelstein, , *The Holocaust Industry: Reflection on the Exploitation of Jewish Suffering*, Verso, London, New York, (2000); and *Beyond Chutzpah: On the Misuse of Anti-semitism and the Abuse of History*, University of California Press, Berkeley, (2005). *See also:* B. Kimmerling, "Israel's Culture of Martyrdom", *The Nation*, (10-17 January 2005), pp. 29-30, 33-34, 36, 38, 40; which is a review of I. Zertal, *Israel's Holocaust and the Politics of Nationhood*, Cambridge University Press, (2005); and Y. Grodzinsky, *In the Shadow of the Holocaust: The Struggle Between Jews and Zionists in the Aftermath of World War II*, Common Courage Press, Monroe, Maine, (2004).

and that the whole nation perish not. 51 And this spake he not of himself: but being high priest that year, he prophesied that Jesus should die for that nation; 52 And not for that nation only, but that also he should gather together in one the children of God that were scattered abroad. 53 Then from that day forth they took counsel together for to put him to death."

The book of *Matthew* 1:21-23 states that "Jesus"—the Jew—was meant to rescue the Jewish Nation,

"21 And she shall bring forth a son, and thou shalt call his name JESUS: for he shall save his people from their sins. 22 Now all this was done, that it might be fulfilled which was spoken of the Lord by the prophet, saying, 23 Behold, a virgin shall be with child, and shall bring forth a son, and they shall call his name Emmanuel, which being interpreted is, God with us."

If the New Testament is a fiction in part, or in whole, it is a fabrication that fixes blame for the destruction of the Temple and of Jerusalem on Jesus—from the Jewish point of view, instead of on the corruption of some leading Jews against the Roman government and the murder of Caligula for defiling the Temple. It also makes Jesus a means by which to preserve and consolidate the Jewish nation—a human sacrifice. If the New Testament is authentic, then Jesus' murder was a ploy, which enabled Jewish leaders to secure the unity of their people. In either event, the founding of Christianity—the story of the crucifixion of Christ—was a nationalistic attempt to unite the Jews of the world through human sacrifice—an alleged unity that some Jews have since sought and continue to seek at all costs to the themselves and without any regard for the rights and interests of others, both selflessly and selfishly willing to lead the world into an apocalyptic war in order to preserve their nationalistic vision.

At this critical time when humanity faces many important decisions and should be planning for the future of the survival of the human race, the tiny and insignificant country of Israel with a population of only six million receives grossly disproportionate attention on the world stage, draining off resources and time that the other six-and-one-half-billion human beings cannot afford to spare. Humanity would be better served to devote its resources to more important problems and simply impose an equitable solution to the problems in the Middle East with overwhelming force, or overwhelming disinterest. Though it seems the racist Jews will never rest until they have murdered off the Gentiles in way or another.

Early Christians inherited their love of martyrdom from the Jews and mostly were Jews. Ancient writers assert that ancient Jews believed that death by martyrdom was a certain means to immortality. One is further reminded of the countless failed attempts to form a Jewish nation and the desperation of the Zionists to find a means to achieve their ends because they believed the "Jewish race" was on the verge of extinction. The political Zionists embraced anti-Semitism as that meanest of means.

It is a fact that the Nazis in their writings and in their speeches promoted the

Zionists, and that the Zionists in their writings and in their speeches promoted anti-Semitism and the Nazis. It is also a fact that the Zionists advocated the racist position that Jews cannot and should not assimilate and were a foreign, disloyal, and combative nation within Germany.

This common interest between Nazis and Zionists includes financial collusion between Zionists and anti-Semites—the type of financial collusion Herzl advocated in his book *The Jewish State*. Herzl, who exhibited a psychopathic personality, held the majority of the Jews in low regard, and was eager to "sacrifice" them for his cause. His philosophical descendants were even more inhumane. Of course, the guilt of the Zionists who fomented the political climate which precipitated the Holocaust in no way abrogates the guilt of the many Germans and Europeans who participated in murdering millions of innocent men, women and children in the Holocaust and the Second World War. It serves as a warning to us all of the power held by those who mold public opinion and the possibilities for good or evil that control over that force holds. It is presently in the hands of the Zionists and has been for centuries.

7.5.1 Nazism is a Stalking Horse for Zionism and Communism

Adolf Hitler was a former Bolshevik with connections to members of the *Thule-Gesellschaft*—a subversive organization founded by crypto-Jewish Zionists on the Illuminati model to foment an anti-Semitic revolution that would force the Jews out of Europe and into Palestine.[607] Hitler was filmed marching in the funeral procession of Jewish Communist Kurt Eisner, who led a short-lived Soviet Republic in Munich at the end of the First World War. Hitler was a Bolshevik and a Zionist with many strong ties to the Jewish community. He surrounded himself with Jews and crypto-Jews throughout his life.

To many of his contemporaries, Hitler appeared to be a Jewish actor, comically spoofing a ranting anti-Semitic demagogue. Many of Hitler's contemporaries knew that Hitler was a Red subversive who was trying to weasel his way into power through Jew-baiting, and pretending to fight Bolshevism, in order to convert Germany into a Zionist Bolshevist nation led by crypto-Jews. This was a common Communist tactic. At the same time, American Jewish Communists were also Jew-baiting and trying to attract a following through the use of anti-Semitic propaganda in an effort to use anti-Jewish prejudice as a means to fulfill Jewish prophecy and put Jews in power around the world. This was an old Frankist trick. One could even say that Christianity served the same purpose.

The Soviet Union tried to subvert Moslem nations with anti-Israeli positions meant to lure Moslem nations into turning Communist and to make it appear that

---

[607]. Rudolf Glandeck Freiherr von Sebottendorf (b. Adam Alfred Rudolf Glauer), *Bevor Hitler kam: Urkundliches aus der Frühzeit der nationalsozialistischen Bewegung*, Grassinger, München, (1933).

Israel was a necessary ally to the United States and Western Europe. If the Moslem nations had gone Communist, Israel, which itself had a Communist Party, would then have had complete control over those nations, which undoubtedly would have been ruled by crypto-Jews or Jewish agents. In a short period of time, the Moslem faith would have been proscribed, and the Moslems, even in the oil rich nations, would have found themselves completely ruined and in abject poverty. The current President of Iran is serving Israel's interests by making anti-Israeli statements which serve as a spurious pretext for war. The Israelis have placed crypto-Jews and Jewish agents in power throughout the Middle East and have organized anti-Israeli terrorist organizations so as to provide Israel with pretexts to attack and dehumanize Moslems.

The Nazi Party's platform of "The 25 Points" published on 24 February 1920 was so obviously Bolshevistic, that Adolf Hitler had to apologize for it on 13 April 1928 in order to appease the German Capitalists who had sponsored Hitler believing he would fight Bolshevism and fatten their pockets with profitable wars against the Bolsheviks,

> "On April 13th, 1928, Adolf Hitler made the following declaration:
> It is necessary to reply to the false interpretation on the part of our opponents of Point 17 of the Programme of the N.S.D.A.P.
> Since the N. S. D. A. P. admits the principle of private property, it is obvious that the expression 'confiscation without compensation' merely refers to possible legal powers to confiscate, if necessary, land illegally acquired, or not administered in accordance with national welfare. It is directed in accordance with national welfare. It is directed in the first instance against the Jewish companies which speculate in land.
> Munich, April 13th, 1928.
> (signed) Adolf Hitler."[608]

The Nazis wanted to eliminate class differences, abolish personal property and make businesses communal. Hitler and others expressly stated that the Nazi agenda was to produce a classless society as can be seen and heard in Leni Riefenstahl's film *Triumph des Willens* or *Triumph of the Will*.

Nazism was in many respects quite Marxist. When the NSDAP began, many in the *Freikorps* believed that Adolf Hitler was a Communist and that the Nazis were "Reds". The term "NAZI" comes from the party's name, *Nationalsozialistische Deutsche Arbeiterpartei*, or "National Socialist German Worker's Party"; which sounded very much like the older Communist parties *Allgemeiner Deutscher Arbeiter-Verein*, or "Universal German Worker's Union"; and *Sozialdemokratische Arbeiterpartei*, or "Social Democrat Worker's

---

[608]. E. T. S. Dugdale, "The 25 Points", *The Programme of the N.S.D.A.P. and its General Conceptions*, Franz Eher Nachf., Munich (1932), pp. 18-20, at 19; reprinted as an Appendix "The Program of the National Socialist German Workers Party" in K. G. W. Ludecke, *I Knew Hitler: The Story of a Nazi AWho Escaped the Blood Purge*, Charles Scribner's Sons, New York, (1937), pp. 793-796, at 795, note 1.

Party". Ernst Röhm, Chief of the *SA*, was very liberal minded and sought support from Communists. Gregor Strasser was another Communist Nazi. When Hitler made it very clear that he would protect the interests of the wealthy capitalists who supported him financially, while stating that he reserved the right to confiscate land for the State, it confused many Socialists who weren't sure whether Hitler embraced Socialism or Capitalism. Though the Nazis sought to distance themselves from Bolshevism, their propaganda of revolution and worker's rights was often directly copied from Bolshevik propaganda.

Indeed, archival footage shows that Hitler marched with a detachment from his regiment in the funeral procession of the Jewish Communist Kurt Eisner in February of 1919.[609] Eisner was shot after the Bavarian Revolution in November, 1918. Hitler's group wore both red and black armbands to sponsor Socialism and to mourn the death of a Socialist revolutionary. Soon thereafter, Bolshevists established a Soviet Republic in Bavaria, and Adolf Hitler became a spokesman for the Soviet Counsel. After the Bolshevik Revolution was suppressed, Hitler began to work for the Right as an anti-Bolshevist and an anti-Jewish propagandist. But he soon showed his true colors as a devoted Communist Red working towards a controlled opposition when he turned the rightist party to the left and converted the DAP into the NSDAP. In the Bolshevik totalitarian tradition, Hitler eventually destroyed all political parties but his. Like other crypto-Jewish Bolshevik tyrants, he wanted a thoroughly homogenous State with only one political party, Socialism.

Hermann Rauschning, who was himself at one time a powerful Nazi leader, wrote several books in the late 1930's and early 1940's, which alleged that many Nazis were essentially Bolshevist revolutionaries and that Hitler was in many respects seen by them as a Marxist revolutionary.[610] Rauschning knew in

---

609. G. Knopp, M. P. Remy, "Hitler: The Private Man", *The Rise and Fall of Adolf Hitler*, Volume 1, Video Documentary, The History Channel, (1995).

610. H. Rauschning, *Germany's Revolution of Destruction*, W. Heinemann, London, Toronto, (1939); **and** *The Revolution of Nihilism: Warning to the West*, Alliance Book Corp., New York, Longmans, Green & Co.; **and** (1939); **and** *The Voice of Destruction*, Putnam, New York, (1939); **and** *Germany's Revolution of Destruction*, W. Heinemann, London, (1939); **and** *Hitler Speaks: A Series of Political Conversations with Adolf Hitler on His Real Aims*, T. Butterworth Ltd., London, (1939); **and** *Hitler's Aims in War and Peace*, W. Heinemann London, Toronto,(1939); **and** *Hitler Could Not Stop*, New York, Council on Foreign Relations, Inc., (1939); **and** *The Revolution of Nihilism: Warning to the West*, New York, Alliance Book Corp., Longmans, Green & Co., (1940); **and** *Hitler and the War*, American Council on Public Affairs, Washington, D.C., (1940); **and** *Verboten! The book that Hitler Fears*, Kelly and Walsh, Shanghai, (1940); **and** *Make and Break with the Nazis: Letters on a Conservative Revolution*, Secker and Warburg, London, (1941); **and** *The Beast from the Abyss*, W. Heinemann, London, (1941); **and** *Hitler Wants the World*, Argus Press, London, (1941); **and** *The Redemption of Democracy: The Coming Atlantic Empire*, Alliance Corp., New York, (1941); **and** *The Conservative Revolution*, G.P. Putnam's Sons, New York, (1941); **and** *Makers of Destruction*, Eyre & Spottiswoode, London, (1942); **and** *Men of Chaos*, G.P. Putnam's Sons, New York, (1942); **and** *Time of Delirium*, D. Appleton-Century company, Inc. New

advance that Hitler would turn on Russia. Rauschning, who made it clear that he believed Hitler hated Bolshevists, stated in 1939,

> "There has been from the beginning in the National Socialist Party a group favoring close alliance with Soviet Russia. [***] It insisted upon the need to create this continental line as the foundation for a new world order — not through war, but through an alliance with Russia. After all, the advocates of this scheme said, it mattered little whether the vast empire was National Socialist or Bolshevik. The differences were, in their opinion, of no importance as against the larger world-revolutionary tasks of rational economic planning, of creating the new social order, and a 'just' redistribution of the world's wealth. It was not of such paramount importance, in the end, whether Germans or Russians would come out on top in this close symbiosis of Germany and Russia. What really mattered was the finish of the democratic order, free economy, and capitalism. Though he did not accept these ideas, Hitler never rejected them."[611]

Though he was considered highly credible for many years, several researchers have discredited Rauschning's claim to have had numerous conversations with Hitler, and he is today disregarded as a historical witness to Hitler's personality.[612] Regardless of these facts and allegations, many of Rauschning's general predictions came true and he was a witness to, and a member of, the inner circles of the Nazi hierarchy. His statements with respect to the redder tones of Nazism are verified by the actions and beliefs of Ernst Röhm and the fact that many sincere Socialists became uneasy and began to leave the Party in the 1930's when it became clear that Capitalism was still king. Adolf Hitler was more devoted to Zionism than to Communism. He was put into power to create a "Jewish State" and for those who put him in power, Communism was a transitory means to achieve that end.

The seemingly paradoxical accusation, that both Capitalism and Marxism are ultimately centralized Jewish movements, can be explained in many ways. It is usually dismissed as paranoia sponsored by *The Protocols of the Learned Elders of Zion*, which presents a plan to pit Liberals against Conservatives, in order to weaken and confuse both groups, and in order to control all government from behind the scenes without detection and regardless of which political persuasion is in power at any particular time in any given place. Jewish financiers had sponsored an arms build up in opposing empires, which led to the

---

York, London, (1946).
611. H. Rauschning, "Hitler Told Me This", *The American Mercury*, Volume 48, Number 192, (December, 1939), pp. 385-393, at 389.
612. H.W. Koch, "1933: The Legality of Hitler's Rise to Power", *Aspects of the Third Reich*, St. Martin's Press, New York, (1985), p. 39. C. Zentner and F. Bedürftig, *The Encyclopedia of the Third Reich*, Volume 2, Macmillan, New York, Toronto, (1991), p. 757. S. W. Mitscham, *Why Hitler?: The Genesis of the Nazi Reich*, Praeger, Westport, Connecticut, (1996), p. 137.

First World War, through their control of news media by direct ownership, tribal loyalty and with advertising dollars. Their media control gave them control over public opinion and control over politicians. Jewish financiers funded the Bolshevik Revolution in Russia in order to free the Jews there from the Pale of Settlement and to seize the reins of power in Russia and rob the nation of its vast wealth. The artificial struggle between Capitalists and Communists weakened peoples and states and left them vulnerable to Jewish exploitation and totalitarian control. The constant war resulting from the battles of Capitalism and Communism suppressed the masses, weakened the nations in preparation for world revolution, and enriched the Jewish arms dealers and bankers.

A large branch of the Nazis, primarily under the leadership of Ernst Röhm, were in greater sympathy with the Communists than the Zionists. The industrialists who sponsored Hitler were vehemently anti-Communist Capitalists. Hitler was more of a Zionist than a Communist. He knew that the primary goal of Communism was to destroy Gentile society. Hitler's primary goal was the establishment of the "Jewish State". Like many top Bolshevik dictators, he did not care about the working class.

The Zionists hated the Capitalism that enabled and sponsored Jewish assimilation and the Zionists hated the anti-nationalistic Communism which led to "Red Assimilation". This is why Bolshevism morphed into Nazism, which destroyed both assimilationist Capitalism and assimilationist Communism. Zionist Berl Katzenelson stated,

> "[...]they enjoy emancipation purchased through assimilation in capitalistic France and communistic Russia[... .]"[613]

Racist political Zionist Theodor Herzl stated in his book *The Jewish State*,

> "When we sink, we become a revolutionary proletariat, the subordinate officers of the revolutionary party; when we rise, there rises also our terrible power of the purse. [***] Again, people will say that I am furnishing the Anti-Semites with weapons. Why so? Because I admit the truth? Because I do not maintain that there are none but excellent men amongst us? Again, people will say that I am showing our enemies the way to injure us. This I absolutely dispute. My proposal could only be carried out with the free consent of a majority of Jews. Individuals or even powerful bodies of Jews might be attacked, but Governments will take no action against the collective nation. The equal rights of Jews before the law cannot be withdrawn where they have once been conceded; for the first attempt at withdrawal would immediately drive all Jews rich and poor alike, into the ranks of the revolutionary party. The first official violation of Jewish liberties invariably brings about economic crisis. Therefore no weapons can

---

613. B. Katzenelson, *Ba-Mivhan*, Tel-Aviv, (1935), pp. 67-70; translated to English in A. Hertzberg, *The Zionist Idea*, Harper Torchbooks, New York, (1959), pp. 390-395, at 395.

be effectually used against us, because these cut the hands that wield them."[614]

The Nazis' attacks on Jews aided the Zionists' agenda of forcing rich assimilationist Jews towards Zionism, and punishing them for their opposition to it, as was accomplished by brutal Nazi persecution; while concurrently eliminating the sanctuary that Marxist nations afforded liberated Jews; as was accomplished by the Nazi invasion of the Soviet Union—these being major objectives of both Theodor Herzl and Chaim Weizmann, as evinced in Herzl's diaries[615] and Weizmann's autobiography.[616] Hitler's strategies in some ways copied those of Napoleon and in some ways were opposites of those of Napoleon. Both Napoleon and Hitler were Zionists who seemingly irrationally attacked Russia in order in part to force, or to enable, Jews to move to Palestine. Both had Zionist allies and both fit the Zionist mold of a dictator. Napoleon sought to fulfill the Zionist dream of the Jews with philo-Semitism. After Napoleon failed, Hitler sought to fulfill the Zionist dream of the Jews with anti-Semitism, and succeeded.

In an article in 1943 in which he acknowledged that the First World War freed the Jews of central and eastern Europe then led to rabid anti-Semitism, and in which he acknowledged that the Zionists had allies in the newspapers of New York and in American Presidents from "Wilson down", Zionist Rabbi Abba Hillel Silver emphasized the importance of America to Zionism and called for "American Israel" to unite behind Zionism. Hillel also recognized the repeated Zionist connections between revolutions, emancipation, assimilation, anti-Semitism and world war,

> "The story of Jewish emancipation in Europe from the day after the French Revolution to the day before the Nazi revolution is the story of political positions captured in the face of stubborn and sullen opposition, which left our emancipated minority in each country encamped within an unbeaten and unreconciled opposition, so that at the slightest provocation, as soon as things got out of order, the opposition returned to the attack and inflicted grievous wounds. And in our day, stirred by the political and economic struggles which have torn nations apart, this never-failing, never-reconciled opposition swept over the Jewish political and economic positions in Europe and completely demolished them. There is a stout black cord which connects the era of Fichte in Germany with its feral cry of *'hep, hep,'* and the era of Hitler with its cry of *'Jude verrecke.'* The Damascus affair of 1840 links up with the widespread reaction after the Revolution of 1848—the Mortara

---

614. T. Herzl, *A Jewish State: An Attempt at a Modern Solution of the Jewish Question*, The Maccabæan Publishing Co., New York, (1904), pp. 23, 99.
615. T. Herzl, *The Complete Diaries of Theodor Herzl*, In Five Volumes, Herzl Press, New York, (1960).
616. C. Weizmann, *Trial and Error: The Autobiography of Chaim Weizmann*, Harper & Brothers, New York, (1949), p. 201.

affair of Italy; the Christian Socialist Movement in the era of Bismarck; the Tisza-Eszlar affair in Hungary; the revival of blood accusations in Bohemia; the pogroms in the eighties in Russia; *La France Juive* and the Dreyfus affair in France; the pogroms of 1903; the Ukrainian blood baths after the last war, and the human slaughter houses of Poland in this war."[617]

Some political Zionists wanted to unite the Jews of the world, gather them together and forcibly expel them to Palestine, punishing those who had abandoned Israel with death as was prophesied in the Old Testament. German Bolshevik movements were often led by Eastern European Jews. Anti-Bolshevik movements were also led by Eastern European Bolshevik Jews, who wanted a controlled opposition they could use to sponsor revolutions which would ultimately place them in power in fulfillment of Jewish Messianic prophecy. This controlled opposition became known as "The Trust". They followed the example of "Judas"—the "Jew", who placed "Jesus"—the "Jew" on the throne of the Messiah by betraying him and fulfilling the Old Testament Jewish prophecy, which saved the Jewish Nation (*Zechariah* 11:12-13. *Matthew* 27:9). Note that "Judas"="Judah"="Yehuda"="Jew". The myth of Judas' betrayal of Jesus may well have been derived from the myth of Judah and Joseph (*Genesis* 37:26-28). Just as Judah's decision to sell Joseph for 20 pieces of silver saved Joseph's life, in the *Gospel of Judas*, Judas' betrayal of Jesus freed Jesus from the prison of flesh and gave the promise of redemption and eternal life to Christians. This became a Frankist theme.

The Jewish bankers may well have been inspired by the story of Judas the "Jew" and Jesus the "Jew" to betray the Jews of Europe into a Holocaust of the Jewish bankers' design, and in so doing to have fulfilled Jewish prophecy. Note that the story of Jesus and Judas is suspiciously similar to the story of Julius Cæsar and Brutus—the story of a man who would be king and his friend turned betrayer who caused his murder and so saved the nation. Note further that the parallel between Judas and Brutus was captured in the story of Dante's *Inferno*, and that John Wilkes Booth, the Jewish actor who assassinated President Lincoln perhaps at the behest of the Jewish bankers, likened himself to Brutus.

Eastern European Jews were the most ardent political Zionists. The Bolsheviks were among the most dogmatic thinkers, the most ruthless and undemocratic tyrants the world has ever known; and, like the Nazis, they had no compunctions about forcing people into acts they would not voluntarily commit. Both the Nazis and the Bolsheviks outlawed all rival political parties in territories under their control. Their rigid dogmatism and totalitarianism were typically Judaic, as were their terror tactics and genocides meant to segregate Jews.

Throughout its existence, the Nazi regime preached revolution by the working class. Like many totalitarian Socialist regimes, National Socialism

---

617. A. H. Silver, *Vision and Victory*, Zionist Organization of America, New York, (1949); as quoted in A. Hertzberg, *The Zionist Idea*, Harper Torchbooks, New York, (1959), pp. 592-600, at 593-594.

punished free thought and banned all political parties other than National Socialism. While preaching the superiority of the "Nordic race", it subverted the intellectual growth of Northern Europe and promoted *Gleichschaltung* and the *Ermächtigungsgesetz*, which enslaved and degraded the German People in the same way Stalin enslaved the Soviets. This resulted in the degradation of German culture and the growth of the decadent mythologies of *Germanenorden*.

Hitler attacked German and European society in the exact way he alleged that Jews sought to undermine it. In the name of rescuing Europe from Jewish Bolshevism, Hitler immediately destroyed the intellectual classes who opposed him or who even had the potential to oppose him. It was obvious that Hitler was an agent of the Bolsheviks and the Zionists, and was accomplishing their goals. Jewish leadership yet again used anti-Semitism as means to put Jewish agents into power who would ruin Gentile nations and segregate Jews.

The hypocrisy of Hitler's attacks on Jews versus his own assumption of dictatorial powers[618] was apparent in an interview he gave to Anne O'Hare McCormick which was published in *The New York Times* on 10 July 1933. As early as 8 April 1933, in the "Topics of the Times" Section of *The New York Times* on page 12, the following statement appeared,

> ""HITLER's chief enemy, over whose prostrate body he has ridden to victory, is 'Marxianism.' But Marxianism and Hitlerism are really brothers. They are both the offspring of the Absolute of Hegelian dialectic. KARL MARX, riding the theory of materialistic determinism to death, and HITLER, setting out to reconstruct Christianity on a purely Aryan basis, are equally good illustrations of what the German mind is likely to do when it gets hold of a formula."

The commonality of the oppression of both Bolshevist and Nazi Socialist dictatorships, and the common totalitarianism, was so obvious to so many that Goebbels protested loudly that Nazism was not Bolshevism—despite the fact that it was. In response to the comparison of Hitlerism to Stalinism in the *London Times*,[619] Goebbels gave a speech in 1935, "Communism with the Mask Off", in which he stated, *inter alia*,

> "In the beginning of August, this year, one of the most authoritative English newspapers published a leading article entitled 'Two Dictatorships', in which a naive and misdirected attempt was made to place before the readers of the paper certain alleged similarities between Russian Bolshevism and German National Socialism. This article gave rise to an extraordinary amount of heated discussion in international centres, which was only another proof of the fact that an astonishing misconception exists among the most prominent West European circles as to the danger which

---

618. "Rule till 1937 Sought", *The New York Times*, (21 March 1933), pp. 1, 11.
619. "Two Dictatorships", *The London Times*, (7 August 1935), p. 11.

communism presents to the life of the individual and of the nation. Such people still cling to their opinion in face of the terrible and devastating experiences of the past eighteen years in Russia.

The author of the article stated that the two symbols which are to-day opposed to one another, namely that of Bolshevism and National Socialism, stand for regimes which 'in essential structure are similar and in many of their laws—their buttresses—are identical. The similarity is moreover increasing'. He went on to say:

'In both countries are the same censorships on art, literature, and of course the Press, the same war on the intelligentsia, the attack on religion, and the massed display of arms, whether in the Red Square or the Tempelhofer Feld.'

'The strange and terrible thing is', he declared, 'that two nations, once so widely different, should have been schooled and driven into patterns so drably similar.'"[620]

The Times truly touched a Nazi nerve. Cesare Santoro wrote in his book *Hitler Germany as Seen by a Foreigner*, Second Edition, Internationaler Verlag, Berlin (1939), page 59,

"A particularly vehement and outspoken speech was delivered on this occasion by the Reich Minister of Propaganda, Dr. Goebhels, who is the most fertile orator in new Germany, a master of the art of polemics and endowed with a rare gift for irony, and whose persuasive eloquence played a decisive part in the development of the party, especially in Berlin. In the speech in question Dr. Goebbels cited an article in a leading London newspaper which pointed out a certain analogy between the Russian and German systems. With the help of extensive statistical and other material, Dr. Goebbels showed that the author of the article had not taken the trouble to study the fundamental and essential principles either of National Socialism or of Bolshevism; and that he was consequently not qualified to appreciate the differences which separate them."

In 1938, Nesta Helen Webster stated that Fascism and Bolshevism were commonly considered to be the same thing, in Chapter 4, "Bolshevism and Fascism", of her book, *Germany and England*, Boswell, London, (1938). She tried to convince her readers that Nazism was not Bolshevism, in spite of the obvious parallels.

Adolf Hitler was *Time* Magazine's "Man of the Year" for 1938. The article on Hitler in the 2 January 1939 issue of *Time* stated,

"The Fascintern, with Hitler in the driver's seat, with Mussolini, Franco and

---

[620]. J. Goebbels, *Communism with the Mask Off. Speech Delivered in Nürnburg on September 13th, 1935 at the Seventh National Socialist Party Congress.*, M. Müller, Berlin, (1935).

the Japanese military cabal riding behind, emerged in 1938 as an international, revolutionary movement. Rant as he might against the machinations of international Communism and international Jewry, or rave as he would that he was just a Pan-German trying to get all the Germans back in one nation, Führer Hitler had himself become the world's No. 1 International Revolutionist—so much so that if the oft-predicted struggle between Fascism and Communism now takes place it will be only because two revolutionist dictators, Hitler and Stalin, are too big to let each other live in the same world. [***] Most cruel joke of all, however, has been played by Hitler & Co. on those German capitalists and small businessmen who once backed National Socialism as a means of saving Germany's bourgeois economic structure from radicalism. [***] Hard-pressed for foodstuffs as well as funds, the Nazi regime has taken over large estates and in many instances collectivized agriculture, a procedure fundamentally similar to Russian Communism."[621]

One of the major mistakes Germany had made in the First World War was to make it easy for England to enter the war. Many have asserted that Goebbels and Hitler thought that England would stay out of the approaching second war as long as England believed that Germany would safeguard Western Europe from Bolshevism. In fact, it did not matter whether England entered the war, or not. Stalin and Hitler would not rest until Eastern Europe came under Bolshevist control. Most of the world's Jews lived in Eastern Europe.

It worried Nazi leadership when they learned that the British public had discovered that Nazism was a twin bother to Bolshevism. German Jewish bankers and German industrialists had financed the Nazis and the Bolshevik Revolution in Russia.[622] Some German industrialists were duped into sponsoring the attack on the Soviets, because they were glad to learn that the Nazis would attack the anti-Capitalistic Bolsheviks (whom the German industrialist had helped to put into power—they were then also the dupes of Jewish bankers, who promised them victory over Pan-Slavism and unlimited access to Russia's vast wealth—the German industrialists did not know that the Bolsheviks would mass murder 30 million Gentiles in the first six years of their reign). They believed that perpetual war would make them rich beyond their wildest dreams.

Gentile German industrialists had become increasingly concerned by the Bolshevik Nazis, who were planning to nationalize industry. In order to dispel their fears, Hitler arrested and murdered the most outspoken Communists in the Nazi Party including Ernst Röhm and Gregor Strasser in the infamous "night of the long knives" on 30 June 1934 and 1 July 1934. After Hitler slaughtered the most obvious Bolsheviks in the Nazi Party, and concurrently killed off any potential rivals less inclined to Zionism than himself, coal magnate Emil Kirdorf reassured his fellow industrialists that Hitler was their man. Some say Emil

---

621. "Man of the Year", *Time*, Volume 33, Number 1, (2 January 1939), pp. 11-14.
622. A. C. Sutton, *Wall Street and the Rise of Hitler*, GSG & Associates, San Pedro, California, (2002).

Kirdorf was half Jewish. He had long financed and promoted Adolf Hitler and even promoted Hitler's little book to his industrialist friends: A. Hitler, *Der Weg zum Wiederaufstieg*, H. Bruckmann, München, (1927).

The Nazis used both the threat of Bolshevism and the alleged need for *Lebensraum* as pretexts to attack Poland and then the Soviets in order to destroy Eastern Europe and ready it for a Communist takeover, and to attack the defenseless Jewish families who lived in the East and segregate them, then, it was planned, force them into a "Jewish State". The Zionist Winston Churchill had issued the same carrot and stick threats at the same time. Churchill helped Zionist Adolf Hitler to turn Eastern Europe into a Communist bloc and to create the State of Israel—all of this vast destruction, communization and the ruin of Gentile nations and peoples, took place in the name of protecting the world from Jewish Bolshevism.

In 1932, Goebbels combined Bolshevist propaganda with anti-Semitic propaganda and misrepresented Marxism in order to mask his advocacy of its ideals. Goebbels adopted Socialism while presenting it as nationalistic racism, as opposed to *international* communism, which the Nazis attributed to "Jews".[623] However, this was exactly what racist Zionist Communist National Socialist Moses Hess had proposed in the mid-Nineteenth Century. In addition, the Nazis called for world revolution as loudly as had Trotsky.

When the Nazis strengthened their hand, the Nazi propaganda, which had initially declared that Nazism differed from Bolshevism in that it was limited to a German revolution, became international, or multinational, and declared itself to be on a "world mission" to stamp out "international Bolshevism". Russian Bolshevism had criminalized anti-Semitism on pain of death, which political Zionists feared would cause the extinction of the "Jewish race" in the East through assimilation. On this point, as with so many others, the Zionists and Nazis supported one another. Santoro continued on page 60,

> "This last argument put forward by Dr. Goebbels reveals one of the main reasons of the hostility to Bolshevism manifested by the new Germany—namely, the predominance in the development of the Bolshevist creed of Jewish elements similar to those which National Socialism considers to have been the chief cause of all the evils that befell Germany after the World War, and which have now been completely eliminated from German public life. Hitler combats Russian Bolshevism for the same motive which dictated his hostility to Marxism in Germany, which was likewise dominated by Jewish influence.

---

623. J. Goebbels and Mjölnir—pseudonym of the caricature artist Hans Schweitzer, *Die verfluchten Hakenkreuzler: Etwas zum Nachdenken*, Verl. F. Eher, München, (1929). *See also*: J. Goebbels, *Die zweite Revolution: Briefe an Zeitgenossen*, Streiter-Verlag Zwickau (Sa.), (1926); **and** "Goldene Worte für einen Diktator und für solche, die es werden wollen", *Der Angriff*, (1 September 1932); reprinted in: *Wetterleuchten: Aufsätze aus der Kampfzeit*, Zentralverlag der NSDAP., Franz Eher Nachf., München, (1939), pp. 325-327.

From an international point of view it is interesting to note that for the first time an allusion was made in the speech of Dr. Goebbels to the 'world mission' of Hitler as champion in the struggle against Bolshevism outside the German frontiers. 'If' (said Dr. Goebbels) 'Germany which has been redeemed and united in the spirit of National Socialism takes the lead, at the head of all those groups which are animated by a similar spirit, in this struggle against international Bolshevism, she is convinced that over and above her national aims she has a world mission to fulfil, on the successful issue of which the fate of all civilised nations will depend.'"

The Bolsheviks were always nihilistic. They wanted to tear down society. They did not care whether Hitler won, or Stalin won, because in either event the revolution won, which is to say humanity lost. Hitler and Stalin initially had a pact which troubled unaware Jewish Communists in America, but under this pact which brought peace, they could not impart the destruction to Europe both men sought. When the time was right, they started the war the Jews had been planning for centuries.

7.5.2 Hitler and Goebbels Reveal Their True Motives at War's End

Joseph Goebbels, who was called the "little rabbi" in school, revealed himself as a Bolshevik yet again at the end of the war when the Nazis and Bolsheviks had crushed the spirit of Eastern Europe and readied it for a Communist takeover. Goebbels rejoiced in Hitler's "Nero Order", which called for the destruction of Germany, for the destruction of "the last so-called achievements of the bourgeois nineteenth century".[624]

Hitler issued the "Nero Order" on 19 March 1945 and demanded the destruction of German infrastructure, industry, etc. in the hopes that the German People would be annihilated—which was his Bolshevik and Zionist goal from the very beginning. Goebbels stated,

> "If the Führer were to meet an honourable death in Berlin, with Europe falling to the Bolsheviks, within five years at the latest, the Führer would become a legendary personality and National Socialism mythic, because he would have been sanctified by this greatest and last act, and all the human frailties which today people criticise him for would be wiped away at one stroke."[625]

---

624. M. Burleigh, *The Third Reich: A New History*, Hill and Wang, New York, (2000), p. 789. Burleigh cites: H. R. Trevor-Roper, *The Last Days of Hitler*, Fourth Edition, Macmillan, London, (1971), p. 51.

625. M. Burleigh, *The Third Reich: A New History*, Hill and Wang, New York, (2000), pp. 790-791. Burleigh cites: "'... warum dann überhaupt noch leben!' Hitlers Legebesprechungen am 23, 25 und 27 April 1945", *Der Spiegel*, Volume 3, Number 20, (10 January 1966), pp. 32-46; **and** S. Behrenbeck, *Der Kult um die toten Helden:*

One might conclude that Goebbels believed that Hitler would be revealed as a Bolshevik who had conquered Europe for the world revolution the Nazis had been preaching in a chorus with the Bolsheviks from the beginnings of the Nazi movement. One might alternatively conclude that Goebbels believed that Hitler would be seen as a hero because he had opposed the Bolsheviks, who would certainly impose terror on a conquered Europe. An eyewitness account of some of Goebbels' last words provides us with a means to determine his intentions—to determine that he was as an *agent provocateur* for the Bolsheviks—and the Zionists,

> "the German people deserved the fate that awaited them.... . [Goebbels] remarked cynically that the German people had after all chosen this fate themselves. 'In the referendum on Germany's quitting the League of Nations they chose in a free vote to reject a policy of subordination and in favour of a bold gamble. Well, the gamble hadn't come off.... . Yes, that may surprise some people, including my colleagues. But have no illusions. I never compelled anybody to work for me, just as we didn't compel the German people. They themselves gave us the job to do. Why did you work with me? Now, you'll have your little throat cut.' Striding towards the door, [Goebbels] turned round once more and shouted: 'but the earth will shake as we leave the scene.'"[626]

Goebbels murdered his wife and children at the end of the war. He was never close to them. He preferred dark-haired Jewish women to his "Aryan" wife.

In the last days of the war on 16 April 1945, Hitler proclaimed,

> "For the last time the Jewish-Bolshevik deadly foe has come forth with his masses to attack. He is seeking to destroy Germany and to exterminate our people. Many of you soldiers from the East already know yourselves what fate threatens above all German women and children. While the elderly, menfolk and children will be murdered, women and girls will be degraded into barrack-room whores. The rest will be marched off to Siberia."[627]

The best means Hitler had to ensure that the Bolsheviks would impose this horrible fate on the Germans was for the Nazis to continue to fight the Soviets and to resist any attempts at a negotiated peace that would end the destruction of

---

Nationalsozialistische Mythen, Riten und Symbole 1923 bis 1945, SH-Verlag, Vierow, (1996), p. 584.
626. M. Burleigh, *The Third Reich: A New History*, Hill and Wang, New York, (2000), p. 791. Burleigh cites: J. Noakes and G. Pridham, Editors, *Nazism, 1919-1945: A Documentary Reader*, Volume 4, Document 1397, University of Exeter Press, Exeter, United Kingdom, (1998), p. 667.
627. M. Burleigh, *The Third Reich: A New History*, Hill and Wang, New York, (2000), p. 789.

Germany and it secure its borders from a Soviet takeover. Nazi leaders Rudolf Hess and Heinrich Himmler sought peace at the beginning, and at the end of the war, and both were silenced by the British. Goebbels relished the fact that the crimes the Nazis committed against the Jews would mean that the Germans would have to fight to the very end and consume themselves.

Hitler continued the war in the knowledge and the hopes that his failure to seek peace terms would lead to the destruction of Germany and the extermination of the German People, and note that he knew that the war was killing off the best of the German's genetic stock,

> "If the war is to be lost, the nation will also perish. This fate is inevitable. There is no need to consider the basis of a most primitive existence any longer. On the contrary it is better to destroy even that, and to destroy it ourselves. The nation will have proved itself the weaker and the future will belong exclusively to the stronger Eastern nation. Those who remain alive after the battles are over are in any case only inferior persons, since the best have fallen."[628]

Hitler stated,

> "That is the decision. To save everything here, and only here, and to deploy the last man, that is our duty."[629]

Hitler, who had once stated that Oliver Cromwell was his hero[630]—Oliver Cromwell who had emancipated the Jews and welcomed them to England—Cromwell the Puritan revolutionary who had declared the Pope in Rome to be the anti-Christ—this Adolf Hitler likened himself to Napoleon, the revolutionary who had emancipated the Jews of Europe—Napoleon who had fought to take Palestine for the Jews—Napoleon who had suicidally attacked Russia in order to emancipate its Jews and bring them to Palestine—Adolf Hitler iterated the nihilistic Bolshevistic mantra:

> "I have been Europe's last hope. She proved incapable of refashioning herself by means of voluntary reforms. She showed herself impervious to charm and persuasion. To take her I had to use violence.

---

628. R. Payne, *The Life and Death of Adolf Hitler*, Praeger Publishers, New York, (1973), p. 541. Payne cites: H. Guderian, *Panzer Leader*, E. P. Dutton, New York, (1952), p. 423.
629. M. Burleigh, *The Third Reich: A New History*, Hill and Wang, New York, (2000), p. 791. Burleigh cites: "'... warum dann überhaupt noch leben!' Hitlers Legebesprechungen am 23, 25 und 27 April 1945", *Der Spiegel*, Volume 3, Number 20, (10 January 1966), pp. 32-46; **and** S. Behrenbeck, *Der Kult um die toten Helden: Nationalsozialistische Mythen, Riten und Symbole 1923 bis 1945*, SH-Verlag, Vierow, (1996), p. 584.
630. A. O'Hare McCormick, "Hitler Seeks Jobs for All Germans", *The new York Times*, (10 July 1933), pp. 1, 6, at 6.

Europe can be built only on a foundation of ruins. Not material ruins, but ruins of vested interests and economic coalitions, of mental rigidity and narrow-mindedness. Europe must be refashioned in the common interest of all and without regard for individuals. Napoleon understood this perfectly.

I, better than anyone else, can well imagine the torments suffered by Napoleon, longing, as he was, for the triumph of peace and yet compelled to continue waging war, without ceasing, and without seeing any prospect of ceasing—and still persisting in the hope eternal of at last achieving peace."[631]

Like Napoleon, Hitler was viewed by his subjects as a messiah.

Hennecke Kardel entertained the possibility of links between Jewish self-hatred among the Nazi hierarchy, Nazism, Bolshevism, Zionism and Jewish financing in his book *Adolf Hitler, Begründer Israels*, Verlag Marva, Genf, (1974); English translation *Adolf Hitler: Founder of Israel*, Modjeskis' Society Dedicated to Preservation of Cultures, San Diego, (1997). Though it is often claimed that Hitler and other Nazi leaders who were of mixed Jewish descent, or in some instances pure Jewish descent, were self-hating Jews; it is more likely that they hated the "Aryans" far more, their eternal enemy Esau, whom they did so much to destroy.

Zionist racist Moses Hess stated in 1862 that the only obstacle to the success of Zionism was the reluctance of cultured Jews to accept their fate and move to Palestine. Hess forecast the Nazi régime in 1862, established most of its tenets, and predicted that the assimilatory aspirations of cultured Jews would "be shattered only by a blow from without," a blow that would "close their ephemeral existence". Hess concluded his racist Zionist treatise *Rome and Jerusalem* with the apocalyptic forecast:

"In contradistinction to orthodoxy, which cannot be destroyed by an external force without at the same time endangering the embryo of Jewish Nationalism that slumbers within it, the hard covering that surrounds the hearts of our cultured Jews will be shattered only by a blow from without, one that world events are already preparing; and which will probably fall in the near future. The old frame-work of European Society, battered so often by the storms of revolution, is cracking and groaning on all sides. It can no longer stand a storm. Those who stand between revolution and reaction, the mediators, who have an appointed purpose to push modern Society on its path of progress, will after society becomes strong and progressive, be swallowed up by it. The nurses of progress, who would undertake to teach the Creator himself wisdom, prudence and economy; those carriers of culture, the saviors of Society, the speculators in politics, philosophy and religion, will not survive the last storm. And along with the other nurses of progress our Jewish reformers will also close their ephemeral existence. On

---

[631]. R. Payne, *The Life and Death of Adolf Hitler*, Praeger Publishers, New York, (1973), p. 542. Payne cites: M. Bormann, H. R. Trevor-Roper, Editor, *The Bormann Letters*, Nicolson and Weidenfeld, London, (1954), pp. 103-104.

the other hand, the Jewish people, along with other historical nations will, after this last catastrophe, the approach of which is attested by unmistakable signs of the times, receive its full rights as a people.

> 'Remember the days of old,
> Consider the years of many generations;
> Ask thy father and he will tell thee,
> Thy elders and they will inform thee,
> When the Most High divided to the nations
>   their inheritance,
> When he separated the sons of Adam,
> He set the bounds of the peoples
> According to the number of the Children of
>   Israel.'[*Footnote:* Deut. xxxii, 7-8.]

Just as after the last catastrophe of organic life, when the historical races came into the world's arena, there came their division into tribes, and the position and rôle of the latter was determined, so after the last catastrophe in social life, when the spirit of humanity shall have reached its maturity, will our people, with the other historical people, find its legitimate place in universal history."[632]

When the pressure from without of Nazism failed to persuade the cultured Jews of Europe to move to Palestine, Hitler set out to fulfill his promise of 1939,

> "[I want to be a prophet again today:] If international finance Jewry in and outside Europe succeeds in plunging the peoples into another world war, then the end result will not be the Bolshevization of the earth and the consequent victory of Jewry but the annihilation of the Jewish race in Europe."[633]
>
> "Ich will heute wieder ein Prophet sein: Wenn es dem internationalen Finanzjudentum in und außerhalb Europas gelingen sollte, die Völker noch einmal in einen Weltkrieg zu stürzen, dann wird das Ergebnis nicht die Bolschewisierung der Erde und damit der Sieg des Judentums sein, sondern die Vernichtung der jüdischen Rasse in Europa!"[634]

---

632. M. Hess, *Rom und Jerusalem: die letzte Nationalitätsfrage*, Eduard Wengler, Leipzig, (1862); English translation by M. Waxman: *Rome and Jerusalem: A Study in Jewish Nationalism*, Bloch, New York, (1918/1943), pp. 177-178.

633. English translation in: R. S. Levy, *Antisemitism in the Modern World: An Anthology of Texts*, D. C. Heath and Company, Toronto, (1991), pp. 222-223, at 223. An alternative translation appears in: "Holocaust", *Encyclopaedia Judaica*, Volume 8, Macmillan, Jerusalem, (1971), col. 852.

634. A. Hitler in M. Domarus, Editor, *Hitler: Reden und Proklamationen, 1932-1945: Kommentiert von einem deutschen Zeitgenossen*, Süddeutscher Verlag, München, (1965), pp. 1057-1058.

The Jewish Nazi tyrant of Poland, Dr. Hans Frank, stated at a Cabinet Session on 16 December 1941,

> "As far as the Jews are concerned, I want to tell you quite frankly, that they must be done away with in one way or another. The Fuehrer said once: should united Jewry again succeed in provoking a world war, the blood of not only the nations which have been forced into the war by them, will be shed, but the Jew will have found his end in Europe"[635]

Did the crypto-Jewish Zionists Adolf Hitler and Hans Frank mean that they would exterminate the Jews of Europe in death camps, or did they mean that they would deport the Jews of Europe to Palestine as a final solution to the Jewish question? Frank was a long-term Zionist who wanted to segregate the Jews in Polish concentration camps and then ship them to Palestine—not to say that he did not intend to kill off a large percentage of his brethren in the process. In the fall of 1933 in Nuremberg on *Reichsparteitag*, Frank stated that his goal was to secure a "Jewish State",

> "Unbeschadet unseres Willens, uns mit den Juden auseinanderzusetzen, ist die Sicherheit und das Leben der Juden in Deutschland staatlich, reichsamtlich und juristisch nicht gefährdet. Die Judenfrage ist rechtlich nur dadurch zu lösen, dass man an die Frage eines jüdischen Staates herangeht."[636]

The Zionists had always viewed wealthy Jewish assimilationists as their arch-enemy in their struggle to force Jews to Palestine against their will. Hitler's last testament states, among other things,

> "But I left no doubt about the fact that if the peoples of Europe were again to be treated as so many packages of shares by these international money and finance conspirators, then the people who bear the real guilt for this murderous struggle would also have to answer for it: the Jews! It also left no doubt that this time we would not permit millions of European children of Aryan descent to die of hunger, or millions of grown-up men to suffer death, or hundreds of thousands of women and children to be burned and

---

635. H. Frank, (16 December 1941), quoted in: *Nazi Conspiracy and Aggression*, Volume 2, United States, Office of Chief of Counsel for the Prosecution of Axis Criminality, Washington, D. C., United States Government Printing Office, (1946), p. 634. ***See also:*** Y. Arad, Yitzhak, I. Gutman, A. Margaliot, Abraham,Editors, *Documents on the Holocaust: Selected Sources on the Destruction of the Jews of Germany and Austria, Poland, and the Soviet Union*, Yad Vashem in cooperation with the Anti-Defamation League and Ktav Pub. House, Jerusalem, (1981).
636. H. Frank quoted in H. Kardel, *Adolf Hitler, Begründer Israels*, Verlag Marva, Genf, (1974).

bombed to death in the cities, without the real culprit suffering his due punishment, though in a more humane way."[637]

Hitler was put into power by political Zionists to create an anti-Semitic Bolshevist revolution in Europe that would destroy the intellectual class, all forms of Monarchy and would place the working class proletariat in the hands of absolute Jewish rule in achievement of the Messianic vision of racist Zionists like Moses Hess and Theodor Herzl. Since it was the goal of political Zionists to eliminate the sanctuary that Marxism afforded Jews, Hitler preached anti-Semitism while concurrently preaching "World Revolution", *i. e.* thinly veiled Bolshevism. Among Adolf Hitler's first anti-Semitic statements after leaving Bolshevism to become an anti-Semitic propagandist was his assertion that the fight against Bolshevism meant the extirpation of the Jews—which was also a goal of the political Zionists.[638] Hitler later inexplicably attacked the Soviet State in which Jews were becoming assimilated. Hitler attempted to create an anti-Semitic Bolshevist tyranny in Europe and to found a Jewish State to provide a homeland for forcibly expelled Jews. Liebenfels, Rosenberg and the other architects of Nazi ideology had always sponsored Zionism as a right of expelled Jews, in full agreement with Theodor Herzl's prescriptions for a final resolution to the Jewish question.

Why did not the Russian Bolsheviks do to the Jews what the Nazis later would, if Jewish leadership controlled both movements? There are several reasons. While some Zionists predicted that assimilation would take place after the revolution emancipated the Jews, there were also many Zionists who hoped that the Russian People were too anti-Semitic and the Jews were of Russia were too racist and tribalistic for assimilation to occur in the East. Another reason is that the Zionists hoped to found a "Jewish State" soon after the war and they wanted to maintain Russia as a source of wealth and power and leverage against the Moslems of the Middle East, or, alternatively, they wanted to found a "Jewish State" in formerly Soviet territory. Yet another reason is that there were far greater numbers of Jews in Russia than in Germany, and the Bolshevik Revolution in Russia would not as easily have succeeded with a Nazi-style party platform, which, given the Jewish propaganda of the time, would not have appeared to have differed greatly from the Czar's platform as depicted in the press. There are several other more obvious reasons.

The Nazis eventually and inevitably lost their perpetual war of revolution on the world. Hitler's posthumously published sequel to his *Mein Kampf*, which sequel was written in 1928, asserted that "eternal war" was a doomed proposition. He must have known that his completely unnecessary declarations of war against the United States and the Soviet Union were suicidal to the

---

637. A. Hitler, "My Political Testament", English translation in: R. Payne, *The Life and Death of Adolf Hitler*, Praeger Publishers, New York, (1973), pp. 589-591, at 589.
638. A. Hitler, "Staatsmänner oder Nationalverbrecher", *Völkischer Beobachter* (formerly Thule's *Münchner Beobachter*), (15 March 1921), p. 2.

German Nation.[639] He knew the history of the First World War. It seems that he was either a complete fool, or was bent on destroying Germany, Communizing Europe and founding a "Jewish State" at the expense of the World. Given that Hitler's régime so exactly fulfilled Jewish Messianic prophecy, and given that Hitler had so many relations with Zionists, and further given that Jews who sought to fulfill those Jewish Messianic prophecies put Adolf Hitler into power, the "coincidences" are too many and too unlikely to have been the products of chance.

7.5.3 Zionists and Communists Delight in Massive Human Sacrifices to the Jewish Messianic Cause

The Second World War ended in 1945 with Albert Einstein's 1915 vision of a divided and destroyed Germany made real. Communism was infinitely stronger than before the war and it looked as if France, Greece, Italy, Germany and even England, in their weakened state, would succumb to it. Zionists used the Nazis' crimes against Jews, which the Zionist Jews intentionally caused, to justify the formation of the State of Israel, and the theft of Palestine, and the perpetual vilification of the Moslems.

Since the ancient Diaspora, all previous attempts to found a State of Israel had failed and the outlook for Jews after the First World War was near total assimilation, and, in the racist minds of political Zionists, the consequent extermination of the "Jewish race". They were, in fact, desperate enough to create the Nazis as a means achieve their ends and they believed Jewish Messianic prophecy fully justified their treachery.

In 1921, political Zionist Jakob Klatzkin stated,

> "[I applaud] the contribution of our enemies in the continuance of Jewry in eastern Europe. [***] We ought to be thankful to our oppressors that they closed the gates of assimilation to us and took care that our people were concentrated and not dispersed, segregatedly united and not diffusedly mixed [***] One ought to investigate in the West and note the great share which antisemitism had in the continuance of Jewry and in all the emotions and movements of our national rebirth . [***] Truly our enemies have done much for the strengthening of Judaism in the diaspora . [***] Experience teaches that the liberals have understood better than the antisemites how to destroy us as a nation. [***] We are, in a word, naturally foreigners; we are an alien nation in your midst and we want to remain one."[640]

---

[639]. A. Hitler, *Hitler's Secret Book*, Grove Press, New York, (1928/1962).

[640]. K. J. Herrmann, "Historical Perspectives on Political Zionism and Antisemitism", *Zionism & Racism: Proceedings of an International Symposium*, International Organization for the Elimination of All Forms of Racial Discrimination, Tripoli, (1977), pp. 197-210, at 204-205. A lengthy quotation from Klatzkin, in English translation, appears in: M. Menuhin, *The Decadence of Judaism in Our Time*, Exposition Press, New York, (1965), pp. 482-483.

In 1898, Nachman Syrkin wrote,

"Nonetheless, the enemy has *always* considered the Jews a nation, and they have always known themselves as such."[641]

In 1945, after the Zionist Nazi atrocities, Albert Einstein callously reminded the world of the Balfour Declaration and the Palestine Mandate in order to exploit the tragedy of the Holocaust the Zionists had deliberately caused. Einstein used the Holocaust to justify the fulfilment of his pre-Nazi political Zionist agenda. Einstein asserted that the Holocaust proved that the world thought of the Jews as a nation. Genocidal human sacrifice had long been a Judaic tradition, and in more recent times, Friedrich Engels made it clear that the Communists were comfortable with human sacrifices amounting to ten million lives lost in order to prepare the way for revolution and Communist world dominance. In 1887, Frederick Engels knew that the First World War was coming and that it would destroy the Empires of Europe and leave them ripe for revolution,

"No other war is now possible for Prussia-Germany than a world war, and indeed a world war of hitherto unimagined sweep and violence. Eight to ten million soldiers will mutually kill each other off, and in the process devour Europe barer than any swarm of locusts ever did. The desolation of the Thirty Years' War compressed into three or four years and spread over the entire continent: famine, plague, general savagery, taking possession both of the armies and of the masses of the people, as a result of universal want; hopeless demoralization of our complex institutions of trade, industry and credit, ending in universal bankruptcy; collapse of the old states and their traditional statecraft, so that crowns will roll over the pavements by the dozens and no one be found to pick them up; absolute impossibility of foreseeing where this will end, or who will emerge victor from the general struggle. Only *one* result is absolutely sure: general exhaustion and the creation of the conditions for the final victory of the working class."[642]

In 1945, Einstein wrote, among other things,

"[The Jews'] status as a uniform political group is proved to be a fact by the behavior of their enemies. Hence in striving toward a stabilization of the

---

[641]. N. Syrkin, under the nom de plume "Ben Elieser", *Die Judenfrage und der socialistische Judenstaat*, Steiger, Bern, (1898); English translation in A. Hertzberg, *The Zionist Idea*, Harper Torchbooks, New York, (1959), pp. 333-350, at 343.

[642]. B. D. Wolfe, *Marxism: One Hundred Years in the Life of a Doctrine*, Dial Press, New York, (1965), p. 67. Wolfe cites: "From Engels's introduction to the reissue of a pamphlet by Sigismund Borkheim. Borkheim's pamphlet, *Zur Errinnerung fuer die deutschen Mordspatrioten 1806-07* [***] The introduction is reproduced in *Werke*, Vol. XXI, pp. 350-351."

international situation they should be considered as though they were a nation in the customary sense of the word. [***] In parts of Europe Jewish life will probably be impossible for years to come. In decades of hard work and voluntary financial aid the Jews have restored the soil of Palestine to fertility. All these sacrifices were made because of trust in the officially sanctioned promise given by the governments in question after the last war, namely that the Jewish people were to be given a secure home in their ancient Palestinian country. To put it mildly, the fulfillment of this promise has been but hesitant and partial. Now that the Jews—especially the Jews in Palestine—have in this war too rendered a valuable contribution, the promise must be forcibly called to mind. The demand must be put forward that Palestine, within the limits of its economic capacity, be thrown open to Jewish immigration. If supranational institutions are to win that confidence that must form the most important buttress for their endurance, then it must be shown above all that those who, trusting to these institutions, have made the heaviest sacrifices are not defrauded."[643]

Einstein's statements prove that the human sacrifice of countless Jewish lives in the Zionist Holocaust had not changed the nationalistic racism of the political Zionists at all, but rather had strengthened their hand—in fulfillment of the Zionists' expressed plans. The racist Zionists had no regrets over their mass murder of Jews and they rejoiced at their slaughter of Gentiles. In the 1890's, Bernard Lazare iterated the Zionist mantra:

"It is because the Jews are a nation that anti-Semitism exists. [***] If the cause of anti-Semitism is the existence of the Jews as a nationality, its effect is to make this nationality more tangible for the Jews, to make them more aware of the fact that they are a people."[644]

Albert Einstein told Peter A. Bucky that the Holocaust had the benefit of uniting "all the Jews in the world":

"But the suffering had not been in vain, in Einstein's view. He felt that the Jews who died in Hitler's pogroms had strengthened the bond uniting all of the Jews in the world."[645]

Einstein was simply repeating the Zionist party line, as expressed by Rabbi Abba Hillel Silver in 1943,

---

[643]. A. Einstein, "Unpublished Preface to a Blackbook", *Out of My Later Years*, Philosophical Library, New York, (1950), pp. 258-259, at 259.
[644]. B. Lazare, "Jewish Nationalism and Emancipation (1897-1899)", in A. Hertzberg, *The Zionist Idea: A Historical Analysis and Reader*, Garden City, N.Y. Doubleday, (1959), pp. 471-476, at 471.
[645]. P. A. Bucky, Einstein, and A. G. Weakland, *The Private Albert Einstein*, Andrews and McMeel, Kansas City, (1992), p. 84.

"Should not, I ask you fellow Jews, ought not, the incalculable and unspeakable suffering of our people and the oceans of blood which we have shed in this war and in all the wars of the centuries; should not the myriad martyrs of our people, as well as the magnificent heroism and the vast sacrifices of our brave soldier sons who are today fighting on all the battle fronts of the world—should not all this be compensated for finally and at long last with the re-establishment of a free Jewish Commonwealth?"[646]

Did it occur to no one that the world, including the Jews, would be far better off if racist Zionism and Jewish tribalism were eradicated, rather than further justified, as a result of yet another massive Jewish tragedy? What, other than Jewish racism, prevented a massive drive for assimilation world-wide after the Holocaust?

---

[646]. A. H. Silver, *Vision and Victory*, Zionist Organization of America, New York, (1949); in A. Hertzberg, *The Zionist Idea*, Harper Torchbooks, New York, (1959), pp. 592-600, at 597.

OTHER TITLES

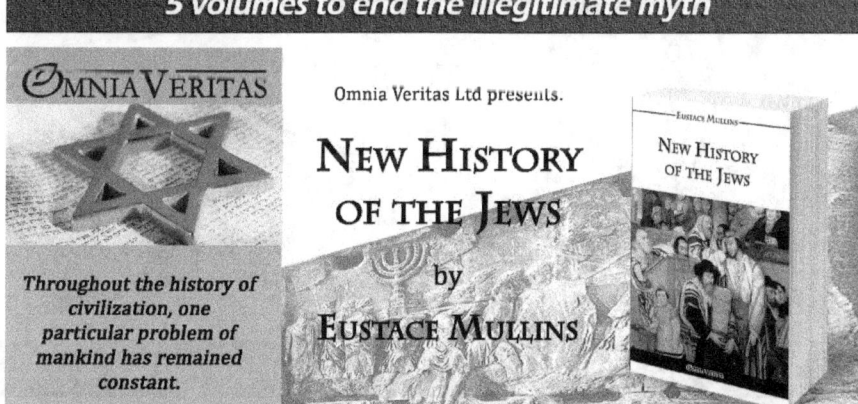

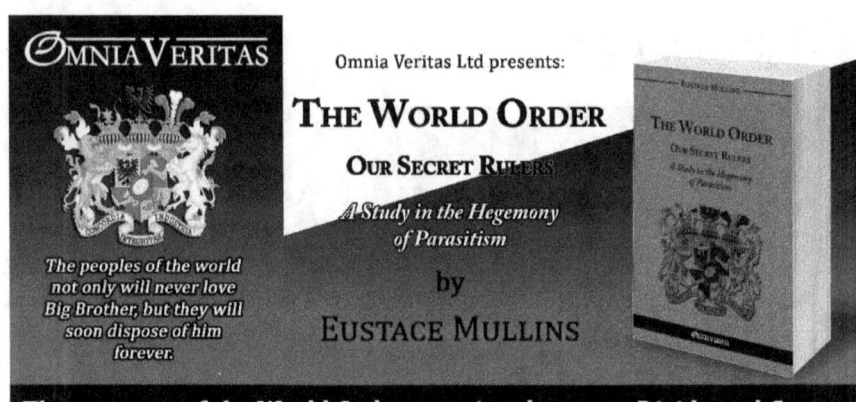

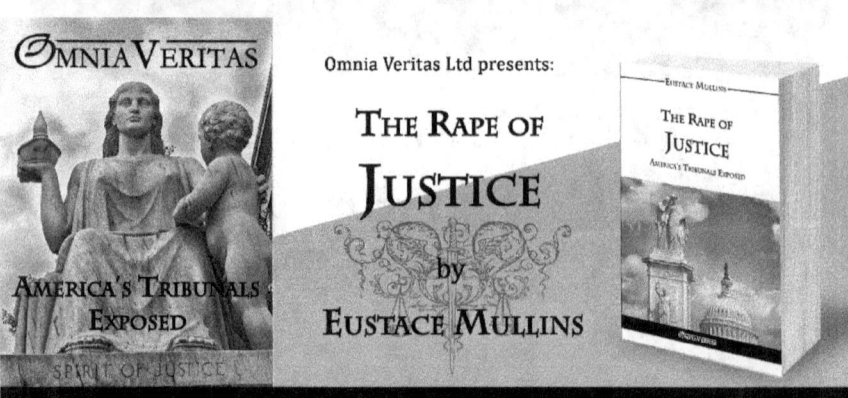

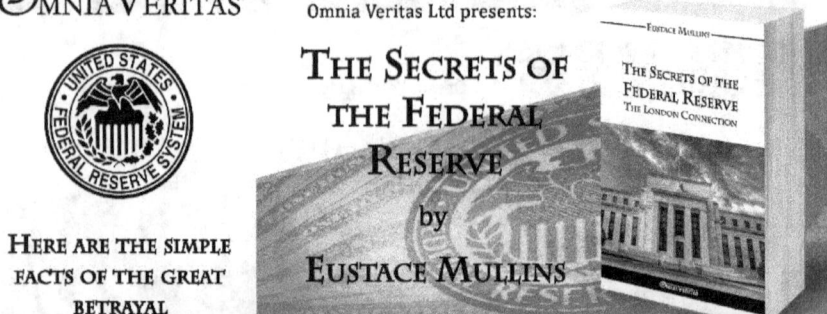

  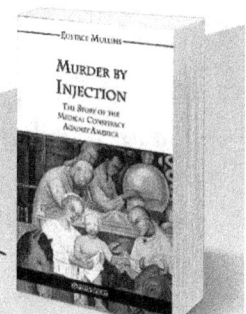

Omnia Veritas Ltd presents:

MURDER BY INJECTION

THE STORY OF THE MEDICAL CONSPIRACY AGAINST AMERICA

by

EUSTACE MULLINS

The cynicism and malice of these conspirators is something beyond the imagination of most Americans.

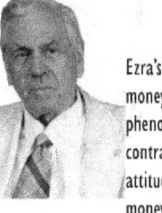 

Omnia Veritas Ltd presents:

EZRA POUND
THIS DIFFICULT INDIVIDUAL

by

EUSTACE MULLINS

Ezra's interest in money as a phenomenon, in contrast to the usual attitude toward money as something to get, is a legitimate one.

An illustration for his own monetary theories...

OMNIA VERITAS LTD PRESENTS:

MY LIFE IN CHRIST

BY

EUSTACE MULLINS

Christ did not wish to be followed by robots and sleepwalkers, He desired man to awaken, and to attain the full use of his earthly powers.

THIS is the story of my life in Christ

I wish to tell of the things which have happened to me in my struggle against the forces of darkness.

Omnia Veritas Ltd presents: COLLECTED ESSAYS — EUSTACE MULLINS

It is my hope that others will be forewarned of what to expect in this fight

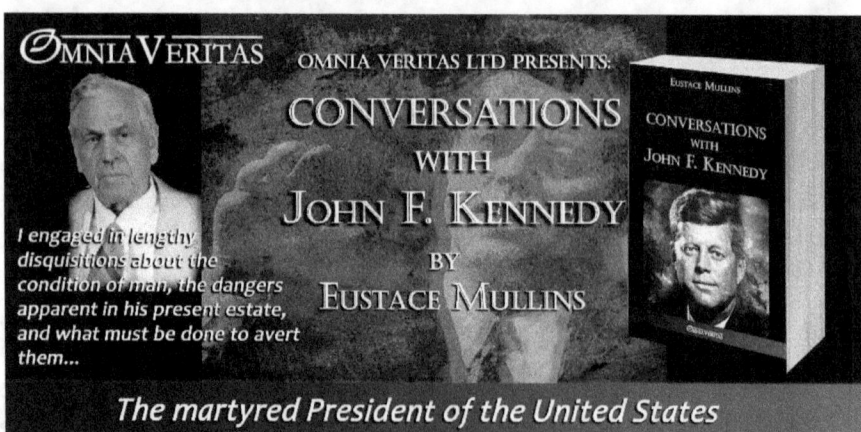

I engaged in lengthy disquisitions about the condition of man, the dangers apparent in his present estate, and what must be done to avert them...

OMNIA VERITAS LTD PRESENTS: CONVERSATIONS WITH JOHN F. KENNEDY BY EUSTACE MULLINS

The martyred President of the United States

My research has revealed that there are two separate and opposing powers in Freemasonry.

OMNIA VERITAS LTD PRESENTS: SCARLET AND THE BEAST — A HISTORY OF THE WAR BETWEEN ENGLISH AND FRENCH FREEMASONRY

One is Scarlet. The other, the Beast.

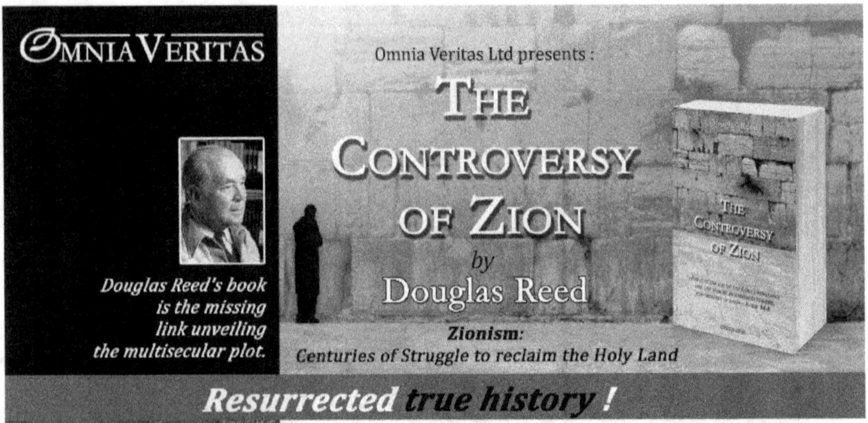
OMNIA VERITAS
Omnia Veritas Ltd presents :
THE CONTROVERSY OF ZION
by
Douglas Reed
Douglas Reed's book is the missing link unveiling the multisecular plot.
Zionism:
Centuries of Struggle to reclaim the Holy Land
Resurrected true history !

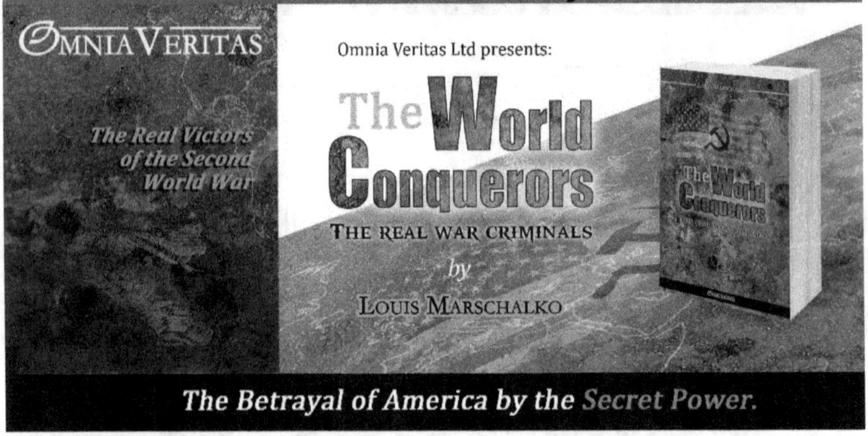
OMNIA VERITAS
Omnia Veritas Ltd presents:
The Real Victors of the Second World War
The World Conquerors
THE REAL WAR CRIMINALS
by
LOUIS MARSCHALKO
The Betrayal of America by the Secret Power.

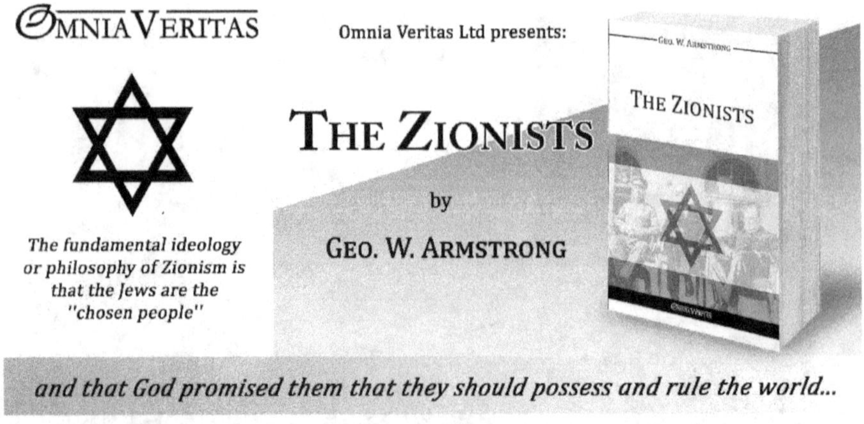
OMNIA VERITAS
Omnia Veritas Ltd presents:
THE ZIONISTS
by
GEO. W. ARMSTRONG
The fundamental ideology or philosophy of Zionism is that the Jews are the "chosen people"
and that God promised them that they should possess and rule the world...

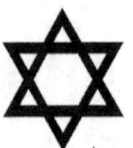

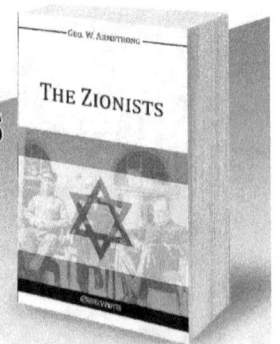

OMNIA VERITAS

www.omnia-veritas.com

https://www.instagram.com/omnia.veritas/

https://twitter.com/OmniaVeritasLtd

www.ingramcontent.com/pod-product-compliance
Lightning Source LLC
Chambersburg PA
CBHW071359230426
43669CB00010B/1390